GENERAL MICROBIOLOGY

General Microbiology

FOURTH EDITION

ROGER Y. STANIER
l'Institut Pasteur, Paris 15e, France

EDWARD A. ADELBERG
Yale University, New Haven, Connecticut

JOHN L. INGRAHAM
University of California, Davis, California

Original American edition published by Prentice-Hall, Inc.,
Englewood Cliffs, New Jersey, U.S.A.

First published in the United Kingdom 1958
Second Edition 1963
Third Edition 1971
Fourth Edition 1977
Reprinted 1978, 1979, 1980

Published by
THE MACMILLAN PRESS LTD
London and Basingstoke
Associated companies in Delhi Dublin
Hong Kong Johannesburg Lagos Melbourne
New York Singapore and Tokyo

ISBN 0 333 22013 7 (hard cover)
      0 333 22014 5 (paper cover)

Printed in Hong Kong

To the memory of Michael Doudoroff (1911–1975).

CONTENTS

1

2

vii

3

THE NATURE OF THE MICROBIAL WORLD   59

4

THE PROTISTS   89

5

THE PROCARYOTES: AN INTRODUCTORY SURVEY   119

6

MICROBIAL METABOLISM: THE GENERATION OF ATP   154

7

8

9

14

THE EXPRESSION OF MUTATION IN VIRUSES, CELLS, AND CELL POPULATIONS 430

15

GENETIC RECOMBINATION 452

16

THE CLASSIFICATION OF BACTERIA 502

17

THE PHOTOSYNTHETIC PROCARYOTES   527

18

GRAM-NEGATIVE BACTERIA: THE CHEMOAUTOTROPHS AND METHYLOTROPHS   564

19

GRAM-NEGATIVE BACTERIA: AEROBIC CHEMOHETEROTROPHS   589

20

THE ENTERIC GROUP AND RELATED ORGANISMS   612

21

GRAM-NEGATIVE BACTERIA: MYXOBACTERIA AND OTHER GLIDING ORGANISMS   630

26

27

28

29

30

THE EXPLOITATION OF MICROORGANISMS BY MAN   831

PREFACE TO THE FOURTH EDITION

The fourth edition of this book is dedicated to the memory of our friend and colleague Michael Doudoroff, one of the original authors of *The Microbial World*. His entire career was spent at the University of California in Berkeley; he was a leading figure in the group which made Berkeley a great center of research in microbial biochemistry and physiology during the decades following the second World War. He entered the Department of Bacteriology in 1940, and organized single-handed the program of instruction in general microbiology, for which he remained solely responsible until two of us (Roger Y. Stanier and Edward A. Adelberg) joined the Department some years later. *The Microbial World*, first published in 1957, grew out of the introductory courses which we then taught together at Berkeley. The design of these courses was almost entirely the creation of Michael Doudoroff, who was thus the real founder of *The Microbial World*.

When the first edition was written, microbiology still remained—as it had been for many decades—a discipline almost completely isolated from the rest of biology. Our goal in 1957 was to present microbiology within the framework of the concepts of general biology, since it was already evident that the period of isolation was coming to an end. By then, microbiology had begun to provide the experimental material for the discovery of the molecular basis of biological functions, even though the molecular biological revolution was only beginning to get under way. Now, almost 20 years later, the intellectual climate of biology in general and microbiology in particular has changed out of all recognition. This has forced us, in preparing successive editions of *The Microbial World*, to reappraise the scope and level of the material to be included, and to modify profoundly the structure of the book. The present revision is less drastic than the preceding ones. We have retained the organization and scope of the third

xvii

edition, although few chapters have escaped a complete rewriting. As always, we shall welcome comments and criticisms from readers.

We should like to express our thanks to the many individuals who helped us in the preparation of this edition. Once again, we are grateful to the many colleagues who responded to our requests for new illustrative material. Dr. Mark Wheelis read the entire manuscript and made many constructive suggestions with respect to style and content. Mrs. Marjorie Ingraham gave us invaluable help both with the typing of the manuscript and with the reading of proofs.

ROGER Y. STANIER

EDWARD A. ADELBERG

JOHN L. INGRAHAM

THE MICROBIAL WORLD

1

THE BEGINNINGS OF MICROBIOLOGY

Microbiology is the study of organisms that are too small to be clearly perceived by the unaided human eye, called *microorganisms*. If an object has a diameter of less than 0.1 mm, the eye cannot perceive it at all, and very little detail can be perceived in an object with a diameter of 1 mm. Roughly speaking, therefore, organisms with a diameter of 1 mm or less are microorganisms and fall into the broad domain of microbiology. Microorganisms have a wide taxonomic distribution; they include some metazoan animals, protozoa, many algae and fungi, bacteria, and viruses. The existence of this microbial world was unknown until the invention of microscopes, optical instruments that serve to magnify objects so small that they cannot be clearly seen by the unaided human eye. Microscopes, invented at the beginning of the seventeenth century, opened the biological realm of the very small to systematic scientific exploration.

Early microscopes were of two kinds. The first were *simple microscopes* with a single lens of very short focal length, consequently capable of a high magnification; such instruments do not differ in optical principle from ordinary magnifying glasses able to increase an image severalfold, which had been known since antiquity. The second were *compound microscopes* with a double lens system consisting of an ocular and objective. The compound microscope has the greater intrinsic power of magnification and eventually displaced completely the simple instrument; all our contemporary microscopes are of the compound type. However, nearly all the great original microscopic discoveries were made with simple microscopes.

THE DISCOVERY OF THE MICROBIAL WORLD

The discoverer of the microbial world was a Dutch merchant, Antony van Leeuwenhoek (Figure 1.1). His scientific activities were fitted into a life well filled with business affairs and civic duties. In this, he was no exception for his time; the great discoveries of this period in all fields of science were made by amateurs who earned their living in other ways, or who were freed from the necessity of earning a living because of their personal wealth. However, Leeuwenhoek differed from his scientific contemporaries in one respect: he had little formal education and never attended a university. This was probably no disadvantage scientifically, since the scientific training then available would have provided little basis for his life's work; more serious handicaps, insofar as the communication of his discoveries went, were his lack of connections in the learned world and his ignorance of any language except Dutch. Nevertheless, through a fortunate chance, his work became widely known in his own lifetime, and its importance was immediately recognized. About the time that Leeuwenhoek began his observations, the Royal Society had been established in England for the communication and publication of scientific work. The Society invited Leeuwenhoek to communicate his observations to its members and a few years later (1680) elected him as a Fellow. For almost 50 years, until his death in 1723, Leeuwenhoek transmitted his discoveries to the Royal Society in the form of a long series of letters written in Dutch. These letters were largely translated and published in English in the *Proceedings of the Royal Society*, so becoming quickly and widely disseminated.

Leeuwenhoek's microscopes (Figure 1.2) bore little resemblance to the instruments with which we are familiar. The almost spherical lens (a) was mounted between two small metal plates. The specimen was placed on the point of a blunt pin (b) attached to the back plate and was brought into focus by manipulating two screws (c) and (d), which varied the position of the pin relative to the lens. During this operation the observer held the instrument with its other face very close to his eye and squinted through the lens. No change of magnification was possible, the magnifying power of each microscope being an intrinsic property of its lens. Despite the simplicity of their construction, Leeuwenhoek's microscopes were able to give clear images at magnifications which ranged, depending on the focal length of the lens, from about 50 to nearly 300 diameters. The highest magnification that he could obtain was consequently somewhat less than one-third of the highest magnification that is obtainable with a modern compound light microscope. Leeuwenhoek constructed hundreds of such instruments, a few of which survive today.

Leeuwenhoek's place in scientific history depends not so much on his skill as a microscope maker, essential though this was, as on the extraordinary range and skill of his microscopic observations. He was endowed with an unusual degree of curiosity and studied almost every conceivable object that could be looked at through a microscope. He made magnificent observations on the microscopic structure of the seeds and embryos of plants and on small invertebrate animals. He discovered the existence of spermatozoa and of red blood cells and was thus

FIGURE 1.1

Antony van Leeuwenhoek (1632–1723). In this portrait, he is holding one of his microscopes. Courtesy of the Rijksmuseum, Amsterdam.

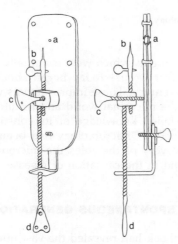

FIGURE 1.2

A drawing to show the construction of one of Leeuwenhoek's microscopes: (a) lens, (b) mounting pin, (c) and (d) focusing screws. After C. E. Dobell, *Antony van Leeuwenhoek and His Little Animals.* New York: Russell and Russell, Inc., 1932.

the founder of animal histology. By discovering and describing capillary circulation he completed the work on the circulation of blood begun by Harvey half a century before. Indeed, it would be easy to fill a page with a mere list of his major discoveries about the structure of higher plants and animals. His greatest claim to fame rests, however, on his discovery of the microbial world: the world of "animalcules," or little animals, as he and his contemporaries called them. A new dimension was thus added to biology. All the main kinds of unicellular microorganisms that we know today—protozoa, algae, yeasts, and bacteria—were first described by Leeuwenhoek, often with such accuracy that it is possible to identify individual species from his accounts of them. In addition to the diversity of this microbial world, Leeuwenhoek emphasized its incredible abundance. For example, in one letter describing for the first time the characteristic bacteria of the human mouth, he wrote:

I have had several gentlewomen in my house, who were keen on seeing the little eels in vinegar; but some of them were so disgusted at the spectacle, that they vowed they'd never use vinegar again. But what if one should tell such people in future that there are more animals living in the scum on the teeth in a man's mouth, than there are men in a whole kingdom?

Although Leeuwenhoek's contemporaries marveled at his scientific discoveries, the microscopic exploration of the microbial world which he had so brilliantly begun was not appreciably extended for over a century after his death. The principal reasons for this long delay seem to have been technical ones. Simple microscopes of high magnification are both difficult and tiring to use, and the manufacture of the very small lenses is an operation that requires great skill. Consequently, most of Leeuwenhoek's contemporaries and immediate successors used compound microscopes. Despite the intrinsic superiority of compound microscopes, the ones available in the seventeenth and eighteenth centuries suffered from serious optical defects, which made them less effective working instruments than Leeuwenhoek's simple microscopes. Thus, Leeuwenhoek's English contemporary, Robert Hooke, who was a very capable and careful observer, could not repeat with his own compound microscope many of the finer observations reported by Leeuwenhoek.

The major optical improvements which were eventually to lead to compound microscopes of the quality that we use today began about 1820 and extended through the succeeding half century. These improvements were closely followed by resumed exploration of the microbial world and resulted, by the end of the nineteenth century, in a detailed knowledge of its constituent groups. In the meantime, however, the science of microbiology had been developing in other ways, which led to the discovery of the roles the microorganisms play in the transformations of matter and in the causation of disease.

THE CONTROVERSY OVER SPONTANEOUS GENERATION

After Leeuwenhoek had revealed the vast numbers of microscopic creatures present in nature, scientists began to wonder about their origin. From the beginning there were two schools of thought. Some believed that the animalcules were formed spontaneously from nonliving materials, whereas others (Leeuwenhoek included) believed that they were formed from the "seeds" or "germs" of these animalcules, which were always present in the air. The belief in the spontaneous formation of living beings from nonliving matter is known as the doctrine of *spontaneous generation,* or *abiogenesis,* and has had a long existence. In ancient times it was considered self-evident that many plants and animals can be generated spontaneously under special conditions. The doctrine of spontaneous generation was accepted without question until the Renaissance.

As knowledge of living organisms accumulated, it gradually became evident that the spontaneous generation of plants and animals simply does not occur. A decisive step in the abandonment of the doctrine as applied to animals took place as the result of experiments performed about 1665 by an Italian physician, Francesco Redi. He showed that the maggots that develop in putrefying meat are the larval stages of flies and will never appear if the meat is protected by placing it in a vessel closed with fine gauze so that flies are unable to deposit their eggs on it. By such experiments, Redi destroyed the myth that maggots develop spontaneously from meat. Consequently, the doctrine of spontaneous generation was already being weakened by studies on the development of plants and animals at the time when Leeuwenhoek discovered the microbial world. For technical reasons, it is far more difficult to show that microorganisms are not generated spontaneously, and as time went on the proponents of the doctrine came to center their claims more and more on the mysterious appearance of these simplest forms of life in organic infusions. Those who did not believe in the spontaneous generation of microorganisms were in the position, always difficult, of having to prove a negative point; in fact, it was not until the middle of the nineteenth century that the cumulative negative evidence became sufficiently abundant to lead to the general abandonment of this doctrine.

One of the first to provide strong evidence that microorganisms do not arise spontaneously in organic infusions was the Italian naturalist Lazzaro Spallanzani, who conducted a long series of experiments on this problem in the middle of the eighteenth century. He could show repeatedly that heating can prevent the appearance of animalcules in infusions, although the duration of the heating necessary is variable. Spallanzani concluded that animalcules can be carried into

5 infusions by air and that this is the explanation for their supposed spontaneous generation in well-heated infusions. Earlier workers had closed their flasks with corks, but Spallanzani was not satisfied that any mechanical plug could completely exclude air, and he resorted to hermetic sealing. He observed that after sealed infusions had remained barren for a long time a tiny crack in the glass would be followed by the development of animalcules. His final conclusion was that to render an infusion *permanently* barren, it must be sealed hermetically and boiled. Animalcules could never appear unless new air somehow entered the flask and came in contact with the infusion.

Spallanzani's beautiful experiments showed clearly all the difficulties of work of this kind. However, faulty experiments continued to be performed, and the results continued to be brought forward as evidence for the occurrence of spontaneous generation. In the meantime, an interesting *practical* application of Spallanzani's discoveries had been made. His experiments had shown that even very perishable plant or animal infusions do not undergo putrefaction or fermentation when they have been rendered free of animalcules, from which it seemed probable that these chemical changes were in some way connected with the development of microbes. In the beginning of the nineteenth century François Appert found that one can preserve foods by enclosing them in airtight containers and heating the containers. He was able in this way to preserve highly perishable foodstuffs indefinitely, and "appertization," as this original canning process was called, came into extensive use for the preservation of foods long before the scientific issue had been finally settled.

In the late eighteenth century the work of Priestley, Cavendish, and Lavoisier laid the foundations of the chemistry of gases. One of the gases first discovered was oxygen, which soon was recognized to be essential for the life of animals. In the light of this knowledge, it seemed possible that the hermetic sealing recommended by Spallanzani and practiced by Appert was effective in preventing the appearance of microbes and the decomposition of organic matter, not because it excluded air carrying germs but because it excluded oxygen, required both for microbial growth and for the initiation of fermentation or putrefaction. Consequently, the influence of oxygen on these processes was a matter of much discussion in the early nineteenth century. It was finally shown that neither growth nor decomposition will occur in an infusion that has been properly heated, even when it is exposed to air, provided that the air entering the infusion has been previously treated so as to remove any germs that it contains.

• **The experiments of Pasteur** By 1860 some scientists had begun to realize that there is a *causal relationship* between the development of microorganisms in organic infusions and the chemical changes that take place in these infusions; *microorganisms are the agents that bring about the chemical changes.* The great pioneer in these studies was Louis Pasteur (Figure 1.3). However, the acceptance of this concept was conditional on the demonstration that spontaneous generation does not occur. Stung by the continued claims of adherents to the doctrine of spontaneous generation, Pasteur finally turned his attention to this problem. His work on the subject was published in 1861 as a *Memoir on the Organized Bodies Which Exist in the Atmosphere.*

Pasteur first demonstrated that air does contain microscopically observable "organized bodies." He aspirated large quantities of air through a tube that contained a plug of guncotton to serve as a filter. The guncotton was then removed and dissolved in a mixture of alcohol and ether, and the sediment was examined

FIGURE 1.3

Louis Pasteur (1822–1895). Courtesy of the Institut Pasteur, Paris.

microscopically. In addition to inorganic matter, it contained considerable numbers of small round or oval bodies, indistinguishable from microorganisms. Pasteur next confirmed the fact that heated air can be supplied to a boiled infusion without giving rise to microbial development. Having established this point, he went on to show that in a closed system the addition of a piece of germ-laden guncotton to a sterile infusion invariably provoked microbial growth. These experiments showed Pasteur how germs can enter infusions and led him to what was perhaps his most elegant experiment on the subject. This was the demonstration that infusions will remain sterile indefinitely in open flasks, provided that the neck of the flask is drawn out and bent down in such a way that the germs from the air cannot ascend it. Pasteur's swan-necked flasks are illustrated in Figure 1.4. If the neck of such a flask was broken off, the infusion rapidly became populated with microbes. The same thing happened if the sterile liquid in the flask was poured into the exposed portion of the bent neck and then poured back.

Pasteur rounded out his study by determining in semiquantitative fashion the distribution of microorganisms in the air and by showing that these living organisms are by no means evenly distributed through the atmosphere.

• **The experiments of Tyndall**

The last proponents of spontaneous generation maintained a stubborn rear-guard action for some years. The English physicist, John Tyndall, an ardent partisan of Pasteur, undertook a series of experiments designed to refute their claims; in the course of them, he established an important fact that had been overlooked by Pasteur, and in part accounted for the conflicting claims of the spontaneous generationists.

In a long series of experiments with infusions prepared from meat and fresh vegetables Tyndall obtained satisfactory sterilization by placing tubes of these infusions for 5 minutes in a bath of boiling brine. However, when he undertook similar experiments with infusions prepared from dried hay, this sterilization procedure proved completely inadequate. Worse still, when he then attempted to repeat his earlier experiments with other types of infusions, he found that they could no longer be sterilized by immersion in boiling brine, even for periods of as long as an hour. After many experiments, Tyndall finally realized what had happened. Dried hay contained spores of bacteria that were many times more

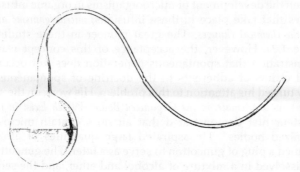

FIGURE 1.4

The swan-necked flask used by Pasteur during his studies on spontaneous generation. The construction of the neck permitted free access of air to the flask contents but prevented entry of microorganisms present in the air.

7 resistant to heat than any microbes with which he had previously dealt, and as a result of the presence of the hay in his laboratory, the air had become thoroughly infected with these spores. Once he had grasped this point, he proceeded to test the actual limits of heat resistance of the spores of hay bacteria and found that boiling infusions for even as long as $5\frac{1}{2}$ hours would not render them sterile with certainty. From these results he concluded that bacteria have phases, one relatively thermolabile (destroyed by boiling for 5 minutes) and one thermoresistant to an almost incredible extent. These conclusions were almost immediately confirmed by a German botanist, Ferdinand Cohn, who demonstrated that the hay bacteria can produce microscopically distinguishable resting bodies (*endospores*), which are highly resistant to heat.

Tyndall then proceeded to develop a method of sterilization by *discontinuous heating*, later called *tyndallization*, which could be used to kill *all* bacteria in infusions. Since growing bacteria are easily killed by brief boiling, all that is necessary is to allow the infusion to stand for a certain period before applying heat to permit germination of the spores with a consequent loss of their heat resistance. A very brief period of boiling can then be used, and repeated, if need be, several times at intervals to catch any spores late in germination. Tyndall found that discontinuous boiling for 1 minute on five successive occasions would make an infusion sterile, whereas a single continuous boiling for 1 hour would not. Recognition of the tremendous heat resistance of bacterial spores was essential to the development of adequate procedures for sterilization.

It has been stated that the work of Pasteur and Tyndall "disproved" the possibility of spontaneous generation, and their experimental findings have been used to support the contention that spontaneous generation has never occurred. This is an unjustifiable extension of their actual findings. The conclusion that we may safely draw is a much more limited one: that at the present time microorganisms do not arise spontaneously in properly sterilized organic infusions. It is probable that the primary origin of life on earth did involve a kind of spontaneous generation, although a far more gradual and subtle one than that envisaged by the proponents of the doctrine during the eighteenth and nineteenth centuries.

THE DISCOVERY OF THE ROLE OF MICROORGANISMS IN TRANSFORMATIONS OF ORGANIC MATTER

During the long controversy over spontaneous generation, a correlation between the growth of microorganisms in organic infusions and the onset of chemical changes in the infusion itself was frequently observed. These chemical changes were designated as "fermentation" and "putrefaction." Putrefaction, a process of decomposition that results in the formation of ill-smelling products, occurs characteristically in meat and is a consequence of the breakdown of proteins, the principal organic constituents in such natural materials. Fermentation, a process that results in the formation of alcohols or organic acids, occurs characteristically in plant materials as a consequence of the breakdown of carbohydrates, the predominant organic compounds in plant tissues.

• **Fermentation as a biological process** In 1837 three men, C. Cagniard-Latour, Th. Schwann, and F. Kützing, independently proposed that the yeast which appears during alcoholic fermentation is a microscopic plant and that the conversion of sugars to ethyl alcohol and carbon dioxide characteristic of the alcoholic fermentation is a physiological function of the yeast cell. This theory was bitterly attacked by such leading chemists of the time as J. J. Berzelius, J. Liebig, and F. Wohler, who held the view that fermentation and putrefaction are purely chemical processes. The science of chemistry had made great advances during the first decades of the nineteenth century and in 1828 the whole field of synthetic organic chemistry had been opened up by the first synthesis of an organic compound, urea, from inorganic materials. With the demonstration that organic compounds, until that time known exclusively as products of living activity, could be made in the laboratory, the chemists rightly felt that a large body of natural phenomena had now become amenable to analysis in physicochemical terms. The conversion of sugars to alcohol and carbon dioxide appeared to be a relatively simple chemical process. Accordingly, the chemists did not look with favor on the attempt to interpret this process as the result of the action of a living organism.

Ironically enough it was Pasteur, himself a chemist by training, who eventually convinced the scientific world that *all fermentative processes are the results of microbial activity*. Pasteur's work on fermentation extended with minor interruptions from 1857 to 1876. This work had a practical origin. The distillers of Lille, where the manufacture of alcohol from beet sugar was an important local industry, had encountered difficulties and called on Pasteur for assistance. Pasteur found that their troubles were caused by the fact that the alcoholic fermentation had been in part replaced by another kind of fermentative process, which resulted in the conversion of the sugar to lactic acid. When he examined microscopically the contents of fermentation vats in which lactic acid was being formed, he found that the cells of yeast characteristic of the alcoholic fermentation had been replaced by much smaller rods and spheres. If a trace of this material was placed in a sugar solution containing some chalk, a vigorous lactic fermentation ensued, and eventually a grayish deposit was formed, which again proved on microscopic examination to consist of the small spherical and rod-shaped organisms. Successive transfers of minute amounts of material to fresh flasks of the same medium always resulted in the production of a lactic fermentation and an increase in the amount of the formed bodies. Pasteur argued that the active agent, or new "yeast," was a microorganism that specifically converted sugar to lactic acid during its growth.*

Using similar methods, Pasteur studied a considerable number of fermentative processes during the following 20 years. He was able to show that fermentation is invariably accompanied by the development of microorganisms. Furthermore, he showed that each particular chemical type of fermentation, as defined by its principal organic end products (for example, the lactic, the alcoholic, and the butyric fermentations), is accompanied by the development of a *specific type of microorganism*. Many of these specific microbial types could be recognized and differentiated microscopically by their characteristic size and shape. In addition, they could be distinguished by the specific environmental conditions which favored their development. To cite one example of such physiological specificity, Pasteur observed very early that whereas the agent of alcoholic fermentation can

---

*The agents of the lactic acid fermentation are in fact bacteria, but in Pasteur's time, the different taxonomic groups of microorganisms had not yet been clearly distinguished.

flourish in an acid medium, the agents of the lactic fermentation grow best in a neutral medium. It was for this reason that he added chalk (calcium carbonate) to his medium for the cultivation of the lactic organisms; this substance serves as a neutralizing agent and prevents too strong an acidification of the medium that would otherwise occur as a result of the formation of lactic acid.

• The discovery of anaerobic life

During his studies on the butyric fermentation Pasteur discovered another fundamental biological phenomenon: the *existence of forms of life which can live only in the absence of free oxygen*. While examining microscopically fluids that were undergoing a butyric fermentation, Pasteur observed that the bacteria at the margin of a flattened drop, in close contact with the air, became immotile, whereas those in the center of the drop remained motile. This observation suggested that air had an inhibitory effect on the microorganisms in question, an inference which Pasteur quickly confirmed by showing that passage of a current of air through the fermenting fluid could retard, and sometimes completely arrest, the butyric fermentation. He thus concluded that some microorganisms can live only in the absence of oxygen, a gas previously considered essential for the maintenance of all life. He introduced the terms *aerobic* and *anaerobic* to designate, respectively, life in the presence and in the absence of oxygen.

• The physiological significance of fermentation

The discovery of the anaerobic nature of the butyric fermentation provided Pasteur with an important clue for the understanding of the role that fermentations play in the life of the microorganisms that bring them about. Free oxygen is essential for most organisms as an agent for the oxidation of organic compounds to carbon dioxide. Such oxygen-linked biological oxidations, known collectively as *aerobic respirations*, provide the energy that is required for maintenance and growth.

Pasteur was the first to realize that the breakdown of organic compounds in the absence of oxygen can also be used by some organisms as a means of obtaining energy; as he put it, *fermentation is life without air*. Some strictly anaerobic microorganisms, such as the butyric acid bacteria, are dependent on fermentative mechanisms to obtain energy. Many other microorganisms, including certain yeasts, are *facultative anaerobes*, which have two alternative energy-yielding mechanisms at their disposal. In the presence of oxygen they employ aerobic respiration, but they can employ fermentation if no free oxygen is present in their environment. This was beautifully demonstrated by Pasteur, who showed that sugar is converted to alcohol and carbon dioxide by yeast in the absence of air but that in the presence of air little or no alcohol is formed; carbon dioxide is the principal end product of this aerobic reaction.

The amount of growth that can occur at the expense of an organic compound is determined primarily by the amount of energy that can be obtained by the breakdown of that compound. Fermentation is a less efficient energy-yielding process than aerobic respiration, because part of the energy present in the substance decomposed is still present in the organic end products (for example, alcohol or lactic acid) characteristically formed by fermentative processes. As Pasteur was the first to show, the breakdown of a given weight of sugar results in substantially less growth of yeast under anaerobic conditions than under aerobic ones, thus establishing the relative inefficiency of fermentation as an energy source.

Pasteur's work showed that fermentations are "vital processes," which play a role of basic physiological importance in the life of many cells. The further development of knowledge about the nature of fermentation resulted from an accidental observation made in 1897 by H. Buchner. In attempting to preserve an extract of yeast, prepared by grinding yeast cells with sand, Buchner added a large quantity of sugar to it and was surprised to observe an evolution of carbon dioxide accompanied by the formation of alcohol. A soluble enzymatic preparation, able to carry out alcoholic fermentation, was thus discovered. Buchner's discovery inaugurated the development of modern biochemistry; the detailed analysis of the mechanism of cell-free alcoholic fermentation was eventually to show that this complex metabolic process can be interpreted as resulting from a succession of chemically intelligible reactions, each catalyzed by a specific enzyme. Today, the belief that even the most complex physiological process can be similarly understood in physicochemical terms is accepted as a matter of course by all biologists. In this sense, the intuition of the nineteenth-century chemists who battled against the biological theory of fermentation has proved to be a correct one.

THE DISCOVERY OF THE ROLE OF MICROORGANISMS IN THE CAUSATION OF DISEASE

During his studies on fermentation Pasteur, ever conscious of the practical applications of his scientific work, devoted considerable attention to the spoilage of beer and wine, which he showed to be caused by the growth of undesirable microorganisms. Pasteur used a peculiar and significant term to describe these microbially induced spoilage processes; he called them "diseases" of beer and wine. In fact, he was already considering the possibility that microorganisms may act as agents of infectious disease in higher organisms. Some evidence in support of this hypothesis already existed. It had been shown in 1813 that specific fungi can cause diseases of wheat and rye, and in 1845 M. J. Berkeley had proved that the great Potato Blight of Ireland, a natural disaster which deeply influenced Irish history, was caused by a fungus. The first recognition that fungi may be specifically associated with a disease of animals came in 1836 through the work of A. Bassi in Italy on a fungal disease of silkworms. A few years later J. L. Schönlein showed that certain skin diseases of man are caused by fungal infections. Despite these indications, very few medical scientists were willing to entertain the notion that the major infectious diseases of man could be caused by microorganisms, and fewer still believed that organisms as small and apparently simple as the bacteria could act as agents of disease.

• Surgical antisepsis The introduction of anesthesia about 1840 made possible a very rapid development of surgical methods. Speed was no longer a primary consideration, and the surgeon was able to undertake operations of a length and complexity that would have been unthinkable previously. However, with the elaboration of surgical technique, a problem that had always existed became more and more serious: *surgical sepsis*, or the infections that followed surgical intervention and often resulted in the death of the patient. Pasteur's studies on the problem of spontaneous generation had shown the presence of microorganisms in the air and at the same time indicated various ways in which their access to and development in organic infusions could be prevented. A young British surgeon, Joseph Lister, who was deeply impressed by Pasteur's work, reasoned that surgical sepsis might well result from microbial infection of the tissues exposed during operation. He decided to develop methods for preventing the access of microorganisms to surgical wounds. By the scrupulous sterilization of surgical instruments, by the use of disinfectant dressings, and by the conduct of surgery under a spray of disinfectant to prevent airborne infection, he succeeded in greatly reducing the incidence of surgical sepsis. Lister's procedures of antiseptic surgery, developed about 1864, were initially greeted with considerable scepticism but, as their striking success in the prevention of surgical sepsis was recognized, gradually became common practice. This work provided powerful *indirect* evidence for the germ theory of disease, even though it did not cast any light on the possible microbial causation of specific human diseases. Just as in the case of Appert's development of canning as a means of food preservation half a century before, so with Lister's introduction of surgical antisepsis, practice had run ahead of theory.

● **The bacterial etiology of anthrax**

The discovery that bacteria can act as specific agents of infectious disease in animals was made through the study of anthrax, a serious infection of domestic animals that is transmissible to man. In the terminal stages of a generalized anthrax infection, the rod-shaped bacteria responsible for the disease occur in enormous numbers in the bloodstream. These objects were first observed as early as 1850, and their presence in the blood of infected animals was reported by a series of investigators during the following 15 years. Particularly careful and detailed studies were carried out between 1863 and 1868 by C. J. Davaine, who showed that the rods are invariably present in diseased animals but are undetectable in healthy ones and that the disease can be transmitted to healthy animals by inoculation with blood containing these rod-shaped elements.

The conclusive demonstration of the bacterial causation, or *etiology*, of anthrax was provided in 1876 by Robert Koch (Figure 1.5), a German country doctor. He had no laboratory, and his experiments were conducted in his home, using very primitive improvised equipment and small experimental animals. He showed that mice could be infected with material from a diseased domestic animal. He transmitted the disease through a series of 20 mice by successive inoculation; at each transfer, the characteristic symptoms were observed. He then proceeded to cultivate the causative bacterium by introducing minute, heavily infected particles of spleen from a diseased animal into drops of sterile serum. Observing hour after hour the growth of the organisms in this culture medium, he saw the rods change into long filaments within which ovoid, refractile bodies eventually appeared. He showed that these bodies were spores, which had not been seen by previous workers (Figure 1.6). When spore-containing material was transferred to a fresh drop of sterile serum, the spores germinated and gave rise once more to typical rods. In this fashion, he transferred cultures of the bacterium eight successive times. The final culture of the series, injected into a healthy animal, again produced the characteristic disease, and from this animal the organisms could again be isolated in culture.

FIGURE 1.5

Robert Koch (1843–1910). Courtesy of VEB George Thieme, Leipzig.

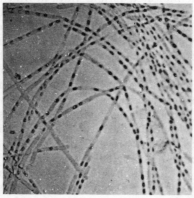
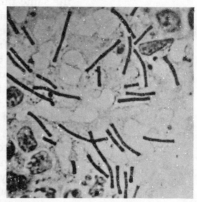

(a)   (b)   (c)

FIGURE 1.6

The first photomicrographs of bacteria, taken by Robert Koch in 1877. (a) Unstained chains of vegetative cells of *Bacillus anthracis*. (b) Unstained chains of *B. anthracis*; the cells contain refractile spores. (c) A stained smear of *B. anthracis* from the spleen of an infected animal. Note the rod-shaped bacilli and the larger tissue cells.

This series of experiments fulfilled the criteria which had been laid down 36 years before by J. Henle as logically necessary to establish the causal relationship between a specific microorganism and a specific disease. In generalized form, these criteria are (1) the microorganism must be present in every case of the disease; (2) the microorganism must be isolated from the diseased host and grown in pure culture; (3) the specific disease must be reproduced when a pure culture of the microorganism is inoculated into a healthy susceptible host; and (4) the microorganism must be recoverable once again from the experimentally infected host. Since Koch was the first to apply these criteria experimentally, they are now generally known as *Koch's postulates*.

Koch carried out another series of experiments which demonstrated the *biological specificity* of disease agents. He showed that another spore-forming bacterium, the hay bacillus, does not cause anthrax upon injection, and he also differentiated bacteria that cause other infections from the anthrax organism. From these studies he concluded that "only one kind of bacillus is able to cause this specific disease process, while other bacteria either do not produce disease following inoculation, or give rise to other kinds of disease."

In the meantime, Pasteur had found a collaborator, J. Joubert, with a knowledge of medical problems. Unaware of Koch's work, Pasteur and Joubert undertook the study of anthrax. They did not add anything new to the conclusions reached by Koch, but they confirmed his work and provided additional demonstrations that the bacillus, and not some other agent, was the specific cause of the disease.

• **The rise of medical bacteriology**

This work on anthrax abruptly ushered in the golden age of medical bacteriology, during which newly established institutes, created in Paris and in Berlin for Pasteur and Koch, respectively, became the world centers of bacteriological science. The German school, led by Koch, concentrated primarily on the isolation, cultivation, and characterization of the causative agents for the major infectious diseases of man. The French school, under the leadership of Pasteur, turned almost immediately to a more subtle and complex problem: the experimental analysis of how infectious disease takes place in the animal body and how recovery and immunity are brought about. Within 25 years most of the major bacterial agents of human disease had been discovered and described, and methods for the prevention of many of these diseases, either by artificial immunization or by the application of hygienic measures, had been developed. It was by far the greatest medical revolution in all human history.

• **The discovery of filterable viruses**

One of the early technical contributions from Pasteur's new institute was the development of filters able to retain bacterial cells and thus to yield bacteria-free filtrates. Infectious fluids were often tested for the presence of disease-producing bacteria by passing them through such filters; if the filtrate was no longer able to produce infection, the presence of a bacterial agent in the original fluid was indicated. In 1892 a Russian scientist, D. Iwanowsky, applied this test using an infectious extract from tobacco plants infected with mosaic disease. He found to his surprise that the filtrate was fully infectious when applied to healthy plants. His specific discovery was soon confirmed, and within a few years other workers found that many major plant and animal diseases are caused by similar, filter-passing, submicroscopic agents. A whole class of infectious entities, much smaller than any previously known organisms, was thus discovered. The true nature of these *viruses*, as they came to be known, remained obscure for many decades, but eventually it was established that they are a distinctive group of biological objects entirely different in structure and mode of development from all cellular organisms (see Chapter 12).

THE DEVELOPMENT OF PURE CULTURE METHODS

Pasteur possessed an intuitive skill in the handling of microorganisms and was able to reach correct conclusions about the specificity of fermentative processes, even when working with cultures that contained a mixture of microbial forms. The classical studies of Koch and Pasteur on anthrax, which firmly established the germ theory of animal disease, were conducted under experimental conditions which did not really permit certainty that rigorously pure cultures of the causative organism had been obtained. There are pitfalls in working with mixed microbial populations, and not all the scientists who began to study microorganisms in the middle of the nineteenth century were as skillful as Pasteur and Koch. It was frequently claimed that microorganisms had a large capacity for variation with respect both to their *morphological form* and to their *physiological function*. This belief became known as the doctrine of *pleomorphism*, while the opposing belief, that microorganisms show constancy and specificity of form and function, became known as the doctrine of *monomorphism*.

• **The origin of the belief in pleomorphism**

Let us consider what happens when a nutrient solution is inoculated with a mixed microbial population. The principle of natural selection at once begins to operate, and the microbe that can grow most rapidly under the conditions provided soon predominates. As a result of its growth and chemical activities, the composition of the medium changes; after some time, conditions no longer permit growth of the originally predominant form. The environment may now be favorable for the growth of a second kind of microorganism, also originally introduced into the medium but hitherto unable to develop, which gradually replaces the first as the predominant form in the culture. In this fashion one may obtain the *successive development of many different microbial types* in a single culture flask seeded with a mixed population. It is often possible to maintain the predominance of the form that first develops by repeated transfer of the mixed population at short intervals into a fresh medium of the same composition; this was essentially the device used by Pasteur in his studies on fermentation.

If one does not recognize the possibility of such microbial successions, it is easy to conclude that the chemical and morphological changes observable over the course of time in a single culture inoculated with a mixed population reflect *transformations undergone by a single kind of microorganism.* Between 1865 and 1885, claims for the extreme variability of microorganisms, based on such observations, were frequently made.

The term *pleomorphism* (derived from the Greek, meaning "doctrine of many shapes") implies that its proponents were concerned primarily with the possibilities of morphological variation. In fact, this was often not the case. Many pleomorphists insisted equally on the variability of function. For them, there was no such thing as a specific microbial agent for alcoholic fermentation or for a particular disease; they considered that it is the nature of the environment that determines both form and function. The widespread persistence of such beliefs represented a threat to the development of microbiology, and they were opposed by such leaders of the new discipline as Pasteur, Koch, and Cohn, who upheld the doctrine of monomorphism, insisting on the constancy of microbial form (and function).

Around 1870 it began to be realized that a sound understanding of the form and function of microorganisms could be obtained only if the complications inherent in the study of mixed microbial populations were avoided by the use of pure cultures. *A pure culture is one that contains only a single kind of microorganism.* The leading advocates of the use of pure cultures were two great mycologists (students of fungi), A. de Bary and O. Brefeld.

• The first pure cultures   Much of the pioneering work on pure culture techniques was done by Brefeld, working with fungi. He introduced the practice of isolating single cells, as well as the cultivation of fungi on solid media, for which purpose he added gelatin to his culture liquids. His methods of obtaining pure cultures worked admirably for the fungi but were found to be unsuitable when applied to the smaller bacteria. Other methods had, therefore, to be devised for bacteria. One of the first to be proposed was the *dilution method.* A fluid containing a mixture of bacteria was diluted with sterile medium in the hope that ultimately a growth could be obtained that took its origin from a single cell. In practice, the method is tedious, difficult, and uncertain; it also has the obvious disadvantage that at best one can only isolate in pure form the particular microorganism that predominates in the original mixture.

Koch realized very early that the development of simple methods for obtaining pure cultures of bacteria was a vital requirement for the growth of the new science.

15   The dilution method was obviously too tedious and uncertain for routine use. A more promising approach had already been suggested by the earlier observations of J. Schroeter, who had noted that on such solid substrates as potato, starch paste, bread, and egg albumen, isolated bacterial growths, or *colonies*, arose. The colonies differed from one another, but within each colony the bacteria were of one type. At first Koch experimented with the use of sterile, cut surfaces of potatoes, which he placed in sterile, covered glass vessels and then inoculated with bacteria. However, potatoes have obvious disadvantages: the cut surface is moist, which allows motile bacteria to spread freely over it; the substrate is opaque, and hence it is often difficult to see the colonies; and most important of all, the potato is not a good nutrient medium for many bacteria. Koch perceived that it would be far better if one could solidify a well-tried liquid medium with some clear substance. In this fashion, a translucent gel could be prepared on which developing bacterial colonies would be clearly visible. At the same time, the differing nutritional requirements of different bacteria could be met by modifying the composition of the liquid base. With this in mind, he added gelatin as a hardening agent. Once set, the gelatin surface was seeded by picking up a minute quantity of bacterial cells (the *inoculum*) on a platinum needle, previously sterilized by passage through a flame, and drawing it several times rapidly and lightly across the surface of the jelly. Different bacterial colonies soon appeared, each of which could be purified by a repetition of the streaking process. This became known as the *streak method* for isolating bacteria. The pure cultures were transferred to tubes containing sterile nutrient gelatin that had been plugged with cotton wool and set in a slanted position. Such cultures became known as *slant cultures*. Shortly thereafter, Koch discovered that instead of streaking the bacteria over the surface of the already solidified gelatin, he could mix them with the melted gelatin. When the gelatin set, the bacteria were immobilized in the jelly and there developed into isolated colonies. This became known as the *pour plate method* for isolating bacteria.

Gelatin, the first solidifying agent used by Koch, has several disadvantages. It is a protein highly susceptible to microbial digestion and liquefaction. Furthermore, it changes from a gel to a liquid at temperatures above 28°C. A new solidifying agent, *agar*, was soon introduced. Agar is a complex polysaccharide, extracted from red algae. A temperature of 100°C is required to melt an agar gel, so it remains solid throughout the entire temperature range over which bacteria are cultivated. However, once melted, it remains a liquid until the temperature falls to about 44°C, a fact that makes possible its use for the preparation of cultures by the pour plate method. It produces a stiff and transparent gel. Finally, it is a complex carbohydrate that is attacked by relatively few bacteria, so the problem of its liquefaction rarely arises. For these reasons, agar rapidly replaced gelatin as the hardening agent of choice for bacteriological work. All modern attempts to find an equally satisfactory synthetic substitute for agar have failed.

• **The development of culture media by Koch and his school**   Pasteur had used simple, transparent liquid media of known chemical composition for the selective cultivation of fermentative microorganisms. For the isolation of the microbial agents of disease, different types of culture media were required, and this was the second major technical problem to which Koch and his collaborators devoted their attention. Disease-producing bacteria develop normally within the tissues of an infected host, so it seemed logical that their cultivation outside the animal body would succeed best if the medium resembled as much as possible the environment of the host tissues. This line of reasoning led Koch to adopt *meat infusions* and *meat extracts* as the basic ingredients in his culture media. *Nutrient broth* and its solid counterpart, *nutrient agar,* which are still the most widely used media in general bacteriological work, were the outcome of Koch's experiments along these lines. Nutrient broth contains 0.5 percent peptone, an enzymatic digest of meat; 0.3 percent meat extract, a concentrate of the water-soluble components of meat; and 0.8 percent NaCl, to provide roughly the same total salt concentration as that found in tissues. For the cultivation of more fastidious disease-producing organisms, this basal medium can be supplemented in various ways (e.g., with sugar, blood, or serum). Considering the specific purposes for which these media were designed, the choice of ingredients may be considered logical, although there is no evidence that the traditional inclusion of NaCl has any real value, for most bacteria are insensitive to changes in the salt concentration of their environment over a very wide range. As time went on, however, many bacteriologists came to consider that these media were universal ones, suitable for the cultivation of nearly all bacteria. This is untrue; bacteria vary greatly in their nutritional requirements, and no single medium is capable of supporting the growth of more than a very small fraction of the bacteria that exist in nature (see Chapter 2).

MICROORGANISMS AS GEOCHEMICAL AGENTS

Although the role played by microorganisms as agents of infectious disease was the central microbiological interest in the last decades of the nineteenth century, some scientists carried forward the work initiated by Pasteur through his early investigations on the role of microorganisms in fermentation. This work had clearly shown that microorganisms can serve as specific agents for large-scale chemical transformations and indicated that the microbial world as a whole might well be responsible for a wide variety of other geochemical changes.

The establishment of the cardinal roles that microorganisms play in the biologically important cycles of matter on earth—the cycles of carbon, nitrogen, and sulfur—was largely the work of two men, S. Winogradsky (Figure 1.7) and M. W. Beijerinck (Figure 1.8). In contrast to plants and animals, microorganisms show an extraordinarily wide range of physiological diversity. Many groups are specialized for carrying out chemical transformations that cannot be performed at all by plants and animals, and thus play vital parts in the turnover of matter on earth.

One example of microbial physiological specialization is provided by the autotrophic bacteria, discovered by Winogradsky. These bacteria can grow in completely inorganic environments, obtaining the energy necessary for their growth by the oxidation of reduced inorganic compounds, and use carbon dioxide as the source of their cellular carbon. Winogradsky found that there are several physiologically distinct groups among the autotrophic bacteria, each characterized by the ability to use a particular inorganic energy source; for example, the sulfur bacteria oxidize inorganic sulfur compounds, the nitrifying bacteria, inorganic nitrogen compounds.

FIGURE 1.7

Sergius Winogradsky (1856–1953). Courtesy of Masson et Cie., Paris. Reprinted with the permission of the *Annales de l'Institut Pasteur.*

19   which play indispensable roles in the metabolism of the cell. These discoveries, spanning the period from 1920 to 1935, demonstrated the fundamental similarities of all living systems at the metabolic level: a doctrine proclaimed by biochemists and microbiologists under the slogan "the unity of biochemistry."

The second great advance of biology in the early twentieth century—the creation of the discipline of genetics, formed through the convergence of cytology and Mendelian analysis—had no immediate impact on microbiology. Indeed, it long seemed doubtful whether the mechanisms of inheritance operative in plants and animals likewise functioned in bacteria. The first important contact between genetics and microbiology occurred in 1941, when Beadle and Tatum succeeded in isolating a series of biochemical mutants from the fungus *Neurospora*. This opened the way to the analysis of the consequences of mutation in biochemical terms, and *Neurospora* joined the fruit fly and the maize plant as a material of choice for genetic research.

In 1943 an analysis by Delbrück and Luria of mutation in bacteria provided the technical and conceptual basis for genetic work on these microorganisms. Soon afterward several mechanisms of genetic transfer were shown to exist in bacteria, all significantly different from the mechanism of sexual recombination in plants and animals. In 1944 the work of Avery, McLeod, and McCarty on the process of bacterial genetic transfer known as *transformation* revealed that it is mediated by free deoxyribonucleic acid (DNA). The chemical nature of the hereditary material was thus discovered.

The confluence of microbiology, genetics, and biochemistry between 1940 and 1945 brought to an end the long isolation of microbiology from the main currents of biological thought. It also set the stage for the second major revolution in biology, to which microbiologists made many contributions of fundamental importance: the advent of molecular biology.

FURTHER READING

Books   BROCK, T. D. (editor and translator), *Milestones in Microbiology*. Englewood Cliffs, N.J.: Prentice-Hall, 1961.

BULLOCH, W., *The History of Bacteriology*. New York: Oxford University Press, 1960.

DOBELL, C., *Antony van Leeuwenhoek and His "Little Animals."* London: Staples Press, 1932.

DUBOS, R., *Louis Pasteur, Free Lance of Science*. Boston: Little, Brown, 1950.

JACOB, F., *The Logic of Living Systems*. London: Allen Lane, 1974.

LARGE, E. C., *Advance of the Fungi*. London: Jonathan Cape, 1940.

STENT, G. (editor), *Phage and the Origins of Molecular Biology*. Cold Spring Harbor, N.Y.: Laboratory of Quantitative Biology, 1966.

WATSON, J. D., *The Double Helix*. New York: Atheneum, 1968.

Review   VAN NIEL, C. B., "Natural Selection in the Microbial World," *J. Gen. Microbial.* 13, 201 (1955).

THE METHODS OF MICROBIOLOGY

2

As a result of the small size of microorganisms, the amount of information that can be obtained about their properties from the examination of *individuals* is limited; for the most part, the microbiologist studies *populations,* containing millions or billions of individuals. Such populations are obtained by growing microorganisms under more or less well-defined conditions, as *cultures.* A culture that contains only one kind of microorganism is known as a *pure* or *axenic culture.* A culture that contains more than one kind of microorganism is known as a *mixed culture;* if it contains only two kinds of microorganisms, deliberately maintained in association with one another, it is known as a *two-membered culture.*

At the heart of microbiology there accordingly lie two kinds of operations: *isolation,* the separation of a particular microorganism from the mixed populations that exist in nature; and *cultivation,* the growth of microbial populations in artificial environments (culture media) under laboratory conditions. These two operations come into play irrespective of the kind of microorganism with which the microbiologist deals; they are basic alike to the study of viruses, bacteria, fungi, algae, protozoa, and even small invertebrate animals. Furthermore, they have been extended in recent years to the study in isolation of cell or tissue lines derived from higher plants and animals (*tissue culture*). The unity of microbiology as a science, despite the biological diversity of the organisms with which it deals, is derived from this common operational base.

PURE CULTURE TECHNIQUE

Microorganisms are ubiquitous, so the preparation of a pure culture involves not only the isolation of a given microorganism from a mixed natural microbial population, but also the maintenance of the isolated individual and its progeny in an artificial environment to which the access of other microorganisms is prevented. Microorganisms do not require much space for development; hence an artificial environment can be created within the confines of a test tube, a flask, or a petri dish, the three kinds of containers most commonly used to cultivate microorganisms. The culture vessel must be rendered initially *sterile* (free of any living microorganism) and, after the introduction of the desired type of microorganism, it must be protected from subsequent external contamination. The primary source of external contamination is the atmosphere, which always contains floating microorganisms. The form of a petri dish, with its overlapping lid, is specifically designed to prevent atmospheric contamination. Contamination of tubes and flasks is prevented by closure of their orifices with an appropriate stopper. This is usually a plug of cotton wool, although metal caps or plastic screw caps are now often employed, particularly for test tubes.

The external surface of a culture vessel is, of course, subject to contamination, and the interior of a flask or tube can become contaminated when it is opened to introduce or withdraw material. This danger is minimized by passing the orifice through a flame, immediately after the stopper has been removed and again just before it is replaced.

The *inoculum* (i.e., the microbial material used to seed or inoculate a culture vessel) is commonly introduced on a metal wire or loop, which is rapidly sterilized just before its use by heating in a flame. Transfers of liquid cultures can also be made by pipette. For this purpose, the mouth end of the pipette is plugged with cotton wool, and the pipette is sterilized in a paper wrapping or in a glass or metal container, which keeps both inner and outer surfaces free of contamination until the time of use.

The risks of accidental contamination may be further reduced by performing transfers in a hood or in a small closed room, the air of which has been specially treated to reduce its microbial content.

• **The isolation of pure cultures by plating methods**

Pure cultures of microorganisms that form discrete colonies on solid media (e.g., most bacteria, yeasts, many fungi and unicellular algae) may be most simply obtained by one of the modifications of the plating method. This method involves the separation and immobilization of individual organisms on or in a nutrient medium solidified with agar or some other appropriate jelling agent. Each viable organism gives rise, through growth, to a colony from which transfers can be readily made.

The *streaked plate* is in general the most useful plating method. A sterilized bent wire is dipped into a suitable diluted suspension of organisms and is then used to make a series of parallel, nonoverlapping streaks on the surface of an already solidified agar plate. The inoculum is progressively diluted with each successive streak, so that even if the initial streaks yield confluent growth, well-isolated colonies develop along the lines of later streaks (Figure 2.1). Alternatively, isolations can be made with *poured plates:* successive dilutions of the inoculum are placed in sterile petri dishes and mixed with the molten but cooled agar medium, which is then allowed to solidify. Colonies subsequently develop embedded in the agar.

The isolation of strictly anaerobic bacteria by plating methods poses special problems. Provided that the organisms in question are not rapidly killed by exposure to oxygen, plates may be prepared in the usual manner and then incubated in closed containers, from which the oxygen is removed either by

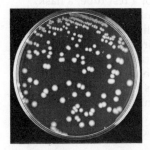

FIGURE 2.1

Isolation of a pure culture by the streak method. A petri dish containing nutrient agar was streaked with a suspension of bacterial cells. As a result of subsequent growth, each cell has given rise to a macroscopically visible colony.

chemical absorption or combustion. For the more oxygen-sensitive anaerobes, a modification of the pour plate method, known as the *dilution shake culture*, is to be preferred. A tube of melted and cooled agar is inoculated and mixed, and approximately one-tenth of its contents is transferred to a second tube, which is then mixed and used to inoculate a third tube in a similar fashion. After 6 to 10 successive dilutions have been prepared, the tubes are rapidly cooled and sealed, by pouring a layer of sterile petroleum jelly and paraffin on the surface, thus preventing access of air to the agar column. In shake cultures the colonies develop deep in the agar column (Figure 2.2), and are thus not easily accessible for transfer. To make a transfer, the petroleum jelly-paraffin seal is removed with a sterile needle, and the agar column is gently blown out of the tube into a sterile petri dish by passing oxygen-free gas through a capillary pipette inserted between the tube wall and the agar. The column is sectioned into discs with a sterile knife to permit examination and transfer of colonies.

In isolating from a mixed natural population it is often possible, provided one's technique is good, to prepare a first plate or dilution shake series in which many of the colonies that develop are well-separated from one another. Can one then pick material from such a colony, transfer it to an appropriate medium, and call it a pure culture? Although this is often done, a culture so isolated may be far from pure. Microorganisms vary greatly in their nutritional requirements, and consequently no single medium and set of growth conditions will permit the growth of all the microorganisms present in a natural population. Indeed, it is probable that only a very small fraction of the microorganisms initially present will be able to form colonies on any given medium. Hence, for every visible colony on a first plate, there may be a thousand cells of other kinds of microorganisms which were also deposited on the agar surface but have failed to give macro-

FIGURE 2.2

Isolation of a pure culture of anaerobic bacteria by the dilution shake method. A complete series of dilution shakes is shown. Note the confluent growth in the more densely seeded tubes (at right), and the well-isolated colonies in the two final tubes of the series (at left). After the agar had solidified, each tube was sealed with a mixture of sterile vaseline and paraffin to prevent the access of atmospheric oxygen, which inhibits the growth of anaerobic bacteria.

scopically visible growth, although they may still be viable. The probability is high that some of these organisms will be picked up and carried over when a transfer is made. *One should never pick from a first plate for the preparation of a pure culture.* Instead, a second plate should be streaked from a cell suspension prepared from a well-isolated colony. If all the colonies on this second plate appear identical, a well-isolated colony can be used to establish a pure culture.

Not all microorganisms able to grow on solid media necessarily give rise to well-isolated colonies. Certain motile flagellated bacteria (*Proteus, Pseudomonas,* for example) can rapidly spread over the slightly moist surface of a freshly poured plate. This can be prevented by the use of plates with well-dried surfaces, on which the cells are immobilized. Spirochetes and organisms that show gliding movement (myxobacteria, blue-green bacteria) can move over or through an agar gel, even when its surface is well dried. In such cases, the movement of the organisms in question is an aid to their purification, since they can move away from other kinds of microorganisms immobilized on the agar. Thus, purification can often be achieved by allowing migration to occur and transferring repeatedly to fresh plates from the advancing edge of the migrating population.

The incorporation into the medium of selectively inhibitory substances is also sometimes helpful in making isolations from nature. Because of their biological specificity, certain antibiotics are particularly useful in this respect. Bacteria vary greatly in their sensitivity to the antibiotic penicillin, which can consequently be used at low concentrations to prevent the development of sensitive bacteria in the initial population. At higher concentrations, penicillin is generally toxic for procaryotic organisms but not for eucaryotic ones. It is thus a very useful agent for the purification of protozoa, fungi, and eucaryotic algae that are contaminated by bacteria. Conversely, procaryotic organisms are insensitive to polyene antibiotics such as nystatin, which are generally toxic for eucaryotic organisms. The incorporation of this kind of antibiotic into the isolation medium can sometimes be used to advantage in the purification of bacteria heavily contaminated by fungi or amebae. Many other variations on the theme of selective toxicity can be used to facilitate isolations by the plating method.

• **The isolation of pure changes in liquid media**

Plating methods are in general satisfactory for the isolation of bacteria and fungi, because the great majority of the representatives of these groups can grow well on solid media. However, some of the larger-celled bacteria have not yet been successfully cultivated on solid media, and many protozoa and algae are also cultivable only in a liquid medium. Although plating methods for the isolation of viruses have been greatly extended in recent years, many of these organisms are most easily isolated by the use of liquid media. In the case of viruses, of course, a pure culture is never obtainable, since these organisms are obligate intracellular parasites; a two-membered culture, consisting of a specific virus and its biological host, represents the goal of purification for this microbial group.

The simplest procedure of isolation in liquid media is the *dilution method.* The inoculum is subjected to serial dilution in a sterile medium, and a large number of tubes of medium are inoculated with aliquots of each successive dilution. The goal of this operation is to inoculate a series of tubes with a microbial suspension so dilute that the probability of introducing even one individual into a given tube is very small: a probability of the order of 0.05. When a large number of tubes is seeded with an inoculum of this size, it can be calculated from probability theory that the fraction of tubes which receives no organisms is 0.95; the fraction receiving one organism is 0.048; the fraction receiving two organisms is 0.0012; the fraction receiving three organisms is 0.00002. As a result, if a tube shows *any* subsequent growth, there is a very high probability that this growth has resulted from the introduction of a *single* organism. The probability is

$$\frac{0.048}{0.048 + 0.0012 + 0.00002} = 0.975$$

The probability that growth has originated from a single organism declines very rapidly as the mean number of organisms in the inoculum increases. It is, therefore, essential to isolate from a series of tubes the great majority of which show *no* growth.

The dilution method has, however, one major disadvantage: it can be used only to isolate the *numerically predominant* member of a mixed microbial population. It can almost never be effectively used for the isolation of larger microorganisms that are incapable of developing on solid media (e.g., protozoa, algae), because in nature these microorganisms are as a rule greatly outnumbered by bacteria. Hence, the usefulness of the dilution method is limited.

When neither plating nor dilution methods can be applied, the only alternative is to resort to the *microscopically controlled isolation of a single cell or organism from the mixed population*, a technique known by the name of *single-cell isolation*. The technical difficulty of single-cell isolation is inversely related to the *size* of the organism which one wishes to isolate: it is relatively easy to use with large-celled microorganisms, such as algae and protozoa, but becomes much harder with bacteria.

In the case of large microorganisms, purification involves the capture of a single individual in a fine capillary pipette and the subsequent transfer of this individual through several washings in relatively large volumes of sterile medium to eliminate microbial contaminants of smaller size. The successive operations can be performed manually, with control by direct microscopic observation at a relatively low magnification, such as that provided by a dissecting microscope.

The technique of the capillary pipette can no longer be applied if the organism which one wishes to isolate is so small that it cannot be readily observed at a magnification of 100 times or less, because one cannot achieve the necessary fineness of control to manipulate a capillary pipette directly at higher magnifications. In this event, a mechanical device known as a *micromanipulator* must be used, in conjunction with specially prepared, very fine glass operating instruments. The essential purpose of a micromanipulator is to *gear down* manual control, so that very slight and precisely controlled movements of the operating instruments can be effected in a small operating area (a microdrop) under continuous microscopic observation at high magnifications (500 to 1,000 ×).

• **Two-membered cultures**   The goal of isolation is normally to obtain a pure culture. However, there are certain situations where this cannot be achieved or where achievement is so difficult as to be impractical. Under such circumstances, the alternative is to obtain the next best degree of purification, in the shape of a *two-membered culture*, which contains only two kinds of microorganisms. As already mentioned, a two-membered culture is in principle the only possible way to maintain viruses, since

25 these organisms are all obligate intracellular parasites of cellular organisms. Obligate intracellular parasitism is also characteristic of several groups of cellular microorganisms. In all these instances a two-membered culture represents the nearest approach to cultivation under controlled laboratory conditions that can be achieved.

Many of the protozoa, which feed in nature on smaller microorganisms, are also most easily maintained in the laboratory as two-membered cultures in association with their smaller microbial prey. This is true, for example, of ciliates, amebas, and slime molds. In such instances the association is probably never an *obligate* one, since careful nutritional studies on a few representatives of these groups have shown that they can be grown in pure culture; however, the nutritional requirements of protozoa are often extremely complex, so that the preparation of media for the maintenance of pure cultures is both difficult and laborious. For purposes of routine maintenance, and also for many experimental purposes, two-membered cultures are satisfactory.

The establishment of a two-membered culture is an operation that is conducted in two phases. First, it is necessary to establish a pure culture of the food organism (the host in the cases of obligate intracellular parasites, the prey in the case of protozoa). Once this has been achieved, the parasite or predator can be isolated by any one of a variety of methods (plating on solid media in the presence of the food organism, dilution in a liquid medium, single-cell isolation) and introduced into the pure culture of the food organism.

The successful maintenance of two-membered cultures requires considerable art, because a reasonably stable biological balance between the two components is essential. The medium must be one that permits sufficient growth of the food organism to meet the needs of the parasite or predator but should not be so rich that the food organism can outgrow its associate or produce metabolic products that are deleterious to it.

THE THEORY AND PRACTICE OF STERILIZATION

Sterilization is a treatment that *frees the treated object of all living organisms*. It can be achieved by exposure to lethal physical or chemical agents or, in the special case of solutions, by filtration.

To understand the basis of sterilization by lethal agents, it is necessary to describe briefly the kinetics of death in a microbial population. The only valid criterion of death in the case of a microorganism is *irreversible loss of the ability to reproduce*; this is usually determined by quantitative plating methods, survivors being detected by colony formation. When a pure microbial population is exposed to a lethal agent, the kinetics of death are nearly always *exponential*: the number of survivors decreases geometrically with time. This reflects the fact that all the members of the population are of similar sensitivity; probability alone determines the actual time of death of any given individual. If the logarithm of the number of survivors is plotted as a function of the time of exposure, a straight line is obtained (Figure 2.3); its negative slope defines the *death rate*.

The death rate tells one only what *fraction* of the initial population survives

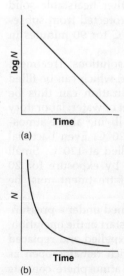

(a)

(b)

FIGURE 2.3

Exponential (logarithmic) order of death of bacteria. The same data are plotted semi-logarithmically in (a) and arithmetically in (b); *N* is the number of surviving bacteria.

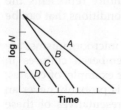

FIGURE 2.4

Relationship of death rate and population size to the time required for the destruction of bacterial cultures; N is the number of surviving bacteria. Cultures B, C, and D have identical death rates. Culture A has a lower death rate.

a given period of treatment. To determine the *actual number* of survivors, one must also know the *initial population size*, as illustrated graphically in Figure 2.4. Accordingly, for the establishment of procedures of sterilization, two factors have to be taken into account: the death rate and the initial population size.

In the practice of sterilization the microbial population to be destroyed is almost always a *mixed* one. Since microorganisms differ widely in their resistance to lethal agents, the significant factors become the initial population size and the death rate of the *most resistant* members of the mixed population. These are almost always the highly resistant endospores of certain bacteria. Consequently, spore suspensions of known resistance are the objects commonly used to assess the reliability of sterilization methods.

Taking into account the kinetics of microbial death, we can formulate the practical goal of sterilization by a lethal agent in a slightly more refined way: *the probability that the object treated contains even one survivor should be infinitesimally small*. For example, if we wish to sterilize a liter of a culture medium, this goal will be achieved for all practical purposes if the treatment is one which will leave no more than one survivor in 10^6 liters; under such circumstances, the probability of failure is very small indeed. Procedures of routine sterilization are always designed to provide a very wide margin of safety.

• Sterilization by heat

Heat is the most widely used lethal agent for purposes of sterilization. Objects may be sterilized by dry heat, applied in an oven in an atmosphere of air, or by moist heat, provided by wet steam. Of the two methods, sterilization by dry heat requires a much greater duration and intensity. Heat conduction is less rapid in air than in steam. In addition, bacteria can survive in a completely desiccated state and, in this state, the intrinsic heat resistance of vegetative bacterial cells is greatly increased, almost to the level characteristic of spores. Consequently, the death rate is much lower for dry cells than for moist ones.

Dry heat is used principally to sterilize glassware or other heatstable solid materials. The objects are wrapped in paper or otherwise protected from subsequent contamination and exposed to a temperature of 170°C for 90 minutes in an oven.

Steam must be used for the heat sterilization of aqueous solutions. Treatment is usually carried out in a metal vessel known as an *autoclave,* which can be filled with steam at a pressure greater than atmospheric. Sterilization can thus be achieved at temperatures considerably above the boiling point of water; laboratory autoclaves are commonly operated at a steam pressure of 15 lb/in.^{2} above atmospheric pressure, which corresponds to a temperature of 120°C. Even bacterial spores that survive several hours of boiling are rapidly killed at 120°C. Small volumes of liquid (up to about 3 liters) can be sterilized by exposure for 20 minutes; if larger volumes are to be sterilized, the time of treatment must be extended.

A temperature of 120°C within the autoclave will be attained under a pressure of 15 lb/in.^{2} *only if the atmosphere consists entirely of steam.* At the start of the operation, accordingly, all the air originally in the chamber must be expelled and replaced by steam; this is achieved by the use of a steam trap, which remains open as long as air is being passed through it but closes when the atmosphere consists of steam. If some air remains in the sterilization chamber, the partial pressure of steam will be lower than that indicated on the pressure gauge, and the temperature will be correspondingly lower. For this reason an autoclave should always be equipped with both a temperature and a pressure gauge. The temperature within the sterilization chamber can be monitored by including, among the objects to be sterilized, special indicator papers, which change color if the heat treatment has been adequate.

• Sterilization by chemical treatment

Many of the substances used in preparing culture media are too heat labile to be sterilized by autoclaving. For such substances, a reliable method of chemical sterilization would be extremely useful. The essential requirement for a chemical sterilizing agent is that it should be *volatile* as well as *toxic*, so that it can be readily eliminated from the object sterilized after treatment. The best available candidate is *ethylene oxide*, a liquid that boils at 10.7°C. It can be added to solutions in liquid form (final concentration of approximately 0.5 to 1.0 percent) at a temperature of 0 to 4°C, or used as a sterilizing gas at temperatures above the boiling point. However, it is chemically unstable, decomposing in aqueous solution to ethylene glycol, which is nonvolatile and may have undesirable effects. Furthermore, ethylene oxide is both explosive and toxic for humans, so special precautions must be taken in its handling. For these reasons, ethylene oxide sterilization has not become a routine laboratory procedure. It is, however, used industrially for the sterilization of plastic petri dishes and other plastic objects which melt at temperatures greater than 100°C.

• Sterilization by filtration

The principal laboratory method used to sterilize solutions of heat-labile materials is *filtration* through filters capable of retaining microorganisms. The action of such filters is almost always complex. Microorganisms are retained in part by the small size of the filter pores and in part by adsorption on the pore walls during their passage through the filter. The importance of adsorption is indicated by the fact that a filter may effectively retain microorganisms even when the average diameter of its pores is somewhat greater than the mean size of the cells that are retained. Sterilization by filtration is subject to one major theoretical limitation. Since the viruses range down in size to the dimensions of large protein molecules, they are not necessarily retained by filters that can hold back even the smallest of cellular microorganisms. Consequently, it is never possible to be certain that filtration procedures which render a solution bacterium-free will also free it of viruses.

THE PRINCIPLES OF MICROBIAL NUTRITION

To grow, organisms must draw from the environment all the substances which they require for the synthesis of their cell materials and for the generation of energy. These substances are termed *nutrients*. A culture medium must therefore contain, in quantities appropriate to the specific requirements of the microorganism for which it is designed, all necessary nutrients. However, microorganisms are extraordinarily diverse in their specific physiological properties, and correspondingly in their specific nutrient requirements. Literally thousands of different media have been proposed for their cultivation, and in the

TABLE 2.1

Approximate elementary
composition of the
microbial cell[a]

ELEMENT	PERCENTAGE OF DRY WEIGHT
Carbon	50
Oxygen	20
Nitrogen	14
Hydrogen	8
Phosphorus	3
Sulfur	1
Potassium	1
Sodium	1
Calcium	0.5
Magnesium	0.5
Chlorine	0.5
Iron	0.2
All others	~0.3

[a] Data for a bacterium,
Escherichia coli, assembled by
S. E. Luria, in *The Bacteria*
(I. C. Gunsalus and R. Y.
Stanier, eds.), Vol. I, Chap. 1
(New York: Academic Press,
1960).

descriptions of these media the reasons for the presence of the various components are often not clearly stated. Nevertheless, the design of a culture medium can and should be based on scientific principles, the *principles of nutrition*, which we shall outline as a preliminary to the description of culture media.

The chemical composition of cells, broadly constant throughout the living world, indicates the major material requirements for growth. *Water* accounts for some 80 to 90 percent of the total weight of cells and is always therefore the major essential nutrient, in quantitative terms. The solid matter of cells (Table 2.1) contains, in addition to *hydrogen* and *oxygen* (derivable metabolically from water), *carbon, nitrogen, phosphorus,* and *sulfur*, in order of decreasing abundance. These six elements account for about 95 percent of the cellular dry weight. Many other elements are included in the residual fraction. Nutritional studies show that *potassium, magnesium, calcium, iron, manganese, cobalt, copper, molybdenum,* and *zinc* are required by nearly all organisms. The known functions in the cell of these 15 elements are summarized in Table 2.2.

All the required metallic elements can be supplied as nutrients in the form of the *cations of inorganic salts*. Potassium, magnesium, calcium, and iron are required in relatively large amounts and should always be included as salts in culture media. The quantitative requirements for manganese, cobalt, copper, molybdenum, and zinc are very small—so small, in fact, that it is often technically difficult to demonstrate their essentiality, since they are present in adequate amounts as contaminants of the major inorganic constituents of media. They are often referred to as *trace elements* or *micronutrients*. One nonmetallic element, phosphorus, can also be used as a nutrient when provided in inorganic form, as phosphate salts.

It should be noted that some biological groups have additional, specific mineral requirements; for example, diatoms and certain other algae synthesize cell walls that are heavily impregnated with silica and consequently have a specific *silicon* requirement, supplied as silicate. Although a requirement for *sodium* cannot be demonstrated for most microorganisms, it is required at relatively high concentrations by certain marine bacteria, by blue-green bacteria, and by photosynthetic bacteria. In these groups it cannot be replaced by other monovalent cations.

The needs for *carbon, nitrogen, sulfur,* and *oxygen* cannot be so simply described, because organisms differ with respect to the *specific chemical form* under which these elements must be provided as nutrients.

**• The
requirements
for carbon**

Organisms that perform photosynthesis and bacteria that obtain energy from the oxidation of inorganic compounds typically use the most oxidized form of carbon, CO_2, as the sole or principal source of cellular carbon. The conversion of CO_2 to organic cell constituents is a reductive process, which requires a net input of energy. In these physiological groups, accordingly, a considerable part of the energy derived from light or from the oxidation of reduced inorganic compounds must be expended for the reduction of CO_2 to the level of organic matter.

All other organisms obtain carbon principally from organic nutrients. Since most organic substrates are at the same general oxidation level as organic cell constituents, they usually do not have to undergo a primary reduction to serve as sources of cell carbon. In addition to meeting the biosynthetic needs of the

TABLE 2.2

General physiological functions of the principal elements

ELEMENT	PHYSIOLOGICAL FUNCTIONS
Hydrogen	Constituent of cellular water, organic cell materials
Oxygen	Constituent of cellular water, organic cell materials; as O_2, electron acceptor in respiration of aerobes
Carbon	Constitutent of organic cell materials
Nitrogen	Constituent of proteins, nucleic acids, coenzymes
Sulfur	Constituent of proteins (as amino acids cysteine and methionine); of some coenzymes (e.g., CoA, cocarboxylase)
Phosphorus	Constituent of nucleic acids, phospholipids, coenzymes
Potassium	One of the principal inorganic cations in cells, cofactor for some enzymes
Magnesium	Important cellular cation; inorganic cofactor for very many enzymatic reactions, including those involving ATP; functions in binding enzymes to substrates; constituent of chlorophylls
Manganese	Inorganic cofactor for some enzymes, sometimes replacing Mg
Calcium	Important cellular cation; cofactor for some enzymes (for example, proteinases)
Iron	Constituent of cytochromes and other heme or nonheme proteins; cofactor for a number of enzymes
Cobalt	Constituent of vitamin B_{12} and its coenzyme derivatives
Copper, zinc, molybdenum	Inorganic constituents of special enzymes

cell for carbon, organic substrates must supply the energetic requirements of the cell. Consequently, much of the carbon present in the organic substrate enters the pathways of energy-yielding metabolism and is eventually excreted again from the cell, as CO_2 (the major product of energy-yielding respiratory metabolism) or as a mixture of CO_2 and organic compounds (the typical end products of fermentative metabolism). Organic substrates thus usually have a *dual nutritional role:* they serve at the same time as a source of carbon and as a source of energy. Many microorganisms can use *a single organic compound* to supply completely both these nutritional needs. Others, however, cannot grow when provided with only one organic compound, and they need a variable number of additional organic compounds as nutrients. These additional organic nutrients have a purely biosynthetic function, being required as precursors of certain organic cell constituents that the organism is unable to synthesize. They are termed *growth factors,* and their roles are described in greater detail below.

Microorganisms are extraordinarily diverse with respect to both the *kind* and the *number* of organic compounds that they can use as a principal source of carbon and energy. This diversity is shown by the fact that *there is no naturally produced organic compound that cannot be used as a source of carbon and energy by some microorganism.* Hence, it is impossible to describe concisely the chemical nature of organic carbon sources for microorganisms. This extraordinary variation with respect to carbon requirements is one of the most fascinating physiological aspects of microbiology.

When the organic carbon requirements of *individual* microorganisms are examined, some show a high degree of versatility, whereas others are extremely specialized. Certain bacteria of the *Pseudomonas* group, for example, can use any one of over 90 different organic compounds as sole carbon and energy source

(Table 2.3). At the other end of the spectrum are methane-oxidizing bacteria, which can use only two organic substrates, methane and methanol, and certain cellulose-decomposing bacteria, which can use only cellulose.

Most (and probably all) organisms that depend on organic carbon sources also require CO_2 as a nutrient in very small amounts, because this compound is utilized in a few biosynthetic reactions. However, as CO_2 is normally produced in large quantities by organisms that use organic compounds, the biosynthetic requirement can be met through the metabolism of the organic carbon and energy source. Nevertheless, the complete removal of CO_2 often either delays or prevents the growth of microorganisms in organic media, and a few bacteria and fungi require

TABLE 2.3

Organic compounds capable of serving as the sole source of carbon and energy
for the bacterium *Pseudomonas cepacia*

1. *Carbohydrates and carbohydrate derivatives (sugar acids and polyalcohols):*	Caprylate	Aspartate
	Pelargonate	Glutamate
	Caprate	Lysine
Ribose		Arginine
Xylose	3. *Dicarboxylic acids:*	Histidine
Arabinose	Malonate	Proline
Fucose	Succinate	Tyrosine
Rhamnose	Fumarate	Phenylalanine
Glucose	Glutarate	Tryptophan
Mannose	Adipate	Kynurenine
Galactose	Pimelate	Kynurenate
Fructose	Suberate	
Sucrose	Azelate	7. *Other nitrogenous compounds:*
Trehalose	Sebacate	Anthranilate
Cellobiose		Benzylamine
Salicin	4. *Other organic acids:*	Putrescine
Gluconate	Citrate	Spermine
2-Ketogluconate	α-Ketoglutarate	Tryptamine
Saccharate	Pyruvate	Butylamine
Mucate	Aconitate	Amylamine
Mannitol	Citraconate	Betaine
Sorbitol	Levulinate	Sarcosine
Inositol	Glycolate	Hippurate
Adonitol	Malate	Acetamide
Glycerol	Tartrate	Nicotinate
Butylene glycol	Hydroxybutyrate	Trigonelline
	Lactate	
2. *Fatty acids:*	Glycerate	8. *Nitrogen-free ring compounds:*
Acetate	Hydroxymethylglutarate	Benzoylformate
Propionate		Benzoate
Butyrate	5. *Primary alcohols:*	o-Hydroxybenzoate
Isobutyrate	Ethanol	m-Hydroxybenzoate
Valerate	Propanol	p-Hydroxybenzoate
Isovalerate	Butanol	Phenylacetate
Caproate	6. *Amino acids:*	Phenol
Heptanoate	Alanine	Quinate
	Serine	Testosterone
	Threonine	

a relatively high concentration of CO_2 in the atmosphere (5 to 10 percent) for satisfactory growth in organic media.

• The requirements for nitrogen and sulfur

Nitrogen and sulfur occur in the organic compounds of the cell principally in reduced form as amino and sulfhydryl groups, respectively. Most photosynthetic organisms assimilate these two elements in the oxidized inorganic state, as nitrates and sulfates; their biosynthetic utilization thus involves a preliminary reduction. Many nonphotosynthetic bacteria and fungi can also meet the needs for nitrogen and sulfur from nitrates and sulfates. Some microorganisms are unable to bring about a reduction of one or both of these anions and must be supplied with the elements in a *reduced form*. The requirement for a reduced nitrogen source is relatively common and can be met by the provision of nitrogen as ammonium salts. A requirement for reduced sulfur is rarer; it can be met by the provision of sulfide or of an organic compound that contains a sulfhydryl group (e.g., cysteine).

The nitrogen and sulfur requirements can often also be met by organic nutrients that contain these two elements in reduced organic combination (amino acids or more complex protein degradation products, such as peptones). Such compounds may also, of course, provide organic carbon and energy sources, meeting simultaneously the cellular requirements for carbon, nitrogen, sulfur, and energy.

Some bacteria can also utilize the most abundant natural nitrogen source, N_2. This process of nitrogen assimilation is termed *nitrogen fixation* and involves a preliminary reduction of N_2 to ammonia.

• Growth factors

Any organic compound that an organism requires as a precursor or constituent of its organic cell material, but which it cannot synthesize from simpler carbon sources, must be provided as a nutrient. Organic nutrients of this type are known collectively as *growth factors*. They fall into three classes, in terms of chemical structure and metabolic function:

1. *Amino acids,* required as constituents of proteins.
2. *Purines* and *pyrimidines,* required as constituents of nucleic acids.
3. *Vitamins,* a diverse collection of organic compounds that form parts of the prosthetic groups or active centers of certain enzymes (Table 2.4).

Because growth factors fulfill specific needs in biosynthesis, they are required in only small amounts, relative to the principal cellular carbon source, which must serve as a general precursor of cell carbon. Some 20 different amino acids enter into the composition of proteins, so the need for any specific amino acid that the cell is unable to synthesize is obviously not large. The same argument applies to specific need for a purine or a pyrimidine: five different compounds of these classes enter into the structure of nucleic acids. The quantitative requirements for vitamins are even smaller, since the various coenzymes of which they are precursors have catalytic roles and consequently are present at levels of a few parts per million in the cell, as shown in Table 2.5.

The biosynthesis of amino acids, purines, pyrimidines, and coenzymes typically involves complex series of individual step reactions, which will be discussed in Chapter 7. The inability to perform *any one* of these step reactions makes an organism dependent on the provision of the end product as a growth factor. However, the growth factor itself may not be absolutely essential; if the blocked

TABLE 2.4

Relation of some water-soluble vitamins to coenzymes

VITAMIN	COENZYME	ENZYMATIC REACTIONS INVOLVING THE COENZYME FORM
Nicotinic acid (niacin)	Pyridine nucleotide coenzymes (NAD and NADP)	Dehydrogenations
Riboflavin (vitamin B_2)	Flavin nucleotides (FAD and FMN)	Some dehydrogenations, electron transport
Thiamin (vitamin B_1)	Thiamin pyrosphosphate (cocarboxylase)	Decarboxylations and some group-transfer reactions
Pyridoxine (vitamin B_6)	Pyridoxal phosphate	Amino acid metabolism Transamination Deamination Decarboxylation
Pantothenic acid	Coenzyme A	Keto-acid oxidation, fatty acid metabolism
Folic acid	Tetrahydrofolic acid	Transfer of one-carbon units
Biotin	Prosthetic group of biotin enzymes	CO_2 fixation, carboxyl transfer
Cobamide (vitamin B_{12})	Cobamide coenzymes	Molecular rearrangement reactions

TABLE 2.5

Concentrations of several water-soluble vitamins[a]
in the cells of bacteria
(in parts per million of dry weight)

VITAMIN	Aerobacter aerogenes	Pseudomonas fluorescens	Clostridium butyricum
Nicotinic acid	240	210	250
Riboflavin	44	67	55
Thiamin	11	26	9
Pyridoxine	7	6	6
Pantothenic acid	140	91	93
Folic acid	14	9	3
Biotin	4	7	Required for growth

[a] In the cell these substances are present in the coenzyme form, but since their quantitative measurement after extraction is dependent on conversion to the corresponding vitamins, the data are presented in this form. Taken from R. C. Thompson, *Texas Univ. Publ.* 4237, 87 (1942).

reaction occurs at an early stage in its biosynthesis, organic precursors that follow the blocked step may be able to satisfy the needs of the cell as specific nutrients. A close analysis of a particular growth-factor requirement shown by a number of different microorganisms usually reveals that they differ in the particular chemical form or forms of the growth factor which they require. This can be illustrated by considering the rather common requirement for vitamin B_1 (thiamin), which has the following structure:

pyrimidine   thiazole

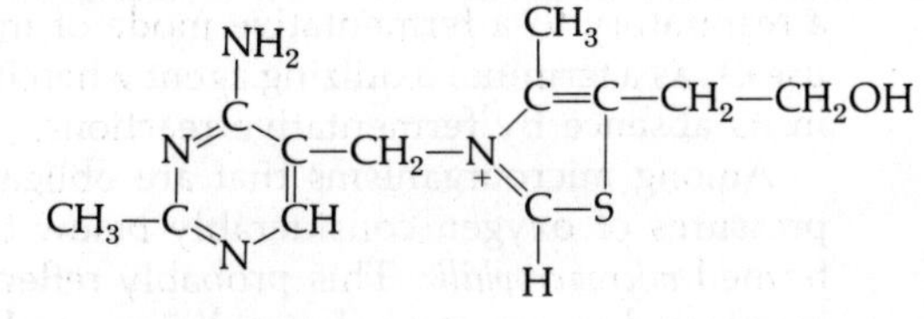

Some microorganisms require the entire molecule as a growth factor. There are, however, some microorganisms that, given the two halves of the molecule as nutrients, can put them together. Others require only the pyrimidine portion because they can synthesize thiazole. Still others need only the thiazole portion, because they can make and add the pyrimidine portion. For each type of organism described above, the *minimal* growth-factor requirement is different. Yet, in every case, what the organism must eventually have is the entire thiamin molecule, and if this compound is provided as a nutrient, it can be used as a growth factor by all the types described. Even the entire thiamin molecule, however, is not the compound that organisms must eventually make as an essential component of their cells. The functional compound is the coenzyme cocarboxylase, which acts as a prosthetic group in several enzymatic reactions. This coenzyme is thiamin pyrophosphate and has the following structure:

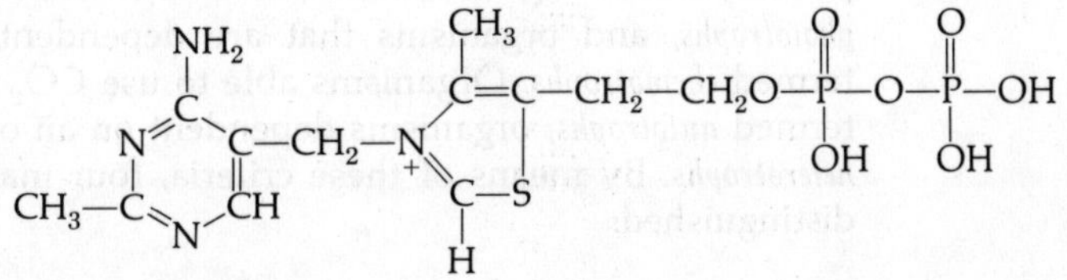

• **The roles of oxygen in nutrition**
As an elemental constituent of water and of organic compounds, oxygen is a universal component of cells and is always provided in large amounts in the major nutrient, water. However, many organisms also require *molecular oxygen* (O_2). These are organisms that are dependent on aerobic respiration for the fulfillment of their energetic needs and for which molecular oxygen functions as a terminal oxidizing agent. Such organisms are termed *obligately aerobic.*

At the other physiological extreme are those microorganisms which obtain energy by means of reactions that do not involve the utilization of molecular oxygen and for which this chemical form of the element is not a nutrient. Indeed, for many of these physiological groups, molecular oxygen is a toxic substance, which either kills them or inhibits their growth. Such organisms are *obligately anaerobic.*

Some microorganisms are *facultative anaerobes,* able to grow either in the presence or in the absence of molecular oxygen. In metabolic terms, facultative anaerobes fall into two subgroups. Some, like the lactic acid bacteria, have an exclusively fermentative energy-yielding metabolism but are not sensitive to the presence of oxygen. Others (e.g., many yeasts, coliform bacteria) can shift from a respiratory to a fermentative mode of metabolism. Such facultative anaerobes use O_2 as a terminal oxidizing agent when it is available but can also obtain energy in its absence by fermentative reactions.

Among microorganisms that are obligate aerobes, some grow best at partial pressures of oxygen considerably below that (0.2 atm) present in air. They are termed *microaerophilic.* This probably reflects the possession of enzymes that are inactivated under strongly oxidizing conditions and can thus be maintained in a functional state only at low partial pressures of O_2. Many bacteria that obtain energy by the oxidation of molecular hydrogen show this behavior, and it is known that hydrogenase, the enzyme involved in hydrogen utilization, is readily inactivated by oxygen.

• **Nutritional categories among microorganisms**

Originally, biologists recognized two principal nutritional classes among organisms: the *autotrophs,* exemplified by plants, which can use completely inorganic nutrients; and the *heterotrophs,* exemplified by animals, which require organic nutrients. Today, these two simple categories are insufficient to encompass the variety of nutritional patterns known to exist in the living world, and various attempts to construct more elaborate systems of nutritional classification have been made. Perhaps the most useful, albeit relatively simple, nutritional classification is that based on two parameters: the nature of the *energy source* and of the *principal carbon source.* With respect to energy source, there is a basic dichotomy between photosynthetic organisms that are able to use light as an energy source, termed *phototrophs,* and organisms that are dependent on a chemical energy source, termed *chemotrophs.* Organisms able to use CO_2 as a principal carbon source are termed *autotrophs;* organisms dependent on an organic carbon source are termed *heterotrophs.* By means of these criteria, four major nutritional categories can be distinguished:

1. *Photoautotrophs,* using light as the energy source and CO_2 as the principal carbon source. They include most photosynthetic organisms: higher plants, algae, and many photosynthetic bacteria.

2. *Photoheterotrophs,* using light as the energy source and an organic compound as the principal carbon source. This category includes certain of the purple and green bacteria.

3. *Chemoautotrophs,* using a chemical energy source and CO_2 as the principal carbon source. Energy is obtained by the oxidation of *reduced inorganic compounds,* such as NH_3, NO_2^-, H_2, reduced forms of sulfur (H_2S, S, $S_2O_3^{2-}$), or ferrous iron. Only members of the bacteria belong to this nutritional category. As a result of their distinctive ability to grow in strictly mineral media in the absence of light, these organisms are sometimes termed *chemolithotrophs* (from the Greek word *lithos,* a rock).

4. *Chemoheterotrophs,* using a chemical energy source and an organic substance as the principal carbon source. The clear-cut distinction between energy source and carbon source, characteristic of the three preceding categories,

loses its clarity in the context of chemoheterotrophy, where *both carbon and energy can usually be derived from the metabolism of a single organic compound*. The chemoheterotrophs include all metazoan animals, protozoa, fungi, and the great majority of bacteria. Certain further subdivisions within this very complex nutritional category can be made. One is based in the *physical state in which organic nutrients enter the cell*. The *osmotrophs* (for example, bacteria and fungi) take up all nutrients in dissolved form; the *phagotrophs* (for example, protozoa) can take up solid food particles by the mechanism termed *phagocytosis* (see p. 73).

It must be emphasized that the marked nutritional versatility of many microorganisms makes the application of this system of nutritional categories to some degree arbitrary. For example, many photoautotrophic algae can also grow in the dark, as chemoheterotrophs. Chemoheterotrophy is likewise an alternate nutritional mode for certain photoheterotrophs and chemoautotrophs. More or less by convention, such organisms are assigned to the category characterized by the simplest nutritional requirements: thus, phototrophy takes precedence over chemotrophy, autotrophy over heterotrophy. The qualifications *obligate* and *facultative* are often used to indicate the absence (or presence) of nutritional versatility. Thus, an *obligate photoautotroph* is strictly dependent on light for its energy source and on CO_2 for its principal carbon source, but a *facultative photoautotroph* is not.

In order to take into account the requirement for growth factors an additional pair of terms, *prototrophy* and *auxotrophy*, are sometimes employed. A prototroph can derive all carbon requirements from the principal carbon source. An auxotroph requires, in addition to the principal carbon source, one or more organic nutrients (growth factors). Both prototrophy and auxotrophy may occur among the organisms assigned to any one of the four major nutritional categories defined in terms of energy requirements and principal carbon sources. For example, auxotrophy, represented by an absolute requirement for one or more vitamins, is characteristic of many photoautotrophic algae and bacteria.

THE CONSTRUCTION OF CULTURE MEDIA

In constructing a culture medium for any microorganism, the primary goal is to provide a balanced mixture of the required nutrients, at concentrations that will permit good growth. It might seem at first sight reasonable to make the medium as rich as possible, by providing all nutrients in great excess. However, this approach is not a wise one. In the first place, many nutrients become growth inhibitory or toxic as the concentration is raised. This is true of organic substrates, such as salts of fatty acids (e.g., acetate) and even of sugars, if the concentration is high enough. Some inorganic constituents may also become inhibitory if provided in excess; many algae are very sensitive to the concentration of inorganic phosphate. Second, even if growth can occur in a concentrated medium, the metabolic activities of the growing microbial population will eventually change the nature of the environment to the point where it becomes highly unfavorable and the population becomes physiologically abnormal or dies. This may be brought about by a drastic change in the hydrogen ion concentration (pH), by the accumulation of toxic organic metabolites, or, in the case of strict aerobes, by the depletion of oxygen. Since the usual goal of the microbiologist is to study the properties and behavior of *healthy* microorganisms, it is wise to limit the total growth of cultures by providing a limiting quantity of one nutrient; in the case of chemoheterotrophs, the principal carbon source is usually selected for this purpose. Examples of the appropriate concentrations of nutrients will be provided in the various media described below.

The rational point of departure for the preparation of media is to compound a *mineral base*, which provides all those nutrients which can be supplied to any organism in inorganic form. This base can then be supplemented, as required, with a carbon source, an energy source, a nitrogen source, and any required growth factors; these supplements will, of course, vary with the nutritional properties of the particular organism that one wishes to grow. A medium composed entirely of chemically defined nutrients is termed a *synthetic medium*. One that contains ingredients of unknown chemical composition is termed a *complex medium*.

We may illustrate these principles by considering the composition of four media of increasing chemical complexity, each of which is suitable for the cultivation of certain kinds of chemotrophic bacteria (Table 2.6). All four media share a common mineral base. Medium 1 is supplemented with NH_4Cl at a concentration of 1 g/liter but has no added source of carbon. However, if it is incubated aerobically, the CO_2 of the atmosphere will be available as a carbon source. In the dark the only organisms that can grow in this medium are chemoautotrophic nitrifying bacteria, such as *Nitrosomonas*, which obtain carbon from CO_2 and energy from the aerobic oxidation of ammonia; the ammonia also provides them with a nitrogen source.

Medium 2 is additionally supplemented with glucose at a concentration of 5 g/liter. Under aerobic conditions, it will support the growth of many bacteria and fungi, since glucose can commonly be used as a carbon and energy source for aerobic growth. If incubated in the absence of oxygen, it can also support the development of many facultatively or strictly anaerobic bacteria, able to derive carbon and energy from the fermentation of glucose. Note, however, that this medium is not a suitable one for any microorganism that requires growth factors; it contains only a single carbon compound.

TABLE 2.6

Four media of increasing complexity

| COMMON INGREDIENTS | ADDITIONAL INGREDIENTS | | | |
	MEDIUM 1	MEDIUM 2	MEDIUM 3	MEDIUM 4
Water, 1 liter	NH_4Cl, 1 g	Glucose,[a] 5 g	Glucose, 5 g	Glucose, 5 g
K_2HPO_4, 1 g		NH_4Cl, 1 g	NH_4Cl, 1 g	Yeast extract,
$MgSO_4 \cdot 7H_2O$, 200 mg			Nicotinic acid,	5 g
$FeSO_4 \cdot 7H_2O$, 10 mg			0.1 mg	
$CaCl_2$, 10 mg				
Trace elements (Mn,				
Mo, Cu, Co, Zn) as				
inorganic salts,				
0.02–0.5 mg of each				

[a] If the media are sterilized by autoclaving, the glucose should be sterilized separately and added aseptically. When sugars are heated in the presence of other ingredients, especially phosphates, they are partially decomposed to substances that are very toxic to some microorganisms.

Medium 3 is additionally supplemented with one vitamin, nicotinic acid. It can therefore support the growth of all those organisms able to develop in medium 2, together with others, such as the bacterium *Proteus vulgaris*, which require nicotinic acid as a growth factor.

For the three media so far described, the chemical nature of every ingredient is known; thus, they are good examples of synthetic media. Medium 4 is a complex medium, in which the NH_4Cl and nicotinic acid of medium 3 have been replaced by a nutrient of unknown composition, yeast extract, at a concentration of 5 g/liter. It can support the growth of a great many chemoheterotrophic microorganisms, both aerobic and anaerobic, having no growth-factor requirements, relatively simple ones, or highly complex ones. The yeast extract provides a variety of organic nitrogenous constituents (partial breakdown products of proteins) which can fulfill the general nitrogen requirements, and it also contains most of the organic growth factors likely to be required by microorganisms.

Complex media are, accordingly, useful for the cultivation of a wide range of microorganisms, including ones whose precise growth-factor requirements are not known. Even when the growth-factor requirements of a microorganism have been precisely determined, it is often more convenient to grow the organism in a complex medium, particularly if the growth-factor requirements are numerous. This point is illustrated in Table 2.7, which describes the composition of a synthetic medium that will support growth of the lactic acid bacterium, *Leuconostoc mesenteroides*. This bacterium can also be cultivated satisfactorily in the complex medium 4. In this particular instance, accordingly, the yeast extract of medium 4 must furnish the following requirements: the organic acid, acetate; 19 amino acids; 4 purines and pyrimidines; and 10 vitamins.

The media described in Table 2.6 can support the development of microorganisms only if certain other requirements for growth are also met. These include a suitable temperature of incubation, favorable osmotic conditions, and a hydrogen ion concentration within the range tolerated by the organism in question. Suitable chemical adjustments may be required to accommodate the osmotic conditions and hydrogen ion concentration to the needs of some microorganisms for which these media are satisfactory with respect to their content of nutrients.

• **The control of pH*** Although a given medium may be suitable for the *initiation* of growth, the subsequent development of a bacterial population may be severely limited by chemical changes that are brought about by the growth and metabolism of the organisms themselves. For example, in glucose-containing media, organic acids that may be produced as a result of fermentation may become inhibitory to growth.

In contrast, the microbial decomposition or utilization of anionic components of a medium tends to make the medium more alkaline. For example, the oxidation of a molecule of sodium succinate liberates two sodium ions in the form of the

---

* The pH scale (Figure 2.5) provides a convenient method of expressing hydrogen ion concentrations in aqueous solutions and is almost invariably used by biologists and biochemists. The pH value of a given solution is the *logarithm of the reciprocal of the hydrogen ion concentration* (expressed in moles per liter), or

$$pH = \log \frac{1}{[H^+]}$$

For example, a solution of acid, which is 0.1 N with respect to hydrogen ions, has a pH value of 1.0. This follows from the fact that log (1/0.1) equals the logarithm of 10, which is 1.0.

TABLE 2.7

Medium for *Leuconostoc mesenteroides*[a]

WATER	1 liter		
ENERGY SOURCE:			
Glucose	25 g		
NITROGEN SOURCE:			
NH$_4$Cl	3 g		
MINERALS:			
KH$_2$PO$_4$	600 mg	FeSO$_4$·7H$_2$O	10 mg
K$_2$HPO$_4$	600 mg	MnSO$_4$·4H$_2$O	20 mg
MgSO$_4$·7H$_2$O	200 mg	NaCl	10 mg
ORGANIC ACID:			
Sodium acetate	20 g		
AMINO ACIDS:			
DL-α-Alanine	200 mg	L-Lysine·HCl	250 mg
L-Arginine	242 mg	DL-Methionine	100 mg
L-Asparagine	400 mg	DL-Phenylalanine	100 mg
L-Aspartic acid	100 mg	L-Proline	100 mg
L-Cysteine	50 mg	DL-Serine	50 mg
L-Glutamic acid	300 mg	DL-Threonine	200 mg
Glycine	100 mg	DL-Tryptophan	40 mg
L-Histidine·HCl	62 mg	L-Tyrosine	100 mg
DL-Isoleucine	250 mg	DL-Valine	250 mg
DL-Leucine	250 mg		
PURINES AND PYRIMIDINES:			
Adenine sulfate·H$_2$O	10 mg	Uracil	10mg
Guanine·HCl·2H$_2$O	10 mg	Xanthine·HCl	10 mg
VITAMINS			
Thiamine·HCl	0.5 mg	Riboflavin	0.5 mg
Pyridoxine·HCl	1.0 mg	Nicotinic acid	1.0 mg
Pyridoxamine·HCl	0.3 mg	p-Aminobenzoic acid	0.1 mg
Pyridoxal·HCl	0.3 mg	Biotin	0.001 mg
Calcium pantothenate	0.5 mg	Folic acid	0.01 mg

[a] From H. E. Sauberlich and C. A. Baumann, "A Factor Required for the Growth of *Leuconostoc citrovorum*," *J. Biol. Chem.* **176**, 166 (1948).

very alkaline salt, sodium carbonate. The decomposition of proteins and amino acids may also make a medium alkaline as a result of ammonia production.

To prevent excessive changes in hydrogen ion concentration either *buffers* or *insoluble carbonates* are often added to the medium.

The phosphate buffers, which consist of mixtures of monohydrogen and dihydrogen phosphates (e.g., K$_2$HPO$_4$ and KH$_2$PO$_4$), are the most useful ones. KH$_2$PO$_4$ is a weakly acidic salt, whereas K$_2$HPO$_4$ is slightly basic, so that an equimolar solution of the two is very nearly neutral, having a pH of 6.8. If a limited amount of strong acid is added to such a solution, part of the basic salt is converted to the weakly acidic one:

$$K_2HPO_4 + HCl \longrightarrow KH_2PO_4 + KCl$$

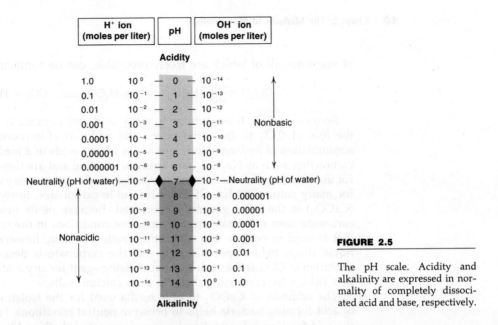

| H⁺ ion (moles per liter) | pH | OH⁻ ion (moles per liter) |

Acidity

1.0	10^0	0	10^{-14}	
0.1	10^{-1}	1	10^{-13}	
0.01	10^{-2}	2	10^{-12}	
0.001	10^{-3}	3	10^{-11}	Nonbasic
0.0001	10^{-4}	4	10^{-10}	
0.00001	10^{-5}	5	10^{-9}	
0.000001	10^{-6}	6	10^{-8}	
Neutrality (pH of water) —	10^{-7}	7	10^{-7}	— Neutrality (pH of water)
	10^{-8}	8	10^{-6}	0.000001
	10^{-9}	9	10^{-5}	0.00001
	10^{-10}	10	10^{-4}	0.0001
Nonacidic	10^{-11}	11	10^{-3}	0.001
	10^{-12}	12	10^{-2}	0.01
	10^{-13}	13	10^{-1}	0.1
	10^{-14}	14	10^0	1.0

Alkalinity

FIGURE 2.5

The pH scale. Acidity and alkalinity are expressed in normality of completely dissociated acid and base, respectively.

If, however, a strong base is added, the opposite conversion occurs:

$$KH_2PO_4 + KOH \longrightarrow K_2HPO_4 + H_2O$$

Thus, the solution acts as a buffer in that it resists radical changes in the hydrogen ion concentration when acid or alkali is produced in the medium. By using different ratios of acidic and basic phosphates, different pH values may be established, ranging from approximately 6.0 to 7.6. Good buffering action, however, is obtained only in the narrower range of pH 6.4 to 7.2 because the capacity of a buffer solution is limited by the amounts of its basic and acidic ingredients. Hence, the more acidic the initial buffer, the less is its ability to prevent increases in hydrogen ion concentration (decreases in pH) and the greater its capacity for reacting with alkali. Conversely, the more alkaline the initial buffer, the less is its ability to prevent increases in pH and the greater its ability to prevent acidification.

The phosphates are used widely in the preparation of media because they are the only inorganic agents that buffer in the physiologically important range around neutrality and that are relatively nontoxic to microorganisms. In addition, they provide a source of phosphorus, which is an essential element for growth. In high concentrations, phosphate becomes inhibitory, so the amount of phosphate buffer that can be used in a medium is limited by the tolerance of the particular organism being cultivated. Generally, about 5 g of potassium phosphates per liter of medium can be tolerated by bacteria and fungi.

When a great deal of acid is produced by a culture, the limited amounts of phosphate buffer that may be used become insufficient for the maintenance of a suitable pH. In such cases, carbonates may be added to media as "reserve alkali" to neutralize the acids as they are formed. In the presence of hydrogen ions, carbonate is transformed to bicarbonate, and bicarbonate is converted further to carbonic acid, which decomposes spontaneously to CO_2 and water. This sequence of reactions, all of which are freely reversible, can be summarized:

$$CO_3^{2-} \underset{-H^+}{\overset{+H^+}{\rightleftharpoons}} HCO_3^- \underset{-H^+}{\overset{+H^+}{\rightleftharpoons}} H_2CO_3 \rightleftharpoons CO_2 + H_2O$$

Because H_2CO_3 is an extremely weak acid and because it decomposes with the loss of CO_2 to the atmosphere, the addition of carbonates prevents the accumulation of hydrogen ions and hence of free acids in a medium. The soluble carbonates, such as Na_2CO_3, are strongly alkaline and are therefore not suitable for use in culture media. In contrast, *insoluble carbonates* are very useful ingredients for many culture media. Of these insoluble carbonates, finely powdered chalk ($CaCO_3$) is the most generally employed. Because of its insolubility, calcium carbonate does not create strongly alkaline conditions in the medium, especially if it is used in conjunction with other buffers. When, however, the pH of the liquid drops below approximately 7.0, the carbonate is decomposed with the evolution of CO_2. It thus acts as a neutralizing agent for any acids that may appear in a culture by converting them to their calcium salts.

The addition of $CaCO_3$ to agar media used for the isolation and cultivation of acid-forming bacteria helps to preserve neutral conditions. Furthermore, since the acid-forming colonies dissolve the precipitated chalk and become surrounded by clear zones, they can be easily recognized against the opaque background of the medium.

In some instances, neither buffers nor insoluble carbonates can be used to maintain a relatively constant pH in a culture medium. Special problems arise, for example, when very large amounts of acid are formed in a medium in which the presence of calcium carbonate is not desired. Even more serious difficulties are encountered in controlling the pH of slightly alkaline media in which basic substances are produced as a result of bacterial growth. This is due to the fact that phosphate buffers are not effective in the pH range 7.2 to 8.5, and no other suitable buffers in this range are available. In certain cases, therefore, it is necessary to adjust the pH of the culture, either periodically or continuously, by the aseptic addition of strong acids or bases. In some laboratories and in industrial plants, elaborate mechanical devices are used for this purpose. With their aid, a continuous titration of the medium is feasible and the pH is kept nearly constant.

The media described in Table 2.6 are all slightly alkaline at the beginning because they contain the alkaline salt K_2HPO_4. Many organisms prefer neutral or slightly acidic conditions, which can be achieved by the use of appropriate buffers.

• The avoidance of mineral precipitates: chelating agents

A troublesome problem often encountered in the preparation of synthetic media is the formation of a precipitate upon sterilization, particularly if the medium has a relatively high phosphate concentration. This results from the formation of insoluble complexes between phosphates and certain cations, particularly calcium and iron. Although it usually does not affect the nutrient value of the medium, it may make the observation or quantitation of microbial growth difficult. The problem can be avoided by sterilizing separately the calcium and iron salts in concentrated solution and adding them to the sterilized and cooled medium. Alternatively, one can incorporate in the medium a small amount of a chelating agent, which will form a soluble complex with these metals and thus prevent them from forming an insoluble complex with phosphates. The chelating agent most commonly used for this purpose is ethylenediaminetetraacetic acid (EDTA), at a concentration of approximately 0.01 percent.

• **The control of oxygen concentration**

Oxygen is an essential nutrient for the obligately aerobic bacteria. Aerobic microorganisms can be grown easily on the surface of agar plates and in shallow layers of liquid medium. In unshaken liquid cultures, growth usually occurs at the surface. Below the surface, however, conditions become anaerobic, and growth is impossible. To obtain large populations in liquid cultures, *it is therefore necessary to aerate the medium*. Various types of shaking machines that constantly agitate, and thus aerate, the medium are available for laboratory use. Another method of aeration is the continuous passage of a stream of air through a culture. To ensure a large surface of contact between gas and liquid, the air may be introduced through a porous "sparger," which delivers it in the form of very fine bubbles.

• **Techniques for cultivation of obligate anaerobes**

Many of the more sensitive, strictly anaerobic microorganisms are rapidly killed by contact with molecular oxygen. The exposure of cultures to air should accordingly be minimized or avoided completely. Furthermore, many strict anaerobes can initiate growth only in media of low oxidation-reduction potential ($\sim$150 mv, or even less). The use of media that are *pre-reduced* by the inclusion of such reducing agents as cysteine, thioglycollate, Na_2S, or sodium ascorbate is therefore a factor of cardinal importance for the cultivation of many anaerobes. Once prepared, such media must, of course, be protected from exposure to air during both storage and use. During use, this can be achieved by passage of a stream of O_2-free CO_2 or N_2 into the orifice of the opened culture vessel.

Liquid cultures of strict anaerobes are usually prepared in tubes or flasks completely filled with medium and closed by rubber stoppers or plastic screw caps. Isolation in solid media can be undertaken by several methods. Organisms able to tolerate a transient exposure to air can be isolated on streaked plates, which are placed after inoculation in sealed anaerobic jars. The atmosphere of the jars is then rendered O_2-free by evacuation and refilling with an inert gas (e.g., N_2); by chemical destruction of oxygen; or by a combination of both methods. For the isolation of more oxygen-sensitive anaerobes it is best to use agar shake cultures, which are individually sealed immediately after inoculation. A modification is the use of a so-called "roll tube," in which the molten agar is distributed in a thin layer over the walls of the tube. The tube is gassed with an oxygen-free mixture during manipulation, and then closed with a rubber stopper.

• **The provision of carbon dioxide**

A problem frequently encountered in the cultivation of photoautotrophs and chemoautotrophs is the provision of carbon dioxide in sufficient amounts. Although the diffusion of carbon dioxide from the atmosphere into the culture medium will permit growth to occur, the carbon dioxide concentration in the atmosphere is very low (0.03 percent in the open atmosphere, somewhat higher inside a building), and the growth rates of autotrophs are often limited by the availability of carbon dioxide under these conditions. The solution is to gas the cultures with air that has been artificially enriched with carbon dioxide and contains from 1 to 5 percent of this gas. The control of pH becomes a problem, for reasons already discussed (p. 40), and if this solution is adopted, care must be taken to modify the buffer composition of the medium. In the case of autotrophs that can be grown under anaerobic conditions in stoppered bottles (e.g., the purple and green sulfur bacteria) the requirement for CO_2 can be met by the incorporation of $NaHCO_3$ in the medium. Soluble carbonates cannot be used in media exposed to air because the rapid loss of CO_2 to the atmosphere causes the medium to become extremely alkaline.

• The provision of light

For the cultivation of phototrophic microorganisms (algae, photosynthetic bacteria), *light* is an essential requirement. The provision of adequate illumination combined with control of temperature is not a simple matter. In the cultivation of nonphotosynthetic organisms, temperature control is provided by the use of incubators, maintained by a thermostatic device at the desired value; however, most commercially available models are not designed with a system of internal illumination and cannot be used for the cultivation of phototrophic organisms. A relatively uncontrolled and discontinuous illumination may be obtained by the exposure of cultures to daylight. Direct exposure to sunlight should be avoided, because the intensity may be too high, and the temperature may rise to a point where growth is prevented. Many phototrophic microorganisms can tolerate continuous illumination, and their growth is much more rapid under these conditions, so artificial light sources are advantageous. The *emission spectrum* of the lamp employed is important. Fluorescent light sources have the practical advantage of producing relatively little heat, so maintenance of a suitable temperature is not difficult. However, their emission spectra are deficient, compared to sunlight, in the longer wavelengths of the visible spectrum and the near infrared region. They are satisfactory for the cultivation of algae and blue-green bacteria, which perform photosynthesis with light of wavelengths shorter than 700 nm, but provide little or no photosynthetically effective light for purple and green bacteria, which use wavelengths in the range 750 to 1,000 nm. The only suitable artificial light sources for the latter photosynthetic bacteria are incandescent lamps, and if high intensities are used, the dissipation of heat becomes a problem. The easiest solution is to immerse culture vessels in a glass or plastic water bath which can be subjected to lateral illumination and maintained at the desired temperature by the circulation of water. The other solution is to construct a light cabinet with internal incandescent illumination, in which the temperature can be controlled by ventilation or refrigeration.

SELECTIVE MEDIA

It is clear that no single medium or set of conditions will support the growth of all the different types of organisms that occur in nature. Conversely, any medium that is suitable for the growth of a specific organism is, to some extent, *selective* for it. In a medium inoculated with a variety of organisms, only those that can grow in it will reproduce, and all others will be suppressed. Further, if the growth requirements of an organism are known, it is possible to devise

43   a set of conditions which will specifically favor the development of this particular organism, thus permitting its isolation from a mixed natural population, even when the organism in question is a minor component of the total population. Microorganisms can be selectively obtained from natural habitats (e.g., soil or water) either by *direct isolation* or by *enrichment*.

• **Direct isolation**   If a mixed microbial population is spread over the surface of a selective medium solidified with agar (or some other gelling agent), every cell in the inoculum capable of development will grow and eventually form a colony. The spatial dispersion of the microbial population on a solid medium considerably reduces the competition for nutrients: under these circumstances, even organisms that grow relatively slowly will be able to produce colonies. Direct plating on a selective medium is the technique of choice when one wishes to isolate a considerable diversity of microorganisms, all able to grow under the conditions of culture employed.

• **Enrichment**   If a mixed microbial population is introduced into a liquid selective medium, there is a direct competition for nutrients among the members of the developing population. Liquid enrichment media therefore tend to select the microorganism of highest growth rate among all the members of the introduced population which are able to grow under the conditions provided.

The selectivity of an enrichment culture is not solely determined by the chemical composition of the medium used. The outcome of enrichment in a given medium can be significantly modified by variation of such other factors as temperature, pH, ionic strength, illumination, aeration, or source of inoculum. For example, medium 2 of Table 2.6 can be used to grow or to enrich enteric bacteria belonging to the genera *Escherichia* and *Enterobacter*. Members of the former genus are normal inhabitants of the intestinal tract of warm-blooded animals, while members of the latter genus are common soil inhabitants. *Escherichia* strains are naturally adapted to growth at somewhat higher temperatures than *Enterobacter* strains, and they are also less sensitive to the toxic effect of bile salts. Consequently, either incubation at an elevated temperature (45°C) or inclusion of bile salts will make medium 2 selective for *Escherichia*.

Temperature selection can also be used with great effectiveness for the isolation of blue-green bacteria, which closely resemble many algae in nutritional and metabolic respects. Both algae and blue-green bacteria can grow in a simple mineral medium, incubated in the light at 25°C. However, development of algae can be almost wholly prevented by incubation at a temperature of 35°C, since algae in general have lower temperature maxima than blue-green bacteria.

An enrichment medium that is not initially highly selective may acquire greatly increased selectivity for a particular type of microorganism, as a result of chemical changes produced by this organism during its development. Thus, fermentative bacteria and yeasts are typically more tolerant than other organisms of the organic end products which they themselves produce from carbohydrates; in a carbohydrate-rich medium their development will therefore tend to suppress competing microorganisms.

In the isolation of spore-forming bacteria (genera *Bacillus* and *Clostridium*), competition from nonsporulating bacteria can be wholly eliminated by a pretreatment of the inoculum. Pasteurization of the inoculum, involving brief exposure to a high temperature (2 to 5 minutes at 80°C) will destroy all vegetative cells, leaving the much more heat-resistant spores unaffected.

• Enrichment methods for some specialized physiological groups

The tool of the enrichment culture is one of the most powerful techniques available to the microbiologist. An almost infinite number of permutations and combinations of the different environment variables, nutritional and physical, can be developed for the specific isolation of microorganisms from nature. Enrichment techniques provide a means for isolating known microbial types at will from nature, by taking advantage of their specific requirements, and can also be indefinitely elaborated, as a means of obtaining for study hitherto undescribed organisms capable of growing in the environments devised by the scientist. Here we shall attempt to summarize a few of the enrichment procedures that can be used to isolate major physiological groups of microorganisms, principally bacteria, from nature.

• Synthetic enrichment media for chemoheterotrophs

The nutritional and environmental conditions necessary for the enrichment of various groups of chemoheterotrophs in synthetic media are outlined in Table 2.8, and the detailed composition of each enrichment medium is described in Table 2.9.

For the enrichment of fermentative organisms, the chemical nature of the organic substrate is important; to be attacked by fermentation, it must be neither too oxidized nor too reduced. Sugars are excellent fermentative substrates, but

TABLE 2.8

Primary environmental factors that determine the outcome of enrichment procedures for chemoheterotrophic bacteria with the use of synthetic media

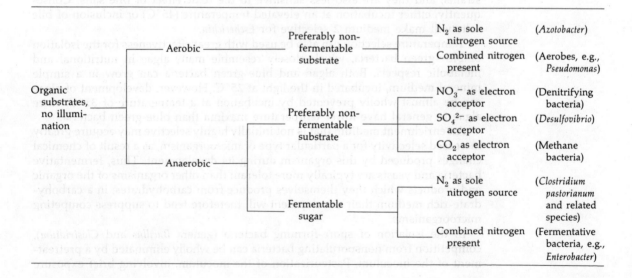

Organic substrates, no illumination	Aerobic	Preferably non-fermentable substrate	N₂ as sole nitrogen source	(*Azotobacter*)
			Combined nitrogen present	(Aerobes, e.g., *Pseudomonas*)
	Anaerobic	Preferably non-fermentable substrate	NO₃⁻ as electron acceptor	(Denitrifying bacteria)
			SO₄²⁻ as electron acceptor	(*Desulfovibrio*)
			CO₂ as electron acceptor	(Methane bacteria)
		Fermentable sugar	N₂ as sole nitrogen source	(*Clostridium pastorianum* and related species)
			Combined nitrogen present	(Fermentative bacteria, e.g., *Enterobacter*)

TABLE 2.9

Enrichment conditions for chemoheterotrophic bacteria[a,b]

ADDITIONS TO BASAL MEDIUM		SPECIAL ENVIRONMENTAL CONDITIONS		
ORGANIC	INORGANIC	ATMOSPHERE	pH	ORGANISMS ENRICHED
Ethyl alcohol, 4.0	None	Air	7.0	*Azotobacter*
Ethyl alcohol, 4.0	NH₄Cl, 1.0	Air	7.0	Aerobes (e.g., *Pseudomonas*)
Ethyl alcohol, 4.0	NaNO₃, 3.0	None (stoppered bottle)	7.0	Denitrifying bacteria (e.g., some species of *Pseudomonas*)
Ethyl alcohol, 4.0	NH₄Cl, 1.0 Na₂SO₄, 5.0	None (stoppered bottle)	7.2	*Desulfovibrio*
Ethyl alcohol, 4.0	NH₄Cl, 1.0 NaHCO₃, 1.0 CaCO₃, 5.0	None (stoppered bottle)	7.4	Methane-producing bacteria
Glucose, 10.0	CaCO₃, 5.0	Pure N₂	7.0	*Clostridium pastorianum* and related species
Glucose, 10.0	NH₄Cl, 1.0	None (stoppered bottle)	7.0	Enteric bacteria (e.g., *Escherichia, Enterobacter*)

[a] The components of the medium are given in grams per liter.
[b] Incubated at 25–30°C without illumination. Basal medium: MgSO₄ · 7H₂O, 0.2; K₂HPO₄, 1.0; FeSO₄ · 7H₂O, 0.05; CaCl₂, 0.02; MnCl₂ · 4H₂O, 0.002; NaMoO₄ · 2H₂O, 0.001.

many other classes of organic compounds on the same approximate oxidation level can also be fermented. The enrichment cultures must be incubated anaerobically, not only because some fermentative organisms are obligate anaerobes, but also to prevent competition from aerobic forms. Nitrate should not be used as a nitrogen source, since it will also allow growth of denitrifying bacteria (see below). Calcium carbonate may be added if the organisms being enriched produce acid and are themselves sensitive to it.

Three special physiological groups of chemoheterotrophic bacteria, which can use inorganic compounds other than molecular oxygen as terminal oxidants for respiratory metabolism, may also be enriched in synthetic media under anaerobic conditions. In these cases, readily fermentable organic substrates such as sugars should be avoided. For the denitrifying bacteria, nitrate is included as a terminal oxidant, and acetate, butyrate, or ethyl alcohol as a carbon and energy source. For the sulfate-reducing bacteria, a relatively large amount of sulfate is included, since it is the specific terminal oxidant; lactate or malate, but not acetate, provides the best source of carbon and energy. For the carbonate-reducing (methane-producing) bacteria, CO_2 must be present as an oxidant and such compounds as formate as the source of carbon and energy. It should also be noted here that in enrichments for sulfate- and carbonate-reducing bacteria, the use of ammonia as a nitrogen source is desirable; addition of nitrate will favor development of the nitrate reducers. Similarly, in enrichments for carbonate and nitrate reducers, the concentration of sulfate should be kept to a minimum to prevent overgrowth by sulfate reducers.

Obviously, aerobic conditions are essential for the enrichment of microorganisms that obtain energy from aerobic respiration. Either fermentable or nonfermentable substrates are satisfactory organic nutrients. If a fermentable substrate is used, however, the culture must be aerated thoroughly and continuously, because the depletion of oxygen by respiration favors the enrichment of anaerobes. Readily fermentable compounds generally stimulate the growth of facultative anaerobes. Thus, when glucose or other sugars are included in enrichment media, *Enterobacter* appears as one of the principal organisms under aerobic as well as under anaerobic conditions.

The use of nonfermentable substrates usually results in the enrichment of organisms that are obligate aerobes, most commonly members of the genus *Pseudomonas*. For instance, benzoate-oxidizing strains of *P. putida* can be obtained by using sodium benzoate (1 g/liter) as the sole organic nutrient. If asparagine (2 g/liter) is used as the sole source of both carbon and nitrogen in the medium, asparagine-oxidizing strains of this or other related species become predominant.

Ammonium salts are generally used as a nitrogen source in synthetic enrichment media for aerobes. If no combined nitrogenous compounds are provided, cultures of the aerobic nitrogen-fixing bacteria of the genus *Azotobacter* can be obtained. Species of *Azotobacter* can use a great variety of organic substrates, including alcohol, butyrate, benzoate, and glucose, as their only organic nutrients.

• **The enrichment of chemoautotrophic and photosynthetic organisms**

For the enrichment of chemoautotrophic and photoautotrophic organisms, organic compounds must be omitted from the medium, and CO_2 or bicarbonate must be used as the only source of carbon (Tables 2.10 to 2.13). Photosynthetic forms require light, whereas the chemolithotrophs should be cultivated in the dark to prevent the development of photosynthetic types. During incubation of cultures, either aerobic or anaerobic conditions must be maintained, depending on whether or not the organisms need oxygen. An exception to this rule is found in the selective cultivation of algae. Since these organisms *produce* oxygen in their metabolism, it makes virtually no difference whether the enrichment cultures are incubated under aerobic or anaerobic conditions.

Media devised for the enrichment and propagation of photosynthetic organisms should contain sodium because this element is known to be required by photosynthetic bacteria.

For the enrichment of purple nonsulfur bacteria, the medium should include a suitable organic substrate and, in some instances, bicarbonate. The substrate should not be a readily fermentable one; acetate, butyrate, or malate is customarily used. Bicarbonate must be added if the substrate (e.g., butyrate) is more reduced than cell material because photosynthesis is accompanied by a net consumption of CO_2. With substrates such as malate, which are metabolized with a net production of CO_2, the addition of bicarbonate is unnecessary. Because the photoheterotrophic bacteria require various growth factors, a small amount of yeast extract is generally added to the enrichment medium.

• **The use of complex media for enrichment**

Some bacteria cannot be enriched in defined media because they have extremely complex nutritional requirements. Nevertheless, such organisms may, in some cases, be obtained from nature by the use of specially designed complex media. The lactic acid bacteria illustrate this point. These organisms are characterized by their remarkable resistance to lactic acid, which they themselves produce in the fermentation of sugar. To enrich for lactic acid bacteria, a poorly buffered

TABLE 2.10

Primary environmental factors that determine the outcome of enrichment procedures for some chemoautotrophic bacteria

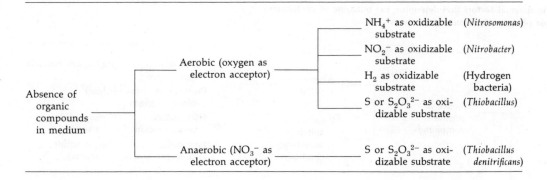

TABLE 2.11

Enrichment conditions for some chemoautotrophic bacteria[a]

| ADDITIONS TO MEDIUM | SPECIAL ENVIRONMENTAL FEATURES | | ORGANISMS ENRICHED |
	ATMOSPHERE	pH	
NH$_4$Cl, 1.5 CaCO$_3$, 5.0	Air	8.5	Nitrosomonas
NaNO$_2$, 3.0	Air	8.5	Nitrobacter
NH$_4$Cl, 1.0	85% H$_2$, 10% O$_2$, 5% CO$_2$	7.0	Hydrogen bacteria
NH$_4$Cl, 1.0 Na$_2$S$_2$O$_3$·7H$_2$O, 7.0	Air	7.0	Thiobacillus
NH$_4$NO$_3$, 3.0 Na$_2$S$_2$O$_3$·7H$_2$O, 7.0 NaHCO$_3$, 5.0	None (stoppered bottle)	7.0	Thiobacillus denitrificans

[a]The components of the medium are given in grams per liter. Basal medium: MgSO$_4$·7H$_2$O, 0.2; K$_2$HPO$_4$, 1.0; FeSO$_4$·7H$_2$O, 0.05; CaCl$_2$, 0.02; MnCl$_2$·4H$_2$O, 0.002; NaMoO$_4$·2H$_2$O, 0.001. Environment: in the dark; temperature, 25°–30°C.

medium containing glucose and a rich source of growth factors is used (e.g., 20 g of glucose and 10 g of yeast extract per liter). After inoculation, preferably with natural materials that are rich in lactic acid bacteria (e.g., vegetable matter, raw milk, sewage), the medium is incubated under anaerobic conditions. The first organisms to develop are usually bacteria such as *Enterobacter* and *Escherichia*. However, as lactic acid gradually accumulates, conditions become less and less favorable for these bacteria, whereas the lactic acid bacteria continue to grow. Eventually, the acidity of the medium becomes so high that the lactic acid bacteria predominate and most other organisms are destroyed.

Another good example of a complex medium that is quite selective for a specific

TABLE 2.12

Primary environmental factors that determine the outcome of enrichment
procedures for photosynthetic microorganisms

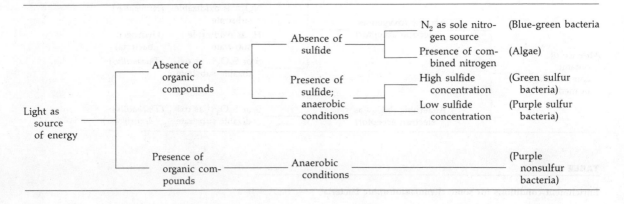

TABLE 2.13

Enrichment conditions for photosynthetic microorganisms[a]

| ADDITIONS TO MEDIUM | | SPECIAL ENVIRONMENTAL FEATURES | | |
ORGANIC	INORGANIC	ATMOSPHERE	pH	ORGANISMS ENRICHED
None	None	Air, or air + 5% CO_2	6.0–8.0	Blue-green bacteria
None	NaNO$_3$ or NH$_4$Cl, 1.0	Air, or air + 5% CO_2	6.0–8.0	Algae
None	NH$_4$Cl, 1.0; Na$_2$S · 9H$_2$O, 2.0; NaHCO$_3$, 5.0	None (stoppered bottle)	7.5	Green sulfur bacteria
None	NH$_4$Cl, 1.0; Na$_2$S · 9H$_2$O, 1.0; NaHCO$_3$, 5.0	None (stoppered bottle)	8.0–8.5	Purple sulfur bacteria
Sodium malate, 5.0; yeast extract, 0.5	NH$_4$Cl, 1.0	None (stoppered bottle)	7.0–7.5	Purple nonsulfur bacteria

[a] The components of the medium are given in grams per liter. Basal medium: MgSO$_4$ · 7H$_2$O, 0.2; K$_2$HPO$_4$, 1.0; FeSO$_4$ · 7H$_2$O, 0.01; CaCl$_2$, 0.02; MnCl$_2$ · 4H$_2$O, 0.002; NaMoO$_4$ · 2H$_2$O, 0.001; NaCl, 0.5. Environment: constant illumination; temperature, 25°–30°C.

group of organisms is one designed for the enrichment of the propionic acid
bacteria. These organisms produce propionic acid, acetic acid, and CO_2 in fermentation. Although they can ferment glucose readily, they cannot compete either
with *Enterobacter* or with the lactic acid bacteria in glucose media because they grow
relatively slowly and do not tolerate acidic conditions. However, the propionic
acid bacteria can also ferment lactic acid, which is not a suitable substrate for

most other fermentative organisms. This capacity is the key to their enrichment. If a neutral medium containing 20 g of sodium lactate and 10 g of yeast extract per liter is inoculated with natural materials containing propionic acid bacteria and incubated at 30°C under anaerobic conditions, an enrichment of these organisms is obtained. Swiss cheese is the best inoculum for the cultures because propionic acid bacteria are the principal agents in its ripening.

Complex media can be used successfully for the selective cultivation of the acetic acid bacteria. These bacteria are especially adapted to environments that contain high concentrations of alcohol. They are also far less susceptible than other bacteria to inhibition by acetic acid, which they produce from alcohol in respiration. To enrich for them, a complex medium containing alcohol is inoculated with materials containing these bacteria and then incubated under aerobic conditions. Fruits, flowers, and unpasteurized (draught) beer are good sources of inoculum. A medium containing 40 ml of alcohol and 10 g of yeast extract per liter, adjusted to pH 6.0, can be used for enrichment. Beer and hard cider are also excellent enrichment media, for these fermented beverages closely resemble the natural environment in which acetic acid bacteria predominate. Although a large surface of contact with air must be ensured, the inoculated medium is usually not aerated vigorously because many acetic acid bacteria grow best in a pellicle that they form at the surface.

Examples of complex media and environmental conditions used for the enrichment of selected organotrophs are given in Table 2.14.

TABLE 2.14

Complex media for enrichment of chemoheterotrophs[a]

ADDITIONS	SPECIAL ENVIRONMENTAL CONDITIONS	PREFERRED CHOICE OF INOCULUM	ORGANISMS ENRICHED
None	pH 7.0; aerobic	Soil	Aerobic amino acid oxidizers
None	pH 7.0; aerobic	Pasteurized soil	*Bacillus* spp.
None	pH 7.0; anaerobic	Pasteurized soil	Amino acid—fermenting clostridia
Urea, 50.0	pH 8.5; aerobic	Pasteurized soil	Alkali-tolerant urea-decomposing bacilli (*Bacillus pasteurii*)
Glucose, 20.0	pH 2.0–3.0; anaerobic	Soil	Anaerobic *Sarcina* spp.
Glucose, 20.0	pH 6.5; anaerobic	Plant materials, milk	Lactic acid bacteria
Glucose, 20.0; $CaCO_3$, 20.0	pH 7.0; aerobic or anaerobic	Soil or sewage	Enteric bacteria
Glucose, 20.0; $CaCO_3$, 20.0	pH 7.0; anaerobic	Pasteurized soil	Sugar-fermenting clostridia
Sodium lactate, 20.0	pH 7.0; anaerobic	Swiss cheese	Propionic acid bacteria
Ethanol, 40.0	pH 6.0; aerobic	Fruits, unpasteurized beer	Acetic acid bacteria

[a] Common components are yeast extract, 10.0; KH_2PO_4 or K_2HPO_4, 1.0; $MgSO_4$, 0.2. (All components are given in grams per liter.) Incubation is generally at 30°C.

LIGHT MICROSCOPY

In order to understand the indispensable role played by the microscope in the study of microorganisms, it is necessary to appreciate the intrinsic limitations of the eye as a magnifying instrument. The apparent size of an object viewed by the human eye is directly related to the angle that the object subtends at the eye: hence, if its distance from the eye is halved, its apparent size is doubled. However, the eye cannot focus on objects brought closer to it than approximately 25 cm; this is, accordingly, the distance of maximal effective magnification. In order to be seen at all, an object must subtend an angle at the eye of 1° or greater; and for a distance of 25 cm, this corresponds to a particle with a diameter of approximately 0.1 mm.

Most cells (and hence most unicellular microorganisms) are too small to be detected by the unaided human eye. In order to detect them, and to observe their form and structure, the use of a microscope is therefore essential. The function of the magnifying lens system of this instrument, interposed between the specimen and the eye, is greatly to increase the apparent angle subtended at the eye by objects within the microscopic field. In addition to this factor of *magnification*, two other factors, *contrast* and *resolution*, are of great importance. In order to be perceived through the microscope, an object must possess a certain *degree of contrast* with its surrounding medium; and in order to produce a clear magnified image, the microscope must possess a *resolving power* sufficient to permit the perception as separate objects of closely adjacent points in the image.

• **The light microscope**

As discussed in Chapter 1, Leeuwenhoek discovered the microbial world through the use of simple microscopes containing a single, biconvex lens of short focal length. The development and improvement of the more complex *compound microscopes* now employed required almost two centuries of research in applied optics.

A modern compound microscope contains three separate lens systems (Figure 2.6). The *condenser*, interposed between the light source and the specimen, collimates the light rays in the plane of the microscopic field. The *objective* produces a magnified image of the microscopic field within the microscope; and the *ocular* further enlarges this image and enables it to be perceived by the eye.

Single lenses have two inherent optical defects. They fail to bring the whole microscopic field into simultaneous focus (*spherical aberration*), and they produce colored fringes around objects in the field (*chromatic aberration*). These defects can be largely eliminated by placing additional, correcting lenses adjacent to a primary magnifying lens. Consequently, both the ocular and objective lenses of a modern compound microscope are multiple ones, designed to minimize these aberrations.

Correct adjustment of the condenser lens is of critical importance in providing a clear image. When a microscope is used at high magnifications, the condenser must be positioned in order to provide *critical illumination* of the microscopic field: rays from the light source must be brought to focus in the plane of observation, and the field of light transmitted must almost fill the objective lens.

51

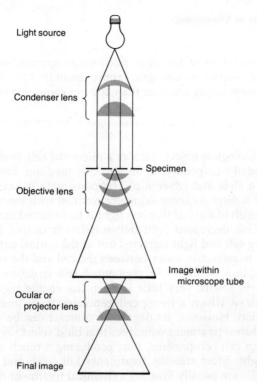

Light source

Condenser lens

Specimen

Objective lens

Image within
microscope tube

Ocular or
projector lens

Final image

FIGURE 2.6

Schematic representation of the compound
bright field microscope, adjusted for Kohler
illumination.

• **Resolving limit**   The physical properties of light set a fixed limit to the effective magnification
obtainable with a light microscope. Because of the wave nature of light, a very
small object appears to be a disc, surrounded by a series of light and dark rings.
Two adjacent points can be distinguished as separate, or *resolved*, only if the rings
surrounding them do not overlap. The distance between two points that can just
be distinguished from one another is known as the *resolving limit*, and it determines
the maximal useful magnification of the light microscope.

The resolving limit (*d*) is defined by the equation:

$$d = \frac{0.5\,\lambda}{N \sin \alpha} \tag{2.1}$$

where λ is the wavelength of the light source employed, α is the half angle of
the objective lens, and N is the refractive index of the medium between the
specimen and the front of the objective lens. The terms of the denominator
($N \sin \alpha$), commonly called the numerical aperture (NA), describe the properties
of the objective lens. Up to a certain limit, an increase in the numerical aperture
of the objective lens increases resolving power. If the medium between the
specimen and the objective is air, a feasible diameter of the objective lens limits
the NA to approximately 0.65. The value of N can be increased by filling the
intervening space with oil, which has a much higher refractive index than air.
With *oil immersion lenses*, the NA can be increased to a value as high as 1.4, although values of 1.25 are more common. Under the best obtainable conditions,

the maximal resolution of the light microscope approaches 200 nm, with light of the shortest visible wavelength (approximately 426 nm). In other words, two adjacent points closer together than 200 nm cannot be resolved into separate images.

• Contrast and its enhancement in the light microscope

When a small biological object, such as a microbial cell, is observed in the living state, it is normally suspended in an aqueous medium, compressed into a thin layer between a slide and cover slip. The perception of such an object depends on the fact that it displays some degree of contrast with the surrounding aqueous medium, as a result of the fact that less light is transmitted through it than through the medium. This decreased light transmission is caused by two factors: light absorbed by the cell and light refracted out of the optical path of the microscope, by a difference in refractive index between the cell and the surrounding medium. With the exception of intensely pigmented cell structures (e.g., chloroplasts) biological objects absorb very little light in the visible region of the spectrum; hence, the contrast which a living cell generates is attributable almost entirely to light refraction. However, its degree of contrast can be greatly increased by staining procedures: treatment with dyes that bind selectively either to the whole cell or to certain cell components, thus producing a much greater absorption of the incident light. Most staining treatments kill cells; and as a preliminary to staining, the cells are usually *fixed*, by a chemical treatment designed to minimize postmortem changes of structure. Commonly used fixatives include osmic acid and aldehydes, notably glutaraldehyde.

It is rarely necessary to stain microbial cells in order to make them visible; the simplest, and for many purposes the most satisfactory, way to observe micro-organisms with a light microscope is in the living state, as *wet mounts*. The principal value of staining procedures is to provide *specific information about the internal structure or the chemical properties of cells*. Thus, staining methods specific for deoxyribonucleic acid can reveal the structure and location of the nucleus; and a variety of special staining methods can be used to demonstrate intracellular deposits of such reserve materials as glycogen, polyphosphate, and poly-β-hydroxybutyrate. The Gram stain (see p. 125) and the acid-fast stain (see p. 168) are used to obtain information on the composition of the wall layers of bacterial cells. So-called *negative stains*, which do not enter the cell, are sometimes useful to reveal surface layers of very low refractive index, such as the capsules and slime layers that often surround microbial cells. These can be made visible by adding India ink to the suspending medium; since the carbon particles of the ink cannot penetrate the capsular layer, it is revealed as a clear zone surrounding the cell (see Figure 5.41).

ENHANCEMENT OF CONTRAST BY PHASE MICROSCOPY   The relatively low contrast of living cells as viewed with a conventional light microscope can be greatly increased by the use of an instrument with a modified optical system, known as the *phase contrast microscope* (Figure 2.7). Phase contrast microscopy is based on the fact that the rate at which light travels through objects is inversely related to their refractive indices. Since the wavelength of light is independent of the medium through which it travels, the phase of a light ray passing through an object of higher refractive

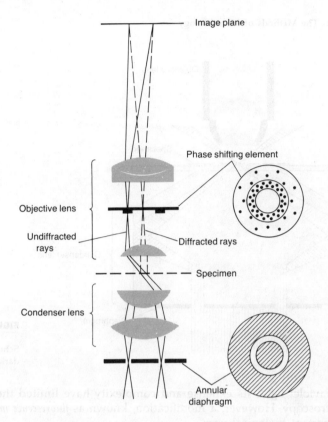

Image plane

Phase shifting element

Objective lens

Undiffracted rays

Diffracted rays

Specimen

Condenser lens

Annular diaphragm

FIGURE 2.7

Schematic representation of the phase contrast microscope.

index than the surrounding medium will be relatively retarded. A system of rings in the condenser and the objective separate the rays diffracted from the specimen from those which are not; the two sets of rays are recombined, after the diffracted rays have been passed through a glass ring, which introduces an additional phase difference. By this optical modification, the degree of contrast of cells or intracellular structures that differ very slightly in refractive index from their surroundings is greatly increased.

DARK-FIELD ILLUMINATION   When light impinges on a small object, some light is scattered, making the object appear luminous and thus visible against a dark background. The technique of *dark-field illumination* exploits this phenomenon and permits the detection of objects so small as to otherwise provide insufficient contrast. Such illumination is achieved by the use of a special kind of condenser, which focuses on the specimen a hollow cone of light (Figure 2.8), the diverging rays of which do not enter the objective. Only light scattered by the specimen enters the objective and is observed.

• Ultraviolet and fluorescence microscopy

Since the resolving power of the light microscope is directly related to the wavelength of light employed, a slight improvement of resolution (about twofold) can be achieved by the use of an ultraviolet light source. Since glass is opaque to ultraviolet light of short wavelengths, the lens system must be composed of quartz, and a camera must be used to record the image, since the eye cannot perceive

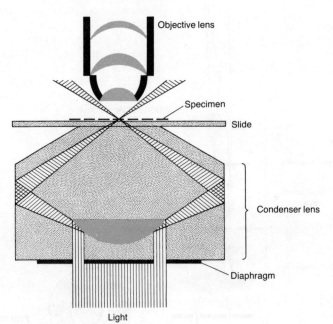

FIGURE 2.8

Schematic representation of dark field illumination.

ultraviolet light. Its expense and complexity have limited the use of ultraviolet microscopy. However, a modification, known as *fluorescence microscopy,* has many important biological uses.

Certain chemical compounds that absorb ultraviolet light reemit part of the radiant energy as light of longer wavelength, situated in the visible region; this phenomenon is termed *fluorescence.* When exposed to ultraviolet light, a fluorescent object can thus be perceived as a brightly colored body against a black background. This is the principle of fluorescence microscopy. Only the condenser lens must be constructed of quartz, since it is the visible light emitted by the fluorescent specimen that is transmitted through the microscope. The major use of fluorescence microscopy in biology involves the technique of *immunofluorescence.* When an animal is immunized with a specific antigen (e.g., a particular type of bacterium), its serum contains antibody proteins, which are capable of binding specifically to the immunizing antigen employed. The antibody proteins can be made intensely fluorescent, by chemical conjugation with a fluorescent dye; and when they combine with the specific antigen, it also becomes fluorescent. By fluorescence microscopy, it is therefore possible to detect specifically the cells of a particular type of bacterium in a mixed microbial population treated with a fluorescent antiserum directed against this bacterium. This technique has also been used to study the mode of growth of the bacterial cell wall (see Chapter 11, p. 331).

ELECTRON MICROSCOPY

The development of the electron microscope is one of the major achievements of applied physics in the twentieth century, and has revolutionized our knowledge of biological structure. It was based on the discovery that an

55 electromagnetic field acts on a beam of electrons in a way analogous to the action of a glass lens on a beam of light. An electron beam has the properties of an electromagnetic wave of very short wavelength; when accelerated through an electric field of 100 kv, its wavelength is only 0.04 nm, about 10,000 times shorter than that of visible light. The resolving limit of the electron microscope is consequently several orders of magnitude lower than that of the light miscroscope [see (Eq.) 2.1], and it thus permits the use of far higher effective magnifications (Table 2.7). As shown in Figure 2.9, the path of electrons through a transmission electron microscope is directed in a manner analogous to the path of light rays through a light microscope (Figure 2.6). A beam of electrons projected from an electron gun is passed through a series of electromagnetic lenses. The condenser lens collimates the electron beam on the specimen, and an enlarged image is produced by a series of magnifying lenses. The image is rendered visible by allowing it to impinge on a phosphorescent screen (similar to the front of a cathode-ray tube in a television set). Since electrons can travel only in a high vacuum, the entire electron path through the instrument must be evacuated; consequently, specimens must be completely dehydrated prior to examination. Furthermore, only very thin specimens (with a thickness of 100 nm or less) can be observed, since the penetrating power of electrons through matter is weak.

In a transmission electron microscope, contrast results from the differential scattering of electrons by the specimen, the degree of scattering being a function of the number and mass of atoms that lie in the electron path. Since most of the constituent elements in biological materials are of low mass, the contrast of these materials is weak. It can be greatly enhanced by "staining" with the salts of various heavy metals (e.g., lead, tungsten, uranium). These may be either fixed on the specimen (positive staining) or used to increase the electron opacity of

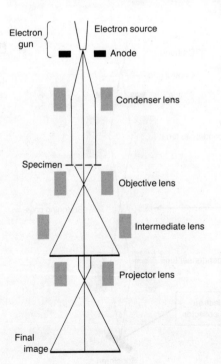

FIGURE 2.9

Schematic representation of the transmission electron microscope (TEM).

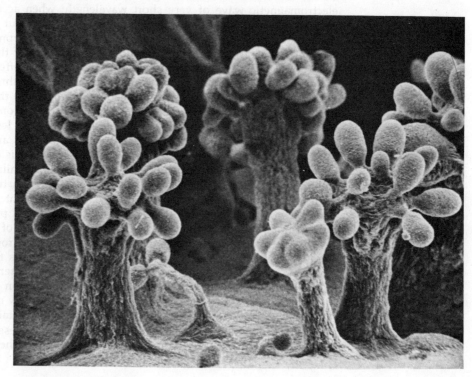

FIGURE 2.10

Scanning electron micrograph of the fruiting bodies of the myxobacterium, *Chrondromyces crocatus* (magnification ×820). Courtesy of J. Pangborn and P. Grilione.

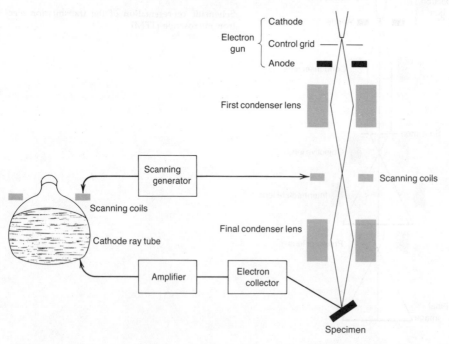

FIGURE 2.11

Schematic representation of the scanning electron microscope (SEM).

the surrounding field (negative staining). Negative staining is particularly valuable for the examination of very small structures such as virus particles, protein molecules, and bacterial flagella (see Figure 12.3). However, cells (even of very small microorganisms) are too thick to be examined satisfactorily in whole mounts. Observation of their internal fine structure requires that they be fixed, dehydrated, embedded in a plastic, and sectioned. Ultrathin sections (not more than 50 nm thick) are then positively stained with heavy metal salts, and are then mounted for examination.

Two other preparatory techniques, *metal shadowing* and *freeze-etching,* are frequently used for the observation of biological specimens with the transmission electron microscope. In metal shadowing the dried specimen is exposed at an acute angle to a directed stream of a heavy metal (platinum, palladium, or gold), thus producing an image which reveals the three-dimensional structure of the object. In the process of freeze-etching, the specimen is frozen; and the frozen block is fractured with a knife, exposing various surfaces on and within the specimen. The fractured surface is shadowed at an acute angle with a heavy metal, and a supporting layer of carbon is evaporated onto the metal surface. The shadowed specimen is then destroyed by chemical treatment, and the replica is examined. Freeze-etching has proved of particular value in the study of the wall and membrane structure of cells.

• **The scanning electron microscope**   Both light microscopes and transmission electron microscopes produce images that are essentially two-dimensional, since at any given setting of the instrument only a small depth of the field is in sharp focus; for example with the oil immersion objective of a light microscope, the depth of focus is only 0.25 μm. A recently developed modification of the electron microscope, known as the *scanning electron microscope*, makes it possible to obtain three-dimensional images of the surface structure of microscopic objects varying widely in size. The depth of focus of this instrument is several millimeters; and its range of effective magnification extends from about 20$\times$ to 10,000$\times$. An example of the type of image which it yields is shown in Figure 2.10.

The specimen to be examined is coated with a thin layer of a heavy metal, and is then scanned by a narrow electron beam (the *electron probe*) in a raster pattern, much like that of a television tube (Figure 2.11). Electrons are emitted from the point on the specimen struck by the probe, collected, and used to produce an image on the face of a cathode-ray tube. Contrast results from the fact that the number of secondary electrons emitted by the specimen is proportional to the angle between the electron beam and the surface of the specimen, the topology of which is thus revealed.

FURTHER READING

Books   MEYNELL, G. G., and E. MEYNELL, *Theory and Practice in Experimental Bacteriology.* New York: Cambridge University Press, 1965.

NORRIS, J. R., and D. W. ROBBINS (editors), *Methods in Microbiology.* London and New York: Academic Press, 1969–1973. A comprehensive reference work in seven volumes.

SCHLEGEL, H. G. (editor), *Anreicherungskultur und Mutantenauslese* (Enrichment Culture and Mutant Selection). Stuttgart: Fischer, 1965. The only systematic description of enrichment methods; many of the articles are in English.

Review LURIA, S. E., "The Bacterial Protoplasm: Composition and Organization," in *The Bacteria*, I. C. Gunsalus and R. Y. Stanier (editors), Vol. I, p. 1. New York: Academic Press, 1960.

THE NATURE OF THE MICROBIAL WORLD

3

The term *microorganism* does not have the precise taxonomic significance of such terms as *vertebrate* or *angiosperm*, each of which defines a restricted biological group, all members of which share numerous common structural and functional properties. In contrast, any organism of microscopic dimensions is by definition a microorganism; and microorganisms occur in a wide diversity of taxonomic groups, some of which (e.g., the algae) also contain members far too large to be assigned to this category. In this and the following two chapters, two major taxonomic groups that consist either in whole or in part of microorganisms—the procaryotes and the protists—will be surveyed and distinguished.

THE COMMON PROPERTIES OF BIOLOGICAL SYSTEMS

Cellular organisms share a *common chemical composition*, their most distinctive chemical attribute being the presence of three classes of complex macromolecules: deoxyribonucleic acid (DNA), ribonucleic acids (RNAs), and proteins. DNA is the constituent that carries in coded form all the genetic information necessary to determine the specific properties of the organism known collectively as its *phenotype*. The genetic information is initially transcribed into complementary RNA sequences, in the form of molecules of RNA known as *messenger* RNA (mRNA). The mRNA molecules subsequently serve as templates for the synthesis of all the specific protein molecules characteristic of the organism; the translation of the transcribed genetic message is mediated on organelles known as *ribosomes*, composed of protein subunits and of special types of RNA, rRNAs. A third class of RNA molecules, the transfer RNAs (tRNAs) also participate in protein synthesis, as carriers of the activated amino acids that are assembled into linear sequence in the primary step of protein synthesis. The proteins of an organism include the enzymes that catalyze its activities, and they also include the subunits from which many classes of proteinaceous cellular microstructures are assembled.

The chemical activities of an organism, catalyzed by its specific array of enzymes, are known collectively as *metabolic* activities. They include (1) the biosynthesis of the macromolecular constituents from the much simpler chemical substances (nutrients), derived from the external environment, and (2) the reactions necessary to generate the energy-rich substances which drive the processes of biosynthesis.

Most organisms share a common physical structure, being organized into microscopic subunits termed *cells*. All cells are enclosed by a thin membrane, the *cytoplasmic membrane*, which retains within its boundary the various molecules, large and small, necessary for the maintenance of biological function, and which at the same time regulates the passage of solutes between the interior of the cell and its external environment. Cells never arise *de novo*: they are always derived from preexisting cells, by the process of growth and cell division.

These generalizations apply to all living objects with the exception of viruses. The general properties of viruses will be described at the end of this chapter.

• Patterns of cellular organization

The simplest cellular organisms consist of a single cell. Since cells are always of microscopic dimensions, such *unicellular organisms* are necessarily small, and thus fall in the general category of microorganisms. Unicellularity is widespread, though not universal, in the microbial groups known as *bacteria, protozoa,* and *algae:* it likewise occurs, though more rarely, in *fungi.* The very considerable differences which exist among the various groups of microorganisms are expressed solely in terms of differences with respect to the *size, form,* and *internal structure* of the cell: sketches of a few unicellular organisms, all drawn to the same scale, are shown in Figure 3.1.

A more complex mode of organization is *multicellularity.* Although a multicellular organism arises initially from a single cell, it consists in the mature state of many cells, attached to one another in a characteristic fashion, which determines the gross external form of the organism. Multicellular organisms composed of a small number of cells may still be of microscopic dimensions; many examples exist among bacteria and algae. Such organisms are usually composed of similar cells, arranged in the form of a thread or filament. However, when the number

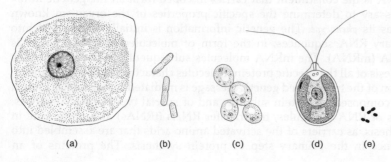

(a)   (b)   (c)   (d)   (e)

FIGURE 3.1

Drawings of several unicellular microorganisms on the same relative scale: (a) An amoeba; (b) a large bacterium; (c) a yeast; (d) a flagellate alga; (e) a small bacterium (×1,000).

of cells composing the organism is larger, the organism acquires a certain degree of structural complexity, simply from the manner in which the constituent cells are arranged. The best illustrations of such simple multicellular organization occur among the larger algae (for example, marine algae) which often have a characteristically plantlike form, even though there is little or no specialization of the component cells. Form is derived by the specific pattern in which the like structural units are arranged.

In metazoan animals and vascular plants, multicellular organization leads to a much higher degree of intrinsic structural complexity, as a consequence of the *differentiation of distinct cell types* during the development of the mature organism. This leads, through cell division, to the emergence of distinct *tissue regions,* each composed of a special type of cell; a further level of internal complexity may be attained by the association of different cell types into functional units known as *organs.* The structural complexity of a vascular plant or a metazoan animal thus proves upon microscopic analysis to be vastly greater than that of large but undifferentiated multicellular organisms such as the marine algae.

In a few biological groups, biological organization assumes a third form, known as *coenocytic structure,* which at first sight seems to contradict the axiom that organisms are composed of cells. A coenocytic organism is not composed of cellular subunits, separated from one another by their bounding membranes; instead, the cytoplasm is continuous throughout the individual organism, which grows in size without undergoing cell division. This type of organization is characteristic of most fungi, and also occurs in many algae.

• **The problem of primary divisions among organisms**

It is a judgment of common sense, as old as mankind, that our planet is populated by two different kinds of organisms, plants and animals. Early in the history of biology this prescientific opinion became formalized in scientific terms: biologists recognized two primary kingdoms of organisms, the *Plantae* and the *Animalia.* The members of the two kingdoms appeared to be readily distinguishable by a whole series of characters, both structural and functional, some of which are summarized in Table 3.1. This traditional bipartite division was in fact a satisfactory one, as long as biologists had to take into account only the more highly differentiated groups of multicellular organisms.

• **The place of microorganisms**

When exploration of the microbial world got under way in the eighteenth and nineteenth centuries, there seemed no reason to doubt that these simple organisms could be distributed between the plant and animal kingdoms. In practice, the assignment was usually made on the basis of the most easily determinable differences between plants and animals: the power of active movement and the ability to photosynthesize. Multicellular algae, which are immotile, photosynthetic, and in some cases plantlike in form, found a natural place in the plant kingdom. Although they are all nonphotosynthetic, the coenocytic fungi were also placed in the plant kingdom on the basis of their general immotility. Microscopic motile forms were lumped together as one group of animals, the Infusoria (Table 3.2).

Following the enunciation and acceptance of the cell theory (about 1840), biologists perceived that the Infusoria were a very heterogeneous group in terms of their cellular organization. Some of these microscopic forms (e.g., the rotifers) are invertebrate animals, with a body plan based on differentiation during multicellular development. Furthermore, the unicellular representatives can be sub-

TABLE 3.1

Some major differences between metazoan animals and
vascular plants

		VASCULAR PLANTS	METAZOAN ANIMALS
Functional characters	Energy source	Light	Chemical energy
	Carbon source	CO_2	Organic compounds
	Growth factor requirements	None	Complex
	Active movement	Absent	Present
Structural characters	Cell walls	Present	Absent
	Chloroplasts	Present	Absent
	Mode of growth	Open[a]	Closed[a]

[a] In animals the individual achieves a more or less fixed size and form
as an adult. In most plants growth continues throughout the life of the individual, and the final size and form are much less rigidly fixed.

divided into two groups: *protozoa*, with relatively large and complex cells, and
bacteria, with much smaller and simpler cells. The old Infusoria was thus split three
ways. Some of its component groups were classified as metazoan (multicellular)
invertebrate animals. Others, the protozoa, were kept in the animal kingdom, but
differentiated from all other animals on the basis of their unicellular structure.
Finally, the bacteria were transferred to the plant kingdom, despite their generally
nonphotosynthetic nature, as a result of the discovery that the blue-green bacteria,
then considered to be algae, were characterized by cells with a comparably simple
structure.

However, subsequent experience showed that the treatment of the protozoa
(a large and complex microbial group) as unicellular animals led to considerable
difficulties. Such protozoa as the ciliates and amebas, phagotrophic organisms
devoid of cell walls, could be fitted quite satisfactorily into the confines of the
animal kingdom, but other protozoa could not. On closer study, the flagellate
protozoa proved to be a very odd assortment of creatures, in some of which
motility by means of flagella was the only "animallike" character. Some possessed
cell walls, others did not. Some were phototrophs, others chemotrophs: and among
the latter, both osmotrophic and phagotrophic representatives occurred. In short,

TABLE 3.2

Early attempts (about 1800) to allocate microorganisms to the
plant and animal kingdoms

PLANTS	ANIMALS
Algae (immotile, photosynthetic)	Infusoria (motile)
Fungi (immotile, nonphotosynthetic)	

TABLE 3.3

Final effort (about 1860) to allocate microorganisms to the plant and animal kingdoms

PLANTS	CONTESTED GROUPS	ANIMALS
		Small metazoans
		Rotifers
		Nematodes (some)
		Arthropods (some)
Algae (photosynthetic)		Protozoa
		Ciliates
Immotile forms ←———	Photosynthetic flagellates ——→	Nonphotosynthetic flagellates
Fungi (nonphotosynthetic)		
True fungi ←———	Slime molds ————→	Ameboid protozoa
Bacteria		

this one microbial group shows all possible combinations of plantlike and animallike characters. The problem of the placement of the flagellates became even more acute when it was recognized that in terms of cellular properties many of the phototrophic flagellates resembled very closely certain of the multicellular, immotile algae. Another protozoan group, the slime molds, also presented difficulties. In the vegetative state these organisms are phagotrophic and ameboid, but they can also form complex fruiting structures, similar in size and form to those characteristic of the true fungi. Should the slime molds be classified with the fungi, as plants, or with the protozoa, as animals?

Consequently, as knowledge of the properties of the various microbial groups deepened, it became apparent that at this biological level a division of the living world into two kingdoms cannot really be maintained on a logical and consistent basis. Some groups (notably the flagellates and the slime molds) were claimed both by botanists as plants and by zoologists as animals (Table 3.3). The problem is easy enough to understand in evolutionary terms. The major microbial groups can be regarded as the descendants of very ancient evolutionary lines, which antedated the emergence of the two great lines that eventually led to the development of plants and animals. Hence, most microbial groups cannot be pigeonholed in terms of the properties that define these two more advanced evolutionary groups.

• **The concept of protists**

Dissatisfaction with existing classification, coupled with a clear understanding of the root of the trouble, led one of Darwin's disciples, E. Haeckel, to propose the obvious way out. In 1866 he suggested that logical difficulties could be avoided by the recognition of a *third* kingdom, the *protists*, to include protozoa, algae, fungi, and bacteria. The protists accordingly include both photosynthetic and nonphotosynthetic organisms, some plantlike, some animallike, some sharing properties specific to both the traditional kingdoms. What distinguished all protists from plants and animals was their *relatively simple biological organization*. Many protists are unicellular or coenocytic; and even the multicellular protists (e.g., the larger algae) lack the internal differentiation into separate cell types and tissue regions

TABLE 3.4

Component groups of the three kingdoms of organisms proposed
by Haeckel (1866)

PROPERTIES	PLANTS	ANIMALS
Multicellular; extensive differentiation of cells and tissues	Seed plants Ferns Mosses and liverworts	Vertebrates Invertebrates
	PROTISTS	
Unicellular, coenocytic, or multicellular; latter with little or no differentiation of cells and tissues	Algae Protozoa Fungi Bacteria	

characteristic of plants and animals. A primary division in the biological world
could accordingly be made in terms of the *degree of complexity of biological organization;*
this could then be followed, for the more highly organized forms, by a secondary
division on the basis of the properties long used to separate plants from animals
(Table 3.4).

EUCARYOTES AND PROCARYOTES

About 1950 the development of the electron microscope and of
associated preparative techniques for biological materials made it possible to
examine the structure of cells with a degree of resolution many times greater
than that previously possible by the use of the light microscope. Within a few
years many hitherto unperceived features of cellular fine structure were revealed.
This led to the recognition of a profoundly important dichotomy among the
various groups of organisms with respect to *the internal architecture of the cell:* two
radically different kinds of cells exist in the contemporary living world. The more
complex *eucaryotic cell* is the unit of structure in plants, metazoan animals, protozoa,
fungi, and all save one of the groups which had been traditionally assigned to
the algae. Despite the extraordinary diversity of the eucaryotic cell as a result
of its evolutionary specialization in these groups, as well as the modifications
which it can undergo during the differentiation of plants and animals, its basic
architecture always has many common denominators. The less complex *procaryotic
cell* is the unit of structure in two microbial groups: bacteria and the organisms
formerly known as blue-green algae, which share with eucaryotic algae a common
mechanism of photosynthesis, housed in a cell of radically different fine structure.
The placement of the so-called blue-green "algae" with other algal groups can
thus no longer be justified, and they will be treated in this book as one group
of photosynthetic bacteria, the blue-green bacteria.

The flood of new information about cellular fine structure provided by electron

TABLE 3.5

The primary subdivisions of cellular organisms which are now recognized

		PLANTS	ANIMALS
Eucaryotes	Multicellular; extensive differentiation of cells and tissues	Seed plants Ferns Mosses Liverworts	Vertebrates Invertebrates
		PROTISTS	
	Unicellular, coenocytic, or mycelia; latter with little or no differentiation of cells and tissues	Algae Protozoa Fungi	
Procaryotes		Bacteria	

microscopy has been paralleled by an equally rapid growth of knowledge about cellular function, in large part a result of the rise of molecular biology. This has made it evident that the structural differences between eucaryotic and procaryotic cells are expressive of highly important differences in the way that universal cell functions are accomplished: notably the transmission and expression of genetic information, the performance of energy-yielding metabolism, and the entry and exit of materials. In summary, it is evident that the line of demarcation between eucaryotic and procaryotic cellular organisms is the largest and most profound single evolutionary discontinuity in the contemporary biological world. Furthermore, it permits a completely unambiguous bipartite division of organisms, exclusively based on their cellular properties.

This newly recognized line of demarcation runs through Haeckel's proposed kingdom of protists. Protozoa, fungi, and algae (with the exclusion of blue-greens) are eucaryotes, indistinguishable in terms of *cellular* attributes from plants and animals. All bacteria (including blue-green bacteria) are procaryotic, and their simple designation as *procaryotes* adequately defines this large biological assemblage. The term *protist* can now be most usefully restricted to the relatively simple eucaryotes: protozoa, fungi, and algae. It becomes a collective name to distinguish them from plants and animals (Table 3.5).

In the following pages the principal features of organization and function that distinguish eucaryotic from procaryotic cells will be summarized.

ORGANIZATION AND FUNCTION IN THE EUCARYOTIC CELL

As already mentioned, all cells are bounded by a surface membrane known as the cytoplasmic membrane. This can be resolved by electron microscopy as a triple-layered structure with a total width of about 8 nm.* Membranes possessing this fine structure are termed *unit membranes*. A characteristic property of the eucaryotic cell is the presence within it of a *multiplicity of unit membrane systems*, many of which are both structurally and topologically distinct from the

---

*The nanometer (nm) is now the standard unit of length used to describe biological fine structure. It equals 10^{-3} micrometer (μm), or 10^{-7} cm. The Angstrom unit (A), frequently used in older descriptions of biological fine structure, equals 10^{-1} nm.

cytoplasmic membrane. These internal unit membrane systems serve to segregate many of the functional components of the eucaryotic cell into specialized and partly isolated regions, between which passage of materials is effected largely through membrane transport.

The most complex internal membrane system, topologically speaking, is the *endoplasmic reticulum* (ER). It consists of an irregular network of interconnected membrane-delimited channels that traverses much of the interior of the cell and is in direct contact with two other major cell components: the nucleus and some of the cytoplasmic ribosomes. Part of the ER surrounds the nucleus, being expanded to form the *nuclear membrane,* which has a distinctive structure, perforated by numerous pores about 40 nm in diameter. In other regions of the ER, known as the *rough endoplasmic reticulum,* the surfaces of the membranes are coated with ribosomes (Figure 3.2). Proteins synthesized on the ribosomes of the rough ER pass into the channels of the ER, through which they are carried to other regions of the cell.

Another characteristic membranous organelle is the Golgi apparatus (Figure 3.3), which consists of a densely packed mass of flattened sacs and vessels of varying size. Its manifold functions include (1) the packaging of material, both proteinaceous and nonproteinaceous, synthesized in the endoplasmic reticulum and (2) their transport to other regions of the cell, or to the cell surface, where Golgi vesicles may coalesce with the cytoplasmic membrane and liberate their contents to the exterior of the cell, a process known as *exocytosis.*

The machinery responsible for the performance of respiration and—in photosynthetic eucaryotes—of photosynthesis is segregated in two functionally distinct

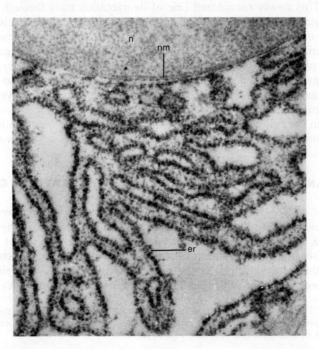

FIGURE 3.2

Electron micrograph of a thin section of a rabbit plasmocyte, showing a portion of the cytoplasm filled with rough endoplasmic reticulum (er); the field also includes a portion of the nucleus (n), surrounded by the nuclear membrane (nm) ($\times$35,000). Courtesy of Dr. L. G. Chevance. Institut Pasteur.

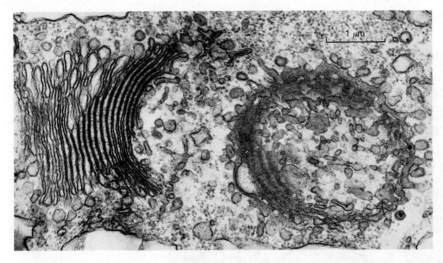

FIGURE 3.3

The Golgi apparatus as seen in an electron micrograph of a thin section of *Euglena gracilis* (×28,000). Two adjacent Golgi bodies have been sectioned in different planes. At left, vertical section through the membrane stack. At right, section parallel to the stack. Courtesy of Gordon F. Leedale.

classes of membrane-bounded organelles, mitochondria (Figure 3.4), and chloroplasts (Figure 3.5). Each type of organelle contains an internal membrane system of characteristic structure and function. The internal membranes of the mitochondrion (*cristae*) house the respiratory electron transport system; the internal membranes of the chloroplast (thylakoids) house the photosynthetic pigments and electron transport system, as well as the photochemical reaction centers.

• **The eucaryotic genome: its replication, transcription, and translation**

In a eucaryotic cell the nucleus is the principal but never the sole repository of genetic information. A quantitatively minor but functionally very important part of the total cellular genome is located in mitochondria and (in photosynthetic organisms) in chloroplasts. The organellar DNA determines some (but by no means all) of the properties of the specific organelle with which it is associated. Furthermore, each type of organelle likewise contains an organelle-specific machinery of transcription and translation. It therefore follows that the replication of the eucaryotic genome, as well as its transcription and translation, occurs at either two or three distinct sites within the cell: in the nucleocytoplasmic region; in the mitochondrion; and in the chloroplast. Furthermore, the machinery of organellar replication, transcription, and translation differs from its counterpart in the nucleocytoplasmic region. The properties of these two systems must therefore be discussed separately.

REPLICATION, TRANSCRIPTION, AND TRANSLATION OF THE NUCLEAR GENOME The genetic information contained in the eucaryotic nucleus is dispersed over a limited number of distinct structural elements known as *chromosomes*. Each chromosome is a threadlike structure containing DNA, a special class of basic proteins known as *histones*, and a group of nonhistone proteins which appear to play a role in the regulation of gene expression. In the nondividing or *interphase* nucleus (Figure 3.6) each chromosome is greatly elongated and is only 20 to 30 nm in width; it therefore cannot be resolved by light microscopy. The interphase nucleus contains an organelle known as the *nucleolus*, rich in RNA and associated with a specific chromosomal region known as the *nucleolar organizer*. The nucleolar organizer

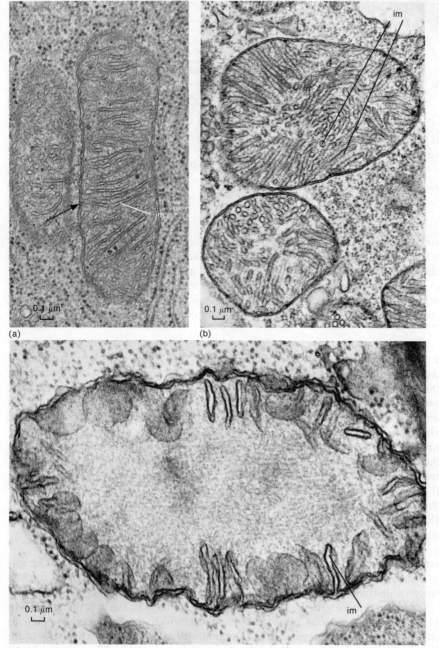

(a)

(b)

(c)

FIGURE 3.4

The structure of mitochondria as seen in electron micrographs of thin sections of eucaryotic cells: (a) Mitochondria in a mammary gland cell of the mouse ($\times$56,100). Numerous flattened internal membranes (im) arise by invagination from the inner enclosing membrane of the organelle (arrow). Courtesy of Dorothy Pitelka. (b) Mitochondria of a ciliate, *Condylostoma* ($\times$56,100). The internal membranes (im) are tubular in cross section and are very abundant. Courtesy of Dorothy Pitelka. (c) Mitochondrion of a photosynthetic flagellate, *Euglena* ($\times$65,450). The flattened internal membrane (im) are less numerous and less extensively intruded than in (a) and (b). Courtesy of Gordon F. Leedale.

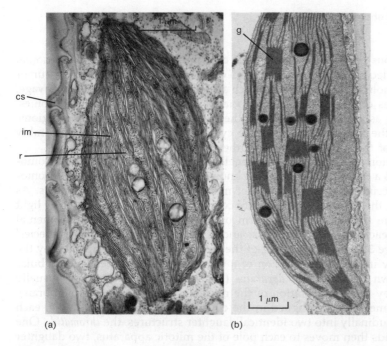

(a)   (b)

FIGURE 3.5

The structure of chloroplasts as revealed in electron micrographs of thin sections of eucaryotic cells. (a) Chloroplast of the unicellular alga *Euglena* (×21,200). The internal membranes (im) are arranged in irregular parallel groups and run in the long axis of the chloroplast. Ribosomes (r) are scattered between the lamellae. The chloroplast lies just below the sculptured cell surface (cs). From G. F. Leedale, B. J. D. Meeuse, and E. G. Pringsheim, "Structure and Physiology of *Euglena spirogyra,*" *Arch. Mikrobiol.* **50,** 68 (1965). (b) Chloroplast of a sugar beet leaf (×14,840). The internal membranes tend to be arranged in dense, regular stacks, termed *grana* (g), in the chloroplasts of plants. Courtesy of W. M. Laetsch.

FIGURE 3.6

An electron micrograph of a freeze-etched preparation of the resting nucleus of a mouse cell. The plane of fracture has passed in part through the nuclear membrane and reveals the nuclear pores both on the inner surface of the membrane (npi) and on its outer surface (npo) (×24,000). Courtesy of Dr. L. G. Chevance, Institut Pasteur.

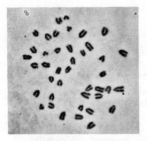

FIGURE 3.7

Stained preparation of a mitotic figure in the nucleus of a mouse cell, showing the chromosome array. Courtesy of Dr. H. Jakob, Institut Pasteur.

carries numerous copies of the genes that determine the structures of ribosomal RNAs; the nucleolus, in fact, functions as a center for the synthesis of a precursor RNA of high molecular weight from which are subsequently derived, by cleavage, the major types of RNA molecules associated with the cytoplasmic ribosomes. These, as well as messenger RNAs synthesized in other chromosomal regions, pass through the nuclear pores into the cytoplasm, where ribosomal assembly and the bulk of the cellular protein synthesis take place.

The replication and partition of the genetic information carried on the chromosomes involves a complex cyclic process known as *mitosis*. Replication of chromosomal DNA occurs prior to the onset of mitosis, in the interphase nucleus. As mitosis begins, the chromosomes coil into compact structures visualizable by light microscopy: the total number and form of the chromosomes are fundamental characters of each eucaryotic species, and collectively determine the species' *karyotype* (Figure 3.7). The shortening of the chromosomes is accompanied by the rapid assembly in the nuclear region of a bipolar, spindle-shaped microtubular structure known as the *mitotic apparatus* (Figure 3.8). Its formation is usually accompanied by the disintegration of the nucleolus and the nuclear membrane. The chromosomes become aligned in the equatorial region of the spindle, each splitting longitudinally into two identical daughter structures, the *chromatids*. One set of chromatids then moves to each pole of the mitotic apparatus, two daughter nuclei being reorganized at the poles, with an accompanying disassembly of the spindle. Cell division normally takes place during the terminal phase of mitosis, the plane of division following the equatorial plane of the spindle.

REPLICATION, TRANSCRIPTION, AND TRANSLATION OF THE ORGANELLAR GENOMES   The DNA in chloroplasts and mitochondria exists as small double helical molecules, usually circular, and not associated with histones (Figure 3.9). The organellar genetic material is thus located in structures which appear to be very similar to procaryotic chromosomes, although of considerably smaller size. Numerous copies (as many as 40 to 50 in the case of some chloroplasts) are present in each organelle. Chloroplasts and mitochondria likewise contain the machinery of transcription and translation including organelle-specific ribosomes, which are smaller than the 80S cytoplasmic ribosomes and resemble in size the 70S ribosomes of procaryotes. Organellar protein synthesis can be inhibited by chloramphenicol and certain other antibiotics which are likewise inhibitors of this process in procaryotes but do not affect eucaryotic cytoplasmic protein synthesis. In a number of important respects, accordingly, chloroplasts and mitochondria show fundamental resemblances to procaryotic cells. Mitochondria also possess a property characteristic of cells, but not of other cell constituents: they are formed by division of preexisting organelles. This has also been demonstrated for many types of algal chloroplasts. In higher plants mature chloroplasts develop from simpler structures known as *proplastids*, and organellar replication takes place at the proplastid stage.

The newly recognized resemblances between procaryotic cells and these two classes of eucaryotic organelles have important evolutionary implications. These resemblances suggest that both chloroplasts and mitochondria may have had evolutionary origins different from that of the rest of the eucaryotic cell, being derived originally from ancestral procaryotes that entered the eucaryotic cell as

71

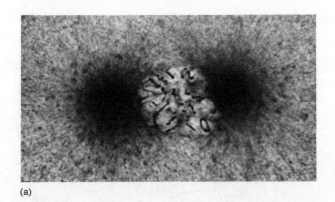

(a)

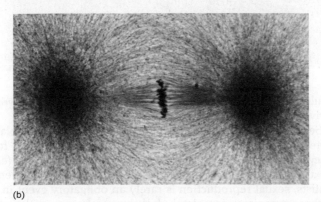

(b)

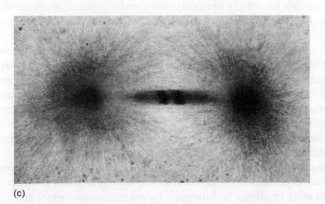

(c)

FIGURE 3.8

Photomicrographs illustrating three successive phases of a mitotic nuclear division. (a) Early organization of the spindle; the nuclear membrane has disappeared, and the chromosomes are already visible in the region of the organizing spindle. (b) The spindle is now fully developed, and the chromosomes are regularly aligned in its equatorial plane; this stage is often referred to as the *metaphase of mitosis*. (c) Separation of the two daughter sets of chromosomes has occurred, and each set is being withdrawn toward one pole of the spindle.

symbionts and eventually lost their capacity to live in independence of the host organism (see Chapters 26–28).

• Eucaryotic sexual processes

Cellular fusion is the first step in the process of *sexual reproduction*. The two cells that participate are known as *gametes* and the resulting fusion cell as a *zygote*. In all eucaryotic organisms, gametic fusion is followed by nuclear fusion, with the

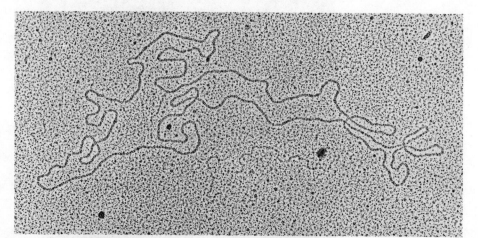

FIGURE 3.9

Electron micrograph of a circular double helical molecule of mitochondrial DNA, isolated from the yeast, *Saccharomyces cerevisiae*. The total length of the molecule is 9.2 nm ($\times$ 160,000). Courtesy of J. Lazowska and P. Slonimski.

result that the zygote nucleus contains *two complete sets of genetic determinants*, one derived from each gametic nucleus.

Sexual reproduction is common in the life cycle of plants and animals. In vertebrates and many invertebrates, it is the *only* method for the production of a new individual. Plants can also be propagated asexually (e.g., by cuttings), and asexual modes of reproduction exist in many groups of invertebrates. Among the protists, sexual reproduction is rarely an obligatory event in the life cycle. Many of these organisms completely lack a sexual stage in their life cycles, and even in species in which sexuality does exist, sexual reproduction may occur infrequently, the formation of new individuals taking place principally by asexual means (for example, by binary fission or the formation of spores).

Sexual fusion results in a *doubling of the number of chromosomes*, since the nuclei of the gametes, each containing N chromosomes, fuse to form the nucleus of the zygote, which consequently contains $2N$ chromosomes. Hence, in passing from one sexual generation to the next, there must at some stage be a *halving of the number of chromosomes*, if the chromosome content of the nucleus is not to increase indefinitely. If fact, the halving of the chromosome number is a universal accompaniment of sexuality. It is brought about by a special process of nuclear division termed *meiosis* (Figure 3.10). In animals, *meiosis takes place immediately prior to the formation of gametes*. In other words, each individual of the species has $2N$ chromosomes in its cells through most of the life cycle. Such an organism is termed *diploid*. This state of affairs is, however, by no means universal among sexually reproducing eucaryotic organisms. In many protists, *meiosis takes place immediately after zygote formation*, with the consequence that the organisms have N chromosomes through most of the life cycle. Such organisms are termed *haploid*. In many algae and plants, as well as in some fungi and protozoa, there is a well-marked *alternation of haploid and diploid generations*. In this type of life cycle, the diploid zygote gives rise to a diploid individual, which forms, by meiosis, haploid *asexual* reproductive cells. Each such haploid cell gives rise to a haploid individual, which eventually forms haploid gametes; gametic fusion, with the formation of a diploid zygote once again, completes the cycle.

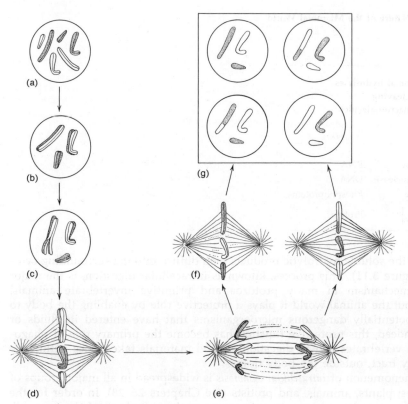

(a)

(b)

(c)

(d)

(e)

(f)

(g)

FIGURE 3.10

Meiosis in a hypothetical diploid plant cell (a) with three pairs of chromosomes. (b) Homologous chromosomes pair. (c) An exchange of segments (crossing over) takes place (shown for one chromosome only). (d) The chromosomes are shown at "first metaphase." (e) The chromosomes of each pair separate. (f) Two nuclei have formed, and within each a metaphase spindle forms. This time, however, the sister chromatids that make up each single chromosome separate. This phase, called the "second metaphase," is thus analogous to a mitotic division. (g) The four haploid nuclei that result from meiosis are shown.

• **The transfer of materials across the eucaryotic cell membrane: endocytosis and exocytosis**

Although small molecules in solution can enter the eucaryotic cell by passage through the cytoplasmic membrane, the entry of other materials can occur by a second quite distinct mechanism: bulk transport of small droplets, enclosed by an infolding of the cytoplasmic membrane to form a membrane-enclosed vacuole. The most familiar example of this phenomenon is the *phagocytosis* of bacteria or other small solid objects by phagotrophic protozoa, or by the phagocytic cells of metazoan animals. Droplets of liquid can enter the eucaryotic cell in a similar fashion, this process being termed *pinocytosis*. Phagocytosis and pinocytosis are known collectively as *endocytosis*. During endocytosis large amounts of surface membrane are taken into the cell to form the membranes of the enclosing vacuoles, the surface area of the cytoplasmic membrane being maintained by synthesis of new membrane material.

Endocytosis is a distinctively eucaryotic process of fundamental biological importance, which initiates both *intracellular digestion* (hydrolysis of biological macromolecules) and the *establishment of endosymbionts*.

One of the products formed in the Golgi apparatus is a membrane bounded vesicle known as the *lysosome*. Lysosomes contain an extensive array of hydrolytic enzymes capable of breaking down most classes of biological macromolecules (Table 3.6). These enzymes do not normally act upon the constituents of the cell in which they are formed, since they are segregated within the lysosomal membrane. However, the lysosomes can fuse with vacuoles formed through endocytosis, thus permitting hydrolysis of the materials (or cells) contained in these

TABLE 3.6

Some lysosomal hydrolases
capable of cleaving
biological macromolecules

ENZYMES	SUBSTRATE
Ribonuclease	RNA
Deoxyribonuclease	DNA
Phosphatase	Phosphoproteins
Cathepsin ⎫ Collagenase ⎰	Proteins

vacuoles: the soluble hydrolytic products then diffuse into the surrounding cytoplasm (Figure 3.11). This process, known as intracellular digestion, is the major feeding mechanism in many protozoa and primitive invertebrate animals. Throughout the animal world it plays a protective role by enabling the body to destroy potentially dangerous microorganisms that have entered its fluids or tissues. Indeed, this protective function has become the primary role of phagocytosis in vertebrates, where digestion of food materials takes place within the alimentary tract, outside the body tissues.

The phenomenon of *intracellular symbiosis* is widespread in all major groups of eucaryotes: plants, animals, and protists (see Chapters 26–28). In order for the cells of future endosymbionts to gain access to the cytoplasm of the host cell, they must pass intact through the host cytoplasmic membrane: this passage invariably occurs through phagocytic engulfment. In this situation, engulfment of the foreign cell is not followed by lysosomal digestion.

Droplets or solid particles of material synthesized within the eucaryotic cell can also pass to the exterior by the converse mechanism, known as *exocytosis*. Here also the Golgi apparatus plays a key role, since materials destined for exocytosis

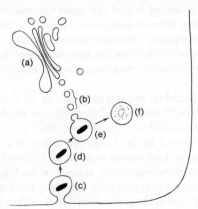

FIGURE 3.11

A diagrammatic representation of the events of intracellular digestion: (a) Golgi apparatus; (b) lysosomes produced from the Golgi apparatus; (c) phagocytic capture of a food particle (a bacterium) at the surface of the cell, during which the particle is almost completely surrounded by the cell membrane; (d) newly formed food vacuole; (e) coalescence of the food vacuole with a lysosome; (f) digestion of the vacuolar contents by hydrolytic enzymes released from the lysosome. Modified from N. Novikoff, E. Essner, and N. Quintana, *Federation Proc.* **23,** 1011 (1964).

are packaged initially in Golgi vesicles. The secretion of enzymes and of hormones by specialized animal cells occurs in this manner, and in algae it has been shown that formation of the cell wall involves the exocytosis of small fragments of the wall fabric, synthesized endogenously and transported to the cell surface in Golgi vesicles.

• **Microtubular systems**

An element of structure that has many functions in the eucaryotic cell is the *microtubule,* an extremely thin cylinder, some 20 to 30 nm in diameter and of indefinite length. The walls of microtubules are composed of globular protein subunits with a molecular weight of 50,000 to 60,000, which can assemble in regular array.

Microtubules provide the structural framework of the mitotic spindle; they also appear to play a role in the establishment and maintenance of the shape of many types of eucaryotic cells. In addition, a regular longitudinal array of microtubules occurs within the eucaryotic locomotor organelles known as *cilia* or *flagella.* The cilium or flagellum is enclosed by an extension of the cytoplasmic membrane and contains a set of nine outer pairs of radially arranged microtubules, in turn surrounding an inner central pair (Figure 3.12). The central microtubules arise from a plate near the surface of the cell, whereas the outer pairs originate from a cylindrical body or *centriole* (Figure 3.13), which is likewise composed of nine pairs of microtubules.

In some eucaryotes the centrioles are also associated with the formation of the microtubular system of the mitotic apparatus, being located at the two poles of the spindle.

• **Osmoregulation in eucaryotes**

Most free-living microorganisms live in environments with a water concentration considerably greater than that inside the cell. Since the cytoplasmic membrane is freely permeable to water, but not to many solutes, there is a tendency for water to enter the cell; unless this tendency is counterbalanced in some manner, the cell swells and eventually undergoes *osmotic lysis.* In many protists (algae, fungi) the danger of osmotic lysis is prevented mechanically by enclosure of the cell in a rigid wall of sufficient tensile strength to counterbalance water pressure, and thus prevent lysis of the cell. However, many protozoa do not possess walls; and in these protists a special type of vacuole, the *contractile vacuole,* functions as a cellular pump to collect water from within the cell and periodically discharge it to the exterior through coalesence with the cell membrane. The operation of the contractile vacuole is accordingly another mode of exocytosis and provides an active mechanism of osmoregulation.

• **Directed intracellular movement in the eucaryotic cell**

Examination of many types of eucaryotic cells in the living state reveals that the cytoplasm is frequently in active movement, a phenomenon known as *cytoplasmic streaming.* Furthermore, it is evident that intracellular movements are often closely directed: the light-induced orientation of chloroplasts, the localized concentration in cells of mitochondria, the roles played by the Golgi apparatus in the intracellular transport of packaged materials, and the movement of chromosomes during

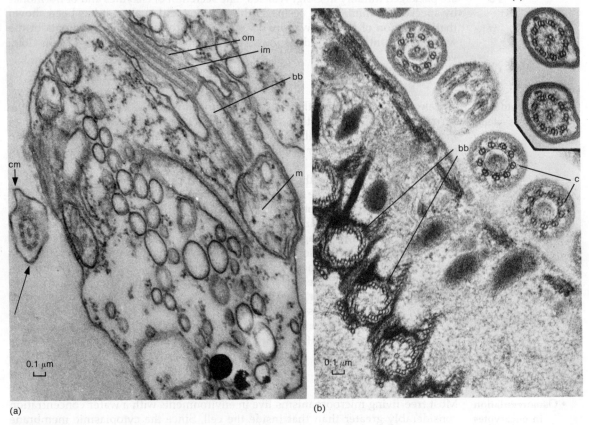

(a)

(b)

(c)

FIGURE 3.12

The fine structure of eucaryotic flagella and cilia, as revealed by electron micrographs of thin sections. (a) Longitudinal section through the cell of *Bodo*, a nonphotosynthetic flagellate (×38,800): cylindrical basal body (bb); outer microtubules (om); inner microtubules (im). Underlying the basal body is a specialized mitochondrion (m). At left (arrow), transverse section of a flagellum external to the cell. Note enclosure by an extension of the cell membrane (cm). (b) Section through the body surface of a ciliate, *Didinium* (×51,800). Within the cell (lower left), basal bodies (bb) have been sectioned transversely; their walls are composed of nine triple rows of microtubules. Just above the cell surface, several cilia (c) have been sectioned transversely; note the nine outer pairs of microtubules and the absence of the inner pair of microtubules. (c) Insert at upper right: section through two cilia at a point some distance from the cell surface. Note the inner pair of microtubules, the nine outer pairs, and the enclosing membrane. Courtesy of Dorothey Pitelka.

mitosis are only some of the phenomena that reveal the precision with which the relative positions of the cell components can be controlled.

In many protists which have cells not enclosed by walls, directed cytoplasmic streaming can be used as a means of cellular translocation over a solid substrate: this is the characteristic mode of movement in ameboid cells (Figure 3.14).

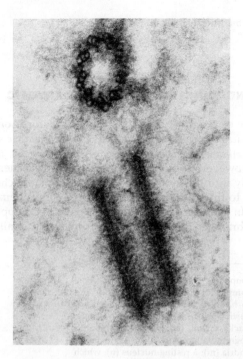

FIGURE 3.13

Electron micrograph of a pair of centrioles in a dividing human lymphosarcoma cell ($\times 104{,}000$). One is sectioned transversely, the other longitudinally, revealing the typical hollow cylindrical structure of this organelle. Courtesy of G. Bernhard, Institut de Recherches sur le Cancer, Villejuif, France.

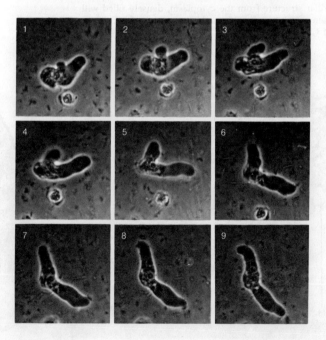

FIGURE 3.14

Amoeboid movement. A series of successive photomicrographs by phase contrast, taken at intervals of 15 seconds, of a small amoeba, *Tetramitus* ($\times 9{,}900$). Courtesy of Jeanne Stove Poindexter.

ORGANIZATION AND FUNCTION IN THE PROCARYOTIC CELL

One of the most striking structural features of the procaryotic cell is the *absence of internal compartmentalization by unit membrane systems* (Figure 3.15). The *cytoplasmic membrane is, in the great majority of procaryotes, the only unit membrane system of the cell.* However, the topology is often complex, membranous infoldings penetrating deeply into the cytoplasm. The blue-green bacteria provide the sole known exception to the rule that there is only one unit membrane system in the procaryotic cell. In these organisms the photosynthetic apparatus is located on a series of membranous flattened sacs or thylacoids similar in structure and

FIGURE 3.15

Electron micrographs of thin sections of two nonphotosynthetic unicellular microorganisms to illustrate the differences in internal complexity between the eucaryotic and procaryotic cell. (a) A protist, *Labyrinthula* which has a relatively undifferentiated eucaryotic cell structure. (b) A bacterium, *Bacillus subtilis,* which has a typical procaryotic cell structure. The *Labyrinthula* cell in part (a) lacks a wall, but it is surrounded by a loose, extracellular slimematrix (sm). Other recognizable structures include the endoplasmic reticulum (er); Golgi bodies (g); mitochondria (m); a resting nucleus (n), which contains a dense, centrally located nucleolus and is surrounded by a nuclear membrane (nm); free 80S ribosomes (r); large lipid droplets (ld); and the cell membrane (cm). The dividing cell of *Bacillus subtilis,* shown in part (b), is surrounded by a relatively dense wall (cw), enclosing the cell membrane (cm). Within the cell, the nucleoplasm (n) is distinguishable by its fibrillar structure from the cytoplasm, densely filled with 70S ribosomes (r). Note the absence of internal unit membrane systems. (a) Courtesy of David Porter; (b) courtesy of C. F. Robinow.

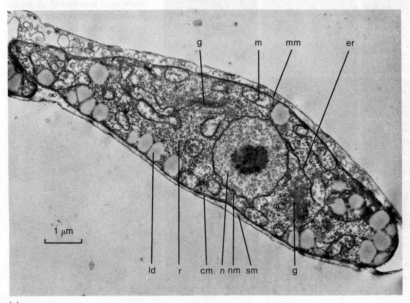

(a)

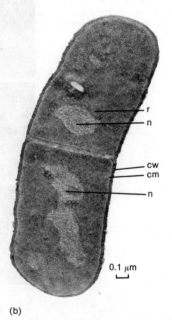

(b)

function to the thylacoids within a chloroplast. However, in blue-green bacteria the thylacoids are not segregated within an organelle, but are dispersed throughout the cytoplasm (Figure 3.16).

Electron microscopy of most procaryotes reveals only two structurally distinguishable internal regions in the cell: cytoplasm and nucleoplasm [see Figure 3.15(b)]. The cytoplasm has a finely granular appearance as a result of its content of ribosomes, each about 10 nm in diameter. These are always of the so-called 70S type, smaller than the cytoplasmic ribosomes of eucaryotes, but similar in size to eucaryotic organellar ribosomes. The nucleoplasm is of irregular contour, sharply segregated from the cytoplasm even though a bounding membrane never separates the two regions. The content of the nucleoplasm is fibrillar; it consists of double helical DNA, the strands of which are some 25 nm wide.

A *cell wall*, considerably thicker than the membrane, encloses the cells of most procaryotes; only the *Mycoplasma* group are devoid of this structure.

• **Functions of the cytoplasmic membrane**

The cytoplasmic membrane of procaryotes constitutes a much more selective barrier between the interior of the cell and the external environment than does the membrane of eucaryotes. The largest objects known to traverse this barrier are of molecular dimensions: fragments of DNA (transforming DNA, see p. 459) and proteins of low molecular weight (exocellular enzymes, secreted by the cell, see p. 302). The phenomena of exocytosis and endocytosis are completely unknown among procaryotes, even those (the *Mycoplasma* group) in which a cell wall is absent and in which there is consequently no mechanical impedance to

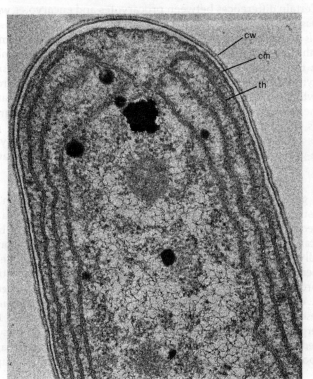

FIGURE 3.16

Electron micrograph of a thin section of a unicellular blue-green bacterium. The cell is enclosed by a cell wall (cw) and cell membrane (cm). The thylacoids (th) which bear the photosynthetic apparatus are located in the cytoplasm, and are thicker than the cell membrane, since they are composed of two closely appressed membranes ($\times$53,000). Courtesy of Dr. G. Cohen-Bazire.

the transfer of particulate objects or liquid droplets across the cell surface. As a result, procaryotes lack biological properties that are dependent on the capacity for endocytosis: notably, the ability to perform intracellular digestion and to harbor cellular endosymbionts.

In many procaryotes the cytoplasmic membrane performs a role in energy-yielding metabolism, a role that it never plays in the eucaryotic cell. Among aerobic bacteria, the respiratory electron transport system is incorporated into the cell membrane. In eucaryotes this part of the machinery of respiration is incorporated into the inner membrane system of the mitochondrion. In one group of photosynthetic procaryotes, the purple bacteria, the centers of photosynthetic activity are likewise incorporated into the cytoplasmic membrane. Both in purple bacteria and in aerobic bacteria which have high rates of respiration, the topology of the membrane is often very complex: it is extensively infolded, lamellar or vesicular intrusions penetrating deep into the cytoplasm. This structure greatly increases the total area of the membrane, and thus permits the cell to accommodate many centers of respiratory (or photosynthetic) function.

There are strong indications that the cytoplasmic membrane also contains specific attachment sites for the DNA of the procaryotic cell, membrane growth being the mechanism responsible for the separation of genomes following completion of their replication. This is, of course, another function never played by the cytoplasmic membrane of the eucaryotic cell in which the separation of genomes is accomplished by mitosis.

Certain differences between procaryotes and eucaryotes with respect to the *lipid composition of cell membranes* deserve mention. Firstly, the class of lipids known as *sterols* are invariably present as components of the cell membrane of eucaryotes, but they are not found in significant amounts in the cell membrane of procaryotes, with the exception of the *Mycoplasma* group. The *Mycoplasma* group are unable to synthesize sterols, but they incorporate exogenous sterols furnished by the growth medium into the cell membrane. Secondly, the fatty acid components of the membrane lipids of all eucaryotes include polyunsaturated types (viz., fatty acids which contain more than one double bond). In most procaryotes the fatty acids are exclusively saturated or monounsaturated; the only exceptions are some blue-green bacteria, which are able to synthesize polyunsaturated fatty acids.

• **The procaryotic genome**   The genetic information of a procaryotic cell is carried in the nucleoplasm on the structure termed the *bacterial chromosome*. It consists of a double helical DNA molecule, never associated with basic proteins, and has been shown in some procaryotes to be circular. The bacterial chromosome is consequently not structurally homologous with the nuclear chromosomes of the eucaryotic cell, but rather with the organellar DNA present in the eucaryotic mitochondria and chloroplasts. It is probable, but not yet certain, that a *single bacterial chromosome* (viz., one very long DNA molecule) *carries all the genetic information necessary to specify the essential properties of the procaryotic cell;* no exceptions to this rule have so far emerged, although the number of organisms in which this fact has been established beyond question still remains small. Many bacteria can also transiently harbor small, extrachromosomal circular DNA molecules capable of autonomous replication, which are known as *plasmids*. The plasmids so far investigated carry the determinants for such phenotypic characters as resistance to drugs and other antibacterial substances, and for the enzymes which mediate certain peripheral metabolic pathways. The amount of DNA in a plasmid is from 0.1 to 5 percent of that in the bacterial chromosome; plasmids can be lost from the cell without impairment of its viability (see Chapter 15).

The monomolecular structure of the procaryotic genome is correlated with the fact that the total amount of genetic information contained in a procaryotic cell is less—in general, by several orders of magnitude—than that contained in a eucaryotic cell. Expressed as lengths of double-stranded DNA, the values for procaryotes extend from 0.25 mm (mycoplasmas) to about 3 mm (certain blue-green bacteria). The lowest value known for the haploid genome size of a eucaryote is 4.6 mm, for the yeast *Saccharomyces cerevisiae*, which has very simple cells; most eucaryotes have genomes at least ten times larger. The haploid genome of *S. cerevisiae* is dispersed over 17 chromosomes; the average DNA content per chromosome in this eucaryote is therefore very low, considerably less than that in the single bacterial chromosome of a procaryote such as *Escherichia coli*.

The segregation of the genome following DNA replication in a procaryote is evidently much simpler than in the nucleus of a eucaryote, and it is achieved by an entirely different mechanism which does not involve the complex sequence of structural events associated with mitosis. During the cell cycle the bacterial chromosome never undergoes changes of length and thickness by coiling, and the separation of daughter chromosomes is not mediated by a microtubular system. In fact, microtubules, which play many roles in the organization and function of the eucaryotic cell, have not so far been detected in procaryotes. The mechanism of chromosomal segregation in procaryotes is not yet fully understood, but the evidence so far available suggests that it involves attachment to specific sites on the membrane, separation of daughter chromosomes being brought about by membrane growth between the two attachment sites.

The separation of daughter genomes and the process of cell division are not as closely linked to one another in the procaryotic cell as they are in the eucaryotic cell in which cell division typically begins in the terminal stages of mitosis, the plane of division coinciding with the equatorial plane of the mitotic spindle. During rapid growth of unicellular procaryotes, nuclear division typically runs ahead of cell division (Figure 3.17). Each daughter cell thus contains two (or more) already separated bacterial chromosomes immediately after the completion of cell division (Figure 3.18), the uninucleate state becoming reestablished only after the cessation of growth.

Most procaryotes (probably all) normally exist and reproduce by asexual means in the haploid state. Consequently, persistent diploidy, which is characteristic of many groups of eucaryotes and which has had profound evolutionary consequences, plays no role in the evolution of procaryotes. A diploid state can arise transiently in procaryotes, as a result of genetic transfer, but full diploidy is rarely attained, as a consequence of the special mechanisms of genetic transfer characteristic of procaryotes.

As will be discussed in Chapter 15, genetic transfer among procaryotes always occurs by a unidirectional passage of DNA from a *donor cell* to a *recipient cell*. This may be mediated either by *conjugation*, involving direct cell-to-cell contact, or by the processes known as *transduction* and *transformation*. Transductional transfer to a recipient cell is mediated by certain bacterial viruses (bacteriophages) which

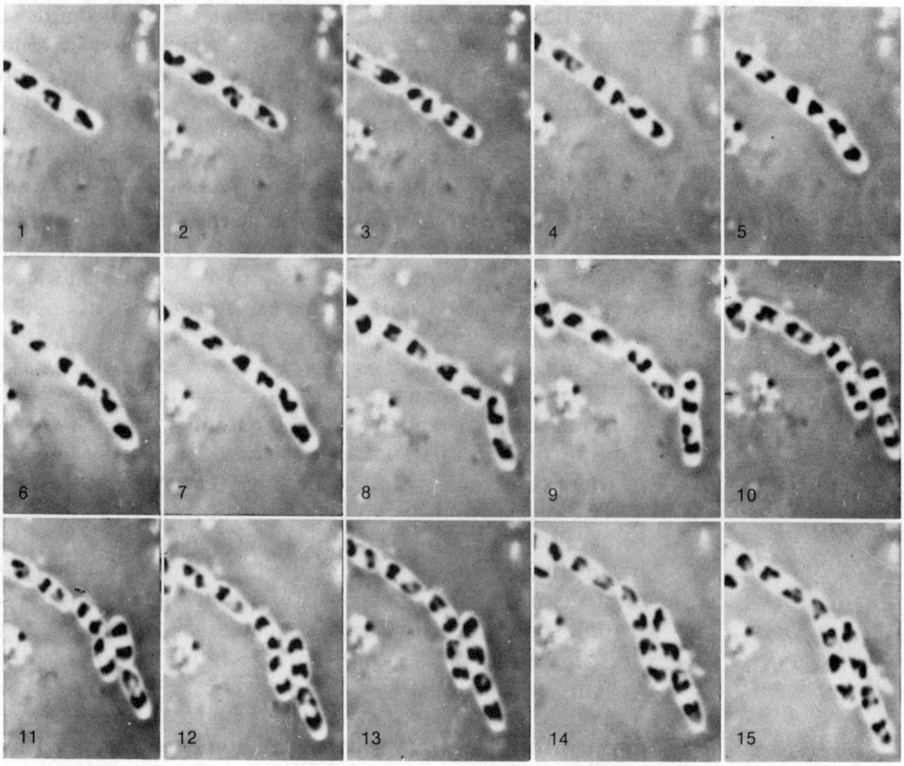

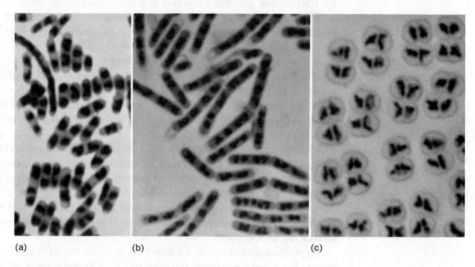

FIGURE 3.17

Successive photomicrographs of growth and nuclear division in a single group of *E. coli* cells suspended in a concentrated protein solution to enhance the contrast between nuclear and cytoplasmic regions (phase contrast, ×975). The sequence was taken over a total period of 78 minutes, equivalent to 2.5 bacterial divisions. Courtesy of D. J. Mason and D. Powelson.

FIGURE 3.18

Stained preparations of the nuclear structures of bacteria photographed through the light microscope: (a) *Proteus vulgaris* (×720); (b) *Bacillus mycoides* (×630); (c) unidentified coccus (×850). Photographs courtesy of C. F. Robinow.

(a)   (b)   (c)

incorporate fragments of the genome of the donor cell. Transformational transfer is mediated by free DNA fragments derived from the donor cell, which pass through the medium and are taken up by the recipient cell. As a rule, only small fragments of the donor genome are transferred by transduction and transformation. Although conjugational transfer can in principle permit transfer of the entire donor chromosome, this rarely occurs. Consequently, the recipient cell usually becomes a *partial diploid* (merodiploid), and subsequent genetic recombination involves exchanges between the complete haploid genome of the recipient and a fraction of the donor genome. The haploid state is usually rapidly restored after recombination, with elimination of supernumerary genes not incorporated into the recipient chromosome. Return to the haploid state does not, accordingly, involve a regular reduction division comparable to the eucaryotic process of meiosis.

Conjugational genetic transfer in procaryotes does not necessarily involve the transfer of chromosomal determinants: the transferred material may be a plasmid. In this event, transfer is not followed by recombination; the plasmid, provided that it is capable of autonomous replication, may instead be maintained by the recipient cell independently of the chromosome. Hence, new genetic elements borne on plasmids, which possess few if any regions homologous with regions of the chromosome, can be introduced into and maintain themselves within the procaryotic cell. For this reason, plasmid transfer can occur between organisms of widely differing (chromosomal) genetic constitution.

• Osmotic control in procaryotes

Procaryotes do not contain organelles with the function of the contractile vacuole, and therefore they are unable to maintain osmotic balance by active means in hypotonic media. Hence, their only device to avoid the danger of osmotic lysis is to synthesize a cell wall possessing the mechanical strength necessary to counterbalance the turgor pressure of the enclosed protoplast. Most procaryotes in fact possess a cell wall. Although the chemical composition of the procaryotic cell wall is complex and varies widely from group to group, it almost invariably contains a distinctive type of wall polymer known as a *peptidoglycan*, which is largely if not entirely responsible for conferring mechanical strength. *The ability to synthesize this distinctive type of wall polymer is confined to procaryotes, and it is thus a chemical attribute which distinguishes these organisms from eucaryotes.* In eucaryotic groups that possess cell walls (fungi, algae, plants), other types of molecular solutions to the problem of conferring mechanical strength on the wall have been found; these solutions are variable, no single class of polymer being present in all types of eucaryotic walls.

The wall is a constant feature of cell structure in all major groups of procaryotes except the members of the *Mycoplasma* group. Most of these organisms are parasites that normally develop in the cells or body fluids of animals or plants; they are osmotically sensitive and can be cultivated only on media of high osmotic strength. There is one exception to the rule that the cell wall of procaryotes contains a peptidoglycan constituent: the extremely halophilic bacteria of the genus *Halobacterium*, which normally live in concentrated brines and salt lakes. Like the *Mycoplasma* group, these bacteria are osmotically sensitive, undergoing rapid lysis in a dilute suspending medium.

• Procaryotic movement

Directed internal cytoplasmic movements which are so conspicuous in most eucaryotic cells do not occur in procaryotic cells. Consequently, ameboid locomotion, characteristic of many eucaryotic protists which lack cell walls, does not occur among procaryotes without walls, namely the members of the *Mycoplasma* group. Many procaryotes which possess cell walls exhibit active movement. One kind of active movement, *gliding motility,* is manifested only when the cell is in contact with a solid substrate; it is not known to be mediated by specific locomotor organelles. Gliding motility is characteristic of many blue-green bacteria, and also occurs in certain groups of nonphotosynthetic bacteria.

The second kind of active movement, *swimming motility,* occurs in cells suspended in a liquid medium and is brought about by locomotor organelles termed *bacterial flagella.* The bacterial flagellum is completely different in fine structure and organization from the eucaryotic flagellum or cilium. It is a proteinaceous filament of molecular dimensions (approximately 12 to 18 nm in diameter) which extends out through the wall and membrane from an anchoring structure located just below the membrane (see Chapter 11, p. 338). A compound structure known as the *axial filament,* comprised of two sets of bacterial flagella which lie within the cell wall, is the organelle responsible for cellular movement in the group of procaryotes known as spirochetes.

• **Intracytoplasmic organelles of procaryotes**

The only intracytoplasmic organelles of procaryotes which bear a structural resemblance to components of a eucaryotic cell are the thylacoids—flattened sacs composed of a unit membrane system—which house the photosynthetic apparatus in one group of photosynthetic procaryotes, the blue-green bacteria. They appear to be homologous in function and structure to the thylacoids of chloroplasts. However, a few other types of distinctively procaryotic organelles do occur in some groups of procaryotes, although none of these structures is of wide distribution. All these organelles are characterized by the *absence of unit membranes:* they are enclosed by a single-layered membrane with a thickness of only some 2 to 3 nm. They include the *chlorobium vesicles* which house the photosynthetic apparatus of green bacteria; the *gas vesicles* which confer bouyancy on the cells of a diversity of aquatic procaryotes; and the *carboxysomes* which contain a key enzyme of reductive CO_2 assimilation, carboxydismutase, in many phototrophs and chemoautotrophs. Their properties will be discussed further in Chapter 11.

TARGETS FOR CERTAIN ANTIBIOTIC AGENTS IN PROCARYOTIC AND EUCARYOTIC CELLS

Antibiotics are organic substances of microbial origin which are either toxic or growth-inhibitory for other organisms. Their great value in the therapy of infectious diseases largely depends on their *selective toxicity,* directed against the infectious agent, but not the infected host. The discovery of antibiotics was empirical, and their specific modes of action against susceptible organisms were established much later. It has now been shown that many classes of antibiotics owe their selective toxicity to the fact that the target is a structure (or function) specific either to the procaryotic or to the eucaryotic cell. Some examples are shown in Table 3.7.

The penicillins and several other classes of antibiotics are selectively toxic for

TABLE 3.7

Classes of antibiotics which act on targets specific to either procaryotic
or eucaryotic cells

ANTIBIOTICS	MODE OF ACTION	ACTIVE AGAINST PROCARYOTES	EUCARYOTES
Penicillins	Block synthesis of peptidoglycan constituent of cell wall	$+^a$	−
Polyene antibiotics	Combine with sterols in cell membrane; affect permeability	$−^b$	+
Glutarimides	Block synthesis of proteins on 80S ribosomes	−	+
Aminoglycosides, tetracyclines, macrolides, chloramphenicol	Block synthesis of proteins on 70S ribosomes	+	$−^c$

[a] Except those (*Mycoplasma* group) which do not produce cell walls.
[b] Some mycoplasmas that incorporate sterols from the growth medium into the membrane are sensitive.
[c] At high concentrations they may affect organellar protein synthesis.

procaryotes because they affect steps in the synthesis of the peptidoglycan component of the cell wall, a molecular component essential for the stability of the cell in most procaryotes, but one that is not synthesized by eucaryotes. The only procaryotes insensitive to these classes of antibiotics are the *Mycoplasma* group, which lack cell walls, and the halophilic bacteria which lack the peptidoglycan component of cell walls.

The polyene antibiotics (exemplified by nystatin, filipin, and amphotericin B) are bound to sterols in the cell membrane of sensitive organisms, resulting in destruction or malfunction of the membrane and the leakage of essential metabolites out of the cell. As previously mentioned (p. 80), a high sterol content is characteristic of the eucaryotic cell membrane, but not of the membranes of most procaryotes. The polyene antibiotics are accordingly agents of selective toxicity for eucaryotic cells; the only procaryotes affected by polyene antibiotics are members of the *Mycoplasma* group, which incorporate exogenous sterols into their membranes.

A very large number of antibiotics interfere with steps in protein synthesis, and many specifically inhibit the function either of 70S ribosomes (characteristic of procaryotes) or 80S ribosomes (characteristic of eucaryotes). The glutarimides (e.g., cycloheximide) specifically inhibit protein synthesis on the 80S ribosome, and are therefore selectively toxic for eucaryotes. The aminoglycosides (e.g., streptomycin), the tetracyclines, chloramphenicol, and macrolide antibiotics (e.g., erythromycin) all specifically inhibit protein synthesis on the 70S ribosome. These antibiotics are thus selectively toxic for procaryotes. However, since the ribosomes of chloroplasts and mitochondria are functionally similar to procaryotic ribosomes, eucaryotic organellar protein synthesis may also be affected. The antibiotic concentrations required to affect significantly the activities of mitochondria or chloroplasts are, however, considerably higher than those necessary to inhibit bacterial growth, probably because the outer membranes of these organelles interpose a partial permeability barrier.

THE DIFFERENCES BETWEEN PROCARYOTES
AND EUCARYOTES: A SUMMING UP

The numerous and profound divergences with respect both to organization and to function between eucaryotic and procaryotic cells have been revealed gradually and have been fully recognized only recently. It is now evident that many universal cellular functions—transmission, transcription, and translation of the genetic message, energy-yielding metabolism, nutrient intake, secretion, movement—are mediated in markedly different ways by the two kinds of cells. This fact raises some major evolutionary questions, notably concerning the primary origins and relationships between the two kinds of cells. Does the relatively simple procaryotic cell, still extant in certain biological groups, represent an early stage in the evolution of the far more complex eucaryotic cell, or did the two kinds of cells have completely independent evolutionary origins? The apparent resemblances between procaryotes and two classes of eucaryotic organelles, chloroplasts and mitochondria, with respect to the structure of the genetic material and the machinery of transcription and translation, suggest another fascinating evolutionary possibility which has been much discussed: namely, that chloroplasts and mitochondria had evolutionary origins distinct from those of the other component parts of the eucaryotic cell. These two classes of organelles might have been derived from free-living procaryotes endowed respectively with respiratory and photosynthetic function, which entered into an endosymbiotic relationship with primitive eucaryotes and gradually became so closely integrated with the machinery of the host cell that they ultimately lost part of their genetic autonomy and became incapable of independent existence.

Some of the major differences between the eucaryotic and the procaryotic cell are summarized in Tables 3.8 to 3.10.

TABLE 3.8

Differences between eucaryotes and procaryotes with respect
to genetic organization

	EUCARYOTES	PROCARYOTES
Nucleoplasm bounded by a membrane	+	−
Number of chromosomes	>1	1[a]
Chromosomes contain histones	+	−
Presence of nucleolus	+	−
Nuclear division by mitosis	+	−
DNA also present in organelles	+	−
Means of genetic recombination:		
fusion of gametes	+	−
formation of partial diploids by unidirectional transfer of DNA	−	+

[a]Some genetic information, nonessential for basic cellular function, may be housed in separate genetic elements (plasmids).

TABLE 3.9

Differences between procaryotes and eucaryotes with respect
to cytoplasmic structure

	EUCARYOTES	PROCARYOTES
Endoplasmic reticulum	+	−
Golgi apparatus	+	−
Lysosomes	+	−
Mitochondria	+	−
Chloroplasts	+ or −	−
Ribosomes	80S (cytoplasmic) 70S (organellar)	70S
Microtubular systems	+	−
Organelles bounded by nonunit membranes	−	+ or −
Presence of cell wall containing peptidoglycan	−	+ or −

TABLE 3.10

Some functional attributes found exclusively
in eucaryotic cells

Phagocytosis
Pinocytosis
Secretions of materials in Golgi vesicles
Intracellular digestion
Maintenance of cellular endosymbionts
Directed cytoplasmic streaming and ameboid
   movement

THE GENERAL PROPERTIES OF VIRUSES

One class of microorganisms, the viruses, are acellular: they differ
from cellular organisms in structure, chemical composition, and mode of growth.

The viruses are obligate parasites, capable of development only within the cells
of susceptible host organisms. Viral hosts include almost all groups of cellular
organisms, both procaryotes and eucaryotes. Viruses are transmitted from cell
to cell in the form of small infectious particles, known as *virions*. The virion of
each type of virus has a fixed form and size. Each virion consists of a core of
nucleic acid, enclosed within a protein coat, or *capsid*, which is composed of a
fixed number of identical protein subunits, the arrangement of which confers in
the virion its external form. Certain virions possess additional structures. Many
of those which infect animals are enclosed in lipoprotein membranes, derived
from the host cell membrane. Certain of those which infect procaryotes have
special proteinaceous tail structures attached to the capsid, which function in the
attachment of the virion to the host cell, and the introduction of viral nucleic
acid into the host.

The core of the virion contains only one kind of nucleic acid; depending on
the virus, it may be double-stranded or single-stranded DNA or double-stranded

or single-stranded RNA, but in all cases it provides the genetic information required for viral synthesis, and it directs the synthesis of new virions by the infected host cell.

Although all viruses are dependent on host cells for their development, the extent and nature of this dependence varies. The simplest viruses contain very little genetic information, sufficient to code at most for three proteins. In such cases, the genetic information and enzymatic machinery of the host cell play the predominant role in viral synthesis. The largest viruses contain genetic information sufficient to code for as many as 500 different proteins, including many enzymes specific to viral synthesis. In all cases, however, the provision of energy and of low molecular weight precursors of proteins and nucleic acids, together with much of the machinery of protein synthesis, is assured by the host cell. As a result of the transcription and translation of the information carried in the viral genome, the activities of the host cell are largely redirected towards the synthesis of viral components, which are then assembled into new virions within the infected cell. Intracellular maturation of the virions is followed by their release from the cell, which is generally killed. For further information about the properties of viruses, see Chapter 12.

FURTHER READING

Books   DuPraw, E. J., *Cell and Molecular Biology*. New York: Academic Press, 1968.

Knight, B. C. J. G., and H. P. Charles (editors), *Organization and Control in Prokaryotic and Eukaryotic Cells*. New York: Cambridge University Press, 1970.

Novikoff, A. B., and E. Holtzmann, *Cells and Organelles*. New York: Holt, Rinehart & Winston, 1970.

Reviews   Lwoff, A., "The Concept of Virus," *J. Gen. Microbiol.* **17**, 239 (1957).

Stanier, R. Y., and C. B. van Niel, "The Concept of a Bacterium," *Arch. Mikrobiol.* **42**, 17 (1962).

4

THE PROTISTS

Among protists, three major groups can be recognized: *algae, protozoa,* and *fungi.* Each of these groups is very large and internally diverse. The more highly specialized representatives—for example, a seaweed, a ciliate, and a mushroom—can be readily assigned to the algae, the protozoa, and the fungi, respectively. However, there are many protists for which the assignment is arbitrary: numerous transitions exist between algae and protozoa and between protozoa and fungi. For this reason, the three major groups of protists cannot be sharply distinguished in terms of simple sets of clear-cut differences. Broadly speaking, the algae may be defined as organisms that perform oxygen-evolving photosynthesis and possess chloroplasts. Some of them are unicellular microorganisms; some are filamentous, colonial, or coenocytic; and some have a plantlike structure that is formed through extensive multicellular development, with little or no differentiation of cells and tissues. In organismal terms, accordingly, the algae are highly diverse, and by no means all fall into the category of microorganisms. The brown algae known as kelps may attain a total length of as much as 50 m. The protozoa and fungi are nonphotosynthetic organisms, and the difference between them is essentially one of organismal structure; protozoa are predominantly unicellular, whereas fungi are predominantly coenocytic and grow in the form of a filamentous, branched structure known as a *mycelium.*

For historical reasons that have been discussed in Chapter 3, the algae and fungi were traditionally regarded as "plants" and have been largely studied by botanists, while the protozoa were traditionally regarded as "animals" and have been largely studied by zoologists. As a result of this specialization, the many interconnections between the three groups have tended to be overlooked. We shall attempt in this chapter to provide a unified account of the properties of protists that emphasizes possible evolutionary interrelationships.

THE ALGAE

The primary classification of algae is based on cellular, not organismal, properties: the chemical nature of the wall, if present; the organic reserve materials produced by the cell; the nature of the photosynthetic pigments; and the nature and arrangement of the flagella borne by motile cells. In terms of these characters, the algae are arranged in a series of divisions, summarized in Table 4.1.

The divisions are not equivalent to one another in terms of the range of organismal structure of their members. For example, the Euglenophyta (euglenid algae) consist entirely of unicellular or simple colonial organisms, while the Phaeophyta (brown algae) consist only of plantlike, multicellular organisms. The largest and most varied group, the Chlorophyta (green algae), from which the higher plants probably originated, spans the full range of organismal diversity, from unicellular organisms to multicellular representatives with a plantlike structure.

The common cellular properties of each algal division suggest that its members, however varied their organismal structure may be, are representatives of a single major evolutionary line. Evolution among the algae thus in general appears to have involved *a progressive increase in organismal complexity in the framework of a particular variety of eucaryotic cellular organization*. Although it is possible to perceive these evolutionary progressions *within* each algal division, the relationships *between* divisions are completely obscure. The primary origin of the algae as a whole is accordingly an unsolved problem.

• **The photosynthetic flagellates**

In many algal divisions, the simplest representatives are motile, unicellular organisms, known collectively as *flagellates*. The cell of a typical flagellate, illustrated by *Euglena* (Figure 4.1), has a very marked polarity: it is elongated and leaf-shaped, the flagella usually being inserted at the anterior end. In the Euglenophyta, to which *Euglena* belongs, there are two flagella of unequal length, which originate from a small cavity at the anterior end of the cell. Many chloroplasts and mitochondria are dispersed throughout the cytoplasm. Near the base of the flagellar apparatus is a specialized organelle, the *eyespot,* which is red, owing to its content of special carotenoid pigments; the eyespot serves as a photoreceptor to govern the active movement of the cell in response to the direction and intensity of illumination. The cell of *Euglena*, unlike that of many other flagellates, is not enclosed within a rigid wall; its outer layer is an elastic *pellicle*, which permits considerable changes of shape. Cell division occurs by *longitudinal fission* [Figure 4.2(a)]. About the time of the onset of mitosis, there is a duplication of the anterior organelles of the cell, including the flagella and their basal apparatus; cleavage subsequently occurs through the long axis, so that the duplicated organelles are equally partitioned between the two daughter cells. This mode of cell division is characteristic of all flagellates except those belonging to the Chlorophyta, such as *Chlamydomonas*, where each cell undergoes *two or more multiple fissions* to produce four smaller daughter cells, liberated by rupture of the parental cell wall [Figure

TABLE 4.1

Major groups of algae

GROUP NAME	PIGMENT SYSTEM		COMPOSITION OF CELL WALL	NATURE OF RESERVE MATERIALS	NUMBERS AND TYPE OF FLAGELLA	RANGE OF STRUCTURE
	CHLOROPHYLLS	OTHER SPECIAL PIGMENTS				
Green algae: division *Chlorophyta*	a + b	—	Cellulose	Starch	Generally two identical flagella per cell	Unicellular, coenocytic, filamentous; plantlike multicellular forms
Euglenids: division *Euglenophyta*	a + b	—	No wall	Paramylum and fats	One, two, or three flagella per cell	All unicellular
Dinoflagellates and related forms: division *Pyrrophyta*	a + c	Special carotenoids	Cellulose	Starch and oils	Two flagella, dissimilar in form and position on cell	Mostly unicellular, a few filamentous forms
Chrysophytes and diatoms: division *Chrysophyta*	a ± c	Special carotenoids	Wall composed of two overlapping halves, often containing silica (some have no walls)	Leucosin and oils	Two flagella, arrangement variable	Unicellular, coenocytic, filamentous
Brown algae: division *Phaeophyta*	a + c	Special carotenoids	Cellulose and algin	Laminarin and fats	Two flagella, of unequal length	Plantlike multicellular forms
Red algae: division *Rhodophyta*	a	Phycobilins	Cellulose	Starch	No flagella	Unicellular; plantlike multicellular forms

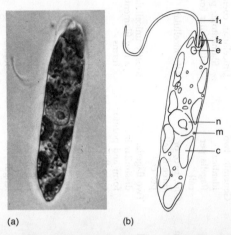

(a)   (b)

FIGURE 4.1

Euglena gracilis. (a) Photomicrograph of fixed cell (×1,000). Courtesy of Gordon F. Leedale. (b) Schematic drawing of the same cell, to show prinicpal structural features: n, nucleus; c, chloroplast; m, mitochondrion; e, eyespot; f$_1$, f$_2$, the two flagella of unequal length, originating within a small cavity of the anterior end of the cell.

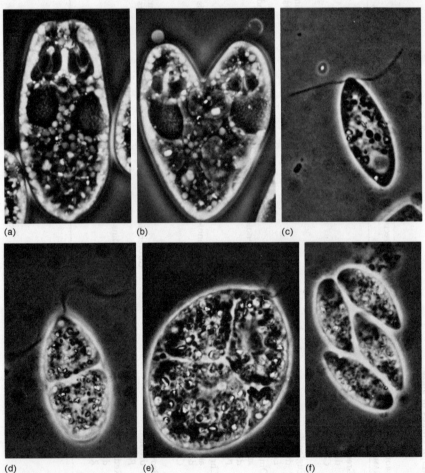

(a)   (b)   (c)

(d)   (e)   (f)

FIGURE 4.2

Longitudinal and multiple fission in flagellate algae. (a, b) Two cells of *Euglena gracilis* in the course of longitudinal fission (phase contrast, ×1,240). [Reproduced from G. F. Leedale, in *The Biology of Euglena,* D. E. Buetow, ed. (New York: Academic Press, 1968).] In (a), division of the nucleus and of the locomotor apparatus at the anterior end of the cell is complete. In (b), cell cleavage has begun. (c, d, e, f) Four steps in the cellular life cycle of *Chlorogonium tetragamum,* a green alga that reproduces by multiple fission (phase contrast, ×1,430). (c) Newly liberated daughter cell. (d) Two-celled stage. (e) Four-celled stage. (f) Four daughter cells just after liberation from the mother cell. Original photomicrographs of material provided by Paul Kugrens, Department of Botany, University of California, Berkeley.

4.2(b)]. Even in such cases, however, the internal divisions take place in the longitudinal plane. As we shall see in a subsequent section, longitudinal division also occurs in the nonphotosynthetic flagellate protozoa and is one of the primary characters that distinguish these organisms from the other major group of protozoa that possess flagellalike locomotor organelles, the ciliates.

Most multicellular algae are immotile in the mature state. However, their reproduction frequently involves the formation and liberation of motile cells, either asexual reproductive cells (*zoospores*) or gametes. Figure 4.3 shows the liberation of zoospores from a cell of a filamentous member of the Chlorophyta, *Ulothrix;* it can be seen that these zoospores have a structure very similar to that of the *Chlorogonium* cell, illustrated in Figure 4.2(c). The structure of the motile reproductive cells of multicellular algae thus often reveals their relatedness to a particular group of unicellular flagellates.

• **The nonflagellate unicellular algae**

By no means all unicellular algae are flagellates; several algal divisions also contain unicellular members which are either immotile or possess other means of movement. Many of these unicellular nonflagellate algae possess strikingly specialized and elaborate cells, which may be illustrated by considering two groups, the *desmids* and the *diatoms*.

The desmids, members of the Chlorophyta, have flattened, relatively large cells, with a characteristic bilateral symmetry (Figure 4.4). Asexual reproduction involves the synthesis of two new half cells in the equatorial plane, followed by cleavage between the new half cells to produce two bilaterally symmetrical daughters, each of which has a cell consisting of an "old" and a "new" half.

The diatoms (Figure 4.5), members of the Chrysophyta, have organic walls impregnated with silica. The architecture of the diatom wall is exceedingly complex; it always consists of two overlapping halves, like the halves of a petri dish. Division is longitudinal, each daughter cell retaining half of the old wall and synthesizing a new half.

Although devoid of flagella, some desmids and diatoms can move slowly over solid substrates. The mechanism of desmid locomotion is not known. The locomotion of diatoms is accomplished by a special modification of ameboid movement. In motile diatoms, there is a narrow longitudinal slot in the wall, known as a *raphe,* through which the protoplast can make direct contact with the substrate. Movement is brought about by directed cytoplasmic streaming in the canal of the raphe, which pushes the cell over the substrate.

Many fossil diatoms are known, because the siliceous skeleton of the wall (Figure 4.6) is practically indestructible, and as diatoms are one of the major groups of algae in the oceans, large fossil deposits of diatom walls have accumu-

FIGURE 4.3

The filamentous green alga, *Ulothrix* (×1,250). At left, the formation and liberation of biflagellate zoospores.

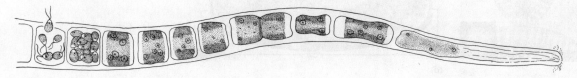

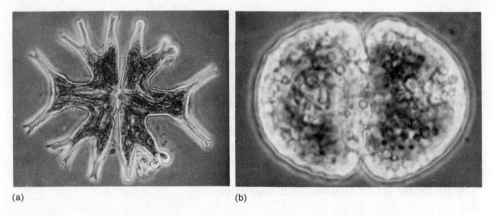

(a)   (b)

FIGURE 4.4

Two phase contrast photomicrographs of living desmid cells: (a) *Micrasterias* (×360); (b) *Cosmarium* (×1,560). Material from the collection of algae of the Department of Botany, University of California, Berkeley, provided by R. Berman.

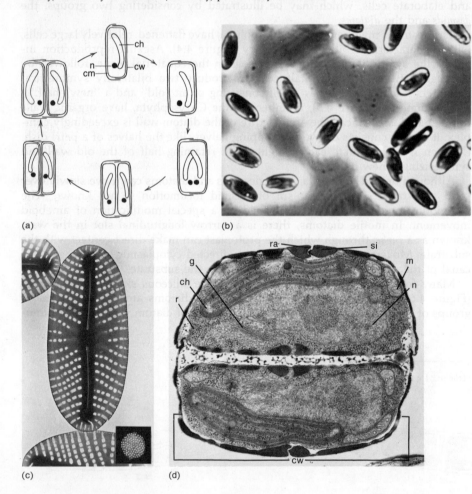

(a)   (b)

(c)   (d)

FIGURE 4.5

The diatom *Navicula pelliculosa*. (a) Diagrammatic representation of the division cycle. (b) Living cells, phase contrast illumination (×1,320). (c) Electron micrograph of the wall (×9,800). Insert depicts fine structure of one of the wall pores (×56,000). (d) Transverse section of a dividing cell (×23,800): ch, chloroplast; g, Golgi apparatus; n, nucleus; m, mitochondrion; r, ribosomes; ra, raphe; si, silica in wall; cw, cell wall; cm, cell membrane, Courtesy of M. L. Chiappino and B. E. Volcani, University of California, San Diego.

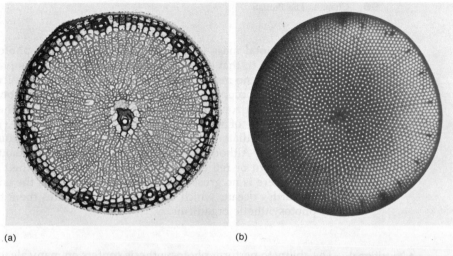

(a)   (b)

(c)   (d)

FIGURE 4.6

The structural complexity of the wall in diatoms, illustrated by electron micrographs of isolated walls. (a) Wall of *Cyclotella nana* ($\times$ 12,100). (b) Wall of *Coscinodiscus granii* ($\times$ 1,370). (c) Part of (b) at a higher magnification ($\times$ 3,630). (d) Part of (b) at still higher magnification, showing detailed fine structure of the pores ($\times$ 22,600). Courtesy of M. L. Chiappino and B. E. Volcani, University of California, San Diego.

lated in many areas. These deposits, known as *diatomaceous earth*, have industrial uses as abrasives and filtering agents.

• **The natural distribution of algae**

Most algae are aquatic organisms that inhabit either fresh water or the oceans. These aquatic forms are principally free-living, but certain unicellular marine algae have established durable symbiotic relationships with specific marine invertebrate animals (e.g., sponges, corals, various groups of marine worms) and grow within the cells of the host animal. Some terrestrial algae grow in soil or on the bark of trees. Others have established symbiotic relationships with fungi, to produce the curious, two-membered natural associations termed *lichens,* which form slowly growing colonies in many arid and inhospitable environments, notably on the surface of rocks. Many of the symbiotic relationships into which algae have entered will be described in Chapters 26 and 27.

The marine algae play a very important role in the cycles of matter on earth, since their total mass (and consequently their gross photosynthetic activity) is at least equal to that of all land plants combined and is probably much greater. This role is by no means evident, because the most conspicuous of marine algae, the seaweeds, occupy a very limited area of the oceans, being attached to rocks in the intertidal zone and the shallow coastal waters of the continental shelves. The great bulk of marine algae are unicellular floating (*planktonic*) organisms, predominantly diatoms and dinoflagellates, distributed through the surface waters of the oceans. Although they sometimes become abundant enough to impart a definite brown or red color to local areas of the sea, their density is usually so low that there is no gross sign of their presence. It is the enormous total volume of the earth's oceans which they occupy that makes them the most abundant of all photosynthetic organisms.

• **Nutritional versatility of algae**   The ability to perform photosynthesis confers on many algae very simple nutrient requirements; in the light they can grow in a completely inorganic medium. This is not always true, however, because many algae have *specific vitamin requirements*, a requirement for vitamin B_{12} being particularly common. In nature the source of these vitamins is probably bacteria that inhabit the same environment. The ability to perform photosynthesis does not necessarily preclude the utilization of organic compounds as the principal source of carbon and energy, and many algae have *a mixed type of metabolism*.

Even when growing in the light, certain algae (e.g., the green alga *Chlamydobotrys*) cannot use CO_2 as their principal carbon source and are therefore dependent on the presence of acetate or some other suitable organic compound to fulfill their carbon requirements. This is caused by a defective photosynthetic machinery: although these algae can obtain energy from their photosynthetic activity, they cannot obtain the reducing power to convert CO_2 to organic cell materials.

Many algae that perform normal photosynthesis in the light, using CO_2 as the carbon source, can grow well in the dark at the expense of a variety of organic compounds; such forms can thus shift from photosynthetic to respiratory metabolism, the shift being determined primarily by the presence or absence of light. Algae completely enclosed by cell walls are osmotrophic and dependent on dissolved organic substrates as energy sources for dark growth. However, a considerable number of unicellular algae which lack a cell wall, or are not completely enclosed by it, can phagocytize bacteria or other smaller microorganisms and thus employ a phagotrophic mode of nutrition as well. It is not correct, accordingly, to regard the algae as an *exclusively* photosynthetic group; on the contrary, many of their unicellular members possess and can use the nutritional capacities characteristic of the two major subgroups of nonphotosynthetic eucaryotic protists, the protozoa and fungi.

• **The leucophytic algae**   Loss of the chloroplast from a eucaryotic cell is an *irreversible event*, which results in a *permanent loss of photosynthetic ability*. Such a change appears to have taken place many times among unicellular algal groups with a mixed mode of nutrition,

FIGURE 4.7

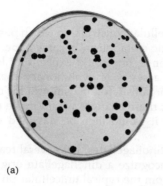

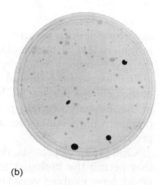

(a)   (b)

The loss of chloroplasts in *Euglena gracilis* as a result of ultraviolet irradiation. (a) A light-grown plate culture of *E. gracilis.* (b) A light-grown culture of the same organism, after exposure to brief ultraviolet irradiation. Most of the cells have given rise to clones devoid of chloroplasts (pale colonies). Courtesy of Jerome A. Schiff.

to yield nonpigmented counterparts, which can be clearly recognized on the basis of other cellular characters as *nonphotosynthetic derivatives of algae.* Such organisms, known collectively as *leucophytes,* exist in many flagellate groups and also in diatoms and in nonmotile groups among the green algae. The recognition of leucophytes is often easy, since they may have preserved a virtually complete structural identity with a particular photosynthetic counterpart. In some cases, this structural near-identity may include the preservation of vestigial, nonpigmented chloroplasts, as well as a pigmented eyespot. There can be little doubt accordingly that these nonphotosynthetic organisms are close relatives of their structural counterparts among the algae and have arisen from them by a loss of photosynthetic ability in the recent evolutionary past. Indeed, the transition can be demonstrated experimentally in certain strains of *Euglena,* which yield stable, colorless races when treated with the antibiotic streptomycin or when exposed to small doses of ultraviolet irradiation or to high temperatures (Figure 4.7). These colorless races cannot be distinguished from the naturally occurring nonphotosynthetic euglenid flagellates of the genus *Astasia.*

The classification of the leucophytes raises a difficult problem. In terms of cell structure, they can be easily assigned to a particular division of algae, as nonphotosynthetic representatives, and this classification is no doubt the most satisfactory one. However, since they are nonphotosynthetic unicellular eucaryotic protists, they can alternatively be regarded as protozoa, and they are, in fact, included among the protozoa by zoologists. The leucophytes accordingly provide the first and by far the most striking case of a group, or rather a whole series of groups, which are clearly transitional between two major assemblages among the eucaryotic protists.

• **The origins of the protozoa**

The protozoa are a highly diverse group of unicellular, nonphotosynthetic protists, most of which show no obvious resemblances to the various divisions of algae. Nevertheless, the various kinds of leucophytes, which are recognizably of algal origin, provide a plausible clue concerning the evolutionary origin of many groups among the protozoa. The loss of photosynthetic function abruptly reduces the nutritional potentialities of an organism; leucophytes are therefore immediately confined to a more restricted range of environments than their photosynthetic

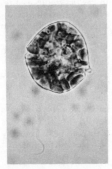

(a)

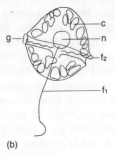

(b)

FIGURE 4.8

A photosynthetic dinoflagellate, *Glenodinium foliaceum*. (a) A living cell (× 1,000). (b) A diagrammatic drawing of the cell: c, chloroplast; n, nucleus; g, girdle; f_1 and f_2, flagella.

ancestors. Specific features of cellular construction which possessed adaptive value in the context of photosynthetic metabolism become superfluous; the eyespot is the most obvious example. Hence, one could expect that loss of photosynthetic ability would be followed by a series of evolutionary changes in the structure of the cell which better fit the organism for an osmotrophic or phagotrophic mode of life. Beyond a certain point, these changes would make the algal origin of the organism unrecognizable, and it would then be classified without question as a protozoon.

One group of protists, the dinoflagellates, has several features of cell structure that permit the biologist to recognize a dinoflagellate origin even in organisms which have evolved very far from the typical unicellular photosynthetic flagellate members of this algal group (Figure 4.8). The motile cell of a dinoflagellate has two flagella, which differ in structure and arrangement. One lies in a groove or girdle around the equator of the cell; the other extends away from the cell in a posterior direction. The dinoflagellate nucleus is also unusual; its division is highly specialized, and the chromosomes remain visible in interphase.

Most photosynthetic dinoflagellates are unicellular planktonic organisms, widely distributed in the oceans, and characteristically brown or yellow in color as a result of the possession of a distinctive set of photosynthetic pigments. Many (the so-called "armored" dinoflagellates) possess very elaborate cell walls, composed of a series of plates, which do not completely enclose the protoplast. There is a very pronounced tendency to phagotrophic nutrition among these photosynthetic members of the group, because the wall structure permits pseudopodial extension and the engulfment of small prey. A few filamentous algae, completely enclosed by walls, can be recognized as of dinoflagellate origin, since they form zoospores with the characteristic flagellar arrangement.

A much more extensive series of specialized forms can be traced among the nonphotosynthetic members of this flagellate group. Many of the free-living unicellular dinoflagellates are nonphotosynthetic phagotrophic organisms. Some preserve close structural similarities to photosynthetic members of the group; others, such as the large marine organism, *Noctiluca*, have a highly specialized cellular organization not found in any photosynthetic member of the group. However, the most far-reaching modifications of cell structure within the donoflagellates are to be found among its parasitic members, most of which occur in marine invertebrates. *Hematodinium*, which occurs in the blood of certain crabs, is completely devoid of flagella. *Chytriodinium*, which parasitizes the eggs of copepods, develops as a large, saclike structure within the egg, subsequently giving rise by multiple internal cleavage to numerous motile spores with a typical dinoflagellate structure. Were it not for the retention of the distinctive nuclear organization (and, in the case of *Chytriodinium*, the flagellar structure of the spores), neither of these parasitic protists could be recognized as belonging to the same group as the photosynthetic dinoflagellates. *Hematodinium* could be classified with the sporozoan protozoa and *Chytriodinium* with the primitive group of fungi known as chytrids.

Accordingly, within this one small flagellate group, it is possible to reconstruct some major patterns of evolutionary radiation that were probably characteristic of protists as a whole (Figure 4.9).

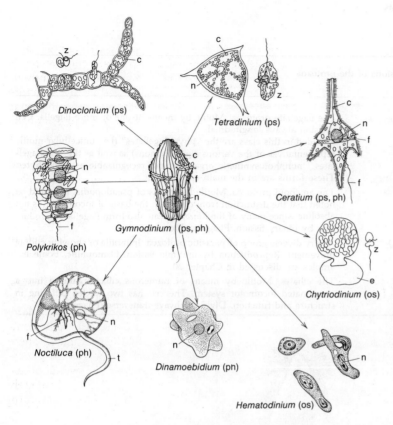

FIGURE 4.9

The different evolutionary trends that are represented among dinoflagellates. *Gymnodinium* is a relatively unspecialized photosynthetic dinoflagellate, which is both photosynthetic (ps) and phagotrophic (ph). *Ceratium* is a more specialized photosynthetic dinoflagellate, characterized by a very complex wall with spiny extensions, comprised of many plates. *Tetradinium* and *Dinoclonium* are nonmotile, strictly photosynthetic organisms, which reproduce by multiple cleavage to form typical dinoflagellate zoospores. *Polykrikos, Noctiluca,* and *Dinamoebidium* are three free-living phagotrophic dinoflagellates. *Polykrikos* is a coenocytic, multinucleate organism, the cell of which bears a series of pairs of flagella. *Noctiluca* has one small flagellum, and bears a large and conspicuous tentacle. *Dinamoebidium* is an ameboid organism. *Chytriodinium* and *Hematodinium* are parasitic dinoflagellates whose nutrition is osmotrophic (os). *Chytriodinium* parasitizes invertebrate eggs and reproduces by cleavage of a large saclike structure into dinoflagellate zoospores. *Hematodinium* is a blood parasite in crabs: n, nucleus; f, flagellum; c, chloroplast; z, zoospore; e, parasitized invertebrate egg; t, tentacle.

THE PROTOZOA

In the light of the preceding discussion, the protozoa can best be regarded as comprising a number of groups of nonphotosynthetic, typically motile, unicellular protists, which have probably derived at various times in the evolutionary past from one or another group among the unicellular algae (see Table 4.2).

• **The flagellate protozoa: the Mastigophora**

The Mastigophora are protozoa that always bear flagella as the locomotor organelles. In contrast to the ciliates, in which cell division is transverse, flagellate protozoa undergo longitudinal division, preceded by duplication of the flagellar apparatus at the anterior end of the cell. This mode of division has already been described for a photosynthetic flagellate, *Euglena*. In addition to leucophytes, this protozoan group includes many representatives that show no resemblance to photosynthetic flagellates and are for the most part parasites of animals.

The trypanosomes are frequently parasitic in vertebrates, where they develop in the bloodstream, being transmitted from host to host by the bite of insects. They include important agents of disease, such as the agent of African sleeping sickness, transmitted by the tsetse fly. The cell is slender and leaf-shaped, its single flagellum being directed posteriorly and attached through part of its length to the body of the cell, to form an undulating membrane [Figure 4.10(a)]. The

TABLE 4.2

Primary subdivisions of the protozoa

I. *Class Mastigophora:*	The flagellate protozoa. Motile by means of one or more flagella. Cell division always longitudinal. Included in this class are the "phytoflagellates" (i.e., unicellular motile representatives of the various algal divisions) as well as the "zooflagellates," nonphotosynthetic organisms not recognizable as leucophytes. These forms are in the main osmophilic.
II. *Class Rhizopoda:*	The ameboid protozoa. Motile by means of pseudopodia. It should be noted that the distinction from class I on the basis of locomotion is not absolute, since many of the *Rhizopoda* can also form flagella. Reproduction by binary fission. Phagotrophic.
III. *Class Sporozoa:*	A very diverse group of parasitic protozoa. Immotile or showing gliding movement. Reproduction by multiple fission. Osmophilic. Some examples are discussed in Chapter 30.
IV. *Class Ciliata:*	The ciliates. Motile by means of numerous cilia, organized into a coordinated locomotor system. The cell has two nuclei, differing in structure and function. Division always transverse. Phagotrophic.

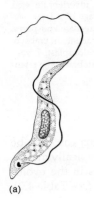

(a)

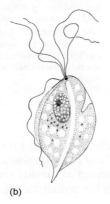

(b)

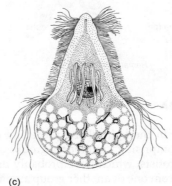

(c)

FIGURE 4.10

Some nonphotosynthetic flagellate protozoa (Mastigophora). (a) A trypanosome. The leaf-shaped cell and the long undulating membrane, to which the flagellum is attached, are characteristic of this organism. (b) *Trichomonas.* (c) *Trichonympha.*

trypanosomes are osmotrophic protozoa, which absorb their nutrients from the blood of the host.

Other parasitic flagellates inhabit the gut of vertebrates or invertebrates. The trichomonads, which have 4 to 6 flagella [Figure 4.10(b)] are harmless inhabitants of the gut of vertebrates. Several very highly specialized groups of flagellate protozoa inhabit the gut of termites; one of the most striking of these organisms, *Trichonympha*, is illustrated in Figure 4.10(c).

• **The ameboid protozoa: the Rhizopoda**

The Rhizopoda are protozoa in which ameboid locomotion is the predominant mode of cell movement, although some of them are able to produce flagella as well. The simplest members of this group are amebas, which have characteristically amorphous cells as a result of the continuous changes of shape brought

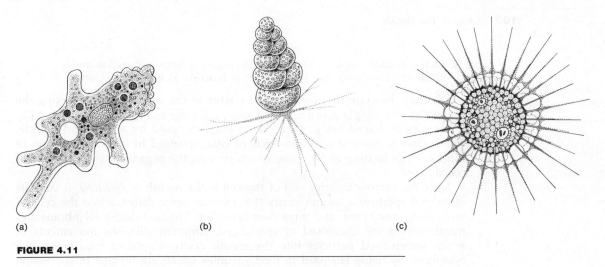

FIGURE 4.11

Some ameboid protozoa (*Sarcodina*). (a) An ameba. (b) A foraminiferan. Note the many-chambered shell, from which the pseudopodia extend. (c) A heliozoan.

about by the extension of pseudopodia. Most amebas are free-living soil or water organisms which phagocytize smaller prey. A few inhabit the animal gut, including forms that cause disease (amebic dysentery). Other members of the Rhizopoda have a well-defined cell form, as the result of the formation of an exoskeleton or shell (typical of the foraminifera) or an endoskeleton (typical of the heliozoa and radiolaria). Several members of the Rhizopoda are illustrated in Figure 4.11.

• **The ciliate protozoa: the Ciliophora**

The ciliate protozoa are a very large and varied group of aquatic, phagotrophic organisms that are particularly widely distributed in fresh water. The ciliates share a number of fundamental cellular characters which distinguish them sharply from all other protists. This suggests that despite the very great internal diversity of this group, it is the one class of protozoa that may have had a single common evolutionary origin.

The common characters of ciliates can be summarized as follows:

1. At some time in the life history, the cell is motile by means of numerous short, hairlike projections, structurally homologous with flagella, which are termed *cilia*.

2. Each cilium arises from a basal structure, the kinetosome, which is homologous with the kinetosome of a flagellum; however, in ciliates the kinetosomes are interconnected by rows of fibrils called *kinetodesmata* to form very elaborate compound locomotor structures termed *kineties*. This internal system persists, even when the cell is devoid of cilia.

3. Cell division is transverse, not longitudinal, as in flagellates. Ciliates show a marked polarity, with posterior and anterior differentiation of the cell, so the transverse mode of cell division necessarily entails an elaborate process of morphogenesis each time division occurs, during which the anterior daughter cell resynthesizes posterior structures, while the posterior daughter cell resynthesizes anterior structures. The morphogenetic transformations are generally almost complete when the two daughter cells separate.

4. Each individual contains two dissimilar nuclei, a large *macronucleus* and a much smaller *micronucleus,* which differ in function as well as in structure.

We may illustrate the distinctive character of the ciliates by considering the properties of a simple member of the group, *Tetrahymena pyriformis* (Figure 4.12). It has a pear-shaped body about 50 µm long, enclosed by a semirigid pellicle. The surface is covered with hundreds of cilia, arranged in longitudinal rows of kineties. The beating of the cilia, which propels the organism, is rhythmic and coordinated.

Near the narrow anterior end of the cell is the mouth or *cytostome.* It consists of an oral aperture, a mouth cavity that extends some distance into the cell, an undulating membrane, and three membranelles. The undulating membrane and membranelles are composed of specialized, adherent cilia, the movements of which sweep food particles into the mouth cavity. Captured food enters the cytoplasm by being enclosed in food vacuoles which are formed in succession at the base of the mouth cavity. These food vacuoles then circulate within the cell as a result of cytoplasmic streaming until the food material has been digested and the soluble products absorbed; undigested material is ejected from the cell

FIGURE 4.12

The ciliate protozoon, *Tetrahymena.* (a) A general view, showing external appearance. (b) Diagrammatic cross section, showing main structural features of the cell.

(a)   (b)

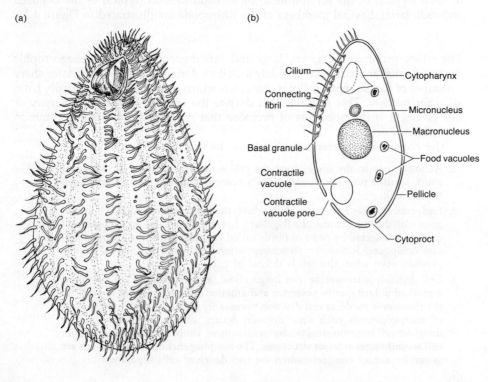

Cilium

Connecting fibril

Basal granule

Contractile vacuole

Contractile vacuole pore

Cytopharynx

Micronucleus

Macronucleus

Food vacuoles

Pellicle

Cytoproct

by a posteriorly located pore known as the *cytoproct*. In nature *Tetrahymena* is normally a predator and feeds on smaller microorganisms. However, in the laboratory it can be grown in pure culture on a medium that contains only soluble nutrients. Under such conditions, the liquid nutrients must still be taken in through the mouth, in the form of vacuoles.

Although its natural environment is a dilute one, with an osmotic pressure far below that of the contents of the cell, *Tetrahymena* is able to maintain water balance by the operation of a *contractile vacuole*. This structure, located near the posterior end of the cell, is formed by the coalescence of smaller vacuoles in the cytoplasm; when it reaches a certain critical size, it discharges its liquid contents into the environment through a pore in the pellicle and then starts to grow in volume again. As mentioned above, typical ciliates have two dissimilar nuclei in the cell. The larger *macronucleus,* which is polyploid, is necessary for normal cell division and growth and is therefore sometimes referred to as the "vegetative nucleus." Some strains of *Tetrahymena* have only this kind of nucleus; they can reproduce indefinitely by binary fission but cannot undergo sexual reproduction. Other strains possess also a small, diploid *micronucleus*, which plays an essential role in sexual reproduction. As will be described presently, the macronucleus can be derived after conjugation from a micronucleus; hence, strains of *Tetrahymena*, having only a macronucleus can be regarded as deficient cell lines which have probably lost their micronucleus by an accident of vegetative growth. In *Tetrahymena* the first step in cell division is an elongation of the macronucleus parallel to the long axis of the cell. At the same time, a structural reorganization of the cytoplasm begins. Its principal feature is the formation of a *second cytostome* just posterior to the future plane of cell division. A furrow then develops across the center of the cell, which becomes dumbbell-shaped. If a micronucleus is present, it divides mitotically, and the two daughter nuclei migrate respectively to the anterior and posterior portions of the cell. Finally, the elongated macronucleus divides, and the two daughter cells separate.

Sexual reproduction in *Tetrahymena* (Figure 4.13) takes place only between cells of two different strains of compatible mating type. The mating cells fuse in the region of the cytostome, and their micronuclei undergo meiotic division, each giving rise to four haploid nuclei. Three of the nuclei in each cell disintegrate, while the fourth undergoes another mitotic division. A nuclear exchange between the mating partners then takes place, each partner receiving one of the two haploid nuclei derived from the micronucleus of the other. This nuclear exchange is followed, in each cell of the pair, by a fusion between the haploid nucleus that was not transferred and the haploid nucleus received from the partner. The resulting diploid zygote nuclei undergo two successive mitotic divisions, producing four diploid nuclei in each cell. Two of these then proceed to develop into macronuclei, a process that involves polyploidization. The two others remain as micronuclei. At the same time, the old macronucleus in each cell disintegrates. The two partners then separate from one another, each now containing two new macronuclei and two new micronuclei, all derived from the zygote nucleus. One of the micronuclei in each cell disintegrates, and the remaining one divides mitotically, after which the cell undergoes binary fission to yield two daughter cells, each containing a single macronucleus and a single micronucleus. The organisms continue to reproduce by binary fission until the next cycle of conjugation.

A curious feature of sexuality in ciliates is that the disintegration of three of the four haploid daughter nuclei in each cell which follows meiotic division of

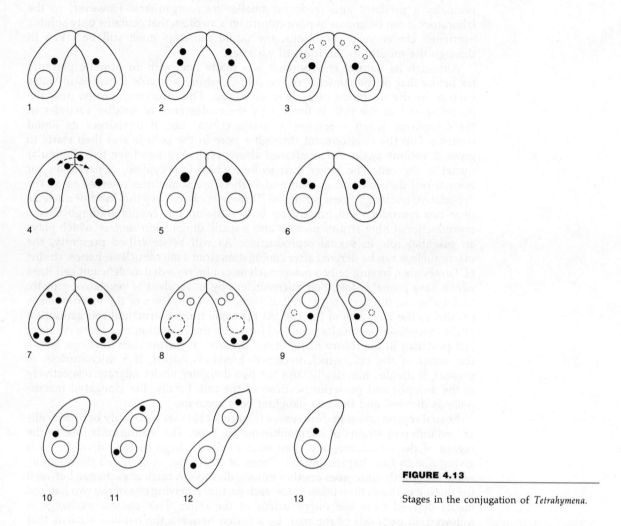

FIGURE 4.13

Stages in the conjugation of *Tetrahymena*.

the micronucleus *results in the complete elimination from the cell of half its previous diploid genome*. When the remaining haploid nucleus divides, each member of the conjugating pair thus contains two genetically identical haploid nuclei. The following reciprocal nuclear exchange and zygotic nuclear fusion therefore makes the two mating cells into *identical diploid twins*; after their separation from one another, they transfer identical genotypes to all their respective progeny during subsequent vegetative growth.

Tetrahymena is among the simplest of ciliates. The foregoing account suffices to show what an extraordinarily elaborate and complex biological organization has been evolved in this protozoan group within the framework of unicellularity. The ciliates represent the apex of biological differentiation on the unicellular level, but they appear to be a terminal evolutionary group. The development of more complex biological systems took place through the establishment of multicellularity and involved the differentiation of specialized cell types during the growth of the individual organism, characteristic of all plants and animals.

THE FUNGI

Like the protozoa, the fungi are nonphotosynthetic. Although some of the more primitive aquatic fungi show resemblances to flagellate protozoa, the fungi as a whole have developed a highly distinctive biological organization that can be regarded as an adaptation to life in their most common habitat, the soil. We shall start out by considering the main features of this type of biological organization.

Most fungi are coenocytic organisms and have a vegetative structure known as a *mycelium* (Figure 4.14). The mycelium consists of a multinucleate mass of cytoplasm enclosed within a rigid, much-branched system of tubes, which are fairly uniform in diameter. The enclosing tubes represent a protective structure that is homologous with the cell wall of a unicellular organism. A mycelium normally arises by the germination and outgrowth of a single reproductive cell, or spore. Upon germination, the fungal spore puts out a long thread, or *hypha*, which branches repeatedly as it elongates to form a ramifying system of hyphae which constitutes the mycelium. Fungal growth is characteristically confined to the tips of the hyphae; as the mycelium extends, the cytoplasmic contents may disappear from the older, central regions. The size of a single mycelium is not fixed; as long as nutrients are available, outward growth by hyphal extension can continue, and in some of the basidiomycetes a single mycelium may be as much as 50 ft in diameter. Usually, asexual reproduction occurs by the formation of uninucleate or multinucleate spores which are pinched off at the tips of the hyphae. Neither the spores nor the mycelium of higher fungi are capable of movement. However, the internal contents of a mycelium show streaming movements, which cannot be translated into progression over the substrate, because the cytoplasm is completely enclosed within its wall. In fact, the simplest brief definition of the structure of a higher fungus is: *a multinucleate mass of cytoplasm, mobile within a much-branched enclosing system of tubes.*

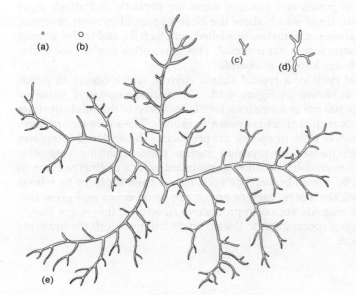

FIGURE 4.14

Successive stages in the development of a fungal mycelium from a reproductive cell or conidium (×85). After C. T. Ingold, *The Biology of Fungi*. London: Hutchinson, 1961.

Since a mycelium is capable of almost indefinite growth, it frequently attains macroscopic dimensions. In nature, however, the vegetative mycelium of fungi is rarely seen, because it is normally embedded in soil or other opaque substrates. Many fungi (the mushrooms) form specialized, spore-bearing fruiting structures, however, which project above soil level and are readily visible as macroscopic objects. Such structures were known long before the beginning of scientific biology, although their nature and mode of formation were not clearly understood until the nineteenth century. The superficial resemblance of these fruiting structures to plants was undoubtedly a very important factor in the decision of the early biologists to assign fungi to the plant kingdom, despite their nonphotosynthetic nature.

Since fungi are always enclosed by a rigid wall, they are unable to engulf smaller microorganisms. Most fungi are free-living in soil or water and obtain their energy by the respiration or fermentation of soluble organic materials present in these environments. Some are parasitic on plants or animals. A number of soil forms are predators and have developed ingenious traps and snares, composed of specialized hyphae, which permit them to capture and kill protozoa and small invertebrate animals such as the soil-inhabiting nematode worms. After the death of their prey, such fungi invade the body of the animal by hyphal growth and absorb the nutrients contained in it.

The fungi comprise three major groups: a lower group, the Phycomycetes, and two higher groups, the Ascomycetes and the Basidiomycetes. A fourth group, the Fungi Imperfecti, has been set aside to include those species for which the sexual stage, and hence the correct classification, is not yet known.

• **The primitive fungi: aquatic Phycomycetes**

Although soil is by far the most common habitat of the fungi as a whole, many of the primitive fungal groups are aquatic. These fungi are known collectively as *water molds* or *aquatic Phycomycetes*. They occur on the surface of decaying plant or animal materials in ponds and streams; some are parasitic and attack algae or protozoa. It is these fungi which show the closest resemblances to protozoa; they produce motile spores or gametes, furnished with flagella, and in the simpler forms the vegetative structure is not mycelial. This description applies, for example, to many of the fungi known as *chytrids*.

The developmental cycle of a typical simple chytrid, which occurs in ponds on decaying leaves, is shown in Figure 4.15. The mature vegetative structure consists of a sac about 100 μm in diameter which is anchored to the solid substrate by a number of fine, branched threads known as *rhizoids*. The sac is a *sporangium*, within which reproductive cells, or spores, are produced. The enclosed cytoplasm contains many nuclei, formed by repeated nuclear division during vegetative growth. Each nucleus eventually becomes surrounded with a distinct volume of cytoplasm, bounded by a membrane. The sporangium then ruptures to release uninucleate flagellated zoospores, each of which can settle down and grow into a new organism. The rhizoids are not reproductive structures; they serve purely to anchor the developing sporangium to the substrate and to absorb the nutrients required for its growth.

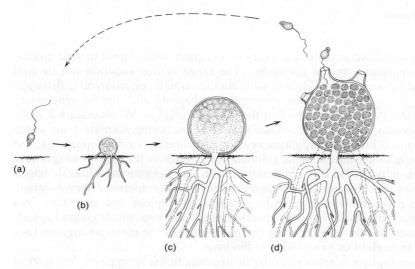

FIGURE 4.15

The life cycle of a primitive fungus, a chytrid. The flagellated zoospore (a) settles down on a solid surface. As development begins (b), a branching system of rhizoids is formed, anchoring the fungus to the surface. Growth results in the formation of a spherical zoosporangium, which cleaves internally to produce many zoospores (c). The zoosporangium ruptures to liberate a fresh crop of zoospores (d).

FIGURE 4.16

The life cycle of *Allomyces*, an aquatic phycomycete with a well-marked alternation of haploid and diploid generations. From a drawing made by Raphael Rodriguez and reprinted by permission of Arthur T. Brice.

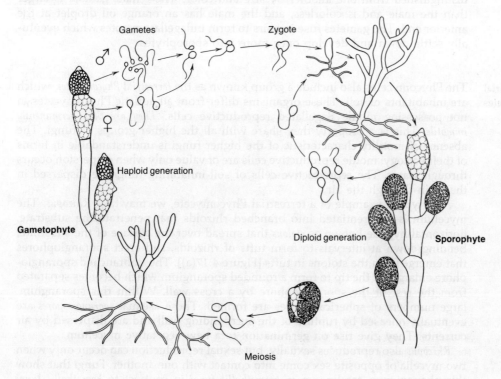

The aquatic Phycomycetes are a very varied group with respect to their mechanisms of reproduction and life cycles. The range of this variation can be well illustrated by comparing a chytrid with another aquatic phycomycete, *Allomyces*. *Allomyces* shows a well-marked alternation of haploid and diploid generations (Figure 4.16). We shall describe first the diploid *sporophyte*. When mature it looks like a microscopic tree, with a basal system of anchoring rhizoids from which springs a much-branched mycelium bearing two different kinds of sporangia. The mitosporangia have thin, smooth, colorless walls, whereas the meiosporangia have thick, dark-pitted walls. Upon maturation, both kinds of sporangia liberate flagellated spores, but the subsequent development of these spores is very different. The mitospores derived from mitosporangia are diploid and germinate into sporophytic individuals. The meiospores derived from meiosporangia are haploid, because meiosis takes place during the maturation of the meiosporangium; they give rise to haploid or *gametophytic* individuals.

The gametophyte is grossly similar in structure to the sporophyte, but instead of bearing meiosporangia and mitosporangia, it produces male and female gametangia, which are generally borne in pairs. The female gametangium looks very much like a mitosporangium, whereas the male gametangium is distinguished by its brilliant orange color. The gametangia rupture to liberate male and female gametes, a considerable number arising from each gametangium. Both male and female gametes are motile, moving by means of flagella, but they can be readily distinguished from one another by size and color. The female gamete is larger than the male and is colorless, and the male has an orange oil droplet at the anterior end. The gametes fuse in pairs to form biflagellate zygotes which eventually settle down and develop once more into sporophytes.

• **The terrestrial Phycomycetes**

The Phycomycetes also include a group known as the *terrestrial Phycomycetes*, which are inhabitants of soil. These organisms differ from all aquatic Phycomycetes in not possessing motile flagellated reproductive cells. *They are thus permanently immotile.* This is a property they share with all the higher groups of fungi. The absence of motility characteristic of the higher fungi is understandable in terms of their ecology: motile reproductive cells are of value only when dispersion occurs through water. The reproductive cells of soil-inhabiting fungi are dispersed in the main through the air.

As a typical example of a terrestrial Phycomycete, we may take *Rhizopus*. The mycelium is differentiated into branched rhizoids that penetrate the substrate, horizontal hyphae known as *stolons* that spread over the surface of the substrate, bending down at intervals to form tufts of rhizoids, and erect sporangiophores that emerge from the stolons in tufts [Figure 4.17(a)]. The unbranched sporangiophore enlarges at the tip to form a rounded sporangium which becomes separated from the rest of the sporangiophore by a cross wall. Within this sporangium, large numbers of spherical spores are formed. These asexual *sporangiospores* are eventually released by rupture of the surrounding wall and are dispersed by air currents. They give rise on germination to a new vegetative mycelium.

Rhizopus also reproduces sexually, but sexual reproduction can occur only when two mycelia of opposite sex come into contact with one another. Fungi that show this phenomenon are known as *heterothallic fungi* in contrast to *homothallic fungi* (such as *Allomyces*) that can produce both kinds of sex cells on a single mycelium.

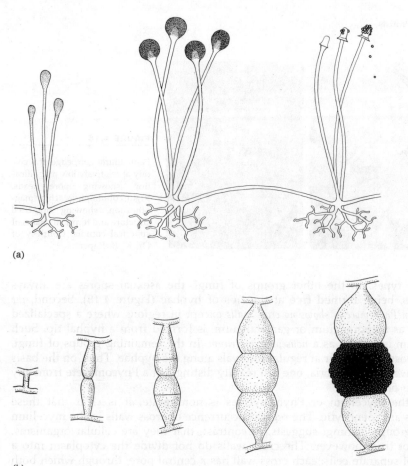

(a)

(b)

FIGURE 4.17

(a) The vegetative stage of *Rhizopus*, a terrestrial phycomycete. (b) Sexuality in *Rhizopus*. Successive stages of sexual fusion and the formation of a zygospore.

In *Rhizopus* the two kinds of mycelia between which sexual reproduction can take place are known as + and − strains, because there are no morphological indications of maleness and femaleness. As the hyphae from a + and a − mycelium meet, each produces a short side branch at the point of contact. This side branch then divides to form a special cell, the *gametangium*. The two gametangia, which are in direct contact with one another, fuse to form a large zygospore, surrounded by a thick, dark wall. The whole sequence of events is shown in Figure 4.17(b). It can be seen that the behavior of both partners in the sexual act is identical; hence, there is no basis for a designation as "male" or "female." Upon germination of the zygospore, meiosis occurs, and a hypha emerges and produces a sporangium. The haploid spores from this sporangium in turn develop into the typical vegetative mycelium.

• Distinctions between Phycomycetes and higher fungi

Despite the considerable differences among them, all Phycomycetes share two properties that readily distinguish them from the remaining classes of fungi (Ascomycetes, Basidiomycetes, and Fungi Imperfecti). First, *their asexual spores are always endogenous*, formed inside a saclike structure, the zoosporangium of the aquatic types or the sporangium containing immotile sporangiospores of the

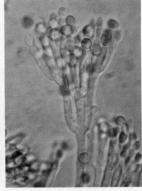

FIGURE 4.18

Penicillium. Left, edge of a colony at relatively low magnification, showing spore heads. Right, conidiophore at high magnification, showing branched structure and terminal chains of spherical conidia. Courtesy of Dr. K. B. Raper.

terrestrial types. In the other groups of fungi, the asexual spores are always exogenous, being formed free at the tips of hyphae (Figure 4.18). Second, *the mycelium in Phycomycetes shows no cross walls* except in regions where a specialized cell, such as a sporangium or gametangium, is formed from a hyphal tip. Such a mycelium is known as a *nonseptate mycelium*. In the remaining groups of fungi, distinct cross walls occur at regular intervals along the hyphae. Thus, on the basis of these two simple criteria, one can readily distinguish a Phycomycete from any other type of fungus.

Since the mycelium of Phycomycetes is nonseptate, it is clear that these organisms are coenocytic. The regular occurrence of cross walls in the mycelium of other groups of fungi suggests, in contrast, that they are cellular organisms. This is not true, however. The cross walls do not divide the cytoplasm into a number of separate cells: each cross wall has a central pore, through which both cytoplasm and nuclei can move freely. There is thus just as much cytoplasmic continuity in the septate fungi as in the Phycomycetes, and both groups are, in fact, coenocytic.

• The Ascomycetes and Basidiomycetes

The higher fungi, with septate mycelia and exogenous asexual spores, are broadly classified into two groups, Ascomycetes and Basidiomycetes, on the basis of their sexual development. Following zygote formation in these fungi, there is an immediate reduction division followed by the formation of four or eight haploid sexual spores, which are borne in or on structures known as *asci* and *basidia*. The formation of an ascus is characteristic of fungi of the class Ascomycetes, and the formation of a basidium is characteristic of fungi of the class Basidiomycetes. In Ascomycetes, the zygote develops into a saclike structure, the ascus, while the nucleus undergoes two meiotic divisions, often followed by one or more mitotic divisions. A wall is formed around each daughter nucleus and the neighboring cytoplasm to produce four, eight, or more ascospores within the ascus (Figure 4.19). Eventually the ascus ruptures, and the enclosed spores are liberated.

In Basidiomycetes, the zygote enlarges to form a club-shaped cell, the basidium; at the same time, the diploid nucleus undergoes meiosis. The subsequent course of events is strikingly different from that which occurs in an ascus. No spores

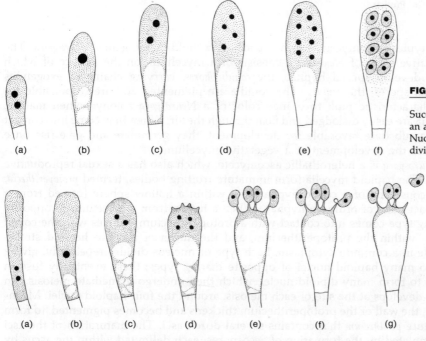

FIGURE 4.19

Successive stages in the formation of an ascus. (a) Binucleate fusion cell. (b) Nuclear fusion. (c), (d), (e), Nuclear divisions. (f) Ascospore formation.

FIGURE 4.20

Successive stages in basidium formation and basidiospore discharge. (a) Binucleate cell. (b) Nuclear fusion. (c), (d) Nuclear division. (e) Formation of basidiospores. (f), (g) Basidiospore discharge.

are formed within the basidium; instead, a slender projection known as a *sterigma* develops at its upper end, and a nucleus migrates into this sterigma as the latter enlarges. Eventually, a cross wall is formed near the base of the sterigma, the cell thus cut off being a basidiospore. The same process is repeated for the remaining three nuclei in the basidium, so that a mature basidium bears on its surface four basidiospores (Figure 4.20). Basidiospore discharge is a remarkable phenomenon. After the basidiospore has matured, a minute droplet of liquid appears at the point of its attachment to the basidium. This droplet grows rapidly until it is about one-fifth the size of the spore, and then, quite suddenly, both spore and droplet are shot away from the basidium.

• **The Fungi Imperfecti**

The classification of the septate fungi into Ascomycetes and Basidiomycetes has one practical disadvantage. Obviously, the assignment of a fungus to its correct class is possible only if one has observed the sexual stage of its life cycle. If one happens to deal with a fungus that is incapable of sexual reproduction, or in which the sexual stage is unknown, it cannot be assigned either to the Ascomycetes or to the Basidiomycetes. Since heterothallism is very common in the higher fungi, it often happens that a single isolate of an ascomycete or basidiomycete will never undergo sexual reproduction, which requires the presence of another strain of opposite mating type. Accordingly, it has been necessary to create a third class, the Fungi Imperfecti, for those kinds in which a sexual stage has not so far been observed. It should be realized that the Fungi Imperfecti is essentially a provisional taxonomic group; from time to time the sexual stage is discovered in a fungus originally assigned to this group, and the organism in question is then transferred to either the Ascomycetes or the Basidiomycetes.

111

• **The development
of an ascomycete**

As a typical ascomycete, we may consider a mold of the genus *Neurospora*. The vegetative stage of *Neurospora* consists of a mycelium, on the surface of which there develop special hyphae, the *conidiophores*, carrying chains of exogenous asexual spores, the *conidia*. The conidia are pigmented and are responsible for the characteristic pink to orange color of a *Neurospora* colony. When mature, conidia are easily dislodged and float through the air. When they come into contact with a substrate favorable for development, they germinate and give rise once more to the development of a vegetative mycelium.

Neurospora is a heterothallic ascomycete, which also has a sexual reproductive cycle. The haploid mycelia form immature fruiting bodies, termed *protoperithecia*, each consisting of a coiled hypha lying within a hollow sphere formed from a compact mass of ordinary hyphae. When a hypha from a mycelium of opposite mating type comes into contact with a protoperithecium, it fuses with the coiled hypha within the protoperithecium, and the nuclei of the two haploid strains mingle in a common cytoplasm. Each type of nucleus divides repeatedly, giving rise to many haploid nuclei of opposite mating types; these eventually fuse in pairs, to form many diploid nuclei, which then undergo immediate meiosis. An ascus develops at the site of each meiosis, around the four haploid nuclei. Meanwhile, the wall of the protoperithecium thickens and becomes pigmented, to form a mature *perithecium* that contains several dozen asci. The maturation of the asci is completed by the formation of ascospores, each delimited within the ascus by a resistant spore wall that surrounds one of the haploid nuclei and the adjacent cytoplasm. A mature ascus may contain four or eight ascospores, the number depending on whether or not meiosis is followed by a further mitotic division of the four haploid nuclei.

The mature perithecium is roughly spherical in shape, with a short protruding neck. At maturity, a pore forms at the tip of the neck, through which the ascospores are violently discharged. Upon germination, ascospores (like conidia) produce a haploid mycelium. This life cycle is shown in Figure 4.21.

FIGURE 4.21

The life cycle of *Neurospora*, a heterothallic ascomycete. Asexual reproduction occurs by the formation of conidia from a haploid mycelium of each mating type, A and a. These mycelia also bear protoperithecia which, when fertilized by conidia or hyphae of the opposite mating type, develop into perithecia, within which numerous zygotes are formed. Each zygote undergoes two meiotic and one mitotic division to form an ascus containing eight ascospores, four of mating type A and four of mating type a. Germination of the ascospores gives rise once more to haploid mycelia.

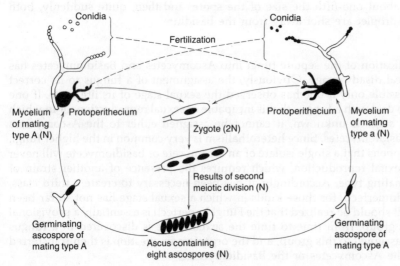

Conidia

Conidia

Fertilization

Mycelium of mating type A (N)

Protoperithecium

Zygote (2N)

Results of second meiotic division (N)

Protoperithecium

Mycelium of mating type a (N)

Germinating ascospore of mating type A

Ascus containing eight ascospores (N)

Germinating ascospore of mating type A

FIGURE 4.22

Cross section of a mushroom, showing the subterranean vegetative mycelium and the fruiting structure. In the mature fruiting body at left, large numbers of basidiospores are being discharged from the gills on the underside of the cap and are being dispersed by the wind. A second, immature fruiting body is shown at right, just emerging from the soil. From A. H. R. Buller, *Researches on Fungi*, Vol. 1, p. 219. New York: Longmans Green, 1909.

• **The development of a basidiomycete**

The most conspicuous members of the Basidiomycetes are the mushrooms. The portion of a mushroom that is seen by the casual observer is a small part of the whole organism, being the specialized fruiting structure that bears the basidia. The vegetative portion of the organism is entirely concealed from view and consists of a loose mycelium spreading, often for many meters, under the soil (Figure 4.22). The vegetative mycelium grows more or less continuously. When conditions are favorable (generally following a spell of wet, warm weather), fruiting bodies are formed at various points on its surface and push up through the soil to become the parts of the fungus visible to the observer.

In the common field mushroom (Figure 4.22) the fruiting body consists of a stalk surmounted by a cap, both composed of closely packed hyphae. The underside of the cap consists of rows of radiating gills, each gill being lined with thousands of basidia. The basidia project horizontally from the vertical walls of the gill, and consequently when the basidiospores are ejected, they pass into the air space between adjacent gills, and from there fall to the ground below. When a mushroom is mature, basidiospore discharge is a massive phenomenon. If one places a ripe cap on a piece of paper for a few hours, a "negative" of the gill structure will be formed on the paper by the deposition of millions of basidiospores.

Heterothallism is widespread among the Basidiomycetes. Consequently, isolated haploid basidiospores give rise on germination to haploid mycelia that are incapable of fructifying. However, if two such haploid mycelia of compatible mating types come into contact, hyphal fusion followed by nuclear exchange takes place, to produce a still haploid *dicaryon;* the two kinds of nuclei become associated in a very regular fashion, one pair occurring in each compartment of the septate mycelium. During growth of the dicaryon, the two kinds of nuclei divide synchronously.

A dicaryotic mycelium may continue to grow vegetatively for a long time, and during such vegetative growth, fusion between the paired nuclei never occurs. Fusion takes place only at the time of fructification, when the basidia are produced and is followed in each basidium by immediate meiosis to produce the four haploid nuclei destined to enter the basidiospores.

• **The yeasts**

Among the Ascomycetes, Basidiomycetes, and Fungi Imperfecti, the characteristic vegetative structure is the coenocytic mycelium. Nonetheless, there are a few groups in these classes that have largely lost the mycelial habit of growth and have become unicellular. Such organisms are known collectively as *yeasts:* A typical yeast consists of small, oval cells that multiply by forming buds. The buds enlarge until they are almost equal in size to the mother cell, nuclear division occurs, and then a cross wall is formed between the two cells (Figure 4.23). Although the yeasts constitute a minor branch of the higher fungi in terms of number of species, they are very important microbiologically. Most yeasts do not live in soil but have instead become adapted to environments with a high sugar content, such as the nectar of flowers and the surface of fruits. Many yeasts (the fermentative yeasts) perform an alcoholic fermentation of sugars and have been long exploited by man (see Chapter 31).

Yeasts are classified in all three classes of higher fungi: Ascomycetes, Basidiomycetes, and Fungi Imperfecti. The principal agent of alcoholic fermentation, *Saccharomyces cerevisiae*, is an ascomycetous yeast. Budding ceases at a certain stage of its growth, and the vegetative cells become transformed into asci, each containing four ascospores. For a long time it was believed that ascospore formation in *S. cerevisiae* was not preceded by zygote formation because pairing of vegetative cells prior to the formation of ascospores could never be observed. Eventually, however, it was discovered that zygote formation takes place at an unexpected stage of life cycle—immediately after the germination of the haploid ascospores. Pairs of germinating ascospores, or the first vegetative cells produced from them, fuse to form diploid vegetative cells. Diploidy is then maintained throughout the entire subsequent period of vegetative development, and meiosis occurs immediately prior to the formation of ascospores. Thus, *S. cerevisiae* exists predominantly in the diplophase. Other ascomycetous yeasts do not share this pattern of behavior

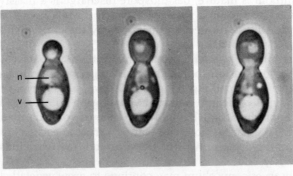

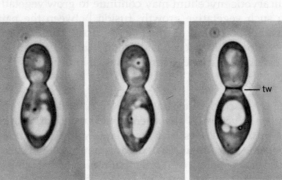

FIGURE 4.23

A sequence of photomicrographs of a budding cell of the ascomycetous yeast, *Wickerhamia*, showing nuclear division and transverse wall formation (phase contrast, ×1,770): n, nucleus; v, vacuole; tw, transverse wall. From P. Matile, H. Moore, and C. F. Robinow, p. 219 in *The Yeasts*, Vol. 1, A. N. Rose and J. S. Harrison, eds. New York: Academic Press, 1969.

FIGURE 4.24

The formation of a mirror image of a colony of *Sporobolomyces* by basidiospore discharge in a petri dish incubated in the inverted position: (top) the colony on the agar surface, streaked in the form of an S; (bottom) the deposit of basidiospores formed on the lid of the petri dish as a result of spore discharge from the colony. From A. H. R. Buller, *Researches on Fungi*, Vol. 5, p. 175. New York: Longmans Green, 1933.

but form zygotes by fusion between vegetative cells immediately before ascospore formation. The germinating ascospores then gives rise to haploid vegetative progeny.

Although budding is the predominant mode of multiplication in yeasts, there are a few that multiply by binary fission, much like bacteria; these are placed in a special genus, *Schizosaccharomyces*.

In ascomycetous yeasts, the vegetative cell or zygote becomes entirely transformed into an ascus at the time of ascospore formation. Yeasts of the genus *Sporobolomyces* form basidiospores, and in this case the entire vegetative cell becomes transformed into a basidium. Just as in the mushrooms, basidiospore discharge in *Sporobolomyces* is a violent process, and the colonies of this yeast are readily detectable on plates that have been incubated in an inverted position because the portion of the glass cover underlying a *Sporobolomyces* colony becomes covered with a deposit of discharged spores that form a mirror image of the colony above (Figure 4.24).

THE SLIME MOLDS

We shall conclude this survey of the protists by discussing the *slime molds*, which are not classified as true fungi, although they possess certain characteristics that resemble those of the fungi. The best-known representatives of the slime molds are the Myxomycetes, organisms that are found most commonly growing on decaying logs and stumps in damp woods. The vegetative structure, known as a *plasmodium*, is a multinucleate mass of cytoplasm unbounded by rigid walls, which flows in ameboid fashion over the surface of the substrate, ingesting smaller microorganisms and fragments of decaying plant material. An actively moving plasmodium is characteristically fan-shaped, with thickened ridges of cytoplasm running back from the edge of the fan; it resembles a spreading layer of thin, colored slime (Figure 4.25). As long as conditions are favorable for vegetative development, the plasmodium continues to increase in bulk with accompanying repeated nuclear divisions. Eventually, the organism may become a mass of cytoplasm containing thousands of nuclei and weighing several hundred grams. Fruiting occurs when a plasmodium migrates to a relatively dry region of the substrate. Out of the undifferentiated plasmodium there is then produced a fruiting structure that is often of remarkable complexity and beauty [as illustrated by the case of *Ceratiomyxa* (Figure 4.26)]. As this fruiting body develops, small, uninucleate sections of the plasmodium become surrounded by walls to form large numbers of uninucleate spores, borne on the fruiting structure. After

FIGURE 4.25

The plasmodium of a myxomycete, *Didymium*, growing at the expense of bacteria on the surface of an agar plate. Courtesy of Dr. K. B. Raper.

liberation, the spores germinate to produce uniflagellate ameboid gametes which fuse in pairs to form biflagellate zygotes. After some time, a zygote loses its flagella and develops into a new plasmodium. The vegetative nuclei in a growing plasmodium are diploid, meiosis taking place just prior to the formation of spores in the fruiting body.

It is, of course, the fruiting stage of a myxomycete that at once reminds one of a true fungus; at first sight, the amorphous, plasmodial vegetative stage appears to resemble little, if at all, the branched, mycelial vegetative stage of the fungi but suggests, rather, a relationship to the ameboid protozoa. In fact, the plasmodium and the mycelium are basically similar structures. Both are coenocytic, and in both the cytoplasm can flow, although in the mycelium cytoplasmic streaming is confined within the walls of branched tubes. The superficial difference between a plasmodium and a mycelium is essentially caused by the fact that in a plasmodium the cytoplasm is not bounded by rigid walls and is thus free to flow in any direction.

The slime molds also include a small group, the Acrasieae (Figure 4.27), which show far greater resemblances to the unicellular ameboid protozoa than do the true Myxomycetes. The vegetative stage of an acrasian consists of small, uninucleate amebas, which multiply by binary fission and can in no way be distinguished, at this stage of their life history, from other small ameboid protozoa. Nevertheless, when conditions are favorable, thousands of these isolated amebas

FIGURE 4.26

Fruiting bodies of a myxomycete, *Ceratiomyxa*, on a piece of wood. From C. M. Wilson and I. K. Ross, "Meiosis in the Myxomycetes," *Am. J. Botany* **42**, 743 (1955).

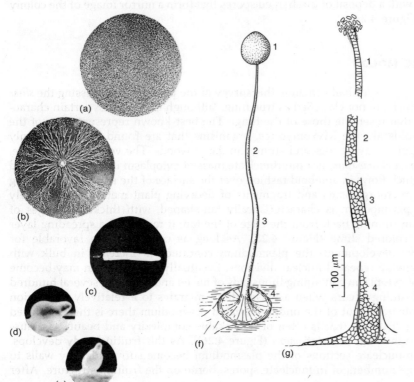

100 μm

FIGURE 4.27

The life cycle of *Dictyostelium*, a representative of the Acrasieae. (a) A uniform mass of vegetative amebae. (b) Aggregation of the amebae to a fruiting center. (c) Motile mass of aggregated cells. (d), (e) Early stages in formation of the fruiting body. (f) A mature fruiting body. (g) Magnified sections through various regions of the fruiting body. From K. B. Raper, "Isolation, Cultivation, and Conservation of Simple Slime Molds," *Quart. Rev. Biol.* **26**, 169 (1951).

are capable of aggregating and cooperating, without ever losing their cellular distinctness, in the construction of an elaborate fruiting body. The first sign of approaching fructification is the aggregation of the vegetative cells to form a macroscopically visible heap. This heap of cells gradually differentiates into a tall stalked structure that bears a rounded head of asexual spores. At all stages in the formation of this fruiting body, the cells remain separate; some individuals form the stalk, which is surrounded and given rigidity by a cellulose sheath, while others are carried up the outside of the rising stalk to form the spore head. As this matures, each ameba in it rounds up and becomes surrounded by a wall. These spores, following their release, germinate and give rise to individual ameboid vegetative cells once more. This remarkable kind of life cycle, where a communal process of fructification is imposed on a unicellular phase of vegetative development, occurs in one procaryotic group, the myxobacteria (described in Chapter 5).

THE PROTISTS: SUMMING UP

It is not possible in one chapter to do justice to the extraordinary profusion and biological variety of the protists; only a few representatives of each major subgroup have been somewhat summarily described. For more detailed information about these organisms, the reader should consult specialized books dealing with the algae, the protozoa, or the fungi (see the bibliography at the end of the chapter). There is, unfortunately, no single book that provides a more extended survey of the entire biological group. Comprehension of the comparative biology of protists is further impeded by major terminological difficulties, because botanists and zoologists have applied entirely different names to structures common to all three subgroups. Moreover, the taxonomic treatments adopted for flagellate algae and leucophytes by zoologists and botanists are widely different. Here we have tried to bridge these differences and to provide a broader account than is customary of the protists, in terms of their possible evolutionary interrelationships. These suggested interrelationships are summarized in Figure 4.28.

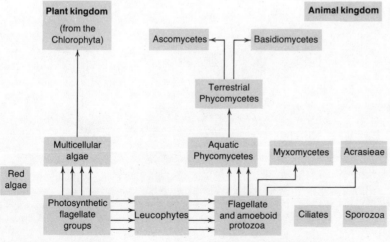

FIGURE 4.28

Possible evolutionary relationships between eucaryotic biological groups. The arrows connecting boxes indicate directions of evolution; where several arrows connect two boxes, this indicates probable multiple evolutionary lines connecting the two groups in question. Isolated boxes contain groups of which the evolutionary relationships are uncertain.

FURTHER READING

Books   *Algae:*

FRITSCH, F. E., *Structure and Reproduction of the Algae*. New York: Cambridge University Press, 1935, Vol I; 1945, Vol II.

LEWIN, R. (editor), *Physiology and Biochemistry of Algae*. New York: Academic Press, 1962.

SMITH, G. M., *Cryptogamic Botany*, 2nd ed., Vol. I. New York: McGraw-Hill, 1955.

Protozoa:

GRELL, K. G., *Protozoology*. Heidelberg: Springer-Verlag, 1973.

HALL, R. P., *Protozoology*. Englewood Cliffs, N.J.: Prentice-Hall, 1953.

MACKINNON, D. L., and R. S. J. HAWES, *An Introduction to the Study of Protozoa*. New York: Oxford University Press, 1961.

SLEIGH, M., *The Biology of Protozoa*. Amsterdam: Elsevier, 1973.

Fungi:

ALEXOPOULOS, C. J., *Introductory Mycology*, 2nd ed. New York: Wiley, 1962.

BURNETT, J. H., *Fundamentals of Mycology*. New York: St. Martin's Press, 1968.

THE PROCARYOTES:
AN INTRODUCTORY SURVEY

Most procaryotes are unicellular organisms; and although some of them display a certain degree of differentiation, it never approaches the remarkable specialization within the confines of a single cell that is characteristic of such protists as dinoflagellates and ciliates. One group of unicellular procaryotes, the myxobacteria, form many-celled fruiting bodies through aggregation of vegetative cells, a developmental cycle analogous to that of the cellular slime molds (Acrasieae). Many blue-green bacteria and some nonphotosynthetic procaryotes are multicellular, filamentous organisms; and the nonphotosynthetic Actinomycetes have a coenocytic, mycelial structure and reproduce by spore formation. However, none of the filamentous and mycelial procaryotes has attained the size and organismal complexity characteristic of such protists as the red and brown algae and the basidiomycetous fungi. It is nevertheless an interesting fact that such developmental patterns as the formation of many-celled fruiting bodies, the mycelial habit of growth, and simple versions of multicellularity, all of which occur among the protists, have counterparts among the procaryotes. Evolution from unicellularity toward more complex forms of development has evidently proceeded along parallel lines in the eucaryotes and in the procaryotes.

Although a certain degree of structural diversity exists among procaryotes, their diversity is expressed primarily in metabolic terms, particularly with respect to the mechanisms of energy-yielding metabolism. The principal versions of energy-yielding metabolism characteristic of protists—photosynthesis, aerobic respiration, and fermentation—all occur in procaryotes. Furthermore, certain procaryotes perform types of energy-yielding metabolism that do not exist in eucaryotes. These include special versions of photosynthesis, anaerobic respiration, a wide variety of fermentations, and the oxidation of reduced inorganic compounds. In addition, many groups of procaryotes can fix molecular nitrogen and use it as a nitrogen source, a capacity which does not exist in eucaryotes.

• **The range of cell size in procaryotes**

Both among protists and among procaryotes the dimensions of the cell cover a substantial range. The largest unicellular bacteria are considerably larger than the smallest unicellular protists, so that cell size cannot serve as an absolute character to distinguish between the members of these two great microbial assemblages. Nevertheless, the average cell size of procaryotes is considerably less than that of protists; and the smallest bacteria have cells far below the lower limit of cell size in eucaryotic groups. Some bacteria have cells only just resolvable with the light microscope, and they overlap in size with the larger viruses.

Because cells can differ so greatly in shape, the only satisfactory basis on which their size can be defined is cellular volume. In Table 5.1 data on the volume of the cell for a number of procaryotes and protists, as well as for the structural unit (virion) of some representative viruses, have been assembled. From the individual data, size range for each of the three groups can be established: these values are also given in Table 5.1.

The smallest known protist is a flagellate alga, *Micromonas*. This interesting organism has a cell that contains a single, small chloroplast and one mitochondrion. Together with the nucleus, they occupy a large part of the total volume of the cell. Therefore, any further reduction of size in such an organism could be achieved only *by elimination of a cellular organelle that is vital to metabolic function.* The lower size limit possible for this kind of cell is accordingly set by *structural limitations.*

In principle, the lower size for a procaryotic cell is set by *molecular limitations.* To reproduce itself, any cell requires a large number of *different enzymes* and hence of *different kinds of proteins.* We do not know the precise minimum, but the number is probably of the order of several hundred. Furthermore, *many molecules* of most enzymes are needed. To these must be added the cellular nucleic acids and the

TABLE 5.1

Size of the unit of structure of unicellular protists and of viruses

NATURE OF STRUCTURAL UNIT	BIOLOGICAL GROUP	VOLUME OF STRUCTURAL UNIT, μm^3		LIMITS FOR MAJOR GROUP
		NORMAL RANGE	EXTREME LIMITS	
Eucaryotic cell	Unicellular algae	5,000–15,000	5–100,000	5–150,000,000
	Protozoa	10,000–50,000	20–150,000,000	
	Yeasts	20–50	20–50	
Procaryotic cell	Photosynthetic bacteria	5–50	0.1–5,000	0.01–5,000
	Spirochetes	0.1–2	0.05–1,000	
	Mycoplasmas	0.01–0.1	0.01–0.1	
Virion	Pox group	0.01	Fixed for each group	0.00001–0.01
	Rabies virus	0.0015		
	Flu virus	0.0005		
	Polio virus	0.00001		

various other organic constituents of the cell such as lipids and carbohydrates. All in all, it seems likely that the smallest bacteria, which can just be seen with a light microscope, have a size very close to this *molecular limit* for the maintenance of cellular function.

The minimal size of the unit of structure of a virus, the virion, is also determined by molecular considerations. Virions are composed of only one kind of nucleic acid and of one (or a few) kinds of protein; their minimal size is, therefore, defined by the volume occupied by a small number of molecules of protein and nucleic acid.

The factors that operate to determine minimal size adequately explain why the cells of some bacteria can be more than two orders of magnitude smaller than the smallest eucaryotic cell. It is not obvious, however, why the cells of bacteria cannot attain the largest cell sizes found among protozoa and algae, although the data assembled in Table 5.1 clearly show that they do not. Several factors probably operate to keep procaryotic cells small. In the first place, it is a general biological rule, valid at all levels of biological complexity, that *the rate of metabolism is inversely related to the size of the organism*. Since the growth rate is primarily determined by the overall rate of metabolism, small size is an essential condition for rapid growth. Many bacteria are characterized by extraordinarily rapid growth, with doubling times of less than an hour under optimal conditions. Such growth rates are far higher than those of most protists. This property undoubtedly confers a decisive biological advantage in many situations where bacteria compete for nutrients with protists, and it is probably the most important single factor in the survival of many bacteria in nature.

A second factor can be surmised from the properties of very large bacteria, such as the chemoheterotroph *Spirillum volutans* and the photoautotrophs *Chromatium okenii* and *Thiospirillum jenense*. In contrast to the smaller-celled members of the same physiological groups, these giant bacteria are extremely difficult to cultivate, primarily because the physicochemical conditions necessary for their development are very narrowly defined. In other words, they lack adaptive

TABLE 5.2

Genome sizes of procaryotes, expressed in daltons ($\times 10^{-9}$)

GRAM-POSITIVE BACTERIA:	
Spore-formers	2.3–2.8
Cocci	1.2–2.8
GRAM-NEGATIVE BACTERIA:	
Enteric bacteria	2.1–2.9
Other nonphotosynthetic groups	1.0–2.9
Unicellular blue-green bacteria	1.6–4.0
Filamentous blue-green bacteria	3.5–5.9
MYCOPLASMAS	0.45–1.1

flexibility and cannot readily accommodate to minor changes in the environment. This suggests that the power to *regulate metabolic function*, essential to achieve adaptive flexibility, is very limited in such organisms. The upper size limit for the procaryotic cell may, accordingly, be determined by the difficulty of maintaining a satisfactory regulation and coordination of metabolic activity in a large cell with the procaryotic type of organization.

• **The range of genome size in procaryotes**

Recently developed physical techniques have made it relatively easy to measure the size of the genome, and many values for procaryotic organisms are now available. Some of the values are summarized in Table 5.2. In general, they lie in a rather narrow range, extending from about 1 to 3×10^9 daltons. Significantly higher values occur in some filamentous blue-green bacteria, and significantly lower ones occur in some members of the *Mycoplasma* group.

MAJOR TAXONOMIC SUBDIVISIONS AMONG PROCARYOTES

The problem of recognizing and distinguishing major constituent groups among the procaryotes is extraordinarily difficult for two reasons. First, as a result of the simplicity of cell structure, the number of structural characters available as useful taxonomic markers is very limited. Second, there is rarely a close correlation between the groups distinguishable in terms of shared *structural* properties and those distinguishable in terms of shared *metabolic* properties. The recently published eighth edition of *Bergey's Manual of Determinative Bacteriology** divides the procaryotes into no less than 20 major assemblages, many of which are distinguished by somewhat arbitrary criteria. Although this taxonomic scheme will be discussed at a later place (see Chapter 16), it is too complex to provide a satisfactory framework for the purposes of this introductory survey. Here, a simpler subdivision of the procaryotes will be presented, primary separations being based on a structural character: *the nature of the bounding layer of the cell.*

This property distinguishes three major groups: the Mycoplasmas, which do not synthesize a cell wall, the membrane serving as the outer bounding layer; the Gram-positive bacteria, which synthesize a monolayered cell wall; and the Gram-negative bacteria, which synthesize a cell wall composed of at least two structurally distinct layers.

In both Gram-positive and Gram-negative bacteria, the distinctive heteropolymers known as *peptidoglycans* are the components primarily responsible for the mechanical strength of the wall. However, the cell wall always contains other classes of polymers, which are of different chemical nature in the walls of Gram-positive and Gram-negative bacteria. Furthermore, the location of peptidoglycan within the wall is different in Gram-positive and Gram-negative bacteria.

Electron micrographs of cross sections of the two types of cell wall reveal the

---

*R. E. Buchanan, and N. E. Gibbons, eds., *Bergey's Manual of Determinative Bacteriology*, 8th ed. (Baltimore: Williams and Wilkins, 1974).

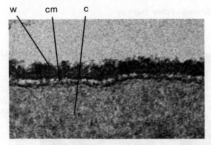

(a)

(b)

FIGURE 5.1

Electron micrographs of sections of the surface layers of (a) a Gram-positive bacterium and (b) a Gram-negative bacterium, illustrating the differences in cell wall profiles: c, cytoplasm; cm, cell membrane; w, wall. In the Gram-positive bacterium the wall consists of a single, thick, continuous layer. In the Gram-negative bacterium it is multilayered; in this particular organism there is a thin inner (peptidoglycan) layer; an intermediate layer, similar in profile to the cell membrane; and a loose outer layer.

wall of a Gram-positive bacterium as a structure of uniform fabric, some 20 to 80 nm in width, whereas the wall of a Gram-negative bacterium is composed of at least two readily distinguishable layers, each considerably thinner than the wall of a Gram-positive organism (Figure 5.1).

Chemical analyses reveal that the walls of Gram-positive bacteria contain peptidoglycan as a major polymer (generally accounting for 50 to 80 percent of the weight of the wall), with which are associated polysaccharides and a special class of polymers, the teichoic acids. The teichoic acids, which occur as wall constituents only in Gram-positive bacteria, are linear polymers derived from either glycerol phosphate or ribitol phosphate. The manner in which peptidoglycan, polysaccharide, and teichoic acid is distributed through the structurally uniform wall of a Gram-positive bacterium is not yet known; however, there are indications that the three types of polymer may be in part covalently linked to one another, to produce a wall fabric of enormous size and molecular complexity. Although proteins are not invariably associated with the wall of Gram-positive bacteria, they are sometimes present on its outer surface, conferring on it a very regular superficial structure.

The walls of Gram-negative bacteria contain a dense, uniform inner layer, only 2 to 3 nm wide (the so-called *rigid layer*). This is overlain by a somewhat thicker (8 to 10 nm) outer layer, which has a fine structure virtually indistinguishable from that of a unit membrane, and for this reason it is sometimes termed the *outer membrane*. It must be emphasized, however, that this wall layer differs both in chemical composition and in function from the cytoplasmic membrane. The surface of this outer wall layer is highly irregular (corresponding to its wavy profile in thin sections), as revealed in electron micrographs of negatively stained cells (Figure 5.2). The walls of some Gram-negative bacteria include an additional outer layer, frequently displaying a remarkably regular fine structure (Figure 5.3).

The peptidoglycan content of the walls of Gram-negative bacteria is relatively

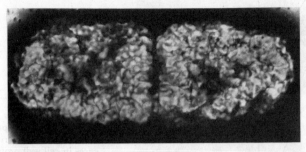

FIGURE 5.2

Electron micrograph of a cell of *E. coli,* negatively stained with silicotungstate to reveal the highly irregular surface structure of the outer layer of the cell wall ($\times$53,000). From M. E. Bayer and T. F. Anderson, "The Surface Structure of *Escherichia coli,*" *Proc. Natl. Acad. Sci. U.S.* **54,** 1592 (1965).

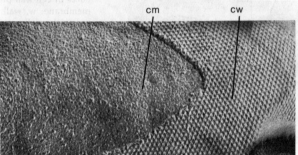

cm   cw

FIGURE 5.3

Electron micrograph of a freeze-etched preparation of the cell surface of a marine *Nitrosomonas,* cut tangentially so as to reveal the regular surface structure of the cell wall (cw) and the surface structure of the underlying cell membrane (cm) ($\times$101,000.) Courtesy of S. W. Watson.

low (as little as 1 to 10 percent of the total weight of the wall, in many species). Chemical fractionation shows that the peptidoglycan is confined to the innermost rigid layer, which consists in effect of a peptidoglycan sac enclosing the protoplast. This sac sometimes bears particles of lipoprotein attached to its outer surface. The so-called *outer membrane layer* of the wall contains proteins, lipopolysaccharides, and lipoproteins. The external layers of regular structure which occur on certain Gram-negative bacteria appear to consist only of protein.

The principal chemical differences between the two types of cell wall characteristic of procaryotes are summarized in Table 5.3.

TABLE 5.3

Principal chemical components of the bacterial cell wall

| | | GRAM-NEGATIVE BACTERIA | |
| | | RIGID WALL | OUTER WALL |
COMPONENT	GRAM-POSITIVE BACTERIA	LAYER	LAYER(S)
Peptidoglycan	+	+	−
Teichoic acid	+	−	−
Polysaccharide	+	−	−
Protein	+ or −	−	+
Lipopolysaccharide	−	−	+
Lipoprotein	−	+ or −	+

• Relation between
wall structure and
staining properties
of the bacterial
cell

A differential stain of great practical value for the identification of bacteria was discovered empirically in 1884 by Christian Gram, and it subsequently became known as the *Gram stain*. Many modifications exist, but they all embody the following essential steps: a heat-fixed smear of cells is stained successively with the basic dye, crystal violet, and with a dilute iodine solution. The preparation is then briefly treated with an organic solvent (alcohol or acetone). *Gram-positive* bacteria resist decolorization and remain stained a deep blue-black; *Gram-negative* bacteria are rapidly and completely decolorized. In performing this staining procedure, it is essential to examine growing cells, since certain Gram-positive bacteria (for example, *Bacillus spp.*) rapidly lose the ability to retain the crystal violet-iodine complex after active growth ceases. The outcome of the Gram reaction is thus to some extent conditioned by the physiological state of the cell.

Despite many attempts to elucidate its mechanism, the precise nature of the Gram reaction is not yet understood. The reaction is given only by intact cells; broken cells of a Gram-positive bacterium are Gram negative. The presence of the cell wall is essential, even though the dye complex is not located in the wall itself. This can be shown by staining a Gram-positive bacterium and then stripping off the cell wall by an appropriate enzymatic treatment in an isotonic medium. Under these conditions, the enclosed protoplasts are liberated intact, and contain the dye complex. However, they are no longer resistant to decolorization with alcohol; the attribute of Gram positiveness is lost upon removal of the surrounding wall. This experiment suggests that the wall of a Gram-positive bacterium constitutes a barrier to the extraction of the dye complex from the enclosed cell by organic solvents, but why this should be so is still not known.

The widespread application of the Gram staining procedure to bacteria soon revealed that most readily recognizable taxonomic subgroups among the bacteria are uniform with respect to their Gram reaction. The evident taxonomic value of the Gram stain, accordingly, led to the proposal that it is correlated with a basic chemical difference between Gram-positive and Gram-negative cells. Initially, attempts were made to detect a substance in Gram-positive cells that combines with and retains the crystal violet-iodine complex, but no such cellular constituent has ever been demonstrated. The true chemical significance of the Gram reaction only began to emerge about 20 years ago, when M. Salton developed procedures for the isolation and chemical analysis of bacterial cell walls. Much subsequent work has established that the outcome of the Gram reaction is systematically correlated with the major differences in the chemical composition and ultrastructure of procaryotic cell walls that have been outlined above. The Gram reaction is, accordingly, an empirical staining method that is easy to apply and that distinguishes, although not always with complete reliability, two major subgroups of procaryotes that differ with respect to the nature of the cell wall. Unfortunately, the differentiation of these two subgroups by other and more reliable methods is not easy: it requires either electron microscopic examination of wall structure in thin sections of cells or chemical detection of the group-specific polymers. We shall designate these two large subgroups as *Gram-positive bacteria* and *Gram-negative bacteria,* since no other simple group designations are in common use, while recognizing that the Gram staining procedure is not always a wholly reliable method to differentiate between them. In fact, the limitations of the staining procedure occur almost entirely in the context of the Gram-positive group: some bacteria clearly assignable to this group by virtue of wall structure and composition may give a Gram-negative or Gram-variable staining reaction. One

125

other point concerning the Gram reaction should be emphasized: *it is of diagnostic value only when applied to procaryotes that possess cell walls,* and it does not yield taxonomically useful information when applied to the *Mycoplasma* group or to the eucaryotic protists.

THE MYCOPLASMAS

Recent studies show that the mycoplasmas (members of the genus *Mycoplasma* and related organisms) are a large and widespread group of procaryotes, although knowledge of their properties has grown slowly. The absence of a cell wall makes the cells plastic, readily deformable, and liable to damage, which has impeded study of their structure and development. Furthermore, the mycoplasmas are chemoheterotrophs with complex nutritional requirements, which makes their cultivation in artificial media difficult.

The first member of this group to be described was the agent of bovine pleuropneumonia, and as a result mycoplasmas were long designated as the PPLO ("pleuropneumonialike organisms") group. Many species have been isolated from man and other vertebrates, where they occur as parasites on moist mucosal surfaces; and a number of additional pathogenic mycoplasmas have been described. Mycoplasmas are frequent contaminants in tissue cultures, grown in media that are likewise very favorable for these procaryotes. Other members of the *Mycoplasma* group have been detected in the vascular tissues of plants, and there is increasing evidence that they are the agents of a large number of plant diseases of previously undetermined etiology. Free-living mycoplasmas have been found in hot springs and other thermal environments.

The structure of the *Mycoplasma* cell is both irregular and variable (Figure 5.4). Cells may be coccoid or pear-shaped, sometimes with filamentous extensions, or grow as thin filaments, branched or unbranched. One plant *Mycoplasma* has the form of a helical filament and is reported to be motile. Other mycoplasmas are immotile. The cells are typically small, equivalent to a sphere with a diameter between 0.3 and 0.9 μm. Mycoplasmas are therefore the smallest known cellular organisms, close to the limit of resolution with the light microscope. The mode of reproduction, long controversial, is now known to be binary fission.

The colonies of mycoplasmas on solid media have a characteristic "fried-egg" structure, consisting of an opaque central area partly embedded in the substrate and a translucent periphery (Figure 5.5).

Two principal genera are distinguished on nutritional grounds. The members of the genus *Mycoplasma* specifically require a sterol, cholesterol. This substance is incorporated in large amounts into the cell membrane, and it appears to contribute to its stability. Members of the genus *Acholeplasma* do not require cholesterol, but they will incorporate it into the membrane if it is furnished in the medium. The mycoplasmas also require fatty acids and proteins. Both the protein and the lipid requirements can be furnished by serum, and media for the cultivation of these organisms are commonly supplemented with 20 percent horse serum.

Since they do not synthesize peptidoglycans, the mycoplasmas are insensitive to penicillins and other types of antibiotics which specifically affect peptidoglycan

FIGURE 5.4

Electron micrograph of cells of a member of the *Mycoplasma* group, the agent of bronchopneumonia in the rat ($\times$ 1,230). From E. Klieneberger-Nobel and F. W. Cuckow, "A Study of Organisms of the Pleuropneumonia Group by Electron Microscopy," *J. Gen. Microbiol.* **12**, 99 (1955).

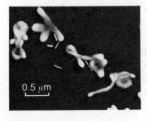

0.5 μm

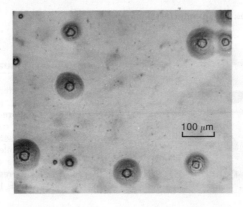

100 μm

FIGURE 5.5

Characteristic colony structure of organisms of the *Mycoplasma* group (×79). Courtesy of M. Shifrine.

synthesis. They are the only procaryotes resistant to penicillin, with the exception of the so-called *L forms* derived from certain Gram-positive and Gram-negative bacteria. L forms are mutants that have lost the ability (or have an impaired ability) to synthesize peptidoglycans; they can be selected from the parental bacteria by penicillin treatment. As a result of their defective walls, L forms are osmotically sensitive, and they can grow only in media of high osmotic strength. The possible relations of bacterial L forms to the mycoplasmas have been much discussed, but it is now generally recognized that the resemblances between them are superficial and do not express evolutionary relationship. L forms do not have the distinctive nutritional requirements of the mycoplasmas, still retain a defective cell wall, and are, of course, genetically nearly identical with the parental bacteria from which they have been derived.

Measurements of the genome size of procaryotes (see Table 5.2) have revealed that the *Mycoplasma* species stand apart from all other procaryotes by virtue of their very small content of genetic information, one-half (or less) that of all other procaryotes. This character makes them a group of exceptional biological interest, since it is probable that their information content is close to the lower limit necessary to specify the properties of a cellular organism.

THE GRAM-POSITIVE BACTERIA

Nearly all Gram-positive bacteria are chemoheterotrophs, dependent either on aerobic respiration or on fermentation for their supply of energy. One small subgroup, the methanogenic bacteria, are highly specialized anaerobic chemoautotrophs, which obtain energy by coupling the oxidation of molecular hydrogen with the reduction of CO_2 to methane.

In terms of structural properties, the Gram-positive bacteria can be divided into four main groups (Table 5.4). Some are unicellular organisms with rod-shaped or spherical cells, which reproduce by binary transverse fission. This category includes a special subgroup known as the sporeformers, which form distinctive resting cells known as *endospores*. An endospore is formed within a vegetative cell by a special process of division that will be described in detail in Chapter 11. As a rule, only one endospore is produced in each vegetative cell; after maturation, it is liberated by lysis of the surrounding mother cell. Endospores are recognizable by their extreme refractility, caused by the presence of a thick, specialized wall,

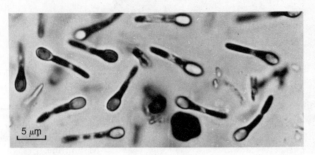

FIGURE 5.6

Bacterial endospores. Stained wet mount of an anaerobic spore-former (*Clostridium*) in the course of sporulation. The rod-shaped vegetative cells are swollen at one end by the presence of oval, highly refractile spores. Photo courtesy of C. F. Robinow.

and by the very low water content of the enclosed resting cell (Figure 5.6). They can be stained only with difficulty, and they are extremely resistant to heat, radiation, and toxic chemical agents. Once produced, endospores can remain dormant but viable for many years. The germination of an endospore occurs by rupture of the enclosing spore wall and emergence of a vegetative cell (Figure 5.7).

In rod-shaped unicellular Gram-positive bacteria, cell division always occurs in the equatorial plane, at right angles to the long axis of the cell. Some spherical unicellular Gram-positive bacteria likewise divide regularly in a single plane. This mode of cell division frequently gives rise to the formation of short chains of coccoid cells, as a result of continued adhesion between daughter cells following division (Figure 5.8). However, some bacteria with spherical cells divide regularly in either two or three successive planes at right angles to one another, thus giving rise to regular tetrads or cubical packets of cells (Figure 5.9).

FIGURE 5.7

Spore germination in (a) *Bacillus polymyxa* and (b) *B. circulans* (stained preparations, ×3,040). Courtesy of C. F. Robinow and C. L. Hannay.

Many Gram-positive unicellular bacteria are permanently immotile. Motility, in species in which it does occur, is always of the swimming type, mediated by bacterial flagella. In most Gram-positive bacteria, each cell bears numerous flagella, inserted at a number of different points over the surface; this flagellar arrangement is known as *peritrichous flagellation*, in order to distinguish it from insertion of the flagella at one site near the pole of the cell, termed *polar flagellation*. Polar flagellar insertion is very rare among Gram-positive bacteria. A few Gram-positive bacteria have a permanently filamentous structure. The very large motile

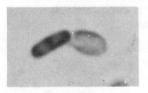

(a)

(b)

TABLE 5.4

Major constituent groups of Gram-positive bacteria distinguishable by structural characters

I. Unicellular bacteria with rod-shaped or spherical cells, reproducing by binary transverse fission: some produce endospores.

II. Unicellular bacteria with a tendency to structural irregularity. Cells spherical, rod-shaped, or branched: coryneform bacteria.

III. Bacteria which develop vegetatively as a mycelium and reproduce by mycelial fragmentation: the Proactinomycetes.

IV. Bacteria which form a persistent vegetative mycelium and reproduce by the formation of spores, formed in various ways at the tips of the hyphae: the Euactinomycetes.

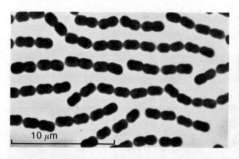

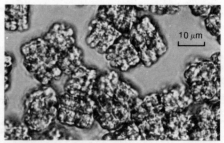

FIGURE 5.8

A chain-forming coccus, *Streptococcus* (×2,710). Courtesy of Daisy Kuhn and Patricia Edlemann.

FIGURE 5.9

A coccus that forms cubical packets of cells, *Sarcina* (×630). From E. Canale-Perola and R. S. Wolfe, "Studies on *Sarcina ventriculi*," *J. Bacteriol.* **79**, 887 (1960).

organism, *Caryophanon* (Figure 5.10), develops in the form of filaments about 4 μm wide and up to 40 μm in length, bearing many flagella. Each filament is composed of as many as 30 cells, which divide at right angles to the long axis of the filament. Reproduction occurs by binary fission of the filament, giving rise to two shorter filaments of cells.

One subgroup of Gram-positive bacteria, the *coryneform bacteria*, is characterized by a tendency to produce cells of variable and irregular form: spherical, rod-shaped, or even branched. These organisms, exemplified by *Arthrobacter* (Figures 5.11 and 5.12), have a limited potential for mycelial growth and form a group transitional between the unicellular Gram-positive bacteria and the other principal assemblage of Gram-positive organisms, the *Actinomycetes*, in which vegetative development characteristically involves the formation of a much-branched mycelium largely devoid of cross walls (Figure 5.13), superficially similar to the non-septate mycelium of certain phycomycetes (see p. 110).

Reproduction among the antinomycetes can occur by two different mechanisms. In the *Proactinomycetes*, the mycelium is undifferentiated and eventually undergoes a more or less massive fragmentation into rod-shaped or spherical mycelial fragments: at this stage of development, a culture can easily be confused with that of a Gram-positive unicellular bacterium.

In the *Euactinomycetes* the mycelium is permanent; and reproduction occurs by

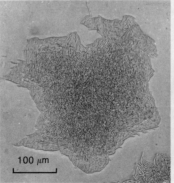

(a)

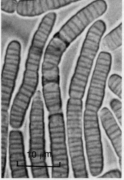

(b)

FIGURE 5.10

Caryophanon. (a) A small colony on agar (×110). (b) Living unstained filaments (×1,110) From E. G. Pringsheim and C. F. Robinow, "Observations on Two Very Large Bacteria. *Caryophanon latum Peshkoff* and *Lineola longa* (nomen provisorium," *J. Gen. Microbiol.* **1**, 278 (1947).

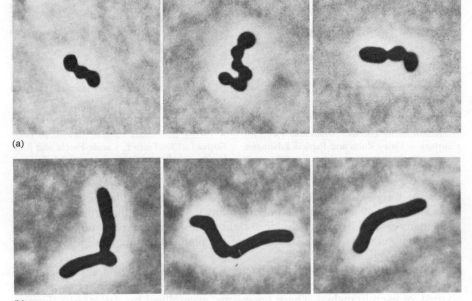

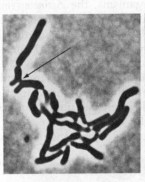

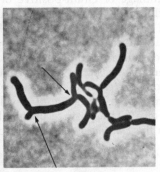

(a)

(b)

FIGURE 5.11

(a) Coccoid cells of *Arthrobacter* from a broth culture in the stationary phase (phase contrast, ×2,100. (b) Rod-shaped cells of *Arthrobacter* from a broth culture in the early exponential phase of growth (phase contrast is ×2,100. From M. P. Starr and D. A. Kuhn, "On the Origin of V-Forms in *Arthrobacter atrocyaneus*," *Arch. Mikrobiol.* **42**, 289 (1962).

FIGURE 5.12

Two microcolonies of *Arthrobacter* on agar, showing angular growth (arrows) (phase contrast, ×1,035). From M. P. Starr and D. A. Kuhn, "On the Origin of V-Forms in *Arthrobacter atrocyaneus*," *Arch. Mikrobiol.* **42**, 289 (1962).

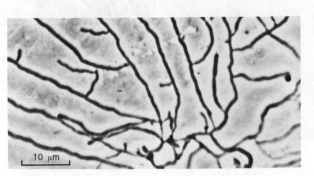

10 μm

FIGURE 5.13

The mycelium of an actinomycete, *Streptomyces* (phase contrast, ×1,160). From D. A. Hopwood, "Phase Contrast Observations on *Streptomyces coelicolor*," *J. Gen. Microbiol.* **22**, 295 (1960).

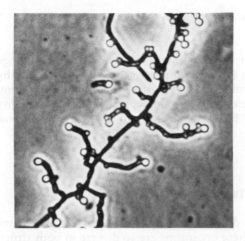

FIGURE 5.14

Micromonospora chalcea. showing spherical conidiospores borne singly at the tips of hyphae (phase contrast, ×1,440). Courtesy of G. M. Luedemann and the Schering Corporation.

(a)   (b)

FIGURE 5.15

The surface of a *Streptomyces* colony, as observed with the scanning electron microscope. (a) general view of the aerial mycelium (×1,740). (b) A helically wound chain of conidia. The individual conidia (not distinguishable in this figure) bear spiny appendages (×5,800). Reproduced with permission from S. Kimoto and J. C. Russ, "The Characteristics and Applications of the Scanning Electron Microscope," *Am. Scientist* **57**, 112 (1969).

the formation of specialized spores. In the simplest case, exemplified by *Micromonospora*, the spores are formed singly at the tips of the hyphae of the vegetative mycelium (Figure 5.14). In other Euactinomycetes, exemplified by *Streptomyces*, the spores are never formed directly in the vegetative mycelium. Instead, a much looser aerial mycelium develops on the surface of the dense, leathery vegetative mycelium, and chains of spores are subsequently differentiated at the ends of the projecting hyphae of the aerial mycelium (Figure 5.15). Finally, certain Euactinomycetes (exemplified by *Streptosporangium*) produce endogenous spores within terminal swellings (sporangia) produced at the tips of the hyphae.

Although immotile in the vegetative phase of growth, some Proactinomycetes and Euactinomycetes produce motile, flagellated spores. This is characteristic of certain sporangial Euactinomycetes and of the genus *Dermatophilus* among Pro-actinomycetes.

THE GRAM-NEGATIVE BACTERIA

It is difficult to give a succinct general account of Gram-negative bacteria because these organisms are so diverse in both structural and functional respects. Furthermore, subdivisions based on structural properties do not necessarily coincide with subdivisions based on functional properties, such as the mode of energy-yielding metabolism. In this chapter we shall be largely concerned with structural properties, although some discussion of major metabolic properties is unavoidable. It may be noted that the Gram-negative bacteria include all photosynthetic procaryotes, assignable to three quite distinct taxonomic groups; most of the chemoautotrophic bacteria; and many groups (some highly specialized) of chemoheterotrophs.

• **Mechanisms of movement**   Many groups of Gram-negative bacteria are characterized by swimming motility, mediated by bacterial flagella. The flagellar insertion is usually either polar or peritrichous. Some polarly flagellated bacteria are monotrichous, bearing a single flagellum at one pole of the cell (Figure 5.16). Others are multitrichous and bear a polar tuft composed of as many as 20 to 30 flagella (Figure 5.17). An individual bacterial flagellum is far too thin to be visualized by light microscopy, unless the cell has been subjected to a special staining procedure that artificially increases the apparent width of the flagella (see Figure 5.16). However, the flagellar tufts of some multitrichous bacteria are sufficiently thick to be detectable on living

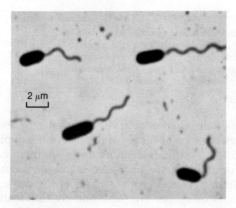

FIGURE 5.16

Stained flagella of a *Pseudomonas* that has a single polar flagellum (×1,890). Courtesy of George Hageage and C. F. Robinow.

2 μm

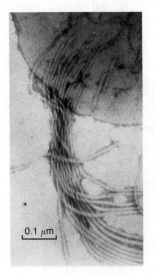

cells by phase contrast microscopy (Figure 5.18), and such organisms have provided valuable experimental objects for observations of flagellar movement.

The site of flagellar insertion (viz., polar or peritrichous) is a character to which considerable importance has been traditionally attributed in the classification of Gram-negative bacteria. However, it is sometimes difficult to determine on stained preparations, particularly of bacteria with a small number of flagella per cell, since the site of insertion can be misconstrued. In such cases, it is necessary to examine a large number of cells and to deduce the site of insertion from frequency distributions (Table 5.5). In addition to this strictly determinative problem, a more fundamental criticism of the use of flagellar insertion as a major taxonomic character can be made. Some Gram-negative bacteria bear two different sets of flagella, one polar and the other peritrichous. This fact can be ascertained by electron microscopy if (as is often the case) the polar and peritrichous flagella differ in wavelength or thickness (Figure 5.19). Such "mixed" flagellation implies that the cell contains two distinct sets of genetic determinants governing flagella formation. A further experimental complication is introduced by the fact that one set of determinants may not be phenotypically expressed under all conditions: for example, some bacteria show mixed flagellation only when grown at low temperatures, the formation of peritrichous flagella being suppressed at higher temperatures.

Other types of cellular movement occur in certain groups of Gram-negative bacteria. Swimming movement of the spirochetes is mediated by a system of axial fibrils, closely wound about the cylindrical cell, and enclosed by the cell wall (Figure 5.20). One set of fibrils is inserted at each pole of the cell, and the two sets overlap one another near the center of the cell. In terms of fine structure, the axial fibrils appear to be homologous with bacterial flagella. The swimming movement of spirochetes can therefore be interpreted as a special modification of flagellar movement.

A different type of locomotion known as *gliding movement* occurs in many groups of Gram-negative bacteria. Its mechanism remains obscure, since no locomotor

FIGURE 5.17

The fine structure of bacterial flagella. Part of the pole of a cell of *Spirillum serpens*, showing numerous bacterial flagella emerging at various points from the cell surface. Electron micrograph of phosphotungstate-stained material (×63,800). Courtesy of R. G. E. Murray.

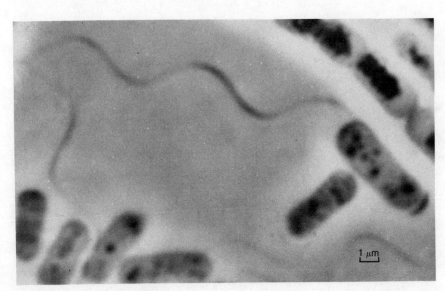

FIGURE 5.18

Polar tufts of flagella on living cells of *Sphaerotilus natans*, demonstrated by phase contrast illumination (×5,670). From J. L. Stokes, "Studies on the Filamentous Sheathed Iron Bacterium *Sphaerotilus natans*," *J. Bacteriol.* **67**, 285 (1954).

133

TABLE 5.5

An analysis of flagellar insertion in several small, rod-shaped
Gram-negative bacteria[a]

| | NUMBER OF FLAGELLA PER CELL | | | | APPARENT ORIGIN OF EACH FLAGELLUM (PERCENTAGE OF ALL FLAGELLA COUNTED) | | | |
| | AVERAGE VALUE | PERCENTAGE OF CELLS WITH: | | | | | | INFERRED MODE OF |
ORGANISM		1	2	3	POLAR	SUBPOLAR	LATERAL	FLAGELLAR INSERTION
A	1.02	98	2	0	98	1	1	Polar monotrichous
B	1.06	94	6	0	69	24	7	Polar or subpolar monotrichous
C	1.55	65	20	15	30	51	19	Sparse peritrichous
D	2.70	20	29	51	30	45	25	Peritrichous

[a]The number and apparent origin of flagella were determined on 100 to 150 cells in stained preparations of each organism. A: *Pseudomonas facilis*; B: *Pseudomonas flava*; C: *Alcaligenes paradoxus*; D: *Alcaligenes eutrophas*. Data from Davis et al., *Int. Jour. Systematic Bacteriol.* **19**, 375 (1969).

organelles have so far been detected in gliding bacteria. Such movement requires contact with a solid substrate, and it is much slower than swimming locomotion. Gliding bacteria typically form flat, spreading colonies on solid media; the periphery of the colony has a highly irregular structure, as a result of the outward migration of groups of cells (Figure 5.21). The path of movement of individual cells is often revealed by a slime track, deposited by the cells as they move over the substrate (Figure 5.22).

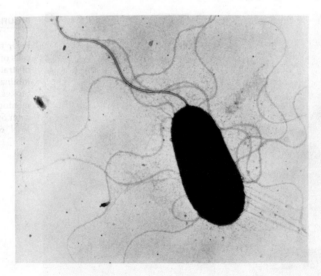

FIGURE 5.19

Electron micrograph of a bacterium with "mixed" polar peritrichous flagellation. The cell bears a single, relatively thick polar flagellum, together with numerous laterally inserted flagella ($\times$16,600). From R. D. Allen and P. Baumann, *J. Bacteriol.* **107**, 295 (1971).

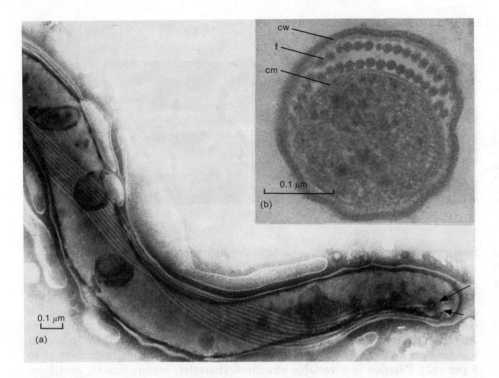

cw

f

cm

0.1 μm

(b)

0.1 μm

(a)

FIGURE 5.20

The structure of the spirochetal cell as shown by electron micrographs of spirochetes from the mouth. (a) End of a cell, negatively stained with phosphotungstic acid, showing the relationship of the multifibrillar axial filament to the protoplast (×51,000). The insertion points of two fibrils from the axial filament are just visible at the pole of the cell (arrows). (b) Cross section of a large spirochete, showing the location of the fibrils (f) of the axial filament between the cell membrane (cm) and the cell wall (cw) (×183,000). From M. A. Listgarten and S. S. Socransky, "Electron Microscopy of Axial Fibrils, Outer Envelope and Cell Division of Certain Oral Spirochetes," *J. Bacteriol.* **88**, 1087 (1964).

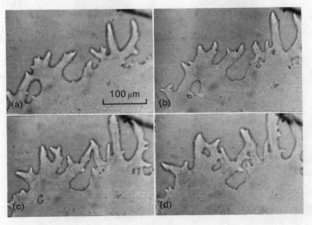

100 μm

(a)

(b)

(c)

(d)

FIGURE 5.21

Expanding edge of myxobacterial colony on agar, photographed at intervals as follows: (a) initial shape; (b) after 7 minutes; (c) after 15 minutes; (d) after 25 minutes. From R. Y. Stanier, "Studies of Nonfruiting Myxobacteria, 1. *Cytophaga johnsonae, n. sp.,* a Chitin-decomposing Myxobacterium," *J. Bacteriol.* **53**, 310 (1947).

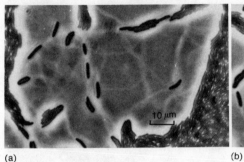
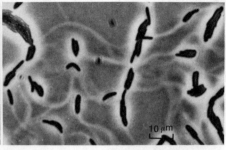

(a)   (b)

FIGURE 5.22

Edge of the growth of *Myxococcus fulvus* on agar. Note slime tracks behind gliding cells. (a) Phase contrast, ×630. (b) Phase contrast ×700. Courtesy of M. Dworkin and H. Reichenbach.

• Special cellular appendages: pili and prosthecae

Some unicellular Gram-negative bacteria produce special cellular appendages that do not occur in Gram-positive bacteria. They are of two distinct types, termed *pili* and *prosthecae*.

Pili are rigid cylindrical rods, composed (like flagella) of protein, which are anchored in the cell wall and extend out to a considerable distance beyond it (Figure 5.23). Many types of pili have been described, differing in width from 3 to as much as 30 nm. Some types occur in large numbers, ranging up to several hundred per cell; others, notably sex pili, usually occur in small numbers (1 to 5 per cell). Piliation is a variable structural character, readily lost by mutation; consequently, bacterial strains of a given species may exist either in the piliated or in the unpiliated state. Pili are not locomotor organelles, and their only known general function is to confer the property of adhesiveness; piliated bacteria tend to stick to one another, and they produce coherent pellicles on the surface of unagitated liquid cultures. However, a special subclass, the *sex pili*, have an important specific property: *they confer on the cell the ability to act as a genetic donor*

FIGURE 5.23

Electron micrograph of a metal-shadowed preparation of a cell of *E. coli* surrounded by pili (×11,200). Courtesy of C. Brinton and E. Kellenberger.

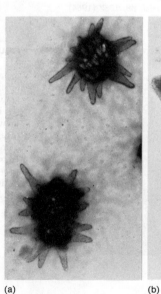

(a)   (b)

FIGURE 5.24

Electron micrographs of two prosthecate freshwater bacteria. (a) *Prosthecomicrobium pneumaticum* (×11,200). Note gas vesicles (light areas) within the cells. (b) Unidentified dividing, star-shaped bacterium. From J. T. Staley, "*Prosthecomicrobium* and *Ancalomicrobium*: New Prosthecate Freshwater Bacteria," *J. Bacteriol.* **95,** 1921 (1968).

in conjugation. Conjugational gene transfer among Gram-negative bacteria is strictly dependent on the possession of a sex pilus by the donor cell. The character of sex piliation is itself a transmissible one, since it is genetically determined by plasmids which are readily transferable by conjugation (see Chapter 15).

Prosthecae are organelles of entirely different structure: they are localized protuberances, usually filiform, of the cell wall and membrane, which occur at one or more sites on the surface of the cell (Figure 5.24). These structures occur in many groups of Gram-negative bacteria.

CONSTITUENT GROUPS OF GRAM-NEGATIVE BACTERIA

Despite its limitations, which have already been discussed, the mechanism of cellular movement appears to be the best character available for the subdivision of Gram-negative bacteria, permitting the recognition of three major assemblages: gliding bacteria; Gram-negative eubacteria, motile by flagella; and spirochetes. Of the organisms assigned to these three assemblages, only spirochetes are universally motile. The assignment of nonmotile Gram-negative organisms to the gliding bacteria or to the eubacteria is arbitrary and must be based on resemblances in other respects to motile representatives of one of these assemblages.

• **The gliding bacteria** The gliding bacteria include two groups of phototrophic organisms, the blue-green bacteria and the green bacteria, as well as several large groups of nonphotosynthetic bacteria: the cytophagas, the fruiting myxobacteria, and the filamentous gliding bacteria.

PHOTOTROPHIC GLIDING BACTERIA   The blue-green bacteria are by far the largest and most widespread group of photosynthetic procaryotes. They are photoautotrophs and are the only procaryotic organisms that perform photosynthesis accompanied by oxygen formation. The mechanism of photosynthesis in these organisms is accordingly identical with that of photosynthetic eucaryotes. They are the predominant photosynthetic inhabitants of hot springs, from which nearly all eucaryotic algae are excluded because of their low temperature maxima. Since many blue-green bacteria can fix nitrogen, a capacity not possessed by any eucaryotes, they are abundant in nitrogen-poor soils and waters.

Although uniform in nutritional and metabolic respects, blue-green bacteria are structurally diverse. Some are immotile, unicellular rods or cocci which reproduce by binary fission (Figure 5.25). Certain of these forms develop as colonial aggregates, held together by thin multilayered sheaths formed by detachment of the outer layer of the cell wall (Figure 5.26).

A mode of reproduction characteristic of some blue-green bacteria that is rare in other procaryotes is *multiple fission.* It is the sole mode of reproduction in the unicellular organism, *Dermocarpa* (Figure 5.27). The cell increases many times in size, and then it rapidly undergoes many cell divisions, after which the daughter cells are released by breakage of the wall of the mother cell. Each daughter cell is motile for a short period, motility being lost as it begins to grow. Other blue-green bacteria, exemplified by *Myxosarcina* (Figure 5.28), can undergo both binary and multiple fission: binary fission initially produces cubical packets of cells, each of which then produces many small daughter cells by multiple cleavage.

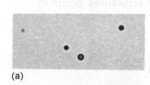

(a)

(b)

(c)

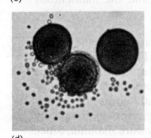

(d)

FIGURE 5.27

Dermocarpa, a unicellular blue-green bacterium which reproduces exclusively by multiple fission. Successive photomicrographs, taken over a period of 240 hours, of the development of three cells. (a–c) Cell enlargement; (d) liberation of small daughter cells produced by multiple fission of a vegetative cell, (×400). Courtesy of John Waterbury.

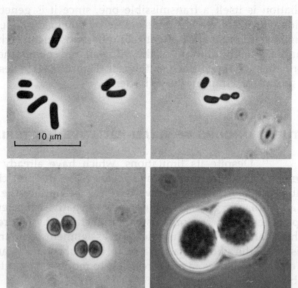

10 μm

FIGURE 5.25

Some unicellular blue-green bacteria. Courtesy of Rosmarie Rippka and Riyo Kunisawa.

Many blue-green bacteria are filamentous organisms, which reproduce by fragmentation of the filament. Some develop as simple filaments; others can form branched filamentous structures (Figures 5.29 and 5.30). Among certain of the filamentous blue-green bacteria, there is some degree of cellular differentiation. Within the filament, vegetative cells may develop into larger, heavy-walled resting cells or *akinetes;* or into cells with a specialized metabolic function, known as *heterocysts* (Figure 5.31). Heterocysts play a role in nitrogen fixation, which will be discussed in Chapter 17.

The green bacteria are a small but distinctive group of phototrophic procaryotes, readily distinguishable from blue-green bacteria both by their pigment system and by their mechanism of photosynthesis, which does not result in oxygen

FIGURE 5.26 (right)

A large unicellular blue-green bacterium, the cells of which are held together in groups by a multilayered sheath (phase contrast ×935). Courtesy of Rosmarie Rippka and Riyo Kunisawa.

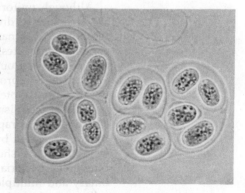

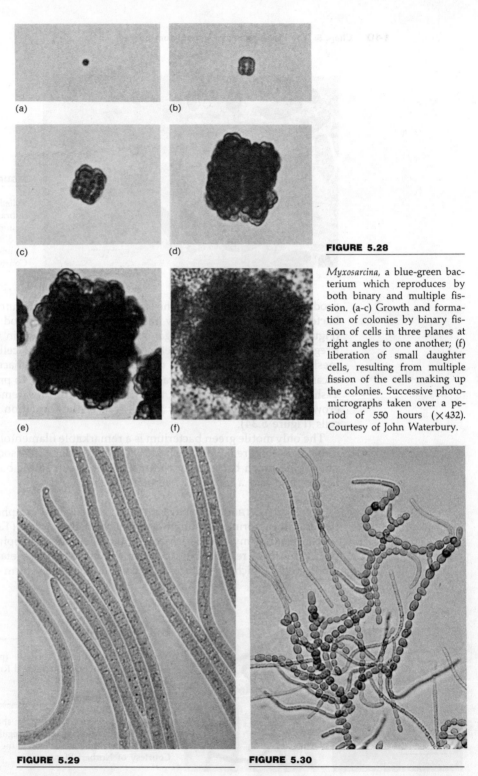

(a)   (b)

(c)   (d)

(e)   (f)

FIGURE 5.28

Myxosarcina, a blue-green bacterium which reproduces by both binary and multiple fission. (a-c) Growth and formation of colonies by binary fission of cells in three planes at right angles to one another; (f) liberation of small daughter cells, resulting from multiple fission of the cells making up the colonies. Successive photomicrographs taken over a period of 550 hours (×432). Courtesy of John Waterbury.

FIGURE 5.29

Oscillatoria, a blue-green bacterium which grows as unbranched filaments (×748). Courtesy of R. Rippka.

FIGURE 5.30

Fischerella, blue-green bacterium which grows as branched filaments (×467). Courtesy of R. Rippka.

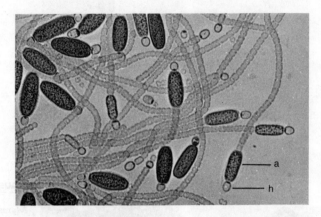

FIGURE 5.31

Cylindrospermum, a filamentous, unbranched blue-green bacterium which forms two types of differentiated cells, akinetes (a) and heterocysts (h) (×339). Courtesy of R. Rippka.

production (see Chapter 17). The strictly anaerobic photoautotrophic green sulfur bacteria are small, permanently immotile rods, which reproduce by binary fission (Figure 5.32). One of these organisms, *Pelodictyon*, grows in the form of a loose, irregular, three-dimensional net, composed of chains of cells (Figure 5.33). The meshes in the net arise as a result of the fact that adjacent cells in a chain occasionally undergo ternary, rather than binary fission, to produce two apposed, Y-shaped structures. Provided that the cells so formed remain in contact, their growth and subsequent binary fission lead to the formation of a closed chain of cells (Figure 5.34).

The only motile green bacterium is a remarkable filamentous, gliding organism, *Chloroflexus* (Figure 5.35), which grows in hot springs in association with thermophilic blue-green bacteria. In contrast to the green sulfur bacteria, it is a photoheterotroph and a facultative aerobe.

NONPHOTOSYNTHETIC GLIDING BACTERIA   Among the nonphotosynthetic gliding bacteria, three principal subgroups can be distinguished (Table 5.6).

The fruiting myxobacteria are aerobic chemoheterotrophs distinguished primarily by their remarkable development cycle. The vegetative cells are small, motile rods, which reproduce by binary fission. They form coherent, spreading

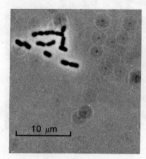

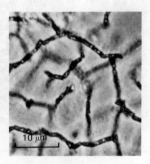

FIGURE 5.32 (left)

The green bacterium, *Chlorobium* (phase contrast, ×1,420). Courtesy of Rosmarie Rippka and Riyo Kunisawa.

FIGURE 5.33 (right)

The green bacterium, *Pelodictyon*, showing the characteristic growth in the form of loose, irregular nets (phase contrast, ×1,420). The light areas within the cells are gas vacuoles. Courtesy of Norbert Pfenning.

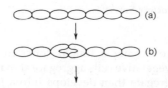

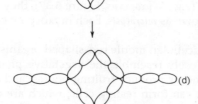

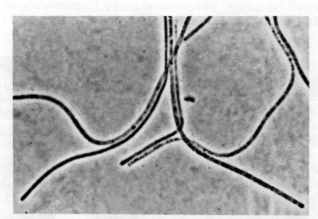

(a)

(b)

(c)

(d)

FIGURE 5.34

The mechanism of mesh formation in the green bacterium *Pelodictyon*. (a) Chain of cells produced by successive binary fissions; (b,c) ternary fission of two apposed cells in the chain, which initiates formation of a mesh; (d) enlargement of the mesh by subsequent binary fissions of its constituent cells.

FIGURE 5.35

Chloroflexus aurantiacus, a filamentous, gliding green bacterium (phase contrast ×1,040). Courtesy of B. K. Pierson and R. W. Castenholz.

TABLE 5.6

The principal groups of nonphotosynthetic gliding bacteria

	CYTOPHAGAS	FRUITING MYXOBACTERIA	FILAMENTOUS GLIDERS
Vegetative structure	Unicellular rods	Unicellular rods	Filaments
Mode of reproduction	Binary fission	Binary fission	Single cells or short chains
Resting cells			
Type	Usually absent (microcysts in one species)	Myxospores	None
Borne on fruiting bodies	−	+	−
Produce carotenoid pigments	+	+	−

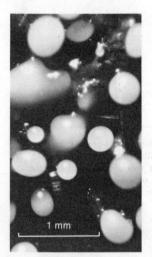

FIGURE 5.36

A fruiting myxobacte-rium, *Myxococcus*. Mature fruiting bodies on a dung particle (×16). From A. T. Henrici, and E. J. Ordal, The Biology of Bacteria, 3rd ed. Boston: Heath, 1948. Courtesy of Mrs. N. A. Woods, E. J. Ordal, and the publisher.

colonies, embedded in or overlying a tough, coherent layer of slime. Under appropriate conditions, the vegetative cells aggregate at many points on the surface of the colony, and each aggregate then develops into a *fruiting body* constructed of slime and cells, which is brightly colored and of macoscopic size (Figure 5.36). The cells within a fruiting body become converted to resting cells, known as *myxospores*. In some genera, the myxospores are not easily distinguished from vegetative cells; in others (e.g., *Myxococcus*, Figure 5.37), they are refractile, spherical, or oval structures, known as *microcysts*. Each myxospore can germinate to form a vegetative cell.

The cytophagas are unicellular, motile rod-shaped organisms, which reproduce by binary fission, and closely resemble the vegetative phase of fruiting myxo-bacteria. However, they never produce fruiting bodies, and only one member of the group, *Sporocytophaga*, can form resting cells, which are called *microcysts*. Like the fruiting myxobacteria, the cytophagas are aerobic chemoheterotrophs.

The third group of gliding, nonphotosynthetic bacteria is composed of filamentous organisms, which reproduce by fragmentation of the filament (Figure 5.38). Their properties are further discussed in Chapter 21.

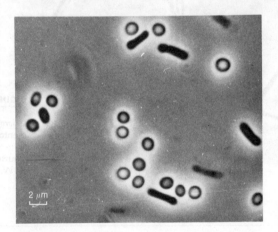

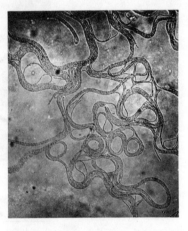

FIGURE 5.37

Microcysts and shortened rods from a fruiting body of *Myxococcus fulvus* (phase contrast, ×2,210). Courtesy of M. Dworkin and H. Reichenbach.

FIGURE 5.38

Vitreoscilla filaments growing on the surface of an agar plate. Courtesy of E. G. Pringsheim.

• The Gram-negative eubacteria

These organisms comprise the many groups of Gram-negative procaryotes which swim by means of bacterial flagella, together with related permanently immotile organisms. They include one group of phototrophs, the purple bacteria, together with numerous assemblages of chemoautotrophic and chemoheterotrophic pro-caryotes, and are without exception unicellular. The cells may be spherical, oval, cylindrical, or helical in shape. Most forms with spherical cells (the Gram-negative cocci) are permanently immotile, whereas forms with helical cells (the spirilla) are invariablly motile by means of polar tufts of flagella. Forms with oval or cylindrical cells include permanently immotile organisms, as well as organisms with polar flagella, with peritrichous flagella and with "mixed" flagellation.

As a general rule, the cells separate soon after division; consequently, chains or aggregates of cells are rarely formed. The principal exceptions are bacteria that form chains or aggregates of cells held together by a common outer sheath, wall layer, or slime layer and bacteria that form colonies as a result of the secretion of stalks. Two examples are to discussed in the following paragraphs.

Lampropedia is an immotile Gram-negative coccus that grows in the form of flat, rectangular plates of cells (Figure 5.39). Cell division occurs in two successive planes at right angles to one another, and the orderly arrays are maintained by enclosure within the outer layers of a very complex, multilayered cell wall.

Many Gram-negative eubacteria produce discrete extracellular slime layers, known as *capsules,* demonstrable by negative staining (Figure 5.40). Normally, capsule formation does not cause adhesion of the cells after division. However, *Zoogloea* synthesizes a coherent, gelatinous slime layer within which the cells are entrapped, and it develops as many-celled colonies which are often of macroscopic size (Figures 5.41 and 5.42).

DIVISION PATTERNS AND CELLULAR LIFE CYCLES AMONG GRAM-NEGATIVE EUBACTERIA   As a general rule, the eubacteria reproduce by binary transverse fission, with the formation of two identical daughter cells. There are, however, a number of exceptions to this rule.

One variation on binary fission is the process known as *budding*. Budding occurs by the formation of a small protuberance on the surface of the mother cell. The protuberance grows until it attains a volume similar to that of the mother cell, and then becomes detached from it.

Two prosthecate bacteria which reproduce by budding are the purple bacterium

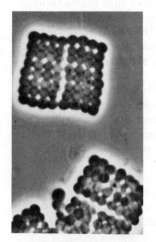

FIGURE 5.39

A coccus that forms flat, rectangular plates of cells, *Lampropedia* (×1,725). Courtesy of Daisy Kuhn.

FIGURE 5.40

Bacterial capsules, demonstrated by dispersing the cells in India ink. The organism shown is *Bacillus megaterium* (×2,160). Courtesy of C. F. Robinow.

FIGURE 5.41

Star-shaped flocs of *Zoogloea ramigera* in a broth (×23). From K. Crabtree and E. McCoy, "*Zoogloea ramigera Itzigsohn,* Identification and Description," *Intern. J. Syst. Bacteriol.* **17,** 1 (1967).

FIGURE 5.42

Phase contrast photomicrograph of projections at the periphery of a floc of *Zoogloea ramigera,* showing cells contained in slime layer (×1,110). From K. Crabtree and E. McCoy, "*Zoogloea ramigera Itzigsohn,* Identification and Description," *Intern. J. Syst. Bacteriol.* **17,** 1 (1967).

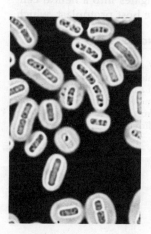

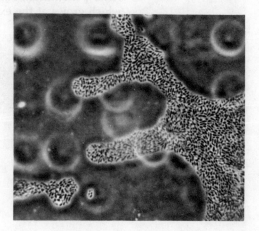

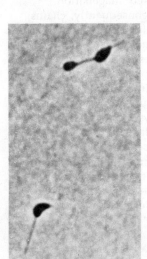

FIGURE 5.43

Hyphomicrobium vulgare. Phase contrast micrograph of two budding cells (×2,450). Micrograph courtesy of Peter Hirsch.

Rhodomicrobium and the aerobic chemoheterotroph *Hyphomicrobium* (Figures 5.43 and 5.44). The small ovoid, immotile mother cell bears a very thin filamentous prostheca, about 0.3 μm in width and frequently several times as long as the cell. Buds, initially spherical, form at the tips of the prosthecae. They enlarge, develop flagella, and swim away after detachment; the flagella are shed as these swarmers begin to grow. Lateral branches, which can also bear buds, are often formed on the prosthecae.

However, not all prosthecate bacteria that reproduce by budding develop daughter cells at the tips of the proshtecae. In *Ancalomicrobium,* a permanently immotile prosthecate bacterium, the cells of which bear from 2 to 8 prominent prosthecae (Figure 5.45), buds develop directly on the surface of the cell. As the bud grows, it synthesizes a new set of prosthecae, which are fully formed at the time of its liberation (Figure 5.46).

A less conspicuous kind of budding occurs in several species of polarly flagellated purple bacteria belonging to the genus *Rhodopseudomonas* (Figure 5.47). Outgrowth of buds occurs at the nonflagellated pole; a new flagellum forms on the free pole of the bud, which detaches from the mother cell after it has attained full size.

In the prosthecate caulobacters, cell division occurs by binary fission, but invariably it results in the formation of two *dissimilar* daughter cells. The reproductive cell of a caulobacter is an immotile rod which bears a filiform polar or subpolar prostheca (Figure 5.48). Just before division, a flagellum is formed at the corresponding site on the opposite pole; binary fission thus produces an immotile, prosthecate cell and a flatellated, nonprosthecate *swarmer*. The prosthecate cell can divide again in the same manner. However, the swarmer is incapable of division until it has undergone differentiation into a prosthecate cell; this occurs by loss of the flagellum and outgrowth of a prostheca at the same site (Figure 5.49).

Another unusual developmental cycle, involving multiple fission, occurs in the bdellovibrios, parasitic eubacteria which attack and destroy other bacteria (see Chapter 28). The parasite is a very small, curved rod with a single polar flagellum (Figure 5.50), which penetrates the cell wall of its host and grows between the wall and the protoplast of the host cell, within the periplasmic space. After penetration, *Bdellovibrio* loses its flagellum and elongates into a helical cell, which grows at the expense of nutrients derived from the host protoplast, which

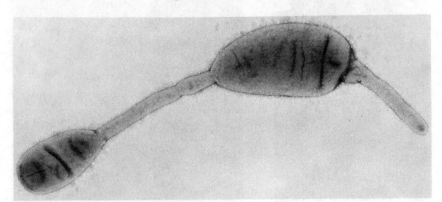

FIGURE 5.44

Hyphomicrobium vulgare. Electron micrograph of a whole mount of a budding cell, showing continuity between cell wall and prosthecae. Courtesy of Peter Hirsch.

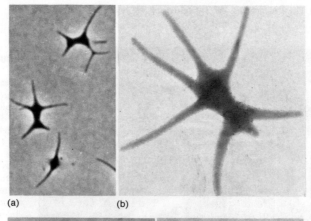

(a)　　　　　　(b)

FIGURE 5.45

A prosthecate, freshwater bacterium, *Ancalomicrobium adetum*. (a) Phase contrast micrograph of a group of cells (×2,180). (b) Electron micrograph of a single cell, showing continuity of prosthecae with the body of the cell (×8,530). From J. T. Staley, "*Prosthecomicrobium* and *Ancalomicrobium*: New prosthecate Freshwater Bacteria," *J. Bacteriol.* **95**, 1921 (1968).

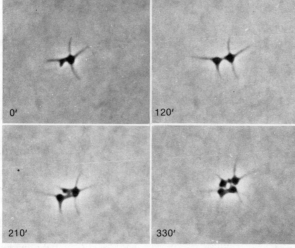

FIGURE 5.46

Budding growth of *Ancalomicrobium adetum*, followed by phase contrast photomicrography over a period of 330 minutes. The cell at 0 time has started to form a bud, which already bears one prostheca. After 120 minutes, the bud has attained the same size as the mother cell, and it has formed a second prostheca. Both cells then start to bud again (210 minutes), the daughter buds being mature after 330 minutes. Courtesy of J. T. Staley.

FIGURE 5.47

Photomicrographs of budding growth of two cells in a slide culture of a *Rhodopseudomonas* species, *R. acidophila*, taken at intervals of 30 minutes to show the outgrowth of buds at the poles of the rod-shaped cells (×1,500). Courtesy of Norbert Pfennig and Heather Johnston. From N. Pfennig, "*Rhodopseudomonas acidophila*, a New Species of the Budding Nonsulfur Purple Bacteria," *J. Bacteriol.* **99**, 597 (1969).

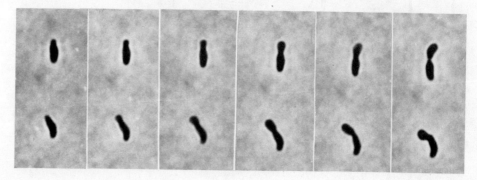

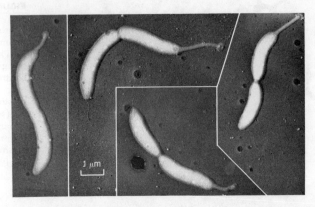

FIGURE 5.48

Metal-shadowed electron micrograph of stalked cells of Caulobacter, several of which are dividing (×8,850). Note the flagellum on the distal daughter cell in the divisional stages and the continuity between the body of the cell and the stalk. From A. E. Houwink, "Caulobacter, its Morphogenesis, Taxonomy, and Parasitism," *Antonie van Leeuwenhoek,* **21,** 1, 54 (1955).

gradually disintegrates. Its growth completed, the helical cell undergoes multiple fission, to produce several vibrioid, flagellated cells which emerge from the carcass of the host to seek other prey (Figure 5.51).

FORMATION OF HOLDFASTS Some aquatic Gram-negative eubacteria are sessile during part of the cellular life cycle. They attach to solid surface by means of

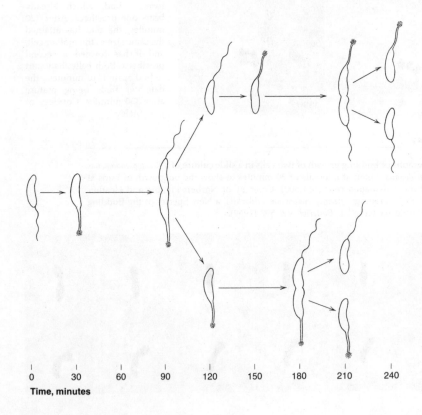

FIGURE 5.49

A diagrammatic representation of clonal growth in *Caulobacter,* based on continuous microscopic observations. Note that the time required for the division of a swarmer cell is considerably longer than that required for the division of its stalked sibling. After J. L. S. Poindexter. "Biological Properties and Classification of the Caulobacter group," *Bacteriol. Rev.* **28,** 231 (1962).

0   30   60   90   120   150   180   210   240

Time, minutes

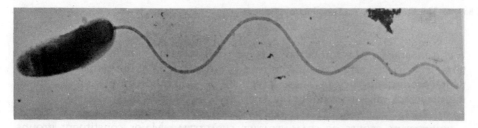

FIGURE 5.50

Electron micrograph of a cell of *Bdellovibrio*, showing the unusually thick single polar flagellum (uranyl acetate stain; ×29,000). From R. J. Seidler and M. P. Starr, "Structure of the Flagellum of *Bdellovibrio bacteriovorus*," *J. Bacteriol.* **95**, 1952 (1968).

holdfasts: small discs of adhesive material, secreted at one point on the cell surface.

The production of holdfasts is characteristic of the caulobacters, the holdfast being formed during division at the distal pole of the swarmer cell. In the genus *Caulobacter*, where the flagellum and the prostheca are also polar, the holdfast becomes attached to the tip of the prostheca after the outgrowth of this organelle. In the genus *Asticcacaulis*, the flagellum and prostheca are subpolar, and consequently the polarly located holdfast remains on the surface of the cell, and is not attached to the tip of the prostheca (Figure 5.52). In pure cultures of bacteria with holdfasts, collisions between motile cells lead to the formation of rosettes, by common adhesion through the holdfasts (Figure 5.53).

FIGURE 5.51

The growth cycle of *Bdellovibrio* in the cell of its bacterial host. (a,b) Attachment and entry of *Bdellovibrio* into the periplasmic space of the host; (c–e) growth and elongation of *Bdellovibrio* in the host cell, accompanied by progressive destruction of the host protoplast; (f) division of *Bdellovibrio*; (g) released *Bdellovibrio* cells.

FIGURE 5.52

A diagrammatic representation of cellular differentiation and division in two genera of stalked bacteria, *Caulobacter* and *Asticcacaulis*: h, holdfast; s, stalk; f, flagellum. From J. M. Schmidt and R. Y. Stanier, "The Development of Cellular Stalks in Bacteria," *J. Cell Biol.* **28**, 423 (1966).

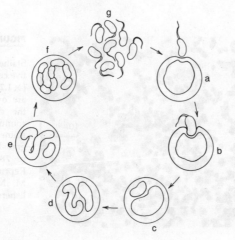

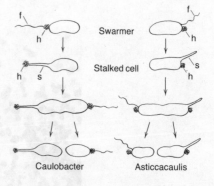

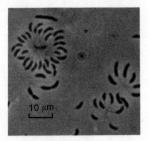

FIGURE 5.53

Caulobacter rosettes are formed by the adherence of cells to one another through holdfasts located at the base of the stalks (phase contrast, ×1,050). Courtesy of Jeanne Stove Poindexter.

RESTING STAGES The formation of resting cells is rare in Gram-negative eubacteria. It occurs in certain aerobic, nitrogen-fixing bacteria of the genus *Azotobacter*, which produce microcysts. These structures develop by the formation of a thick, multilayered wall surrounding the vegetative cell (Figure 5.54).

CONSTITUENT GROUPS OF GRAM-NEGATIVE EUBACTERIA Major constituent groups among the eubacteria will be described in detail in later chapters. However, two special groups, the rickettsias and chlamydias, merit more detailed discussion at this point, because their very unusual biological properties once appeared to raise a fundamental determinative problem: the distinction between cellular organisms and viruses.

Like the viruses, most rickettsias and chlamydias appear to be obligate intracellular parasites: with very few exceptions, all attempts to cultivate them in artificial media—or even to maintain them in a viable state outside the host cell—have so far failed. Since they are also very small organisms, it was at one time believed that they might represent a biological group transitional between bacteria and viruses, or even in some cases actually be large viruses. However, study of their fine structure by electron microscopy has shown that they are cellular organisms, of typically procaryotic structure; and a few of these organisms have recently been grown outside host cells, on complex artificial media. There is, accordingly, good reason to believe that they are members of the bacteria.

THE RICKETTSIAS The rickettsias are obligate intracellular parasites in certain groups of arthropods (notably fleas, lice, and ticks). They do not appear to produce symptoms of disease in their arthropod hosts, but if they are transmitted by bite to a vertebrate host, a severe and often fatal infection may result. The major rickettsial disease of man is epidemic typhus, the biography of which has been recounted in popular form in one of the few literary classics written by a microbiologist, Hans Zinsser's *Rats, Lice and History*. Other human rickettsial diseases of minor importance are Rocky Mountain spotted fever, transmitted by ticks, and scrub typhus, normally transmitted by mites to field mice, but also transmissible to man. The laboratory study of the rickettsias has been greatly facilitated by

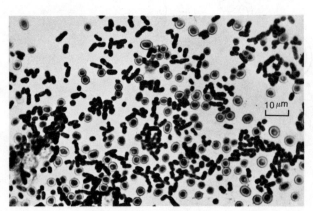

FIGURE 5.54

Stained preparation of vegetative cells and cysts of *Azotobacter* (×1,210). The vegetative cells are oval, deeply stained rods; the cysts are spherical, surrounded by thick, lightly stained walls. From S. Winogradsky, *Microbiologie du sol*, p. 780. Paris: Masson, 1949. Reprinted with permission of M. Manigault and the publishers.

the discovery that they can be grown in the yolk sacs of chick embryos, a much more convenient laboratory material than the animal hosts.

The rickettsias are short rods, about 0.3 by 1.0 μm (Figure 5.55), which multiply by binary transverse fission. Their cellular nature is clearly established by the fact that they contain both DNA and RNA, are enclosed by a cell wall, and can perform certain metabolic activities. The structure of the cell, as revealed by the electron microscopy of thin sections, is clearly procaryotic, and this has been confirmed chemically, by the detection of components of peptidoglycan as cell constituents.

The explanation of their obligate intracellular parasitism is still far from clear. Isolated cell suspensions can oxidize the amino acid glutamate through the reactions of the tricarboxylic acid cycle, which suggests that rickettsias have an autonomous system of energy-yielding metabolism. However, isolated cells rapidly become inviable. The loss of viability can be retarded by adding certain

FIGURE 5.55

Electron micrographs of thin sections through the tissues of chicken embryos infected with Rickettsia. (a) A single embryonic cell in the later stage of rickettsial infection ($\times$10,026). The whole cytoplasmic region is filled with the rod-shaped rickettsial cells (r); the centrally located nucleus (n), although abnormal in appearance, is not infected. (b) Portion of an infected embryonic cell at a high magnification ($\times$57,031). Several rickettsial cells (r) are to be seen in longitudinal or transverse section; also a mitochondrion (m), which is roughly of the same dimensions as the rickettsial cells. From S. L. Wissig, L. G. Caro, E. B. Jackson, and J. E. Smadel, "Electron Microscopic Observations on Intracellular Rickettsiae," *Am. J. Pathol.* **32**, 1117 (1956).

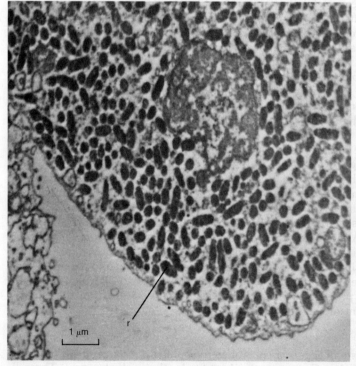

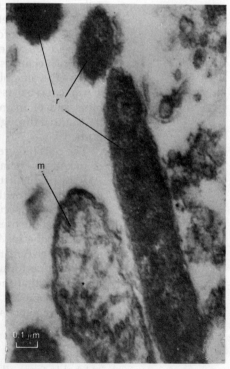

(a)   (b)

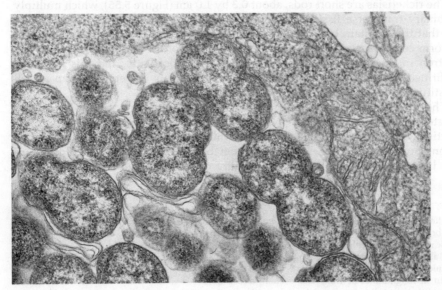

FIGURE 5.56

Electron micrograph of a thin section of part of a mammalian cell growing in tissue culture, and infected by many chlamydia cells (×31,185). Each of the small round-to-ovoid chlamydia cells is surrounded by a unit membrane; many are in the course of binary fission. Courtesy of R. R. Friis, Department of Microbiology, University of Chicago.

coenzymes to the suspending medium. Since these are substances to which intact cells are normally completely impermeable, it has been suggested that the adaptation of the rickettsias to an intracellular mode of life has caused far-reaching changes in the properties of the cell membrane, which enable them to absorb and use coenzymes or other chemically complex metabolites produced by the host. Such changes in membrane properties would, however, make the rickettsias extremely vulnerable in an extracellular environment.

THE CHLAMYDIAS   The chlamydias are the agents of a number of diseases of birds and of mammals, including man. In contrast to rickettsias, they are not transmitted through invertebrate hosts, but pass directly between their vertebrate hosts.

The cells (Figure 5.56) are more or less spherical and slightly smaller than those of rickettsias, being 0.2 to 0.7 μm in diameter.

As a result of their very small size, it is difficult to establish with certainty the mode of growth within the host cell; and there is some dispute about whether multiplication occurs by binary fission or by budding. Chemical analyses have established the presence in the cells of both DNA and RNA, as well as components of peptidoglycan. Although biochemical studies show that the chlamydias possess extensive enzymatic capacities, it has not yet been demonstrated that they are capable of performing energy-yielding metabolism. This has led to the interesting hypothesis that they may be "energy parasites," the growth of which is strictly dependent on the provision of energy-rich materials derived from the energy-yielding metabolism of the host cell.

• **The spirochetes**   The spirochetes, a small group of chemoheterotrophic bacteria, are readily distinguishable from all other Gram-negative bacteria by their cell structure. The cell,

151 always very long relative to its width, is helical in form and is surrounded by a delicate, flexible wall. All spirochetes can swim actively in liquid media by means of their system of axial fibrils, described on p. 133; and during their movement, the cells often bend into loops or coils, while retaining at all times their intrinsic helical structure (Figure 5.57). They reproduce by binary fission, and no resting stages are known.

The number of fibrils that surround the protoplast varies widely in different members of the group, and it is more or less directly related to the size of the cell. In the smallest spirochetes there is only a single fibril inserted at each pole. In the very large forms, such as *Cristispira*, there are several hundred. Cell division in spirochetes (Figure 5.58) involves a primary division of the protoplast within the enclosing wall. This is followed by division of the wall and separation of the two daughter individuals, each individual bearing one of the two sets of fibrils that compose the axial filament. A new initiation point and a new set of fibrils are synthesized at the newly formed pole of each daughter cell.

Free-living spirochetes (*Spirochaeta*) are inhabitants of mud and water; most of these organisms appear to be anaerobes, found characteristically in the oxygen-free regions of these environments. The members of the genus *Cristispira* are parasites of clams and other mollusks. They occur, frequently in enormous numbers, in a special region of the host, a gelatinous rod known as the *crystalline style*, which lies in a sac connected to the digestive tract. Many of the smaller spirochetes are parasites of man or of other vertebrates and include agents

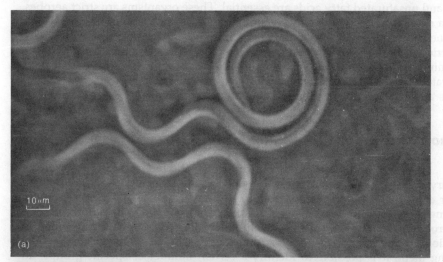
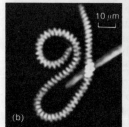
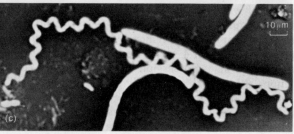

FIGURE 5.57

Some of the larger spirochetes. (a) Phase contrast of living *Cristispira*, a form found in the crystalline style of clams (×380). Courtesy of S. W. Watson. (b) Nigrosin mount of an unidentified spirochete from water (×341). (c) Nigrosin mount of *Spirochaeta plicatilis*, a large spirochete common in water (×341). The preparation also contains cells of rod-shaped bacteria. (b) and (c) courtesy of C. F. Robinow.

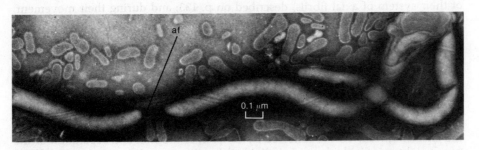

FIGURE 5.58

Electron micrograph of negatively stained dividing cell of the spirochete *Treponema microdentium* (×46,200). The daughter cells are beginning to separate, but are still connected by fibrils of the axial filament (af). From M. A. Listgarten and S. S. Socransky, "Electron Microscopy of Axial Fibrils, Outer Envelope and Cell Division of Certain Oral Spirochetes," *J. Bacteriol.* **88**, 1087 (1964).

responsible for several human diseases: syphilis and yaws (*Treponema*), relapsing fever (*Borrelia*), and one type of infectious jaundice (*Leptospira*).

The members of the genus *Leptospira* are readily cultivable, although their nutrient requirements, which are complex, are not fully elucidated. Several vitamins and unsaturated fatty acids are essential. These organisms are strict aerobes, with a respiratory mode of energy-yielding metabolism. Many of the pathogenic treponemas and borrelias have not yet been grown in culture; those which have are strict anaerobes, which obtain energy by the fermentation of carbohydrates or amino acids. Several free-living aquatic *Spirochaeta* species have also been cultivated; they are either strict or facultative anaerobes, which ferment carbohydrates and have relatively complex nutritional requirements.

THE PROCARYOTES: A SUMMING UP

As the preceding account indicates, the biological diversity of the procaryotes is very great, in fact, so great that the recognition of well-defined major constituent groups, comparable to those which exist among the protists, becomes very difficult. It seems probable that the contemporary procaryotes are the products of numerous distinct evolutionary lines, which have developed in isolation from one another throughout much of the course of biological evolution. Among them, it is possible to identify a certain number of specific groups, all the members of which share sufficient common properties to suggest a common origin: two examples are the blue-green bacteria and the fruiting myxobacteria. However, the attempt to discern *intergroup* evolutionary relationships, as was done for the protists in Chapter 4, seems doomed to failure. The only attribute common to all these organisms is possession of the procaryotic cell as the unit of structure and function; although this certainly suggests a common primary evolutionary origin, the later divergences have been too great, and too remote, to permit reconstruction of evolutionary relationships on the basis of the properties of contemporary representatives.

FURTHER READING

Books Buchanan, R. E., and N. E. Gibbons (editors), *Bergey's Manual of Determinative Bacteriology,* 8th ed. Baltimore: Williams & Wilkins, 1974.

Fogg, G. E., W. D. P. Stewart, P. Fay, and A. F. Walsby, *The Blue-Green Algae.* London and New York: Academic Press, 1973.

Hayflick, L. (editor), *The Mycoplasmatales and the L-Phase of Bacteria.* New York: Appleton-Century-Crofts, 1969.

Smith, P. F., *The Biology of Mycoplasmas.* New York: Academic Press. 1971.

Reviews Dworkin, M., "Biology of the Myxobacteria," *Ann. Rev. Microbiol.* **20,** 75 (1966).

Lechevalier, H. A., and M. P. Lechevalier, "Biology of Actinomycetes," *Ann. Rev. Microbiol.* **21,** 71 (1967).

Maniloff, J., and H. J. Morowitz, "Cell Biology of the Mycoplasmas," *Bacteriol. Revs.* **36,** 263 (1972).

Moulder, J. W., "The Relation of the Psittacosis Group (Chlamydiae) to Bacteria and Viruses," *Ann. Rev. Microbiol.* **20,** 107 (1966).

Pfennig, N., "Photosynthetic Bacteria," *Ann. Rev. Microbiol.* **21,** 285 (1967).

Schmidt, J. L., "Prosthecate Bacteria," *Ann. Rev. Microbiol.* **25,** 93 (1972).

Starr, M. P., and V. B. D. Skerman, "Bacterial Diversity: The Natural History of Selected Morphologically Unusual Bacteria," *Ann. Rev. Microbiol.* **19,** 407 (1965).

6

MICROBIAL METABOLISM: THE GENERATION OF ATP

The sum total of all the chemical transformations that occur in cells is termed *metabolism*. Certain of these transformations are involved in the synthesis of macromolecules, which constitute the major portion of cellular mass, from the simpler compounds present in the extracellular environment. This portion of metabolism is termed *biosynthesis*.* Many of the reactions in biosynthesis require an activation of their reactants to an increased energy state (i.e., an input of energy is required for biosynthesis). Hence, an important part of the metabolism of any organism involves the mobilization of chemical energy to drive biosynthesis. The usual chemical form of energy is the highly reactive compound, ATP.

Although the mechanisms of biosynthesis are quite uniform among organisms, mechanisms of generating ATP are diverse. By the process of *photosynthesis*, light energy can be converted to chemical energy to drive biosynthesis. Chemical energy in the form of inorganic or organic compounds can also be used; such processes, involving the direct utilization of chemical energy, are termed *catabolism* or degradative metabolism.

SOME THERMODYNAMIC CONSIDERATIONS

The second law of thermodynamics states that only part of the energy released in a chemical reaction, called *free energy* (ΔG), is available for the performance of work; the rest is lost through the increase in *entropy* (randomness of the system). Chemical reactions occur only if the free energy change, ΔG, is negative. Since ΔG is an index of the tendency of a reaction to occur, its numerical value is dependent on the *concentration* of the reactants and products in the reaction

*Sometimes termed *anabolism*.

mixture. If all reactants are present at a concentration of one molar, the value of ΔG is called the *standard free energy change*, $\Delta G°$, which is related to the *equilibrium constant*, K_{eq}, of the reaction by the equation,

$$\Delta G° = -RT \ln K_{eq} \tag{6.1}$$

where R is the gas constant and T is the absolute temperature. Thus, if $\Delta G°$ of a reaction is a negative value, the equilibrium constant (K_{eq}) is greater than 1.0 and formation of products is favored; if it is positive, K_{eq} is less than 1.0 and the reaction, as written, tends to proceed in the reverse direction.

The value of $\Delta G°$ of a reaction can be calculated from the equilibrium constant of the reaction (Eq. 6.1). It can also be calculated as the *difference between the standard free energy of the products and the standard free energy of the reactants*, each term being adjusted to the stoichiometry of the reaction equation, i.e.,

$$\Delta G° = \Sigma\, G° \text{ products} - \Sigma\, G° \text{ reactants} \tag{6.2}$$

THE ROLE OF ATP IN BIOSYNTHESIS

All biosynthetic pathways require the participation of adenosine triphosphate, the chemical structure of which is shown in Figure 6.1. It is a derivative of adenosine monophosphate (AMP) to which two additional phosphate groups are attached through an *anhydride linkage*. The two bonds indicated by the symbol, $\sim$, are high-energy bonds* and thus are particularly reactive. Hence, ATP is able to donate phosphate groups to a number of metabolic intermediates, thereby converting them to activated forms. Their standard free energy is increased to a level that allows the phosphorylated intermediate to participate in biosynthetic reactions which are thermodynamically favorable ($-\Delta G°$), while the comparable reaction of which the unphosphorylated form is a reactant would be thermodynamically unfavorable ($+\Delta G°$). Thus, ATP generation is required in order for biosynthetic pathways to function.

The special reactivity of the high-energy bonds of ATP is apparent when the $\Delta G°$ of their hydrolysis is compared with the $\Delta G°$ of hydrolysis of the phosphate of AMP (attached to adenosine by an ester linkage, therefore less reactive and

*The term *high-energy bond* should not be confused with the term *bond energy* that is used by the physical chemist to denote the energy *required to break a bond between two atoms.*

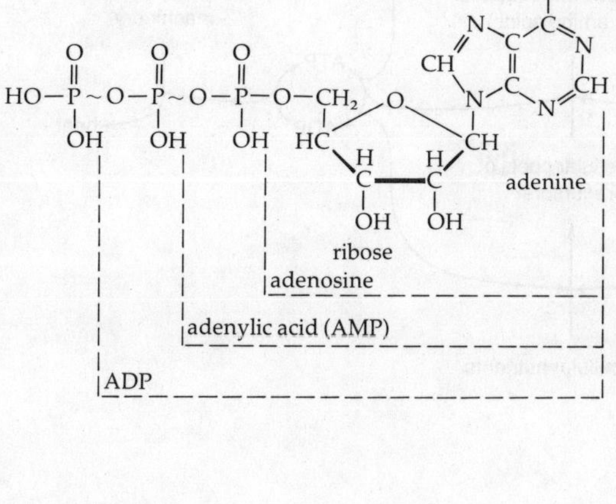

FIGURE 6.1

The structure of ATP (adenosine triphosphate), showing the various components of the molecule that can be obtained by hydrolysis.

termed a *low-energy bond*), i.e.,

$$\text{adenosine—}\textcircled{P}\sim\textcircled{P}\sim\textcircled{P} + H_2O \longrightarrow \text{adenosine—}\textcircled{P}\sim\textcircled{P} + \textcircled{P} \qquad \Delta G° = -7.3 \text{ Kcal}$$
$$\text{(ATP)} \qquad\qquad\qquad\qquad\qquad \text{(ADP)}$$

$$\text{adenosine—}\textcircled{P}\sim\textcircled{P} + H_2O \longrightarrow \text{adenosine—}\textcircled{P} \qquad \Delta G° = -7.3 \text{ Kcal}$$
$$\text{(ADP)} \qquad\qquad\qquad\qquad \text{(AMP)}$$

$$\text{adenosine—}\textcircled{P} + H_2O \longrightarrow \text{adenosine} + \textcircled{P} \qquad \Delta G° = -3.4 \text{ Kcal}$$
$$\text{(AMP)}$$

• **The generation**
of ATP

In cellular metabolism, ATP is generated by two fundamentally different biochemical mechanisms: *substrate level phosphorylation* and *electron transport*. In substrate level phosphorylation, ATP is formed from ADP by transfer of a high-energy phosphate group from an intermediate of a dissimilatory pathway. The following reactions serve as an example:

$$
\begin{array}{ccccc}
\text{CH}_2\text{OH} & & \text{CH}_2 & & \text{CH}_3 \\
| & \xrightarrow{\;H_2O\;} & \| & \xrightarrow{\;\text{adenosine—}\textcircled{P}\sim\textcircled{P}\;(ADP)\;} & | \\
\text{CHO—}\textcircled{P} & & \text{C—O}\sim\textcircled{P} & & \text{C=O} \quad + \text{ adenosine—}\textcircled{P}\sim\textcircled{P}\sim\textcircled{P} \\
| & & | & & | \qquad\qquad\qquad\qquad\quad \text{(ATP)} \\
\text{COOH} & & \text{COOH} & & \text{COOH} \\
\text{2-phosphoglyceric} & & \text{phosphoenol} & & \text{pyruvic} \\
\text{acid} & & \text{pyruvic acid} & & \text{acid}
\end{array}
$$

As a consequence of the removal of a molecule of water, the low-energy ester linkage of phosphate in 2-phosphoglyceric acid is converted to the high-energy enol linkage in phosphoenol pyruvic acid. This high-energy linked phosphate group can then be transferred to ADP, the consequence of which is the generation of a molecule of ATP.

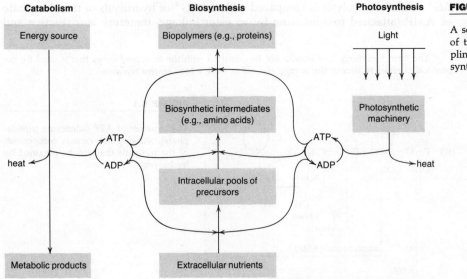

FIGURE 6.2

A schematic representation of the role of ATP in coupling catabolism and photosynthesis to biosynthesis.

All other mechanisms of generating ATP occur in membranes, as a consequence of the transfer of electrons between certain carrier molecules with fixed orientation in the membrane. The mechanism of ATP generation by electron transport will be discussed in a later section of this chapter.

Generation of ATP is the fundamental mechanism by which some free energy can be trapped; in fact, most is dissipated in the form of heat. The role of ATP in coupling energy to biosynthesis is summarized in Figure 6.2.

• **Other compounds with high-energy bonds**

We have emphasized the primary role of ATP in "trapping" a portion of the free energy made available through catabolic reactions, and its role in "driving" biosynthetic reactions by phosphorylating and thereby activating certain intermediary metabolites of biosynthesis, some of which also contain high-energy bonds. Although ATP is directly involved in the majority of such activation reactions, a number of other highly reactive metabolites that also contain high-energy bonds enter into specific activation steps in certain pathways of biosynthesis. All these high-energy compounds can be formed at the expense of one or more of the high-energy bonds of ATP, but sometimes they are formed directly in catabolic reactions. These compounds, along with some representative activation steps in which they participate, are listed in Table 6.1.

TABLE 6.1

High-energy compounds other than ATP that activate metabolic intermediates and thereby drive certain reactions of biosynthesis

HIGH-ENERGY COMPOUND[a]	CAUSE ACTIVATION IN THE BIOSYNTHESIS OF
Guanine-(P)~(P)~(P) GTP	Proteins (ribosome function)
Uridine-(P)~(P)~(P) UTP	Peptidoglycan layer of the bacterial wall
Cytidine-(P)~(P)~(P) CTP	Phospholipids
Deoxythymidine-(P)~(P)~(P) dTTP	Lipopolysaccharide of bacterial wall
Acyl~SCoA Acyl coenzyme A	Fatty acids

[a] High-energy bonds are indicated by the symbol "~."

THE ROLE OF PYRIDINE NUCLEOTIDES IN METABOLISM

Like all oxidations, the biological oxidation of organic metabolites is the removal of electrons. In most cases, each step of oxidation of a metabolite involves the removal of two electrons and thus the simultaneous loss of two protons; this is equivalent to the removal of two hydrogen atoms and is called *dehydrogenation.* Conversely, the *reduction* of a metabolite involves the addition of two electrons and two protons and can therefore be considered a *hydrogenation.*

For example, the oxidation of lactic acid to pyruvic acid and the reduction of pyruvic acid to lactic acid can be expressed:

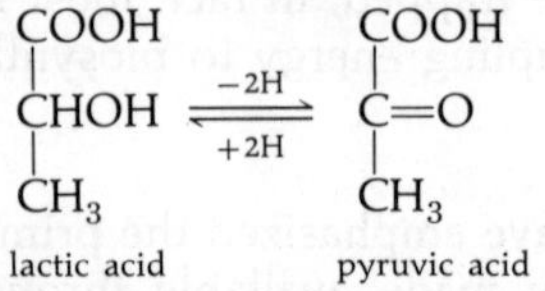

$$
\begin{array}{ccc}
\text{COOH} & & \text{COOH} \\
| & & | \\
\text{CHOH} & \underset{+2H}{\overset{-2H}{\rightleftharpoons}} & \text{C=O} \\
| & & | \\
\text{CH}_3 & & \text{CH}_3 \\
\text{lactic acid} & & \text{pyruvic acid}
\end{array}
$$

The compounds which most often mediate biological oxidations and reductions (i.e., which serve as acceptors for hydrogen atoms released by dehydrogenation reactions, and as donors of hydrogen atoms required for hydrogenation reactions) are two pyridine nucleotides: *nicotinamide adenine dinucleotide* (NAD) and *nicotinamide adenine dinucleotide phosphate* (NADP), the structures of which are shown in Figure 6.3.

Both of these pyridine nucleotides can readily undergo reversible oxidation and reduction, the site of which is the nicotinamide group (Figure 6.4). It can be seen that the oxidized form of the pyridine nucleotides carries one hydrogen atom less than the reduced form; in addition, it has a positive charge on the nitrogen atom, which enables it to accept a second electron upon reduction. The reversible oxidation–reduction of NAD and NADP can thus be symbolized:

$$NAD^+ + 2H \rightleftharpoons NADH + H^+$$

and

$$NADP^+ + 2H \rightleftharpoons NADPH + H^+$$

The pyridine nucleotides function in metabolism either in the *transfer of reducing power* or in the *net generation of reducing power*. With respect to carbon, the overall level of oxidation of organic components of the cell is approximately that of carbohydrate (it can be symbolized as $[CH_2O]$). Hence, chemoheterotrophic organisms that use glucose or other carbon sources on this oxidation level for

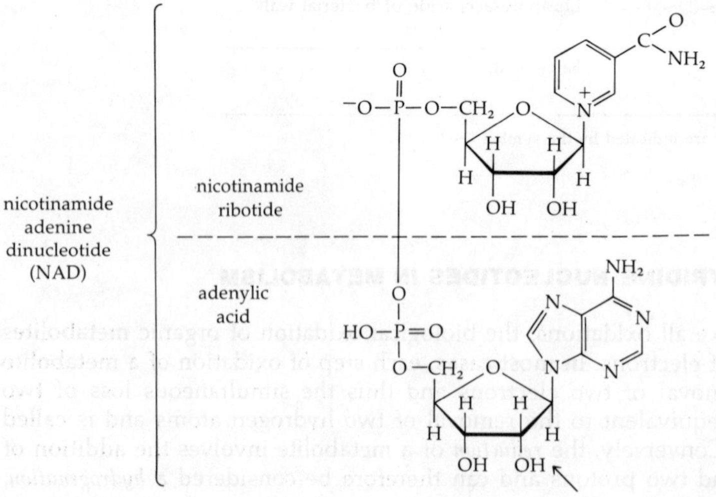

FIGURE 6.3

Structure of NAD (nicotinamide adenine dinucleotide). NADP (nicotinamide adenine dinucleotide phosphate) has an additional phosphate group which is esterified to the 2′ position of the ribose group in the adenyl moiety of the molecule (indicated by the arrow).

nicotinamide adenine dinucleotide (NAD)

nicotinamide ribotide

adenylic acid

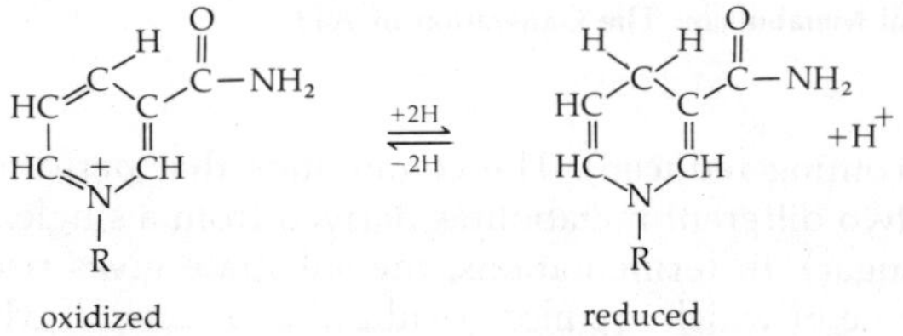

FIGURE 6.4

Oxidized and reduced forms of the nicotinamide moiety of pyridine nucleotides.

the synthesis of cell material require little or no net input of reducing power for the assimilation of carbon. In these organisms the role of pyridine nucleotides is largely in the transfer of reducing power. However, an organism that uses CO_2 as its sole source of carbon (e.g., a plant or an autotrophic bacterium) must have available a large amount of reducing power for the purposes of carbon assimilation. In these organisms a major role of the pyridine nucleotides is in the net generation of reducing power. Roughly speaking, a net input of four hydrogen atoms is required for the assimilation of each molecule of CO_2, to reduce it to the oxidation level of cell material:

$$CO_2 + 4H \longrightarrow (CH_2O) + H_2O$$

In the transfer of reducing power, pyridine nucleotides serve as couplers between the dehydrogenation of substrates and electron transport. They also couple the oxidation of one metabolic intermediate with the reduction of another (e.g., NAD couples the oxidation of glyceraldehyde phosphate with the reduction of pyruvic acid in lactic acid bacteria).

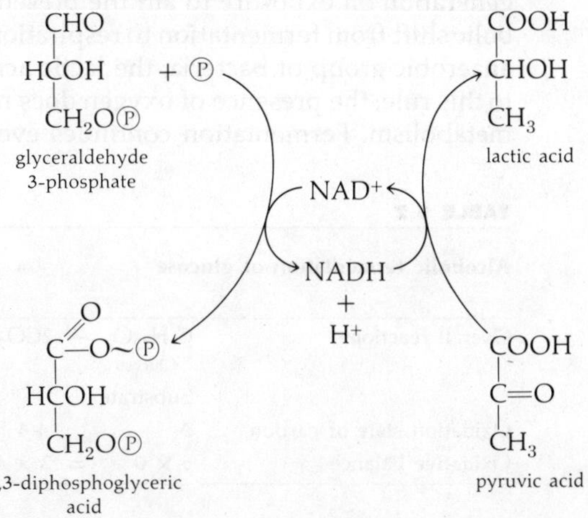

In the net generation of reducing power, electrons arising from the oxidation of inorganic compounds or from the photosynthetic machinery directly reduce pyridine nucleotides.

MODES OF ATP-GENERATING METABOLISM

• The definition and nature of fermentation

Of the three principal modes of ATP-generating metabolism, the simplest in terms of mechanism is fermentation. It can be defined as an *ATP-generating metabolic process in which organic compounds serve both as electron donors* (becoming oxidized) *and electron acceptors* (becoming reduced). The compounds that perform these two functions are usually two different metabolites derived from a single fermentable substrate (such as a sugar). In fermentations, the substrate gives rise to a mixture of end products, some of which are more oxidized and some of which are more reduced (i.e., fermentative processes always maintain a strict oxidation–reduction balance). The *average oxidation level of the end products is identical to that of the substrate*; this can be readily seen in the case of alcoholic fermentation of glucose (Table 6.2). The requirement for a precise oxidation balance limits the kinds of organic compounds that can be degraded by fermentation: they must be neither highly oxidized nor highly reduced. *Carbohydrates* are the principle substrates of fermentation. Among the bacteria, some compounds belonging to other chemical classes can also be fermented: organic acids, amino acids, purines, and pyrimidines.

Substrate-level phosphorylation is the only mode of ATP synthesis possible as a result of fermentation.

Pasteur, who first recognized the physiological role of fermentation, called it "the consequence of life without air." This statement is still correct; all fermentations can proceed under strictly anaerobic conditions. Many of the organisms that generate ATP by fermentation are strict anaerobes (the physiological basis of their inability to grow in the presence of air will be considered in Chapter 10). Others are facultative anaerobes, being able to grow either in the presence or absence of air. As a rule, facultative anaerobes change their mode of ATP generation on exposure to air: the presence of molecular oxygen induces a metabolic shift from fermentation to respiration (see below). However, one facultatively anaerobic group of bacteria, the lactic acid bacteria, provides a notable exception to this rule; the presence of oxygen does not modify their mode of ATP-generating metabolism. Fermentation continues even in the presence of air.

TABLE 6.2

Alcoholic fermentation of glucose

Overall reaction	$C_6H_{12}O_6 \rightarrow 2CO_2 + 2C_2H_5OH$		
	Glucose		Ethanol
	Substrate		End products
Oxidation state of carbon	0	+4	−2
Oxidative balance	6 × 0	= 2 × 4 +	2(2 × −2)

• **The definition and nature of respiration**   Respiration can be defined as an *ATP-generating metabolic* process in which either organic or inorganic compounds serve as electron donors (becoming oxidized) and inorganic compounds serve as the ultimate acceptors (becoming reduced). Usually the ultimate electron acceptor is molecular oxygen. However, in the *anaerobic respirations,* a special class of respiratory processes characteristic of a few bacteria, an inorganic compound other than oxygen serves as the ultimate electron acceptor; the compounds that can so act are sulfates, nitrates, and carbonates. To distinguish oxygen-linked respiratory processes from these anaerobic respirations, it is useful to qualify the former as *aerobic respiration.*

Many microorganisms that perform aerobic respiration are strict aerobes. Some, however, are facultative anaerobes since they can also generate ATP, either by fermentation (as stated above) or by anaerobic respiration with nitrate as the terminal electron acceptor. The bacteria that perform anaerobic respiration utilizing sulfate or carbonate as electron acceptors are strict anaerobes; they cannot use aerobic respiration as an alternative means of generating ATP.

The limits to the kinds of organic compounds that can be degraded by respiration are quite broad; they need only be at an oxidation state less than that of CO_2. Accordingly, all naturally occurring organic compounds can be respired by some microorganism. However, certain man-made compounds (see Chapter 25) appear to be remarkably resistant to respiratory microbial metabolism and, consequently, they accumulate in the environment, often with undesirable ecological impact.

A distinctive feature of most respiratory processes is the presence in the cell of a special set of compounds capable of being reversibly oxidized and reduced, i.e., they can accept electrons from one compound and donate them to another. They comprise the *electron transport chain*. Electrons removed from the substrate enter the chain, passing from one carrier compound to the next, ultimately to the terminal inorganic electron acceptor (O_2, NO_3^-, SO_4^{2-}, or CO_3^{2-}). During this process, ATP is generated by mechanisms to be discussed later in this chapter. This mode of generating ATP is called *oxidative phosphorylation*.

Respiration usually permits a complete oxidation of organic compounds to CO_2. However, certain bacteria, notably, the acetic acid bacteria, only partially oxidize certain substrates; an example of such an *incomplete oxidation* is the formation of acetic acid from ethanol.

$$\underset{\text{ethanol}}{CH_3-CH_2OH} + O_2 \longrightarrow \underset{\text{acetic acid}}{CH_3-COOH} + H_2O$$

The free energy change for the complete oxidation of an organic compound is very much greater than that for its fermentations. For example, the complete oxidation of 1 mole of glucose liberates 688 Kcal, whereas most fermentations of this sugar liberate only about one-tenth as much energy. The substrate-level phosphorylations characteristic of the fermentation of organic compounds also operate in the respiration of organic compounds, but a considerably greater quantity of ATP is generated by oxidative phosphorylation. The gross yield of ATP per mole of substrate respired is, accordingly, much greater than that obtainable by the fermentative metabolism of the same compound.

• The definition and nature of photosynthesis

The third (and mechanistically most complex) mode of ATP generation is *photosynthesis:* the use of light as a source of energy. Historically, the term photosynthesis was used to describe the overall metabolism of plants, algae, and blue-green bacteria, represented by the following reaction:

$$CO_2 + H_2O \xrightarrow{\text{light}} (CH_2O) + O_2$$

where the term (CH_2O) represents organic compounds at the average oxidation state found in cells.

This reaction does not describe a process by which ATP is generated, but rather its biosynthetic consequence: the light-mediated conversion of CO_2 to organic cell materials. ATP is generated by a transfer of the light energy absorbed by the photosynthetic pigment system; this process is called *photophosphorylation*. It is mechanistically analogous to oxidative phosphorylation; ATP is formed during the passage of electrons through an electron transport chain. Substrate-level phosphorylations, which provide all the ATP in fermentations and some of the ATP in respirations, do not occur in photosynthetic metabolism.

Since most photosynthetic organisms use CO_2 as their principal carbon source, they have a considerable metabolic requirement for reducing power. The majority of photosynthetic organisms (plants, algae, and blue-green bacteria) use water as an ultimate source of reductant, the concomitant oxidation of water leading to the formation of O_2 as a photosynthetic product. This kind of photosynthesis will be termed *oxygenic photosynthesis*.

Certain photosynthetic procaryotes (purple and green bacteria) are unable to use water as an ultimate reductant, and their photosynthetic metabolism is never accompanied by the formation of O_2. Instead, other reduced inorganic compounds (e.g., H_2S and H_2) are used by some of these organisms as reductants. Some purple and green bacteria use organic compounds in place of CO_2 as the principal carbon source; in these groups, the net biosynthetic requirement for reducing power becomes negligible. These modes of photosynthesis will be termed collectively *anoxygenic photosynthesis*.

Molecular oxygen (O_2) does not play a role in ATP-generating reactions in any of these forms of photosynthesis. Consequently, all photosyntheses can, in principle, occur under strictly anaerobic conditions. However, all organisms which carry out oxygenic photosynthesis are aerobes in the limited sense that they must be able to tolerate oxygen. The fundamentally anaerobic nature of photosynthesis is evident among the organisms that carry out anoxygenic photosynthesis. Most of these organisms are strict anaerobes; in the minority that are facultative aerobes, photosynthetic generation of ATP is suppressed by the presence of oxygen, being replaced by the respiratory generation of ATP.

THE BIOCHEMISTRY OF ATP GENERATION BY HETEROTROPHS

• **The pathways of formation of pyruvate**

Certain catabolic reaction pathways are common to both respiratory and fermentative metabolism. Among these are the three pathways of conversion of sugars to the key metabolic intermediate, pyruvic acid. They are the *Embden-Meyerhof pathway* (also called the *glycolytic pathway*), the *pentose phosphate pathway* (also called the *hexose monophosphate shunt*), and the *Entner-Doudoroff pathway*. The first two occur in many organisms, including both procaryotes and eucaryotes. The third is restricted to certain groups of procaryotes.

In the Embden-Meyerhof pathway (Figure 6.5), 2 molecules of ATP are expended in the initial reactions to produce fructose-1,6-diphosphate. This is cleaved to yield the triose phosphates, glyceraldehyde phosphate and dihydroxyacetone phosphate, which are freely interconvertible. The oxidation of triose phosphate, coupled with a reduction of NAD, is accompanied by an esterification of inorganic phosphate to yield, from each C_3 moiety, a molecule of 1,3-diphosphoglyceric acid. The subsequent steps in the conversion of this compound to pyruvic acid permit a transfer of both phosphate groups to ADP (substrate-level phosphorylation), so that a total of 4 moles of ATP are formed per mole of glucose used. Since 2 moles of ATP are expended in the initial activation

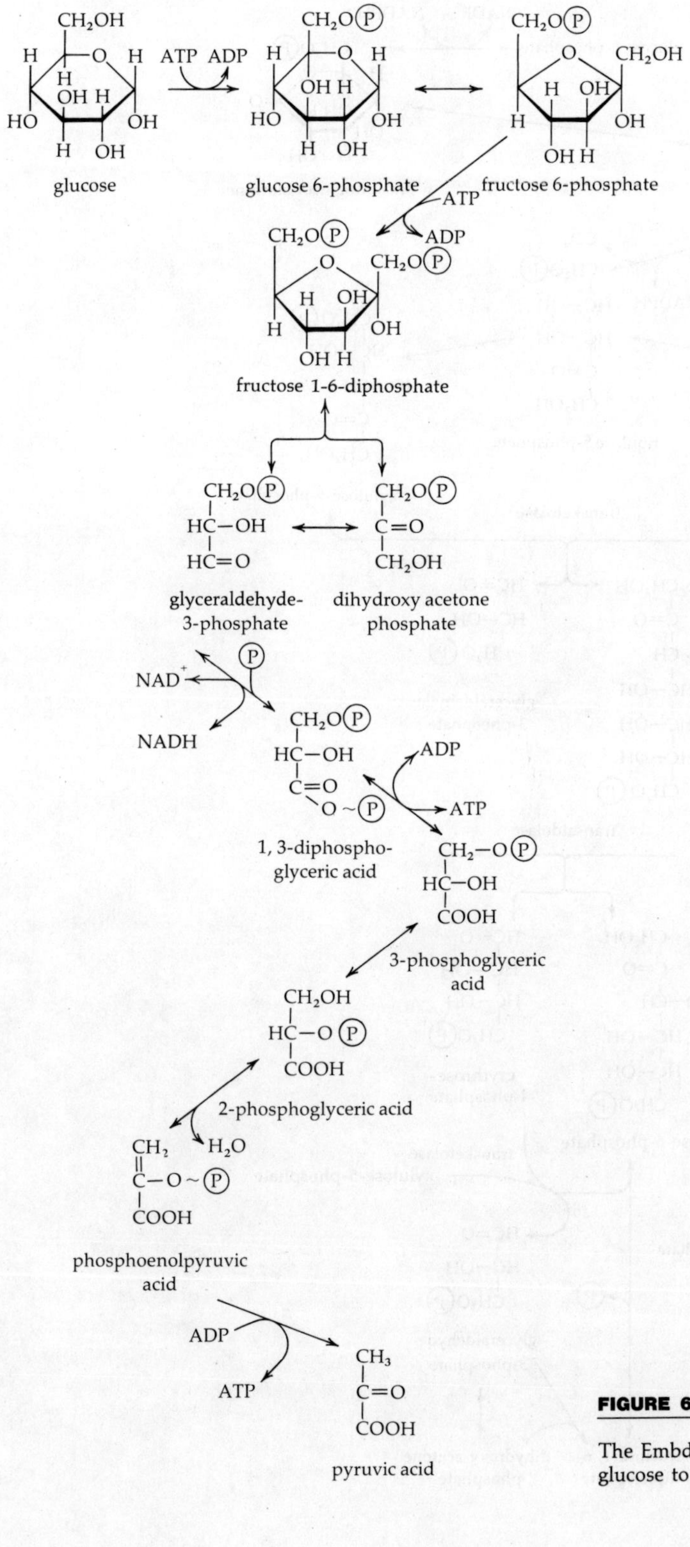

FIGURE 6.5

The Embden-Meyerhof (glycolytic) pathway of conversion of glucose to pyruvic acid.

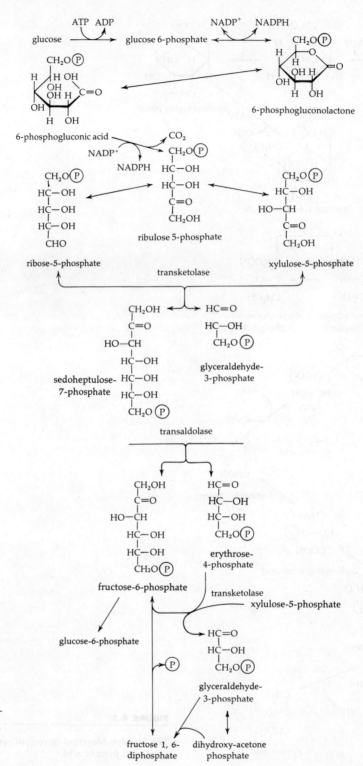

FIGURE 6.6

The pentose phosphate pathway of glucose oxidation.

steps, the net yield is 2 moles of ATP per mole of glucose fermented; 2 moles of NADH are also produced.

The pentose phosphate pathway does not lead directly to pyruvate; rather it provides only for the oxidation of one of the carbon atoms of the substrate (Figure 6.6). It involves the initial phosphorylation of glucose followed by an oxidation of the product, glucose-6-phosphate, coupled with a reduction of NADP, to yield 6-phosphogluconic acid. Then, glucose-6-phosphate enters into a series of reactions involving 2 NADP-linked oxidations and a decarboxylation to yield the pentose phosphate D-ribulose 5-phosphate. By epimerization, D-xylose-5-phosphate and ribose-5-phosphate are formed; D-ribose-5-phosphate and D-xylulose-5-phosphate are the starting point for a series of transketolase reactions (transfer of a 2-carbon glycoaldehyde group, $CH_2OH—CO—$) and trans-aldolase reactions (transfer of a 3-carbon dihydroxy-acetone group, $CH_2OH—CO—CHOH—$) leading eventually to the initial compound of the pathway, glucose-6-phosphate. The pathway is thus cyclic in nature. Passage of 6 molecules through the cycle results in the complete oxidation of 1 molecule of glucose-6-phosphate to CO_2 and the reduction of 6 molecules of $NADP^+$ to NADPH.

The pentose phosphate pathway serves two vital functions in metabolism: it produces the ribose-5-phosphate required for the synthesis of nucleic acids (see Chapter 7) and it produces much of the NADPH required in the various biosynthetic reactions of the cell.

Glucose-6-phosphate is the first intermediate of the Entner-Doudoroff pathway (Figure 6.7). Following oxidation to 6-phosphogluconic acid the unique intermediate of the pathway, 2-keto-3-deoxy-6-phosphogluconic acid (KDPG), is formed by a dehydration step. KDPG is cleaved to 1 molecule of pyruvic acid and 1 molecule of glyceraldehyde-3-phosphate which is metabolized by enzymes of the Embden-Meyerhof pathway to produce a second molecule of pyruvic acid. The net yield of the metabolism of 1 molecule of glucose through the Entner-Doudoroff pathway is 1 molecule of ATP and 2 molecules of NADH.

• Pathways of metabolism of pyruvate In addition to the specific reactions which pyruvate undergoes in certain fermentative processes, pyruvate is oxidized in a cyclic manner through a pathway known as the *tricarboxylic acid (TCA) cycle* (Figure 6.8). This cycle is the major route of ATP generation in aerobes (by oxidative phosphorylation through the transport of electrons from NADH to the terminal electron acceptor). The TCA cycle generates certain intermediates which are required in biosynthetic pathways, so that even strict anaerobes possess most of the enzymes of the cycle, lacking only the step between α-ketoglutaric acid and succinic acid. Thus, by reverse flow from oxalacetic acid to succinic acid and forward flow from citric acid to α-ketoglutaric acid, they are able to synthesize all intermediates even under anaerobic conditions. The TCA cycle effects the complete oxidation of 1 molecule of acetic acid to CO_2 and generates 3 molecules of reduced pyridine nucleotides, 1 molecule of GTP, and 1 pair of electrons which enter the electron transport chain independent of pyridine nucleotide.

If the TCA cycle operated exclusively for the terminal oxidation of acetic acid (acetyl residues) derived from primary substrates, it could be maintained without a net input of oxalacetic acid, the role of this compound being purely catalytic. However, as has been stated, the TCA cycle also generates intermediates which are utilized in biosyntheses. Hence, in a growing organism, the cycle is never

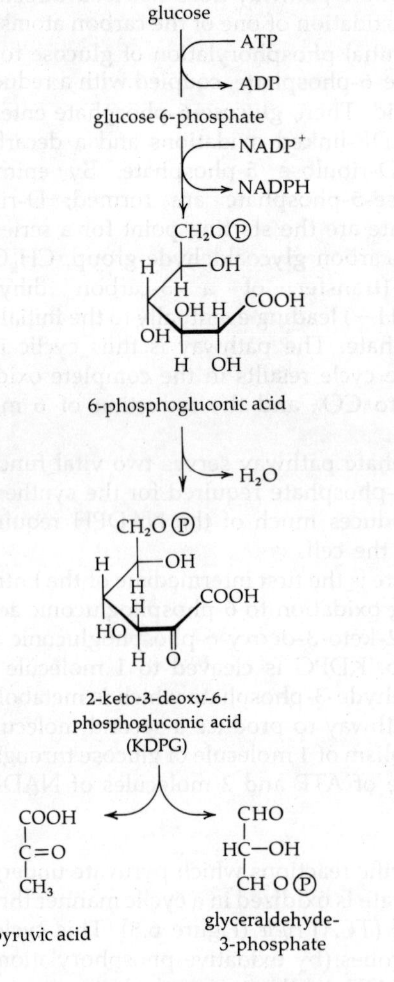

glucose

$\downarrow$ ATP

$\searrow$ ADP

glucose 6-phosphate

$\downarrow$ NADP$^+$

$\searrow$ NADPH

6-phosphogluconic acid

$\downarrow$ $\searrow$ H$_2$O

2-keto-3-deoxy-6-
phosphogluconic acid
(KDPG)

pyruvic acid

COOH
|
C=O
|
CH$_3$

glyceraldehyde-
3-phosphate

CHO
|
HC—OH
|
CH$_2$O Ⓟ

FIGURE 6.7

The Entner-Doudoroff pathway of conversion of glucose to pyruvate and glyceraldehyde-3-phosphate.

in fact closed, and its maintenance requires a considerable net synthesis of oxalacetic acid. This is usually formed by the carboxylation of either pyruvate or phosphoenolpyruvic acid:

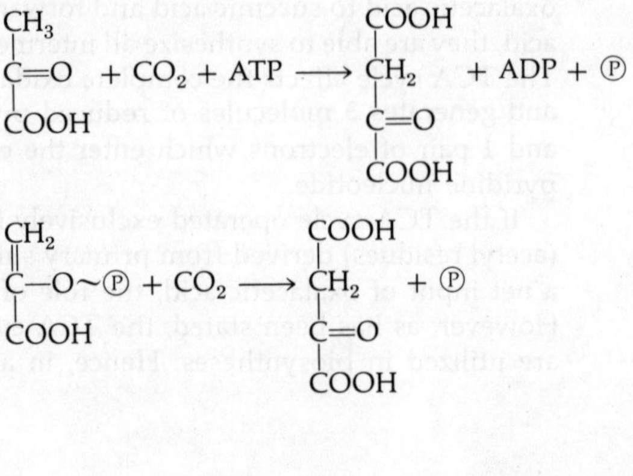

$$
\begin{array}{c}
\text{CH}_3 \\
| \\
\text{C=O} \\
| \\
\text{COOH}
\end{array}
+ \text{CO}_2 + \text{ATP} \longrightarrow
\begin{array}{c}
\text{COOH} \\
| \\
\text{CH}_2 \\
| \\
\text{C=O} \\
| \\
\text{COOH}
\end{array}
+ \text{ADP} + \text{Ⓟ}
$$

$$
\begin{array}{c}
\text{CH}_2 \\
\| \\
\text{C—O}\sim\text{Ⓟ} \\
| \\
\text{COOH}
\end{array}
+ \text{CO}_2 \longrightarrow
\begin{array}{c}
\text{COOH} \\
| \\
\text{CH}_2 \\
| \\
\text{C=O} \\
| \\
\text{COOH}
\end{array}
+ \text{Ⓟ}
$$

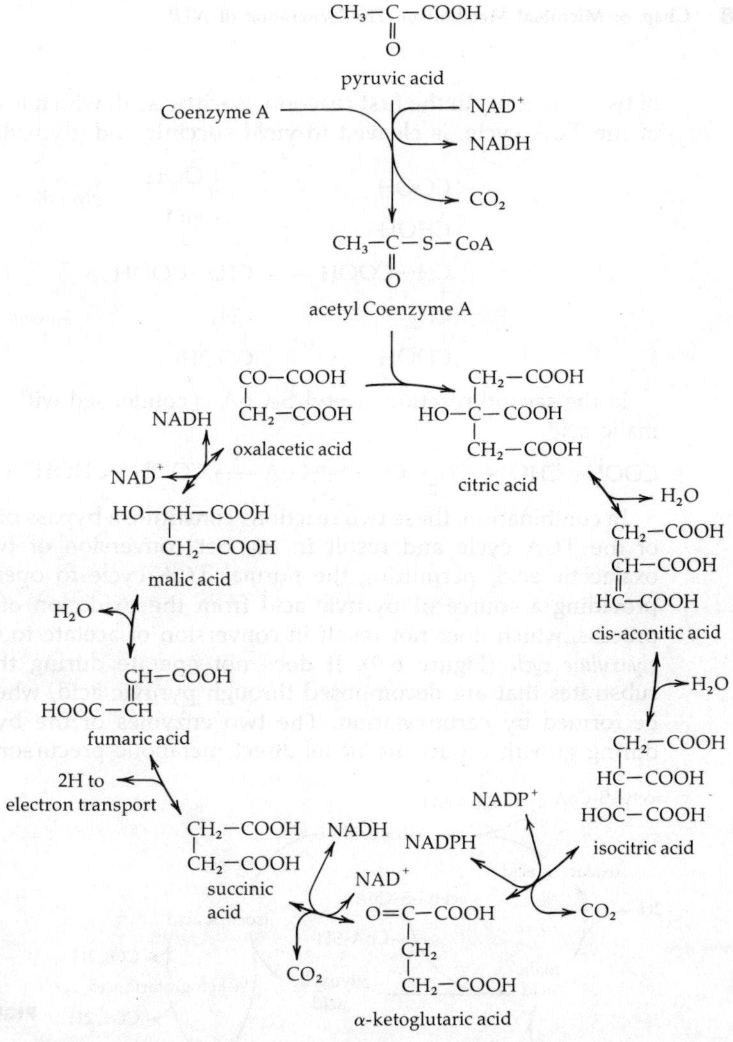

$$CH_3-\overset{\displaystyle\underset{\displaystyle O}{\|}}{C}-COOH$$

pyruvic acid

Coenzyme A —

NAD⁺ → NADH

CO₂

$$CH_3-\overset{\displaystyle\underset{\displaystyle O}{\|}}{C}-S-CoA$$

acetyl Coenzyme A

$$\begin{array}{l} CO-COOH \\ | \\ CH_2-COOH \end{array}$$

oxalacetic acid

NADH

NAD⁺

$$\begin{array}{l} HO-CH-COOH \\ | \\ CH_2-COOH \end{array}$$

malic acid

H₂O

$$\begin{array}{l} HOOC-CH \\ \quad\quad\| \\ \quad\quad CH-COOH \end{array}$$

fumaric acid

2H to electron transport

$$\begin{array}{l} CH_2-COOH \\ | \\ CH_2-COOH \end{array}$$

succinic acid

NADH

NAD⁺

CO₂

$$\begin{array}{l} O=C-COOH \\ | \\ CH_2 \\ | \\ CH_2-COOH \end{array}$$

α-ketoglutaric acid

$$\begin{array}{l} CH_2-COOH \\ | \\ HO-C-COOH \\ | \\ CH_2-COOH \end{array}$$

citric acid

H₂O

$$\begin{array}{l} CH_2-COOH \\ | \\ CH-COOH \\ \quad\quad\| \\ HC-COOH \end{array}$$

cis-aconitic acid

H₂O

$$\begin{array}{l} CH_2-COOH \\ | \\ HC-COOH \\ | \\ HOC-COOH \end{array}$$

isocitric acid

NADP⁺ → NADPH

CO₂

FIGURE 6.8

The tricarboxylic acid (TCA) cycle by which pyruvic acid is oxidized.

As a result, carbon from pyruvic acid enters the cycle by two routes: via oxalacetic acid and via acetyl-S-CoA.

• **The role of the glyoxylate cycle in acetic acid oxidation**

A special modification of the TCA cycle, known as the *glyoxylate cycle*, comes into play during oxidation of acetic acid or of primary substrates (such as higher fatty acids), which are converted to acetyl-S-CoA without the intermediate formation of pyruvic acid. Under these circumstances, oxalacetic acid cannot be generated by the carboxylation of pyruvic or phosphoenolpyruvic acid, since in aerobic microorganisms there is no mechanism for synthesizing pyruvic acid directly from acetic acid: the oxidation of pyruvic acid to acetyl-S-CoA and CO_2 is a completely irreversible reaction.

The oxalacetic acid required for acetic acid oxidation is replenished by the oxidation of succinic and malic acids, which are produced through a sequence of two reactions. In the first reaction, isocitric acid, which is a normal intermediate of the TCA cycle, is cleaved to yield succinic and glyoxylic acids:

$$
\begin{array}{ccc}
\text{COOH} & & \text{COOH} \\
| & & | \\
\text{CHOH} & & \text{CHO} \qquad \text{glyoxylic acid} \\
| & & + \\
\text{CH—COOH} & \longrightarrow & \text{CH}_2\text{—COOH} \\
| & & | \\
\text{CH}_2 & & \text{CH}_2 \\
| & & | \qquad \text{succinic acid} \\
\text{COOH} & & \text{COOH}
\end{array}
$$

In the second reaction, acetyl-S-CoA is condensed with glyoxylic acid to yield malic acid:

$$COOH\text{—}CHO + CH_3\text{—}CO\text{—}S\text{—}CoA \longrightarrow COOH\text{—}CHOH\text{—}CH_2\text{—}COOH + CoASH$$

In combination, these two reactions constitute a bypass of some of the reactions of the TCA cycle and result in the net conversion of two acetyl residues to oxalacetic acid, permitting the normal TCA cycle to operate and, in addition, providing a source of pyruvic acid from the oxidation of C_4 acids. The cyclic process, which does not result in conversion of acetate to CO_2, is known as the *glyoxylate cycle* (Figure 6.9). It does not operate during the attack on primary substrates that are decomposed through pyruvic acid, when oxalacetic acid can be formed by carboxylation. The two enzymes of the bypass are synthesized during growth on acetate or its direct metabolic precursors.

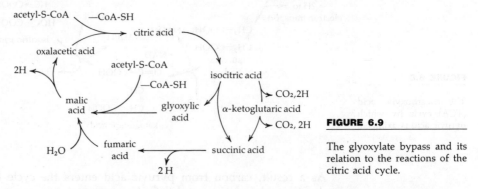

FIGURE 6.9

The glyoxylate bypass and its relation to the reactions of the citric acid cycle.

THE BIOCHEMISTRY OF FERMENTATIONS

As stated earlier, fermentations involve a strict balance between oxidation and reduction; pyridine nucleotides reduced in one step of the process are subsequently oxidized in another. This general principle is illustrated by two fermentations: alcoholic fermentation (typical of the anaerobic metabolism of yeasts) and the homolactic acid fermentation (typical of the metabolism of certain lactic acid bacteria in the presence and absence of air). Both of these fermentative processes (Figure 6.10) are slight modifications of the Embden-Meyerhof pathway, whereby the 2 molecules of NADH oxidized by this pathway are reduced in

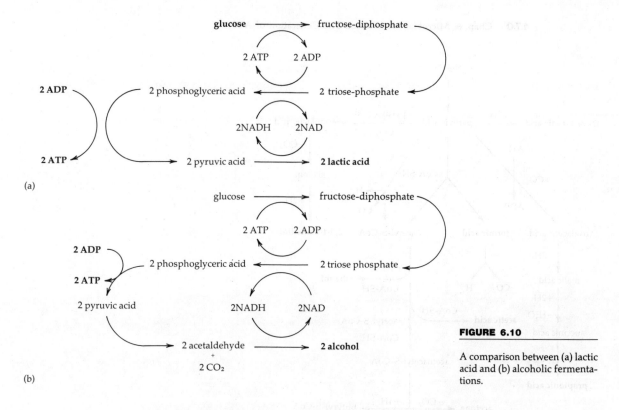

(a)

(b)

FIGURE 6.10

A comparison between (a) lactic acid and (b) alcoholic fermentations.

reactions involving the subsequent metabolism of pyruvic acid. In the case of the homolactic acid fermentation, this oxidation occurs as a direct consequence of the reduction of pyruvic acid to lactic acid. In the case of the alcoholic fermentation, pyruvic acid is first decarboxylated to form acetaldehyde; the oxidation of NADH occurs concomitantly with the reduction of acetaldehyde to form ethanol.

The Embden-Meyerhof pathway is the most widespread mechanism for the fermentative conversion of glucose to pyruvic acid, and it is employed by many groups of bacteria that produce fermentative end products different from those characteristic of the alcoholic and homolactic fermentations. These differences reflect exclusively *differences with respect to pyruvic acid metabolism.*

The pathways leading from pyruvic acid to the various end products of bacterial fermentations are summarized in Figure 6.11. Most bacterial fermentations produce several end products; however, no single fermentation produces all of the end products shown in Figure 6.11.

Not all fermentative mechanisms involve the Embden-Meyerhof pathway. Certain fermentative conversions of glucose involve reactions of the pentose phosphate pathway, and others involve reactions of the Entner-Doudoroff pathway. Fermentative conversions of substrate other than glucose (e.g., amino acids) involve highly specific pathways.

The end products of bacterial fermentations and the pathway by which they are formed are group specific. The specific pathway of fermentation will be considered in subsequent chapters along with the other characteristic features of the various physiological groups of bacteria.

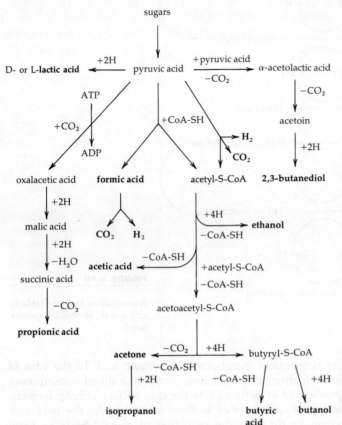

FIGURE 6.11

Derivations of some major end products of the bacterial fermentations of sugars from pyruvic acid. The end products are shown in boldface type.

SPECIAL PATHWAYS FOR THE PRIMARY ATTACK ON ORGANIC COMPOUNDS BY MICROORGANISMS

As stated earlier, there is probably no naturally occurring organic compound that cannot be used as a substrate for respiratory metabolism by some microorganism. However complex the structure of the primary substrate may be, its utilization as a source of energy always involves the same basic principle: a stepwise degradation to yield eventually one or more small fragments capable of entering the reactions of the TCA cycle. As a specific example of the many such specialized microbial pathways, we shall describe those involved in the utilization of aromatic compounds.

• **Oxidations of aromatic compounds through the β-ketoadipate pathway**   Most aerobic bacteria that use aromatic compounds as respiratory substrates attack them through one or another of the two convergent branches of the β-ketoadipate pathway (Figure 6.12). Through these reactions, the six carbon atoms of the aromatic nucleus in the primary substrate are converted to the six carbon atoms of an aliphatic acid, β-ketoadipic acid. This is in turn cleaved to acetyl-S-CoA and succinic acid, both of which can immediately enter the TCA cycle.

A number of other structurally related compounds are metabolized to intermediates of the β-ketoadipate pathway and are further metabolized through it. Some of these compounds and their points of convergence with the β-ketoadipate pathway are shown in Figure 6.13.

THE OXIDATION OF INORGANIC COMPOUNDS

The use of inorganic compounds as substrates for respiratory metabolism is confined to bacteria, and is characteristic of a number of special physiological groups known collectively as the *chemoautotrophs*. The substrates that can so serve as energy sources are H_2, CO, NH_3, NO_2^-, Fe^{2+}, and reduced sulfur compounds (H_2S, S, $S_2O_3^{2-}$). In this mode of respiratory metabolism, the sole function of substrate oxidation is to provide ATP through oxidative phosphorylation and to provide reducing power. Oxidation of certain inorganic substrates (H_2 and reduced sulfur compounds) is characterized by transfer of electrons to NAD^+, thereby directly generating reducing power. However, such direct transfer of electrons is thermodynamically unfavorable with other inorganic substrates (NO_2^- or Fe^{2+}). In the latter cases, NADH is generated by *reverse electron transport*, the ATP generated by oxidative phosphorylation being used to drive electrons back through the electron transport chain, thereby reducing NAD^+.

PHOTOSYNTHESIS

Photosynthetic energy conversion is a process of considerable complexity and occurs in a membrane system containing pigments, electron carriers, lipids, and proteins. This is termed the *photosynthetic apparatus*. In eucaryotic organisms, it is contained within the chloroplast. Among procaryotes, the location within the cell of the photosynthetic apparatus is different in the three major photosynthetic groups (see Chapter 11 for further discussion).

• Light absorption   Radiant energy is always transferred in discrete packets known as *photons*; the energy content of a photon is inversely related to its wavelength (Figure 6.14). When radiant energy is absorbed by matter, its possible effects are a function of the energy content of the photon and hence the wavelength of the radiation. Infrared light of wavelengths longer than 1,200 nm has an energy content so small that the absorbed energy in this spectral range is immediately converted to heat; it cannot mediate chemical change. So-called *ionizing* radiations of very short wavelength (X rays, α particles, cosmic rays) have such a high energy content that molecules in their path are immediately ionized. Between these two extremes, radiations ranging in wavelengths from 200 to 1,200 nm (ultraviolet, visible, and near-infrared light) have an energy content such that their absorption is capable of producing a chemical change in the absorbing molecule; it is this portion of the electromagnetic spectrum that can serve for the performance of photosynthesis.

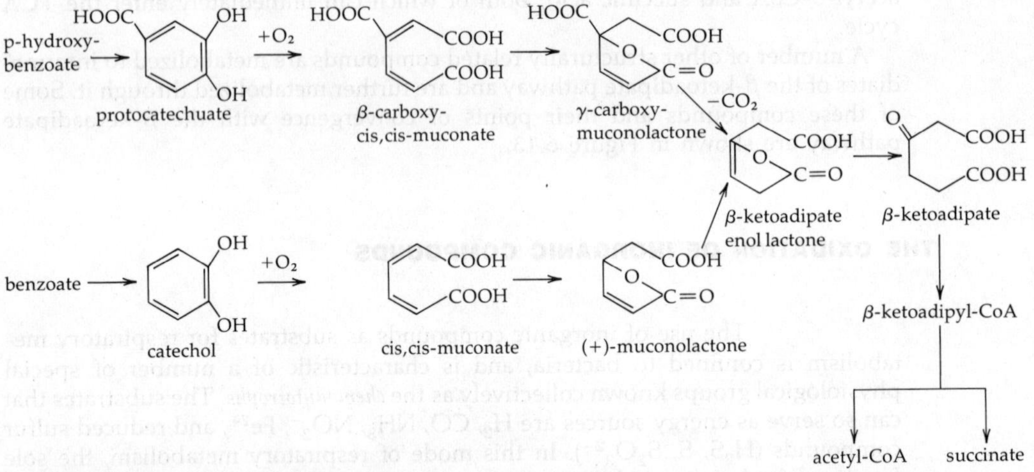

FIGURE 6.12

The chemistry of the β-ketoadipate pathway.

FIGURE 6.13

Certain compounds which are metabolized through the β-ketoadipate pathway: their structures and points of convergence with the pathway. Intermediates of the β-keto-adipate pathway (see Figure 6.12) are shown here in boldface type.

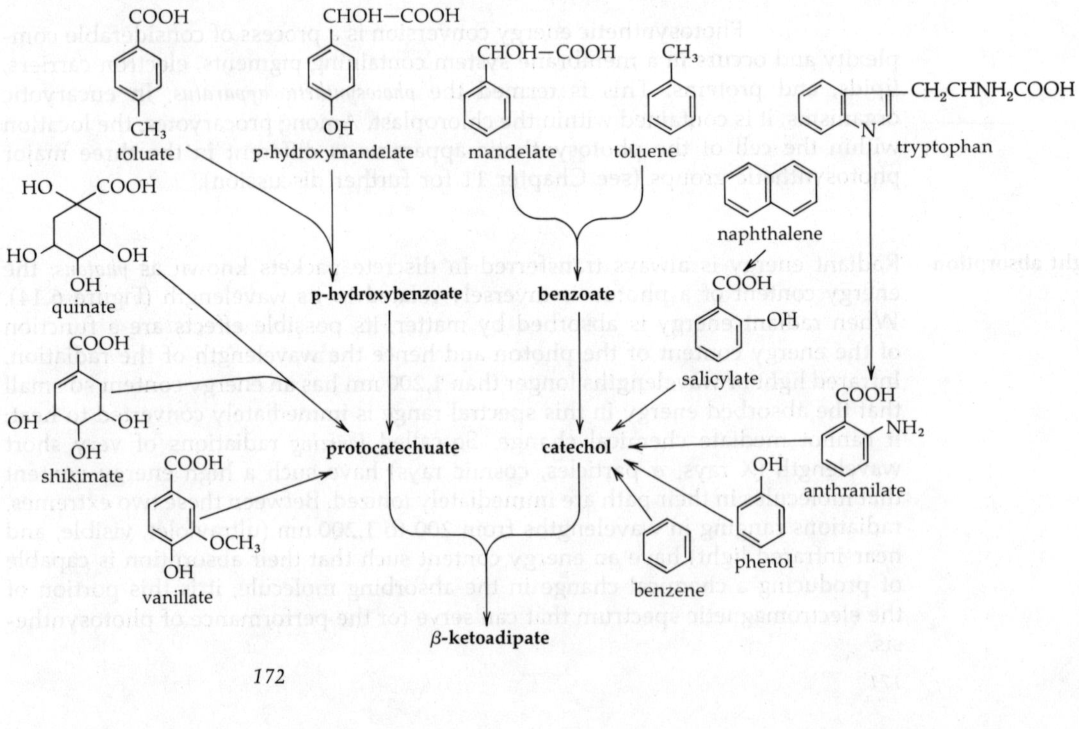

172

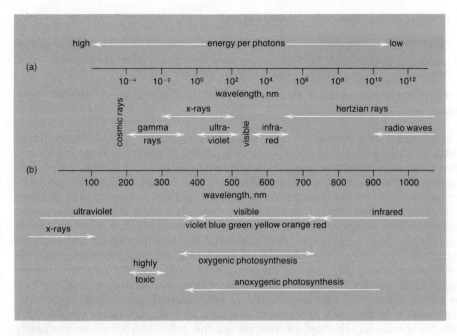

FIGURE 6.14

The electromagnetic spectrum. (a) The entire spectrum is plotted on an exponential scale. (b) The ultraviolet, visible, and near-infrared regions are greatly expanded and plotted on an arithmetic scale.

• The spectral quality of sunlight

Solar radiation is profoundly modified by its passage through the atmosphere which effectively filters out much of the shorter, high-energy radiation. At sea level about 75 percent of the total energy of sunlight is contained in the light of wavelengths between 400 and 1,100 nm (the visible and near-infrared portions of the spectrum), and it is within these limits that the pigments responsible for light capture in photosynthesis have their effective absorption bands.

• The photosynthetic apparatus

The photosynthetic apparatus of all organisms capable of carrying out photosynthesis consists of three essential components:

1. *An antenna of light-harvesting pigments.* These pigments can include chlorophylls, carotenoids, and (in some groups) phycobiliproteins. The particular light-harvesting pigments which comprise the antenna system are group specific, and their cumulative light-absorptive properties determine the range of wavelength of light over which photosynthesis occurs.

2. *A photosynthetic reaction center.* Light energy absorbed by the antenna causes a photosynthetic energy conversion at the photosynthetic reaction center which contains chlorophyll molecules in a special state. Light energy causes an ejection of an electron from a molecule of chlorophyll.

3. *An electron transport chain.* Electrons generated at the reaction center are passed through an electron transport chain thereby generating ATP.

The three components of photosynthetic apparatus will be considered separately.

COMPONENTS OF ANTENNA OF LIGHT-HARVESTING PIGMENTS   The chlorophylls, of which at least seven kinds occur in various groups of photosynthetic organisms, have a common molecular ground plan shown in Figure 6.15. These compounds are related structurally and biosynthetically to the hemes, which serve as prosthetic groups in carriers (cytochromes), in electron transport chains, and in many respiratory enzymes; both hemes and chlorophylls have a central tetrapyrrolic nucleus, within which a metallic ion is chelated. In hemes, this metal is iron; in chlorophylls, it is magnesium. Two additional features distinguish chlorophylls chemically from hemes: the existence of a fifth ring, the pentanone ring, and the esterification of one side chain of the tetrapyrrolic nucleus with an alcohol, either phytol or farnesol. Chlorophylls absorb light intensely in two spectral regions: the violet, around 400 nm; and the red or near-infrared, around 600 to 800 nm. The position of the peak absorption at longer wavelengths varies considerably with the particular chlorophyll species. In the cell, chlorophyll is intimately associated with proteins which may considerably modify its spectral properties.

The carotenoids, of which a large number of different kinds occur in photosynthetic organisms, have the basic structure of long, unsaturated hydrocarbons with projecting methyl groups, containing a total of 40 carbon atoms (Figure 6.16). In particular members of the class, this basic structure can be modified in several ways: by terminal ring closure to form six-membered alicyclic or aromatic rings, and by the addition of oxygenated substituents, notably hydroxyl, ethoxyl, or keto groups. Carotenoids have a single broad region of light absorption between 450 and 550 nm.

The phycobiliproteins are water-soluble chromoproteins containing linear tetrapyrroles. They absorb light in a broad region from 550 to 650 nm, between the major absorption regions of chlorophyll.

PHOTOCHEMICAL REACTION CENTERS AND ELECTRON TRANSPORT CHAINS   The energy contained in the photons absorbed by the photosynthetic pigment system are transferred to the reaction centers in which the chlorophyll molecules are in a

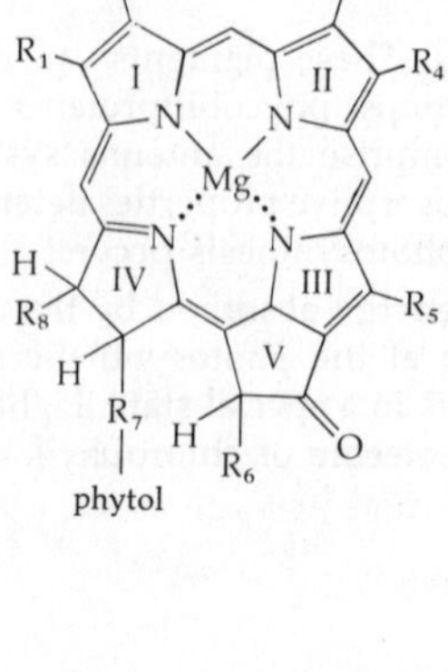

FIGURE 6.15

The molecular ground plan of the chlorophylls. The tetrapyrrolic nucleus (rings I, II, III, and IV) has the same derivation as that of the hemes, but it is chelated with magnesium. In chlorophylls, one or more of the pyrrole rings are reduced; in this diagram, ring IV is shown reduced, as is characteristic of chlorophyll a; R_1, R_2, and so forth, designate aliphatic side chains attached to the tetrapyrrolic nucleus. The presence of ring V, the pentanone ring, and the substitution of R_7 by a long-chain alcohol (phytol or farnesol) are characteristic features of the chlorophylls that do not occur in hemes.

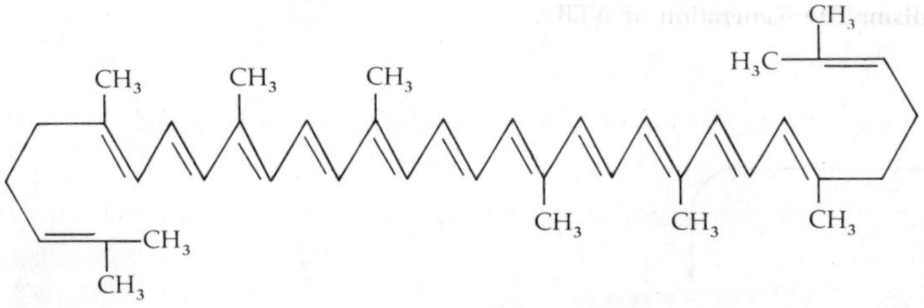

FIGURE 6.16

The molecular ground plan of the carotenoids, illustrated by an open-chain carotenoid which does not contain oxygen. This basic structure may be modified, in the different kinds of carotenoids, by terminal ring closure at one or both ends of the molecule and by the introduction of hydroxyl (—OH), methoxyl (—OCH$_3$), or ketone (=O) groups.

special state, closely associated with the components of the photosynthetic electron transport chain. The first identified electron acceptor in this chain is the nonheme iron protein, ferredoxin, which also plays roles in the metabolism of anaerobic nonphotosynthetic bacteria. Also present in the photosynthetic electron chains are components which are found in respiratory electron transport chains.

The process of photosynthetic energy conversion is initiated when light energy is absorbed by a molecule of reaction center chlorophyll. This chlorophyll molecule is oxidized by ejection of an electron, which is accepted by ferredoxin:

$$\text{chlorophyll} + \text{light energy} \longrightarrow \text{chlorophyll}^\oplus + e^-$$

$$\text{ferredoxin} + e^- \longrightarrow \text{reduced ferredoxin}$$

The reoxidation of reduced ferredoxin makes available approximately as much energy as the oxidation of molecular hydrogen. This energy is harnessed through the intermediacy of other electron carriers of the photosynthetic electron transport system to generate ATP. The mechanism by which electron transport generates ATP will be discussed in a later section of this chapter.

Chlorophyll thus plays a dual role in photosynthetic energy conversion: as a light-harvesting pigment and as the site of the initial photochemical event. Carotenoids and phycobiliproteins, however, function uniquely as light-harvesting pigments, passing the light energy they absorb to the reaction center (i.e., to the chlorophyll).

In all photosynthetic organisms the energy of the electrons ejected from the chlorophyll can be used to generate ATP by the process of *cyclic photophosphorylation,* the pathway of which is shown schematically in Figure 6.17. The electrons removed from chlorophyll by light energy are passed through the photosynthetic electron transport chain, the terminal component of which reduces chlorophyll; the electrons thus flow through a closed circuit, the flow being triggered by the absorption of light energy. Part of the light energy is captured in the transport chain by the synthesis of ATP.

The electrons ejected by light from reaction center chlorophyll can alternatively be used for the reduction of pyridine nucleotides (Figure 6.18). In this event, the flow of electrons become open or noncyclic, because chlorophyll$^\oplus$ must be reduced by electrons derived from an appropriate chemical donor, via the transport system of the photosynthetic apparatus. Since this transfer of electrons can also generate ATP, the process is termed *noncyclic photophosphorylation.*

175

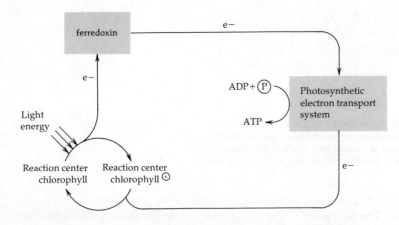

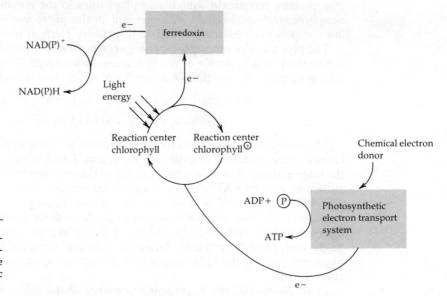

FIGURE 6.17

A schematic diagram of cyclic photophosphorylation.

FIGURE 6.18

The photosynthetic reduction of pyridine nucleotides through the reactions of noncyclic photophosphorylation.

• **The differences between oxygenic ("plant") photosynthesis and anoxygenic ("bacterial") photosynthesis**

The most obvious gross difference between oxygenic and anoxygenic photosynthesis is the evolution of oxygen. In oxygenic photosynthesis, the formation of oxygen results from the oxidation of water, which is coupled, through the reaction of noncyclic photophosphorylation, with the reduction of NAD(P):

$$2NAD(P)^+ + 2H_2O + 2ADP + 2(P) \longrightarrow 2NAD(P)H + O_2 + 2ATP + 2H^+$$

The oxidation of water to O_2 does not proceed spontaneously: on the contrary, it requires energy. How, then, are plants, algae, and blue-green bacteria able to use this thermodynamically unfavorable electron donor? A partial answer to the question can now be given because of the discovery that the photosynthetic apparatus of these organisms contains *two distinct kinds of photochemical reaction centers*, which can be distinguished by their responses to certain wavelengths of light and to inhibitors of the photosynthetic process. One type of reaction center (Type I) mediates both cyclic and noncyclic photophosphorylation. Light absorption by these reaction centers alone is, however, insufficient to couple the reduction of NADP with the oxidation of water. Noncyclic photophosphorylation of the oxygenic type requires simultaneous light absorption by another type of reaction center (Type II) at which photochemically driven *oxidation of water* takes place. The electrons from this oxidation pass through the photosynthetic electron transport chain and reduce the oxidized chlorophyll by passage of electrons to NAD(P) at reaction centers of Type I. This coupling of the two light reactions characteristic of oxygenic photosynthesis is shown in Figure 6.19.

All current evidence suggests that the inability of bacteria to bring about a photosynthetic production of oxygen is a consequence of the absence of Type II reaction centers from the anoxygenic photosynthetic apparatus. Anoxygenic photochemical reaction centers, exclusively of Type I, mediate cyclic photophosphorylation. In terms of photochemistry, anoxygenic photosynthesis is a simpler process than oxygenic photosynthesis.

We have described how light energy absorbed by the pigments of the photosynthetic apparatus is used to mediate the synthesis of ATP and reduced pyridine nucleotides. It is these reactions of energy conversions (i.e., the conversion of light energy to chemical energy) that are unique to photosynthesis. The particular manner in which the chemical energy so produced is used in biosynthetic reactions is neither unique to photosynthesis nor is it, strictly speaking, a part of the photosynthetic process. The ensuing biosynthetic processes involve so-called *dark reactions*, catalyzed by enzymes in the absence of light. It may nevertheless be useful to discuss the coupling between photosynthetic energy generation and the assimilation of CO_2, the primary source of carbon for most photosynthetic organisms.

In all organisms, photosynthetic and nonphotosynthetic, which use CO_2 as a principal carbon source, CO_2 assimilation occurs through a cyclic series of reac-

FIGURE 6.19

The coupling of the two light reactions characteristic of oxygenic photosynthesis, accompanied by the oxidation of water to O_2.

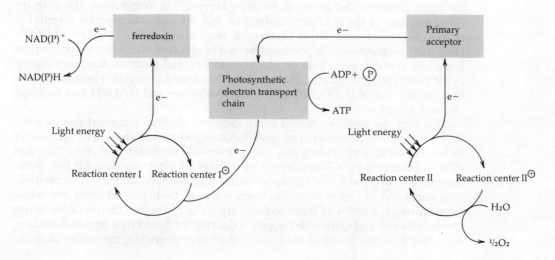

tions (to be discussed in Chapter 7), which result in conversion of CO_2 to cellular components (CH_2O). Grossly, this conversion involves an expenditure of two molecules of reduced pyridine nucleotides and three molecules of ATP:

$$CO_2 \xrightarrow[\underset{2NAD(P)H}{}]{\overset{3ATP}{}} \xleftarrow[\underset{2NAD(P)^+}{}]{\overset{3ADP + 3\,\circled{P}}{}} (CH_2O)$$

The reactions of noncyclic photophosphorylation in oxygenic photosynthesis yield equimolar quantities of reduced pyridine nucleotides and of ATP. Consequently, noncyclic photophosphorylation cannot supply all the ATP needed for the assimilation of CO_2 and must be supplemented by an additional synthesis of ATP, derived from the reactions of cyclic photophosphorylation. Both cyclic and noncyclic photochemical processes must therefore operate in oxygenic photosynthesis to permit the assimilation of CO_2.

The situation in an anoxygenic photosynthesis is less clear, since at least some of the electron donors used by these organisms (e.g., H_2) can serve directly for the generation of reduced pyridine nucleotides by "dark" reactions, thus making unnecessary the intervention of a photochemical reaction for the production of reducing power. Therefore, it seems likely that noncyclic photophosphorylation plays a much less important role in anoxygenic photosynthesis than in oxygenic photosynthesis. The most important function of the photochemical events in anoxygenic photosynthesis is the provision of ATP through the reactions of cyclic photophosphorylation.

THE MECHANISM OF ELECTRON TRANSPORT

In aerobic respiration of organic and inorganic compounds, in anaerobic respiration, and in photosynthesis, ATP is generated as a consequence of passage of electrons from the primary electron donor through an electron transport chain to the terminal electron acceptor. In respiration, the primary electron donor is the oxidizable substrate, and the terminal electron acceptor is an inorganic compound, either O_2 (aerobic respiration) or NO_3^-, SO_4^{2-}, or CO_3^{2-} (anaerobic respirations). In photosynthesis, the chlorophyll molecules of the type I reaction centers serve both as electron donors and electron acceptors during cyclic photophosphorylation. However, the process of noncyclic phosphorylation involves the use of H_2O as a primary electron donor and NADPH as a terminal electron acceptor.

Although the complexity and components of electron transport chains vary, they have certain common features: the components of the chain are carrier molecules capable of undergoing freely reversible oxidation and reduction, and ATP is generated as a consequence of passage of electrons through the chain. An example of an electron transport chain, that involved in aerobic respiration, is schematized in Figure 6.20. Electrons removed from the substrate are transferred through a series of intermediate carriers until the last carrier of the series reacts in its reduced state with oxygen, a reaction mediated by a terminal oxidase. As a result of this terminal oxidation, which is irreversible, the whole chain of

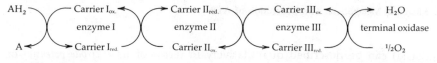

FIGURE 6.20

A schematic representation of a respiratory electron transport system, linking the dehydrogenation of an oxidizable substrate, AH_2, with the reduction of molecular oxygen to water.

carriers is reoxidized and oxygen is reduced to water. Other electron transport chains operate similarly, differing only in the length, the primary electron donor, and the terminal electron acceptor.

The electron transport chain can also operate in the reverse direction. Rather than generating ATP as a consequence of the flow of electrons from a reduced compound such as NADH, utilization of ATP can cause the reverse flow of electrons through the chain, thereby generating reducing power in the form of NADH. Such reverse electron transport is necessary in the metabolism of certain chemoautotrophs such as those that oxidize Fe^{+2} or NO_2^-, because the direct generation of reducing power by transfer of electrons from the inorganic substrate to NAD^+ is thermodynamically unfavorable. Some reducing power in photosynthetic organisms is probably also derived by reverse electron transport.

• Intracellular location of electron transport chains

Electron transport chains are very complex. They comprise a large number of electron carriers and distinct enzymes, physically associated in a rigid matrix having a high lipid content. The multicomponent system is *always contained in membranes*. In eucaryotic cells, these membranes are located within mitochondria or chloroplasts; in procaryotic cells, they are usually the cell membrane itself or invaginations thereof.

The biochemical analysis of electron transport chains is difficult because of their physical structure. Since most of their components are insoluble in water, they cannot be easily separated from one another in a functional state. Most of our current knowledge on this subject has been derived from work with the respiratory electron transport chain of mitochondria, which is remarkably uniform in all eucaryotic groups. The nature of the respiratory electron transport chains found in different bacteria is much more varied, and the components of the chains often differ markedly from those characteristic of mitochondrial systems.

• The physiological function of electron transport chains

When the removal of a pair of electrons or hydrogen atoms from an organic substrate is coupled with the reduction of oxygen to water, there is a large free energy change ($\Delta G°$) approximately equivalent to the combustion of a molecule of hydrogen gas. The passage of these through the transport chain permits a stepwise release of this energy, some of which can be converted into the high-energy bonds of ATP. In order for an electron transport chain to function, each component must be capable of being reduced by the reduced form of the previous component and oxidized by the oxidized form of the subsequent component in the chain (i.e., there must be a gradient of susceptibility to oxidation in an electron transport chain).

The tendency of a substance to donate or accept electrons (i.e., to be oxidized

179

or reduced) can be described quantitatively in terms of its *electrode potential* or *oxidation-reduction potential;* it is the relative voltage required to remove an electron from a compound. It is measured relative to a *standard potential,* that of the hydrogen electrode,

$$\tfrac{1}{2}H_2 \rightleftharpoons H^+ + e^-$$

assigned an arbitrary value of 0.0 volts (V) under standard conditions, [i.e., all reactants at 1 molar (pH 0.0)]. At pH 7.0, near which most biological reactions occur, the standard potential (25°C, all reactants at 1.0 molar, pH 7.0) of the hydrogen electrode is -0.42 V. The symbol E_0' designates electrode potentials in volts measured under these conditions.

When E_0' values of the two half reactions are known, the free energy change resulting from the coupled oxidation–reduction reaction can be calculated from the relationship:

$$\Delta G_0' \rightleftharpoons nF \Delta E_0'$$

where $\Delta G_0'$ is the standard free energy at pH 7.0; n is the number of electrons transferred; F is the faraday (a physical constant equal to 23,000 cal/volt); and $\Delta E_0'$ is the algebraic difference between the potentials of the two half reactions. Thus, the combustion of hydrogen (oxidation of hydrogen and reduction of oxygen) involves the two half reactions:

$$H_2 \rightleftharpoons 2H^+ + 2e^- \qquad (E_0' = -0.42 \text{ V})$$

and

$$\tfrac{1}{2}O_2 \rightleftharpoons O^{2-} - 2e^- \qquad (E_0' = +0.82 \text{ V})$$

It can readily be calculated that the free energy change is:

$$-2 \times 23,000[0.82 - (-0.42)] = -57,040 \text{ cal}$$

For a typical biological oxidation [e.g., the oxygen-linked oxidation of malate to pyruvate and CO_2 ($E_0' = -0.33$)], the analogous calculation shows a free energy change of $-53,000$ cal, not significantly different from that for the oxidation of hydrogen.

The carriers in an electron transport chain participate in a series of reactions

TABLE 6.3

Summary of primary electron donors and terminal electron acceptors in the various modes of ATP generation involving electron transport

MODE OF ATP GENERATION	PRIMARY ELECTRON DONOR	TERMINAL ELECTRON ACCEPTOR
Aerobic respiration	Organic or inorganic compound	O_2
Anaerobic respiration	Organic compound	NO_3^-, SO_4^{2-}, or CO_3^{2-}
Cyclic photophosphorylation	Reaction center chlorophyll	Oxidized reaction center chlorophyll
Noncyclic photophosphorylation	Chemical electron donor	Oxidized reaction center chlorophyll

of gradually increasing E_0' values, between the primary electron donor and the terminal electron acceptor. These are summarized for the various modes of ATP generation involving electron transport in Table 6.3. The position on the E_0' scale of several typical electron carriers, primary electron donors, and terminal electron acceptors are shown in Figure 6.21.

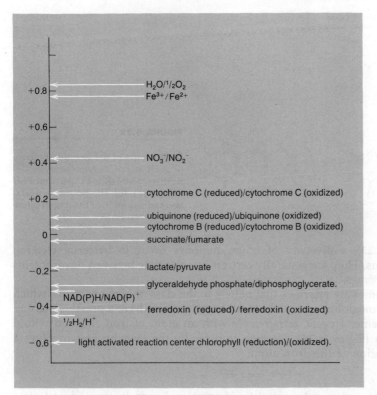

FIGURE 6.21

The E_0' values of various half reactions of primary electron donors, electron carriers, and terminal electron acceptors.

• The major components of electron transport chains

The electron transport chains of respiratory metabolism are the most thoroughly studied. Those involved in the respiration of organic compounds always contain three different classes of molecules. Two classes consist of enzymes with firmly bound prosthetic groups capable of undergoing reversible oxidation and reduction: the *flavoproteins* and the *cytochromes*. The third class consists of nonprotein carriers of relatively low molecular weight, the *quinones*.

Flavoproteins are enzymes with a yellow-colored prosthetic group, derived biosynthetically from the vitamin riboflavin [Figure 6.22(a)]. The prosthetic group may be either flavin mononucleotide (FMN) or flavin adenine dinucleotide (FAD); both possess the same active site, capable of undergoing reversible oxidation and reduction [Figure 6.22(b)]. The flavoproteins are a large class of enzymes which differ widely with respect to their E_0' values. Some are active in the primary dehydrogenation of organic substrates (e.g., succinic dehydrogenases); others transfer electrons from reduced pyridine nucleotides to subsequent members of the electron transport chain; and still others transfer electrons directly to molecular oxygen with the formation of H_2O_2.

A major component of electron transport systems of mitochondria is ubiqui-

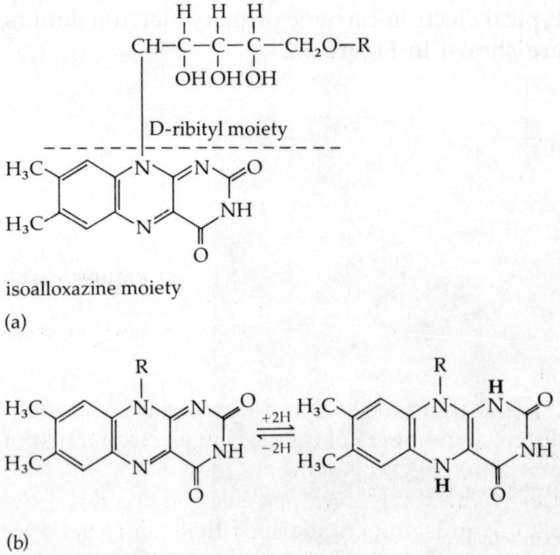

D-ribityl moiety

isoalloxazine moiety

(a)

(b)

FIGURE 6.22

(a) Structures of the vitamin riboflavin (R is H) and of its two coenzyme derivatives, FMN (R is PO_3H_2) and FAD (R is ADP). (b) The reversible oxidation and reduction of the ring structure of FMN and FAD.

none (Figure 6.23); a diversity of other quinones occur in bacterial electron transport systems. They are generally carriers intermediate between flavoproteins and the cytochromes.

The cytochromes are enzymes that belong to the class of heme proteins, which also includes hemoglobin and catalase. All heme proteins have a prosthetic group derived from heme, a cyclic tetrapyrrole with an atom of iron chelated within the ring system (Figure 6.24). Electron transfer by the cytochromes involves a reversible oxidation of this iron atom:

$$Fe^{2+} \rightleftharpoons Fe^{3+} + e^-$$

The cytochromes have characteristic absorption bands in the reduced state

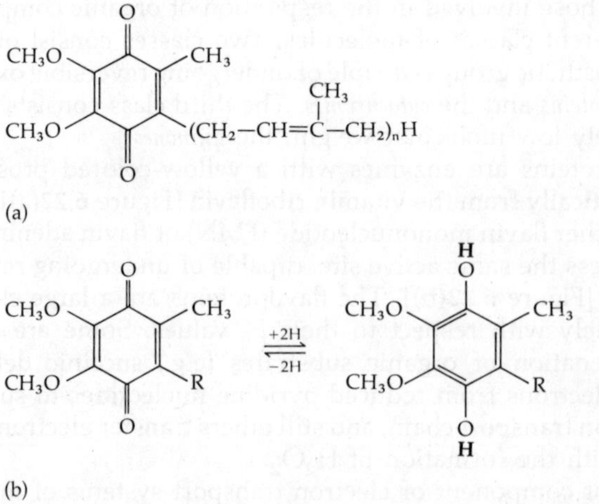

(a)

(b)

FIGURE 6.23

(a) Structure of ubiquinone. The number of isoprenoid units

$$-(CH_2-CH=\overset{\overset{\displaystyle CH_3}{|}}{C}-CH_2)-$$

in the side chain varies in different organisms from 6 to 10. (b) Reversible oxidation and reduction of a quinone.

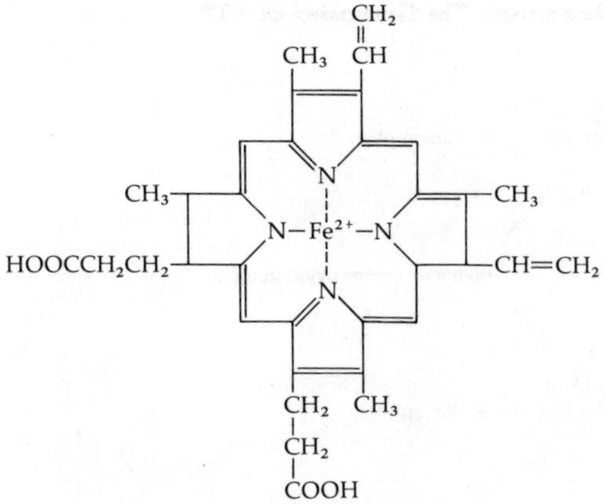

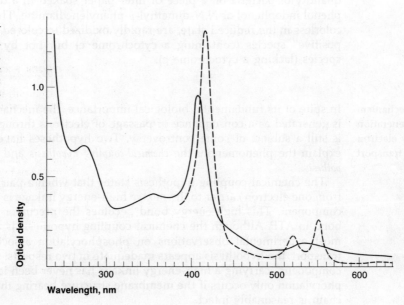

FIGURE 6.24

Structure of heme.

(Figure 6.25), and their spectral and functional properties permit the recognition of several different members of the class in electron transport chains, distinguished by a terminal letter (e.g., cytochrome c).

The carrier chain of a respiratory electron chain can be schematized as shown in Figure 6.26. It must be emphasized that many variations on this theme exist among the bacteria. For example, in the nitrifying bacteria and sulfur-oxidizing bacteria that derive energy from oxidations with relatively low E_0' values, electrons enter the chain at the level of cytochrome c.

• **Bacterial cytochrome systems**

The remarkable diversity of bacterial respiratory electron transport systems, in contrast to the essential uniformity of those in mitochondria, is revealed by the comparison of the cytochrome absorption spectra of intact bacterial cells and mitochondria (Table 6.4). By measuring the absorption spectrum of the reduced

FIGURE 6.25

The absorption spectrum of cytochrome c. Solid line: oxidized cytochrome c; dotted line: reduced cytochrome c.

Optical density

1.0

0.5

300   400   500   600

Wavelength, nm

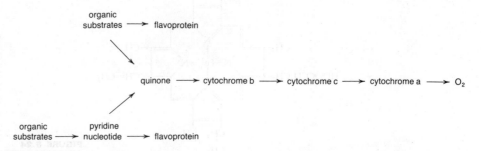

system (anaerobic suspension) relative to that of the oxidized system (aerobic suspension), it is possible to identify the characteristic absorption bands of the various cytochrome components. The so-called α bands at longest wavelengths are particularly characteristic; the shorter β bands and the so-called Soret bands (named for their discoverer) are generally composite ones, to which several pigments contribute.

As Table 6.4 shows, the mitochondria of mammalian heart muscle and of yeasts (at opposite ends of the scale of eucaryotic biological complexity) have identical cytochrome spectra. The cytochrome spectra of a few bacteria (e.g., *Bacillus subtilis*) are similar to those of mitochondria. However, in most bacteria, few if any of the main cytochrome peaks coincide with those of mitochondria.

It will be noted that a number of the bacterial species shown in Table 6.4 show no peak at 552 to 554 nm, characteristic of cytochromes of the c type. The lack of a c-type cytochrome in such organisms is correlated with the outcome of an empirical procedure, the *oxidase test*, which has considerable diagnostic importance in the identification of aerobic bacteria. The test is performed by putting a small quantity of bacteria on a piece of filter paper soaked in a dye: either dichlorophenol indophenol or *N,N*-dimethyl-*p*-phenylenediamine. These dyes, which are colorless in the reduced state, are rapidly oxidized to colored forms by "oxidase-positive" species (containing a cytochrome c) but not by "oxidase-negative" species (lacking a cytochrome c).

• **The mechanism of ATP generation by electron transport**

In spite of its fundamental biological importance, the mechanism by which ATP is generated as a consequence of passage of electrons through a transport chain is still a subject of some controversy. Two hypotheses have been advanced to explain the phenomenon: the *chemical-coupling hypothesis* and the *chemiosmotic hypothesis*.

The chemical-coupling hypothesis states that when a pair of electrons passes from one electron carrier to the next, a high-energy linkage is generated in a third component. This high-energy bond becomes the precursor of the high-energy bond in ATP. Although the chemical-coupling hypothesis is adequate to explain most experimental observations on phosphorylation associated with electron transport, the hypothesis appears inadequate in two respects: the proposed "third component" carrying a high-energy linkage has never been identified, and phosphorylation only occurs if the membrane structure bearing the electron transport chain is reasonably intact.

TABLE 6.4

Wavelengths (nm) of the main cytochrome absorption bands in mitochondrial and bacterial electron transport systems

	α BANDS				PRINCIPAL β BANDS	PRINCIPAL SORET BANDS
	a^a		b^a	c^a		
Mitochondrial systems						
Yeast	605	—	—	563	552	525	426
Mammalian heart muscle	605	—	—	563	552	525	426
Bacterial systems						
Bacillus subtilis	604	—	—	564	552	523	422
Micrococcus luteus	605	—	—	562	552	523	430
Micrococcus lysodeikticus	—	600	—	—	552	520	432
Pseudomonas acidovorans	—	600	—	—	553	530	425
Pseudomonas aeruginosa	—	—	—	560	552	530	428
Pseudomonas putida	—	—	—	560	552	530	427
Gluconobacter suboxydans	—	—	—	—	554	523	428
Acetobacter pasteurianum	—	—	588	—	554	523	428
Enterobacter aerogenes	628	—	592	560	—	530	430
Escherichia coli	630	—	593	560	—	533	433
Proteus vulgaris	630	—	595	560	—	533	430
Pseudomonas maltophilia	628	—	597	558	—	530	430

[a] Designates type of cytochrome in mitochondrial system to which each α band corresponds.

The chemiosmotic hypothesis, formulated by P. Mitchell, does not assume the existence of the high-energy third components and accounts for the requirement for an intact membrane structure. It states that the electron transport chain is oriented across the membrane so that oxidation of the electron carriers is accompanied by stoichiometric translocation of protons in an outward direction (Figure 6.27). It further states that membranes are impermeable to protons. Hence, as a consequence of electron transport, a proton gradient (ΔpH) is created between the outside and inside of the membrane. Protons can only reenter the membrane structure at specific sites. At certain of these sites, specific proteins (ATPase) are located which catalyze the following reaction (at physiological values of pH):

$$ADP + ⓟ + H^+ = ATP + H_2O$$

The pH gradient drives this reaction resulting in the generation of ATP.

Certain experimental evidence supports the chemiosmotic hypothesis: pH gradients have been shown to result from electron transport, and ATPase activities are associated with membranes.

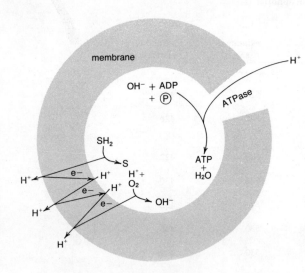

FIGURE 6.27

The chemiosmotic hypothesis in principle, showing the mechanism of extrusion of protons by a respiratory electron transport chain. A pH gradient (ΔpH) between the interior and exterior of the membrane is established as a consequence of the passage of electrons (e^-) from the electron donor (SH_2) to the ultimate electron acceptor (in this case, O_2). Protons can reenter the cell only at specific sites (ATPase) where they drive the reaction which synthesizes ATP. After F. M. Harold, "Chemiosmotic Interpretation of Active Transport in Bacteria," *Ann. N.Y. Acad. of Sci.* **227**, 297 (1974).

FURTHER READING

Books   GUNSALUS, I. C., and R. Y. STANIER (editors), *The Bacteria*, Vol. II. New York: Academic Press, 1961.

LEHNINGER, A. L., *Bioenergetics.* New York: Benjamin, 1965.

————, *Biochemistry.* New York: Worth Publishers, 1970.

MANDELSTAM, J., and K. McQUILLEN (editors), *Biochemistry of Bacterial Growth*, 2nd ed. New York: Wiley, 1973.

SOKATCH, J. R., *Bacterial Physiology and Metabolism.* New York: Academic Press, 1969.

7

MICROBIAL METABOLISM: BIOSYNTHESIS

Despite their mechanistic diversity, all the metabolic pathways discussed in Chapter 6 have the same common function: the provision of ATP (and reduced pyridine nucleotides, where there is a net requirement for reducing power) to drive the reactions of biosynthesis. In this sense, there is a fundamental unity underlying the superficial diversity of catabolic metabolism. This unity of biochemistry, a concept first emphasized by the microbiologist, A. J. Kluyver, in 1926, becomes even more evident when we analyze the ways in which ATP is employed in biosynthesis. In all cells the major end products of biosynthesis are proteins and nucleic acids, and the biochemical reactions leading to their formation show little variation from group to group among procaryotes and even between procaryotes and eucaryotes. There is, accordingly, a *central core of biosynthetic reactions that are similar in all organisms*. A greater degree of diversity occurs in the synthesis of certain other classes of cell constituents, in particular polysaccharides and lipids, since the chemical composition of these substances is often group specific.

In this chapter we will focus attention on the reactions of biosynthesis that are common to most or all organisms. Some more specialized biosynthetic processes, distinctive of procaryotic organisms, will also be discussed.

METHODS OF STUDYING BIOSYNTHESIS

Biochemistry was initially concerned with the elucidation of ATP-generating reactions, many of which (e.g., the fermentations) are chemically fairly simple. The principal technique employed was the direct study of the enzymes involved; from the reactants and products of the individual reactions, the complete reaction sequence was deduced. The technique of *sequential induction* (Chapter 8) has also been used to advantage for the elucidation of inducible pathways. By comparing cells grown on the inducer substrate of the pathway under investigation with cells grown on a substrate which is metabolized through alternate pathways, deductions concerning probable intermediates of the inducible pathway can be made. Because enzymes of an inducible pathway are not synthesized in the absence of the primary inducer substrate and because, through sequential induction, all enzymes of a pathway are synthesized in its presence, probable intermediates of a pathway are identified as those that are immediately metabolized by cells grown on the primary inducer substrate, but are metabolized only after a lag period by cells grown on substrates of alternate pathways. For example, benzoate is the primary inducer and catechol is an intermediate of the β-ketoadipate pathway (Chapter 6). If catechol is added to a suspension of cells of *Pseudomonas putida* which were grown on benzoate, immediate oxidation of catechol ensues; if it is added to cells of the same organism which were grown on asparagine, oxidation begins only after a lag period of about 40 minutes.

The elucidation of biosynthetic mechanisms has come more recently, largely through studies on bacteria. The information gained through these studies, however, was later shown to hold for other organisms. Work on this problem could not even be initiated until the role of ATP as an energetic coupling agent between catabolism and biosynthesis was established. Furthermore, the unraveling of biosynthetic pathways required the development of new techniques which, although helpful, are rarely essential for the analysis of catabolism. The most important of these is the *use of mutants* and the *use of isotopic labeling*.

• **Use of biochemical mutants**   Biochemical mutants (see Chapter 14) became an important tool for the study of biosynthesis after the demonstration in 1940 by G. Beadle and E. Tatum that it is possible to isolate so-called *auxotrophic* mutants. Such mutants require as growth factors biosynthetic intermediates which the parental strain can synthesize *de novo*. Such requirements are caused by the genetic loss of the ability to synthesize, in a functional form, one enzyme mediating a specific step in the affected pathway. The early studies with biochemical mutants led to the hypothesis that each individual enzyme is encoded by a specific gene, which became known as the *one-gene-one-enzyme* hypothesis. Now is it known that there are exceptions: certain genes play exclusively regulatory roles; others encode RNA which is not translated into protein; and some enzymes are composed of dissimilar subunits each of which is encoded by a distinct gene. Still the hypothesis remains a valid and useful generalization.

Biochemical mutants can be utilized in the following ways to determine the sequence of reactions in a biosynthetic pathway:

1. By determining the number of different genes which can undergo mutation resulting in a nutritional requirement for the same growth factor, the number of different enzymatically catalyzed reactions in the pathway of biosynthesis of that growth factor can be determined. For example, mutations in eight different genes lead to a requirement (auxotrophy) for the amino acid arginine; hence, there are eight different enzymatically catalyzed reactions in the arginine pathway (Figure 7.1).

2. Genetic blockades in a pathway tend to cause the accumulation and excretion into the medium of metabolic intermediates prior to the blockade. These

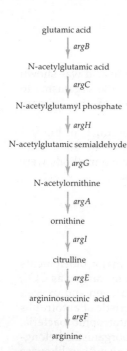

glutamic acid

↓ argB

N-acetylglutamic acid

↓ argC

N-acetylglutamyl phosphate

↓ argH

N-acetylglutamic semialdehyde

↓ argG

N-acetylornithine

↓ argA

ornithine

↓ argI

citrulline

↓ argE

argininosuccinic acid

↓ argF

arginine

FIGURE 7.1

Reaction sequence leading to the biosynthesis of arginine in *Salmonella typhimurium*. The designations of the genes which encode the various enzymes are written to the right of the arrows.

intermediates sometimes allow the growth of other mutant strains blocked in the same pathway at an earlier step; thus, the sequence of blockades in a series of mutant strains can be determined. For example, strains with mutations in the gene *argI* excrete ornithine (Figure 7.2) which can be utilized by strains blocked at the earlier steps under the control of genes *B* and *H*. In addition, the intermediates excreted by the mutant strains can be chemically isolated and identified.

3. Information on the sequence of reactions in a biosynthetic pathway can also be obtained by testing the growth response of mutant strains to suspected intermediates of the pathway being investigated. For example, an *argI* mutant strain will grow if arginine or citrulline is added to the medium; an *argA* strain will grow if arginine, citrulline, or ornithine is added. From such experiments citrulline and ornithine would appear to be probable intermediates of the arginine pathway.

arg I

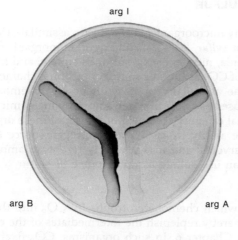

arg B   arg A

FIGURE 7.2

Strains of *Salmonella typhimurium* carrying mutations in genes *argB, argA,* and *argI* (Figure 7.1) were streaked adjacent to one another on a plate lacking arginine. Since all three strains are genetically incapable of synthesizing arginine, they would be unable to grow if streaked alone on such a plate; the *argI* strain excretes ornithine into the medium, however, which allows the *argB* and *argA* strains to grow in that region. From such an experiment one can conclude that *argB* and *argA* encode enzymes which catalyze steps of the arginine pathway prior to that encoded by *argI*. Similarly, the *argA* strain excretes an intermediate allowing growth of the *argB* strain.

• Use of isotopic labeling

When a biosynthetic intermediate (e.g., an amino acid) is added to a growing population of cells, it will often prevent its own endogenous synthesis (the mechanism by which this control is effected is discussed in Chapter 8). The exogenously furnished compound is, therefore, preferentially incorporated by the cell into biosynthetic end products. If the exogenously furnished compound is labeled with a radioisotope, chemical fractionation of the labeled cells can reveal the ultimate location of radioactivity in the various cell constituents. Such experiments show, for example, that ^{14}C-labeled glutamic acid is incorporated into protein not only as glutamic acid residues but also as residues of two other amino acids, arginine and proline. This result demonstrates that glutamic acid is a biosynthetic precursor of arginine and proline.

Another valuable technique employing radioisotopes is *pulse labeling*. A growing culture is briefly exposed to a radioactive biosynthetic precursor. During this exposure a small quantity of the precursor enters the cell and starts to be distributed through the various pathways in which it participates. If samples of cell material are subjected to chemical fractionation at various times after pulse labeling, the sequence of chemical transformations in pathways leading from the radioactive precursor is revealed. The pathway for the conversion of CO_2 to organic compounds by photosynthetic organisms and chemoautotrophs was largely established by experiments of this kind; phosphoglyceric acid was shown to be the first compound to be labeled following exposure of the organisms to radioactive CO_2; hence, it was established as the product of the primary CO_2-utilizing reaction.

Radioisotopic methods are also valuable for detecting the products of biosynthetic reactions in cell-free systems, in which the reaction rates are sometimes too low for ordinary chemical methods to be feasible. Radioactive methods were indispensable in the early studies on the synthesis of protein.

THE ASSIMILATION OF INORGANIC CARBON, NITROGEN, AND SULFUR

Many microorganisms are able to assimilate the major bioelements (i.e., *carbon, nitrogen, sulfur, hydrogen,* and *oxygen*) in inorganic form: carbon as CO_2; nitrogen as ammonia, nitrate, or N_2; sulfur as sulfate; and hydrogen and oxygen as water. The use of CO_2 as the sole source of carbon is characteristic of autotrophs including the algae and many photosynthetic and chemoautotrophic bacteria. Ammonia can be used as a sole nitrogen source by many microorganisms belonging to all nutritional categories; some, but not all, of these organisms can likewise utilize nitrate. The ability to use N_2 as a nitrogen source (nitrogen fixation) is restricted to procaryotes, and it is relatively rare even among this group. Most microorganisms can use sulfate as their source of sulfur.

• **The assimilation of CO_2**

The reactions by which chemoheterotrophs fix CO_2 into the carboxyl group of oxalacetate, and thereby replenish the intermediates of the citric acid cycle, have been described in Chapter 6. In such organisms, CO_2 fixation contributes only a minor fraction of the cellular carbon; most is supplied from the organic carbon source. Autotrophs, for which CO_2 serves as the sole or principal source of cellular carbon, fix CO_2 by a different reaction catalyzed by *ribulose diphosphate carboxylase* (Figure 7.3). The primary product of CO_2 fixation is glyceric acid 3-phosphate, from which all other organic molecules of the cell are synthesized. However, CO_2 fixation is dependent on a supply of the second substrate (ribulose diphosphate)

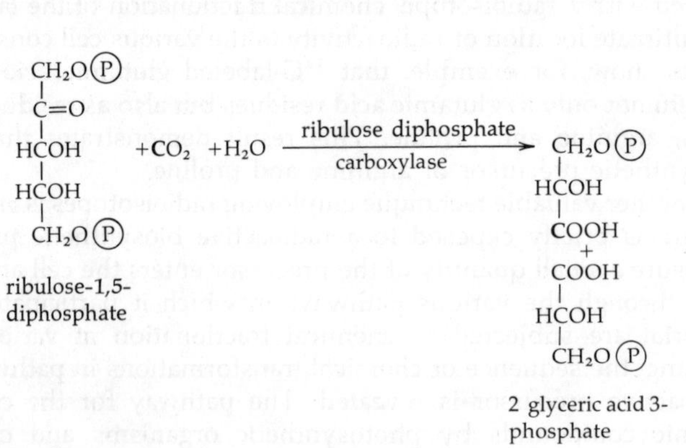

FIGURE 7.3

The CO_2-fixing reaction of autotrophs. The reaction is catalyzed by *ribulose diphosphate carboxylase*. The phosphorylated pentose, ribulose-1,5-diphosphate, accepts 1 molecule of CO_2 and is simultaneously cleaved yielding 2 molecules of glyceric acid-3-phosphate; the carboxyl group of one glyceric acid-3-phosphate molecule is thus derived from CO_2.

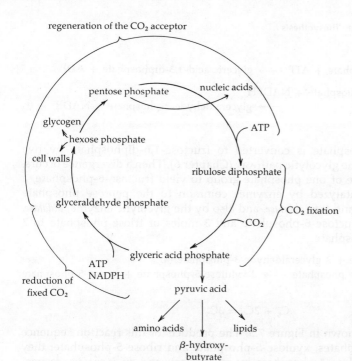

regeneration of the CO₂ acceptor

pentose phosphate — nucleic acids

glycogen

hexose phosphate

cell walls

ATP

ribulose diphosphate

glyceraldehyde phosphate

CO₂ fixation

CO₂

ATP
NADPH

glyceric acid phosphate

reduction of
fixed CO₂

pyruvic acid

amino acids · lipids

β-hydroxy-
butyrate

FIGURE 7.4

The cyclic mechanism of the Calvin cycle, illustrating the sources of major cellular constituents and the three phases of the cycle: CO_2 fixation, reduction of fixed CO_2, and regeneration of the CO_2 acceptor.

of the fixation reaction. Consequently, most of the glyceric acid phosphate must be utilized to regenerate ribulose diphosphate. Thus, the process of CO_2 fixation is cyclic, each turn of the cycle resulting in the fixation of one molecule of CO_2. Various intermediates of the cycle are drawn off and enter various biosynthetic pathways.

This mechanism of CO_2 fixation was first elucidated in a green alga by M. Calvin, A. Benson, and J. Bassham and is sometimes called the *Calvin cycle*. Subsequent work has shown it to be universal among autotrophs.

The Calvin cycle is complex, sharing certain reactions of the glycolytic and pentose phosphate pathways (Chapter 6). Only two reactions are specific to the cycle: the CO_2 fixation reaction itself, and the reaction which generates the CO_2 acceptor (ribulose 1,5-diphosphate) from its immediate precursor, ribulose-5-phosphate.

For simplicity of analysis, the Calvin cycle can be divided into three phases: *CO_2 fixation, reduction of fixed CO_2, and regeneration of the CO_2 acceptor* (Figure 7.4).

• **CO₂ fixation**　As discussed, CO_2 fixation is a consequence of a single reaction producing 2 molecules of glyceric acid phosphate (Figure 7.3).

• **Reduction of fixed CO₂**　In order to be utilized in biosynthesis, CO_2 fixed in the carboxylic acid group of glyceric acid phosphate must first be reduced to the oxidation level of the major cellular constituents. This occurs in two steps by reactions common to glycolysis (Chapter 6) but operating in the reverse direction.

191

glyceric acid-3-phosphate + ATP $\longrightarrow$ glyceric acid-1,3-diphosphate + ADP

glyceric acid-1,3-diphosphate + NADPH

$\longrightarrow$ glyceraldehyde-3-phosphate + NADP$^+$ + $\circled{P}$

• Regeneration of
the CO_2 acceptor

Glyceraldehyde-3-phosphate is converted to fructose-1,6-diphosphate by two enzymes common to the glycolytic pathway (Chapter 6). Then a divergence occurs. Following the cleavage of one phosphate group to yield fructose-6-phosphate, a series of reactions catalyzed by enzymes common to the pentose phosphate pathway, *transketolase* and *transaldolase,* and also by the glycolytic enzyme, *aldolase,* converts 1 mole of fructose-6-phosphate and 3 moles of triose-phosphate to 3 moles of pentose-phosphate.

Fructose-6-phosphate + 2 glyceraldehyde-3-phosphate +

dihydroxyacetone phosphate $\longrightarrow$ 2 xylulose-5-phosphate + ribose-5-phosphate

Schematically,

$$C_6 + 3C_3 = 3C_5$$

These reactions are shown in Figure 7.5. The products of this reaction sequence are the pentose phosphates, xylulose-5-phosphate and ribose-5-phosphate; they are converted to ribulose-5-phosphate by an *epimerase* and an *isomerase,* respectively. Ribulose-5-phosphate is then phosphorylated by the second enzyme spe-

FIGURE 7.5

The interconversion of sugar phosphates by transketolase (TK), transaldolase (TA), and aldolase (A), which results in the formation of 3 moles of pentose phosphates from 1 mole of fructose-6-phosphate and 3 moles of triosephosphates. Reactants of the sequence are in boldfaced type; products are in boxes.

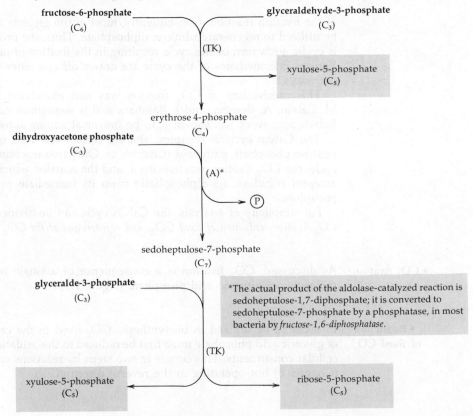

fructose-6-phosphate
(C$_6$)

glyceraldehyde-3-phosphate
(C$_3$)

(TK)

xylulose-5-phosphate
(C$_5$)

erythrose 4-phosphate
(C$_4$)

dihydroxyacetone phosphate
(C$_3$)

(A)*

$\circled{P}$

sedoheptulose-7-phosphate
(C$_7$)

glyceralde-3-phosphate
(C$_3$)

*The actual product of the aldolase-catalyzed reaction is sedoheptulose-1,7-diphosphate; it is converted to sedoheptulose-7-phosphate by a phosphatase, in most bacteria by *fructose-1,6-diphosphatase.*

(TK)

xylulose-5-phosphate
(C$_5$)

ribose-5-phosphate
(C$_5$)

cific to the Calvin cycle, *phosphoribulokinase;* the acceptor compound of the CO_2-fixing reaction is thus regenerated.

$$\text{ribulose-5-phosphate} + \text{ATP} \longrightarrow \text{ribulose-1,5-diphosphate} + \text{ADP}$$

One molecule of ribulose-1,5-diphosphate must be regenerated for each molecule of CO_2 fixed. Although ATP is not required in the reaction in which CO_2 is fixed, 3 molecules of ATP are required to regenerate ribulose-1,5-diphosphate: 2 molecules to phosphorylate glyceric acid 3-phosphate and 1 molecule to phosphorylate ribulose-5-phosphate. In addition, 2 molecules of NADPH are required to reduce the 2 molecules of glyceric acid 1,3-diphosphate in the reductive steps of the cycle. Consequently, the fixation of a molecule of CO_2 requires the expenditure of 3 molecules of ATP and 2 molecules of NADPH (see Figure 7.4).

• The assimilation of ammonia

The nitrogen atom of ammonia (valence of -3) is at the same oxidation level as the nitrogen atoms in the organic constituents of the cell. The assimilation of ammonia does not, therefore, necessitate oxidation or reduction. There are 3 NH_3 fixation reactions: one forming the amino group of glutamic acid,

$$\text{HOOC—(CH}_2)_2\text{—C—COOH} + \text{NH}_3 + \text{NADH} + \text{H}^+$$
(α-ketoglutaric acid)

$$\xrightarrow{\text{glutamate dehydrogenase}} \text{HOOC—(CH}_2)_2\text{—C—COOH} + \text{NAD}^+ + \text{H}_2\text{O}$$
(glutamic acid)

and two others forming the amido group of asparagine and glutamine,

$$\text{HOOC—CH}_2\text{—CHNH}_2\text{—COOH} + \text{NH}_3 + \text{ATP}$$
(aspartic acid)

$$\xrightarrow{\text{asparagine synthetase}} \underset{\text{NH}_2}{\text{C}}\text{—CH}_2\text{—CHNH}_2\text{—COOH} + \text{AMP} + \text{P—P}$$
(asparagine)

and,

$$\text{HOOC—(CH}_2)_2\text{—CHNH}_2\text{—COOH} + \text{ATP} + \text{NH}_3$$
(glutamic acid)

$$\xrightarrow{\text{glutamine synthetase}} \underset{\text{NH}_2}{\text{C}}\text{—(CH}_2)_2\text{—CHNH}_2\text{—COOH} + \text{ADP} + \text{P}$$
(glutamine)

All three products of NH_3 fixation (glutamic acid, asparagine, and glutamine) are direct precursors of proteins, and asparagine serves only in this role. However, both glutamic acid and glutamine play additional roles as agents for the transfer of amino and amido groups to all other nitrogenous precursors of cellular macromolecules. For example, the amino acids alanine, aspartic acid, and phenylalanine are formed by transamination between glutamic acid and non-nitrogenous metabolites, i.e.,

$$\text{L-glutamic acid} + \text{pyruvic acid} \longrightarrow \alpha\text{-ketoglutaric acid} + \text{L-alanine}$$

$$\text{L-glutamic acid} + \text{oxalacetic acid} \longrightarrow \alpha\text{-ketoglutaric acid} + \text{L-aspartic acid}$$

$$\text{L-glutamic acid} + \text{phenylpyruvic acid} \longrightarrow \alpha\text{-ketoglutaric acid} + \text{L-phenylalanine}$$

and the amido group of glutamine is the source of the amino groups of cytidine triphosphate, carbamyl phosphate, NAD, and guanosine triphosphate, among others; e.g.,

$$\text{uridine triphosphate} + \text{glutamine} + \text{ATP}$$
$$\longrightarrow \text{cytidine triphosphate} + \text{glutamic acid} + \text{ADP} + \textcircled{P}$$

The pathways of synthesis of glutamic acid and glutamine depend on the concentration of NH_3 available in the cell. At high concentrations of NH_3, the two sequential reactions, catalyzed by a dehydrogenase and glutamine synthetase, lead to the synthesis of these two compounds:

$$\alpha\text{-ketoglutaric acid} \xrightarrow{NH_3} \text{glutamic acid} \xrightarrow{NH_3} \text{glutamine}$$

However, the substrate affinity of α-ketoglutaric dehydrogenase for NH_3 is relatively low; consequently, this enzyme ceases to function effectively at low concentrations of NH_3, and the above pathway becomes inoperative. Under these conditions, a new enzyme, *glutamate synthase*, sometimes called GOGAT (an acronym for the alternate name glutamine-oxyglutarate amino transferase) is induced which catalyzes the reaction:

$$\text{glutamine} + \alpha\text{-ketoglutaric acid} \rightleftharpoons 2 \text{ glutamic acid}$$

Under these conditions, the glutamine synthetase reaction becomes the major route of NH_3 assimilation, i.e., instead of being synthesized by glutamate dehydrogenase, glutamic acid is synthesized by the reaction sequence:

$$\text{glutamic acid} + NH_3 + \text{ATP} \xrightarrow{\text{glutamine synthetase}} \text{glutamine} + \text{ADP} + \textcircled{P}$$

$$\text{glutamine} + \alpha\text{-ketoglutaric acid} \xrightarrow{\text{GOGAT}} 2 \text{ glutamic acid}$$

NET REACTION: $\alpha\text{-ketoglutaric acid} + NH_3 + \text{ATP} \longrightarrow \text{glutamic acid} + \text{ADP} + \textcircled{P}$

• The assimilation of nitrate   Nitrate ion (NO_3^-) is used by many microorganisms as a source of nitrogen. The valence of the nitrogen atom in NO_3^- is $+5$; consequently, its assimilation involves reduction to the level of -3, by preliminary conversion to ammonia.

As was discussed in Chapter 6, nitrate is also reduced when, in certain microorganisms, it serves as a terminal electron acceptor for anaerobic respiration. Some microorganisms that use nitrate as a nitrogen source cannot use it in anaerobic respiration, because these two processes which reduce nitrate are catalyzed by different enzyme systems.

The process of *assimilatory nitrate reduction* is mediated by two enzyme complexes called *nitrate reductase* and *nitrite reductase*:

$$NO_3^- \xrightarrow[\text{reductase}]{\text{nitrate}} NO_2^- \xrightarrow[\text{reductase}]{\text{nitrite}} NH_3$$

Assimilatory nitrate reductases from bacteria, as well as those from plants, specifically require NADH as the source of electrons; those from fungi preferentially utilize NADPH. However, except for this difference, the nitrate reductases from all organisms are similar. They are high molecular weight complexes of two separable proteins. The first is an FAD-dependent protein which accepts electrons from reduced pyridine nucleotides and transfers them to the second protein which contains molybdenum (Mo), thereby reducing it from Mo^{6+} to Mo^{5+}. The second protein effects the reduction of nitrate to nitrite. The pathway of electron transfer can be schematically represented as:

Assimilatory nitrite reductase from bacteria also utilizes NADH as the electron donor and requires FAD for activity. In algae and higher plants this enzyme functions with ferredoxin instead of FAD. The reduction of nitrite to ammonia involves the addition of six electrons, and it has frequently been claimed that reduction proceeds by a series of two-electron transfers with the production of discrete intermediates, notably hydroxylamine. However, current evidence indicates that nitrite is directly reduced to NH_3, no free intermediates being formed.

The reduction of nitrate to ammonia is thus a relatively complex process, involving the action of several proteins. This explains why many microorganisms that grow readily with ammonia cannot use nitrate as an alternative inorganic nitrogen source. It should be noted that molybdenum plays an essential role in assimilatory nitrate reduction. This fact was suspected long before the biochemical mechanism of nitrate reduction was known, as a result of the nutritional finding that molybdenum is essential for microbial growth with nitrate but not with ammonia. This element also plays an important role in the fixation of N_2, and is therefore an essential nutrient for all organisms growing at the expense of the oxidized forms of nitrogen: nitrate or N_2.

• The assimilation of molecular nitrogen

Gaseous nitrogen (N_2) with a valence of 0 must also be reduced to ammonia prior to incorporation into nitrogenous components of the cell. This process, called *nitrogen fixation*, is limited to procaryotes.

Although the ability of certain bacteria, both free-living and symbiotic, to fix N_2 has been recognized for about 100 years, attempts to elucidate the biochemical mechanism of N_2 fixation were long frustrated by the difficulty of preparing active cell-free extracts. This was accomplished by L. Mortenson and his associates who first established the peculiar properties now known to be common to all N_2-fixing enzyme systems: (1) their extreme sensitivity to irreversible inactivation by low concentrations of O_2 and (2) their requirements for ATP, which must be supplied continuously by an ATP-generating system, because the enzyme is inhibited by high concentrations of ATP.

The enzyme system responsible for N_2 fixation, called *nitrogenase*, also requires a source of electrons. Irrespective of their ultimate source, they must be furnished to the nitrogenase system through a low potential reductant, the nonheme iron electron carrier, ferredoxin (Fd):

$$\text{source of electrons} \longrightarrow \text{Fd} \cdot 2e^- \overset{\displaystyle\frown}{\underset{\displaystyle\smile}{}} \text{(AzoFd)}_2 \cdot 2e^- \overset{\displaystyle\frown}{\underset{\displaystyle\smile}{}} \text{MoFd} \cdot 2e^- \overset{\displaystyle\frown}{\underset{\displaystyle\smile}{}} 2\text{NH}_3$$

The electron transport chain of ferredoxin (Fd), azoferredoxin (AzoFd), and molybdoferredoxin (MoFd) transfers only two electrons at each step (the final transfer requiring the expenditure of 1 molecule of ATP); nevertheless, six electrons are required to reduce N_2 to ammonia. Therefore, the reaction must involve three sequential two-electron steps. However, the partially reduced intermediates have never been detected in the reaction mixture. It is likely that these intermediates remain bound to the enzyme and that they are:

$$\text{enzyme-N}\equiv\text{N} \xrightarrow[2\text{H}^+]{2e^-} \text{enzyme-N}\overset{\text{H}}{=}\overset{\text{H}}{\text{N}} \xrightarrow[2\text{H}^+]{2e^-} \text{enzyme-}\overset{\text{H}\ \ \text{H}}{\underset{\text{H}\ \ \text{H}}{\text{N--N}}} \xrightarrow[2\text{H}^+]{2e^-} 2\text{NH}_3 + \text{enzyme}$$

The substrate specificity of nitrogenase is relatively low; a number of other compounds, including N_3^-, N_2O, HCN, CH_3NC, CH_2CHCN, and C_2H_2 are also reduced by it. Some of these reductions involve the transfer of only two electrons rather than the six required to reduce N_2. The proposed mechanism of the reaction suggests that such two-electron reductions should proceed at three times the rate of the reduction of N_2, and in most cases this is true.

The study of biological nitrogen fixation both in whole cells and in extracts has been greatly aided by the introduction of an assay method using the substrate, acetylene, which is reduced to ethylene

$$\text{CH}\equiv\text{CH} \xrightarrow[2\text{H}^+]{2e^-} \text{CH}_2{=}\text{CH}_2$$

The product can be easily quantitated by gas chromotography, and the reaction is a highly specific one since no enzyme system other than nitrogenase can affect this reduction.

• The assimilation of sulfate

The great majority of microorganisms can fulfill their sulfur requirements from sulfate. Sulfate with a valence of +6 is reduced to sulfide (valence −2) prior to its incorporation into cellular organic compounds. Chemically, this is equivalent to the reduction of sulfate by the sulfate-reducing bacteria which use it as the terminal electron acceptor in anaerobic respiration, as discussed in Chapter 6. The enzymatic mechanisms are different, however; the reduction of sulfate for use as a sulfur source is termed *assimilatory sulfate reduction* (by analogy with assimilatory nitrate reduction) to distinguish it from *dissimilatory sulfate reduction*, the use of sulfate as a terminal electron acceptor.

The pathway of assimilatory sulfate reduction to H_2S is outlined in Figure 7.6. The initial two-electron reduction of sulfate occurs only after it has been converted to an activated form, adenylylsulfate, by a series of three enzymatic steps requiring the expenditure of three high-energy phosphate bonds. The final six-electron reduction is catalyzed by a huge, complex flavometallo-protein, *sulfite reductase*. Sulfite reductase from *E. coli* has a molecular weight of 750,000 and contains 4 FAD, 4 FMN, and 12 Fe prosthetic groups.

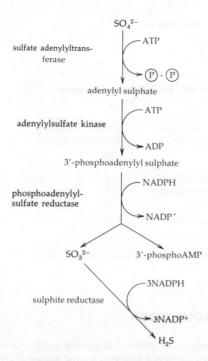

FIGURE 7.6

The assimilatory reduction of sulfate to produce H_2S for use in biosynthetic reactions.

THE STRATEGY OF BIOSYNTHESIS

On a weight basis most of the organic matter of the cell consists of macromolecules which belong to four classes: nucleic acids, proteins, polysaccharides, and complex lipids. These macromolecules are polymers of lower molecular weight organic precursors. Each class of macromolecules is defined by the chemical type of precursors that are polymerized to form it: nucleotides in the case of nucleic acids, amino acids in the case of proteins, and simple sugars (monosaccharides) in the case of polysaccharides. Complex lipids are more variable and heterogeneous in composition; their precursors include fatty acids, polyalcohols, simple sugars, amines, and amino acids. As shown in Table 7.1, approximately 70 different kinds of precursors are required to synthesize the four major classes of macromolecules.

In addition to the precursors of macromolecules, the cell must synthesize a number of compounds that play catalytic roles. These include about 20 coenzymes and electron carriers.

In all, about 150 different small molecules are required to produce a new cell. These small molecules are, in turn synthesized from a smaller number of central intermediary metabolites formed in the course of catabolism by heterotrophs (Chapter 6) or of CO_2 assimilation by autotrophs. The most important of these intermediates are sugar phosphates, pyruvate, acetate, oxalacetate, succinate, and α-ketoglutarate.

In the following pages we shall trace the pathways of biosynthesis of small molecules from central intermediary metabolites. In the concluding section of the chapter we shall discuss the processes by which they are polymerized into macromolecules.

TABLE 7.1

Classes of macromolecules of the cell and their component building blocks

MACROMOLECULE	CHEMICAL NATURE OF BUILDING BLOCKS	NUMBER OF KINDS OF BUILDING BLOCKS
Nucleic acids		
RNA	Ribonucleotides	4
DNA	Deoxyribonucleotides	4
Proteins	Amino acids	20
Polysaccharides	Monosaccharides	~15[a]
Complex lipids	Variable	~20[a]

[a] The number of building blocks in any particular representative of these macromolecules is usually much smaller.

THE SYNTHESIS OF NUCLEOTIDES

The precursors of nucleic acids are purine and pyrimidine nucleoside triphosphates, all of which have the same general structure. A purine or pyrimidine base is attached through nitrogen atoms to a pentose; this combination is called a *nucleoside*. Phosphate groups are attached to the 5' position of the nucleoside (to distinguish between the base and pentose moieties of a nucleoside, positions on the pentose are assigned a prime following the number). This combination is called a *nucleotide*. The general structure of nucleoside triphosphates is shown in Figure 7.7. The names and structures of specific nucleosides are shown in Figure 7.8. Nucleotides are symbolized by letters, A, G, U, C, or T, to indicate the purine or pyrimidine base they contain; MP, DP, or TP indicates whether they are mono-, di- or triphosphates. Deoxynucleotides are indicated by a d (e.g., CDP symbolizes cytidine diphosphate, and dGTP symbolizes 2'deoxyguanosine triphosphate). The two purine (dATP and dGTP) and two pyrimidine (dCTP and dTTP) nucleoside triphosphates containing deoxyribose are the specific precursors of DNA; the two purine (ATP and GTP) and two pyrimidine (CTP and UTP) nucleoside triphosphates containing ribose are specific precursors of RNA. Some of these nucleoside triphosphates also serve as activators (Chapter 6, Table 6.1) and thus play dual roles.

Deoxyribonucleotides are formed by the reduction of the corresponding ribonucleotides. The pathways of synthesis of ribonucleotides will, therefore, be considered first; later the manner by which ribonucleotides are converted to deoxyribonucleotides will be considered.

• **Synthesis of ribonucleotides**

The ribose-phosphate moiety of all ribonucleotides is derived from the same precursor, 5-phosphoribosyl-1-pyrophosphate (PRPP) which, in turn, is synthesized from ribose-5-phosphate (an intermediate of the pentose phosphate pathway) and ATP:

$$\text{ribose-5-phosphate} + \text{ATP} \xrightarrow[\text{synthetase}]{\text{PRPP}} \text{PRPP} + \text{AMP}$$

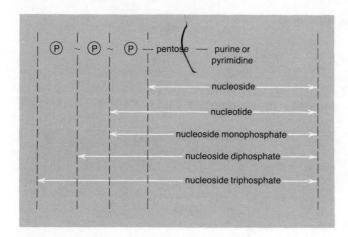

FIGURE 7.7

The general structure of nucleoside triphosphates. High-energy (anhydride) phosphate bonds are symbolized by a wavy line (∼); low-energy (ester) phosphate bonds are symbolized by a straight line (−).

FIGURE 7.8

Names and composition of nucleoside triphosphates. Purines at the 9 position, and pyrimidines at the 3 position, are attached to the 1 position of pentoses to form nucleosides.

BASE		RIBONUCLEOSIDES		2′DEOXYRIBONUCLEOSIDES	
Name	Base structure	Pentose structure	Name	Pentose structure	Name
purines					
adenine	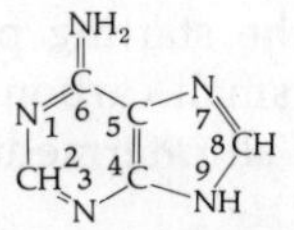		adenosine		2′-deoxyadenosine
guanine	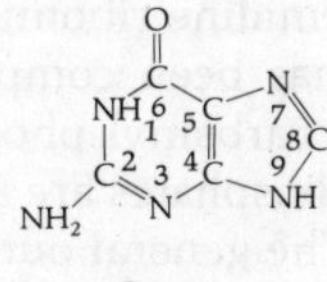	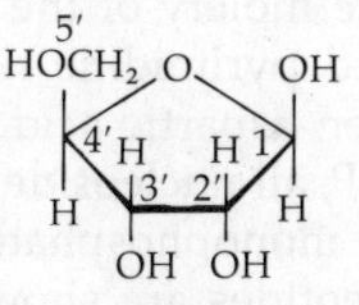	guanosine	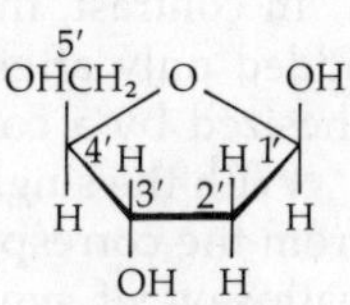	2′-deoxyguanosine
		ribose		2′-deoxyribose	
pyrimidines					
uracil	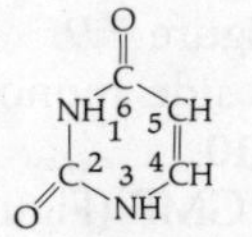		uridine		————
cytosine	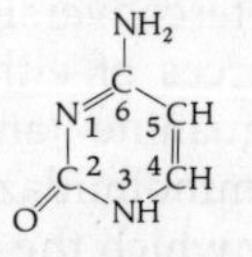		cytidine		2′-deoxycytidine
thymine	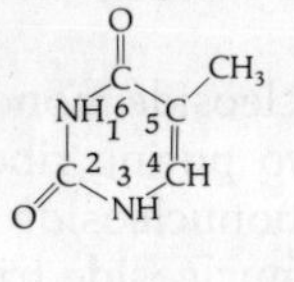				2′-deoxythymidine

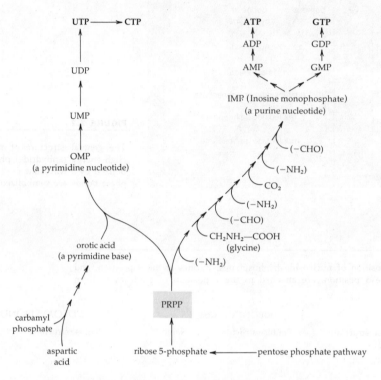

FIGURE 7.9

The general outlines of the pathways of synthesis of purine and pyrimidine ribonucleoside triphosphates.

In the case of the purine ribonucleotides, PRPP is the starting point of the pathway. By successive additions of amino groups and small carbon-containing groups, the nine-membered purine ring is synthesized, all intermediates of the pathway being ribonucleotides (Figure 7.9).

In contrast, the ribose-phosphate moiety of the pyrimidine ribonucleotides is added only after the six-membered pyrimidine ring has been completely synthesized by a condensation between aspartic acid and carbamyl phosphate.

With the single exception of CTP, all nucleoside triphosphates are synthesized from the corresponding nucleoside monophosphates. The general outlines of the pathways of synthesis of ribonucleotides are shown in Figure 7.9.

The detailed reactions by which the purine ribonucleoside monophosphates (AMP and GMP) are synthesized are shown in Figure 7.10.

Although the reactions leading from IMP to AMP and GMP (Figure 7.10) are irreversible, ancillary pathways exist which permit the interconversion of GMP and AMP through IMP (Figure 7.11). Thus, external sources of either guanine or adenine can satisfy the cell's requirement for both guanine- and adenine-containing nucleotides. The pathway between ATP and aminoimidazole carboxamide ribotide (AICAR) is also common to the pathway by which the amino acid, histidine, is synthesized (Figure 7.26).

The detailed reactions by which the pyrimidine ribonucleoside monophosphate, UMP, is synthesized are shown in Figure 7.12. The two purine ribonucleoside monophosphates, AMP and GMP, and the pyrimidine ribonucleoside monophosphate, UMP, are the precursors of the four essential ribonucleoside triphosphates

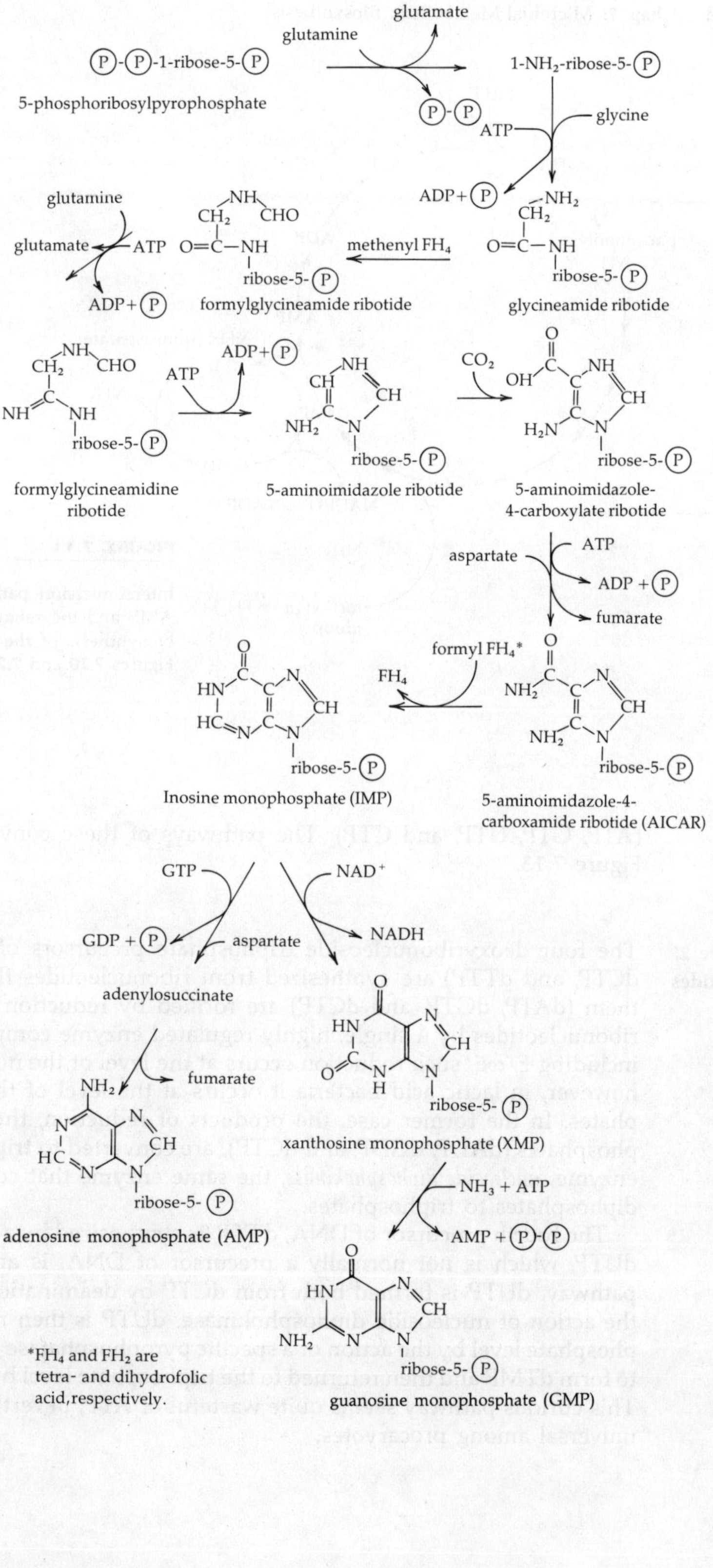

FIGURE 7.10

The biosynthesis of the purine ribonucleotides, AMP and GMP.

*FH₄ and FH₂ are tetra- and dihydrofolic acid, respectively.

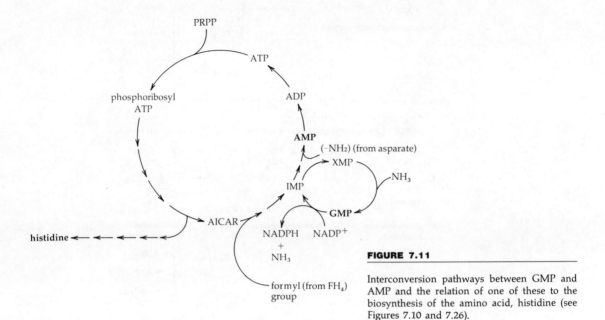

FIGURE 7.11

Interconversion pathways between GMP and AMP and the relation of one of these to the biosynthesis of the amino acid, histidine (see Figures 7.10 and 7.26).

(ATP, GTP, UTP, and CTP). The pathways of these conversions are shown in Figure 7.13.

• **Synthesis of the 2′ deoxyribonucleotides**

The four deoxyribonucleoside triphosphate precursors of DNA (dATP, dGTP, dCTP, and dTTP) are synthesized from ribonucleotides (Figure 7.14). Three of them (dATP, dGTP, and dCTP) are formed by reduction of the corresponding ribonucleotides by a single, highly regulated enzyme complex. In most bacteria, including *E. coli,* such reduction occurs at the level of the nucleoside diphosphate; however, in lactic acid bacteria it occurs at the level of the nucleoside triphosphates. In the former case, the products of reduction, the deoxynucleoside diphosphates (dADP, dGDP, and dCTP), are converted to triphosphates by a single enzyme, *nucleoside diphosphokinase,* the same enzyme that converts ribonucleoside diphosphates to triphosphates.

The fourth precursor of DNA, dTTP, is synthesized by a more circuitous route; dUTP, which is not normally a precursor of DNA, is an intermediate of this pathway. dUTP is formed both from dCTP by deamination and from dUDP by the action of nucleoside diphosphokinase. dUTP is then returned to the monophosphate level by the action of a specific pyrophosphatase before it is methylated to form dTMP and then returned to the triphosphate level by two kinase reactions. This curious pathway seems quite wasteful of ATP; nevertheless, it is apparently universal among procaryotes.

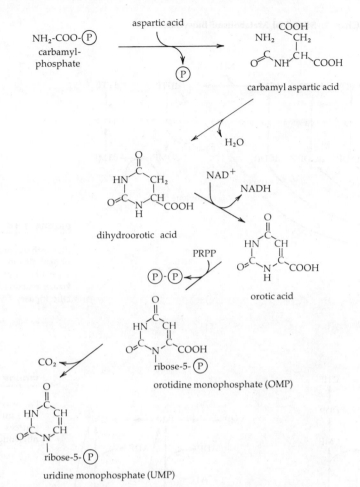

FIGURE 7.12

Biosynthesis of the pyrimidine ribonucleotide, UMP.

• **Utilization of exogenous purine and pyrimidine bases and nucleosides**

Most, but not all, bacteria are able to carry out the synthesis of all nucleoside triphosphates by the pathways outlined in Figures 7.10, 7.11, 7.12, 7.13, and 7.14. They are also able to utilize purines and pyrimidines in the form of free bases as well as nucleosides, when these compounds are supplied in the medium. The pathways by which these compounds are utilized when supplied exogenously have

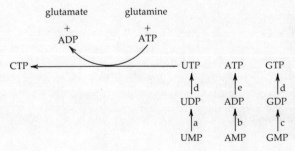

FIGURE 7.13

Biosynthesis of ribonucleoside triphosphates from UMP, AMP, and GMP. Reactions a, b, and c are catalyzed by three specific kinases; reactions labeled d are catalyzed by a nonspecific kinase, nuleoside diphosphokinase. Reaction e symbolizes the many ATP-yielding reactions discussed in Chapter 6.

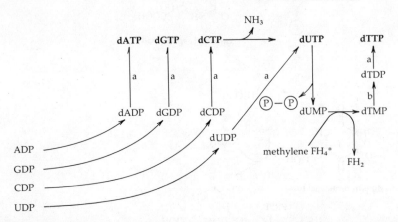

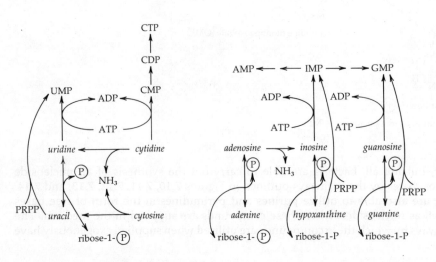

*FH₄ and FH₂ are tetra- and dihydrofolic acid, respectively.

FIGURE 7.14

Biosynthesis of deoxyribonucleoside triphosphates in *E. coli*. Reactions labeled a are all catalyzed by *nucleoside diphosphokinase;* reaction b is catalyzed by a specific kinase, *TMP kinase.*

FIGURE 7.15

Pathways in enteric bacteria for the utilization of exogenous sources of purine and pyrimidine nucleotides.

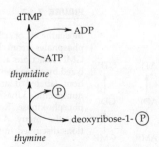

been called *salvage pathways*. Although there are only minor variations among bacteria with respect to the *de novo* pathways of nucleotide biosynthesis, there are considerable variations with respect to the salvage pathways. The nucleotide salvage pathways found in enteric bacteria are shown in Figure 7.15.

THE SYNTHESIS OF AMINO ACIDS AND OTHER NITROGENOUS CELL CONSTITUENTS

Twenty amino acids are required for the biosynthesis of proteins. Only one amino acid, histidine, has a completely isolated biosynthetic origin. The other 19 are derived through branched pathways from a relatively small number of central intermediary metabolites. They can be grouped, in terms of biosynthetic origin, into a total of five "families," as shown in Table 7.2. In addition, certain other nitrogenous cell constituents that do not enter into the synthesis of protein are also derived from these pathways (Table 7.3). We shall describe in a summary manner the pathways involved.

• The glutamate family

We have already discussed the origin of two members of the glutamate family (glutamic acid and glutamine) in the context of ammonia assimilation. The other two members of the glutamate family, proline and arginine, are synthesized from glutamic acid by separate pathways (Figure 7.16).

TABLE 7.2

Biosynthetic derivations of amino acids

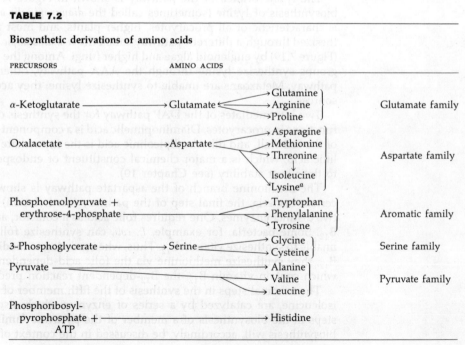

PRECURSORS		AMINO ACIDS	
α-Ketoglutarate ⟶	Glutamate ⟵	Glutamine → Arginine → Proline	Glutamate family
Oxalacetate ⟶	Aspartate ⟨	Asparagine → Methionine → Threonine ↓ Isoleucine Lysine[a]	Aspartate family
Phosphoenolpyruvate + erythrose-4-phosphate ⟶		Tryptophan → Phenylalanine → Tyrosine	Aromatic family
3-Phosphoglycerate ⟶	Serine ⟨	Glycine → Cysteine	Serine family
Pyruvate ⟶		Alanine → Valine → Leucine	Pyruvate family
Phosphoribosyl- pyrophosphate + ATP ⟶		Histidine	

[a] In certain algae and fungi, lysine is synthesized from α-ketoglutarate (see text).

TABLE 7.3

Derivation of other nitrogenous cell constituents from the
pathways of amino acid biosynthesis

PATHWAY (FAMILY)	OTHER NITROGENOUS PRODUCTS
Glutamate[a]	Polyamines
Aspartate[a]	Diaminopimelic acid, dipicolinic acid
Aromatic	p-Hydroxybenzoic acid, p-aminobenzoic acid
Serine	Purines, porphyrins
Pyruvate	Pantothenic acid

[a] In addition, glutamate, glutamine, and aspartate serve as amino donors
in a number of biosynthetic pathways.

• The aspartate family

The parent amino acid of the aspartate family, aspartic acid, arises by transamination of oxalacetate and can be further aminated to yield the amide asparagine, in a reaction analogous to the formation of glutamine from glutamate. The other amino acids belonging to this family are formed through a branched pathway. The common core of the pathway leading to the synthesis of threonine, and the branch points leading to lysine, methionine, and isoleucine, are shown in Figure 7.17.

The lysine branch of the pathway is shown in Figure 7.18. This pathway of biosynthesis of lysine (sometimes called the *diaminopimelic acid* or *DAP pathway*) is characteristic of all procaryotes, higher plants, and most algae. Lysine is synthesized through a different pathway called the *α-aminoadipic acid* or *AAA pathway* (Figure 7.19) by euglenoid algae and higher fungi. Among the phycomycetes, some groups synthesize lysine through the AAA pathway, others through the DAP pathway. Metazoans are unable to synthesize lysine; they acquire it from dietary sources.

Two intermediates of the DAP pathway for the synthesis of lysine have special functions in procaryotes. Diaminopimelic acid is a component of the peptidoglycan of the cell wall, and dihydrodipicolinic acid is the immediate precursor of dipicolinic acid, which is a major chemical constituent of endospores and contributes to their heat stability (see Chapter 10).

The methionine branch of the aspartate pathway is shown in Figure 7.20. In certain bacteria, the final step of the pathway (methylation) can be catalyzed by two distinct enzymes. One requires folic acid as a cofactor, and the other vitamin B_{12}. Some bacteria, for example, *E. coli*, can synthesize folic acid, but they are unable to synthesize vitamin B_{12}. Thus, when growing in media which lack vitamin B_{12}, they synthesize methionine via the folic acid-dependent reaction. In media which contain vitamin B_{12}, the B_{12}-dependent reaction predominates.

The terminal steps in the synthesis of the fifth member of the aspartate family, isoleucine, are catalyzed by a series of enzymes which also catalyze analogous steps in the biosynthesis of a member of the pyruvate family, valine. Isoleucine biosynthesis will, accordingly, be discussed in the context of valine biosynthesis.

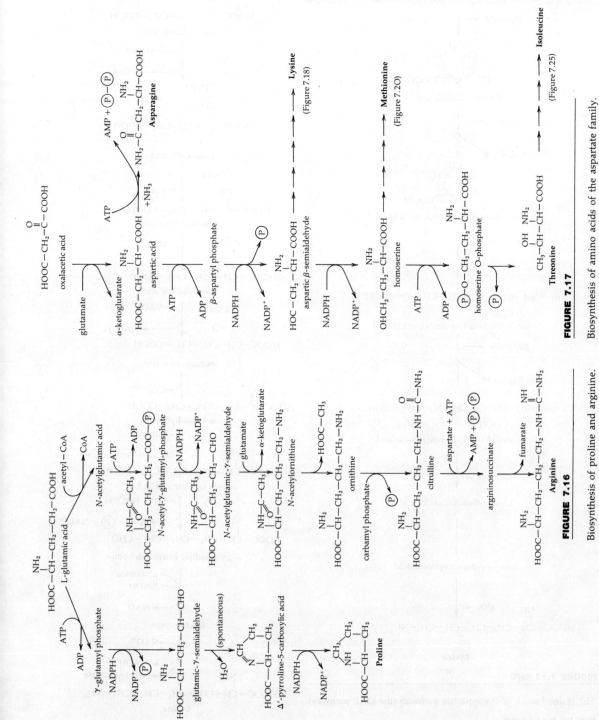

FIGURE 7.17 Biosynthesis of amino acids of the aspartate family.

FIGURE 7.16 Biosynthesis of proline and arginine.

207

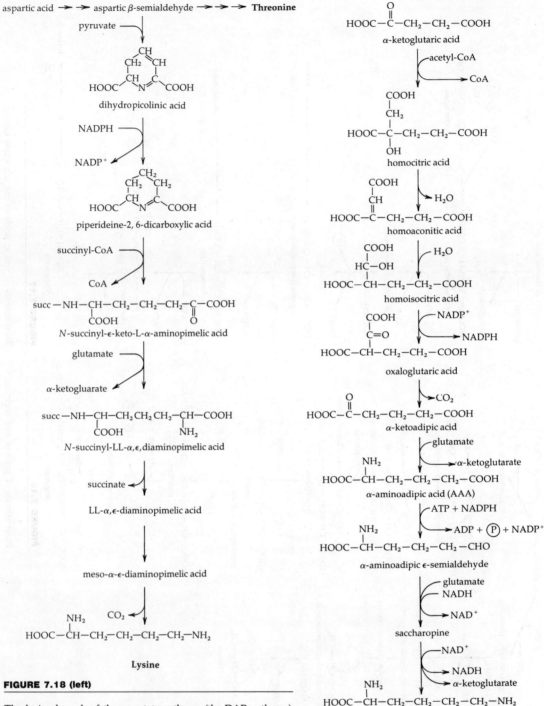

aspartic acid $\rightarrow \rightarrow$ aspartic β-semialdehyde $\rightarrow \rightarrow$ **Threonine**

pyruvate

dihydropicolinic acid

NADPH

NADP$^+$

piperideine-2, 6-dicarboxylic acid

succinyl-CoA

CoA

succ—NH—CH—CH$_2$—CH$_2$—CH$_2$—C—COOH
　　　　｜　　　　　　　　　　　　　║
　　　　COOH　　　　　　　　　　　O

N-succinyl-ϵ-keto-L-α-aminopimelic acid

glutamate

α-ketogluarate

succ—NH—CH—CH$_2$CH$_2$CH$_2$—CH—COOH
　　　　｜　　　　　　　　　　　　　｜
　　　　COOH　　　　　　　　　　NH$_2$

N-succinyl-LL-α,ϵ,diaminopimelic acid

succinate

LL-α,ϵ-diaminopimelic acid

meso-α-ϵ-diaminopimelic acid

CO$_2$

　　　　NH$_2$
　　　　｜
HOOC—CH—CH$_2$—CH$_2$—CH$_2$—CH$_2$—NH$_2$

Lysine

FIGURE 7.18 (left)

The lysine branch of the aspartate pathway (the DAP pathway).

FIGURE 7.19 (right)

The AAA pathway of lysine biosynthesis.

　　　　　O
　　　　　║
HOOC—C—CH$_2$—CH$_2$—COOH

α-ketoglutaric acid

acetyl-CoA

CoA

　　　　COOH
　　　　｜
　　　　CH$_2$
　　　　｜
HOOC—C—CH$_2$—CH$_2$—COOH
　　　　｜
　　　　OH

homocitric acid

　　　　COOH
　　　　｜
　　　　CH
　　　　‖
HOOC—C—CH$_2$—CH$_2$—COOH

H$_2$O

homoaconitic acid

　　　　COOH
　　　　｜
　　　　HC—OH
　　　　｜
HOOC—CH—CH$_2$—CH$_2$—COOH

H$_2$O

homoisocitric acid

　　　　COOH
　　　　｜
　　　　C=O
　　　　｜
HOOC—CH—CH$_2$—CH$_2$—COOH

NADP$^+$

NADPH

oxaloglutaric acid

　O
　║
HOOC—C—CH$_2$—CH$_2$—CH$_2$—COOH

CO$_2$

α-ketoadipic acid

glutamate

α-ketoglutarate

　　　　NH$_2$
　　　　｜
HOOC—CH—CH$_2$—CH$_2$—CH$_2$—COOH

α-aminoadipic acid (AAA)

ATP + NADPH

ADP + (P) + NADP$^+$

　　　　NH$_2$
　　　　｜
HOOC—CH—CH$_2$—CH$_2$—CH$_2$—CHO

α-aminoadipic ϵ-semialdehyde

glutamate
NADH

NAD$^+$

saccharopine

NAD$^+$

NADH
α-ketoglutarate

　　　　NH$_2$
　　　　｜
HOOC—CH—CH$_2$—CH$_2$—CH$_2$—CH$_2$—NH$_2$

Lysine

208

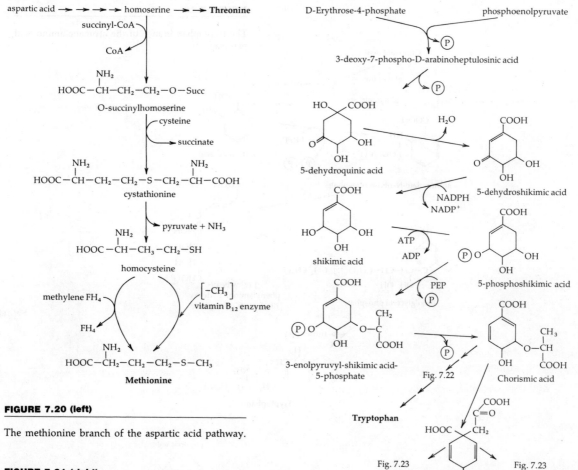

FIGURE 7.20 (left)

The methionine branch of the aspartic acid pathway.

FIGURE 7.21 (right)

Biosynthesis of amino acids of the aromatic family.

• **The aromatic family** The products of this pathway include the three amino acids: tyrosine, phenylalanine, and tryptophan. The first reaction of this pathway is a condensation between an intermediate of the pentose-phosphate cycle, erythrose-4-phosphate, and an intermediate of the glycolytic pathway, phosphoenolpyruvate. Early steps of this pathway leading to the formation of chorismic acid and prephenic acid, both situated at major metabolic branch points, are shown in Figure 7.21. The tryptophan branch of the pathway is shown in Figure 7.22. The phenylalanine and tyrosine branches are shown in Figure 7.23. The aromatic pathway also furnishes, via chorismic acid, p-aminobenzoic acid (one precursor of folic acid), and p-hydroxybenzoic acid (a precursor of the quinones, which are members of certain electron transport chains).

• **The serine and pyruvate families** The pathway for the formation of the amino acids of the serine family (serine, glycine, and cysteine) is shown in Figure 7.24.

FIGURE 7.22

The tryptophan branch of the aromatic amino acid pathway.

Chorismic acid

glutamine ⟶ pyruvate

glutamate ⟵

COOH
NH
ribose-5-(P)

PRPP

(P)-(P)

COOH
NH₂

anthranilic acid

anthranilate-N-ribose phosphate

COOH
N
H
CH
C—OH
CHOH
CHOH
CHO(P)

CO₂ + H₂O

1'(o-carboxy-phenylamino)-1'-deoxy-ribulose-5'-phosphate

H
N
CH
C—CH—CH—CH₂O(P)
OH OH

Indolglycerol phosphate

serine

triose-(P)

H
N
CH
NH₂
CH₂—CH—COOH

Tryptophan

prephenic acid

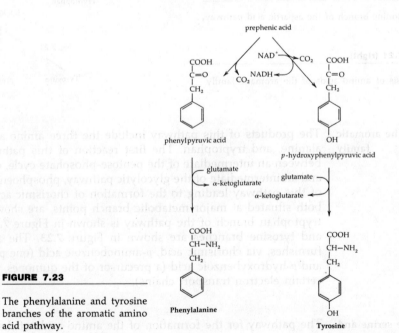

COOH
C=O
CH₂

NAD⁺ ⟶ CO₂
CO₂ ⟵ NADH

COOH
C=O
CH₂
OH

phenylpyruvic acid

p-hydroxyphenylpyruvic acid

glutamate
α-ketoglutarate

glutamate
α-ketoglutarate

COOH
CH—NH₂
CH₂

Phenylalanine

COOH
CH—NH₂
CH₂
OH

Tyrosine

FIGURE 7.23

The phenylalanine and tyrosine branches of the aromatic amino acid pathway.

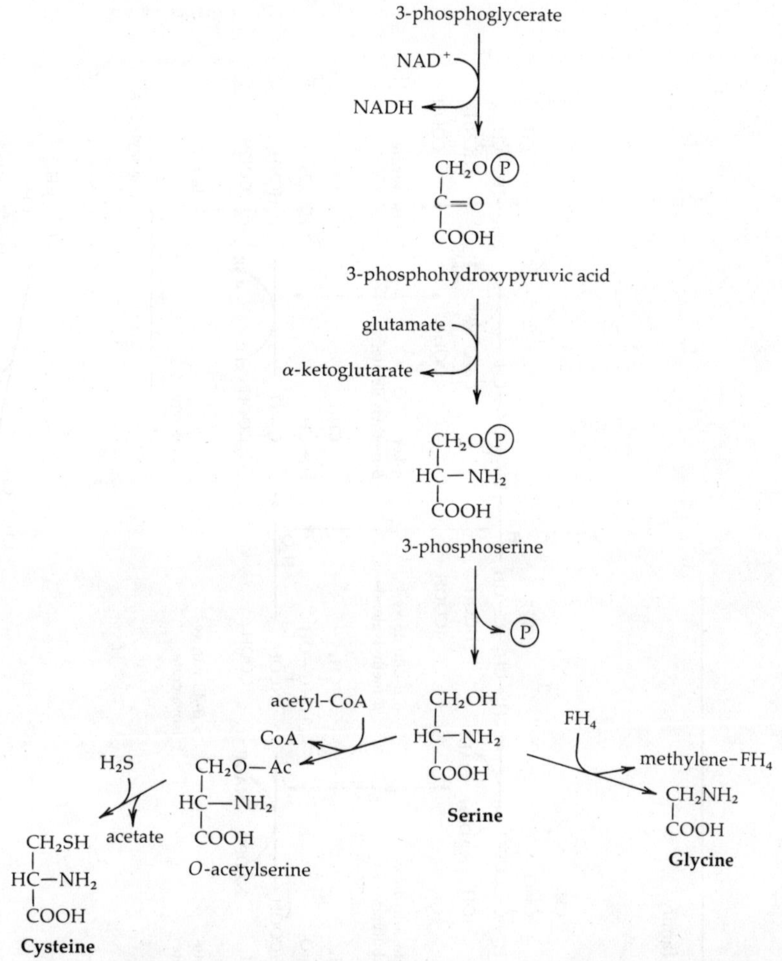

3-phosphoglycerate

NAD$^+$
NADH

$$\begin{array}{l} CH_2O\,\circled{P} \\ | \\ C=O \\ | \\ COOH \end{array}$$

3-phosphohydroxypyruvic acid

glutamate
α-ketoglutarate

$$\begin{array}{l} CH_2O\,\circled{P} \\ | \\ HC-NH_2 \\ | \\ COOH \end{array}$$

3-phosphoserine

$\circled{P}$

acetyl–CoA
CoA

$$\begin{array}{l} CH_2OH \\ | \\ HC-NH_2 \\ | \\ COOH \end{array}$$
Serine

FH$_4$
methylene–FH$_4$

$$\begin{array}{l} CH_2NH_2 \\ | \\ COOH \end{array}$$
Glycine

H$_2$S

$$\begin{array}{l} CH_2O-Ac \\ | \\ HC-NH_2 \\ | \\ COOH \end{array}$$
O-acetylserine

acetate

$$\begin{array}{l} CH_2SH \\ | \\ HC-NH_2 \\ | \\ COOH \end{array}$$
Cysteine

FIGURE 7.24

Biosynthesis of the amino acids of the serine family.

The pathway for the formation of the amino acids of the pyruvate family (alanine, valine, and leucine), as well as isoleucine, which is synthesized by common enzymes, is shown in Figure 7.25. Pantothenate is synthesized from an intermediate in the biosynthesis of valine.

• **Histidine synthesis**

The isolated pathway of histidine biosynthesis is shown in Figure 7.26. The chain of five carbon atoms in the skeleton of this amino acid is derived from PRPP; two of these atoms contribute to the five-membered imidazole ring and the rest give rise to the three-carbon side chain. The remaining three atoms of the imidazole ring have a curious origin: a C-N fragment is contributed from the purine nucleus of ATP, and the other N atom from glutamine. This utilization of ATP as a donor of two atoms of the purine nucleus is unique. Its physiological rationale lies in the fact that cleavage of the purine nucleus of ATP leads to the formation of another biosynthetic intermediate, aminoimidazole carboxamide ribotide (AICAR), which is itself a precursor of purines (Figure 7.10). This intimate connection between the biosynthesis of histidine and purines has been discussed previously (Figure 7.11).

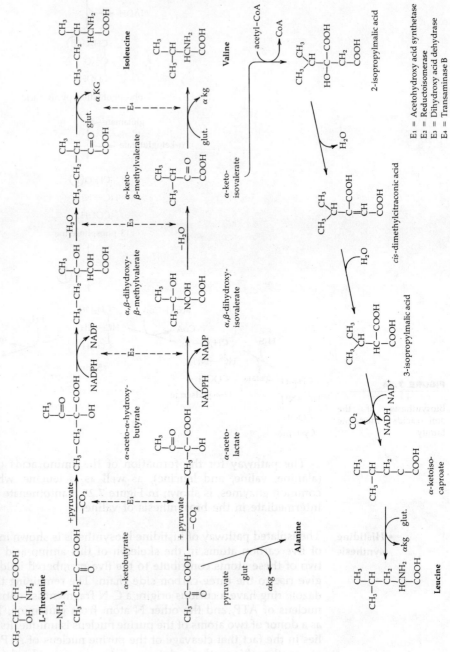

FIGURE 7.25 Biosynthesis of amino acids of the pyruvate family.

E_1 = Acetohydroxy acid synthetase
E_2 = Reductoisomerase
E_3 = Dihydroxy acid dehydrase
E_4 = Transaminase B

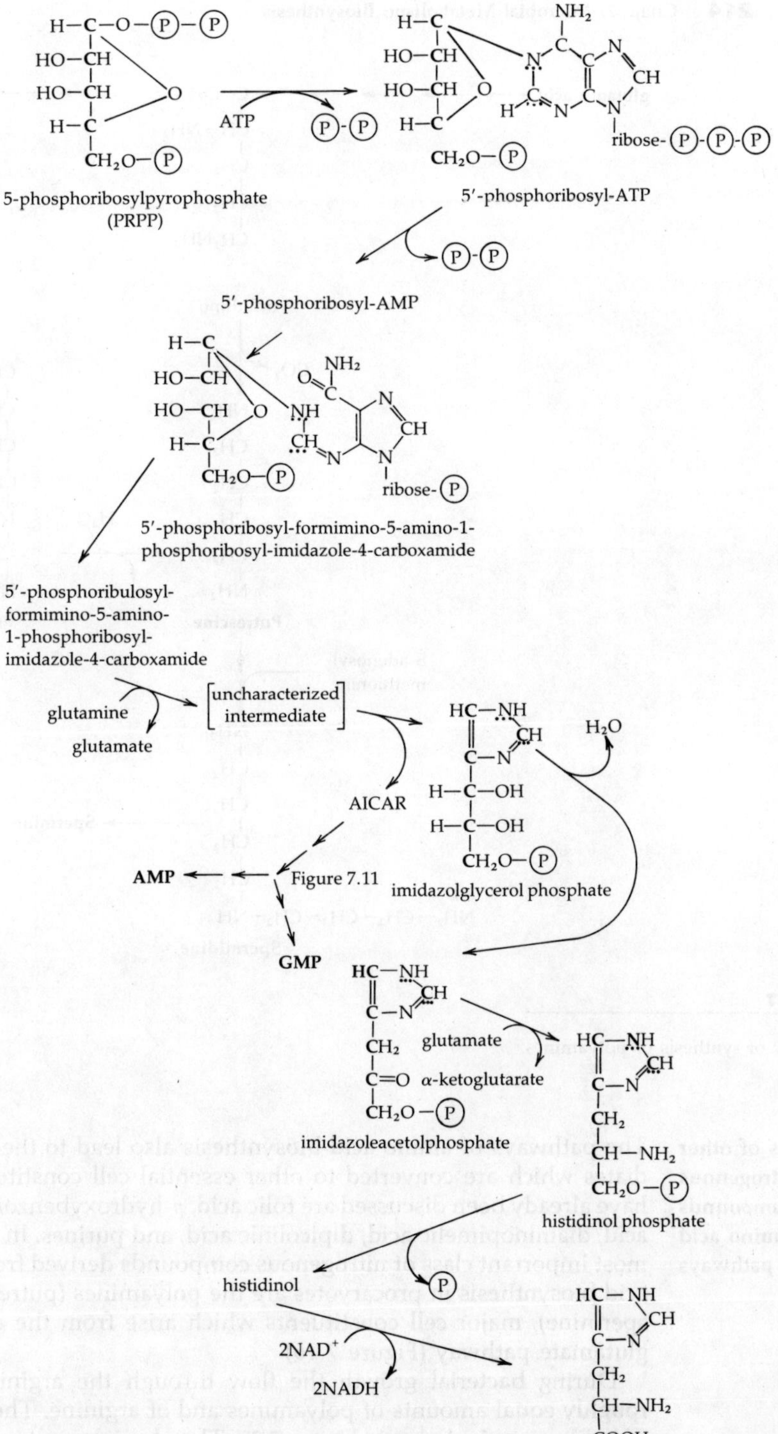

213

FIGURE 7.26

The biosynthesis of histidine.

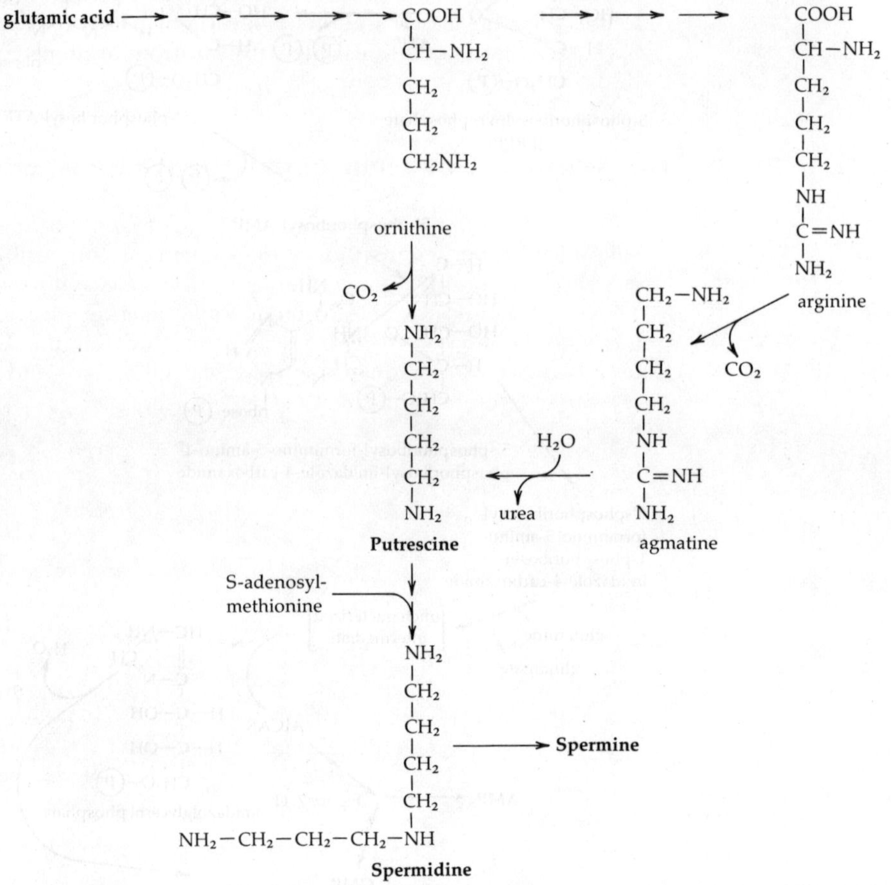

glutamic acid $\longrightarrow$ $\longrightarrow$ $\longrightarrow$ $\longrightarrow$ $\longrightarrow$ COOH $\longrightarrow$ $\longrightarrow$ $\longrightarrow$ COOH

$$\begin{array}{cc}
\text{COOH} & \text{COOH} \\
| & | \\
\text{CH}-\text{NH}_2 & \text{CH}-\text{NH}_2 \\
| & | \\
\text{CH}_2 & \text{CH}_2 \\
| & | \\
\text{CH}_2 & \text{CH}_2 \\
| & | \\
\text{CH}_2\text{NH}_2 & \text{NH} \\
 & | \\
 & \text{C}=\text{NH} \\
 & | \\
 & \text{NH}_2
\end{array}$$

ornithine   arginine

CO_2

$$\begin{array}{cc}
\text{NH}_2 & \text{CH}_2-\text{NH}_2 \\
| & | \\
\text{CH}_2 & \text{CH}_2 \\
| & | \\
\text{CH}_2 & \text{CH}_2 \\
| & | \\
\text{CH}_2 & \text{NH} \\
| & | \\
\text{CH}_2 & \text{C}=\text{NH} \\
| & | \\
\text{NH}_2 & \text{NH}_2
\end{array}$$

H_2O CO_2

Putrescine   urea   agmatine

S-adenosyl-
methionine

$$\begin{array}{c}
\text{NH}_2 \\
| \\
\text{CH}_2 \\
| \\
\text{CH}_2 \\
| \\
\text{CH}_2 \\
| \\
\text{CH}_2 \\
| \\
\text{NH}_2-\text{CH}_2-\text{CH}_2-\text{CH}_2-\text{NH}
\end{array}$$

$\longrightarrow$ **Spermine**

Spermidine

FIGURE 7.27

The pathway of synthesis of polyamines.

• **Synthesis of other nitrogenous compounds via amino acid pathways**

The pathways of amino acid biosynthesis also lead to the formation of intermediates which are converted to other essential cell constituents. Examples which have already been discussed are folic acid, p-hydroxybenzoic acid, p-aminobenzoic acid, diaminopimelic acid, dipicolinic acid, and purines. In quantitative terms, the most important class of nitrogenous compounds derived from a pathway of amino acid biosynthesis in procaryotes are the polyamines (putrescine, spermidine, and spermine), major cell constituents which arise from the arginine branch of the glutamate pathway (Figure 7.16).

During bacterial growth the flow through the arginine pathway produces roughly equal amounts of polyamines and of arginine. The pathway of synthesis of polyamines is shown in Figure 7.27. The physiological importance of putrescine as an osmotic regulator will be discussed in Chapter 10.

Putrescine can be synthesized either from an intermediate of the arginine pathway, ornithine, or directly from arginine. The ornithine route predominates in cells growing in the absence of exogenous arginine. When arginine is supplied to the cells, the *de novo* arginine biosynthetic pathway ceases to function (Chapter 8). Under these conditions, the pathway from arginine comes into operation for the synthesis of polyamines.

THE SYNTHESIS OF LIPID CONSTITUENTS FROM ACETATE

Lipids are a class of cell constituents defined on the basis of their solubility properties instead of their chemical composition. They are insoluble in water and soluble in nonpolar solvents such as ether, chloroform, and benzene. They are chemically heterogeneous and include fats, phospholipids, steroids, isoprenoids, and poly-β-hydroxybutyrate. However, they can be grouped into two broad classes: those containing esterified fatty acids and those which consist of repeating C_5 units with the structure of isoprene:

$$\begin{array}{c} CH_3 \\ | \\ -CH_2-C=CH-CH_2- \end{array}$$

Certain lipids and their functions are listed in Table 7.4.

The *phospholipids* are universal membrane components. Their general structure is shown in Figure 7.28. The chemical nature of the residue (X) attached to the phosphate group defines the class of phospholipid. In *E. coli* and *Salmonella typhimurium*, in which the most careful measurements have been made, the major phospholipid of the membrane is phosphatidylethanolamine ($\sim$75 percent). Lesser amounts of phosphatidylglycerol ($\sim$18 percent) and cardiolipin ($\sim$5 percent) and only traces of phosphatidylserine ($\sim$1 percent) are found.

FIGURE 7.28

General structure of phospholipids.

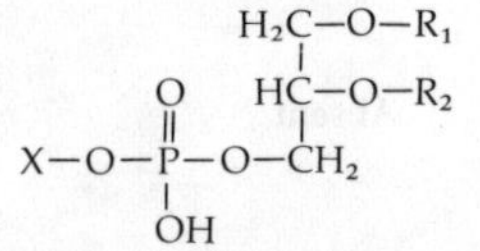

X	Name of phospholipid
$-CH_2-CH_2-NH_2$	phosphotidylethanolamine
$-CH_2-CHNH_2-COOH$	phosphotidylserine
$-CH_2-CHOH-CH_2OH$	phosphotidyglycerol

• The synthesis of fatty acids   Fatty acids are synthesized separately and then esterified to form complex lipids. Scores of different kinds of fatty acids are found in bacteria: they contain different numbers of carbon atoms; they are straight chained or branched; they may or may not contain double bonds; they may or may not contain —OH groups; and they may or may not contain cyclopropane rings. In any particular bacterial species the number of fatty acids is limited. *E. coli*, for example, contains only six while *Bacillus subtilis* contains eight; only two fatty acids are common to both species (Table 7.5).

Certain generalizations can be drawn about the types of fatty acids found in bacteria. Like almost all fatty acids, most fatty acids found in bacteria contain an even number of carbon atoms. Although polyunsaturated fatty acids (more than one double bond) are common constituents of the lipids of eucaryotes, they are confined among procaryotes to the blue-green bacteria.

Saturated fatty acids are synthesized by the general pathway outlined in Figure 7.29. A special protein known as *acyl carrier protein* (ACP) plays a vital role. The formation of long-chain fatty acids starts with the transfer of an acetyl group from

TABLE 7.4

Lipids and their function

	FUNCTION IN	
	PROCARYOTES	EUCARYOTES
I. Lipids containing esterified fatty acids		
A. Simple (a single monomeric unit)		
1. Poly-β-hydroxybutyrate	Reserve material	Absent
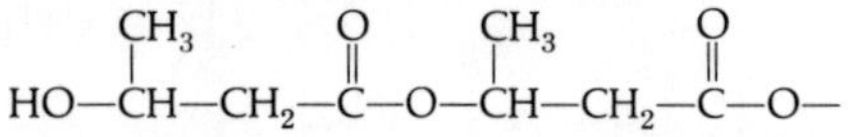 $\underset{\text{HO}}{}\text{—CH—CH}_2\text{—C—O—CH—CH}_2\text{—C—O—}$ (CH$_3$, O, CH$_3$, O)		
B. Complex (fatty acids esterified to other compounds)		
1. Esterified to glycerol		
a. Neutral fats	Absent	Reserve material
CH$_2$—O—R$_1$ R1,2,3=fatty acyl CH—O—R$_2$ $\left(\text{R—C—}\right)$ residue CH$_2$—O—R$_3$		
b. Phospholipid (see Figure 7.28)	Membrane constituent	Membrane constituent
c. Glycolipid CH$_2$—O—R$_1$ CH—O—R$_2$ CH$_2$—O—sugar residue	Membrane constituent of blue-green bacteria	Membrane constituent of chloroplasts
2. Esterified to an amino sugar Lipid A (See Figure 11.15.)	Component of lipopolysaccharide wall layer	Absent

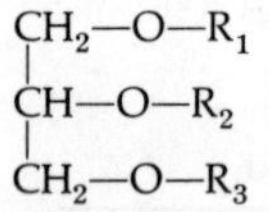

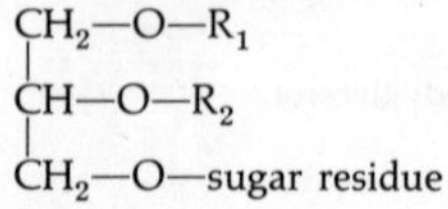

TABLE 7.4 cont.

	FUNCTION IN	
	PROCARYOTES	EUCARYOTES

II. *Lipids containing isoprene units*

A. Polyisoprenoids $\left(-CH_2-\overset{\overset{\displaystyle CH_3}{|}}{C}=CH-CH_2-\right)$

1. Carotenoids $C_{40}(8 \times C_5)$	Photoprotection and light-harvesting	Light-harvesting pigments
2. Sterols $C_{30}(6 \times C_5)$	Absent in most	Membrane constituent
3. Bactoprenol $C_{55}(11 \times C_5)$	Component upon which wall constituents are synthesized	Absent

B. Compounds with isoprenoid components

1. Chlorophyll	Component of photosynthetic apparatus	Component of photosynthetic apparatus
2. Quinones	Component of electron transport chains	Component of electron transport chains

---

TABLE 7.5

Fatty acid composition of lipids in *Escherichia coli* and *Bacillus subtilis*

Number of carbon atoms	14	14	14	15	15	16
Number of double bonds	0	0	0	0	0	0
Number of hydroxyl groups	0	1	0	0	0	0
Structure[a]	Normal	Normal	Iso	Antiiso	Iso	Normal
Common name	Myristic	β-hydroxy myristic				Palmitic
Percent in *E. coli*[b]	6.1	4.8	0	0	0	37.1
Percent in *B. subtilis*[c]	Trace	0	3.9	36.6	12.1	6

Number of carbon atoms	16	16	17	17	17	18
Number of double bonds	1	0	0	0	0	1
Number of hydroxyl groups	0	0	0	0	0	0
Structure[a]	Normal	Iso	Antiiso	Iso	Cyclo	Normal
Common name	Palmitoleic					cis-vaccinic
Percent in *E. coli*[b]	28	0	0	0	3.2	20.8
Percent in *B. subtilis*[c]	0	11.1	14.4	15.9	0	0

[a] *Normal* indicates a straight chain fatty acid; *iso* indicates a methyl group branch at the penultimate carbon; *antiiso* indicates a methyl group branch at the antipenultimate carbon; *cyclo* indicates that the fatty acid contains an internal cyclopropane ring.

[b] From A. G. Marr and J. L. Ingraham, "Effect of Temperature on the Composition of Fatty Acids in *Escherichia coli*," *J. Bacteriol.* **84**, 1260 (1962).

[c] Recalculated from T. Kaneda, "Fatty Acids in the Genus *Bacillus*," *J. Bacteriol.* **93**, 894 (1967).

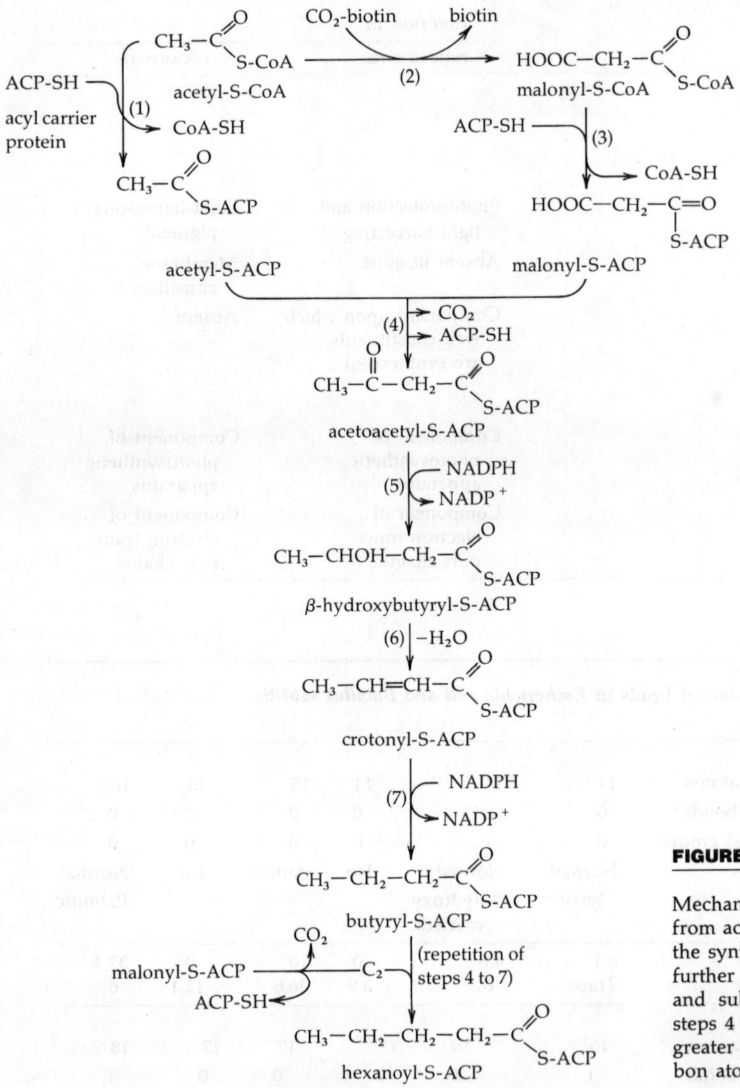

FIGURE 7.29

Mechanism for the synthesis of saturated fatty acids from acetyl-S-CoA in *E. coli*. The reactions leading to the synthesis of hexanoyl-(C_6)-S-ACP are shown. By further transfers of acetyl units from malonyl-S-ACP and subsequent reductions (repetitions of reaction steps 4 to 7), unbranched fatty acids of progressively greater chain length, containing even numbers of carbon atoms, are formed.

acetyl CoA to ACP. This complex serves as an acceptor to which successive C_2 units are transferred. The C_2 donor is malonyl ACP which is formed, in turn, by carboxylation of acetyl CoA; during transfer of the C_2 unit, CO_2 is released and free ACP is regenerated. The product of C_2 transfer carries a terminal acetyl group, in which in subsequent reactions is sequentially reduced, dehydrated, and reduced again, yielding an unsaturated acyl-ACP complex with two additional carbon atoms. Repetitions of this set of reactions progressively lengthen the fatty acid chain until the length characteristic of the particular bacterium (usually between C_{14} and C_{18}) is reached.

Monounsaturated fatty acids are formed in various bacteria by one of two

TABLE 7.6

Biological distribution of mechanisms for the
synthesis of monounsaturated fatty acids

ANAEROBIC PATHWAY	AEROBIC PATHWAY
Clostridium spp.	*Mycobacterium spp.*
Lactobacillus spp.	*Corynebacterium spp.*
Escherichia coli	*Micrococcus lysodeikticus*
Pseudomonas spp.	*Bacillus spp.*
Photosynthetic bacteria	Fungi
Blue-green bacteria	Protozoa
	Animals

different pathways (Table 7.6), the *aerobic pathway* and the *anaerobic pathway* (which occurs in aerobes as well as anaerobes).

The aerobic pathway intervenes as a subsequent modification of fully synthesized saturated fatty acids, while in the anaerobic pathway unsaturation takes place during elongation of the fatty acid chain. The aerobic pathway requires the direct intervention of molecular oxygen (Figure 7.30).

The mechanisms of the anaerobic pathway are outlined in Figure 7.31. The C_{10} hydroxyacyl intermediate, β-OH-decanoyl-ACP, can undergo normal desaturation, leading to the formation of longer chain saturated fatty acids, or it can undergo a dehydration, leading to the homologous monounsaturated fatty acids. Note that in the anaerobic pathway, the position of the double bond in the carbon chain of the eventual end products is determined by the point in biosynthesis at which it is introduced. Subsequent chain elongation leads to its location between carbon atoms 9 and 10 in the C_{16} product (palmitoleic acid). In the C_{18} product, however, the double bond becomes located between carbon atoms 11 and 12. Hence, bacteria that employ the anerobic pathway contain cis-vaccenic acid as their monounsaturated C_{18} fatty acid, rather than oleic acid, the product of direct desaturation of stearic acid by the aerobic pathway.

• Synthesis of phospholipids   Phospholipids are synthesized from fatty acids and an intermediate of glycolysis, dihydroxyacetone-phosphate, by the pathway outlined in Figure 7.32. Dihydroxyacetone-phosphate is reduced to 3-glycerophosphate, which is subsequently esterified by two fatty acid residues. The resulting diglyceride, phosphatidic acid, is then activated by CTP to form CDP-diglyceride, which undergoes transfer reactions with serine and α-glycerophosphate, releasing CMP. The reaction product with serine, phosphatidylserine, itself constitutes a minor class of phos-

FIGURE 7.30

The formation of the monounsaturated fatty acid, palmitoleic acid, from the corresponding saturated fatty acid, palmitic acid, by the aerobic pathway.

$$CH_3-(CH_2)_{14}-\overset{\overset{\text{O}}{\|}}{C}-S-ACP + \tfrac{1}{2}O_2 \longrightarrow CH_3-(CH_2)_5-CH=CH(CH_2)_7-COOH + H_2O$$

ACP derivative of palmitic acid   ACP derivative of palmitoleic acid

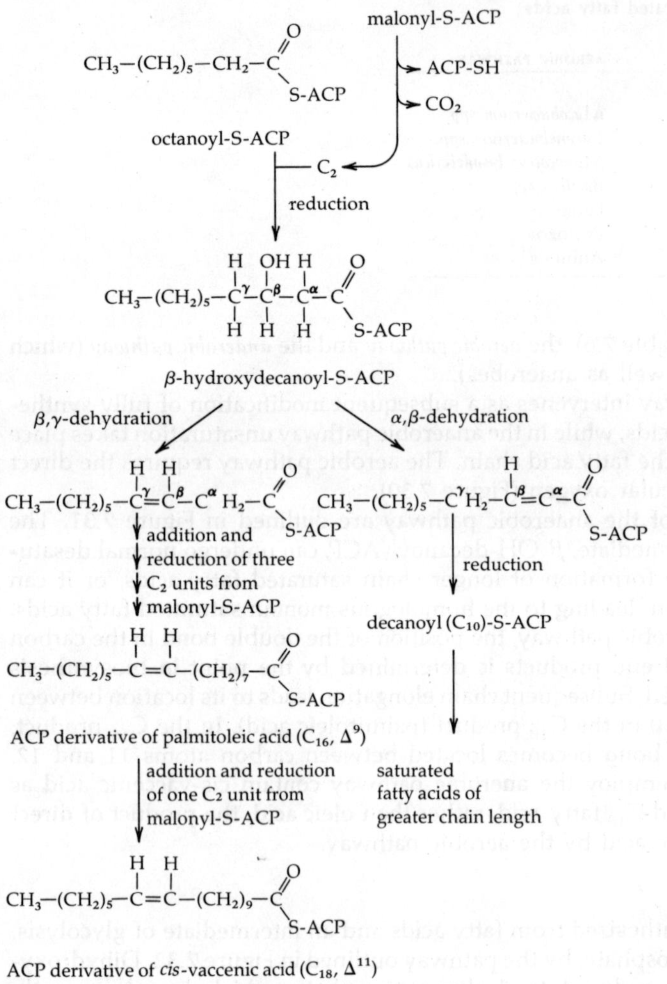

$$CH_3-(CH_2)_5-CH_2-C\overset{O}{\underset{S-ACP}{\big\langle}}$$

octanoyl-S-ACP

malonyl-S-ACP

ACP-SH

CO_2

C_2

reduction

$$CH_3-(CH_2)_5-\underset{\underset{H}{|}}{\overset{\overset{H}{|}}{C^\gamma}}-\underset{\underset{H}{|}}{\overset{\overset{OH}{|}}{C^\beta}}-\underset{\underset{H}{|}}{\overset{\overset{H}{|}}{C^\alpha}}-C\overset{O}{\underset{S-ACP}{\big\langle}}$$

β-hydroxydecanoyl-S-ACP

β,γ-dehydration

α,β-dehydration

$$CH_3-(CH_2)_5-\underset{\underset{H}{|}}{\overset{\overset{H}{|}}{C^\gamma}}=\overset{H}{\overset{|}{C^\beta}}-C^\alpha H_2-C\overset{O}{\underset{S-ACP}{\big\langle}}$$

$$CH_3-(CH_2)_5-C^\gamma H_2-\overset{\overset{H}{|}}{\underset{\underset{}{}}{C^\beta}}=\overset{H}{\overset{|}{C^\alpha}}-C\overset{O}{\underset{S-ACP}{\big\langle}}$$

addition and
reduction of three
C_2 units from
malonyl-S-ACP

reduction

decanoyl (C_{10})-S-ACP

$$CH_3-(CH_2)_5-\overset{H}{\overset{|}{C}}=\overset{H}{\overset{|}{C}}-(CH_2)_7-C\overset{O}{\underset{S-ACP}{\big\langle}}$$

ACP derivative of palmitoleic acid (C_{16}, Δ^9)

addition and reduction
of one C_2 unit from
malonyl-S-ACP

saturated
fatty acids of
greater chain length

$$CH_3-(CH_2)_5-\overset{H}{\overset{|}{C}}=\overset{H}{\overset{|}{C}}-(CH_2)_9-C\overset{O}{\underset{S-ACP}{\big\langle}}$$

ACP derivative of *cis*-vaccenic acid (C_{18}, Δ^{11})

FIGURE 7.31

The anaerobic pathway to monounsaturated fatty acids, characteristic of many bacteria, showing its relationship to the pathway for saturated fatty acid synthesis.

pholipids. The major phospholipid class is the decarboxylation product, phosphatidylethanolamine. The reaction between CDP-diglyceride and α-glycerophosphate leads to the other phospholipid classes, phosphatidylglycerol and cardiolipin.

• **The synthesis of polyisoprenoid compounds**

A large number of different cell constituents have carbon skeletons which consist of repeating C_5 units with the structure of isoprene. These *polyisoprenoid compounds* are synthesized exclusively from acetyl units; however, the mechanism of chain elongation differs markedly from that characteristic of fatty acid synthesis, diverging at the C_4 level (Figure 7.33). Acetoacetyl CoA undergoes a "head-to-head" condensation with acetyl CoA, to yield, after rearrangement, *mevalonic acid*, a branched C_6 acid. This is, in turn, converted, by two successive phosphorylations

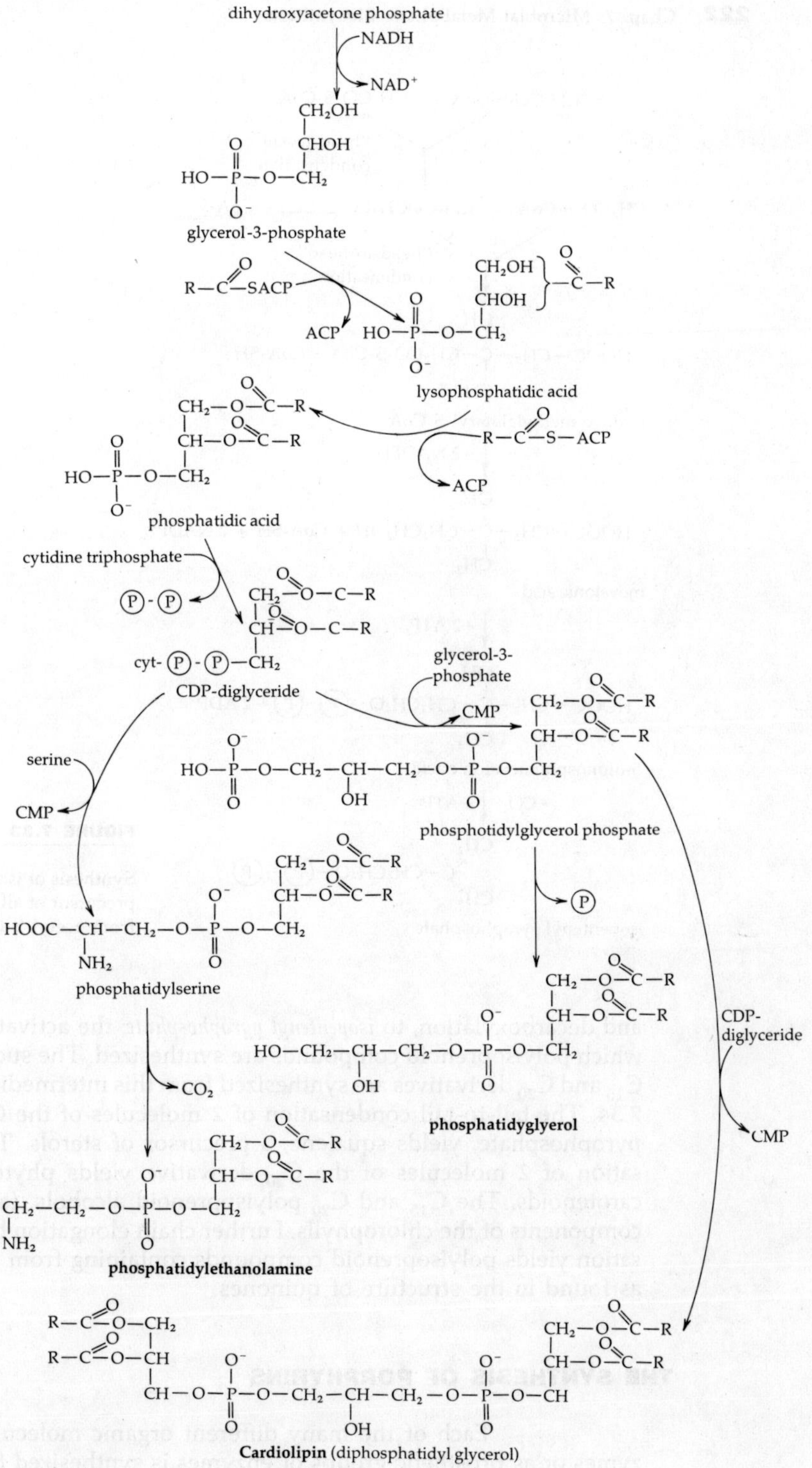

221

FIGURE 7.32

Pathway of formation of the major phospholipid classes found in *E. coli.*

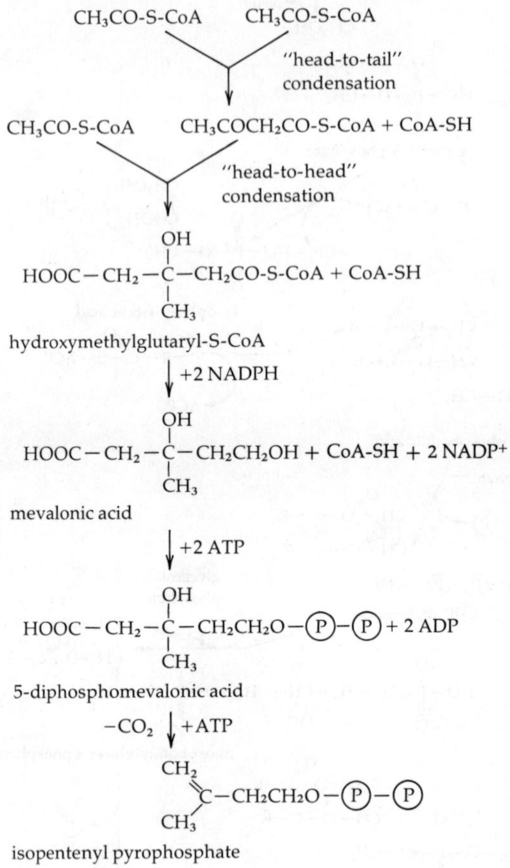

$$CH_3CO\text{-}S\text{-}CoA \qquad CH_3CO\text{-}S\text{-}CoA$$

"head-to-tail"
condensation

$$CH_3CO\text{-}S\text{-}CoA \qquad CH_3COCH_2CO\text{-}S\text{-}CoA + CoA\text{-}SH$$

"head-to-head"
condensation

$$\underset{\text{hydroxymethylglutaryl-S-CoA}}{HOOC - CH_2 - \underset{\underset{CH_3}{|}}{\overset{\overset{OH}{|}}{C}} - CH_2CO\text{-}S\text{-}CoA + CoA\text{-}SH}$$

$\downarrow$ +2 NADPH

$$\underset{\text{mevalonic acid}}{HOOC - CH_2 - \underset{\underset{CH_3}{|}}{\overset{\overset{OH}{|}}{C}} - CH_2CH_2OH + CoA\text{-}SH + 2\,NADP^+}$$

$\downarrow$ +2 ATP

$$\underset{\text{5-diphosphomevalonic acid}}{HOOC - CH_2 - \underset{\underset{CH_3}{|}}{\overset{\overset{OH}{|}}{C}} - CH_2CH_2O - \textcircled{P} - \textcircled{P} + 2\,ADP}$$

$-CO_2 \; \downarrow$ +ATP

$$\underset{\text{isopentenyl pyrophosphate}}{\overset{CH_2}{\underset{CH_3}{\diagdown}}\!\!\!\!C - CH_2CH_2O - \textcircled{P} - \textcircled{P}}$$

FIGURE 7.33

Synthesis of isopentenylpyrophosphate, the precursor of all polyisoprenoid compounds, from acetyl-S-CoA.

and decarboxylation, to *isopentenyl pyrophosphate,* the activated C_5 compound from which polyisoprenoid compounds are synthesized. The successive steps by which C_{15} and C_{20} derivatives are synthesized from this intermediate are shown in Figure 7.34. The tail-to-tail condensation of 2 molecules of the C_{15} derivative, farnesyl pyrophosphate, yields squalene, a precursor of sterols. The analogous condensation of 2 molecules of the C_{20} derivative yields phytoene, the precursor of carotenoids. The C_{15} and C_{20} polyisoprenoid alcohols, farnesol and phytol, are components of the chlorophylls. Further chain elongation by head-to-tail condensation yields polyisoprenoid compounds containing from 50 to 60 carbon atoms, as found in the structure of quinones.

THE SYNTHESIS OF PORPHYRINS

Each of the many different organic molecules that serve as coenzymes or as prosthetic groups of enzymes is synthesized through a special path-

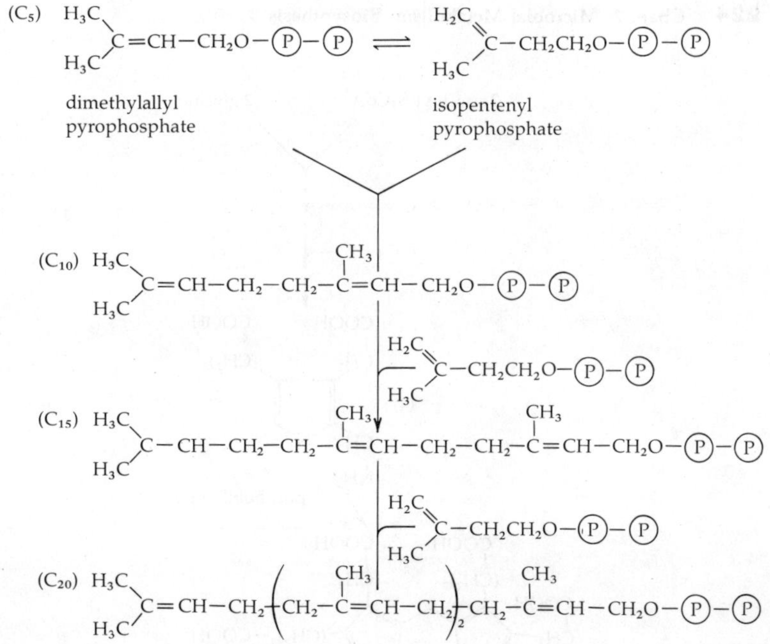

(C5) dimethylallyl pyrophosphate ⇌ isopentenyl pyrophosphate

(C10)

(C15)

(C20)

FIGURE 7.34

Chain elongation in polyisoprenoid biosynthesis.

way. As one illustration, we shall describe the synthesis of *porphyrins*. They fall into two major groups: the iron-containing *hemes*, which serve as prosthetic groups of cytochromes and many other enzymes, known collectively as *heme proteins*; and the magnesium-containing chlorophylls. Vitamin B_{12}, which is a precursor of the prosthetic group for certain enzymes that catalyze the transfer of C_1 groups, is synthesized from an intermediate in the biosynthetic pathway leading to the synthesis of porphyrins.

The synthesis of porphyrins is initiated by a condensation of the amino acid glycine with succinyl CoA; this gives rise in three steps to porphobilinogen (Figure 7.35). The condensation of 4 molecules of this intermediate forms the tetrapyrrolic nucleus of uroporphyrinogen III; subsequent modifications and eventual oxidation yield protoporphryn IX. The insertion of iron as a chelate in this molecule leads directly to the formation of a heme. Alternatively, if magnesium is chelated with the ring, a long series of subsequent steps leads to the formation of the chlorophylls characteristic of the various groups of photosynthetic organisms. Most of these steps in chlorophyll synthesis are common ones: the divergences that give rise to the various specific plant chlorophylls and the bacteriochlorophylls occur near the end of the biosynthetic sequence.

INTERCONNECTIONS BETWEEN CATABOLIC AND BIOSYNTHETIC PATHWAYS

In the preceding pages the pathways of synthesis of many of the biosynthetic intermediates of the cell have been considered. Before continuing, it may be useful to summarize this in the form of a metabolic map (Figure 7.36).

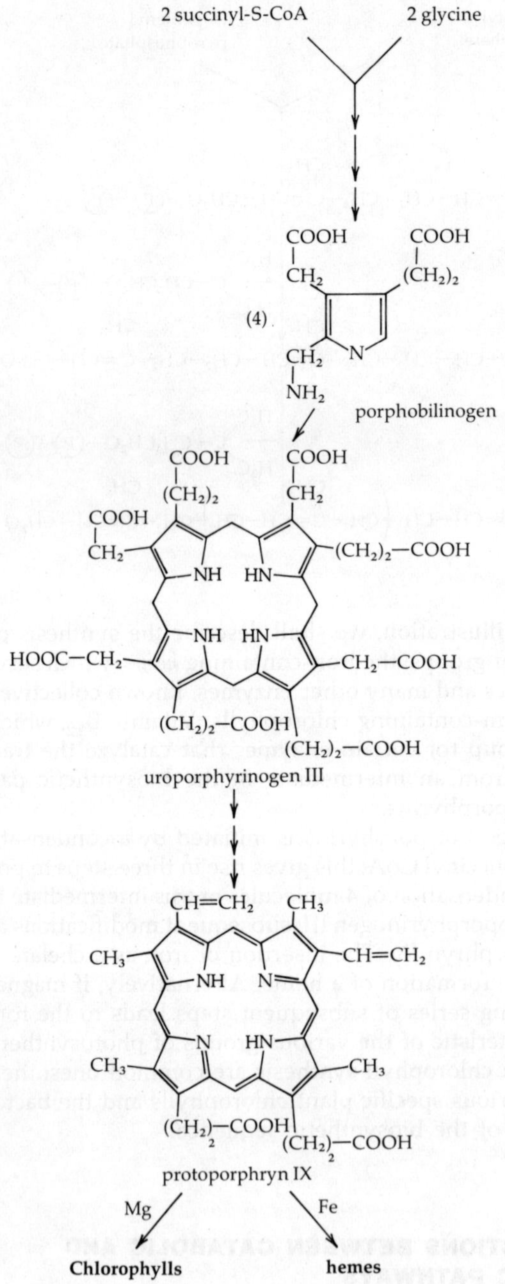

2 succinyl-S-CoA   2 glycine

porphobilinogen

uroporphyrinogen III

protoporphryn IX

Mg   Fe

Chlorophylls   **hemes**

FIGURE 7.35

The general scheme of the synthesis of porphyrins.

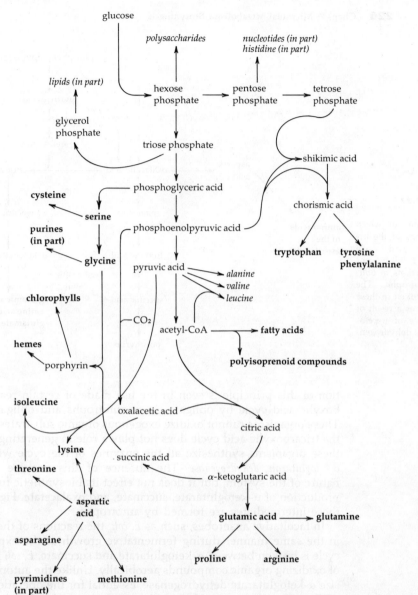

FIGURE 7.36

A generalized metabolic map, showing the derivations of major organic cell constituents from some key products of the intermediary metabolism of glucose.

This makes more evident a point, already alluded to, that the special pathways of biosynthesis are interconnected through reaction sequences (e.g., the reactions of the Embden-Meyerhof pathway and the tricarboxylic acid cycle) which also play important roles in the ATP-generating metabolism of organic substrates.

In heterotrophic organisms there is a dual flow of carbon: part emerges as catabolic end products, and part is drawn off in the form of intermediates of biosynthetic pathways. In autotrophic organisms, however, carbon flow leads exclusively to biosynthesis, producing certain biosynthetic intermediates by sequences of reactions that generate ATP for heterotrophs. An interesting illustra-

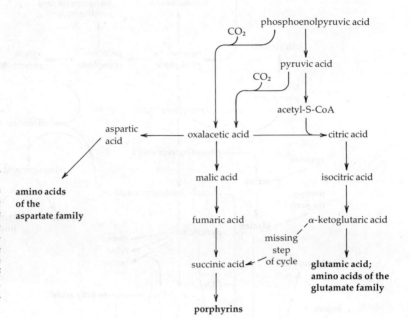

FIGURE 7.37

The manner in which the reactions of the tricarboxylic acid cycle operate to fulfill biosynthetic functions in obligate autotrophs. The cycle is broken in these organisms as a result of the absence of α-ketoglutarate dehydrogenase.

tion of this principle is seen in the use made of certain reactions of the tricarboxylic acid cycle by obligate chemoautotrophs and obligate photoautotrophs. These organisms cannot oxidize exogenous organic substrates, and, consequently, the tricarboxylic acid cycle does not play a role in generating ATP. Nevertheless, these organisms synthesize all the enzymes of the cycle with the exception of *α-ketoglutarate dehydrogenase*. The absence of this enzyme destroys the cyclic nature of the system, but it does not effect its biosynthetic function, which is the production of α-ketoglutarate, succinate, and oxalacetate. Figure 7.37 shows how these intermediates are formed by autotrophs.

In facultative anaerobes, such as *E. coli*, the reactions of the TCA cycle operate in the same manner during fermentative growth at the expense of sugars: the cycle is broken between α-ketoglutarate and succinate. *E. coli* is, however, capable of oxidizing organic compounds aerobically. Unlike the autotrophs, it can synthesize α-ketoglutarate dehydrogenase, essential for the operation of the cycle as an ATP-generating system. Hence, when growing aerobically, the full cycle operates, serving at one time a respiratory and a biosynthetic function.

THE BIOSYNTHESIS OF MACROMOLECULES: GENERAL PRINCIPLES

Proteins and nucleic acids are biopolymers composed of subunits (monomers) linked together by bonds which are characteristic of each class of macromolecule (Figure 7.38). The subunits of all biopolymers can be liberated in free form by hydrolysis. Thus, the biosynthesis of biopolymers involves the

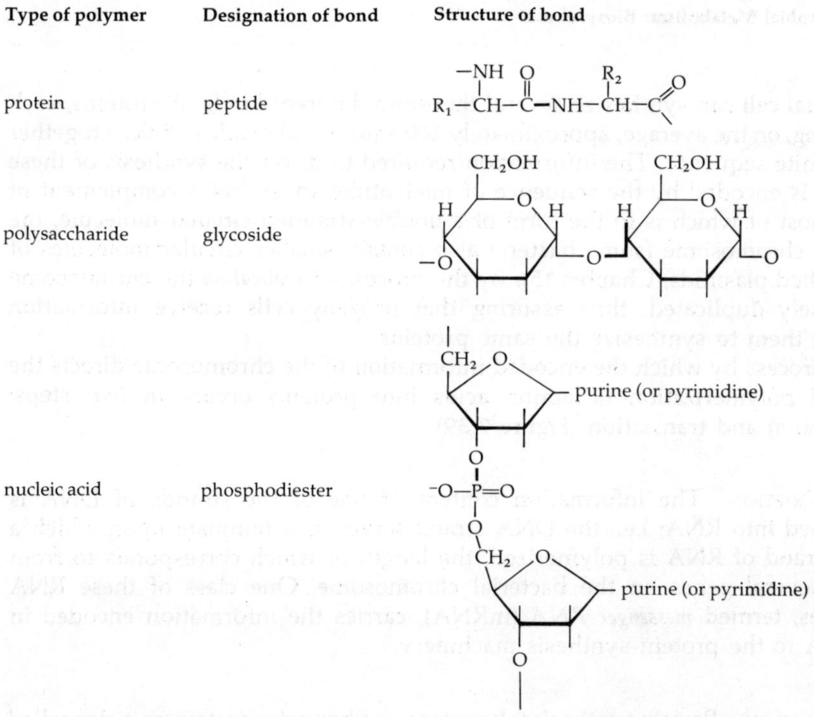

Type of polymer	Designation of bond	Structure of bond
protein	peptide	
polysaccharide	glycoside	
nucleic acid	phosphodiester	

(protein, peptide)

$-NH \quad O \qquad R_2 \quad O$
$R_1-CH-C-NH-CH-C$

(polysaccharide, glycoside)

CH$_2$OH ... CH$_2$OH structures

(nucleic acid, phosphodiester)

CH$_2$ — O — purine (or pyrimidine)

O — P = O

CH$_2$ — O — purine (or pyrimidine)

O

FIGURE 7.38

Nature of the bonds that link to-gether the subunits in the major classes of biological polymers.

joining of subunits through reactions which are, in a formal chemical sense, the reverse of hydrolysis: namely, *dehydration*.

Biopolymers can be hydrolyzed to their subunits by either chemical or en-zymatic means. Thus, their biosynthesis by simple dehydration is thermodynami-cally unfavorable; in the aqueous intracellular environment, breakdown through hydrolysis predominates over synthesis through dehydration. The net synthesis of all biopolymers is therefore accomplished by a preliminary *chemical activation* of the monomer. Such activation requires the expenditure of ATP, and involves the attachment of the monomer to a carrier molecule. Polymerization then occurs by transfer of the monomer from the carrier to the growing polymer chain, a thermodynamically favorable reaction. The activated forms of monomers of the major classes of biopolymers are shown in Table 7.7.

TABLE 7.7

Biopolymers and their monomeric constituents, showing the activated forms of the monomers

BIOPOLYMER	CONSTITUENT MONOMER[a]	ACTIVATED FORM OF MONOMER
Protein	Amino acids	Aminoacyl tRNAs
Nucleic acid	Nucleoside monophosphates	Nucleoside triphosphates
Polysaccharide	Sugars	Sugar-nucleoside diphosphates

[a] Product formed by hydrolysis.

227

• **The general plan
of synthesis
of nucleic acids
and proteins**

A bacterial cell can synthesize several thousand different kinds of proteins, each containing, on the average, approximately 200 amino acid residues linked together in a definite sequence. The information required to direct the synthesis of these proteins is encoded by the sequence of nucleotides in the cell's complement of DNA, most of which is in the form of a double-stranded circular molecule, the bacterial chromosome (some bacteria also contain smaller circular molecules of DNA called plasmids, Chapter 15). By the process of *replication* the chromosome is precisely duplicated, thus assuring that progeny cells receive information enabling them to synthesize the same proteins.

The process by which the encoded information of the chromosome directs the order of polymerization of amino acids into proteins occurs in two steps: transcription and translation (Figure 7.39).

TRANSCRIPTION   The information content of one of the strands of DNA is transcribed into RNA; i.e., the DNA strand serves as a template upon which a single strand of RNA is polymerized, the length of which corresponds to from one to several genes on the bacterial chromosome. One class of these RNA molecules, termed *messenger RNA* (mRNA), carries the information encoded in the DNA to the protein-synthesis machinery.

TRANSLATION   Protein synthesis takes place on ribonucleoprotein particles called *ribosomes* [composed of ribosomal RNA (rRNA) and protein], which attach themselves to the molecule of mRNA. The information carried by the mRNA molecules

FIGURE 7.39

The general plan of synthesis of nucleic acids and proteins.

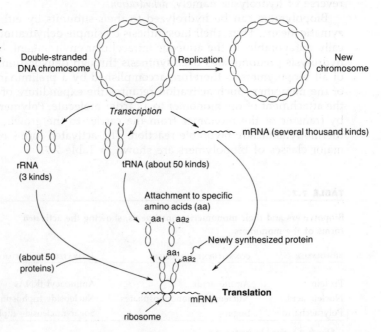

is translated into protein molecules by a special class of RNA molecules called *transfer RNA* (tRNA). These molecules are multifunctional: they are able to bind to the ribosome, to be attached to specific amino acids, and to recognize specific nucleotide sequences of the mRNA. Each molecular species of tRNA recognizes a specific sequence of three nucleotides (a *codon*) on the mRNA molecule, and can be attached to a specific amino acid. Thus, the various amino acids are brought by their cognate tRNA molecules to the ribosome, where they are polymerized into protein in the sequence encoded by the mRNA.

The details of these processes will be discussed in subsequent sections.

THE SYNTHESIS OF DNA

The structure of the DNA molecule, elucidated by Watson and Crick in 1953, immediately suggests how it can be accurately replicated. It is a double helix, each strand of which consists of 2'-deoxyribose molecules linked together by phosphodiester bonds between the 3' hydroxyl group of one and the 5' hydroxyl group of the next. The purine and pyrimidine bases (attached to the 1 position of deoxyribose) project toward the center of the molecule, holding the two strands together by hydrogen bonding between specific purine-pyrimidine pairs. Guanine is paired with cytosine (G—C) and adenine is paired with thymine (A—T) (Figure 7.40). When the bases are present in their energetically most favorable forms (the keto, rather than the enol, form of the oxygenated bases,

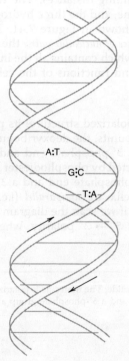

FIGURE 7.40

Schematic representation of the DNA double helix. The outer ribbons represent the two deoxyribosephosphate strands. The parallel lines between them represent the pairs of purine and pyrimidine bases held together by hydrogen bonds. Specific examples of such bonding is shown in the center section, each dot between the pairs of bases representing a single hydrogen bond. The direction of the arrows correspond to the 3' to 5' direction of the phosphodiester bonds between adjacent molecules of 2'deoxyribose. After J. Mandelstam and K. McQuillen, *Biochemistry of Bacterial Growth,* 2nd ed. New York: Wiley, 1973.

A:T

G:C

T:A

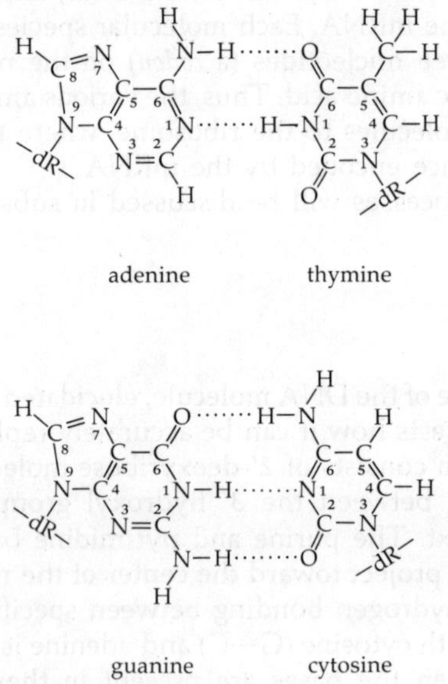

adenine   thymine

FIGURE 7.41

The pairing of adenine with thymine and guanine with cytosine by hydrogen bonding. The symbol —dR— represents the deoxyribose moieties of the sugar-phosphate backbones of the double helix. Hydrogen bonds are shown as dotted lines.

guanine   cytosine

and the amino, rather than the imino form of the aminated bases), only these pairs can fit within the hydrogen-bonding distances. The two hydrogen bonds that form between adenine and thymine, and the three hydrogen bonds that form between guanine and cytosine, are shown in Figure 7.41. The entire molecule can thus be described as a *linear sequence of nucleotide pairs*; the exact order of these pairs constitutes the genetic message which contains all the information necessary to determine the specific structures and functions of the cell.

• **The antiparallel structure of the DNA double helix**

Each strand of the double helix is a polarized structure; its polarity results from the sequential linkage of polarized subunits, the deoxyribonucleotides. As shown in Figure 7.42, each nucleotide has a 5'-phosphate end and a 3'-hydroxyl end; when a series of nucleotides are connected by phosphodiester linkages, a polarized chain is formed which also has a 5'-phosphate end and a 3'-hydroxyl end.

The two strands of the double helix are *antiparallel* (i.e., they have *opposite polarity*). This is shown in Figure 7.43; if we scan the diagram from top to bottom, we see that the left-hand strand has 5' → 3' polarity, whereas the right-hand

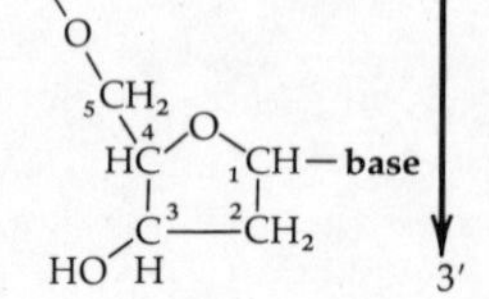

FIGURE 7.42

A deoxyribonucleotide. The molecule is polarized, having a 3'-hydroxyl group at one end and a 5'-phosphate group at the other.

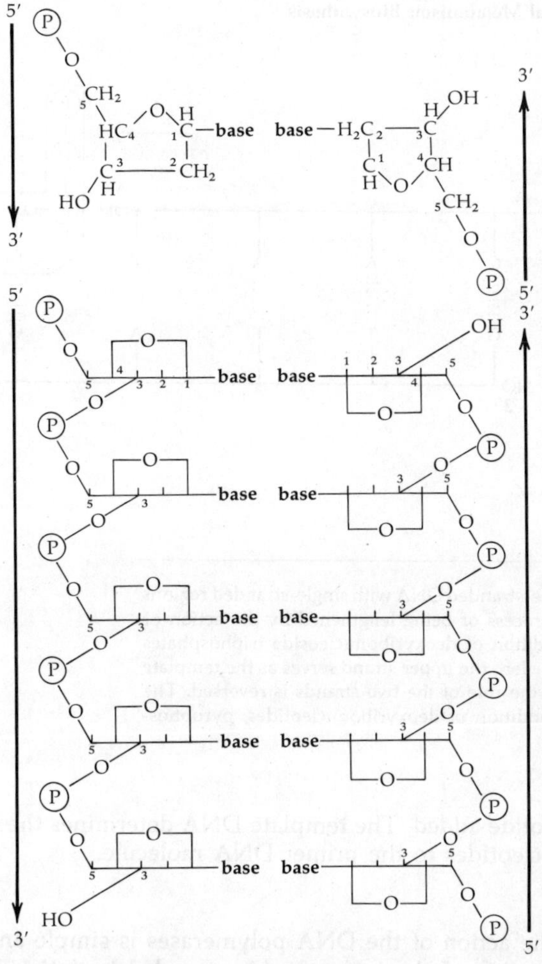

FIGURE 7.43

The antiparallel nature of the double helix. Above, the complete structural formula of one base pair is shown. Below, a segment of duplex is shown diagrammatically. Note that the left-hand strand runs from 5' to 3', reading from top to bottom, while the right-hand strand runs from 3' to 5'.

strand has 3' → 5' polarity. The significance of the antiparallel structure of DNA will become apparent when we consider the process of DNA replication.

• **DNA polymerases** The polymerization of DNA is catalyzed by enzymes called *DNA polymerases*. In addition to the four deoxynucleoside triphosphates (dATP, dGTP, dCTP, and dTTP), which are substrates for the reaction, 2 molecules of DNA are required: one is the *template* to which the substrate deoxynucleoside triphosphate molecules pair according to the rules of hydrogen bonding (G with C and A with T); and a second is the primer, to which the nucleotides are attached as a consequence of the polymerization (Figure 7.44). DNA synthesis proceeds in the 5' to 3' direction by the sequential formation of phosphodiester bonds between the α-phosphates of the 5'-deoxynucleoside triphosphates and the terminal 3'-hydroxy group of the primer, with the release of one pyrophosphate molecule (Ⓟ – Ⓟ) for each

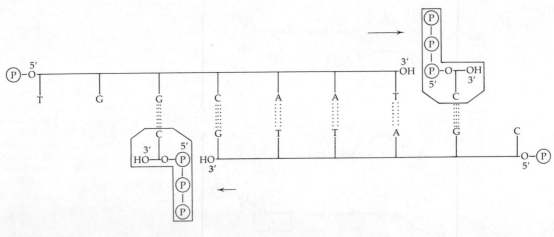

FIGURE 7.44

Diagram showing a short fragment of double-stranded DNA with single-stranded regions at each end. The two strands are in the process of being lengthened by the action of DNA polymerase, which catalyzes the addition of deoxyribonucleoside triphosphates to the 3'-hydroxyl ends of the chains. At the left, the upper strand serves as the template and the lower as the primer. At the right, the role of the two strands is reversed. The arrows show the direction of sequential addition of deoxyribonucleotides; pyrophosphate (Ⓟ—Ⓟ) is split off in the process.

deoxynucleotide added. The template DNA determines the sequence of addition of deoxynucleotides to the primer DNA molecule.

• Replication

Although the action of the DNA polymerases is simple and thoroughly understood, replication of the intact double-stranded bacterial chromosome is much more complicated, and many questions remain to be answered. As previously suggested, the DNA polymerases require a single-stranded template; thus, by still unknown mechanisms, the double-stranded chromosome must be separated into single strands, at least locally, before replication can occur. The requirement for the initial primer is met by the synthesis of a short complementary piece of RNA on each of the separated DNA strands (RNA polymerase does not require a primer). Then, polymerization occurs using the single-stranded DNA template and the RNA primer (Figure 7.45). From the point of initiation, replication occurs simultaneously in both directions around the chromosome (i.e., a "bubble" is formed in the chromosome), creating two "forks" where replication occurs. Since polymerization catalyzed by a DNA polymerase can occur only in the 5' to 3' direction and since the DNA helix is antiparallel, the replication of each strand of DNA at a replicating fork takes place in the opposite direction. After about 100 nucleotides have been polymerized onto the RNA primer by DNA polymerase III, thus creating short strands of DNA attached to RNA (sometimes called *Okazaki fragments* for their discoverer R. Okazaki), another type of polymerase, DNA

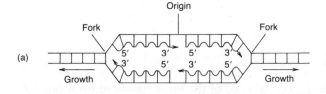

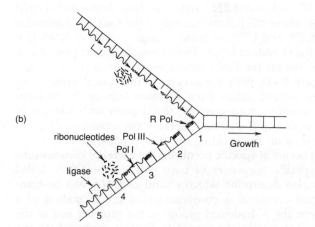

FIGURE 7.45

Schematic representation of the steps of replication of the bacterial chromosome. Part (a) represents a portion of a replicating bacterial chromosome at a stage shortly after replication has begun at the origin. The newly polymerized strands of DNA (wavy lines) are synthesized in the 5' to 3' direction (indicated by the arrows) using the preexisting DNA strands (solid lines) as a template. The process creates two replication forks which travel in opposite directions until they meet on the opposite side of the circular chromosome, completing the replication process. Part (b) represents a more detailed view of one of the replicating forks and shows the process by which short lengths of DNA are synthesized and eventually joined to produce a continuous new strand of DNA. For purposes of illustration, four short segments of nucleic acid are illustrated at various stages. In the first (1) primer RNA (thickened area) is being synthesized by an RNA polymerase (R Pol). Then, sucessively in (2) DNA is being polymerized to it by DNA polymerase III (Pol III); in (3) a preceding primer RNA is being hydrolyzed while DNA is being polymerized in its place by the exonuclease and polymerase activities of DNA polymerase I (Pol I); finally, the completed short segment of DNA (4) is joined to the continuous strand (5) by the action of DNA ligase (ligase).

polymerase I, with exonuclease activity, hydrolyzes the RNA fragment and simultaneously polymerizes more DNA in its place. Finally, the short pieces of DNA thus synthesized are linked together to form continuous complementary copies of the original DNA strands by the action of an enzyme, DNA ligase, which catalyzes the two successive reactions:

$$NAD^+ + enzyme \longrightarrow enzyme\text{-}AMP + nicotinic\ mononucleotide$$

$$enzyme\text{-}AMP + 5'\text{-phosphate end of DNA} + 3'\text{-hydroxyl end of DNA} \longrightarrow$$
$$5'\text{-}3'\ phosphodiester\ bond + enzyme + AMP$$

• Enzymes involved in replication of DNA

The formation of RNA primer is catalyzed by a still uncharacterized DNA-dependent RNA polymerase; it is probably a different enzyme from that involved in transcription, as suggested by the differing sensitivity of the two processes to the antibiotic, rifampin.

Three different DNA polymerase enzymes are present in *E. coli*: polymerase I (Pol I), polymerase II (Pol II), and polymerase III (Pol III). Pol III catalyzes the addition of nucleotides to the RNA primer; Pol I subsequently hydrolyzes it and polymerizes DNA in its place. The role of polymerase II remains unknown.

Much more remains to be learned about the details of the replication process. A hint of its complexity comes from the fact that mutations in at least seven different genes (the *dna genes*) prevent replication; the functions of the products of many of these genes are still not known. Some might unwind the DNA double helix in preparation for replication; others might attach it to the cell membrane where replication is thought to occur (Chapter 11).

THE SYNTHESIS OF RNA

Although the functional roles of the three classes of RNA (mRNA, tRNA, and rRNA) differ markedly, their synthesis is mechanistically identical. A highly complex enzyme, *DNA-dependent RNA polymerase* [commonly called *RNA polymerase* (Table 7.8)], brings about the polymerization of the four ribonucleoside triphosphates (ATP, GTP, CTP, and UTP) to form a single strand of RNA that is complementary to one of the strands of DNA. The ribonucleoside triphosphates pair with complementary bases on the DNA strand according to the base-pair rules discussed earlier (Figure 7.41), with the exception that uracil rather than thymine pairs with adenine. (Uracil differs from thymine only in that it lacks the methyl group in the 5 position; consequently, uracil pairs with adenine in the same manner as thymine does.) The ribonucleotides are subsequently polymerized by RNA polymerase, with the splitting off of pyrophosphates.

Initiation of transcription occurs at specific points on the bacterial chromosome that are defined by short specific sequences of base pairs in the DNA, called *promoters*. These sequences also determine which strand of DNA will be transcribed. Since polymerization occurs as a consequence of the formation of a phosphodiester bond between the 3'-hydroxyl group on the growing end of the RNA strand and the α-phosphate group of a free 5'-nucleoside triphosphate, the direction of transcription along the double-stranded DNA is set by the determination of the DNA strand to be copied. Polymerization continues in this manner until one or more genes have been transcribed. Then, when a sequence of base pairs (in the DNA) is encountered that causes the RNA polymerase molecule to be released, the synthesis of 1 molecule of RNA is completed. Although only one strand of DNA is transcribed, it is not the same strand in all regions of the bacterial chromosome. The strand copied, and hence the direction of transcription along the chromosome, varies without a discernible pattern. The product of transcription is single-stranded RNA. However, in certain cases (notably in tRNA), the strand always folds back on itself with the formation of hydrogen bonds between certain sequences of complementary bases (uracil with adenine and cytosine with guanine) to produce double-stranded regions. A schematic summary of the process of transcription is shown in Figure 7.46.

TABLE 7.8

Subunit structure of DNA-dependent RNA polymerase from *E. coli*

SUBUNIT	MOLECULAR WEIGHT	NUMBER PER POLYMERASE		
Alpha	41,000	2		
Beta	155,000	1	Core enzyme	Holoenzyme[a]
Beta-prime	165,000	1		
Sigma	86,000	1		

[a] The holoenzyme initiates the transcription process; the core enzyme catalyzes polymerization.

RNA + ATP + GTP + CTP + UTP

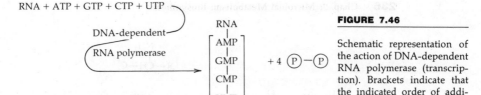

DNA-dependent

RNA polymerase

$$\begin{bmatrix} RNA \\ | \\ AMP \\ | \\ GMP \\ | \\ CMP \\ | \\ UMP \end{bmatrix} + 4 \; (P)-(P)$$

FIGURE 7.46

Schematic representation of the action of DNA-dependent RNA polymerase (transcription). Brackets indicate that the indicated order of addition of nucleotides is arbitrary.

Another enzyme, *polynucleotide phosphorylase*, present in procaryotes but not found in eucaryotes, can synthesize RNA. It catalyzes a DNA-independent reversible synthesis of RNA from nucleoside diphosphates, with the release of orthophosphate. In spite of its apparent ubiquity among procaryotes, its physiological function remains unknown.

THE SYNTHESIS OF PROTEINS

The products of transcription, mRNA, tRNA, and rRNA, all function in protein synthesis. rRNA is a component of *ribosomes*, which are the sites of protein synthesis. The functional form of the procaryotic ribosome, which has a sedimentation velocity of 70S, can be dissociated reversibly into a 30S and a 50S subunit by changing the concentration of Mg^{2+} in the suspending buffer. The molecular composition of procaryotic ribosomes is shown in Figure 7.47.

The many biochemical steps involved in the synthesis of a single protein will be considered separately.

• **Amino acid activation**
The activated forms of amino acids which are polymerized to form proteins are aminoacyl-tRNAs. They are synthesized in two steps catalyzed by a group of

FIGURE 7.47

The composition of procaryotic ribosomes.

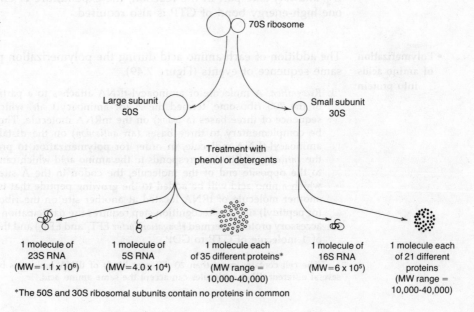

70S ribosome

Large subunit 50S

Small subunit 30S

Treatment with phenol or detergents

| 1 molecule of 23S RNA (MW=1.1 x 10⁶) | 1 molecule of 5S RNA (MW=4.0 x 10⁴) | 1 molecule each of 35 different proteins* (MW range = 10,000-40,000) | 1 molecule of 16S RNA (MW=6 x 10⁵) | 1 molecule each of 21 different proteins (MW range = 10,000-40,000) |

*The 50S and 30S ribosomal subunits contain no proteins in common

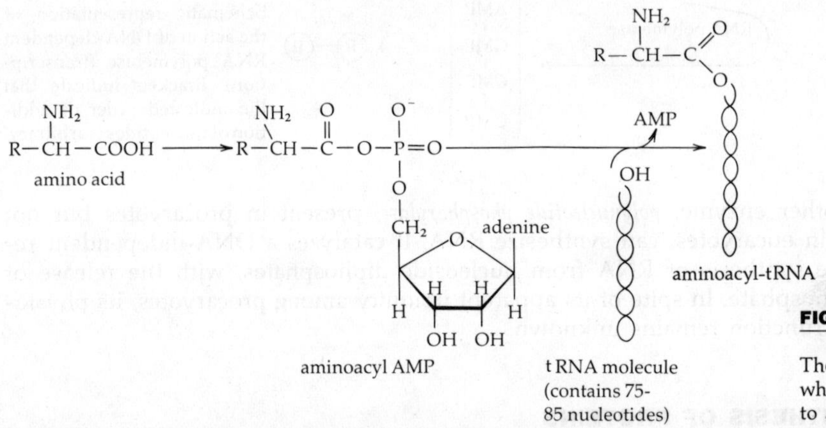

FIGURE 7.48

The process of amino acid activation, in which an amino acid becomes attached to a specific tRNA molecule.

enzymes, *aminoacyl-tRNA synthetases*. There are 20 such enzymes, each specific for a particular amino acid and particular molecules of tRNA.* The amino acid reacts with ATP to form an enzyme-bound intermediate, *aminoacyl adenylic acid;* then the aminoacyl group is transferred to a free hydroxyl group of the terminal AMP residue which all tRNA molecules contain (Figure 7.48).

• Initiation of transcription

The 30S ribosomal subunit, along with a special aminoacyl-tRNA (a methionyl-tRNA, to the amino group of which a formyl residue is attached), combine at a specific site on a molecule of mRNA, forming the *initiation complex.* Then, a 50S ribosomal subunit attaches to the complex to form the 70S ribosome, which mediates polymerization. Three accessory proteins (termed *initiation factors*, IF$_1$, IF$_2$, and IF$_3$) take part in the reaction; the expenditure of energy in the form of one high-energy bond of GTP is also required.

• Polymerization of amino acids into protein

The addition of each amino acid during the polymerization process involves the same sequence of events (Figure 7.49).

1. *Recognition.* A molecule of aminoacyl-tRNA attaches to a particular region of the 70S ribosome, termed the *A* (or aminoacyl) *site*, which exposes a sequence of three bases (a *codon*) on the mRNA molecule. The bases must be complementary to three bases (an *anticodon*) on the distal end of the aminoacyl-tRNA molecule, in order for polymerization to proceed. Since the *anticodon* of tRNA corresponds to the amino acid which can be attached to the opposite end of the molecule, the codon in the A site determines which amino acid will be added to the growing peptide that is attached to another molecule of tRNA located at another site on the ribosome, the *P* (or peptidyl) *site*. The recognition step requires the participation of two more accessory proteins (termed *elongation factor* EFT$_u$ and EFT$_s$) and the conversion of 1 molecule of GTP to GDP.

*The cell contains more than 20 different kinds of tRNA molecules because, in certain cases, several different tRNA molecules can accept the same amino acid.

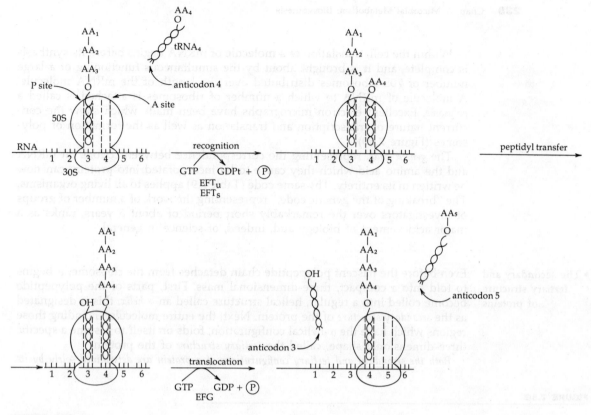

FIGURE 7.49

The sequence of events in the lengthening of the peptide chain. Amino acids (AA) are numbered by their order of addition to the peptide; numbered trios of lines symbolize the codons on the mRNA molecule.

2. *Peptidyl transfer.* The peptide at the P site is then transferred to the amino acid attached to the tRNA molecule at the A site, forming a new peptide bond and thereby lengthening the peptide chain by one aminoacyl residue. Peptidyl transfer is catalyzed by the 50S ribosomal subunit itself; no accessory proteins are required.

3. *Translocation.* Following peptidyl transfer, the peptide-bearing tRNA moves to the P site and the mRNA moves with it. The free tRNA molecule is thus displaced and the next codon is brought into the A site. Translocation requires one accessory protein (termed *elongation factor* EFG) and the hydrolysis of an additional molecule of GTP.

4. *Chain termination.* By repetition of the recognition, peptidyl transfer, and translocation steps, successive aminoacyl residues are added to the peptide chain in the order encoded by the sequence of codons in the mRNA molecule. The process continues until a codon is reached (UAG, UAA, or UGA) which causes the release of the completed peptide from the 70S ribosome. This process requires the intervention of a protein *release factor* (RF). Then, through the action of IF_3, the 70S ribosome dissociates into its 30S and 50S subunits.

Within the cell, translation of a molecule of mRNA begins before its synthesis is complete, and it is brought about by the simultaneous functioning of a large number of 70S ribosomes distributed over the length of the mRNA molecule. A molecule of mRNA to which a number of ribosomes is attached is called a *polysome*. Excellent electron micrographs have been made which show the concurrent nature of transcription and translation as well as the formation of polysomes (Figure 7.50).

The *genetic code*, representing the correspondence between codons on mRNA and the amino acid which they cause to be incorporated into protein, can now be written in its entirety. The same code (Table 7.9) applies to all living organisms. The "breaking of the genetic code," representing the work of a number of groups of investigators over the remarkably short period of about 5 years, ranks as a major achievement of biology and, indeed, of science in general.

• **The secondary and tertiary structure of proteins**

Even before the nascent polypeptide chain detaches from the ribosome, it begins to fold into a compact, three-dimensional mass. First, parts of the polypeptide become coiled into a regular, helical structure called an *α-helix*: this is designated as the *secondary structure* of the protein. Next, the entire molecule, including those regions which have the α-helical configuration, folds on itself to assume a specific three-dimensional shape, called the *tertiary structure* of the protein.

Both the secondary and tertiary configurations of a protein are determined solely by its

FIGURE 7.50

Photomicrograph of the simultaneous transcription and translation of a fragment of the chromosome of *E. coli*. The central horizontal line is DNA. The more wavy lines extending from it are molecules of mRNA to which a number of ribosomes are attached (polysomes). The gradual increase in length of the mRNA from left to right of the picture indicates that transcription by DNA-dependent RNA polymerase molecules (barely discernible at the junction of the DNA and mRNA) was proceeding in that direction, and that transcription began near the left edge of the picture. (×62,350) From B. Hamkalo and O. Miller, Jr., "Electronmicroscopy of Genetic Material," *Ann. Revs. Biochem.* **42**, 379 (1973).

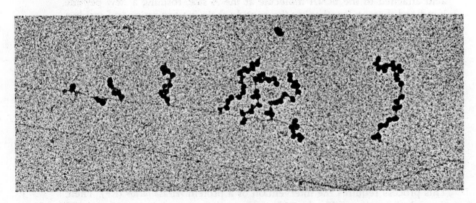

TABLE 7.9

The genetic code

FIRST LETTER	SECOND LETTERS							
	U		C		A		G	
U	UUU	phe[a]	UCU	ser	UAU	tyr	UGU	cys
	UUC	phe	UCC	ser	UAC	tyr	UGC	cys
	UUA	leu	UCA	ser	UAA	(none)[b]	UGA	(none)[b]
	UUG	leu	UCG	ser	UAG	(none)[b]	UGG	try
C	CUU	leu	CCU	pro	CAU	his	CGU	arg
	CUC	leu	CCC	pro	CAC	his	CGC	arg
	CUA	leu	CCA	pro	CAA	glu-N	CGA	arg
	CUG	leu	CCG	pro	CAG	glu-N	CGG	arg
A	AUU	ileu	ACU	thr	AAU	asp-N	AGU	ser
	AUC	ileu	ACC	thr	AAC	asp-N	AGC	ser
	AUA	ileu	ACA	thr	AAA	lys	AGA	arg
	AUG	met	ACG	thr	AAG	lys	AGG	arg
G	GUU	val	GCU	ala	GAU	asp	GGU	gly
	GUC	val	GCC	ala	GAC	asp	GGC	gly
	GUA	val	GCA	ala	GAA	glu	GGA	gly
	GUG	val	GCG	ala	GAG	glu	GGG	gly

[a] Amino acids are abbreviated as the first three letters in each case, except for glutamine (glu-N), asparagine (asp-N), and isoleucine (ileu).

[b] The codons UAA, UAG, and UGA are nonsense codons (see Chapter 13); UAA and UAG are called the ochre codon and the amber codon, respectively.

primary structure; a polypeptide with a given amino acid sequence will ultimately assume one particular form which represents its most stable state. This can be demonstrated experimentally. If a protein, such as an enzyme, is *denatured* (i.e., caused to unfold) under special conditions, removal of the denaturing agent permits the protein to refold into its native state, regaining full enzymatic activity.

The fraction of the molecule which exists in the α-helical configuration varies from zero to nearly 100 percent in different proteins. The α-helix is maintained by hydrogen bonds between the carboxyl oxygen of one peptide bond and the amide nitrogen of another peptide bond three residues farther along the chain. The resulting configuration is a helix in which a complete 360° turn is made once every 3.6 residues.

The tertiary structures of several different proteins have been completely elucidated by X-ray diffraction analysis. The first to be determined were those of myoglobin, the oxygen-carrying heme protein of muscle, and hemoglobin, the oxygen-carrying heme protein of blood. The complete solutions of these structures required many years of work. The complete three-dimensional structures of two enzymes, lysozyme and ribonuclease, have since been determined. A model of ribonuclease is shown in Figure 7.51.

The tertiary structure of most proteins is maintained by several types of bonds, of which the most important are *disulfide bridges* and *hydrophobic bonds*. Lysozyme, for example, contains four disulfide bridges; these are shown more clearly in Figure 7.52. Hydrophobic bonds result in the folding of the molecule such that the hydrophobic (nonpolar) amino acid side chains are closely packed together

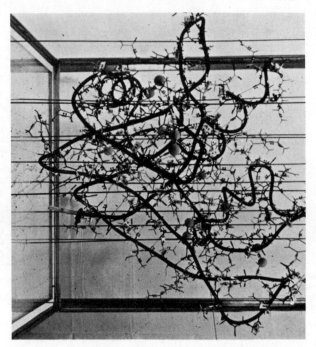

FIGURE 7.51

Photograph of a three-dimensional model of the protein ribonuclease S. The polypeptide backbone of the molecule is represented by the continuous black tube; the amino acyl side chains are represented by thinner stick models. Sulfur atoms are represented by large balls; four disulfide bridges are visible. Courtesy of F. Richards.

on the inside of the three-dimensional mass, most of the polarized groups projecting toward the outside. The configuration of the molecule is also maintained by interpeptide hydrogen bonds, similar to those of the α-helix; by side-chain hydrogen bonds, such as a hydrogen bond between a tyrosine hydroxyl group and a free carboxyl group; and ionic bonds, between free carboxyl and amino groups of acidic and basic amino acids, respectively.

• The quaternary structure of proteins

The quaternary structure of a protein is formed by the noncovalent association of several polypeptides, each of which has its own primary, secondary, and tertiary structure. Glutamic dehydrogenase (GDH), for example, is composed of between 24 and 30 identical subunits, each with a molecular weight of about 40,000. These subunits show two levels of aggregation: first, they associate to form monomers with a molecular weight of 250,000 to 350,000; second, four of these monomers associate to form an enzyme "molecule" (oligomer). The oligomer and monomer, but not the subunit, are catalytically active.

An extremely important property of many oligomeric enzymes is their susceptibility to regulation by *effectors*, small molecules that are unrelated to their substrates. Such enzymes are *allosteric proteins*: in addition to its catalytic sites, an allosteric protein possesses one or more binding sites for specific effectors.

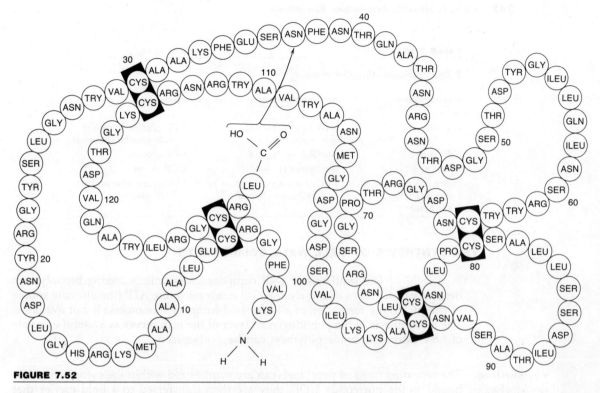

FIGURE 7.52

The primary structure of hen egg-white lysozyme, showing the four disulfide bridges between cysteine residues. The amino acids are abbreviated as the first three letters in all cases except isoleucine (ILEU), asparagine (ASN), and glutamine (GLN). From R. E. Canfield and A. L. Liu, "The Disulfide Bonds of Egg-white Lysozyme (Muramidase)," *J. Biol. Chem.* **240**, 1997 (1965).

The binding of an effector by an allosteric enzyme changes its affinity for its substrate so that its catalytic action is either stimulated or inhibited, as discussed in Chapter 8.

THE SYNTHESIS OF POLYSACCHARIDES

The properties of several systems for the synthesis of polysaccharides are described in Table 7.10. A characteristic feature of polysaccharide synthesis, like that of DNA synthesis, is the requirement for a primer. In the case of polysaccharide synthesis it is a short segment of the polysaccharide in question which acts as an acceptor of additional monomer units. In the synthesis of glycogen, where the function of the primer has been studied in detail, it has been found that the primer must contain more than four sugar units to function effectively (Figure 7.53). The molecular branching characteristic of glycogen is produced by a special *branching enzyme* that cleaves off small fragments from the end of the 1,4 linked linear polysaccharide chain, and inserts them in 1,6 linkage at another point.

TABLE 7.10

Polysaccharide-synthesizing systems

POLYSACCHARIDE	REPEATING UNIT	PRECURSOR
Glycogen	α-D-Glucose (1 → 4)	UPD-glucose (animals), ADP-glucose (bacteria)
Cellulose	β-D-Glucose (1 → 4)	GDP-glucose
Xylan	β-D-Xylose (1 → 4)	UDP-xylose
Pneumococcus type III capsular polysaccharide	β-D-Glucuronic acid (1 → 4) β-D-Glucose (1 → 3)	UDP-glucuronic acid, UDP-glucose

THE SYNTHESIS OF CELL WALL COMPONENTS

The synthesis of wall components is unique among biopolymers because *polymerization occurs outside the cell membrane* where ATP (the ultimate source of energy for the formation of all activated forms of monomers) is not available. The biosynthesis of the peptidoglycan layer of the wall serves as a useful example of how extra-membrane polymers can be synthesized.

• Synthesis of peptidoglycan

The repeating units of peptidoglycan are synthesized within the cytoplasm while bound to the nucleotide UDP; they are then transferred to a lipid carrier that facilitates their movement across the membrane. Finally, they are polymerized into peptidoglycan on the outside of the membrane by enzymes located on the membrane's outer surface.

Peptidoglycan differs from all biopolymers considered so far in that it is a two-dimensional network rather than a strand of molecules; it surrounds the cell like a sac. Therefore, its synthesis requires that the repeating units be chemically bonded in two dimensions (Figure 7.54). The chemical composition of the peptidoglycan sac is similar in all procaryotes, differing only in the amino acid composition of the tetrapeptide chain and the nature and frequency of bonding between tetrapeptide chains.

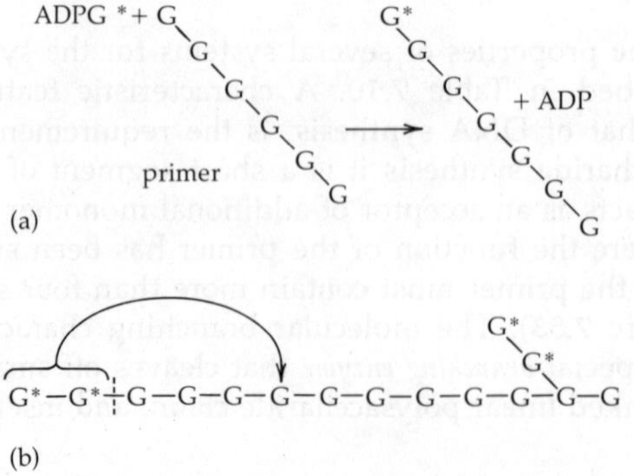

FIGURE 7.53

Chain elongation and branching in the enzymatic synthesis of the polysaccharide glycogen: G denotes glucosyl units. (a) Transfer of a glycosyl unit from ADPG to a primer molecule in glycogen synthesis. (b) Reaction catalyzed by the branching enzyme in glycogen synthesis.

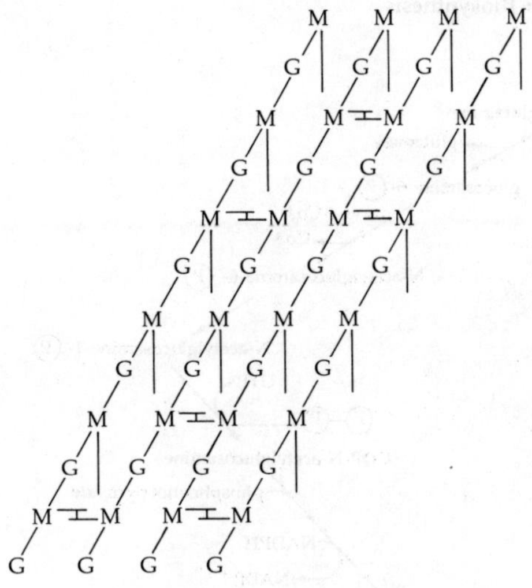

(a)

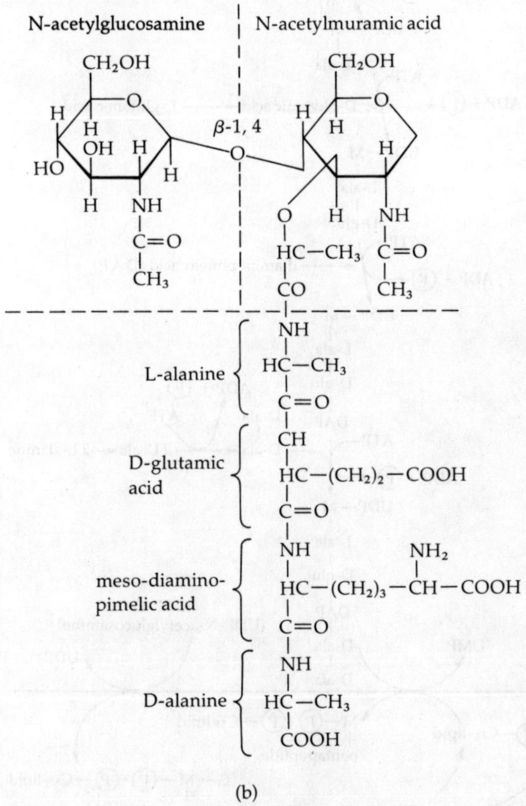

N-acetylglucosamine | N-acetylmuramic acid

β-1, 4

L-alanine

D-glutamic acid

meso-diamino-pimelic acid

D-alanine

(b)

FIGURE 7.54

Part (a) shows a schematic representation of the organization of the intact peptidoglycan sac of *E. coli*: G and M designate residues of N-acetylglucosamine and N-acetylmuramic acid, respectively. The lines extending from M represent tetrapeptides attached to muramic acid residues. In part (b), the repeating units are polymerized in one dimension by β-1,4-glycosididic bonds and in the other dimension by a peptide bond (⊥) between the carboxyl group of a D-alanine residue of one tetrapeptide and an ε-amino group of another tetrapeptide. Not all pairs of tetrapeptides are so joined. From J. M. Ghuysen, "Bacteriolytic Enzymes in the Determination of Wall Structure," *Bacteriol. Rev.* **32,** 425 (1968).

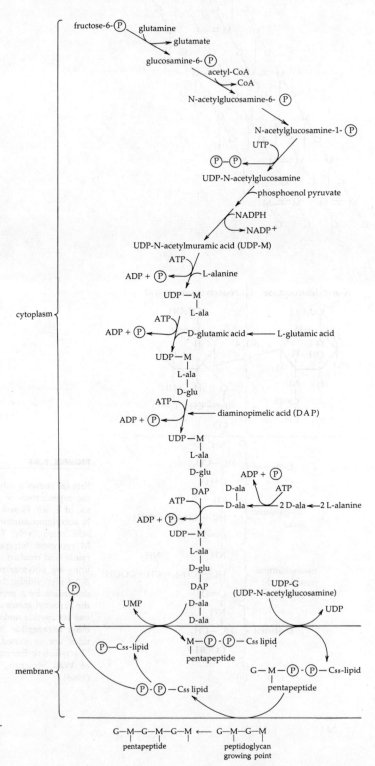

FIGURE 7.55

Pathway of peptidoglycan synthesis in *E. coli*.

The steps of biosynthesis of peptidoglycan by *E. coli* and their cellular location are summarized in Figure 7.55. The N-acetylmuramic acid is synthesized stepwise in the cytoplasm while attached to UDP; it is then transferred to a C_{55} isosprenoid carrier lipid (bactoprenol) in the membrane. In this form, an N-acetylglucosamine residue is added, completing the monomeric subunit of peptidoglycan. Attached to the long-chain lipid, the subunit can traverse the membrane to the outer face where it is polymerized by 1,4 glycosidic bonds to a growth point in the peptidoglycan sac. Finally, if the new monomeric subunit is to be involved in a cross link between peptide strands, an enzyme in the outer portion of the membrane catalyzes a *transpeptidation* reaction breaking the peptide bond between the two terminal D-alanine residues while forming a peptide bond between the subterminal D-alanine residue and the free amino group of diaminopimelic acid on an adjacent peptide strand. If the peptide is not to be involved in a cross link, the same enzyme removes the terminal D-alanine group without forming a new peptide bond.

Since the peptidoglycan layer is responsible for the structural strength that resists the internal osmotic pressure typical of bacteria growing in most environments, it must remain largely intact as the cell grows. However, the peptidoglycan layer can be likened to a mesh. Severing a mesh at one point does not reduce significantly its structural strength. During growth the peptidoglycan sac is opened at points by highly controlled autolytic enzymes, thus allowing enlargement through the insertion of new monomeric units.

• Lipopolysaccharides

The outermost layer of the wall of Gram-negative bacteria appears in thin-section electron micrographs to be similar to the cell membrane. However, its chemical structure is quite different (see Figure 7.56). It contains a complex lipid material, *lipid A*, which contains pairs of glucosamine residues that are substituted with long-chain fatty acids and phosphate (see Figure 7.56). Lipid A is attached to a complex carbohydrate (*core polysaccharide*) which is composed of a variety of sugars including 2-keto-3-deoxyoctonate (KDO), a compound that is only found in the lipopolysaccharide layer. Finally, on the outside of the layer strands of polysaccharides are found which protrude into the medium; these are the *O-specific chains*, which confer strain-specific antigenic properties on certain Gram-negative bacteria.

The pathway of synthesis of the O-specific chain is now understood in some detail. It bears certain resemblances to the synthesis of peptidoglycan, for during synthesis it is attached to a C_{55} isoprenoid lipid. However, it differs in that its entire synthesis occurs while it is attached to the C_{55} lipid.

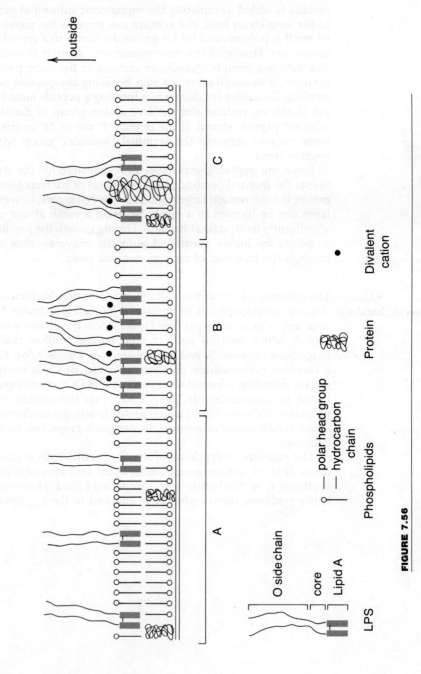

FIGURE 7.56

A possible arrangement of the major constituents in the outer layer of the Gram-negative cell wall. After H. Nikkaido, "Biosynthesis and Assembly of Lipopolysaccharide," in *Bacterial Membranes and Walls*, L. Leive (editor) New York: Marcel Dekker, 1973.

outside

Divalent cation

Protein

— polar head group
— hydrocarbon chain

Phospholipids

O side chain

core

Lipid A

LPS

8

247

FURTHER READING

Books   Kornberg, A., *DNA Synthesis.* San Francisco: W. H. Freeman, 1974.

Lehninger, A. L., *Biochemistry.* New York: Worth Publishers, 1970.

Mandelstam, J., and K. McQuillen (editors), *Biochemistry of Bacterial Growth,* 2nd ed. New York: Wiley, 1973.

Review   Dalton, H., and L. E. Mortenson, "Dinitrogen (N_2) Fixation (with a Biochemical Emphasis)," *Bact. Revs.* **36,** 231 (1972).

REGULATION

The overall metabolic activity of a growing microorganism reflects the simultaneous operation of a large number of interconnected pathways, both energy-yielding and biosynthetic. Each specific pathway comprises a number of individual reactions catalyzed by specific enzymes. The product of these many reactions is a new cell. Successful competition in nature demands that the growth process be both rapid and efficient. The cell must therefore possess appropriate means of balancing the rates of the constituent reactions in each metabolic pathway as well as the overall rates of flow through the various pathways. The cell is also able to make qualitative adjustments in its metabolic machinery in response to changing needs imposed by environmental changes. Microorganisms have evolved a variety of *regulatory mechanisms* which accomplish this.

The fact that microbial metabolism is highly regulated can be inferred from several observations. One concerns the changes in macromolecular composition of bacteria that occur in response to the kinds of nutrients available to the cells. The first systematic study of this aspect of regulation was made almost 20 years ago by O. Maaløe and his associates with enteric bacteria, principally *Salmonella typhimurium*. Enteric bacteria can synthesize all cellular components from a single carbon source and inorganic salts, and they can utilize a number of organic compounds as carbon sources. Certain carbon sources such as acetate are metabolized slowly and support growth at a low rate; others, such as glucose, are metabolized more rapidly and support more rapid growth. If precursors of macromolecules (e.g., amino acids) are added to the medium, growth is even more rapid. Thus, by changing the composition of the medium while maintaining temperature constant, one can obtain growth of *S. typhimurium* with doubling times that vary between 20 minutes and several hours (at 37°C). Furthermore, the size and composition of the cells vary systematically with growth rate (Table 8.1). Cells growing at high rates are richer in RNA, poorer in DNA, and larger than cells growing at lower rates. These changes in cellular composition are dependent on growth rate alone, provided that temperature is not changed.

The variation of the macromolecular composition of cells with growth rate can be interpreted as follows: a cell growing at a high rate must synthesize protein much more rapidly than does a slowly growing cell. This higher rate of protein synthesis requires that the cell contain more ribosomes, since the rate of protein synthesis per ribosome is constant. The ability of a bacterium to modulate its content of ribosomes is of great importance for the maintenance of high growth rates under changing environmental conditions. An insufficient complement of ribosomes would clearly restrict growth rate; an excess of ribosomes would also do so because the cell would be engaged in nonproductive synthesis of ribosomal protein and RNA.

The biosynthesis and degradation of small molecules by bacteria are similarly subject to close regulation, as shown by the following observations on enteric bacteria:

1. When growing in synthetic media containing a single organic compound as source of energy, bacteria synthesize all the monomeric precursors (e.g., amino acids) of macromolecules at rates which are precisely coordinated with the rates of macromolecule synthesis.

2. The endogenous synthesis of any one of these monomeric precursors is immediately arrested when the same compound is added to the medium, provided that the exogenously furnished monomer can enter the cell.

3. Formation of the enzymes that mediate biosynthesis of the monomers in question is also arrested.

4. Bacteria frequently synthesize enzymes responsible for the dissimilation of certain organic substrates only if the compounds in question are present in the medium.

5. When presented with two organic substrates, a bacterium first synthesizes the enzymes required to dissimilate the compound which supports the more rapid growth; only after this compound has been completely utilized are the enzymes required to dissimilate the second compound synthesized.

TABLE 8.1

Size and composition of *Salmonella typhimurium* growing exponentially at various rates

DOUBLING TIME (MIN)	AVERAGE WEIGHT OF A CELL (μg)	CONTENT OF			70S RIBOSOMES (NUMBER/CELL)
		DNA (%)[a]	TOTAL RNA(%)[a]	rRNA[b] (%)[a]	
25	0.77	3.0	31	25	69,800
50	0.32	3.5	22	14	16,300
100	0.21	3.7	18	9	7,100
300	0.16	4.0	12	4	2,000

[a] Calculated as a percentage of the dry weight of the cell.
[b] RNA contained in ribosomes.

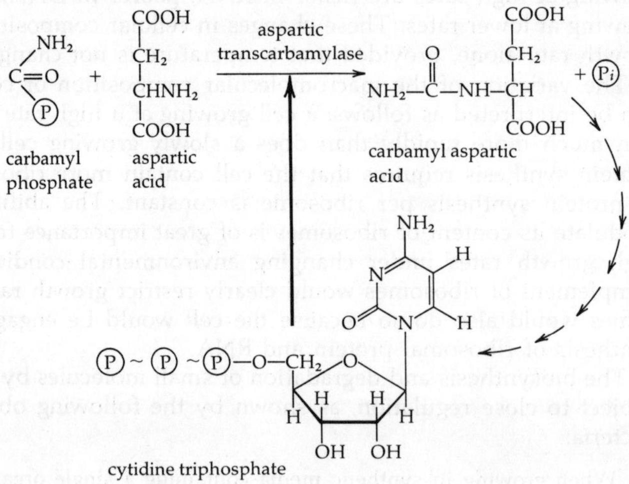

FIGURE 8.1

The allosteric control of the first step in pyrimidine biosynthesis (condensation of carbamyl phosphate and aspartic acid to form carbamyl aspartic acid). The enzyme responsible, aspartic transcarbamylase, is allosterically inhibited (bold arrow) by cytidine triphosphate, the eventual product of the biosynthetic sequence.

THE BIOCHEMICAL BASIS OF REGULATION

Two different regulatory mechanisms operate in the cell: the regulation of *enzyme synthesis*, and the regulation of *enzyme activity*. Both are mediated by compounds of low molecular weight, which are either formed in the cell as intermediary metabolites or enter it from the environment. Both regulatory mechanisms involve the operation of a special class of proteins, called *allosteric proteins*.* The fundamental importance of allosteric proteins as the key components of regulatory systems was first perceived by J. Monod in 1963.

Allosteric proteins are proteins whose *properties change* if certain specific small molecules, *effectors*, are bound to them. Hence, *allosteric proteins are mediators of metabolic change which is directed by changes in concentration of the small effector molecules*.

There are two classes of allosteric proteins: *allosteric enzymes*, whose activities are either enhanced or inhibited when combined with their effectors, and *regulatory allosteric proteins*, devoid of catalytic activity, which modulate the synthesis of specific enzymes.

Regulatory allosteric proteins attach to the bacterial chromosome near the specific structural genes whose repression they control. This attachment can be modified by the binding of small effector molecules to the regulatory proteins, thereby changing the rate at which specific messenger RNAs are synthesized.

• **The regulation of enzymatic activity**

The most thoroughly studied allosteric proteins are the allosteric enzymes, exemplified by aspartic transcarbamylase (ATCase). ATCase catalyzes the first reaction in the pathway of biosynthesis of pyrimidines (Figure 8.1). Its activity is inhibited by an end product of the pathway, cytidine triphosphate (CTP). Thus,

*The word allosteric means *differently shaped*, and it alludes to the fact that the effectors that regulate the activity of an allosteric enzyme have a structure different from that of the substrate of the enzyme.

elevated intracellular concentrations of CTP inhibit the functioning of ATCase and consequently the formation of more CTP until its concentration decreases to an optimal level. ATP, a second effector of ATCase, *activates* the enzyme, and thus serves to coordinate the synthesis of purine and pyrimidine nucleotides.

The rate of the reaction catalyzed by ATCase is a sigmoid function of substrate concentration (Figure 8.2), rather than a hyperbolic function which is typical of nonallosteric enzymes (Figure 8.3). Such kinetics are frequently associated with allosteric enzymes. The specific action of the allosteric inhibitor (also shown in Figure 8.2) is to increase the sigmoid nature of the curve, and hence to reduce the rate of reaction at low concentrations of substrate; such inhibition is largely reversed by increasing the concentration of the substrate. The sigmoid nature of the curve relating activity to substrate concentration also shows that the enzyme has more than one site able to bind substrate molecules (*catalytic sites*). The attachment of a substrate molecule to one of these sites increases the ability of the enzyme to bind additional molecules of substrate at other catalytic sites; i.e., there is a *cooperative interaction* of substrate molecules with the enzyme. Thus, as substrate concentrations are increased, the *rate of acceleration* of the velocity of the reaction increases. The same relationship obtains for effector molecules; they too interact cooperatively with the enzyme at the specific *effector sites*.

As might be suggested by the complexity of the curves shown in Figure 8.2, allosteric enzymes are always proteins of relatively high molecular weight, and are composed of multiple subunits. As a rule, these subunits are identical, each possessing both a catalytic and an allosteric site. However, ATCase is composed of two different kinds of subunits, one with catalytic function and the other with regulatory functions. This fact makes it particularly easy to show that the allosteric and catalytic sites are topologically distinct. Upon mild chemical treatment (e.g., with *p*-Cl-mercuribenzoate), ATCase dissociates into subunits. One (the catalytic subunit) possesses all the catalytic activity of the intact enzyme, but is insensitive to allosteric inhibition by cytidine triphosphate or to allosteric activation by ATP. The other (the regulatory subunit) has no catalytic activity, but has the capacity to bind either cytidine triphosphate or ATP (Figure 8.4). Thus, binding cytidine triphosphate to one subunit inhibits the specific, enzyme-catalyzed reaction, which

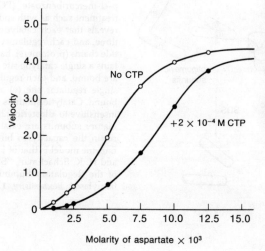

FIGURE 8.2

The rate of reaction of aspartic transcarbamylase as a function of the concentration of one of its substrates, aspartic acid. Note the sigmoid nature of the curve. The effect of the allosteric inhibitor, CTP, on aspartic transcarbamylase activity is also shown. Redrawn from J. C. Gerhart and A. B. Pardee, "The Enzymology of Control by Feedback Inhibition," *J. Biol. Chem.* **237**, 891 (1962).

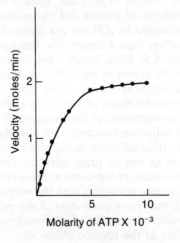

FIGURE 8.3

The rate of reaction of a typical nonallosteric enzyme (nucleoside diphosphokinase) as a function of the concentration of one of its substrates, ATP. Note the hyperbolic nature of the curve. Redrawn from C. L. Ginther and J. L. Ingraham, "Nucleoside Diphosphokinase of *Salmonella typhimurium*," *J. Biol. Chem.* **249**, 3406 (1974).

occurs on the other. The precise mechanism of such allosteric inhibition remains unexplained, but it evidently involves a conformational change in the enzyme. When the concentration in the cell of the end product (effector) of a given biosynthetic pathway rises, the catalytic activity of the allosteric enzyme with which it combines is reduced. Since the activity of this enzyme in turn controls the rate of biosynthesis of the end product (effector), the formation of the latter is also reduced and its intracellular concentration begins to fall. Therefore, allosteric inhibition also decreases. Through this device of *feedback regulation*, termed *end-*

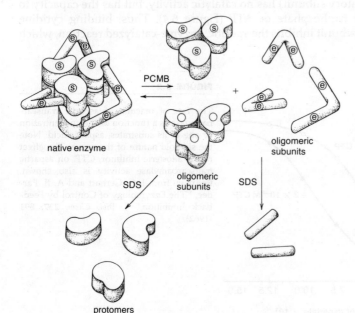

native enzyme

oligomeric subunits

oligomeric subunits

SDS

SDS

protomers

FIGURE 8.4

Dissociation of native aspartic transcarbamylase into two catalytic and three regulatory subunits (oligomers) by mild chemical treatment such as with p-cl-mercuribenzoate (PCMB). Stronger chemical treatment such as with sodium dodecylsulfate (SDS) reveals that each catalytic subunit is comprised of three, and each regulatory subunit of two, polypeptide chains (protomers). Each catalytic protomer contains a single catalytic site (s) at which the substrates are bound, and each regulatory protomer contains a single regulator site (e) at which the effectors are bound. Catalytic subunits are catalytically active but insensitive to allosteric inhibition or activation. Regulatory subunits are catalytically inactive, but they retain the capacity to bind the allosteric effectors. Enzyme model is that of J. A. Cohlberg, V. R. Pigiet, and H. K. Schachman, "Structure and Arrangement of the Regulatory Subunits in Aspartate Transcarbamylase," *Biochemistry*, **11**, 3393 (1972).

product inhibition, the intracellular concentrations of biosynthetic intermediates are very closely controlled. Typically, the enzyme that mediates the first reaction of a given biosynthetic pathway is the specific target of inhibition by the end product (or products) of that pathway. It is evident that when the first enzyme of a specific pathway is the target of regulation, neither the end product nor the intervening intermediates in its formation can accumulate in the cell. By such regulation, the rate of generation of metabolic intermediates also regulates the rates of operation of catabolic pathways. By this means, the primary rate of carbon flow into all biosynthetic pathways is controlled, as well as the overall rate of ATP synthesis.

REGULATION OF ENZYME SYNTHESIS

End-product inhibition mediated by allosteric enzymes is in large part sufficient to assure that all biosynthetic and catabolic pathways operate in balance with one another. However, when the product of a pathway is not required, the enzymes that catalyze the reactions of that pathway become super-fluous. The regulation of microbial metabolism also includes mechanisms which modulate the enzymatic composition of the cell; this regulation is effected at the level of gene expression, and its genetic aspects are considered in Chapter 13.

• Induction of
enzyme synthesis

Many bacteria can use a wide range of different organic compounds as carbon and energy sources, but at any given time, only one of these compounds may be present in the environment. Although the genetic information necessary to synthesize the relevant enzymes is always present, its phenotypic expression is environmentally determined, a given enzyme being synthesized in response to the presence of its substrate.

Induction of enzyme synthesis is mediated by noncatalytic allosteric proteins. These substances are the products of specific regulatory genes; they control enzyme synthesis is a negative manner, by binding to the bacterial chromosome at a site near the structural genes which determine enzyme synthesis, thus preventing their transcription. They are termed *repressors.* When a repressor is bound to its specific allosteric effector, called an *inducer,* its ability to block transcription is lost, and specific enzyme synthesis is initiated.

This kind of regulation of enzyme synthesis was first elucidated by J. Monod and his collaborators through study of the induction of the enzymes which govern the dissimilation of lactose by *Escherichia coli.*

Although *E. coli* contains the enzymes necessary for the metabolism of glucose under all growth conditions (such enzymes are termed *constitutive*), glucose-grown cells contain only barely detectable levels of the enzymes necessary for the initial steps in the metabolism of lactose. These enzymes are, however, present at high levels in lactose-grown cells. Since lactose induces their formation, they are termed *inducible enzymes;* they include *galactoside permease,* an enzyme that mediates entry of lactose into the cell, and *β-galactosidase,* an enzyme that catalyzes the hydrolytic cleavage of lactose to its constituent monosaccharides, glucose and galactose. Galactose formed by the action of *β*-galactosidase induces, in turn, a sequence of enzymes responsible for the metabolism of galactose. Thus, exposure of the cell to lactose leads to a direct induction of the enzymes responsible for the cleavage of lactose into its constituent monosaccharides and an indirect (secondary) induction of the enzymes responsible for galactose metabolism. Such a complex inductive event is known as *sequential induction*, since the metabolism of the primary substrate inducer (in this case, lactose) leads to the intracellular formation of a metabolite (in this case, galactose) with inductive properties different from those of the primary substrate (Figure 8.5).

If the only source of carbon and energy available to a culture is the inducer substrate itself, the kinetics of enzyme synthesis is complex because the enzyme is required to metabolize the substrate and to produce the ATP and metabolic intermediates required to synthesize more enzyme. The rate of enzyme synthesis is thus directly related to the amount of enzyme already synthesized. Such complexity can be avoided by conditions of *gratuity*: enzyme synthesis unaffected by metabolic events which normally result from the activity of the enzymes being induced. Gratuity can be attained by employing a *gratuitous inducer* (an analogue of the inducer substrate which can cause induction but which cannot be metabolized by the induced enzyme); the culture is grown at the expense of a metabolically unrelated carbon and energy source (e.g., glycerol).

Gratuitous induction of β-galactosidase follows a very characteristic time course. Addition of inducer is followed by a brief lag period, after which enzyme synthesis begins, and continues (as long as the inducer is present) at a constant *differential rate*, i.e., β-galactosidase comprises a constant fraction of all newly

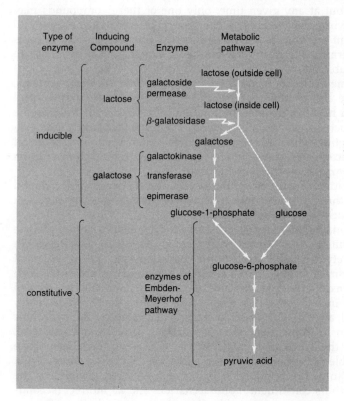

FIGURE 8.5

The changes in the enzymatic composition of *Escherichia coli* that result from exposure to lactose. Two inducible enzymes are synthesized coordinately in direct response to lactose. One product of this pathway (glucose) is futher metabolized by the constitutive enzymes of the Embden-Meyerhof pathway. The other (galactose) induces the coordinate synthesis of three additional enzymes that catalyze its conversion to an intermediate of the Embden-Meyerhof pathway (glucose-1-phosphate).

synthesized cell protein. The constancy of the differential rate of enzyme synthesis following induction can be demonstrated by plotting enzyme activity as a function of cell mass (Figure 8.6). The straight line obtained shows that a given increment of total cell mass is always accompanied by a fixed increment of β-galactosidase. Each inducer elicits a characteristic differential rate of enzyme synthesis; measurements of differential rates allows the effectiveness of different inducers to be compared in quantitative terms. Although the conditions are not gratuitous if lactose is the inducer, estimations of its effectiveness can also be made (Figure 8.6). Curiously enough, the natural inducer of β-galactosidase, lactose, is less effective than many synthetic analogues, which do not occur in nature. In fact, a refined study of the role of lactose as an inducer has revealed that in order to initiate enzyme synthesis, small amounts of lactose must first be converted (through a different catalytic activity of the very low levels of β-galactosidase present in uninduced cells) to another galactoside which is, in reality, the inducing compound. Therefore, lactose is not a direct inducer of enzyme synthesis, whereas the more active synthetic analogues (e.g., isopropyl β-thio-D galactoside) are. This appears to be an exceptional situation; as a rule, the substrates of inducible enzymes or their major metabolic products act directly as inducers.

Any compound capable of inducing β-galactosidase also induces the synthesis of galactoside permease, and the ratio of the differential rates of synthesis of these two enzymes is always constant. When the synthesis of two or more enzymes shows such a tight physiological linkage, it is said to be *coordinate*. Coordinacy is usually, but not always, the consequence of a close association on the chromosome of the structural genes encoding the enzymes in question; i.e., it is the phenotypic expression of common genetic control (see Chapter 13).

When sequential induction occurs as a consequence of exposure of the cells to a primary inducer (e.g., lactose, Figure 8.5), the secondary inductive event (e.g., induction of the galactose-catabolizing enzymes), is never coordinate with the primary one (e.g., induction of β-galactosidase and galactoside permease), since there is no common *genetic* control governing the two processes. This can be demonstrated physiologically: direct exposure of the cells to a metabolite inducer normally formed from the primary substrate does not result in synthesis of the enzymes which dissimilate the primary substrate. Thus, cells of *E. coli* grown at the expense of galactose contain only low levels of β-galactosidase.

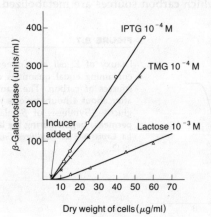

FIGURE 8.6

The differential rate of β-galactosidase synthesis by *E. coli* growing in a mineral salts medium with glycerol as the carbon source and exposed to three different inducers: the natural inducer, lactose, and two lactose analogues, isopropyl thiogalactoside (IPTG) and thiomethylgalactoside (TMG). Although the analogues do not occur in nature, and cannot be hydrolyzed by β-galactosidase, they are both considerably more effective than lactose as inducers.

• Positive control of enzyme induction   Although most systems of enzymatic induction so far analyzed have been shown to operate solely through negative control, like the lactose system, the induction of the enzymes of arabinose metabolism in *E. coli* involves a type of control by the repressor which has both a positive and a negative component. The arabinose repressor is always bound to the chromosome, its influence on the transcription process being changed by binding of the inducer (arabinose). When arabinose is not present, the repressor blocks transcription just as does the lactose repressor. However, when bound to arabinose, the repressor undergoes a conformational change converting it into an activator, which triggers transcription. In terms of their physiological consequences, positive and negative control cannot be distinguished. They can be distinguished only by a refined analysis of the properties of mutations that affect the regulatory gene.

• Catabolite repression   About 35 years ago, J. Monod discovered a phenomenon known as *diauxy*. He noted that a culture of *E. coli* in a medium containing certain pairs of compounds as carbon sources (e.g., glucose and lactose) underwent two distinct growth cycles, characterized by two exponential phases of growth separated by a distinct lag phase (Figure 8.7). Glucose is utilized during the first growth cycle and lactose is utilized during the second. The enzymes necessary for the metabolism of lactose are not synthesized (even though the inducer is always present) until the glucose in the medium has been exhausted. While glucose is being metabolized, the induction of β-galactosidase and galactoside permease is prevented. Growth curves of similar form are obtained in media containing glucose and various other carbon sources which are metabolized by inducible enzymes. This kind of repression of enzyme induction was at first thought to be restricted to glucose and was termed the *glucose effect*. Subsequent work has shown that all rapidly metabolizable energy sources repress the formation of enzymes necessary for the dissimilation of energy sources that are more slowly attacked, and the phenomenon is now known as *catabolite repression*.

As a result of catabolite repression, the cell always first utilizes the substrate that supports the most rapid rate of growth. However, catabolite repression also regulates the rate at which carbon sources are metabolized. Even when growing

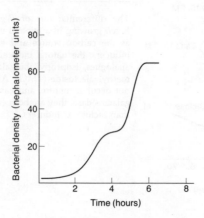

FIGURE 8.7

Diauxy of *E. coli* in a mineral salts medium initially containing equal quantities of glucose and lactose as sources of carbon. The transient cessation of growth after about 4 hours reflects the complete utilization of glucose. Synthesis of β-galactosidase and galactoside permease begins during the lag period. After J. Monod, *La Croissance des Cultures Bacteriennes*. Paris: Hermann, 1942.

on a single carbon source, the levels of enzymes required to generate ATP are regulated through catabolite repression.

The extent of catabolite repression is proportional to the intracellular concentration of ATP. A single allosteric protein, termed CAP (*catabolite activator protein*) is responsible for the regulation of all enzymes under this form of control. Its effector is a cyclic nucleotide, 3',5'-cyclic adenosine monophosphate (cyclic AMP), which is synthesized from ATP. By mechanisms that are still incompletely understood, the intracellular concentration of cyclic AMP varies inversely with ATP supply. Hence, when the cell is growing on rapidly metabolizable substrates, intracellular concentrations of cyclic AMP are low; when the cell is growing on slowly metabolizable substrates, the concentrations are high. Cyclic AMP has been found in every bacterium so far examined. Since it is not an intermediate of any known metabolic pathway, its only physiological role in bacteria is presumably a regulatory one.* Binding of cyclic AMP to CAP causes it to undergo an allosteric transition which allows it to bind to the chromosome near the genes encoding catabolite repressible enzymes and, through such binding, to stimulate transcription of these genes. When unbound to cyclic AMP, the CAP protein is inactive. Hence, both the CAP protein and relatively high internal concentrations of cyclic AMP are necessary for the synthesis of enzymes under control of catabolite repression. Mutant strains that are unable to synthesize either CAP or cyclic AMP are, as a consequence, unable to synthesize a wide variety of enzymes.

It should be noted that the effectiveness with which an energy source can act as a catabolite repressor depends not on its specific chemical structure, but solely on its effectiveness as a source of carbon and energy. Since chemoheterotrophic bacteria differ widely with respect to the relative effectiveness of different organic compounds as carbon and energy sources, the compounds which are most effective as catabolite repressors in one organism may be relatively ineffective in another. Thus, glucose is far more effective than succinate as a catabolite repressor for *E. coli*, whereas the relative effectiveness of these two compounds is reversed for *Pseudomonas putida*.

• End-product repression   In many bacteria the addition of a compound which is the end product of a biosynthetic pathway (e.g., an amino acid) to the growth medium of a bacterium results in immediate slowing or arrest of the *synthesis* of the enzymes of that particular pathway. This is called *end-product repression*.

Enzymes normally subject to end-product repression can be *derepressed* (i.e., synthesized at greater than normal rates) when the intracellular concentration of the end product falls to a very low level. For example, the enzymes which mediate arginine biosynthesis can often be derepressed in bacteria as a result of growth in a medium that is supplemented with many other amino acids. Under these conditions, the rate of protein synthesis is essentially limited by the rate of arginine synthesis. The intracellular concentration of arginine therefore falls, triggering derepression.

Thus, biosynthetic pathways are subject to two types of feedback regulation: end-product inhibition, which regulates enzyme activity, and end-product repression, which regulates enzyme synthesis. Although the effect of end-product

---

*Cyclic AMP also acts as a regulator in eucaryotes where it has been shown to play a role not solely in enzyme expression, but also in cellular differentiation.

repression is immediate in regulating the rate of enzyme synthesis, were it to act alone, the pathway would continue to function until preexisting enzymes were diluted to low levels as a consequence of further growth; end-product inhibition, however, brings about an immediate cessation of the operation of the pathway. Thus, end-product repression and end-product inhibition are complementary mechanisms, which in combination achieve highly efficient regulation of biosynthetic pathways. All biosynthetic pathways are regulated by end-product repression, and usually all enzymes of each specific pathway are so regulated.

End-product repression of the enzymes catalyzing the biosynthetic pathway for tryptophan in E. coli has been explored in considerable detail. A specific gene (*trpR*) governs the synthesis of an allosteric protein called the *tryptophan repressor*, whose only function is to regulate the biosynthesis of the enzymes of this pathway. In its free state the repressor is inactive, but when bound to tryptophan (the *corepressor*), it undergoes an allosteric change that enables it to bind to a region of the chromosome near the structural genes, thus preventing their transcription. Mutant strains that are unable to produce the repressor become insensitive to the presence of the end product; they produce high levels of the tryptophan biosynthetic enzymes under all conditions (i.e., they are *constitutive* for these enzymes).

Enzymes of the arginine biosynthetic pathway in enteric bacteria are regulated by a mechanism quite analogous to that which regulates the tryptophan pathway; their synthesis is also controlled by a specific repressor protein, encoded by the gene *argR*. However, the enzymes of the histidine pathway are regulated in a different manner. Although the detailed mechanism still remains unknown, it is clear that no specific repressor protein for the histidine pathway exists. It is possible that one of the enzymes of the pathway also acts as a repressor.

The generalizations drawn in this chapter come largely from studies on enteric bacteria; they might apply to other bacteria as well.

• **Regulatory mechanisms: a summary**

All known regulatory mechanisms are mediated through allosteric proteins, the activity of which is altered by the binding of small molecules. Hence, they serve as sensitive detectors of the intracellular concentration of key metabolites and modulate the overall metabolic activity of the cell in such a manner as to maximize the rate of growth, by ensuring the conversion of nutrients into cell material with maximal efficiency. These mechanisms are summarized and compared in Table 8.2.

PATTERNS OF REGULATION

Certain generalizations can be made about the regulation of metabolic pathways. In unbranched biosynthetic pathways, the first enzyme of the pathway is regulated by end-product inhibition and the biosynthesis of all enzymes of the pathway are subject to end-product repression. In catabolic pathways involving the metabolism of carbon sources that are commonly present in the environment of the organism, flow through the pathway is regulated largely by allosteric enzymes; the enzymes of the pathway are constitutively synthesized and are subject to minimal catabolite repression. Pathways of catabolism of substrates

TABLE 8.2

Key regulatory mechanisms operative in bacteria

MECHANISM	ALLOSTERIC PROTEIN	EFFECTOR	ACTIVITY OF ALLOSTERIC PROTEIN		PHYSIOLOGICAL RESULT
			ALONE	BOUND TO EFFECTOR	
End-product inhibition (feedback control of pyrimidine synthesis)[a]	First enzyme of a pathway (aspartic transcarbamylase)	End product of pathway (CTP)	Catalyzes first step of pathway	Has decreased catalytic activity	Regulates biosynthesis of small molecules (CTP)
Enzyme induction: negative control (induction of β-galactosidase)	The repressor (product of *lacI* gene)	The inducer (lactose)[b]	Binds to chromosome; prevents enzyme synthesis	Cannot bind to chromosome; enzyme synthesis occurs	Enzymes synthesized only when substrate is present in medium
Enzyme induction: positive control (induction of enzymes that metabolize arabinose)	Repressor-activator (product of *araC* gene)	The inducer (arabinose)	Repressor form; binds to chromosome and prevents enzyme synthesis	Activator form: binds to chromosome and permits enzyme synthesis	Enzymes synthesized only when substrate is present in medium
Catabolite repression (glucose repression of β-galactosidase synthesis)	CAP protein	Cyclic AMP	Cannot bind to chromosome	Binds to chrosome and stimulates enzyme synthesis	Allows cell to use most favorable source of carbon; regulates catabolic rate
End-product repression (regulation of enzymes necessary for synthesis of tryptophan)	Repressor (product of *trpR* gene)	End product of the pathway (tryptophan)	Cannot bind to chromosome	Binds to chromosome and prevents enzyme synthesis	Regulates synthesis of biosynthetic enzymes (enzymes of tryptophan pathway)

[a] Note that specific examples are shown in parentheses, for each of the key regulatory mechanisms.

[b] Actual inducer is not lactose; see text.

that occur more rarely in the environment of the organism are regulated primarily by enzyme induction and by catabolite repression, rather than by allosteric enzymes.

• End-product inhibition in branched pathways

Many biosynthetic pathways have two or more end products. End-product inhibition in such pathways is more complex than it is in simple, unbranched pathways. In a branched pathway leading, for example, to two different amino acids (Figure 8.8), it is obvious that feedback inhibition exerted by an end product (e.g., amino acid I) on the enzyme catalyzing the first step of the pathway (enzyme **a**) would likewise prevent synthesis of other end products (i.e., amino acid II). Consequently, the presence of amino acid I in the medium would effectively prevent endogenous synthesis of amino acid II, and growth would cease. In fact, feedback inhibition by the end product of a branched biosynthetic pathway is often exerted specifically on the enzyme which catalyzes the initial step following a metabolic branch point in the chain of reactions which leads specifically to its synthesis. Thus, amino acid I exerts feedback control on enzyme **d,** and amino acid II on enzyme **g.** The effective regulation of a branched biosynthetic pathway nevertheless requires feedback control of the initial enzyme, **a,** which catalyzes the first step of the common pathway (e.g., A → B → C). A number of different feedback mechanisms which permit this are known.

1. *Isofunctional enzymes.* The cell synthesizes two enzymes (**a** and **a′**) which have the same catalytic activity but are subject to feedback inhibition by different end products (Figure 8.9). When neither end product is present in the environment, the combined activities of enzymes **a** and **a′** produce a sufficient quantity of the intermediate A to meet the cellular demands for both end products. If one end product is present in the environment, synthesis of A is reduced as a result of the specific feedback inhibition of either **a** or **a′**. If both end products are present in the environment, the pathway ceases to function, since both **a** and **a′** are inhibited.
2. *Concerted feedback inhibition.* The reaction subject to control is catalyzed by a single enzyme, **a,** with two different allosteric sites, each of which binds one of the specific end products of the pathway (Figure 8.10). When only one of these sites is occupied by an effector, activity of the enzyme is not affected. However, when both effectors are bound to the enzyme, it becomes inactive. Concerted feedback inhibition exerts a somewhat imprecise control,

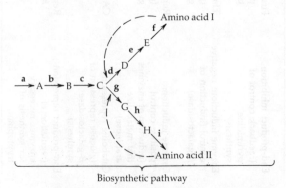

FIGURE 8.8

Generalized scheme of a branched biosynthetic pathway leading to two essential metabolites (in this case, amino acids). Arrows indicate reactions catalyzed by enzymes (lower-case letters) producing biosynthetic intermediates (capital letters). Dashed lines lead from end-product inhibitors to susceptible allosteric enzymes.

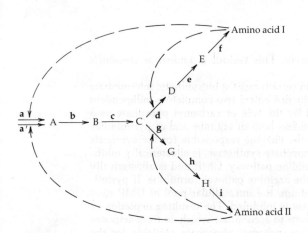

FIGURE 8.9

Scheme of regulation of a branched biosynthetic pathway by end-product inhibition of isofunctional enzymes. (Symbols are the same as in Figure 8.8.)

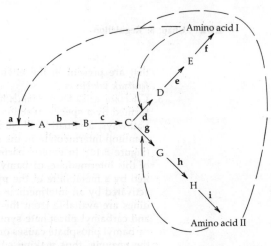

FIGURE 8.10

Scheme of regulation of a branched biosynthetic pathway by concerted feedback inhibition. (Symbols are the same as in Figure 8.8.)

since the rate of the reaction is unchanged if only one end product is present. However, it does prevent operation of the pathway when both end products are present.

3. *Sequential feedback inhibition.* The reaction subject to control is catalyzed by a single enzyme, **a,** for which the effector is not an end product of the pathway, but the intermediate (C) immediately preceding a metabolic branch point. Elevated concentrations of the end product (amino acids I or II) inhibit enzymes **d** and **g,** and thus they cause a rise in the intracellular concentration of C. This, in turn, inhibits the activity of enzyme **a** (Figure 8.11).

4. *Cumulative feedback inhibition.* In certain branched pathways leading to multiple end products, a single allosteric enzyme has effector sites for all end products. Each end product (even at a high concentration) causes only partial inhibition of the enzyme, and inhibitory effects of the different end products are additive. Thus, the rate of the reaction mediated by enzyme **c** is determined by the number (and concentration) of different end products of the pathway

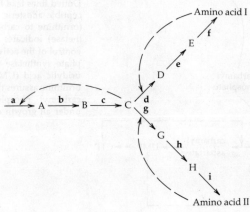

FIGURE 8.11

Scheme of regulation of a branched biosynthetic pathway by sequential feedback inhibition. (Symbols are the same as in Figure 8.8.)

261

that are present in the environment. This control is known as *cumulative feedback inhibition.*

5. *Combined activation and inhibition.* In certain cases, a biosynthetic intermediate formed by a specific reaction sequence enters two completely independent pathways, a situation illustrated by the role of carbamyl phosphate as a common intermediate in the synthesis both of arginine and of pyrimidines (Figure 8.12). In enteric bacteria the enzyme responsible for the synthesis of this intermediate, carbamyl phosphate synthetase, is allosterically inhibited by a metabolite of the pyrimidine pathway, UMP, and is allosterically activated by an intermediate of the arginine pathway, ornithine. If pyrimidines are available from the medium, the intracellular pool of UMP rises and carbamyl phosphate synthetase is inhibited. The resulting depletion of carbamyl phosphate causes ornithine to accumulate which, in turn, activates the enzyme, thus making sufficient carbamyl phosphate available for the synthesis of the alternate end product, arginine. Alternatively, if arginine is available from the medium, the biosynthesis of ornithine is arrested as a consequence of the feedback inhibition by arginine of *N*-acetylglutamic acid synthetase. As a result, the intracellular concentration of ornithine falls, and the activity of carbamyl phosphate synthetase is decreased.

• **End-product repression in branched biosynthetic pathways**

The repression of enzyme synthesis in branched biosynthetic pathways is, like feedback inhibition, both complex and variable in mechanism. For instance, the synthesis of carbamyl phosphate synthetase by *E. coli* is partially repressed either by arginine or by cytidine triphosphate (CTP), and is fully repressed by both together. Thus, the synthesis of this key allosteric enzyme is regulated independently by two end products, neither of which affects the activity of the enzyme.

In the case of isofunctional enzymes subject to independent allosteric control by different end products, the synthesis of each enzyme is frequently controlled by the end product that inhibits its activity; an illustration of this will be described below.

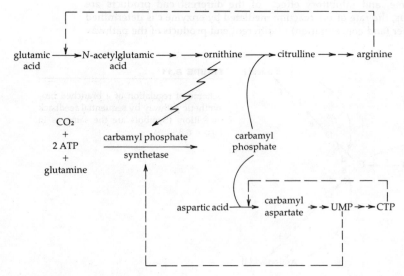

FIGURE 8.12

Regulation of carbamyl phosphate synthetase activity in enteric bacteria. Dotted lines lead from inhibitors to susceptible allosteric enzymes. Jagged line (ornithine to carbamyl phosphate synthetase) indicates activation. The dual control of the activity of carbamyl phosphate synthetase through inhibition of uridylic acid (UMP) and activation by ornithine assures that proper amounts of carbamyl phosphate are synthesized under all growth conditions.

- **Examples of regulation of complex pathways**

In *E. coli* the conversion of aspartic acid to aspartyl phosphate is mediated by three isofunctional enzymes of which two (designed as **a** and **c** in Figure 8.13) possess two catalytic functions, also mediating the conversion of aspartic acid semialdehyde to homoserine. Enzyme **a,** possessing both these functions, is feedback inhibited and its synthesis is repressed by threonine. Enzyme **c,** which similarly possesses both functions, is inhibited and repressed by lysine. The third aspartokinase (enzyme **b**), is not subject to end-product inhibition, but its synthesis is repressed by methionine (Table 8.3).

The enzymes of the L-lysine branch (**m-q**) and the L-methionine branch (**r-v**) catalyze reactions leading in each case to a single end product and are subject to specific repression by that end product (L-lysine and L-methionine, respectively).

The third branch of the aspartate pathway is subject to much more complex regulation, for two reasons. First, L-threonine, formed through this branch, is both a component of proteins and an intermediate in the synthesis of another amino

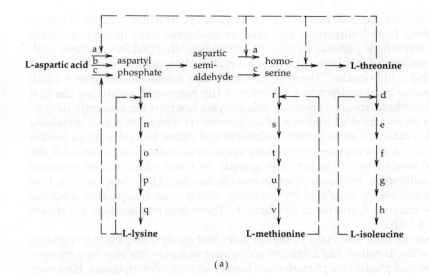

(a)

(b)

263

FIGURE 8.13

A simplified diagram of the aspartate pathway in *E. coli*. Each solid arrow designates a reaction catalyzed by one enzyme. The biosynthetic products of the pathway (in boldface) are all allosteric inhibitors of one or more reactions. Aspartyl-phosphate is synthesized by three isofunctional enzymes, and homoserine by two. Careful study of this diagram reveals that with a single exception (the inhibition exerted by valine, see text) the inhibition imposed by one amino acid does not cause starvation for a different amino acid. End-product repression is described in the text. Part (a) shows the regulatory interrelationships of the L-lysine, L-methionine, and L-isoleucine branches of the pathway. Part (b) shows the regulatory interrelationships of the L-isoleucine, L-valine, and L-leucine branches.

TABLE 8.3

Control of the first step of the aspartate pathway, mediated by three different aspartokinases, in the bacterium *Escherichia coli*

ENZYME	REPRESSED BY	INHIBITED ALLOSTERICALLY BY
Aspartokinase I	Threonine and isoleucine	Threonine
Aspartokinase II	Methionine	No allosteric control
Aspartokinase III	Lysine	Lysine

acid, L-isoleucine. Second, four of the five enzymes (**e-h**) which catalyze L-isoleucine synthesis from L-threonine, also catalyze analogous steps in the completely separate biosynthetic pathway by which L-valine is synthesized from pyruvic acid. The intermediate of this latter pathway, α-ketoisovaleric acid, is also a precursor of the amino acid L-leucine. These interrelationships are shown in Figure 8.13(a).

L-isoleucine is an end-product inhibitor of the enzyme, **d**, catalyzing the first step in its synthesis from L-threonine; this enzyme has no other biosynthetic role. L-valine is an end-product inhibitor of an enzyme (**e**) which has a dual metabolic role, since it catalyzes steps in both isoleucine and valine biosynthesis. In certain strains of *E. coli*, this enzyme is extremely sensitive to valine inhibition, with the result that exogenous valine prevents growth, an effect which can be reversed by the simultaneous provision of exogenous isoleucine. The L-leucine branch of the valine pathway is regulated by L-leucine, which is an end-product inhibitor of the first enzyme, **i**, specific to this branch. These interrelationships are shown in Figure 8.13(b).

As shown in Table 8.4, many of the enzymes that catalyze steps in the synthesis of L-isoleucine, L-valine, and L-leucine are subject to repression only by a mixture of the three end products, a phenomenon known as *multivalent repression*. However, the five enzymes specific to L-leucine synthesis are specifically repressed by this amino acid alone.

As previously mentioned, catabolic pathways mediated by constitutive enzymes are regulated exclusively by allosteric modulation of enzyme activity. This form of pathway control is schematized in Figure 8.14 for the pathway of glucose

TABLE 8.4

Repressive control of the enzymes of the isoleucine-valine-leucine pathway
(see Figure 8.13)

ENZYMES	REPRESSED BY
d, e, f, g, h	Isoleucine + valine + leucine
i, j, jj, k, l	Leucine

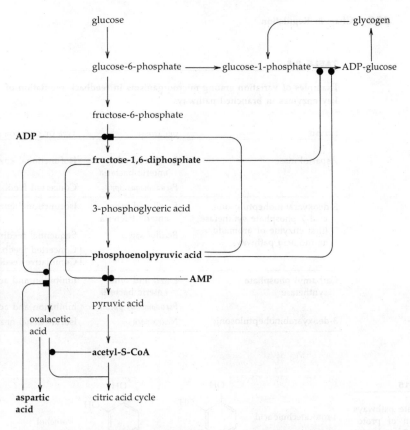

FIGURE 8.14

Some sites of allosteric activation and inhibition in the synthesis of the reserve material glycogen and in the breakdown of glucose by *E. coli*. Metabolites that exert allosteric effects are shown in boldface; the reactions they affect are indicated thus: ——●, an allosteric activation; ──■, an allosteric inhibition. Note, for example, that accumulation of fructose-1,6-diphosphate and of phosphoenolpyruvic acid promotes conversion of glucose to glycogen by allosteric activation of one of the enzymes of glycogen synthesis.

metabolism and glycogen synthesis in *E. coli*. Excess concentrations of the catabolic intermediates, fructose-1,6-diphosphate and phosphenolpyruvate, signal the diversion of carbon flow into glycogen.

• **The diversity of bacterial regulatory mechanisms**

With very few exceptions, biosynthetic pathways are biochemically identical in all microorganisms. However, a given pathway may be subject, in different organisms, to markedly different modes of regulation which tend to be group specific and probably indicate evolutionary relatedness. Certain examples of regulatory diversity in end-product inhibition are summarized in Table 8.5.

Although the members of a given bacterial group may all possess the same mechanism for the regulation of a particular enzyme (e.g., the isofunctional aspartokinases characteristic of the bacteria of the enteric group), a different kind of regulation may operate in the same group with respect to the key enzyme of another pathway (e.g., the regulation by multiple allosteric controls of a single carbamyl phosphate synthetase in the enteric bacteria).

Many catabolic pathways are likewise biochemically identical in a wide diversity of bacterial groups. Here, also, the patterns of regulation are both diverse and group-specific. The β-ketoadipate pathway, for example, serves for the oxidation of aromatic substrates in several groups of bacteria. It is mediated by strictly inducible enzymes, the synthesis of which is regulated in a markedly different way in the *Pseudomonas* and *Acinetobacter* groups (see Figure 8.15 and

TABLE 8.5

Examples of variation among microorganisms in feedback regulation of certain key enzymes in branched pathways

ENZYME	ORGANISM	TYPE OF CONTROL
Aspartokinase	E. coli and other enteric bacteria	Isofunctional enzymes
	Pseudomonas spp.	Concerted feedback inhibition
3-deoxyarabinoheptulosonic acid-7-phosphate synthetase (first enzyme of aromatic amino acid pathway)	E. coli and other enteric bacteria	Isofunctional enzymes
	Bacillus spp.	Sequential feedback inhibition
	Pseudomonas spp.	Concerted feedback inhibition / Cumulative feedback inhibition
Carbamyl phosphate synthetase	E. coli and other enteric bacteria	Inhibition and activation
	Pseudomonas putida	Inhibition and activation
3-deoxyarabinoheptulosonic	Neurospora	Isofunctional enzymes

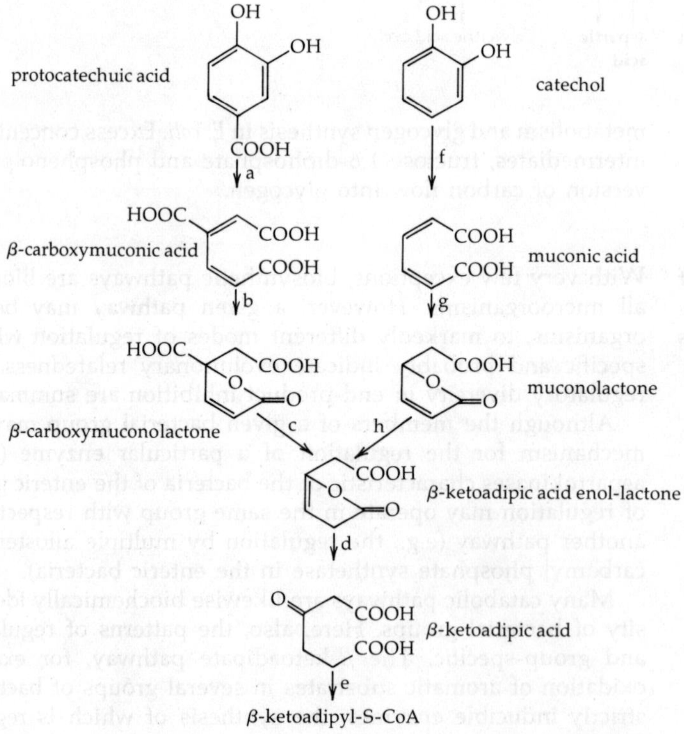

FIGURE 8.15

β-ketoadipate pathways of oxidation of proto-catechuic acid and cate-chol. The pattern of in-duction of the enzymes (a—h) of these pathways in *Acinetobacter* and *Pseu-domonas* are summarized in Table 8.6.

TABLE 8.6

Control of the enzymes of the β-ketoadipate pathways of oxidation of protocatechuic acid and catechol by *Acinetobacter* and *Pseudomonas*[a]

| PSEUDOMONAS | | ACINETOBACTER | |
INDUCER	ENZYME	ENZYME	INDUCER
Protocatechuic acid	a	a ⎫	
β-ketoadipic acid or β-ketoadipyl-S-CoA	b c d e	b c d e ⎬	Protocatechuic acid
Muconic acid	f	f	Muconic acid
Muconic acid	g h	g h d′ e′ ⎬	Muconic acid

[a] Brackets indicate groups of enzymes subject to coordinate induction. Lower-case letters designating enzymes are those shown in Figure 8.15; d′ and e′ indicate enzymes isofunctional with **d** and **e** but subject to independent regulation.

Table 8.6). The differences are expressed in the chemical nature of the metabolite inducers, in the degree of coordinate regulation, and in the presence or absence of isofunctional enzymes.

• **Covalent modification of enzymes**

As has been discussed, the activity of allosteric enzymes is modified by conformational changes in the enzyme induced by binding to the small effector molecule; formation of covalent chemical bonds is not involved in the transition between the two states of an allosteric enzyme. However, many enzymes of mammals are known to exist in active and inactive forms, between which the transition occurs by the formation or breakage of covalent bonds. In some cases, the two forms of the enzyme differ in the number of amino acid residues which they contain; i.e., the transition involves the hydrolysis of a peptide bond. In other cases, active and inactive forms differ with respect to the presence or absence of other chemical substances covalently bonded to the protein. Since the transition between active and inactive forms is effected by rupture or formation of covalent bonds, it is enzyme catalyzed.

One bacterial enzyme which is subject to covalent modification with a resulting change of activity is the glutamine synthetase of *E. coli* (Figure 8.16). The two forms of the enzyme differ with respect to the presence or absence of an adenyl group which is added to the protein or removed from it enzymatically. The substituted form of the enzyme is considerably less active than the unsubstituted form.

It should be emphasized that the existence of regulation through covalent modification of enzyme structure does not contradict the generalization that regulation is mediated by allosteric proteins. In fact, the enzymes that catalyze covalent modification of the enzymes are themselves allosterically regulated. In this sense, covalent modification can be regarded as a mechanism that modulates the expression of allosteric regulation.

FIGURE 8.16

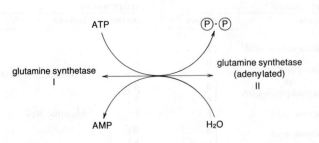

Covalent modification of glutamine synthetase through adenylation is stimulated by high concentrations of ammonia and results in a reduction of activity of the enzyme. The adenylated form of glutamine synthetase is sensitive to inhibition by the end products of glutamine metabolism, whereas the unsubstituted form is considerably less sensitive.

REGULATION OF DNA SYNTHESIS AND CELL DIVISION

In Chapter 7 the synthesis of DNA was described as a process in which the two strands of the double helix separate to produce two single strands, each of which serves as a template for the polymerization of deoxyribonucleotides. It will be recalled that the replication process creates a *fork* in the DNA molecule; as the DNA separates, the fork progresses, accompanied by replication of the two branches (Figure 7.43).

Regulation of DNA synthesis must be precisely controlled since at division each daughter cell receives a full complement of genetic material. Evidence that DNA synthesis and cell division are intimately interconnected comes from the observation that a wide variety of chemical treatments or mutational changes which inhibit DNA synthesis also inhibit cell division. Cells in which DNA synthesis has been inhibited elongate without division, eventually forming very long cells (Figure 8.17).

Data obtained largely through studies on *E. coli* indicate that segregation of DNA into daughter cells is a simple consequence of cell growth. The circular chromosome is attached to the cytoplasmic membrane. Following replication, the newly formed chromosome becomes attached to an adjacent site on the membrane. These sites are separated by interstitial membrane growth, and a cross wall is then formed between the two chromosomes (Figure 8.18). If, for any reason, chromosomal replication is not completed, cell division does not occur. The relationship between completion of replication of the chromosome and subsequent cell division appears to be causal; when the chromosomal replication is completed, a series of metabolic events are initiated which eventually result in cell division. *Regardless of growth rate, cell division in E. coli always occurs about 20 minutes after completion of chromosomal replication* (at 37°C).

Figure 8.18 also shows that chromosomal replication is bidirectional (i.e., two replication forks are formed which travel in opposite directions around the chromosome; when they meet, chromosomal replication is completed). In *E. coli*, the time required for a complete doubling of DNA (i.e., the time required for one of the pairs of forks to travel one-half the length of the molecule) is approximately 40 minutes at 37°C. If the organism grows in a medium in which the cycle of growth and cell division takes longer than 40 minutes, the cycle of DNA

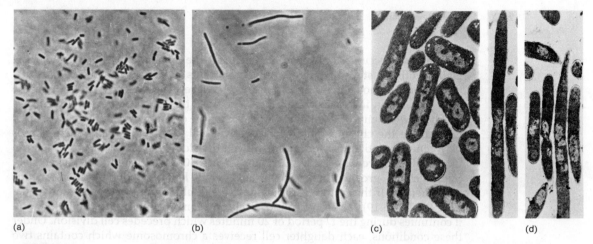

(a)   (b)   (c)   (d)

FIGURE 8.17

Phase (a, b) and electron (c, d) micrographs of *E. coli* B illustrating the consequences of
inhibition of DNA synthesis for cell division. Specific inhibition of DNA synthesis (in
this case effected by addition of mitomycin C to an exponentially growing culture) allows
cell growth to continue, but stops further cell division. Thus, the normal short rods (a)
become highly elongated forms. Electron micrographs before (c) and 3 hours after (d)
addition of mitomycin C show in fact that little or no increase of the nuclear material
(light central regions of the cells) occurs in the presence of the drug. From H. Suzuki,
J. Pangborn, and W. W. Kilgore, "Filamentous Cells of *Escherichia coli* Formed in the
Presence of Mitomycin," *J. Bacteriol.* **93**, 684 (1967).

FIGURE 8.18

Diagram showing relationship between DNA replication, nuclear segregation, physiological division, and cell separation in *E. coli* growing with a doubling time of 45 minutes.
Dotted lines (.....) indicate area of wall and membrane growth prior to duplication of
attachment points. Dashed lines (---) indicate newly replicated portion of the chromosome. Wavy lines (⌒⌒) indicate area of wall and membrane growth after formation
of new attachment. Septum formation is indicated by a vertical dashed line (¦) at 30
minutes. After D. J. Clark, "The Regulation of DNA Replication and Cell Division in
E. coli B/r," *Cold Spring Harbor Symp. Quant. Biol.* **33**, 825 (1968).

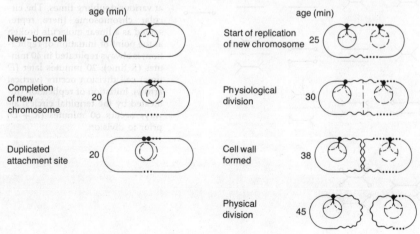

synthesis still occupies only 40 minutes; during the remainder of the division cycle, DNA synthesis stops. Therefore, DNA synthesis in slowly growing cells is discontinuous. In nonsynchronous cultures, however, the quantity of DNA in the total population increases continuously because the cell cycles are not in phase.

Thus, two events of fixed duration make up the division cycle: the time required to replicate the chromosome, which we will call R, and the time which elapses between completion of replication and cell division, which we will call D. In the case of *E. coli* growing at 37°C, R and D are 40 and 20 minutes, respectively (regardless of growth rate). As we have seen, R and D can be simultaneous or sequential. If the culture is growing with a doubling time approximately equal to R (Figure 8.19), a second round of chromosomal replication is initiated, and it continues during the D period of 20 minutes which precedes cell division. Under these conditions, each daughter cell receives a chromosome which contains two replication forks. At doubling times greater than 40 minutes, the portion of the D period during which no chromosome replication occurs increases. As a consequence, at the time of division the daughter cells receive chromosomes with smaller and smaller replication forks. At generation times greater than $R + D$ (viz., greater than 60 minutes), daughter cells receive chromosomes without replication forks, because reinitiation of chromosome replication always begins 60 minutes prior to the next cell division.

Thus, in cells growing with doubling times equal to or greater than R, the coordination of DNA synthesis with cell division is achieved by two means (1) chromosomal replication is initiated at time intervals equal to the doubling time of the culture and (2) initiation occurs 60 minutes prior to cell division. The only variable in this process is the time at which initiation of chromosomal replication occurs; R and D are invariant. In rich media in which *E. coli* is capable

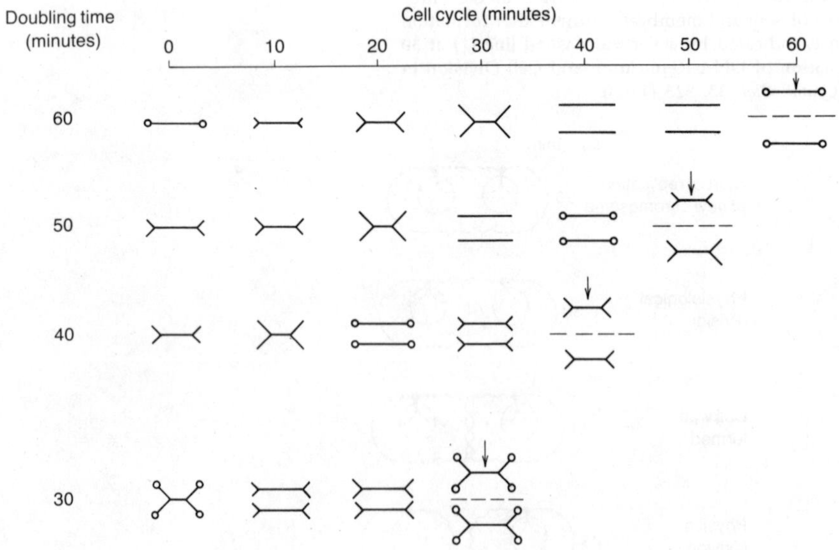

FIGURE 8.19

Schematic diagram of the coupling between chromosome replication and cell division of *E. coli* growing at various doubling times. The circular chromosome (here represented as a linear molecule broken at the point of initiation of replication) is always replicated in 40 minutes (R time); 20 minutes later (D time) cell division occurs (vertical arrow). Initiation of replication (indicated by the terminal circles) always occurs 60 minutes ($R + D$) prior to division.

Doubling time (minutes) / Cell cycle (minutes) — 0, 10, 20, 30, 40, 50, 60 — rows: 60, 50, 40, 30

of growing at doubling times less than R, conditions 1 and 2 still hold. Since reinitiation begins at intervals equal to doubling time which is less than R, *multiple forks* are generated, thus allowing a rate of DNA synthesis commensurate with the higher growth rate.

The relation between initiation of chromosomal replication and cell division at various growth rates is summarized in Figure 8.19. It can be seen that the critical factor is the *time* and *frequency* of initiation of a new round of DNA replication.

The initiation process requires that protein be synthesized; initiated cycles of DNA synthesis can be completed if protein synthesis is blocked, but *a new cycle cannot be initiated*. This has led to the hypothesis that initiation is under the positive control of a specific regulatory protein, the *initiator*. When the concentration of this initiator rises to a certain threshold level, initiation occurs, after which the initiator is destroyed. The time required for effective concentrations of the initiator to be synthesized is precisely equal to the doubling time of the culture. The initiator hypothesis is consistent with all known facts, but so far an initiator protein has not been identified.

REGULATION OF RNA SYNTHESIS

The regulation of the synthesis of the various species of RNA involves several different mechanisms. Regulation of the synthesis of messenger RNA (mRNA) is the fundamental mode of control of synthesis of proteins. As we have seen, the processes of enzyme induction, end-product repression, and catabolite repression all act by regulating transcription, i.e., the synthesis of mRNA.

The regulation of the synthesis of the stable molecular forms of RNA [i.e., ribosomal RNA (rRNA) and transfer RNA (tRNA)] is effected by mechanisms still only very incompletely understood. Their regulation appears to be coordinate because under a wide variety of environmental conditions that greatly affect the total RNA concentration (see Table 8.1), the ratio of the intracellular concentrations of the two types of stable RNA remains constant. No satisfactory explanation is yet available that accounts for the changes in the concentration of stable RNA which occur as a result either of changes in growth rate or of sudden changes in the composition of the medium. Two facts are, however, well established. (1) Regulation is not effected by the concentration of the immediate precursors of RNA, i.e., the ribonucleoside triphosphates. (2) Changes in the rate of RNA synthesis are brought about by changes in the number of sites of RNA polymerization; in other words, the number of RNA polymerase molecules actively engaged in polymerization determines the rate of synthesis.

• **The dependence of stable RNA synthesis on protein synthesis**

About 20 years ago, A. Pardee observed that cultures of *E. coli* which were deprived of a required amino acid not only stopped synthesizing protein, but also abruptly stopped synthesizing stable RNA. Mutant strains were subsequently discovered which lack this *stringent* dependence of RNA synthesis on the ability of the cell to make protein. The mutant strains, termed *relaxed*, differ gentically from their parents, termed *stringent*, by a single mutation in a specific gene, the *rel* gene. Relaxed strains continue to synthesize appreciable quantities of stable RNA for some time after they are deprived of an essential amino acid (Figure 8.20).

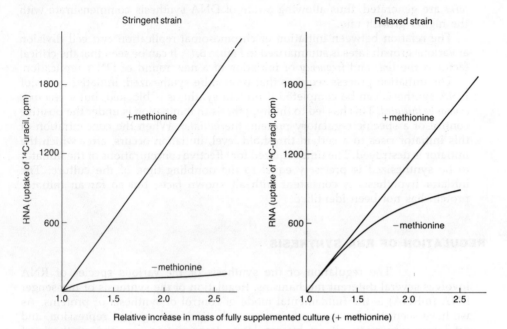

FIGURE 8.20

Effect of methionine starvation on the synthesis of a stringent and a relaxed strain of
E. coli. Both strains are methionine auxotrophs and hence incapable of protein synthesis
in the absence of the amino acid, but unlike the stringent strain which is virtually
incapable of synthesizing stable RNA in the absence of methionine, the relaxed strain
synthesizes significant amounts of RNA under these conditions, initially at full rate.
Redrawn from G. S. Stent and S. Brenner, "A Genetic Locus for the Regulation of
Ribonucleic Acid Synthesis," *Proc. Natl. Acad. Sci. U.S.* **47**, 2005 (1961).

At first, the existence of relaxed mutants seemed to imply that the *rel* gene
encodes the essential regulator of stable RNA synthesis. However, later work has
shown that it controls only one aspect of RNA synthesis, because the modulation
of RNA synthesis as a function of growth rate remains normal in relaxed strains.
The *rel* gene product appears only to prevent RNA synthesis when protein
synthesis is arrested. Continued RNA synthesis in the absence of protein synthesis
is detrimental to the cell, as shown by the fact that relaxed strains which have
been allowed to synthesize RNA in the absence of a required amino acid undergo
an extensive lag before resuming growth when that amino acid is added to the
medium. Amino acid starvation causes a rapid intracellular accumulation of two
unusual nucleotides, guanosine-3′,5′-di (diphosphate) (ppGpp) and guanosine-
3′-diphosphate-5′-triphosphate (pppGpp), in stringent strains, but not in relaxed
strains. These compounds are synthesized by ribosomes from GDP and GTP,
respectively, in the presence of ATP. Although their precise functions are not
known, there are some indications that they play important roles in regulating
RNA synthesis.

The regulation of the syntheses of proteins and nucleic acids has one striking common feature. *The rate of all these syntheses is regulated by varying the number of sites of polymerization and not by varying the concentration of substrates.* The number of replicating forks determines the rate of synthesis of DNA; the number of ribosomes determines the rate of synthesis of protein; and the number of active RNA polymerase molecules determines the rate of synthesis of RNA.

THE TELEONOMIC NATURE OF BIOLOGICAL SYSTEMS

It will not have escaped the attention of the thoughtful reader that the regulatory mechanisms which operate in microorganisms reveal a seemingly strange fact: *microorganisms behave as if they were goal-directed.* Such behavior is characteristic of all living organisms, and it is known as *teleonomic behavior.* The sum total of their activities seems to be directed to the fulfillment of a plan that is preordained. In the case of a unicellular microorganism, the plan that directs cellular activity, and that is so strikingly revealed by the analysis of metabolic regulation, is a simple one: to use the nutrients immediately available to the cell in order to make two cells from one, with the greatest possible rapidity.

From the time of Aristotle, biologists have speculated about the source of the plan. Some have sought it in a supernatural agency, external to the biological world; others have attributed it to an innate, though mysterious internal force possessed by living systems: the *entelechy* of Aristotle; the *elan vital,* or *life force,* of Bergson.

It was Charles Darwin who first attempted to explain in purely scientific terms the teleonomic properties of living organisms, and the Darwinian explanation outlined below is now generally accepted by biologists.

The plan is created and then progressively refined through time by the agency of natural selection which acts upon small, inherited differences that arise through mutations within an initially uniform population. Only those inherited changes which increase the fitness of an organism to its environment are conserved by natural selection; and every aspect of the structure and function of an organism is continuously subjected to scrutiny by natural selection.

The allosteric proteins, which are the key elements of metabolic regulation, provide one of the most remarkable examples of the results of natural selection, operating at a molecular level. It can scarcely be doubted that the initial function of these proteins was purely catalytic, their allosteric properties being acquired secondarily. Through mutations which modified the structure of an initially catalytic protein molecule, additional combining sites, capable of binding small molecules other than the substrate, arose by chance on the enzyme surface. In the very rare instances in which these structural modifications increased the physiological effectiveness of the enzyme, and thereby increased the fitness of the organism in which they occurred, natural selection assured their preservation; all other structural modifications that decreased physiological effectiveness were eliminated by natural selection. What the scientist perceives today when he examines the properties of allosteric proteins are the individual end products of such molecular evolution which have been continuously refined and perfected by the play of natural selection.

FURTHER READING

Books   MAALØE, O., and N. KJELDGAARD, *Control of Macromolecule Synthesis.* New York: Benjamin, 1965.

MANDELSTRAM, J., and K. McQUILLEN, *Biochemistry of Bacterial Growth,* 2nd ed. New York: Wiley, 1973.

MONOD, J., *Chance and Necessity, An Essay on the Natural Philosophy of Modern Biology* (translated from the French). New York: Knopf, 1971.

Review   EDLIN, G., and R. BRODA, "Physiology and Genetics of the Ribonucleic Acid Control Locus in *Escherichia coli,*" *Bact. Revs.* **32,** 206 (1968).

9

MICROBIAL GROWTH

This chapter will describe the methods used for the measurement of microbial growth, together with the various phases and modes of growth. The discussion will focus on the growth of unicellular bacteria, since they are ideal objects for study of the growth process and have largely served for the elucidation of its nature.

THE DEFINITION OF GROWTH

In any biological system, growth can be defined as the orderly increase of all chemical components. Increase of mass might not really reflect growth because the cells could be simply increasing their content of storage products such as glycogen or poly-β-hydroxybutyrate. In an adequate medium to which they have become fully adapted, however, bacteria are in a state of *balanced growth*. During a period of balanced growth, a doubling of the biomass is accompanied by a doubling of all other measurable properties of the population, e.g., protein, RNA, DNA, and intracellular water. In other words, cultures undergoing balanced growth maintain a constant chemical composition. The phenomenon of balanced growth simplifies the task of measuring the rate of growth of a bacterial culture; since the rate of increase of *all* components of the population is the same, measurements of *any* component suffice to determine the growth rate.

THE MATHEMATICAL NATURE AND EXPRESSION OF GROWTH

A bacterial culture undergoing balanced growth mimics a first-order autocatalytic chemical reaction; i.e., the rate of increase in bacteria at any particular time is proportional to the number or mass of bacteria present at that time.

$$\text{rate of increase of cells} = \mu(\text{number or mass of cells}) \tag{9.1}$$

The constant of proportionality, μ, is an index of the rate of growth and is called the *growth rate constant*. Since we assume growth to be balanced, μ also relates the rate of increase of any given cellular component to the amount of that cellular component, or in mathematical terms,

$$\frac{dN}{dt} = \mu N, \qquad \frac{dX}{dt} = \mu X, \qquad \frac{dZ}{dt} = \mu Z \tag{9.2}$$

where N is the number of cells/ml, X is the mass of cells/ml, Z is the amount of any cellular component/ml, t is time, and μ is the growth rate constant. These equations, in fact, accurately describe the growth of most unicellular bacterial cultures. Other (nondifferential) forms of these equations are more useful in practice. Upon integration Eq. (9.2), yields:

$$\ln Z - \ln Z_0 = \mu(t - t_0) \tag{9.3}$$

and on converting natural logarithms to logarithms to the base 10,

$$\log_{10} Z - \log_{10} Z_0 = \frac{\mu}{2.303}(t - t_0) \tag{9.4}$$

where the values of Z and Z_0 correspond to the amount of any bacterial component of the culture at times t and t_0, respectively. By measuring Z and Z_0, one can compute the value of μ, the growth rate constant of the culture. Thus, if the culture contains 10^4 cells/ml at t_0 and 10^8 cells/ml 4 hours later, the specific growth rate of the culture is:

$$\mu = \frac{(8 - 4)2.303}{4} = 2.303 \text{ hours}^{-1} \tag{9.5}$$

The value of μ suffices to define the rate of growth of a culture. However, certain other parameters are also commonly used. One is the mean doubling time or generation time (g) defined as the time required for all components of the culture to increase by a factor of 2.* The relationship between g and μ can be derived from Eq. (9.3), since if the time interval considered $(t - t_0)$ is equal to g, then Z will be twice Z_0. Making these substitutions, one obtains:

$$\mu = \frac{\ln 2}{g} = \frac{0.693}{g} \tag{9.6}$$

In the case of the example we have chosen, the mean doubling time, g, of the

---

*Sometimes the reciprocal of doubling time is used as an index of growth rate. This index (k) is numerically similar to μ and can be confusing; its use should be avoided.

culture is $g = 0.693/2.303 = 0.42$ hour or 25 minutes. This is a relatively high growth rate for a bacterium, as shown by the representative examples assembled in Table 9.1.

The above mathematical expressions for bacterial growth rate have been developed from the premise that the rate of increase is proportional to the number (or mass) present at any given time. From this premise, it was shown [Eq. (9.6)] that doubling time (g) is constant during a period of balanced growth. The same equations can be derived from the premise that mean doubling time is constant, and lead to the conclusion that the rate of increase of number (or mass) is proportional to the number (or mass) at any given time.

• The growth curve

Equation (9.4) predicts a straight-line relationship between the logarithm of cell number (or any other measurable property of the population) and time [Figure 9.1(a)] with a slope equal to $\mu/2.303$ and an ordinate intercept of $\log N_0$. By taking the antilogarithm, Eq. (9.4) can be written in the exponential form:

$$Z = Z_0 10^{\mu(t-t_0)/2.303} \tag{9.7}$$

which predicts an exponential relationship between the number of cells in the population (or any other measurable property) and time [Figure 9.1(b)]. Populations of bacteria growing in a manner that obeys these equations are said to be in the *exponential phase* of growth.

Microbial populations seldom maintain exponential growth at high rates for long. The reason is obvious if one considers the consequences of exponential growth. After 48 hours of exponential growth, a single bacterium with a doubling time of 20 minutes would produce a progeny of 2.2×10^{31} g, or roughly 4,000 times the weight of the earth.

The growth of bacterial populations is normally limited either by the exhaustion of available nutrients or by the accumulation of toxic products of metabolism. As a consequence, the rate of growth declines and growth eventually stops. At this point a culture is said to be in the *stationary phase* (Figure 9.2). The transition between the exponential phase and the stationary phase involves a period of

TABLE 9.1

Maximal recorded growth rates for certain bacteria, measured at or near their temperature optimum, in complex media unless otherwise noted

ORGANISM	TEMPERATURE (°C)	DOUBLING TIME (HOURS)
Beneckea natriegens	37	0.16
Bacillus stearothermophilus	60	0.14
Escherichia coli	40	0.35
Bacillus subtilis	40	0.43
Pseudomonas putida	30	0.75[a]
Vibrio marinus	15	1.35
Rhodopseudomonas sphaeroides	30	2.2
Mycobacterium tuberculosis	37	~6
Nitrobacter agilis	27	~20[a]

[a] Grown in synthetic medium.

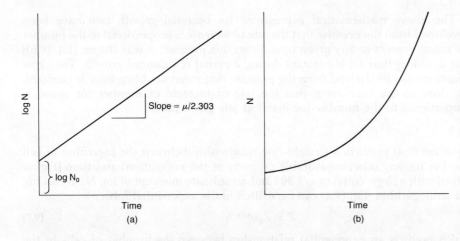

FIGURE 9.1

Comparison of methods of plotting growth data. Plotting the logarithm of cell density (number of cells/ml, N) of a culture undergoing balanced growth as a function of time yields a straight line (a); the slope of the line is the growth rate constant (μ) divided by 2.303, and the intercept is log N_0. Plotting the cell density directly as a function of time yields an exponential curve (b).

unbalanced growth during which the various cellular components are synthesized at unequal rates. Consequently, cells in the stationary phase have a chemical composition that is different from that of cells in the exponential phase. The cellular composition of cells in the stationary phase depends on the specific growth-limiting factor. Despite this, certain generalizations hold; cells in the stationary phase are small relative to cells in the exponential phase (since cell division continues after increase in mass has stopped), and they are more resistant to adverse physical (heat, cold, radiation) and chemical agents.

• **The death phase**   Bacterial cells held in a nongrowing state eventually die. Death results from a number of factors; an important one is depletion of the cellular reserves of energy. Like growth, death is an exponential function and hence in a logarithmic plot (Figure 9.2) the death phase is a linear decrease in number of *viable* cells with time. The death rate of bacteria is highly variable, being dependent on the environment as well as on the particular organism (e.g., enteric bacteria die very slowly, while certain *Bacillus spp.* die rapidly).

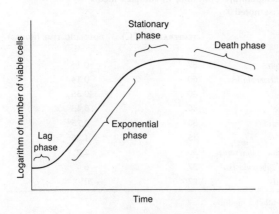

FIGURE 9.2

Generalized growth curve of a bacterial culture.

• The lag phase Cells transferred from a culture in the stationary phase to a fresh medium of the same composition undergo a change of chemical composition before they are capable of initiating growth. This period of adjustment, called the *lag phase* (Figure 9.2), is extremely variable in duration; in general, its length is directly related to the duration of the preceding stationary phase.

• Arithmetic growth In certain situations, the kinetics of bacterial growth is arithmetic rather than exponential. Under this circumstance, the rate of increase is a constant, i.e.,

$$\frac{dN}{dt} = C \tag{9.8}$$

and on integration:

$$N = Ct \tag{9.9}$$

The number of cells (*N*), rather than the logarithm of the number of cells, is a linear function of time (*t*).

A number of conditions can lead to arithmetic growth. For example, if a bacterium that requires nicotinic acid as a growth factor is deprived of this nutrient, it is unable to synthesize pyridine nucleotides, of which nicotinic acid is a specific biosynthetic precursor. As a result of the functions of pyridine nucleotides in electron transport, their levels in the cell determine the overall rate of metabolism and hence of growth. When no further synthesis of these compounds can occur, the growth rate of the population becomes directly proportional to the supply of pyridine nucleotides in the cells; the total catalytic power can no longer increase. Hence, the rate of increase of cells remains constant. The addition of *p*-fluorophenylalanine, an analogue of the natural amino acid, phenylalanine, to a culture also results in arithmetic growth (Figure 9.3). The analogue is sufficiently similar to the natural amino acid that it can become incorporated into newly synthesized proteins in place of phenylalanine. However, proteins containing *p*-fluorophenylalanine are largely nonfunctional, and hence they do not increase the catalytic capacity of the cell. As a consequence, the growth rate cannot increase beyond that determined by the cells' catalytic capacity at the time of addition of the analogue to the culture.

Certain mutations which preclude further synthesis of an essential protein in a particular environment lead to arithmetic growth when the culture is exposed to that environment.

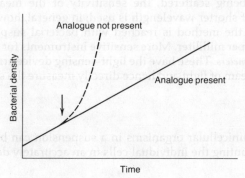

FIGURE 9.3

Linear growth of *E. coli* in a medium containing an amino acid analogue, *p*-fluorophenylalanine (solid line). The analogue was added at the time indicated by the vertical arrow. The dashed line shows the growth of a parallel culture not containing *p*-fluorophenylalanine.

THE MEASUREMENT OF GROWTH

To follow the course of growth, it is necessary to make quantitative measurements. As discussed earlier, exponential growth is usually balanced so any property of the biomass can be measured to determine growth rate. As a matter of convenience, the properties measured are usually cell mass or cell number.

• **Measurement of cell mass**

The only direct way to measure cell mass is to determine the dry weight of cell material in a fixed volume of culture by removing the cells from the medium, drying them, and then weighing them. Such determinations are time consuming and relatively insensitive. With ordinary equipment it is difficult to weigh with accuracy less than 1 mg, yet this dry weight may represent as many as 5 billion bacteria.

The method of choice for measuring the cell mass of unicellular microorganisms is an optical one: the determination of the amount of light scattered by a suspension of cells. This technique is based on the fact that small particles scatter light proportionally, within certain limits, to their concentration. When a beam of light is passed through a suspension of bacteria, the reduction in the amount of light transmitted as a consequence of scattering is thus a measure of the cell density. Such measurements are usually made in a spectrophotometer. This instrument reads in *absorbancy* (A) units; absorbancy is defined as the logarithm of the ratio of intensity of light striking the suspension (I_0) to that transmitted by the suspension (I):

$$A = \log \frac{I_0}{I} \tag{9.10}$$

The instrument is convenient for estimating cell density, and when calibrated against bacterial suspensions of known density (Figure 9.4), it becomes an accurate and rapid way to estimate the dry weight of bacteria per unit volume of culture. It should be emphasized that measurements of cell density are meaningful only when used in conjunction with a standard curve such as that shown in Figure 9.4. Since scattering is inversely proportional to the fourth power of the wavelength of light being scattered, the sensitivity of the measurements increases sharply if light of shorter wavelength is used; in general, however, the lower limit of sensitivity of the method is reached with bacterial suspensions that contain about 10 million per milliliter. More sensitive instruments for measuring scattering are called *nephelometers*. These have the light-sensing device arranged at right angles to the incident beam of light and hence directly measure the scattered light (Figure 9.5).

• **The measurement of cell number**

The number of unicellular organisms in a suspension can be determined microscopically by counting the individual cells in an accurately determined very small

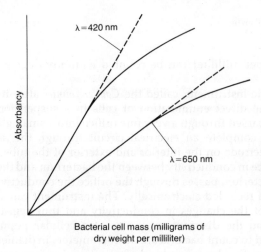

λ = 420 nm

λ = 650 nm

Absorbancy

Bacterial cell mass (milligrams of
dry weight per milliliter)

FIGURE 9.4

The relationship between absorbancy of a suspension of bacteria and bacterial cell mass. Note that proportionality is strict only at low values of absorbancy and deviates from strict proportionality (dashed line) at higher absorbancy values; also note that the measurements are more sensitive with light of shorter wavelength (λ).

volume. Such counting is usually done with special microscope slides known as *counting chambers*. These are ruled with squares of known area and are so constructed that a film of liquid of known depth can be introduced between the slide and the cover slip. Consequently, the volume of liquid overlying each square is accurately known. Such a direct count is known as the *total cell count*. It includes both viable and nonviable cells, since, at least in the case of bacteria, these cannot be distinguished by microscopic examination.

The principal limitation of the direct microscopic enumeration of bacterial populations is the relatively high concentrations of cells that must be present in the suspension. The high magnification required for seeing the bacteria limits the volume of liquid that can be examined carefully with the microscope; yet, a sufficient number of cells must be found in a known volume to make the count statistically significant. As a consequence, only suspensions that contain 10 million

FIGURE 9.5

Comparison of arrangement of the optical components of a photometer and nephelometer. The greater sensitivity of the nephelometer depends on its measuring scattered light rather than residual unscattered light.

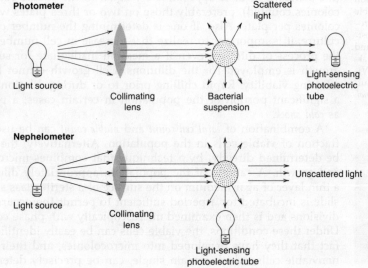

Photometer

Light source — Collimating lens — Bacterial suspension — Scattered light — Light-sensing photoelectric tube

Nephelometer

Light source — Collimating lens — Unscattered light — Light-sensing photoelectric tube

or more cells per milliliter can be counted with any degree of accuracy by this technique.

An electronic instrument, called the *Coulter counter* after its inventor, can also be used for the direct enumeration of cells in a suspension. A portion of the suspension is passed through a very fine orifice into a small glass tube. The orifice also serves to complete an electrical circuit through the suspending medium between an electrode on the interior and exterior of the tube. Detection depends on the difference in conductivity between the bacterium and the suspending liquid. Each time a bacterium passes through the orifice the conductivity drops; this event is detected and recorded electronically. The instrument can score the magnitude and duration of the changes in conductivity and thus register and record both the number and the distribution of size of a cellular population. The orifices commonly used to count bacteria are 30 micrometers in diameter. The suspending liquid must therefore be scrupulously free of inanimate particles (e.g., dust) since smaller ones will score as cells and larger ones will plug the orifice.

The enumeration of unicellular organisms can also be made by *plate count*, because single viable cells separated from one another in space by dispersion on or in an agar medium give rise through growth to separate, macroscopically visible colonies. Hence, by preparing appropriate dilutions of a bacterial population and using them to seed an appropriate medium, one can ascertain the number of viable cells in the initial population by counting the number of colonies that develop after incubation of the plates, and multiplying this figure by the dilution factor. This method of enumeration is often termed a *viable count:* in contrast to direct microscopic enumeration and electronic counting, it measures only those cells that are capable of growth on the plating medium used. The viable count is by far the most sensitive method of estimating bacterial number, since even a single viable cell in a suspension can be detected. Its accuracy depends on observing certain precautions. Significant numbers of colonies must be counted (the standard error is approximately equal to the square root of the number of colonies counted), preferably those on two or three plates with several hundred colonies per plate. Also, if one is determining the number of cells in a growing culture, it is important to realize that increase in cell numbers continues during the process of dilution, even if a medium inadequate for supporting continued growth is employed for the dilutions. Cell growth cannot be stopped without affecting viability. Rapid chilling prior to or during dilution can cause death of a significant portion of the population in certain cases, a phenomenon known as *cold shock*.

A combination of *total cell count* and *viable count* can be used to determine the fraction of viable cells in the population. Alternatively, the viable fraction can be determined directly, by a technique that combines microscopic examination and growth. A sample of the population, appropriately diluted, is spread over a thin layer of agar medium on the surface of a sterile glass slide. The inoculated slide is incubated for a period sufficient to permit the occurrence of several cell divisions and is then examined microscopically with phase contrast illumination. Under these conditions, the viable cells can be easily identified as a result of the fact that they have developed into microcolonies, and their number relative to nonviable cells, which remain single, can be precisely determined (Figure 9.6).

FIGURE 9.6

Determination of viability by a combination of growth and microscopic examination. A sample of a culture was spread on a thin layer of agar medium and was photographed after incubation for 3.5 hours. Viable cells have formed microcolonies. From J. R. Postgate, J. E. Crumpton, and J. R. Hunter, "The Measurement of Bacterial Viabilities by Slide Culture," *J. Gen. Microbiol.* **24**, 15 (1961).

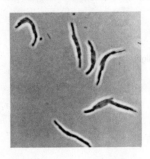

THE EFFICIENCY OF GROWTH: GROWTH YIELDS

The net amount of growth of a bacterial culture is the difference between the cell mass (number of cells) used as an inoculum and the cell mass (number of cells) present in the culture when it enters the stationary phase. When growth is limited by a particular nutrient, there is a fixed linear relationship between the concentration of that limiting nutrient initially present in the medium and the net growth which results, as shown in Figure 9.7. The mass of cells produced per unit of limiting nutrient is, accordingly, a constant, the *growth yield* (Y). The value of Y can be calculated from single measurements of total growth by the equation:

$$Y = \frac{X - X_0}{C} \tag{9.11}$$

where X is the dry weight/ml of cells present when culture enters the stationary phase, X_0 is the dry weight/ml of cells immediately after inoculation, and C is the concentration of limiting nutrient.

The growth yield can be measured for any required nutrient and once determined can then be used to calculate the concentration of that nutrient in an unknown mixture simply by measuring how much growth a sample of the unknown mixture supports when added to a medium complete in all respects except for the limiting nutrient. Such a determination is called a *bioassay*. In the past, bioassays were extensively used for the determination of concentration of amino acids and vitamins in foodstuffs. Now, chemical and physical methods have come into more common use, but the principle of bioassay remains an important research tool for detecting and quantitating compounds with growth-promoting activities. To perform a bioassay one requires only a microbial strain for which the substance to be assayed is an essential nutrient (Chapter 31).

In the case of a chemoheterotrophic bacterium, the growth yield measured in terms of organic substrate utilized becomes an index of the efficiency of conversion of substrate into bacterial mass. The data shown in Figure 9.7 were obtained with an obligately aerobic chemoheterotrophic pseudomonad growing in a synthetic

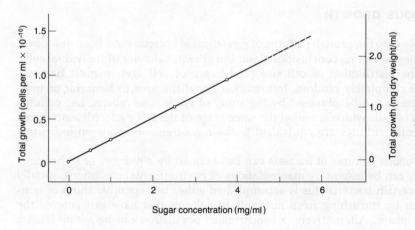

FIGURE 9.7

The relationship between total growth of an aerobic bacterium (*Pseudomonas sp.*) and the initial concentration of the limiting nutrient (fructose). The experiments were done in a synthetic medium with fructose as the sole source of carbon and energy. The slope of the line is the growth yield (Y) of the bacterium on fructose (see text).

medium with fructose as the sole source of carbon and energy. Inspection of the graph reveals a growth yield of approximately 0.4. Considering that the carbon content of fructose and cell material is 40 and 50 percent, respectively, the fraction of fructose carbon converted to cell carbon can be calculated to be about 0.5. Accordingly, this microorganism uses about one-half the carbon of fructose to make cells and oxidizes the other one-half to CO_2. Analogous experiments with other aerobic chemoheterotrophs utilizing sugars as the sole source of carbon reveal that the efficiency of conversion of carbon from the sugars to cellular carbon varies between about 20 and 50 percent. These differences probably reflect differences in efficiency of generating ATP through catabolism of the substrate; evidence to support this inference is described below.

When certain obligately fermentative microorganisms are grown in rich media, radioactive tracer experiments show that little or no carbon from the fermentable substrate is converted into cellular material. Under these conditions, cellular material is derived from the other medium components (amino acids, purines, pyrimidines, etc.), and the fermentable substrate serves only as a source of energy. The ATP yield from many fermentations is known, and the yield of cells (grams dry weight) per mole of ATP produced (Y_{ATP}) approximates 10 grams of cell material/mole of ATP, suggesting that the energy to polymerize carbon building blocks into macromolecules does not vary much among microorganisms. Such constancy is not seen in terms of yield of cells per *mole of substrate fermented* because different microorganisms may ferment a given substrate by pathways which differ in yields of ATP. For example, *Zymomonas mobilis* ferments glucose to produce ethyl alcohol via the Entner-Doudoroff pathway, with formation of 1 mole of ATP per mole of glucose; yeasts, in contrast, ferment glucose to ethyl alcohol via the Embden-Meyerhof pathway, yielding 2 moles of ATP per mole of glucose. Yeasts produce considerably more cells than *Zymomonas* per mole of glucose fermented but approximately the same amount per mole of ATP produced by the fermentation (Table 9.2). The approximate constancy of Y_{ATP} can be used to deduce the yield of an unknown dissimilatory pathway.

SYNCHRONOUS GROWTH

So far, growth patterns of *populations* of bacteria have been described. Such studies permit no conclusions about the growth behavior of individual cells, because the distribution of cell size (and hence of cell age) in most bacterial cultures is completely random. Information about the growth behavior of individual bacteria can be obtained by the study of *synchronous cultures*, i.e., cultures composed of cells which are all at the same stage of the cell cycle. Measurements made on such cultures are equivalent to the measurements made on individual cells.

Synchronous cultures of bacteria can be obtained by a number of techniques. Synchrony can be *induced* by manipulations of environmental conditions, usually cyclic. In certain bacteria this is accomplished either by repetitive shifts of temperature or by furnishing fresh nutrients to cultures that have just entered the stationary phase. Alternatively, a synchronous population can be *selected* from a

TABLE 9.2

Growth yields of fermentative microorganisms, measured in terms of glucose
fermented or ATP produced

ORGANISM	FERMENTATION AND PATHWAY	MOLES ATP FORMED PER MOLE OF GLUCOSE FERMENTED	MOLAR GROWTH YIELD EXPRESSED AS GRAMS OF CELLS PRODUCED PER MOLE OF:	
			GLUCOSE FERMENTED	ATP PRODUCED
Saccharomyces cerevisiae (yeast)	Alcoholic, Embden-Meyerhof	2	21	10.5
Streptococcus faecalis	Homolactic, Embden-Meyerhof	2	22	11.0
Lactobacillus delbruckii	Homolactic, Embden-Meyerhof	2	21	10.5
Zymomonas mobilis	Alcoholic, Entner-Doudoroff	1	8.6	8.6

FIGURE 9.8

Helmstetter - Cummings
technique of obtaining
synchronous cultures.

random population by physical separation of cells that are at the same stage of
the cell cycle. This can be accomplished by differential filtration or by centrifugation. For physiological studies, techniques based on selection are preferable
to those based on induction because the technique of induction may cause cyclic
changes that are not typical of the normal cell cycle.

An excellent selective method for obtaining synchronous cultures is the Helmstetter-Cummings technique, which is based on the fact that certain bacteria stick
tightly to cellulose nitrate (Millipore) filters. The technique involves filtering an
unsynchronized culture of bacteria through a Millipore filter, then inverting the
filter and allowing fresh medium to flow through it (Figure 9.8). After loosely
associated bacteria have been washed from the filter, the only bacterial cells in
the effluent stream of medium are those which arise through division. Hence,
all cells in the effluent are newly formed and are therefore at the same stage of
the cell cycle.

The growth of a culture of *E. coli* so synchronized is shown in Figure 9.9. The
number of cells in the culture remains approximately constant for about one hour
while the newly formed cells grow in size. Then, rather abruptly, the number
of cells doubles. In the second division cycle, the plateau is less distinct and
the population rise extends over a longer period, indicating that synchrony is
already being lost. In the third division cycle, almost no indication of synchrony
remains.

Synchronous cultures rapidly lose synchrony because various cells of a population do not all divide at the same size (age, or time following the previous
division). From the rate of loss of synchrony one can calculate *the distribution of
the age of cells at division*. Using electronic counting techniques, one can determine
the distribution of cell size in an exponentially growing culture. These two distributions
allow the calculation of growth rate as a function of cell size (Figure 9.10). The
relationship is complex: very small cells grow slowly, cells of intermediate size
grow more rapidly, and very large cells again grow slowly.

Similar computations permit an estimate of the distribution of cells of various
ages (time elapsed since the division which produced them) in an exponential

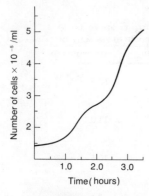

FIGURE 9.9

Synchronous growth of *E. coli* in glucose minimal medium. The effluent from a membrane culture (Helmstetter-Cummings) technique) was collected for 3 minutes and incubated at 30°C. After A. G. Marr, P. R. Painter, and E. H. Nilson, "Growth and Division of Individual Bacteria," in *Microbial Growth*, 19th Symposium of the Society of General Microbiology. Cambridge, England: Cambridge University Press, 1969.

culture (Figure 9.11). Again, the relationship is somewhat complex, reflecting the relationship between growth rate and cell size. Figure 9.11 illustrates a general property of any expanding population: the numerical predominance of young individuals.

Perhaps the most clear illustration of the fact that the kinetics of growth of individual cells cannot be deduced from the kinetics of growth of the overall population is provided by the growth of *Caulobacter*. A population of this type of unicellular bacterium always consists of two structurally differentiated cell types, stalked cells and swarmer cells. The stalked cells always grow significantly faster than the swarmer cells (see page 146) even though the whole population grows exponentially at a constant rate.

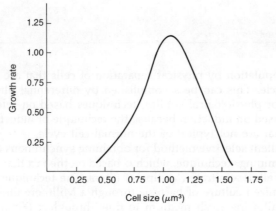

FIGURE 9.10

Growth rate (μm³/hour) of individual cells of *E. coli* growing in a synthetic medium as a function of the size of the cell (μm³). After A. G. Marr, P. R. Painter, and E. H. Nilson, "Growth and Division of Individual Bacteria," in *Microbial Growth*, 19th Symposium of the Society of General Microbiology. Cambridge, England: Cambridge University Press, 1969.

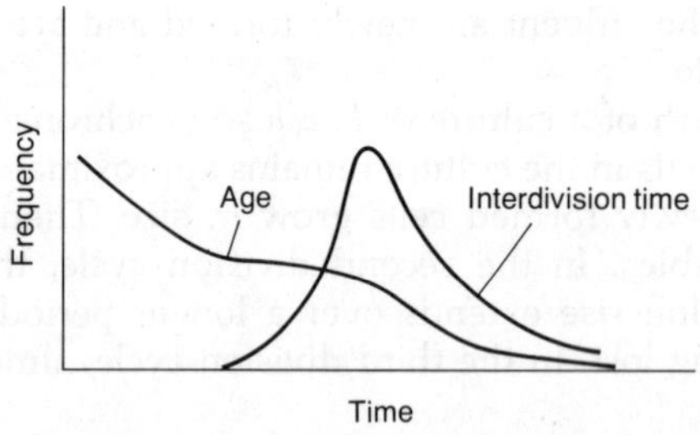

FIGURE 9.11

Comparison of the distribution of interdivision times and age of cells in an exponentially growing population of *E. coli*. After A. G. Marr, P. R. Painter, and E. H. Nilson, "Growth and Division of Individual Bacteria," in *Microbial Growth*, 19th Symposium of the Society of General Microbiology. Cambridge, England: University of Cambridge, 1969.

EFFECT OF NUTRIENT CONCENTRATION ON GROWTH RATE

In many respects, the bacterial growth process can be likened to a chemical reaction in which the components of the medium (the reactants) produce more cells (the product of the reaction), a process catalyzed by the bacterial population. The velocity of chemical reactions is determined by the concentration of reactants, but as we have seen, bacterial growth rate remains

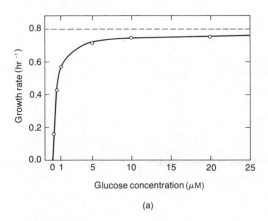

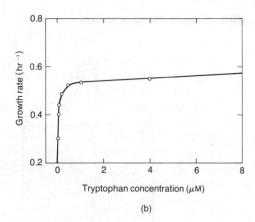

(a)

(b)

FIGURE 9.12

The effect of nutrient concentration on the specific growth rate of *E. coli*. (a) Effect of glucose concentration, and (b) effect of tryptophan concentration (for a tryptophan requiring mutant). From T. E. Shehata and A. G. Marr, "Effect of Nutrient Concentration on the Growth of *Escherichia coli*," *J. Bacteriol.* **107,** 210 (1971).

constant until the medium is almost exhausted of the limiting nutrient. This seeming paradox is explained by the action of permeases which are capable of maintaining saturating intracellular concentrations of nutrients over a wide range of external concentrations (Chapter 10). Nevertheless, at extremely low concentrations of external nutrients the permease systems are no longer able to maintain saturating intracellular concentrations, and the growth rate falls.

The curves relating growth rate to nutrient concentration are typically hyperbolic (Figure 9.12), and fit the equation:

$$\mu = \mu_{max} \frac{C}{K_s + C} \tag{9.12}$$

where μ is the specific growth rate at limiting nutrient concentration (C), μ_{max} is the growth rate at saturating concentration of nutrient, and K_s is a constant analogous to the Michaelis-Menten constant of enzyme kinetics, being numerically equal to the substrate concentration supporting a growth rate equal to $\frac{1}{2}\mu_{max}$. Values of K_s for glucose and tryptophan utilization by *E. coli* (Figure 9.12) are 1×10^{-6} and $2 \times 10^{-7} M$, respectively, or 0.18 and 0.03 microgram per milliliter. These very low values are attributable to the high affinities characteristic of many bacterial permeases, which can be construed as an evolutionary adaptation to growth in extremely dilute solutions. In this respect, conventional laboratory media are very different from many natural environments.

CONTINUOUS CULTURE OF MICROORGANISMS

Cultures of the type so far discussed are called *batch cultures;* nutrients are not renewed and hence growth remains exponential for only a few generations. Microbial populations can be maintained in a state of exponential growth over a long period of time by using a system of continuous culture (Figure 9.13). The growth chamber is connected to a reservoir of sterile medium. Once

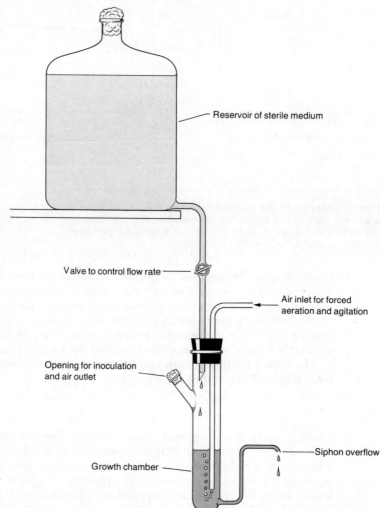

Reservoir of sterile medium

Valve to control flow rate

Air inlet for forced
aeration and agitation

Opening for inoculation
and air outlet

Siphon overflow

Growth chamber

FIGURE 9.13

Simplified diagram of a
continuous culture system.

growth has been initiated, fresh medium is continuously supplied from the
reservoir. The volume of liquid in the growth chamber is maintained constant
by allowing the excess volume to be removed continuously through a siphon
overflow.

If the fresh medium enters at a constant rate, the density of bacteria in the
growth chamber remains constant after an initial period of adjustment. In other
words, the bacteria in the growth chamber grow just fast enough to replace those
lost through the siphon overflow. If the rate of entry of fresh medium is changed,
another adjustment period occurs followed by maintenance of a constant population at a new density; the growth rate changes to match the new rate of loss of

cells through the overflow. A continuous culture system responds in this manner to a wide variation in the rate of addition of fresh medium. However, no matter the rate of inflow of medium, bacteria cannot grow faster than they would in batch culture.

The question posed by the observation that culture densities in continuous culture systems remain constant is the following: how does the rate of addition of fresh medium to the culture vessel determine the growth rate of the culture? The explanation lies in the fact that the rate of growth of bacteria in continuous culture devices is always limited by the concentration of one nutrient. Consequently, the rate of addition of fresh medium determines the rate of growth of the culture: the system is self-regulating. Consider a continuous culture device that is operating at a constant rate of addition of fresh medium. After inoculation, the culture will at first grow at maximum rate (μ_{max}). As the culture density increases, the rate of utilization of nutrients will increase until the depletion of one nutrient begins to limit the growth rate. As long as the growth rate exceeds the rate of loss through the siphon overflow, density will continue to increase, the steady-state concentration of limiting nutrient in the growth vessel will continue to decrease and, as a consequence, the growth rate will decrease until the rate of increase of cells through growth will just equal the rate of loss of cells through the overflow. Were the growth rate transiently to become lower than the rate of loss of cells, cell density would decrease, limiting nutrient concentration would increase, and the growth rate would increase until the balance between the growth rate and the loss of cells is again reached.

Using the fact that cell density in a continuous culture system remains constant and is self-regulating, we can describe the system in mathematical terms:

$$\begin{pmatrix} \text{rate of production of} \\ \text{cells through growth} \end{pmatrix} = \begin{pmatrix} \text{rate of loss of cells} \\ \text{through the overflow} \end{pmatrix}$$

Previously we described the rate of production of bacterial mass (dX/dt) by Eq. (9.2):

$$\frac{dX}{dt} = \mu X$$

The rate of loss of cells through the overflow (dX/dt) can be stated as:

$$\frac{dX}{dt} = \frac{f}{V} = DX \tag{9.13}$$

where flow rate (f) is measured in culture volumes (V) per hour. The expression (f/VX) is called the *dilution rate*, D. Thus,

$$\mu X = DX \tag{9.14}$$

or

$$\mu = D \tag{9.15}$$

which states that the growth rate equals the dilution rate in a stabilized continuous culture device. Substituting Eq. (9.12), we have:

$$\mu_{max} \frac{C}{K_s + C} = D \tag{9.16}$$

and solving for C, we have:

$$C = K_s \frac{D}{\mu_{max} - D} \tag{9.17}$$

which states the fundamental relationship between substrate concentration (C) in the growth vessel and dilution rate (D).

In steady-state operation of a continuous culture device, concentration of the limiting nutrient (C) also remains constant. Thus, the rate of addition of the nutrient must equal the rate at which it is utilized by the culture together with that lost through the overflow.

$$\begin{pmatrix} \text{substrate added} \\ \text{from reservoir} \end{pmatrix} = \begin{pmatrix} \text{substrate used} \\ \text{for growth} \end{pmatrix} + \begin{pmatrix} \text{substrate lost} \\ \text{through overflow} \end{pmatrix} \qquad (9.18)$$

or

$$DC_r = \frac{dc}{dt} + DC$$

where C_r is the concentration of limiting nutrient in the reservoir and dc/dt is the rate of utilization of limiting nutrient for growth. Substituting $(dX/dt)\,(dc/dX)$ for dc/dt, and since $dX/dt = \mu X$ and $dc/dt = Y$, we have:

$$DC_r = \frac{\mu X}{Y} + DC \qquad (9.19)$$

In the steady state, $D = \mu$; hence, solving for X yields:

$$X = Y(C_r - C) \qquad (9.20)$$

the fundamental relation between cell density (X) and the concentration of limiting nutrient (C) in the growth vessel. Together, Eqs. (9.17) and (9.20) allow us to see the relationship between cell density, limiting nutrient concentration, and the dilution rate (Figure 9.14). Cell density and the concentration of limiting nutrient change little at low dilution rates. As the dilution rate approaches μ_{max}, cell density drops rapidly to zero, and the concentration of the limiting nutrient approaches its concentration in the reservoir (C_r).

• **Chemostats and turbidostats**

Continuous culture systems can be operated as *chemostats* or as turbidostats. In a chemostat the flow rate is set at a particular value and the rate of growth of the culture adjusts to this flow rate. In a turbidostat the system includes an optical-sensing device which measures the absorbancy of the culture (culture density) in the growth vessel; the electrical signal from this device regulates the flow rate. Thus, the absorbancy of the culture controls the flow rate and the rate of growth of the culture adjusts to this flow rate.

As a practical matter, chemostats and turbidostats are usually operated at different dilution rates. In the chemostat, maximum stability is attained within a range of dilution rates over which cell density changes only slightly with changes in dilution rate, i.e., at low dilution rates. In contrast, in the turbidostat, maximum sensitivity and stability are achieved at high dilution rates, within a range over which culture density changes rapidly with dilution rate (Figure 9.14).

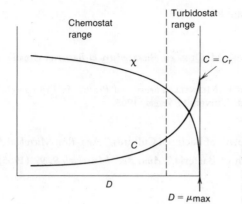

FIGURE 9.14

Relationship between cell density (X), limiting nutrient concentration (C), and dilution rate (D) in a continuous culture system.

• **Use of continuous culture systems**   Continuous culture systems offer two valuable features for the study of microorganisms. They provide a constant source of cells in an exponential phase of growth, and they allow cultures to be grown continuously at extremely low concentrations of substrate. The practical advantages of the former feature are obvious. Growth at low substrate concentrations is valuable in studies on regulation of synthesis or catabolism of the limiting substrate, in selection of various classes of mutants, and in ecological studies.

MAINTENANCE ENERGY

It will be noted in Figure 9.14 that the curves for cell density and concentration of limiting nutrient are not extended into the range of very low dilution rates. This is because the response of a culture at very low dilution rates depends on the nature of the limiting nutrient. If the limiting nutrient is the energy source for the culture, growth ceases at very low dilution rates because a certain amount of energy, termed *maintenance energy*, is always used for purposes other than growth; when the rate of entry of the source of energy is insufficient to supply more energy than that required for maintenance, growth cannot occur. Maintenance energy can be stated mathematically as follows:

$$\frac{dX}{dt} = \mu x - ax \qquad (9.21)$$

where a is the *specific maintenance rate constant*. Equation (9.21) is a variation of the growth equation [Eq. (9.2)] which applies if the growth rate is limited by the supply of energy. Since values of a are typically low (e.g., 0.018 hr^{-1} for *E. coli* strain ML30, growing on glucose), the effect of maintenance energy on the growth rate is usually negligible.

FURTHER READING

Books   Kubitschek, H. E., *Introduction to Research with Continuous Cultures*. Englewood Cliffs, N.J.: Prentice-Hall, 1970.

MANDELSTAM, J., and K. McQUILLEN, *Biochemistry of Bacterial Growth*, 2nd ed. New York: Wiley, 1973.

MEYNELL, G. G., and E. MEYNELL, *Theory and Practice in Experimental Bacteriology*. Cambridge, England: University Press, 1965.

Reviews   MONOD, J., "The Growth of Bacterial Cultures," *Ann. Rev. Microbiol.* **3**, 371 (1949).

NOVICK, A., "Growth of Bacteria," *Ann. Rev. Microbiol.* **9**, 97 (1955).

10

THE EFFECT OF ENVIRONMENT ON MICROBIAL GROWTH

This chapter will consider the interactions between the microbial cell and its chemical and physical environment.

FUNCTIONS OF THE CELL MEMBRANE

The concentration of solutes in a microbial cell is typically much higher than the concentration of solutes in the extracellular environment. This is true both of the natural environment of microorganisms and of most culture media. Furthermore, the solutes present in the cell and the external environment differ qualitatively. The major barrier governing the passage of solutes between the cell and the external environment is the cell membrane, a structure of fixed width (approximately 10 nm) which is composed of a complex mixture of proteins, lipids, and glycoproteins. Although the molecular organization of the membrane constituents is not yet known, a considerable body of evidence suggests that the cell membrane is made up of a phospholipid bilayer in which are imbedded a variety of proteins (see Figure 11.4).

The roles of the bacterial membrane in respiratory metabolism and in chromosomal segregation have been discussed in Chapters 6 and 7. Here we shall consider the function of the membrane in the maintenance of the distinctive internal chemical composition of the cell.

In Gram-negative bacteria, the outer layer of the cell wall (sometimes called the *outer membrane*, because in transverse sections it has a fine structure that resembles that of the cell membrane) also plays a role, although a subordinate one, in regulating the passage of certain molecules between the cell and its environment. The region contained between the cell membrane and the outer wall layers of Gram-negative bacteria, known as the *periplasmic space*, has a chemical composition that is distinct from that of the cell interior and from that of the external environment.

In this respect, membrane function is multifaceted: it keeps essential metabolites and macromolecules inside the cell; it pumps certain nutrients into the cell against a gradient of concentration; it permits other nutrients to flow freely across it; and it excludes the entry of certain solutes present in the environment.

Since the cell membrane contains a lipid bilayer, chemical considerations alone would suggest that it should offer an absolute barrier to the passage of all polar molecules. In fact, however, most nutrients are polar molecules. Hence, the membrane must contain certain chemically modified regions through which those essential polar molecules flow; the regions in question are, therefore, operative in *membrane transport*.

ENTRY OF NUTRIENTS IN THE CELL

Membrane transport occurs by a variety of mechanisms, the simplest of which is *passive diffusion*. Net flow of a compound by passive diffusion occurs only in response to a difference in its concentration across the membrane (a concentration gradient) and as a result of such flow the difference diminishes. The rate of flow is a direct function of the magnitude of the gradient, and it does not approach a limiting value even when the concentration difference is great. Passive diffusion implies that there are regions of the membranes through which certain compounds can pass freely, much as small molecules pass through the artificial membranes used for dialysis. Water is the principal nutrient which enters and leaves the cell by passive diffusion.

• **Facilitated diffusion**   Passive diffusion exhibits a minimum of substrate specificity. For example, passive diffusion cannot discriminate between enantiomorphs of optically active compounds. A related transport mechanism, known as *facilitated diffusion*, exhibits the specificity typical of enzyme-catalyzed reactions. Facilitated diffusion is mediated by specific membrane proteins, which in some cases are induced by their substrates. These membrane proteins, collectively known as *permeases*, bind to the substrate molecule on the exterior of the membrane and, by mechanisms still largely unknown, allow its passage through the membrane. Dissociation of the complex at the interior membrane surface completes the process of transport. The simplest interpretation of the function of permeases is that they are able to pass across the membrane barrier either with or without an attached substrate molecule. They, therefore, catalyze the general reaction:

substrate (outside the cell) $\longleftrightarrow$ substrate (inside the cell)

Facilitated diffusion is similar to simple diffusion in the sense that the substrate moves through a concentration gradient in the thermodynamically favorable direction, i.e., from a higher to a lower concentration. Hence, *this process does not require the expenditure of metabolic energy.* It differs from passive diffusion in that it is mediated by a specific protein catalyst which has many properties in common with enzymes. The rate of the process is rapid (more rapid than that predicted by passive diffusion), and it exhibits considerable substrate specificity. The catalysts are often inducible. Finally, the rate of the reaction approaches a limiting value with increasing concentrations of substrate; i.e., it obeys normal enzyme (Michaelis-Menten) kinetics. The velocity of entry of substrate can be described as:

$$V^{entry} = V_{max}^{entry} \frac{[S]_{ex}}{K_m^{entry} + [S]_{ex}} \tag{10.1}$$

and the rate of exit as:

$$V^{exit} = V_{max}^{exit} \frac{[S]_{in}}{K_m^{exit} + [S]_{in}} \tag{10.2}$$

where $[S]_{ex}$ and $[S]_{in}$ are the concentration of the substrate outside and inside the cell, respectively, and V_{max}^{entry} and V_{max}^{exit} are the velocities of entry and exit at saturating concentrations of substrate. K_m^{exit} and K_m^{entry} are the Michaelis constants (inversely proportional to the affinity of the permease for its substrates) for the entry and exit processes. In facilitated diffusion, the values of V_{max} for the entry and exit reactions are equal, as are the values of K_m^{entry} and K_m^{exit}. As a consequence, *at equilibrium, the internal and external concentrations of a nutrient transported by facilitated diffusion are equal.*

Although facilitated diffusion is a common mechanism of transport in eucaryotes, it appears to be relatively rare in procaryotes. For example, sugars, which characteristically enter eucaryotic cells by facilitated diffusion, enter procaryotic cells by another group of mechanisms, *active transport*, which are described below. One process of transport in a procaryote which is mediated by facilitated diffusion is the entry of glycerol into the cells of bacteria of the enteric group.

• **Active transport**   The mechanisms of transport known collectively as *active transport* permit a solute to enter the cell against a thermodynamically unfavorable gradient of concentration. At equilibrium, active transport systems create concentrations of solutes within the cell which may be several thousand times as great as those outside the cell. The prevalence of active transport mechanisms in bacteria can be correlated with the facts that bacteria frequently occur in dilute chemical environments and nevertheless exhibit rapid rates of metabolism. Since active transport concentrates solutes within the cell against a thermodynamically unfavorable gradient, *it requires an expenditure of metabolic energy.* The mechanism by which metabolic energy is coupled to active transport remains one of the most intriguing and important questions of cell biology.

As one illustration of the many active transport systems of bacteria, we shall consider the system by which the disaccharide, lactose, is concentrated within the cell of *E. coli.* This system is called the *β-galactoside permease system.* The transport of β-galactosides is mediated by a specific permease, the reaction being coupled with the expenditure of metabolic energy. The metabolic energy is used to decrease the affinity of the permease for lactose (or other β-galactosides) on the internal side of the membrane, relative to its affinity on the external side of the membrane. As a result K_m^{exit} becomes greater than K_m^{entry}, and the velocity of exit becomes less than the velocity of entry [Eqs. (10.1) and (10.2)]. At equilibrium, therefore, $[S]_{in}$ is greater than $[S]_{out}$. If generation of metabolic energy is blocked by adding a metabolic poison such as sodium azide to the cells, the β-galactoside permease system can no longer mediate active transport. Under these circumstances, the permease system catalyzes a facilitated diffusion of β-galactosides; it shows an equal affinity for β-galactosides on both sides of the membrane, and the rate of entry becomes equal to the rate of exit.

The β-galactoside permease (sometimes called the *M protein*) has been isolated from the cell membrane by treatment of this structure with detergents.

β-galactoside permease is encoded by a specific gene, *lacY*. Certain mutations in this locus make the cell incapable of transporting lactose, and hence incapable of metabolizing it, although all the enzymes necessary for lactose metabolism are within the cell. Such mutants are said to be *cryptic*. Crypticity caused by the absence of a specific transport system is relatively common in wildtype organisms.* For example, the inability of wildtype strains of *E. coli* to utilize citrate as a source of carbon, a property which is important in the classification of enteric bacteria (Chapter 21), reflects crypticity. *E. coli* possess the enzymes necessary to dissimilate citrate, which is a key intermediate of respiratory metabolism.

• **Binding proteins**

Many active transport systems of Gram-negative bacteria are associated with so-called *binding proteins,* which are located in the periplasmic space. They can be released from the cell by a treatment called *cold osmotic shock,* namely, a sudden decrease in the osmotic strength of the suspending liquid at low temperature. The procedure damages the outer layer of the cell wall, and hence releases the proteins of the periplasmic space into the external solution. These proteins include the binding proteins, which have very high affinities for specific nutrients. The dissociation constants lie in the range of 10^{-7} to 10^{-9}. Over 100 different binding proteins have been isolated and studied. They have no catalytic activity, but they form tight complexes with specific nutrients, including amino acids, sugars, and inorganic ions. Binding proteins are essential for the active transport of their substrates, as shown by the fact that mutants which lose the ability to synthesize a specific binding protein simultaneously lose the ability to accumulate its substrate. However, they are not permeases because they are not integrated into the cell membrane but occur as water-soluble proteins in the periplasmic space. Although both genetic and biochemical evidence suggests that they play an essential role in transport, they do so in conjunction with specific permeases that are located in the cell membrane. Mutational loss either of the specific binding protein or of the specific permease which governs the entry of a given nutrient into the cell confers crypticity for that nutrient.

• **Group translocation**

The process of active transport catalyzes the movement of chemically unmodified nutrients across the cell membrane against a concentration gradient. Bacteria possess other transport systems which, during the process of transport, convert the nutrient to a chemically modified form to which the membrane is impermeable. Such systems of group translocation do not mediate active transport since the concentration of the unmodified nutrient inside the cell $[S]_{in}$, is always very low. The overall process of group translocation does, however, resemble active

*The genetic strain of any organism as it is first isolated from nature is designated as the *wildtype* strain.

transport since the concentration of the chemically modified nutrient inside the cell can greatly exceed the concentration of free nutrient in the medium.

One example of a group translocation system, widely distributed in bacteria, is the *phosphotransferase system*. It mediates the transport of many sugars and sugar derivatives, which are phosphorylated during the process, entering the cell as sugar phosphates. Since the membrane is highly impermeable to most phosphorylated compounds, the sugar phosphates, once formed, are trapped within the cell.

The nonspecific components of this system are two soluble proteins, Enzyme I and HPr. Enzyme I catalyzes the transfer of a high-energy phosphate group from phosphoenolpyruvic acid to a histidine residue in HPr. Phospho-HPr acts as a general phosphoryl donor to all nutrients transported by the phosphotransferase system. The phosphorylations of these nutrients are catalyzed by a series of substrate-specific membrane-bound proteins collectively called Enzymes II. The net chemical reaction accompanying transport is, accordingly, the phosphorylation of a nutrient at the expense of the high-energy phosphate group of phosphoenolpyruvic acid (Figure 10.1).

Each bacterial species contains a single molecular form of Enzyme I and of HPr which functions in the transport of many sugars. Mutant strains unable to synthesize either Enzyme I or HPr lose the ability to transport and (hence to utilize) many sugars. For example, mutants of *Salmonella typhimurium* which have lost Enzyme I are unable to utilize glucose, mannose, fructose, glycerol, mannitol, sorbitol, N-acetylglucosamine, maltose, and melibiose. Mutants which have lost one of the specific Enzymes II lose the ability to utilize only one sugar.

The membrane-bound Enzymes II appear to play the role of permeases in addition to their specific catalytic roles as phosphotransferases. In mutant strains which have lost either Enzyme I or HPr, the Enzymes II catalyze facilitated diffusion of these specific substrates.

In the enteric group, most utilizable sugars with the exceptions of β-galactosides and pentoses are transported by the phosphotransferase system.

Estimates of energy requirements for the β-galactoside permease system suggest that the equivalent of one molecule of ATP is expended per molecule of lactose transported. The transport of a sugar mediated by the phosphotransferase system likewise requires the expenditure of a high-energy phosphate bond (in this case from phosphoenolpyruvate, whose high-energy bond is ultimately derived

$$\text{phosphoenolpyruvate}_{(i)} + \text{HPr}_{(i)} \xrightarrow{\text{Enzyme I}} \text{phospho-HPr}_{(i)} + \text{pyruvate}_{(i)}$$

$$\text{phospho-HPr}_{(i)} + \text{sugar}_{(o)} \xrightarrow{\text{Enzyme II}} \text{phospho-sugar}_{(i)} + \text{HPr}_{(i)}$$

Net reaction: $\text{phosphoenolpyruvate}_{(i)} + \text{sugar}_{(o)} \xrightarrow[\text{Enzyme II,} \atop \text{HPr}]{\text{Enzyme I}} \text{phospho-sugar}_{(i)} + \text{pyruvate}_{(i)}$

FIGURE 10.1

Reactions of the phosphotransferase system. HPr and Enzyme I are soluble intracellular proteins which generate phospho-HPr. Phospho-HPr, in turn, serves as the phosphoryl donor for a number of sugar transport systems, each catalyzed by a specific Enzyme II located in the cell membrane. The net reaction catalyzed by the three proteins is phosphorylation of a sugar at the expense of phosphoenolpyruvate and the simultaneous transport of the sugar moiety across the membrane. Subscripts refer to the location of the compound as being either outside (o) or inside (i) the cell membrane.

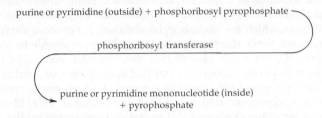

purine or pyrimidine (outside) + phosphoribosyl pyrophosphate

phosphoribosyl transferase

purine or pyrimidine mononucleotide (inside)
+ pyrophosphate

FIGURE 10.2

Reactions involved in the transport of purine and pyrimidine bases across the cell membrane.

from ATP). However, the product of the transfer reaction is a phosphorylated derivative of the sugar; consequently, the first reaction in the catabolism of a free sugar molecule, phosphorylation by a kinase at the expense of ATP, has already been accomplished. Thus, relative to active transport, the transport of sugars mediated by the phosphotransferase system spares ATP. This is particularly important in the fermentative metabolism of sugars, where the net ATP yield per mole of sugar dissimilated is small. This may account for the fact that the phosphotransferase system plays a prominent role in sugar transport in bacteria characterized by fermentative metabolism. Thus, it is invariably operative in the facultatively aerobic enteric bacteria, but is absent from some bacterial groups with a purely respiratory metabolism of carbohydrates such as the aerobic pseudomonads, where active transport is the dominant mechanism for the entry of sugars into the cell.

Group transfer is also involved in the uptake of purine and pyrimidine bases by enteric bacteria. The processes of transport and conversion to nucleotides appear to be intimately linked (Figure 10.2).

•**Summary of membrane transport mechanisms**

The various transport systems known to operate in bacteria are schematized in Figure 10.3. Simple diffusion is the net passage of a nutrient along the thermodynamically favorable concentration gradient through areas of the membrane which offer no barrier to its passage. Facilitated diffusion is mediated by specific

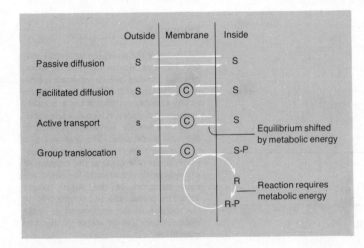

FIGURE 10.3

Comparison of bacterial transport systems. Relative lengths of arrows indicate the direction of equilibria. S and s designate high and low concentrations, respectively, of solutes. C designates a carrier protein (permease). The diagram of group translocation is based on the properties of the phosphotransferase system, where R, R-P, and P represent HPr, phospho-Hpr, and a phosphoryl group, respectively.

carrier proteins or permeases (C) which catalyze passage of a nutrient along a thermodynamically favorable concentration gradient through a membrane which is otherwise impermeable to the nutrient. Active transport resembles facilitated diffusion in that a specific permease (C) mediates passage of a nutrient through the membrane, but it differs in that passage can occur against a concentration gradient, because passage is coupled with the expenditure of metabolic energy. Similarly, group translocation involves the participation of specific permeases, but it is accompanied by a chemical modification of the nutrient undergoing transport.

EFFECTS OF SOLUTES ON GROWTH AND METABOLISM

Transport mechanisms play two essential roles in cellular function. First, they maintain the intracellular concentrations of all primary substrates at levels sufficiently high to ensure operation of both catabolic and biosynthetic pathways at near-maximal rates, even when the concentrations of primary substrates in the external medium are low. This is evidenced by the fact that the exponential growth rates of microbial populations remain constant until one essential nutrient in the medium falls to a very low value, approaching exhaustion. At this limiting nutrient concentration, the growth rate of the population rapidly falls to zero (see Chapter 9). Second, transport mechanisms function in *osmoregulation*, maintaining the solutes (principally small molecules and ions) at levels optimal for metabolic activity, even when the osmolarity of the environment varies over a relatively wide range.*

Most bacteria do not need to regulate their internal osmolarity with precision because they are enclosed by a cell wall capable of withstanding a considerable internal osmotic pressure. Bacteria always maintain their osmolarity well above that of the medium. If the internal osmotic pressure of the cell falls below the external osmotic pressure, water leaves the cell and the volume of the cytoplasm decreases with accompanying damage to the membrane. In Gram-positive bacteria, this causes the cell membrane to pull away from the wall; the cell is said to be *plasmolyzed*. Gram-negative bacteria do not undergo plasmolysis, since the wall retracts with the membrane; this also damages the membrane.

Bacteria vary widely in their osmotic requirements. Some are able to grow in very dilute solutions, and some in solutions saturated with sodium chloride. Microorganisms that can grow in solutions of high osmolarity are called *osmophiles*.

*When a *solution* of any substance (*solute*) is separated from a *solute-free solvent* by a membrane that is freely permeable to solvent molecules, but not to molecules of the solute, the solvent tends to be drawn through the membrane into the solution, thus diluting it. Movement of the solvent across the membrane can be prevented by applying a certain hydrostatic pressure to the solution. This pressure is defined as *osmotic pressure*. A difference in osmotic pressure also exists between two solutions containing different concentrations of any solute.

The osmotic pressure exerted by any solution can be defined in terms of *osmolarity*. An osmolar solution is one that contains one *osmole* per liter of solutes, i.e., a 1.0 molal solution of an ideal nonelectrolyte. An osmolar solution exerts an osmotic pressure of 22.4 atmospheres at 0°C, and depresses the freezing point of the solvent (water) by 1.86°C. If the solute is an electrolyte, its osmolarity is dependent on the degree of its dissociation, since both ions and undissociated molecules contribute to osmolarity. Consequently, the osmolarity and the molarity of a solution of an electrolyte are grossly different. If both the molarity and the dissociation constant of a solution of an electrolyte are known, its osmolarity can be calculated with some degree of approximation, as the sum of the moles of undissociated solute and the mole equivalents of ions. Such a calculation is accurate only if the solution is an ideal one, and if it is extremely dilute. Therefore, it is preferable to determine the osmolarity of a solution experimentally, e.g., by freezing-point depression.

TABLE 10.1

Osmotic tolerance of certain bacteria[a]

PHYSIOLOGICAL CLASS	REPRESENTATIVE ORGANISMS	APPROXIMATE RANGE OF NACL CONCENTRATION TOLERATED FOR GROWTH (%, g/100 ml)
Nonhalophiles	*Spirillum serpens*	0.0–1
	Escherichia coli	0.0–4
Marine forms	*Alteromonas haloplanktis*	0.2–5
	Pseudomonas marina	0.1–5
Moderate halophiles	*Micrococcus halodenitrificans*	2.3–20.5
	Vibrio costicolus	2.3–20.5
	Pediococcus halophilus	0.0–20
Extreme halophiles	*Halobacterium salinarium*	12–36 (saturated)
	Sarcina morrhuae	5–36 (saturated)

[a]Ranges of tolerated salt concentrations are only approximate; they vary with the strain and with the presence of other ions in the medium.

Most natural environments of high osmolarity contain high concentrations of salts, particularly sodium chloride. Microorganisms which grow in this type of environment are called *halophiles*. Bacteria can be divided into four broad categories in terms of their salt tolerance: *nonhalophiles, marine organisms, moderate halophiles,* and *extreme halophiles* (Table 10.1). Some halophiles, for example *Pediococcus halophilus,* can tolerate high concentrations of salt in the growth medium, but they can also grow in media without added NaCl. Other bacteria, including marine bacteria and certain moderate halophiles, as well as all extreme halophiles, require NaCl for growth. The tolerance of environments of high osmolarity and the specific requirement for NaCl are distinct phenomena, each of which has a specific biochemical basis.

•**Osmotic tolerance**

Osmotic tolerance—the ability of an organism to grow in media with widely varying osmolarities—is accomplished in bacteria by an adjustment of the internal osmolarity so that it always exceeds that of the medium. Intracellular accumulation of potassium ions (K^+) seems to play a major role in this adjustment. Many bacteria have been shown to concentrate K^+ to a much greater extent than Na^+ (Table 10.2). Moreover, there is an excellent correlation between the osmotic tolerance of bacteria and their K^+ content. For bacteria as metabolically diverse as Gram-positive cocci, bacilli, and Gram-negative rods, relative osmotic tolerance can be deduced from their relative K^+ contents after growth in a medium of fixed ionic strength and composition. Studies on *E. coli* have shown that the intracellular K^+ concentration increases progressively with increasing osmolarity of the growth medium. Consequently, both the osmolarity and the internal ionic strength of the cell increase.*

*The ionic strength of a solution is defined by the equation: $I = \frac{1}{2} \Sigma M_i Z^2$, where M_i is the molarity of a given ion and Z is the charge, regardless of sign. Since the Z term is squared, the ionic strength of an ion increases exponentially with the magnitude of its charge either positive or negative. The magnitude of ionic charge, however, does not affect osmolarity.

TABLE 10.2

Intracellular concentrations of solutes in various bacteria[a]

ORGANISM	CONCENTRATION (%, w/v) IN GROWTH MEDIUM OF:		RATIO OF INTRACELLULAR TO EXTRACELLULAR CONCENTRATION OF:	
	NaCl	KCl	Na^+	K^+
Nonhalophiles:				
Staphylococcus aureus	0.9	0.19	0.7	27
Salmonella oranienburg	0.9	0.19	0.9	10
Moderate halophiles:				
Micrococcus halodenitrificans	5.9	0.02	0.3	120
Vibrio costicolus	5.9	0.02	0.7	55
Extreme halophiles:				
Sarcina morrhuae	23.4	0.24	0.8	64
Halobacterium salinarium	23.4	0.24	0.3	140

[a] Data from J. H. B. Christian and J. A. Waltho, "Solute Concentrations Within Cells of Halophilic and Non-halophilic Bacteria," *Biochem. Biophys. Acta* **65**, 506 (1962).

The maintenance of a relatively constant ionic strength within the cell is of critical physiological importance, because the stability and behavior of enzymes and other biological macromolecules are strongly dependent on this factor. In bacteria, the diamine putrescine (see Chapter 7, p. 213) probably always plays an important role in assuring the approximate constancy of internal ionic strength. This has been shown through studies on *E. coli*. The concentration of intracellular putrescine varies inversely with the osmolarity of the medium; increases of osmolarity cause rapid excretion of putrescine. An increase in the osmolarity of the medium causes an increase in the internal osmolarity of the cell as a result of uptake of K^+; ionic strength is maintained approximately constant as a result of the excretion of putrescine. This is a consequence of the differing contributions which a multiply-charged ion makes to ionic strength and osmotic strength of a solution; a change of putrescine^{2+} concentration that alters ionic strength by 58 percent alters osmotic strength by only 14 percent.

• **The requirement for Na$^+$ in bacteria**

In most nonhalophilic bacteria it has not been possible to demonstrate a specific Na^+ requirement. In view of the extreme experimental difficulty of preparing a medium that is rigorously free of this very abundant ion, the possibility that nonhalophiles might require a very low concentration of Na^+ cannot be excluded. Among nonhalophiles a Na^+ requirement has been detected only for growth at the expense of certain specific carbon and energy sources: e.g., *E. coli* requires Na^+ for growth at maximal rate with L-glutamate, and *Enterobacter aerogenes* requires Na^+ for growth with citrate. In both cases, the quantitative requirement is small, and absolute growth dependence on Na^+ has not been demonstrated.

In contrast, bacteria of marine origin, moderate halophiles, and extreme halophiles always require Na^+ for growth, at concentrations so high that their absolute dependence on this cation can be demonstrated experimentally without difficulty, even if the basal medium employed has not been prepared from specially purified (Na^+-free) ingredients.

In all these organisms, Na^+ probably plays a number of different roles, all indispensable to the maintenance of cellular function. In marine bacteria, there is good evidence that it assures the correct function of transport mechanisms. In the extreme halophiles, a high concentration of NaCl is essential in order to maintain both the stability and the catalytic activity of enzymes (Figure 10.4).

The walls of extreme halophiles of the genus *Halobacterium* lack a peptidoglycan layer, and are comprised exclusively of protein. Nevertheless, in very concentrated salt solutions (25 to 35 percent w/v), similar to the natural environments in which these organisms live, the proteinaceous wall is sufficiently rigid to confer a cylindrical form on the cell. If the suspending medium is diluted to approximately 15 percent w/v, the cells become round, but they do not lyse. At lower salt concentration, lysis occurs and the wall disaggregates into protein monomers. The cell wall of *Halobacterium* is therefore unique, by virtue of the fact that its structural integrity is assured by ionic bonds, a very high concentration of Na^+ (largely irreplaceable by other monovalent cations) being necessary to maintain the intermolecular association between the protein subunits of the wall.

For marine bacteria and other halophiles, the magnitude of the Na^+ requirement can be substantially reduced (by a factor as great as 2) by increasing the concentrations in the medium of two divalent cations, Mg^{2+} and Ca^{2+}. The quantitative requirements of many halophiles for Mg^{2+} and Ca^{2+} also appear to be much greater than those of nonhalophiles. The influence of the NaCl content of two different media on the growth of a typical marine bacterium is shown in Figure 10.5.

• The formation and excretion of enzymes

Eucaryotic cells devoid of walls are able to engulf particulate matter and liquid droplets from the environment by the processes of phagocytosis and pinocytosis. Bacteria as well as eucaryotic microorganisms completely enclosed by walls cannot do so, but they are able to use as nutrients a wide variety of macromolecules

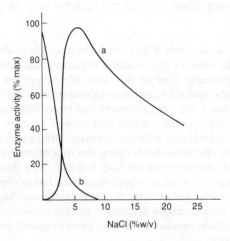

FIGURE 10.4

Effect of NaCl on the activity of the enzyme, malic dehydrogenase, from an extreme halophile (a) and from liver (b). Like most enzymes, the enzyme from liver becomes inactive in high concentrations of NaCl. The enzyme from the extreme halophile requires NaCl for activity. Redrawn from H. Larson, "Biochemical Aspects of Extreme Halophilism," *Advan. Microbiol. Physiol.* **1**, 97 (1967).

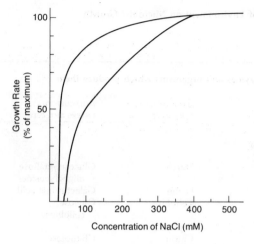

FIGURE 10.5

The influence of NaCl concentration on the growth rate of a typical marine bacterium, *Pseudomonas marina*, growing (a) in a medium with concentrations of Mg^{2+} and Ca^{2+} (2 mM $MgSO_4$ and 0.55 mM $CaCl_2$) typical of a terrestrial environment, and (b) in a medium with concentrations of these ions (50 mM $MgSO_4$ and 10 mM $CaCl_2$) typical of a marine environment; in other respects, the media are the same. It will be noted that the higher concentration of Mg^{+2} and Ca^{+2} found in the marine environment spares the requirement for NaCl. After J. L. Reichelt, and P. Baumann, "Effect of Sodium Chloride on Growth of Heterotrophic Marine Bacteria," *Arch. Microbiol.* **97**, 239 (1974).

including polysaccharides, proteins, and nucleic acids. Most of these macromolecules cannot cross the cell membrane* and are degraded outside the cell by *exoenzymes* synthesized within the cell and excreted into the medium. Exoenzymes are hydrolyases. Most of them catalyze the hydrolysis of macromolecular substrates to soluble end products (usually dimers or monomers) which enter the cell by specific transport mechanisms and serve as carbon and energy sources. The initial hydrolysis of macromolecules is termed *extracellular digestion* in order to distinguish it from *intracellular digestion* which follows phagocytosis and pinocytosis in eucaryotes. When microbial colonies develop on an agar medium containing particles of an insoluble macromolecule which is digestible, each colony is surrounded by an expanding clear area in which the insoluble substrate has been hydrolyzed by the action of exoenzymes. Table 10.3 lists some substrates and products formed by the action of microbial exoenzymes.

Most exoenzymes which have a digestive function are subject to catabolite repression. Their synthesis is consequently repressed if more rapidly metabolized soluble sources of carbon and energy are present in the environment. In addition, many exoenzymes are inducible. Since the macromolecular substrates cannot enter the cell, enzyme synthesis must be triggered by other inducers. The inducers are probably the soluble hydrolytic products, which can enter the cell. Since exoenzymes are always synthesized at a low rate, even in the absence of the specific substrate, the presence of the substrate always leads to the formation of hydrolytic products which can presumably initiate induction.

Not all exoenzymes function to provide the cell with utilizable carbon and energy sources. The 5'-nucleotidases, for example, convert nucleotides to which the membrane is impermeable into nucleosides which can enter the cell and be used directly for biosynthesis. Certain enzymes which destroy antibiotics are also exoenzymes: e.g., the penicillinases of many bacteria, which hydrolyze the four-membered β-lactam ring of penicillins and thereby detoxify them. The extracellular location of this enzyme is of obvious selective advantage to the organism which produces it.

* A notable exception to this generalization is the uptake of DNA of high molecular weight by bacteria that are transformable (Chapter 15).

TABLE 10.3

Examples of exoenzymes and organisms which produce them

EXOENZYME	MACROMOLECULAR SUBSTRATE	MOLECULE THAT ENTERS THE CELL	PRODUCING MICROORGANISM
Polysaccharide-splitting enzymes:			
Amylase	Starch	Glucose, maltose oligoglycosides	*Bacillus subtilis*
Pectinase	Pectin	Galacturonic acid	*Bacillus polymyxa*
Cellulase	Cellulose	Glucose, cellobiose	*Clostridium thermocellum*
Lysozyme	Peptidoglycans		*Staphylococcus aureus*
Chitinase	Chitin	Chitobiose	
Proteinases:			
Peptidase	Peptides	Amino acids	*Bacillus megaterium*
Nucleases:			
Deoxyribonuclease	DNA	Deoxyribonucleosides[a]	*Streptococcus haemolyticus*
Ribonuclease	RNA	Ribonucleosides	*Bacillus subtilis*
Esterases:			
Lipases	Lipids	Glycerols + fatty acids	*Clostridium welchii*
Poly β-hydroxy butyrate depolymerase	Poly β-hydroxy butyrate	β-hydroxy-butyryl-β-hydroxy butyrate	*Pseudomonas spp.*

[a] Although nucleotides are the primary product of hydrolysis, they are further hydrolyzed to nucleosides before entering the cell.

•**Periplasmic enzymes**   The periplasmic space (the space between the cell membrane and the outer layer of the wall) of Gram-negative bacteria contains enzymes called *periplasmic enzymes*. Since they are external to the cell membrane, they are clearly exoenzymes. However, they cannot traverse the outer layer of the cell wall and are trapped in the periplasmic space. This location is usually inferred from the fact that they are released by chemical or physical treatments which damage the outer wall layer but do not release cytoplasmic proteins. These treatments include enzymatic removal of the wall (by treatment with lysozyme in the presence of the chelating agent, EDTA) and cold osmotic shock (rapidly, suspending plasmolyzed cells in dilute media at low temperature). Relatively few of the enzymes found in the periplasmic space of Gram-negative bacteria have digestive functions (Table 10.4). Most of these enzymes are phosphatases which convert phosphorylated compounds, to which the cell membrane is impermeable, into phosphate-free derivatives (Enzymes 3 through 7 in Table 10.4). They also include penicillinase.

It is an interesting fact that many of the enzymes which are periplasmic in Gram-negative bacteria are true exoenzymes in Gram-positive bacteria. This is true of penicillinases, 5'-nucleotidases, and ribonucleases.

In addition to periplasmic enzymes, Gram-negative bacteria produce some extracellular digestive enzymes; it is not known why these traverse the outer wall layers while other enzymes are trapped in the periplasmic space.

TABLE 10.4

Certain enzymes located in the periplasmic space

ENZYME	REACTIONS CATALYZED
1. Ribonuclease I	Hydrolyzes RNA
2. DNA Endonuclease I	Internally cleaves DNA
3. Alkaline phosphotase	Removes phosphate groups from a number of organic compounds
4. 5'-nucleotidase	Converts a number of nucleotides to nucleosides
5. Acid hexose phosphotase	Removes phosphate groups from a number of sugar phosphates
6. Acid phosphatase	Removes phosphate groups from organic compounds
7. Cyclic phosphodiesterase	Converts ribonucleoside-2', 3'-cyclic phosphates to the ribonucleoside-3'-phosphates, and further hydrolyzes the ribonucleoside-3'-phosphates to nucleosides
8. Penicillinase	Hydrolyzes and thereby inactivates penicillin

• **Cell-bound enzymes** A third class of enzymes is exocellular (since they act upon substrates which are external to the cell membrane), but these enzymes always remain tightly bound to the cell surface. Examples are the penicillinase of *Bacillus licheniformis* and the cellulose-hydrolyzing enzymes of the *Cytophaga* group.

• **Mechanism of excretion of exoenzymes** The mechanism by which exoenzymes traverse the cell membrane, which retains other enzymes within the cell, is not yet established. It has been proposed that exoenzymes are synthesized by membrane-bound ribosomes which excrete the unfolded enzyme as synthesis proceeds. However, other studies show that small quantities of exoenzymes exist within the cell, suggesting that they are synthesized normally and then specifically excreted. There are no properties common to exoenzymes which can account for their ability to undergo excretion. Although most are relatively small proteins (molecular weights in the range of 20,000 to 40,000), the exocellular protease of *Pseudomonas myxogenes* has a molecular weight to 77,000. There is no evidence that exoenzymes are conjugated to a common carrier compound which facilitates their excretion.

EFFECT OF TEMPERATURE ON MICROBIAL GROWTH

The rate of chemical reactions is a direct function of temperature and obeys the relationship originally described by Arrhenius:

$$\log_{10} v = \frac{-\Delta H^*}{2.303\,RT} + C \tag{10.3}$$

where v represents the reaction velocity and ΔH^* the activation energy of the reaction, R the gas constant, and T the temperature in degrees Kelvin. Hence, a plot of the velocity of chemical reaction as a function of T^{-1} yields a straight

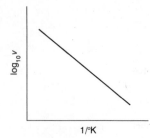

FIGURE 10.6

A generalized Arrhenius plot of the relationship between the velocity of a chemical reaction and temperature.

line with a negative slope (Figure 10.6). Figure 10.7 shows a comparable plot of the rate of growth of *E. coli* as a function of T^{-1}. The curve is linear only over a portion of the temperature range for growth, since the growth rate falls abruptly at both the upper and the lower limits of the temperature range. The abrupt fall in growth rate at high temperatures is caused by the thermal denaturation of proteins and possibly of such cell structures as membranes. The *maximum* temperature for growth is the temperature at which these destructive reactions become overwhelming. This temperature is usually only a few degrees higher than the temperature at which growth rate is maximal, called the *optimum temperature*.

From the effect of temperature on the rate of a chemical reaction one would predict that all bacteria would continue to grow (although at progressively lower rates) as the temperature is reduced, until the system freezes. However, most bacteria stop growing at a temperature (the *minimum temperature of growth*) well above the freezing point of water. Every microorganism has a precise minimum temperature of growth, below which growth will not occur however long the period of incubation.

The numerical values of the *cardinal temperatures* (minimum, optimum, and maximum), and the range of temperature over which growth is possible, vary widely among bacteria. Some bacteria isolated from hot springs are capable of growth at temperatures as high as 95°C; others, isolated from cold environments, can grow at temperatures as low as −10°C if high solute concentrations prevent the medium from freezing. On the basis of the temperature range of growth, bacteria are frequently divided into three broad groups: *thermophiles*, which grow at elevated temperature (above 55°C); *mesophiles*, which grow well in the midrange of temperature (20 to 45°C); and *psychrophiles*, which grow well at 0°C.

As is often true of systems of biological classification, this terminology implies a clearer distinction among types than is found in nature. The tripartite classification of temperature response does not take fully into account the variation among bacteria with respect to the extent of the temperature range over which growth is possible.*

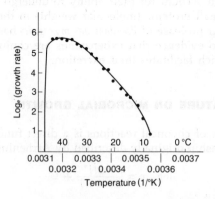

FIGURE 10.7

An Arrhenius plot of the relationship between growth rate and temperature for *E. coli*. After J. L. Ingraham, "Growth of Psychrophilic Bacteria," *J. Bacteriol.* **76,** 75 (1958).

*Differences in temperature range among thermophiles are sometimes indicated by the terms *stenothermophile* (an organism which cannot grow below 37°C), and *eurithermophile* (an organism which can do so). Psychrophiles with temperature ranges that extend above 20°C are termed *facultative psychrophiles*; ones which cannot grow above 20°C are termed *obligate psychrophiles*.

The data describing the temperature ranges of growth of many different bacteria (Table 10.5) show the somewhat arbitrary nature of the designations *thermophile, mesophile,* and *psychrophile.* The range of temperature over which growth is possible is as variable as are the maxima and minima. The temperature range of some bacteria is less than 10 degrees, while for others it is as much as 50 degrees.

•Factors that determine temperature limits for growth

The factors that determine the temperature limits for growth have been revealed by two types of investigations: comparisons of the properties of organisms with widely different temperature ranges; and analyses of the properties of temperature-sensitive mutants, the temperature range of which has been decreased by a single mutational change. Temperature-sensitive mutants are of two types: *heat-sensitive mutants,* with decreased maximum growth temperatures; and *cold-sensitive mutants,* with increased minimum growth temperatures.

Studies on the kinetics of thermal denaturation both of enzymes and of cell

TABLE 10.5

Temperature range of growth of certain procaryotes[a]

ORGANISM	TEMPERATURE (°C)
	−10 0 10 20 30 40 50 60 70 80 90 100
Bacillus globisporus	Psychrophiles
Micrococcus cryophilus	
Marine bacterium (Bedford)	
Candida sp. (Stokes)	
Marine Pseudomonas (Velkamp)	
Vibrio marinus	
Xanthomonas pharmicola	
Pseudomonas alboprecipitans	
Xanthomonas rinicola	Mesophiles
Gaffkya homari	
Neisseria gonorrhoeae	
Escherichia coli	
Acholeplasma blastoclosticum	
Vibrio comma	
Fusobacterium polymorphum	
Haemophilus influenzae	
Lactobacillus lactis	
Bacillus subtilis	
Lactobacillus delbrückii	
Mastigocladus laminosus[b]	Thermophiles
Bacillus coagulans	
Synechococcus lividus[b]	
Bacillus stearothermophilus	
Bacterium in hot springs (Brock)	

[a] Lines terminating in single arrows indicate established temperature limits of growth for at least one strain of the indicated species; variations exist among different strains of some species. Double-headed arrows indicate that the actual temperature limit lies between the arrow points. Solid lines terminating in dotted lines indicate that the minimum growth temperature is not established.

[b] Blue-green bacterium.

structures that contain proteins (e.g., flagella, ribosomes) have shown that many specific proteins of thermophilic bacteria are considerably more heat-stable than their homologues from mesophilic bacteria. It is also possible to make an approximate determination of the overall thermal stability of soluble cell proteins, by measuring the rates at which the protein in a cell-free bacterial extract becomes insoluble as a result of heat denaturation at several different temperatures. Experiments like this (Table 10.6) clearly demonstrate that virtually all the proteins of a thermophilic bacterium remain in the native state after a heat treatment that denatures virtually all the proteins of a related mesophile. It therefore follows that the adaptation of a thermophilic microorganism to its thermal environment can be achieved only through mutational changes affecting the primary structures of most (if not all) proteins of the cell.

Although the evolutionary adaptations which have produced thermophiles must have involved mutations which *increased* the thermal stability of their proteins, most of the mutations which affect the primary structure of a specific protein (e.g., an enzyme) decrease the thermal stability of that protein, even though many of these mutations may have little or no effect on its catalytic properties. Consequently, in the absence of counterselection by a thermal challenge, the maximal temperature for growth of *any* microorganism should decline progressively as a result of random mutations that affect the primary structure of its proteins. This inference is supported by the observation that psychrophilic bacteria isolated from antarctic waters contain a large number of exceptionally heat-labile proteins.

At low temperature, all proteins undergo slight conformational changes, attributable to the weakening of their hydrophobic bonds which play an important role in determining tertiary (three-dimensional) structure. All other types of bonds in proteins become stronger as the temperature is lowered. The importance of precise conformation to the proper function of allosteric proteins and to the self-assembly of ribosomal proteins makes these two classes of proteins particu-

TABLE 10.6

Stability of cytoplasmic proteins from mesophilic and thermophilic bacteria at 60°C[a]

ORGANISM	TEMPERATURE CLASS	% OF PROTEINS DENATURED[b]
Proteus vulgaris	Mesophile	55
Escherichia coli	Mesophile	55
Bacillus megaterium	Mesophile	58
Bacillus subtilis	Mesophile	57
Bacillus stearothermophilus	Thermophile	3
Bacillus sp. (Purdue CD)	Thermophile	0
Bacillus sp. (Texas 11330)	Thermophile	4
Bacillus sp. (Nebraska 1492)	Thermophile	0

[a] Data from H. Koffler and G. O. Gale, "The Relative Thermostability of Cytoplasmic Proteins from Thermophilic Bacteria," *Arch. Biochem. Biophys.* **66**, 249(1957).

[b] Percent of total trichloracetic acid—precipitable material from a sonic extract of cells which is coagulated by an 8-minute heat treatment at 60°C.

TABLE 10.7

Effect of growth temperature on the amounts of major
fatty acids of *E. coli*[a]

FATTY ACID	TEMPERATURE OF GROWTH	
	10°C	43°C
Saturated fatty acids:	%[b]	%
Myristic	3.9	7.7
Palmitic	18.2	48.0
Unsaturated fatty acids:		
Hexadecenoic	26.0	9.2
Octadecenoic	37.9	12.2

[a] Note that at low growth temperatures the proportion of unsaturated fatty acids in the lipids increases dramatically. Data from A. G. Marr and J. L. Ingraham, "Effect of Temperature on the Composition of Fatty Acids in *E. coli.*, *J. Bacteriol.* **8.4**, 1260 (1962).
[b] Percent of total fatty acids in the cell.

lary sensitive to cold inactivation. Therefore, it is not surprising that mutations which increase the minimum temperature for growth usually occur in genes that encode these proteins.

• Effect of growth temperature on lipid composition

The lipid composition of almost all organisms, both procaryotes and eucaryotes, alters with the growth temperature. As the temperature decreases, the relative content of unsaturated fatty acids in the cellular lipids increases.

An illustration of this phenomenon in *E. coli* is shown in Table 10.7. This change in fatty acid composition is a significant component of temperature adaptation in bacteria. The melting point of lipids is directly related to their content of saturated fatty acids. Consequently, the degree of saturation of the fatty acids in membrane lipids determines their degree of fluidity at a given temperature. Since membrane function depends on the fluidity of the lipid components, it is understandable that growth at low temperature should be accompanied by an increase in the degree of unsaturation of fatty acids.

OXYGEN RELATIONS

The present atmosphere of the earth contains about 20 percent (v/v) of the highly reactive gas, oxygen. With the exception of many bacteria and a few protozoa, all organisms are dependent on the availability of molecular oxygen as a nutrient. The responses to O_2 among bacteria are remarkably variable, and this is an important factor in their cultivation (see Chapter 2). The aerobes are dependent on O_2; the facultative anaerobes use O_2 if it is available, but they also can grow in its absence; the anaerobes cannot utilize O_2. Anaerobes are of two types: the *obligate anaerobes,* for which O_2 is toxic, and the *aerotolerant anaerobes,* which are not killed by exposure to O_2. Although the toxicity of O_2 is most strikingly revealed by its effect in the obligate anaerobes, it is, in fact, toxic even for aerobic organisms at high concentration. Many obligate aerobes cannot grow in O_2 concentrations greater than atmospheric (i.e., > 20 percent v/v). Indeed, some obligate aerobes require O_2 concentrations considerably lower than atmospheric (2 to 10 percent v/v) in order to grow. Aerobic bacteria that show such O_2 sensitivity under *all* growth conditions are called *microaerophiles*. Frequently, however, the requirement of a strict aerobe for a reduced O_2 concentration is conditional, being greatly influenced by either the energy source or the nitrogen source. For example, many hydrogen-oxidizing bacteria, which tolerate an atmospheric O_2 concentration when growing with organic substrates, require considerably lower O_2 concentrations when using H_2 as an energy source. Some aerobic nitrogen-fixing bacteria can fix N_2 only in the virtual absence of O_2 (see p. 602).

• **The toxicity of oxygen: chemical mechanisms**   All bacteria contain certain enzymes capable of reacting with O_2; the number and variety of these enzymes determine the physiological relations of the organism to oxygen. The oxidations of flavoproteins by O_2 invariably result in the formation of a toxic compound, H_2O_2, as one major product. In addition, these oxidations (and possibly other enzyme-catalyzed oxidatons or oxygenations) produce small quantities of an even more toxic free radical,* superoxide or $O_2^{\cdot-}$.

In aerobes and aerotolerant anaerobes, the potentially lethal accumulation of superoxide ($O_2^{\cdot-}$) is prevented by the enzyme *superoxide dismutase*, which catalyzes its conversion to oxygen and hydrogen peroxide:

$$2O_2^{\cdot-} + 2H^+ \xrightarrow[\text{dismutase}]{\text{superoxide}} O_2 + H_2O_2$$

Nearly all these organisms also contain the enzyme *catalase*, which decomposes hydrogen peroxide to oxygen and water:

$$2H_2O_2 \xrightarrow{\text{catalase}} 2H_2O + O_2$$

One bacterial group able to grow in the presence of air (lactic acid bacteria, see Chapter 23) do not contain catalase. However, most of these organisms do not accumulate significant quantities of H_2O_2, since they decompose it by means of *peroxidases*, enzymes that catalyze the oxidation of organic compounds by H_2O_2, which is reduced to water.

Superoxide dismutase, catalase, and peroxidase therefore all play roles in protecting the cell from the toxic consequences of oxygen metabolism. The distribution of superoxide dismutase and catalase in bacteria with differing physiological responses to O_2 is shown in Table 10.8. Organisms that can tolerate an exposure to O_2 always contain superoxide dismutase, although not all necessarily contain catalase. However, *all strict anaerobes so far examined lack both superoxide dismutase and catalase*. On present evidence, accordingly, *superoxide dismutase is an indispensable enzyme in any organism which comes in contact with air*. A direct demonstration of this fact has recently been provided by the isolation of mutants which lack this enzyme in a facultative anaerobe, *E. coli*. These mutants have become obligate anaerobes, being rapidly killed by brief exposure to air.

---

* A free radical is a compound with an unpaired electron, indicated by a single dot in the structural formula. Having gained an extra electron, superoxide carries a negative charge.

TABLE 10.8

The distribution of superoxide dismutase and catalase in bacteria with differing physiological response to O_2

| | CONTAINS | |
BACTERIUM	SUPEROXIDE DISMUTASE	CATALASE
Aerobes or facultative anaerobes:		
Escherichia coli	+	+
Pseudomonas spp.	+	+
Micrococcus radiodurans	+	+
Aerotolerant bacteria:		
Butyribacterium rettgeri	+	−
Streptococcus faecalis	+	−
Streptococcus lactis	+	−
Strict anaerobes:		
Clostridium pasteurianum	−	−
Clostridium acetobutylicum	−	−

• The photodynamic effect

The toxicity of O_2 for living organisms can be greatly enhanced if the cells are exposed to light in the presence of air and of certain pigments, known as *photosensitizers*. Light converts the photosensitizer (P) to a highly reactive form, known as the triplet state (P^*):

$$P + hv \longrightarrow P^*$$

A secondary reaction between P^* and O_2 produces singlet-state oxygen ($^{..}O_2$):

$$P^* + O_2 \longrightarrow P + {}^{..}O_2$$

Like the superoxide radical, singlet-state oxygen is a very powerful oxidant, and its formation within the cell is rapidly lethal.

One of the principal biological functions of carotenoid pigments is to act as *quenchers of singlet-state oxygen*, and thus protect the cell from photodynamic death. This function is of particular importance in phototrophs, since chlorophylls are powerful photosensitizers; the photosynthetic apparatus invariably contains carotenoid pigments. Their role in the prevention of lethal chlorophyll-mediated photooxidations was first shown in the purple bacterium *Rhodopseudomonas sphaeroides*. *R. sphaeroides* grows photosynthetically under strictly anaerobic conditions (see p. 551), and can also grow aerobically, either in the light or in the dark. Blue-green mutants, which have lost the ability to synthesize colored carotenoids, can still grow normally either anaerobically in the light or aerobically in the dark, but they are rapidly killed by simultaneous exposure to light and air. In phototrophic organisms which produce O_2 as a photosynthetic product, the loss of colored carotenoids totally abolishes photosynthetic function, since the accumulation of singlet-state oxygen, formed from metabolically generated O_2, occurs as soon as the cells are illuminated.

Many aerobic, nonphotosynthetic microorganisms also synthesize carotenoid pigments, which are incorporated into the cell membrane and function as quenchers of singlet-state oxygen produced by such photosensitizing cellular pigments as the cytochromes. The role of carotenoids in protecting aerobic bacteria against the photodynamic action of sunlight has been demonstrated by studies on the light-sensitivity of nonpigmented mutants of *Micrococcus luteus* and *Halobacterium salinarium*. This protection is probably of ecological importance in all aerobic bacteria which live in environments exposed to high light intensities.

• **Oxygen-sensitive enzymes**   Many enzymes, particularly enzymes of strict anaerobes, are rapidly and irreversibly denatured by exposure to O_2. Their purification and study must therefore be conducted under rigorously anaerobic conditions. A notable example is *nitrogenase*, the enzyme responsible for nitrogen fixation, which catalyzes the reaction:

$$N_2 + 6H^+ + 6e^- \longrightarrow 2NH_3$$

Even the nitrogenases from obligately aerobic nitrogen-fixing bacteria, such as the *Azotobacter* group, exhibit extreme oxygen-sensitivity after extraction from the cell. In intact cells of *Azotobacter*, nitrogenase is evidently protected from inactivation, but the mechanism of protection is not known. The nitrogenases of facultatively anaerobic nitrogen-fixing bacteria (*Enterobacter*, *Bacillus polymyxa*) are not so protected in the intact cell; consequently, these bacteria can fix nitrogen effectively only under anaerobic growth conditions. Most filamentous nitrogen-fixing blue-green bacteria produce specialized cells (heterocysts) lacking photosystem II, in which nitrogenase is protected from oxygen inactivation (see Chapter 17, p. 543).

• **The role of oxygenases in aerobic microorganisms**   Although the primary metabolic function of O_2 in strict aerobes is to serve as a terminal electron acceptor, it also serves a *cosubstrate* for enzymes which catalyze some steps in the dissimilation of aromatic compounds and alkanes. These enzymes are termed *oxygenases*; they mediate a direct addition of either one or two oxygen atoms to the organic substrate. An example is the oxygenative ring cleavage of catechol, an intermediate in the dissimilation of many aromatic compounds:

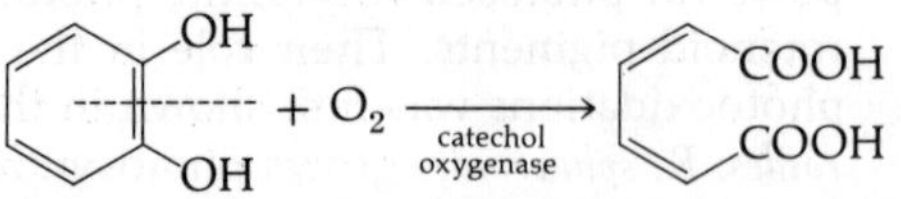

Many aerobic pseudomonads that are able to use aromatic compounds or alkanes as sole sources of carbon and energy are denitrifiers, and hence can grow anaerobically, using nitrate in place of O_2 as a terminal electron acceptor. However, this metabolic option can be exercised only with oxidizable substrates which are catabolized by dehydrogenases. Substrates in the dissimilation of which one or more steps are mediated by oxygenases cannot support anaerobic growth, since nitrate is unable to replace O_2 as a cosubstrate for oxygenases.

In eucaryotes, the biosynthesis of sterols and unsaturated fatty acids involves steps mediated by oxygenases. Consequently, yeasts require sterols and unsaturated fatty acids as growth factors when growing fermentatively under anaerobic conditions, even though they can synthesize these cell components when they are growing aerobically.

FURTHER READING

Book   PRECHT, H. (editor), *Temperature and Life.* Heidelberg: Springer Verlag, 1973.

Reviews and   BROWN, A. D., "Aspects of Bacterial Response to the Ionic Environment," *Bact.*
original papers   *Revs.* **28,** 296 (1964).

HEPPEL, L. A., "The Concept of Periplasmic Enzymes," in *Structure and Function of Biological Membranes,* L. I. Rothfield (editor). New York and London: Academic Press, 1971.

LARSEN, H., "The Halophiles' Confusion to Biology," *Antonie van Leuwenhoek* **39,** 383 (1973).

LIN, E. C. C., "The Molecular Bases of Membrane Transport Systems," in *Structure and Function of Biological Membranes,* L. I. Rothfield, (editor), p. 286. New York and London: Academic Press, 1971.

MacLEOD, R. A., "On the Role of Inorganic Ions in the Physiology of Marine Bacteria," *Advances in Microbiology of the Sea* **1,** 95 (1968).

MORRIS, J. G. "The Physiology of Obligate Anaerobiosis." *Adv. Microbiol. Physiol.* **12,** 169 (1975).

POLLOCK, M. R., "Exoenzymes," in *The Bacteria,* R. Y. Stanier and I. C. Gunsalus (editors), Vol. I. New York and London: Academic Press, 1964.

STANIER, R. Y., "The Relationship between Nitrogen Fixation and Photosynthesis," *Aust. J. Expt. Biol. Med. Sci.* **52,** 3 (1974).

THE RELATIONS BETWEEN STRUCTURE AND FUNCTION IN PROCARYOTIC CELLS

11

The structure of the procaryotic cell has been described in Chapters 3 and 5. In this chapter we shall discuss in greater detail the organization of some of its component parts and the relationships between their structure and function.

SURFACE STRUCTURES OF THE PROCARYOTIC CELL

Very little was known about the composition and functions of the outer layers of bacterial cells until 1952, when M. Salton developed methods for isolating and purifying cell walls (Figure 11.1). The first chemical analyses of such preparations revealed the complexity and diversity of bacterial wall composition, and notably the major compositional differences between the walls of Gram-positive and Gram-negative bacteria, which were discussed in Chapter 5 (p. 122).

Salton showed that the hydrolytic enzyme lysozyme, previously known to lyse many Gram-positive bacteria, can completely destroy the isolated cell walls of such organisms as *Bacillus megaterium*. This observation enabled C. Weibull, in 1953, to perform a simple experiment that clearly revealed the respective functions of the bacterial wall and membrane (Figure 11.2). If cells of *B. megaterium* are suspended in an isotonic sucrose solution prior to lysozyme treatment, the enzymatic dissolution of the wall converts the initially rod-shaped cells into spherical *protoplasts*, which retain full respiratory activity and can synthesize protein and nucleic acid. In media which support good growth of intact cells they are incapable of resynthesizing a cell wall after removal of the lysozyme, but they can increase in size and they have a limited capacity to reproduce. However, under certain special conditions of cultivation, they can be induced to regenerate cell walls and again assume a rod shape.

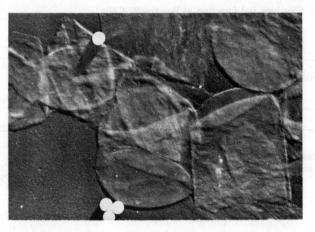

FIGURE 11.1

Isolated and purified cell walls of *Bacillus megaterium*. Electron micrograph. The white spheres are particles of latex exactly 0.25 μm in diameter, which were included in the preparation to show the scale of magnification. From R. Y. Stanier, "Some Singular Features of Bacteria as Dynamic Systems" in *Cellular Metabolism and Infection*, E. Racker (editor). New York: Academic Press, 1954.

When a protoplast suspension is diluted, the protoplasts undergo immediate osmotic lysis. The only structural elements which remain after such lysis are the empty cell membranes or "ghosts," which can be readily isolated and purified by differential centrifugation. Such preparations, derived from lysozyme treatment of Gram-positive bacteria, have provided much information about the properties of the bacterial cell membrane.

Weibull's experiment showed that the cell wall has a vital mechanical function: it protects the cell from osmotic lysis in a hypotonic environment. In addition, it determines cell shape, and it appears to play a role both in movement and in division. However, the wall (at least in Gram-positive bacteria) does not contribute to metabolic activity or constitute part of the osmotic boundary.

In addition to the wall and the membrane, procaryotic cells may be enclosed by a loose outer layer known as a *capsule* or *slime layer*. Two classes of thread-shaped organelles, *flagella* and *pili*, occur on the cell surface of many bacteria.

Table 11.1 summarizes the distinguishing properties of the various surface structures associated with the procaryotic cell.

• **The cell membrane**   The bacterial cell membrane, which bounds the protoplast and is the principal osmotic barrier, can be visualized in electron micrographs of thin sections of cells

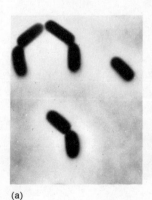
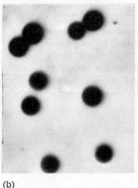
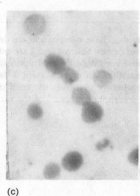

(a)   (b)   (c)

FIGURE 11.2

Bacillus megaterium (phase contrast, ×3,000). (a) The intact cells; (b) the spherical protoplasts, formed by enzymatic dissolution of the cell wall with lysozyme in an isotonic medium; (c) the ghosts (i.e., the empty cytoplasmic membranes), formed by osmotic rupture in a hypotonic medium. Courtesy of C. Weibull.

TABLE 11.1

Surface structures of the procaryotic cell

STRUCTURE	LOCATION	STRUCTURE AND DIMENSIONS	CHEMICAL COMPOSITION
Membrane	Bounding layer of protoplast	Unit membrane, 7.5–8 nm wide	20–30% phospholipid, remainder mostly protein
Wall	Layer immediately external to membrane	Gram-negative groups: Inner single layer 2–3 nm wide Outer unit membranelike layer 7–8 nm wide	Peptidoglycan Phospholipids, proteins, lipopolysaccharides
		Gram-positive groups: Homogeneous layer 10–50 nm wide	Peptidoglycan; teichoic acids; polysaccharides
Capsule or slime layer	Diffuse layer external to wall	Homogeneous structure of low density and very variable width	Diverse; usually a polysaccharide, rarely a polypeptide
Flagella	Anchored in protoplast, traversing membrane and wall	Helical threads, 12–18 nm wide	Protein
Pili	Anchored in protoplast, traversing membrane and wall	Straight threads, 4–35 nm wide	Protein

as a double line, about 8 nm wide: this is the typical fine structure of all so-called *unit membrane* systems. It is made up of a bilayer of phospholipids (Figure 11.3) which are the major lipids of bacterial cell membranes and which account for about 20 to 30 percent of their dry weight. The polar "head" regions of the phospholipids are located at the two outer surfaces of the bilayer, while the hydrophobic fatty acid chains extend into the center of the membrane, perpendicular to its plane (Figure 11.4). The membrane proteins, which account for more than one-half of the dry weight of the membrane, are intercalated into this phospholipid bilayer.

Isolated membranes are highly plastic structures. They can be disaggregated by treatment with detergents, and subsequently reassemble to form new, mem-

FIGURE 11.3

The general structure of phospholipids. These compounds are derivatives of glycerol-3-phosphate. The hydroxyl groups on carbon atoms 1 and 2 are esterified by long-chain fatty acids to yield acyl groups; these constitute the hydrophobic "tail" region of the molecule. The phosphate group esterified on carbon atom 3 can carry a variety of substituents (designated as -R), including glycerol, its aminoacyl derivatives, and ethanolamine. This constitutes the hydrophilic "head" of the molecule.

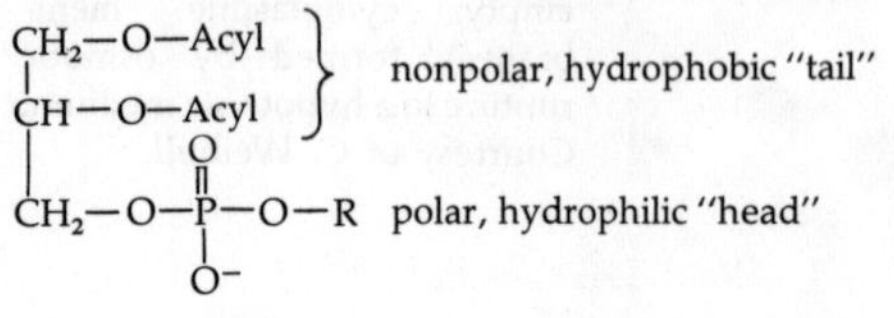

$$\left.\begin{array}{l} CH_2-O-Acyl \\ CH-O-Acyl \end{array}\right\} \text{nonpolar, hydrophobic "tail"}$$

$$CH_2-O-\overset{\overset{\displaystyle O}{\|}}{P}-O-R \quad \text{polar, hydrophilic "head"}$$
$$\underset{O^-}{}$$

FIGURE 11.4

Schematic drawing to show the possible molecular organization of an unit membrane. Folded polypeptide molecules are visualized as being embedded in a phospholipid bilayer, with their hydrophobic regions extending beyond the bilayer of one or both of its surfaces. From S. J. Singer, and A. L. Nicholson, "The Fluid Membrane Model of the Structure of Cell Membranes," *Science* **175**, 720 (1972).

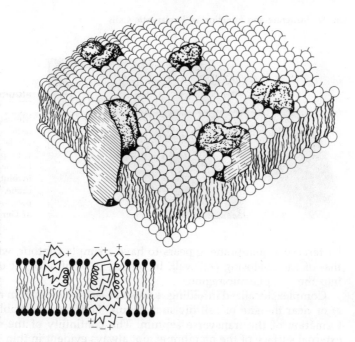

branelike structures. Furthermore, membrane fragments can reseal their edges, to produce closed vesicles with permeability properties similar to those of the cells from which they are derived.

The bacterial cell membrane is an important center of metabolic activity; it contains many different kinds of proteins, each of which probably has a specific catalytic function. Most of these proteins are tightly integrated into the hydrophobic region of the membrane, from which they can be separated only by methods (e.g., detergent treatment) which usually destroy their activity. Major classes of proteins known to be localized in the membrane include (1) the permeases responsible for the transport of many organic and inorganic nutrients into the cell and (2) biosynthetic enzymes that mediate terminal steps in the synthesis of the membrane lipids, and of the various classes of macromolecules that compose the bacterial cell wall (peptidoglycans, teichoic acids, lipopolysaccharides and simple polysaccharides).

In addition, the bacterial membrane often contains important components of the machinery of ATP generation. In aerobic bacteria the carriers of the electron transport chain are membrane-bound. In purple bacteria the entire photosynthetic apparatus (pigments, reaction centers, photosynthetic transport chain) is housed in the membrane. An ATPase is likewise constantly associated with the cell membrane. Lastly, much circumstantial evidence indicates that the procaryotic cell membrane contains specific attachment sites for the chromosome and for plasmids, and that it plays an active role both in the replication and in the subsequent segregation of these genetic elements. In view of its numerous and varied functions, it is not surprising that the bacterial membrane contains from 10 to 20 percent of the total cell protein.

Although the width of the cell membrane is fixed (being determined by the molecular configuration of the phospholipid bilayer), its area is not. In some

317

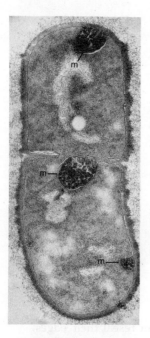

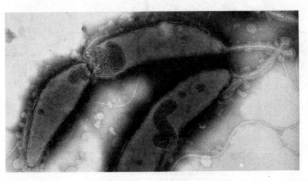

FIGURE 11.6

Mesosomes of *Caulobacter crescentus:* electron micrograph of whole cells negatively stained with phosphotungstate, which has penetrated the mesosomal involutions of the cell membrane, clearly outlining their positions (×22,100). Courtesy of Germaine Cohen-Bazire.

FIGURE 11.5

Electron micrograph of a thin section of a dividing cell of *Bacillus megaterium,* containing three mesosomes (m). One is located in association with the nearly formed transverse septum and wall. From D. J. Ellar, D. Lundgren, and R. A. Slepecky, "Fine Structure of *Bacillus megaterium* During Synchronous Growth," *J. Bact.* **94,** 1189 (1967).

bacteria, the membrane appears to have a simple contour, which closely follows that of the enclosing cell wall. In others, it is infolded, at one or more points, into the cytoplasmic region.

Complex, localized infoldings known as *mesosomes* occur in many bacteria, often at or near the site of cell division (Figure 11.5), and probably participate in the formation of the transverse septum. The continuity of the mesosome with the external surface of the membrane, not always evident in thin sections, is revealed in electron micrographs of whole cells, negatively stained with a heavy metal salt that penetrates through the wall but does not enter the cytoplasm (Figure 11.6).

Membrane infoldings of a different type occur in purple bacteria (Figure 11.7) and in many nonphotosynthetic bacteria that possess a high level of respiratory activity, such as the nitrogen-fixers of the *Azotobacter* group (Figure 11.8) and the nitrifying bacteria (Figure 11.9). It is probable that the greatly enlarged total area of the membrane produced by such intrusions serves to accommodate more centers of respiratory (or photosynthetic) activity than could be housed in a membrane of simple contour. The most convincing evidence in support of this interpretation has come from studies on the membrane structure of certain purple bacteria, where the photosynthetic pigment content (and hence the photosynthetic activity) can vary widely in response to environmental factors (light intensity, presence or absence of oxygen). Here, the extent of the membrane intrusions is directly related to the pigment content and photosynthetic activity of the cells (Figure 11.10).

In most procaryotes, there is a physical continuity between membrane intrusions and the surface region of the membrane. This may not be true, however, of the blue-green bacteria. In these organisms, the photosynthetic apparatus is contained in a system of flattened membranous sacs (thylacoids), which have been very rarely observed in connection with the cell membrane, and may be in large part physically distinct from it (Figure 11.11). If so, the blue-green bacteria constitute the only group of procaryotes that possess a differentiated intracytoplasmic unit membrane system.

• **The bacterial cell wall: its peptidoglycan component**

Of the many classes of macromolecules that may be associated with procaryotic cell walls, only one—the *peptidoglycans*—is of well-nigh universal occurrence. The only procaryotes which possess walls devoid of peptidoglycans are *Halobacterium* and *Halococcus.* Both these organisms are extreme halophiles (see Chapter 10), and

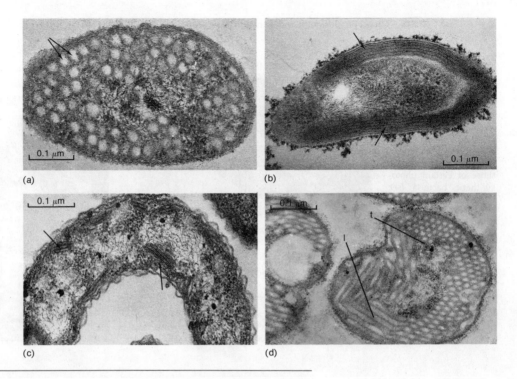

(a)

(b)

(c)

(d)

FIGURE 11.7

Electron micrographs of thin sections of several purple bacteria, illustrating variations in the structure of the internal membranes. (a) *Rhodopseudomonas sphaeroides,* in which the membranes occur as hollow vesicles (arrows) ($\times$5,640). (b) *Rhodopseudomonas palustris,* in which the membranes occur in regular parallel layers in the cortical region of the cell (arrows) ($\times$5,640). (c) *Rhodospirillum fulvum,* in which the membranes occur in small, regular stacks (arrows) ($\times$42,300). (d) *Thiocapsa sp.,* in which the membranes are tubular; some of these tubes are sectioned longitudinally (l), others transversely (t) ($\times$54,400). Micrographs (a), (b), and (c) courtesy of Germaine Cohen-Bazire; (d) courtesy of K. Eimhjellen.

a strong cell wall is unnecessary, since their cell contents are isosmotic with the liquid environment.

Peptidoglycans are heteropolymers of distinctive composition and structure, synthesized uniquely by procaryotes. The monomeric constituents are two acetylated aminosugars, N-acetylglucosamine and N-acetylmuramic acid, and a small number of amino acids, some of which are "unnatural" in the sense that they never occur in proteins (Figure 11.12). The two aminosugars form glycan strands, composed of alternating residues of N-acetylglucosamine (G) and N-acetylmuramic acid (M) in β–1,4 linkage. Each strand contains from 10 to 65 disaccharide units.

Muramic acid is a lactyl ether of glucosamine, and the carboxyl group of the lactyl moiety provides a site to which a peptide chain can be attached. Some of the carboxyl groups of muramic acid in peptidoglycans bear a chain of 4 amino acids. The most common sequence is: muramic acid $\rightarrow$ L-alanine $\rightarrow$ D-glutamic acid $\overset{\gamma}{\rightarrow}$ L-diamino acid $\rightarrow$ D-alanine-COOH. It will be noted that amino acids of L and D configuration alternate in this chain, whereas proteins are composed exclusively of L-amino acids.

FIGURE 11.8

Electron micrograph of thin section of the nitrogen-fixing bacterium, *Azotobacter vinelandii*, showing vesicular intrusions of the membrane similar in structure to those of certain photosynthetic bacteria (see Figure 11.7). Courtesy of J. Pangborn and A. G. Marr.

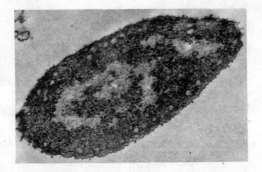

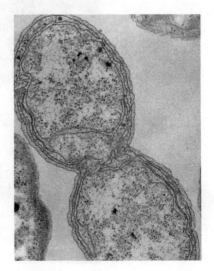

FIGURE 11.9

Electron micrograph of a thin section of *Nitrosomonas europaea,* an obligate chemoautotroph, showing membrane intrusions ($\times$39,100). Courtesy of S. W. Watson.

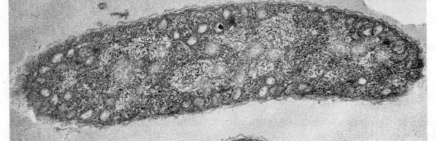

(a)

FIGURE 11.10

Electron micrographs of longitudinal sections of the photosynthetic bacterium, *Rhodospirillum rubrum* ($\times$52,700), showing the effect of the environment on the extent of the intrusion of the cytoplasmic membrane. (a) Cell from a culture grown in bright light (1,000 foot-candles) and having a relatively low chlorophyll content. (b) Cell from a culture grown in dim light (50 foot-candles) and having a high chlorophyll content. Courtesy of Germaine Cohen-Bazire.

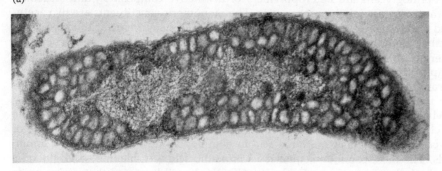

(b)

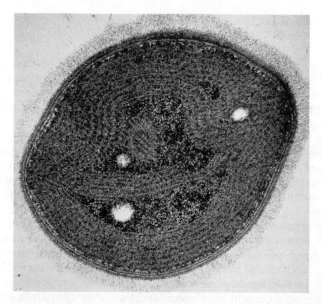

FIGURE 11.11

Electron micrograph of a thin section of an unicellular blue-green
bacterium, showing the extensive
array of internal membranes (thylacoids) characteristic of this procaryotic group (×29,000). Courtesy of
Dr. Germaine Cohen-Bazire.

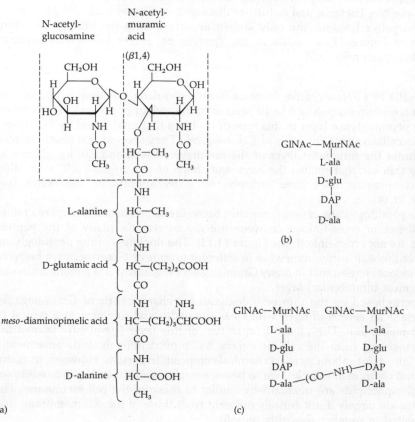

(a)

(b)

(c)

FIGURE 11.12

General structure of a peptidoglycan. (a) Complete structure of a
single subunit, showing the linkage
between the two amino sugars that
make up the glycan strand, and
between muramic acid and the four
amino acids in the short peptide
chain. (b) Schematic, simplified
representation of structure shown
in (a). (c) Representation of the
mode of cross-linking between the
terminal carboxyl group of D-
alanine on one subunit and the
free amino group of the diamino
acid (diaminopimelic acid) on an
adjacent subunit.

Adjacent peptide chains in a peptidoglycan may be *cross-linked*, typically by peptide bonding between the terminal D-alanine in position 4 on one chain, and the free amino group of the diamino acid on another chain. The cross-linking sometimes occurs through an interpeptide bridge, composed of from 1 to 6 amino acid residues.

The two types of linkages (through glycosidic bonds and peptide bonds) that hold together the peptidoglycan subunits confer on these heteropolymers the structure of a *molecular mesh* or *fabric* (Figure 11.13). In fact, the peptidoglycan layer of the bacterial cell wall is a bag-shaped macromolecule of enormous dimensions, which completely encloses and restrains the protoplast, counterbalancing its turgor pressure.

The primary peptidoglycan structure shown in Figure 11.12 is the most widespread one; it occurs in the walls of nearly all Gram-negative bacteria, and in many Gram-positive ones. However, dozens of minor variations on this structure occur, particularly among Gram-positive bacteria. Other diamino acids may replace meso-diaminopimelic acid; interpeptide bridges differing in length and in amino acid content may occur; and the mode of cross-linking between tetrapeptide chains may be different from that shown in Figure 11.12. The primary structure of the cell wall peptidoglycan is an important taxonomic character among Gram-positive bacteria, and is further discussed in Chapters 22 and 23. Among Gram-negative bacteria, the only known departure from the primary structure shown in Figure 11.12 occurs in the spirochetes, where L-ornithine replaces *meso*-diaminopimelic acid.

• **The location of peptidoglycan in the walls of Gram-negative bacteria**
The walls of Gram-negative bacteria have a comparatively low peptidoglycan content, seldom exceeding 5 to 10 percent of the weight of the wall. The location of the peptidoglycan layer in this type of wall was first established by W. Weidel and his collaborators, for walls of *Escherichia coli*. They showed that peptidoglycan constitutes the innermost layer of the multilayered wall and can be isolated as a very thin sac that retains the form and shape of the original cell, after other wall components have been stripped off it by appropriate treatments (see Figure 11.14).

The peptidoglycans of Gram-negative bacteria characteristically display a rather low degree of cross-linkage between the glycan strands: many of the peptide chains are not cross-linked (see Figure 11.13). The thickness of the peptidoglycan layer of the wall varies somewhat in different groups of Gram-negative bacteria. Calculations suggest that in many Gram-negative organisms it is a monomolecular (or at most bimolecular) layer.

Superimposed on the thin peptidoglycan sac characteristic of Gram-negative bacteria is an *outer wall layer*, which has the width and fine structure typical of a unit membrane. The physical separation of this layer (frequently termed the *outer membrane*) from the cell membrane has proved difficult, and consequently some uncertainty about its exact chemical composition persists. However, its major chemical components are known to be *proteins, phospholipids,* and *lipopolysaccharides*. The phospholipids are qualitatively similar to those of the cell membrane. The proteins are largely if not entirely different from those of the cell membrane, and are limited in number (possibly only 5).

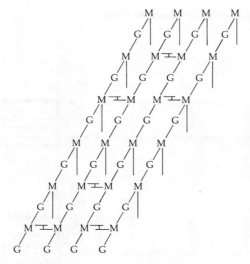

FIGURE 11.13

A schematic representation of the organization of the intact peptidoglycan sac of *E. coli*: G and M designate residues of N-acetylglucosamine and N-acetylmuramic acid, respectively, joined (diagonal lines) by β-1,4-glycosidic bonds. The vertical lines represent free tetrapeptide side chains, attached to muramic acid residues. The symbol ⊥ represents cross-linked tetrapeptide side chains. From J. M. Ghuysen, "Bacteriolytic Enzymes in the Determination of Wall Structure," *Bacteriol. Rev.* **32**, 425 (1968).

The lipopolysaccharides, which are major components of the outer wall layer in most if not all Gram-negative bacteria, are extremely complex molecules, with molecular weights over 10,000; furthermore, these molecules vary widely in chemical composition, both within and between Gram-negative groups. Most of the work on their structure has been conducted with the lipopolysaccharides of the *Salmonella* group, and it is not yet known whether the general features of lipopolysaccharide structure established for these enteric organisms are also characteristic of the lipopolysaccharides of other Gram-negative groups.

The lipopolysaccharides of *Salmonella* spp. are oligomers, which contain on the average about 3 monomeric subunits. Each subunit is composed of three distinct regions: lipid A; the R core region; and the O side-chain. The subunits are linked through pyrophosphate bridges in the lipid A region (Figure 11.15). The lipid A region consists of a phosphorylated glucosamine disaccharide, heavily esterified with long-chain fatty acids; this region of the molecule is consequently hydrophobic. Attached to one of the glucosamine residues of lipid A is the R core

FIGURE 11.14

A diagrammatic representation of successive steps in the fractionation of the cell wall of *E. coli*. Reconstructed from experiments of W. Weidel, H. Frank, and H. H. Martin.

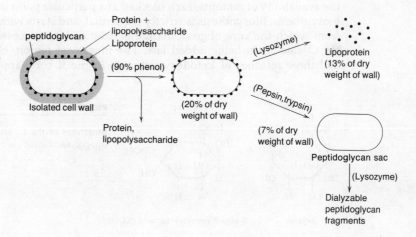

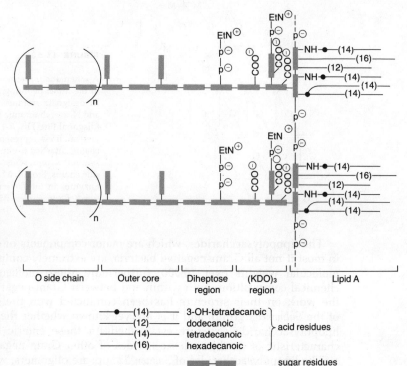

FIGURE 11.15

Schematic representation of a molecule of lipopolysaccharide, containing two cross-linked subunits. The molecular constituents are not drawn to scale. EtN = ethanolamine; KDO = 2-keto-3-deoxyoctonic acid. From H. Nikaido, "Biosynthesis and Assembly of Lipopolysaccharide," in *Bacterial Membranes and Walls*, L. Leive (editor). New York: Marcel Dekker, 1973.

O side chain	Outer core	Diheptose region	(KDO)$_3$ region	Lipid A

——●——(14)—————— 3-OH-tetradecanoic
—————(12)—————— dodecanoic
—————(14)—————— tetradecanoic
—————(16)—————— hexadecanoic
} acid residues

▬▬▬ sugar residues

oligosaccharide: a short chain of sugars, which include two unusual substances, 2-keto-3-deoxyoctonic acid (KDO) and heptose (Figure 11.16). The R core in turn bears the hydrophilic O side-chain, likewise composed of sugars. It is much longer than the R core, being composed of many repeating tetra- or pentasaccharide units. The complete structure of the lipopolysaccharide of *Salmonella typhimurium* is shown in Figure 11.17. The elucidation of this structure was made possible by the availability of mutants, each blocked at a particular point in lipopolysaccharide biosynthesis. Biosynthesis is strictly sequential, and starts with the lipid A portion, from which the core oligosaccharide is extended by successive sugar additions, the O side-chain being added last. The innermost region, consisting of lipid A and three residues of ketodeoxyoctonate in the R core, appears to be essential

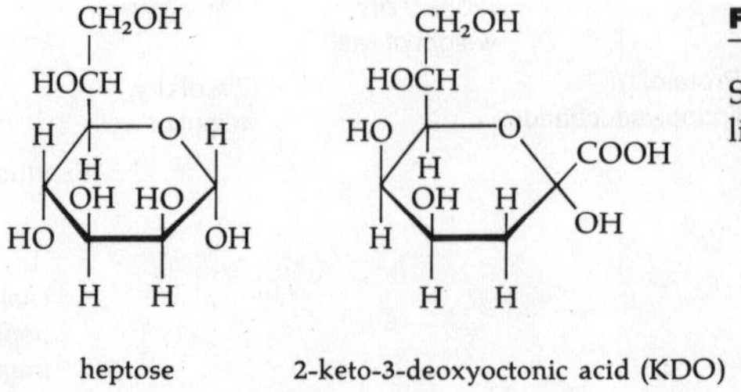

heptose   2-keto-3-deoxyoctonic acid (KDO)

FIGURE 11.16

Structures of the C_7 and C_8 constituents of the lipopolysaccharide core region.

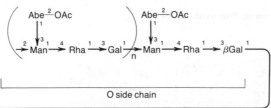

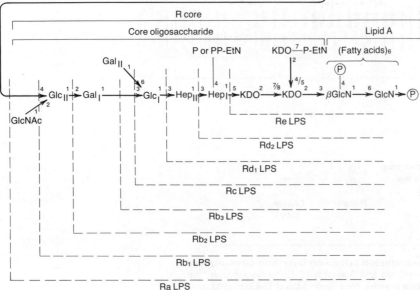

FIGURE 11.17

Structure of one subunit of the lipopolysaccharide of *Salmonella typhimurium*. Abe, abequose; Màn, D-mannose; Rha, L-rhamnose; Gal, D-galactose; GlcNAc, N-acetyl-D-glucosamine; Glc, D-glucose; Hep, heptose; KDO, 2-keto-3-deoxyoctonic acid; EtN, ethanolamine; Ac, acetyl. Biosynthesis starts at the lipid A end, and the molecule is progressively elongated by the addition of sugar residues. The specific points in biosynthesis at which different classes of "rough" mutants are blocked are indicated by dashed lines. From H. Nikkaido, "Biosynthesis and Assembly of Lipopolysaccharide," in *Bacterial Membranes and Walls*, L. Leive (editor). New York: Marcel Dekker, 1973.

to bacterial survival: mutants defective in its synthesis have never been isolated. However, the rest of the core region and the O side-chain are dispensable elements, and mutants blocked at each successive step in their synthesis (indicated by the dotted lines in Figure 11.17) have been isolated.

The lipopolysaccharides are the major antigenic determinant of the wall surface in Gram-negative bacteria, and also serve as the receptors for the adsorption of many bacteriophages. These substances must therefore be located, in part or in whole, on the external surface of the outer wall layer. Furthermore, it has been shown that the long O side-chains extend out to a distance of as much as 30 nm from the wall surface.

The fine structure and width of the outer wall layer suggest that it may consist, like the cell membrane, of a lipid bilayer, in which both phospholipids and lipopolysaccharides are incorporated. A schematic drawing of the possible structure of this layer (which assumes that lipopolysaccharides are concentrated in the outer leaflet) is shown in Figure 11.18.

In enteric bacteria (and probably in other Gram-negative bacteria) the peptidoglycan layer of the wall bears a specific type of lipoprotein on its outer surface. The lipoprotein molecules are covalently linked by a peptide bond to some of the diaminopimelic acid groups in the peptidoglycan mesh. They extend outwards from this layer, the lipid moiety being located at the distal end. The lipoprotein thus serves as a bridge from the peptidoglycan layer to the outer wall layer, to

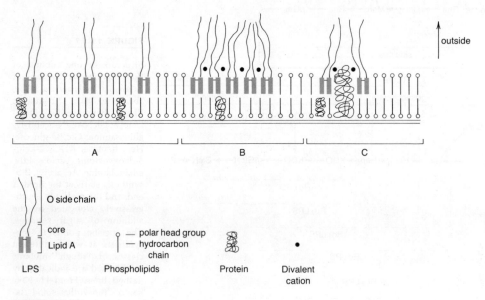

FIGURE 11.18

A possible arrangement of the major constituents in the outer layer of the Gram-negative cell wall. After H. Nikkaido, "Biosynthesis and Assembly of Lipopolysaccharide," in *Bacterial Membranes and Walls*, L. Leive (editor). New York: Marcel Dekker, 1973.

which it is held through hydrophobic interactions between the lipid moiety and the phospholipids of the outer wall layer. A schematic diagram of the probable molecular relations between the outer layers of the Gram-negative cell is shown in Figure 11.19.

• **Peptidoglycan in the walls of Gram-positive bacteria**

In most Gram-positive bacteria, peptidoglycan accounts for some 40 to 90 percent of the dry weight of the cell wall. The wall is homogeneous in structure, and considerably thicker (10–50 nm) than the peptidoglycan wall layer in Gram-negative groups. As already mentioned, there is considerable diversity with respect to peptidoglycan structure and composition among Gram-positive bacteria. The peptidoglycan matrix of the wall is covalently linked to other macromolecular wall constituents which may include a wide variety of polysaccharides and polyolphosphate polymers known as *teichoic acids*. The teichoic acids are water-soluble polymers, containing ribitol or glycerol residues joined through phosphodiester linkages (Figure 11.20). Their exact location in the cell envelope is not certain; most of the teichoic acid remains associated with cell wall material during cell fractionation, and covalent linkage to muramic acid has been demonstrated. However, a small percentage (consisting entirely of glycerol teichoic acids) remains associated with the cell membrane. This material, called *membrane teichoic acid* or *lipoteichoic acid*, has been found to be covalently linked to membrane glycolipid.

The teichoic acids constitute major surface antigens of those Gram-positive

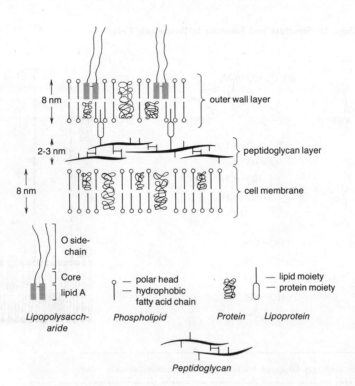

8 nm — outer wall layer

2-3 nm — peptidoglycan layer

8 nm — cell membrane

O side-chain
Core
lipid A

Lipopolysacch-aride

— polar head
— hydrophobic fatty acid chain

Phospholipid

Protein

— lipid moiety
— protein moiety

Lipoprotein

Peptidoglycan

FIGURE 11.19

A schematic drawing to show possible relations between the cell membrane and the inner and outer cell wall layers in a Gram-negative bacterium.

species which possess them, and their accessibility to antibodies has been taken as evidence that they lie on the outside surface of the peptidoglycan layer. Their activity is often increased, however, by partial digestion of the peptidoglycan; thus, much teichoic acid may lie between the cell membrane and the peptido-glycan layer, possibly extending upward through pores in the latter (Figure 11.21).

The repeat units of some teichoic acids are shown in Figure 11.20. The repeat units may be glycerol, joined by 1,3- or 1,2- linkages; ribitol, joined by 1,5-link-ages; or more complex units in which glycerol or ribitol is joined to a sugar residue such as glucose, galactose, or N-acetyl glucosamine. The chains may be 30 or more repeat units in length, although chain lengths of 10 or less are common.

Most teichoic acids contain large amounts of D-alanine, usually attached to position 2 or 3 of glycerol, or position 3 or 4 of ribitol. In some of the more complex teichoic acids, however, D-alanine is attached to one of the sugar residues. In addition to D-alanine, other substituents may be attached to the free hydroxyl groups of glycerol and ribitol: glucose, galactose, N-acetylglucosamine, N-acetylgalactosamine, or succinate. A given species may have more than one type of sugar substituent in addition to D-alanine; in such cases it is not certain whether the different sugars occur on the same or on separate teichoic acid molecules.

The function of the teichoic acids is unknown, but they do provide a high density of regularly oriented charges to the cell envelope, and these must certainly affect the passage of ions through the outer surface layers.

The walls of most Gram-positive bacteria contain no lipid; however, in coryne-bacteria, mycobacteria, and nocardias, a variety of long-chain fatty acids are attached by ester bonds to the wall polysaccharides. Proteins are absent from the

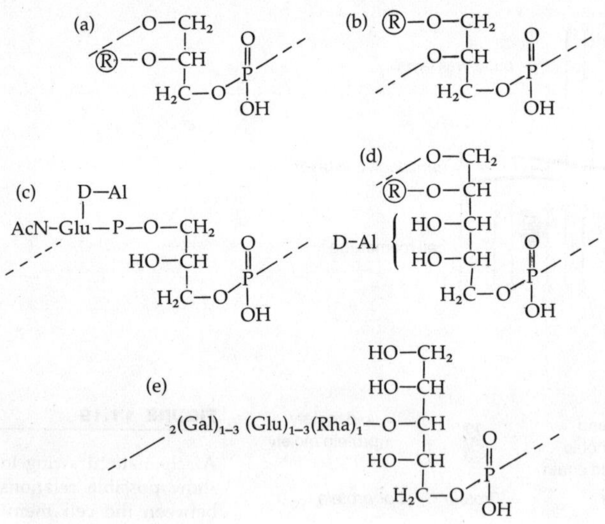

(a)

$$\text{\textcircled{R}}-O-\overset{\displaystyle O-CH_2}{\underset{\displaystyle H_2C-O}{\overset{\displaystyle |}{CH}}}\overset{\displaystyle O}{\underset{\displaystyle OH}{\overset{\displaystyle \|}{P}}}---$$

(b)

$$\text{\textcircled{R}}-O-CH_2 \\ ---\overset{\displaystyle O-CH}{\underset{\displaystyle H_2C-O}{\overset{\displaystyle |}{}}}\overset{\displaystyle O}{\underset{\displaystyle OH}{\overset{\displaystyle \|}{P}}}---$$

(c)

$$\overset{\displaystyle D-Al}{\underset{\displaystyle ---}{AcN-Glu}}-P-O-\overset{\displaystyle CH_2}{\underset{\displaystyle HO-CH}{\underset{\displaystyle H_2C-O}{}}}\overset{\displaystyle O}{\underset{\displaystyle OH}{\overset{\displaystyle \|}{P}}}---$$

(d)

$$\text{\textcircled{R}}-O-\overset{\displaystyle O-CH_2}{CH} \\ D-Al\left\{\begin{array}{l}HO-CH \\ HO-CH \\ H_2C-O\end{array}\right.\overset{\displaystyle O}{\underset{\displaystyle OH}{\overset{\displaystyle \|}{P}}}---$$

(e)

$$_2(Gal)_{1-3}\,(Glu)_{1-3}(Rha)_1-O-\overset{\displaystyle HO-CH_2}{\underset{\displaystyle HO-CH}{\underset{\displaystyle H_2C-O}{\overset{\displaystyle HO-CH}{CH}}}}\overset{\displaystyle O}{\underset{\displaystyle OH}{\overset{\displaystyle \|}{P}}}---$$

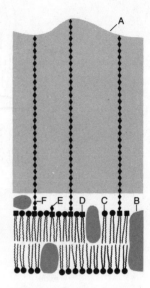

FIGURE 11.20 (left)

Repeats units of some teichoic acids. (a) Glycerol teichoic acid of *Lactobacillus casei* 7469 (R = D-alanine). (b) Glycerol teichoic acid of *Actinomyces antibioticus* (R = D-alanine). (c) Glycerol teichoic acid of *Staphylococcus lactis;* D-alanine occurs in the 6 position of N-acetylglucosamine. (d) Ribitol teichoic acids of *Bacillus subtilis* (R = glucose) and *Actinomyces streptomycini* (R = succinate). (The D-alanine is attached to position 3 or 4 ribitol). (e) Ribitol teichoic acid of the type 6 pneumococcal capsule.

FIGURE 11.21 (right)

A model of the cell wall and membrane of a Gram-positive bacterium, showing lipoteichoic acid molecules extending through the cell wall. The wall teichoic acids, covalently linked to muramic acid residues of the peptidoglycan layer, are not shown. (A) Cell wall, (B) protein, (C) phospholipid, (D) glycolipid, (E) phosphatidyl glycolipid, (F) lipoteichoic acid. From D. van Driel, et al., "Cellular Location of the Lipoteichoic Acids of *Lactobacillus fermenti* NCTC 6991 and *Lactobacillus casei* NCTC 6375," *J. Ultrastruct. Res.* **43**, 483 (1971).

walls of most Gram-positive bacteria. When present, they occur as a separate layer on the outer surface of the wall, often in a very regular ordered array.

• **Function of the peptidoglycan layer**   The well-nigh universal presence of a continuous peptidoglycan layer in the procaryotic cell wall, despite its many other variations of chemical composition, indicates the functional importance of this particular wall polymer. The peptidoglycan layer appears to be the primary determinant of cell shape, as well as the wall component largely responsible for counteracting turgor pressure, and hence preventing osmotic lysis. The action on bacteria of agents which either attack peptidoglycan structure or inhibit peptidoglycan synthesis provide support for these conclusions. Bacterial lysis by lysozyme action, already discussed

(p. 314), is a result of the hydrolytic cleavage of the N-acetylmuramyl-N-acetyl-glucosamine linkage in the glycan strands, which destroys the integrity of the peptidoglycan fabric. The lethal effect of penicillins on growing populations of procaryotes containing peptidoglycans is a consequence of the fact that these antibiotics inhibit the terminal step of peptidoglycan synthesis: cross-linking of the peptide chains. The resultant weakening of the peptidoglycan fabric in the growing cell leads to osmotic lysis.

In this context, the properties of *bacterial L forms* are also highly significant. In 1935 it was observed that the bacterium *Streptobacillus moniliformis* can give rise on rich media to atypical colonies, in which the normal rod-shaped cells are replaced by irregularly shaped, often globular growth forms. It was found that these so-called *L forms* could be propagated indefinitely on serum-enriched complex media. At first, it was believed that the L forms, so different in cell structure from *S. moniliformis*, were symbionts or parasites of the bacterium, but this interpretation had to be abandoned when it was found that they could, on rare occasions, revert to rods. Subsequent observations on L forms of other bacteria showed that the phenomenon was by no means confined to *S. moniliformis*.

After the discovery of penicillin, it was found that L forms are *penicillin resistant* and can in fact be selected by cultivation of bacteria on an osmotically buffered penicillin-containing medium (Figure 11.22). As a rule, removal of penicillin after short exposure leads to immediate reversion to the normal cell form. After longer periods of growth with penicillin, some bacteria may continue to grow in the L form when penicillin is removed. Both the duration of the penicillin exposure necessary to maintain L forms in the absence of penicillin, and the fraction of the population so converted, are variable. Furthermore, the stability of the resulting L forms, even those derived from a single bacterial species, is variable: some ("stable L forms") have never been observed to revert; others ("unstable L forms") do occasionally revert.

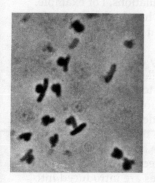

From the known mode of action of penicillin, L forms can be interpreted as bacteria in which the synthesis of the peptidoglycan layer has been severely deranged but which can continue to grow, even though with aberrant cell form, in media sufficiently concentrated to prevent their osmotic lysis.

In certain stable L forms, the components of the peptidoglycan layer are completely absent from the cells. All unstable L forms, and certain stable ones, still contain muramic acid, but the concentration is relatively low (about 10 to 15 percent of its concentration in normal cells). Furthermore, the muramic acid is in an unusual chemical state, being readily extractable with dilute acid, whereas the muramic acid in a normal cell wall is not. From these facts, the following general interpretation of the nature of L forms can be derived. *They are bacteria in which the primer for peptidoglycan synthesis has been either eliminated or modified by penicillin treatment.* In the first case, no new peptidoglycan material can be deposited in the wall; in the second, new subunits are incorporated in an irregular and uncoordinated manner, so that the formation of the normal continuous peptidoglycan layer is prevented. On rare occasions, L forms of the latter type may succeed in resynthesizing a continuous peptidoglycan layer and will then revert to the normal bacterial form.

L forms can be obtained from both Gram-positive and Gram-negative bacteria. In the latter, the outer layer of the cell wall continues to be synthesized in an apparently normal fashion; this is indicated both by electron microscopic examination and by the fact that such L forms retain O antigens and are still susceptible to infection by phages for which the receptors are contained in the outer wall layer.

• Functions of the outer wall layer in Gram-negative bacteria

In Gram-negative bacteria, peptidoglycan is a minor wall polymer, overshadowed quantitatively by the proteins, phospholipids, and lipopolysaccharide contained in the outer wall layer.

There are many indications that the outer wall layer of Gram-negative bacteria interposes a barrier to the passage of some substances from the environment into the cell. These substances include a number of antibiotics, notably penicillins, to which Gram-negative organisms are far less sensitive than Gram-positive ones, even though the target enzyme (transpeptidase), when isolated, is inhibited by similar penicillin concentrations in both groups. Dyes and bile salts likewise appear to be partly excluded by the outer membrane.

The lipopolysaccharides of the outer wall layer seem to play an important role in this barrier function, since rough mutants in which the O chain and part of the core oligosaccharide are eliminated from the wall lipopolysaccharide are much more sensitive than the wild type to penicillin and certain other antibiotics. Treatment of Gram-negative bacteria with the chelating agent, EDTA, which causes the liberation of much lipopolysaccharide from the cell wall, likewise makes the cells more sensitive to many antibiotics.

The outer wall layer also clearly acts as a barrier to the passage of proteins. As discussed in Chapter 10, many enzymes that are exoenzymes in Gram-positive bacteria have a periplasmic location (enclosed by the outer wall layer) in Gram-negative bacteria.

The somatic antigenic specificity of Gram-negative bacteria is largely determined by the lipopolysaccharide in the outer cell wall layer, and more particularly, by the sugars of the O side chains which project from the surface of the wall. As isolated from nature, most Gram-negative bacteria are "smooth," their cell surfaces bearing O side chains. However, "rough" mutants, in which the lipopolysaccharide has lost its O side chains, arise at high frequencies in pure cultures, and they often displace the smooth parental strain. These facts suggest that the preservation of the smooth state (and hence of the O side chain of the lipopolysaccharide) has strong selective advantages in natural environments. At the same time, there is a tremendous diversification of O side-chain structure (and hence of somatic antigenic specificity) within natural bacterial populations. For example, in the enteric bacteria of the *Salmonella* group, there are hundreds of specific serotypes differing in O side-chain antigenic structure, but otherwise they are almost indistinguishable phenotypically.

The selective value of possessing an O side chain is very clear for pathogenic Gram-negative bacteria, such as the members of the *Salmonella* group. Its presence on the cell surface makes the bacterium relatively resistant to engulfment by the phagocytes of the animal host, this resistance being lost only if the host has synthesized antibodies specifically directed against the O side chain. For a pathogenic bacterial species, accordingly, antigenic diversity with respect to the O side chains of its lipopolysaccharide has a considerable selective advantage, since the animal host is unlikely to have high antibody levels directed against a large number of types of O side chains. When lipopolysaccharides are introduced into an animal's bloodstream, they are highly toxic. They cause fever, and at higher levels they cause internal hemorrhage and shock.

• The topology of
wall and membrane
synthesis

During the cell cycle of a bacterium the surface layers of the cell continuously change in form. The increase in the volume of the cell is accompanied by an extension of the area of both wall and membrane. With the onset of division, a vectorial change in wall deposition occurs: the transverse septum begins to grow inward, at right angles to the cell wall, until it forms a complete septum separating the two daughter protoplasts. Thereafter, the transverse septum peels apart into two layers, each of which becomes the newly formed pole of one of the daughter cells.

The terminal steps in the synthesis both of phospholipids and of the monomers destined for incorporation into major wall polymers are all mediated by enzymes localized in the cell membrane, as shown by cell fractionation studies. It is not known, however, if these enzymes are concentrated at specific sites in the membrane or if they are more or less evenly dispersed over its surface.

The increase in the area of the wall and membrane that accompanies cell growth might occur by the insertion of new material at specific growth points or by the intercalation of new material at numerous sites in the preexisting wall and membrane fabric. Many experiments designed to examine this question have been performed, often with conflicting results. It is important to keep in mind that secondary displacement of newly incorporated materials may occur. For example, the plasticity of the cell membrane and the outer wall layer of Gram-negative bacteria could cause an apparently random distribution of newly synthesized components, even if they were initially incorporated at specific points. The same effect would result from a rapid turnover of wall or membrane constituents. Consequently, evidence for localized growth of walls or membranes is incontrovertible, whereas evidence that suggests random incorporation is often ambiguous.

In the streptococci, Gram-positive bacteria with spherical cells, clear evidence for a localized equatorial region of wall growth has been obtained by the use of antisera specifically directed against wall constituents. The antiserum, conjugated with a fluorescent dye, is used to coat the cells, making them intensely and uniformly fluorescent. During subsequent growth in the presence of nonfluorescent antiserum, the poles of the cells remain intensely fluorescent for several generations, new nonfluorescent wall areas being progressively inserted between the "old" wall material as growth proceeds (Figure 11.23). Thus, during exponential growth of streptococci, wall synthesis is highly localized, and the walls, once formed, are not secondarily modified. However, it has also been shown that if the growth of streptococci is prevented by inhibition of protein synthesis (either by specific amino acid deprivation or by chloramphenicol treatment), the synthesis of wall material continues and results in a progressive *thickening* of the wall over its entire surface (Figure 11.24). Under these conditions, accordingly, wall precursors must be incorporated at many points.

Fluorescent antibody labeling experiments with growing cultures of Gram-negative bacteria show no indication of localized wall deposition; the intensity of fluorescence of the cells weakens uniformly and progressively as growth proceeds. In this case, however, the antibodies are directed against components of the outer wall (principally, the lipopolysaccharides), which could be redistributed by the plasticity of the lipid bilayer. One special case of highly localized growth, both of wall and membrane layers, has been demonstrated in a Gram-

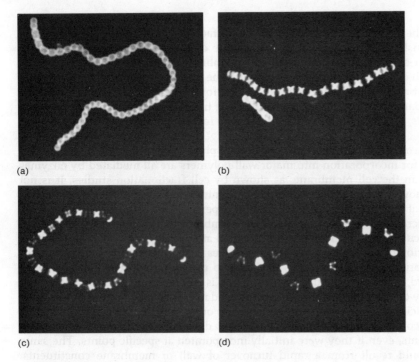

(a)   (b)

(c)   (d)

FIGURE 11.23

Growth of the wall of *Streptococcus pyogenes*, followed by ultraviolet photomicrography of growing chains of cells, in which the wall had been initially coated with a fluorescent anti-body. (a) Immediately after antibody treatment; the cells are evenly fluorescent. (b) After 15 minutes of growth. New (nonfluorescent) wall material has been formed around the equator of each cell; the polar caps of the cells, previously labeled with fluorescent antibody, remain fluorescent. (c), (d) The appearance of cell chains after 30 and 60 minutes of growth, respectively. From R. M. Cole and J. J. Hahn, "Cell Wall Replication in *Streptococcus pyogenes*," *Science* **135,** 722 (1962).

FIGURE 11.24

Electron micrographs of thin sections of *Streptococcus faecalis* to show the manner of wall growth. (a) Longitudinal section of a dividing cell; the circle encloses one of the wall bands that separate the equatorial wall synthesized most recently from that synthesized in the previous generation. The arrow points to a septal mesosome. (b) Section of a cell which has been deprived of an essential amino acid (threonine) for 20 hours, thus arresting protein synthesis, but not wall synthesis. Note the extensive thickening of the wall, compared with (a). The bar indicates 0.1 μm. Reproduced from M. L. Higgins and G. D. Shockman, "Early Changes in the Ultrastructure of *Streptococcus faecalis* After Amino Acid Starvation," *J. Bact.* **103,** 246, (1970).

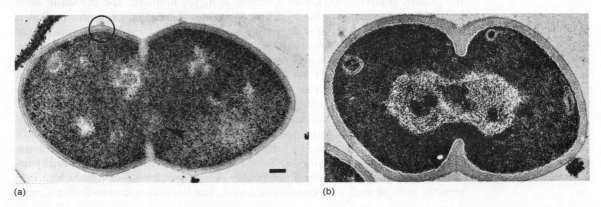

(a)   (b)

negative organism, *Caulobacter*. The filiform prostheca characteristic of caulobacters develops at each cell generation on the initially flagellated pole of the daughter swarmer cell (see p. 146), and normally attains a length of 2 to 3 μm. However, when cells are starved for inorganic phosphate, prosthecal elongation continues, the structure attaining a length of 10 to 15 nm. If swarmer cells are radioactively labeled in a uniform manner by growth with tritiated glucose, and then transferred to a medium containing nonradioactive glucose and a limiting amount of phosphate, the initially short prosthecae formed in the presence of tritiated glucose continue to elongate. Subsequent autoradiography of such cells shows that radioactive material is confined to the body of the cell and the distal end of the prostheca; the proximal part of the prostheca is not radioactive (Figure 11.25). It follows that elongation of the prostheca takes place through highly localized synthesis at the proximal end, close to its attachment to the body of the cell. Since the prostheca is made up of a membranous core, surrounded by both the inner and outer wall layers, it is evident that in this case the synthesis of *all* these structural elements is localized.

In view of the manifold functions of the cell membrane, in particular its postulated role in the segregation of the bacterial genome, the topology of membrane growth is of particular interest. Many studies of membrane growth have been conducted by isotopic or density labeling of precursors of membrane lipids, such as glycerol and fatty acids. This work provides no indication of localized membrane synthesis. However, recent experiments have used as markers of membrane growth certain inducibly synthesized membrane proteins, and the results obtained suggest that membrane growth is in fact highly localized. In *E. coli* the synthesis of β-galactoside permease, located in the membrane, is under the control of the lactose operon; approximately 10^4 permease molecules are

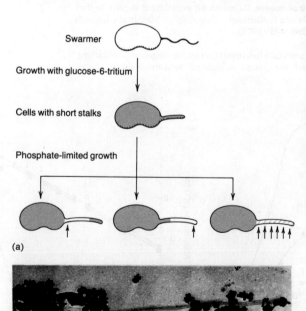

Swarmer

Growth with glucose-6-tritium

Cells with short stalks

Phosphate-limited growth

(a)

(b)

FIGURE 11.25

(a) A diagrammatic drawing of the possible labeling patterns of stalked *Caulobacter* cells resulting from the experiment described in the text. Crosshatching indicates radioactive areas in the cell produced; arrows indicate the corresponding sites of elongation of the prostheca. (b) Radioautograph of a cell with elongated prostheca. From J. Schmidt and R. Y. Stanier, "The Development of Cellular Stalks in Caulobacteria," *J. Cell Biol.* **28,** 423 (1966).

present in the membrane of a fully induced cell. Since the permease is required for lactose uptake, uninduced cells cannot grow with lactose, but induced cells can. When a mixture of induced and uninduced cells is exposed to penicillin in the presence of lactose as the sole carbon source, induced cells start to grow immediately, and undergo rapid lysis as a result of defective wall synthesis. Uninduced cells lyse only somewhat later, after permease synthesis has been again induced by lactose. The heterogeneity of a cell population with respect to permease function can consequently be determined by this method.

When a fully induced population of *E. coli* is transferred to a medium without inducer (glycerol as carbon source), all cells in the population remain subject to rapid penicillin lysis in the presence of lactose for two generations. By the third generation, however, only one-half the population undergoes immediate lysis (Figure 11.26). Thus, a large fraction of the cellular population suddenly becomes permease negative between the second and third generations (Figure 11.27). If

FIGURE 11.26 (left)

Time course of penicillin-induced lysis of *E. coli,* fully induced for galactoside permease (curve GO) and after growth for one, two, and three generations in absence of inducer (G1, G2, G3). Curve NI shows the time course of lysis of a noninduced control population. From A. Képès and F. Autissier, "Topology of Membrane Growth in Bacteria," *Biochem. Biophys. Acta* **265,** 443 (1972).

FIGURE 11.27 (right)

Segregation kinetics of β-galactoside permease, based on an experiment similar to that shown in Figure 11.26. From A. Képès and F. Autissier, "Topology of Membrane Growth in Bacteria," *Biochem. Biophys. Acta* **265,** 443 (1972).
■–■ Absorbance of total population;
●–● Absorbance after 30 minutes of penicillin treatment (permease-negative population);
▲–▲ Absorbance of permease-positive population (calculated by difference).

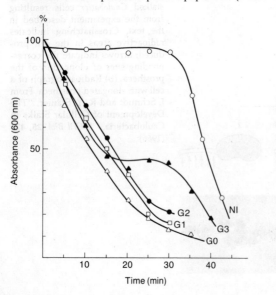

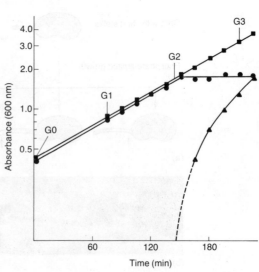

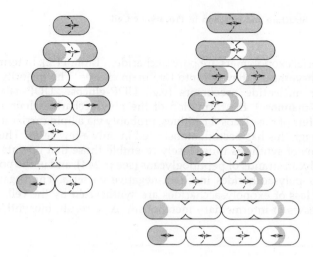

FIGURE 11.28

Distribution of parental membrane (shaded area) in descendants of a bacterium with a single median growing zone. Arrow indicates elongation of new membrane. At left, a newly formed daughter cell; at right, a cell midway between two divisions. From A. Képès and F. Autissier. "Topology of Membrane Growth in Bacteria," *Biochem. Biophys. Acta* **265**, 443 (1972).

one assumes that membrane synthesis by *E. coli* occurs in a central zone, separating the two "old" membrane poles by growth, it could be anticipated that one-half the population would contain no "old" membrane after two cell generations, starting from a newly divided cell. However, exponentially growing populations include cells at all stages of the growth cycle: if a cell half way between two divisions is considered representative of such a nonsynchronized population, it will give rise to descendants with no "old" membrane only after its third division, which will occur after 2.5 generations of growth. These two situations are illustrated schematically in Figure 11.28. The fact that permease negative cells only begin to appear between the second and third generations after removal of the inducer is, therefore, fully compatible with the hypothesis that the membrane is extended by incorporation of newly synthesized material in the central region of the cell. The observations could also be satisfactorily explained on the assumption that the membrane grows at some other localized site (e.g., the cell poles), but they are wholly incompatible with models that assume intercalation of material at many points over the surface of the membrane.

• **Capsules and slime layers**

Many procaryotes synthesize organic polymers which are deposited outside the cell wall, as a loose, more or less amorphous layer called a *capsule*, or *slime layer*. The term *capsule* is usually restricted to a layer that remains attached to the cell wall, as an outer investment of limited extent, clearly revealed by negative staining (see Figure 5.40). However, these *exopolymers* often form much more widely dispersed accumulations, in part detached from the cells that produce them. Such variations in the location and extent of the layer are caused primarily by the abundance with which the exopolymer is formed, and by its degree of water solubility. This layer is clearly not essential to cellular function; many bacteria do not produce it, and those which normally do so can lose the ability as a result of mutation, without any effect on growth.

Exopolymers vary widely in composition. A few *Bacillus* species produce exopolypeptides, made up of only one amino acid, glutamic acid, which is predominantly of D configuration. Glutamyl residues are linked through the γ carboxyl group, as in the side chain of peptidoglycans. With this exception, the bacterial exopolymers are polysaccharides (Table 11.2). In terms of the mechanism of biosynthesis, they fall into two main classes. The majority are synthesized via sugar nucleotide precursors (e.g., UDP-glucose, UDP-galactose, GDP-fucose, GDP-mannose), the formation of the polysaccharide chain involving successive transfers of the glycosyl residues, probably via a lipid carrier in the cell membrane, although this has been demonstrated in only a few cases. The biochemical mechanisms of synthesis thus closely resemble those that operate in the formation of the glycan strands of peptidoglycans (see p. 244), and of the polysaccharide moiety of lipopolysaccharides in Gram-negative bacteria. The sugars incorporated into this class of exopolysaccharides are synthesized by the cell, through the normal processes of intermediary metabolism. As a result, biosynthesis is little, if at all,

TABLE 11.2

Bacterial exopolymers

POLYMER	SUBUNITS	STRUCTURE (IF KNOWN)	ORGANISMS
Polyglutamic acid	Glutamic acid (mainly D isomer)	A. *Exopolypeptides* -γ-glutamyl-γ-glutamyl-	*Bacillus anthracis, B. megaterium*
		B. *Exopolysaccharides synthesized from sugar nucleotides*	
Cellulose	Glucose	β-glu 1 → 4 β-glu	*Acetobacter xylinum*
Glucan	Glucose	β-glu 1 → 2 β-glu	*Agrobacterium tumefaciens*
Colanic acid	Glucose, galactose, fucose, glucuronic acid, pyruvic acid		Enteric bacteria
Polyuronides	Mannuronic acid, glucuronic acid		*Pseudomonas aeruginosa, Azotobacter vinelandii*
Pneumococcal poly-saccharides:			
Type II	Glucose, rhamnose, glucuronic acid		
Type III	Glucose, glucuronic acid	-(-3-β-glucuronyl 1 → 4 β-glucosyl)-	*Streptococcus pneumoniae*
Type XIX	Glucose, rhamnose, N-acetyl-D-mannosamine, phosphate		
Type XXXIV	Glucose, galactose, ribitol, phosphate		
		C. *Exopolysaccharides synthesized from sucrose*	
Dextrans	Glucose (fructose)	α-fru-β-glu 1 → 6 β-glu-	*Leuconostoc spp., Streptococcus spp.*
Levans	Fructose (glucose)	β-glu-α-fru 2 → 6 α-fru-	*Pseudomonas spp., Xanthomonas spp., Bacillus spp., Streptococcus salivarius*

influenced by the nature of the nutrients provided. A wide diversity of mono-saccharides, amino sugars, and uronic acids occur in these compounds; some of the constituent sugars may bear acetyl or pyruvyl substituents. Apart from a few of the simpler, homopolymeric exopolysaccharides (for example, cellulose, synthesized by *Acetobacter xylinum*), the detailed structures have rarely been elucidated.

Two exopolysaccharides, *dextrans* and *levans,* have a different biosynthetic origin: they are formed directly from an exogenous substrate, the disaccharide sucrose, by successive addition of glycosyl units to an acceptor molecule of sucrose. Symbolizing the sucrose molecule as α-glu-β-fru, we can schematize the initial steps of levan synthesis as follows:

$$2 \ \alpha\text{-glu-}\beta\text{-fru} \longrightarrow \alpha\text{-glu-}\beta\text{-fru-}\beta\text{-fru} + \text{glu}$$
$$\alpha\text{-glu-}\beta\text{-fru-}\beta\text{-fru} + \alpha\text{-glu-}\beta\text{-fru} \longrightarrow \alpha\text{-glu-}\beta\text{-fru-}\beta\text{-fru-}\beta\text{-fru} + \text{glu}$$

Thus, the levan molecule has a terminal glucosyl residue, to which is attached a chain of β-fructosyl residues in 2 → 6 linkage. The synthesis of dextrans occurs in an analogous fashion, by successive addition of α-glucosyl units to the fructosyl moiety of an acceptor molecule of sucrose. Therefore, dextrans contain a terminal β-fructosyl residue.

The formation of dextrans and levans without the ATP expenditure necessary to synthesize sugar nucleotide precursors is possible because the glycosidic bond energy in the disaccharide which serves as the substrate is conserved; chain elongation occurs by transglycosylation. For this reason, dextrans and levans cannot be formed at the expense of free monosaccharides: sucrose is the specific substrate for their synthesis. Consequently, dextran- and levan-forming bacteria produce these capsular materials only when they grow on a sucrose-containing medium. The colonies of such organisms thus have an appearance when grown with sucrose entirely different to that when grown with glucose or other sugars (Figure 11.29).

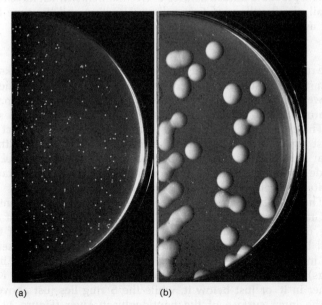

(a)   (b)

FIGURE 11.29

The formation of extracellular polysaccharides by bacteria. Two plates of *Leuconostoc mesenteroides,* streaked on glucose medium (a) and sucrose medium (b). The large size and mucoid appearance of the colonies on sucrose are caused by the massive synthesis and deposition around the cells of dextran.

THE MOLECULAR STRUCTURE OF FLAGELLA AND PILI

Although they differ both in function and in gross form, the two classes of filiform bacterial surface appendages, flagella and pili, share many common structural features. Both originate from the cell membrane and extend outward through the wall to a distance which may be as much as 10 times the diameter of the cell. The external part of these organelles can be detached from the cell by mechanical means (e.g., shearing in a blender), and subsequently isolated and purified. The filaments of flagella and pili are made up of specific proteins, known as *flagellins* and *pilins*. The protein subunits (monomers) can be prepared by treatment of the isolated flagella or pili with heat or acid; they are of relatively low molecular weight (17,000 to 40,000). Studies of isolated flagella and pili by electron microscopy and X-ray diffraction have shown that the protein monomers are assembled in helical chains, wound around a central hollow core. The structure of the filament is consequently a reflection of the properties of the specific type of protein subunit from which it is built; it is determined by the size of the subunit and by the number and pitch of the helical chains into which they aggregate. The flagella of different bacteria differ slightly both in diameter (12 to 18 nm), and in form (i.e., height and wavelength of the helical curvature). Different types of pili differ greatly in width (4 to 35 nm). These minor variations within each class of organelle evidently reflect differences in the assembly properties of different flagellins and pilins. It has been shown that a single mutational change in the amino acid sequence of a flagellin can cause a change in the height and wavelength of the flagellum formed from it.

The probable ultrastructures of two filaments which have been studied in some detail, the flagellum of *Salmonella typhimurium* and the type I pilus of *Escherichia coli*, are shown schematically in Figure 11.30.

• The basal structure of the flagellum

The removal of flagella and pili by mechanical shearing breaks off these organelles near the cell surface; consequently, it does not reveal their basal structures. However, by more gentle methods of cell breakage (osmotic lysis, detergent treatment), it is possible to isolate flagella with their basal structures intact.

The entire flagellar apparatus is made up of three distinct regions. The outermost region is the helical flagellar filament, of constant width, made up flagellin. Near the cell surface this is attached to a slightly wider *hook*, about 45 nm long, made up of a different kind of protein, which is in turn attached to a *basal body*, located entirely within the cell envelope (Figure 11.31).

The basal body consists of a small central rod, inserted into a system of rings. In Gram-negative bacteria the basal body bears two pairs of rings (Figure 11.32). The outer pair (L and P rings) are situated at the level of the outer and inner wall layers, respectively; apparently, their function is to serve as bushings for the insertion of the rod through the two wall layers. The inner pair (S and M rings) are located near the level of the cell membrane: the M ring is embedded either in it, or just below it, while the S ring lies just above, possibly attached to the inner surface of the peptidoglycan layer (Figure 11.33). On flagella of

	Flagellum	Type I pilus
	(*Salmonella typhimurium*)	(*Escherichia coli*)
Form of organelle:	Helical filament, 14 nm in diameter	Straight filament 7 nm in diameter
Protein subunit:	Flagellin (40,000 daltons)	Pilin (17,000 daltons)

Suggested model for the assembly of the subunits (not drawn to same scale):

FIGURE 11.30

Models showing the probable helical arrangement of the protein subunits of a bacterial flagellum (left) and of a pilus (right).

FIGURE 11.31

Electron micrograph of a negatively stained lysate of the purple bacterium, *Rhodospirillum molischianum*, showing the basal structure of an isolated flagellum (×181,000). Note the basal hook and the attached paired discs. The other objects in the field are fragments of the photosynthetic membrane system. From G. Cohen-Bazire and J. London, "Basal Organelles of Bacterial Flagella," *J. Bacteriol.* **94**, 458 (1967).

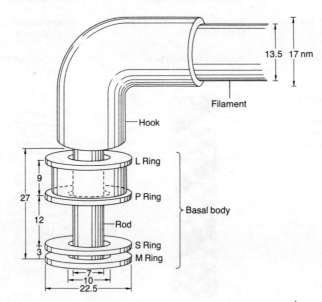

Filament

Hook

L Ring

P Ring

Rod

} Basal body

S Ring
M Ring

FIGURE 11.32

Diagrammatic model of the basal end of the flagellum of *E. coli*, based on electron micrographs of the isolated organelle. From M. L. De Pamphilis and J. Adler, "Fine Structure and Isolation of the Hook-Basal Body Complex of Flagella from *Escherichia coli* and *Bacillus subtilis*," *J. Bact.* **105**, 384 (1971).

Gram-positive bacteria, only the lower (S and M) rings are present; apparently, the upper pair is not required to support the rod as it passes through the relatively thick and homogeneous Gram-positive wall. This difference is significant, since it implies that only the S and M rings are essential for flagellar function.

• **Synthesis of the flagellar filament**

Under appropriate physicochemical conditions, flagellin monomers can reassemble *in vitro* to produce filaments apparently identical with the flagella from which they are derived. This process of assembly requires the presence of seed structures:

FIGURE 11.33

A model showing the possible topological relations between the basal structure of the flagellum and the outer cell layers of *E. coli*. Dimensions in nanometers. From M. L. De Pamphilis and J. Adler, "Attachment of Flagellar Basal Bodies to the Cell Envelope," *J. Bact.* **105**, 396 (1971).

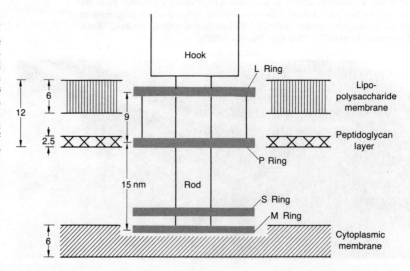

Hook

L Ring

Lipo-polysaccharide membrane

Peptidoglycan layer

P Ring

Rod

S Ring

M Ring

Cytoplasmic membrane

short flagellar fragments to which the flagellin molecules can attach. Such flagellar fragments possess a structural polarity; electron microscopy shows that one end is rounded and the other is indented. By marking the seed fragments with antiflagellar antibody, it can be demonstrated that flagellin subunits add only to the indented end (Figure 11.34). To which end of the intact flagellum, attached to the bacterial cell, does the indented end correspond? This question has been answered by an ingenious experiment on the growth of the flagella of *Salmonella typhimurium*. Addition of the amino acid analogue *p*-fluorophenylalanine to the growth medium causes *S. typhimurium* to synthesize abnormal ("curly") flagella of shorter wavelength than normal. When normal cells are cultivated for 2 to 3 hours in the presence of this analogue, and the distribution of curly waves on their flagella is then examined, it is found that these always occur on the distal parts of the flagella, the basal parts retaining the normal wavelength. Hence, it follows that the elongation of the flagellar filament takes place by the addition of new flagellin subunits to its tip. It is most improbable that flagellins can attain the tip of a growing flagellum by excretion and passage through the medium: they are probably synthesized at the base of the flagellum and move outward through its hollow core to the site of incorporation.

• The mechanism of flagellar movement

The role of flagella as agents of bacterial movement can be demonstrated by a very simple experiment. If cells are mechanically deflagellated by shearing, they become immotile. Regrowth of flagella is rapid, the normal number and length being restored in about one generation. During flagella regrowth the cells at first show only rotatory movement; translational movement begins after the flagella have attained a critical length.

The way by which bacterial flagella propel the cell has been discussed for many decades, but experimental evidence in support of a specific mechanism has been obtained only recently. This evidence, discussed below, indicates that flagella are semi-rigid, helical rotors, each of which spins, either clockwise or counterclockwise, around its long axis. Movement is imparted to the organelle at its base by a flagellar "motor." It has been suggested that the motor operates by causing the S and M rings to rotate relative to each other. As already discussed, the M ring is situated in or just below the cell membrane. In order for relative rotation of the two rings to move the cell, it is essential that the S ring also be inserted into a structural fabric; most likely, it is attached to the inner surface of the cell wall. Such an attachment could account for the fact that complete removal of the bacterial cell wall by lysozyme treatment makes the protoplast immotile, even though it still bears intact flagella.

Evidence in support of the proposed mechanism has been obtained by experiments on mutants of *E. coli*, which are unable to swim as a result of the synthesis

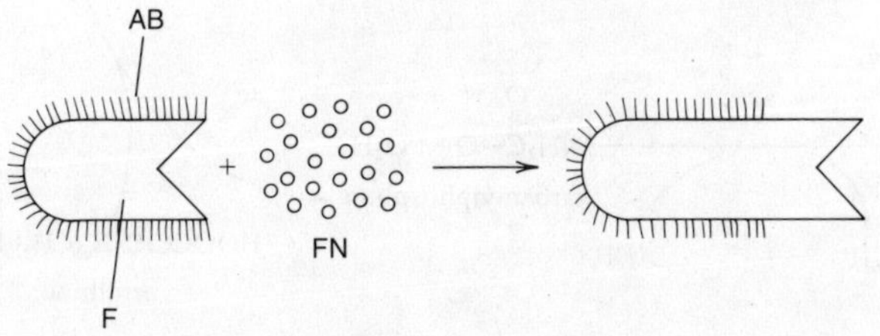

FIGURE 11.34

Schematic representation of experiment demonstrating the unidirectional growth *in vitro* of the bacterial flagellum: F, sheared fragment of a flagellum, coated with antiflagellar antibody (AB); FN, flagellin subunits.

of structurally abnormal flagella. One mutant class produces straight flagellar filaments; a second class produces so-called *polyhook* flagella, which consist of a series of hooks 1 to 2 μm long without an attached filament. When treated with antisera directed against their abnormal flagella, the mutants form clumps, in which the cells are tethered to one another by antibody complexes formed between their respective flagella. Such tethered cells spin about the point of attachment, at rates of 2 to 9 revolutions per second. A spinning cell can modulate its movement in three ways: by continuous spinning in one direction (e.g., clockwise); by stopping, and then restarting in the same direction; and by changing the direction, spinning counterclockwise. Such observations concern the rotation of the cell, a simultaneous rotation of the filament being inferred. In the case of the straight filament mutant, flagellar rotation was directly demonstrated, by coating small latex beads with antiflagellar antibody and adding them to a suspension of the immotile cells. Some beads became attached to the flagella 1 to 2 μm from the surface of the cell, and were observed to rotate very rapidly.

These observations suggest that the flagellar hook is driven in rotary fashion by the underlying motor, resulting in a rotation of the flagellar filament. The cell evidently has the capacity to vary both the speed and the direction of rotation, as well as the frequency of stops and starts. Peritrichously flagellated bacteria swim in straight lines over moderate distances, these runs being interrupted periodically by abrupt, random changes of direction, known as *tumbles*. Recent observations suggest that smooth swimming in a fixed direction is mediated by steady rotation of the flagella in a counterclockwise direction (viewing the flagellum along its rotational axis in the direction of the cell). Tumbles are caused by a reversal of flagellar rotation, to the clockwise sense.

In spirilla, the multitrichous polar tufts of flagella are sufficiently thick to be visualized by phase contrast microscopy. During steady swimming the two polar tufts both rotate in the same sense. Spirilla never tumble; instead, changes in the direction of movement are brought about by a reversal of the sense of rotation of the flagellar tufts, and they result in an exact reversal, by 180°, of the path previously followed.

The mechanochemical basis for the operation of the flagellar motor is not known. However, there is good evidence that it is dependent, either directly (or

FIGURE 11.35

Mechanism of the so-called "arginine dihydrolase" reaction, which permits the generation of ATP by substrate-level phosphorylation.

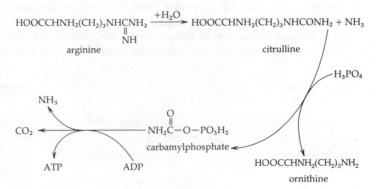

343   more probably indirectly) on the continuous generation of ATP by the cell. Like other strictly aerobic motile bacteria, the fluorescent pseudomonads immediately become immotile when the oxygen supply in wet mounts is exhausted. However, the motility of fluorescent pseudomonads can be maintained, even after oxygen depletion, if they are provided with the amino acid, L-arginine. Anaerobic, arginine-induced motility in a wet mount is less rapid than aerobic motility, and it starts a few seconds after oxygen depletion has caused the cells to become immotile. It is a consequence of the ability of fluorescent pseudomonads to catalyze a nonoxidative conversion of arginine to ornithine, accompanied by synthesis of ATP (Figure 11.35).

THE CHEMOTACTIC BEHAVIOR OF MOTILE BACTERIA

The cells of a suspension of flagellated bacteria are normally in a state of continuous but random active movement. However, if certain chemical gradients are imposed on the population, the cells migrate to and accumulate in that part of the gradient which provides an optimal concentration of the chemical. Some of the substances to which they can respond (for the most part, nutrients) act as *attractants,* in the sense that the cells will accumulate in the more concentrated region of the gradient. Others (for the most part, toxic substances) act as *repellants,* in the sense that the cells avoid regions of high concentration, and accumulate in that part of the gradient in which the concentration is lowest. This behavior is known as *chemotaxis.* The specific substances which can elicit tactic responses differ for different bacteria; a particular chemical spectrum is characteristic for each species.

Molecular oxygen elicits so-called *aerotactic* responses in most motile bacteria. Aerotactic patterns of accumulation can be readily observed in wet mounts in which an oxygen gradient is established by diffusion from the edges of the cover slip. Most strict aerobes accumulate close to the edge of the cover slip; however, the spirilla, which are microaerophils, accumulate in a narrow band some distance from the edge. For motile strict anaerobes, oxygen is a repellant; they accumulate in the center of the wet mount where the oxygen concentration is lowest (see Figure 11.36).

The tactic responses of *Escherichia coli* to organic compounds (sugars and amino acids) have been studied in great detail by J. Adler and his collaborators. Their

FIGURE 11.36

Aerotactic responses of motile bacteria (after Beijerinck). Suspensions of various bacteria were placed on slides under cover slips. (a) Aerobic bacteria accumulate near the edges of the cover slip, where oxygen concentration is greatest. (b) Microaerophilic bacteria accumulate at some distance from the edge. (c) Obligate anaerobes accumulate in the central, almost anaerobic region.

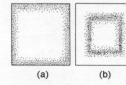

(a)   (b)   (c)

work has shown that by no means all compounds that can serve as nutrients (energy sources) are attractants. For example, the disaccharide maltose is an attractant; the disaccharide lactose (also a good substrate) is not, although its split products (glucose and galactose) are. Conversely, the amino acid serine is a powerful attractant, although pyruvate, the first product formed from it, is not. Compounds that act as attractants are not necessarily metabolizable; for example, D-fucose, a nonmetabolizable analogue of D-galactose, is almost as good an attractant as galactose. Systematic studies on the responses of *E. coli* to sugars and amino acids have involved the use of mutants blocked either in the metabolism of attractants or in the ability to detect them, and have also involved the examination of competition between different attractants. These studies have led to the identification of 11 different chemoreceptors, each capable of eliciting a tactic response to certain specific compounds (Table 11.3). The nature of the receptor molecules for galactose and for maltose is known: they are the specific binding proteins (see p. 296) for these two sugars, located in the periplasmic space of the cell. Mutants that lack one of these binding proteins lose their ability to respond to the sugar in question. The chemoreceptors serve purely as *gradient-sensing devices:* they are not directly associated with locomotion, since specific chemoreceptors can be lost by mutation without any impairment of motility.

Bacterial chemotaxis poses a special problem: how do such small organisms detect the concentration differences in chemical gradients over distances as short as the length of a single cell (2 to 3 μm)? Recent experiments show that the bacterium does not in fact make an instantaneous *spatial* comparison of the attractant concentrations at the two ends of the cell. Instead, it possesses a *temporal*

TABLE 11.3

The chemoreceptors of *Escherichia coli* for sugars and amino acids, with their specificities

NAME OF CHEMORECEPTOR	CHEMICALS DETECTED (IN ORDER OF DECREASING EFFECTIVENESS)	CHEMICALS DETECTED BY MORE THAN ONE RECEPTOR
Aspartate	L-aspartate > L-glutamate > L-methionine }	Asparagine, cysteine
Serine	L-serine > glycine > L-alanine }	
Glucose	D-glucose, D-mannose }	D-glucosamine, 2-deoxy D-glucose, Methyl-α-D-glucoside Methyl-β-D-glucoside
Galactose	D-galactose, D-glucose > D-fucose > L-arabinose > D-xylose > L-sorbose }	
Fructose	D-fructose	
Mannitol	D-mannitol	
Ribose	D-ribose	
Sorbitol	D-sorbitol	
Trehalose	Trehalose	
N-acetylglucosamine	N-acetyl-D-glucosamine	
Maltose	Maltose	

gradient sensing system, i.e., a kind of "memory" device that enables the cell to compare, over a short interval of time, present and past concentrations. The memory system has a decay time of many seconds. Thus, if a bacterium is swimming at 30 μm sec^{-1} and if it possesses a memory with a decay time of 60 seconds, it can compare concentrations over a distance of about 1.8 mm, nearly 1,000 times its body length. The analytical accuracy required is thus several orders of magnitude less than that which would be required if it employed an instantaneous spatial sensing system.

Finally, we must consider how the bacterium uses the information derived from this time-dependent sensing process to migrate toward higher attractant concentrations. The mechanism, at least in *Escherichia coli* and related enteric bacteria, seems to be based on the frequency of tumbles: in other words, on the frequency which the flagellar motor rotates in a clockwise or counterclockwise direction. Bacteria swimming up an attractant gradient sense a positive temporal gradient, and tumble less frequently than normal. However, bacteria swimming down an attractant gradient, and therefore sensing a negative temporal gradient, tumble more frequently than normal. Since each tumble causes a random change of the direction of movement, the net result is that cells placed in a gradient spend more time swimming up the gradient that down it: hence, their characteristic migration to the region of high attractant concentration.

• **The phototactic behavior of purple bacteria**

Motile purple bacteria can respond to a *gradient of light intensity*, a phenomenon known as *phototaxis*. This behavior can be readily demonstrated by projecting a narrow spot of bright light onto an otherwise weakly illuminated suspension of motile purple bacteria, in which the cells are evenly dispersed and moving in a random fashion. Within 10 to 30 minutes most of the population accumulates in the bright spot, which acts as a "light trap". The mechanism of this accumulation is shown in Figure 11.37. Swimming cells enter the light spot by random movement. Once within it, they are prevented from leaving again by a *shock movement* (i.e., an abrupt change in the direction of swimming), which occurs every time they penetrate the sharp gradient of light intensity that separates the brightly illuminated spot from the surrounding dim area.

If a wet mount of motile purple bacteria is illuminated not with white light but with a spectrum produced by focusing light that has passed through a prism on the preparation, the bacteria rapidly accumulate in a series of bands corresponding to the principal absorption bands of their photosynthetic pigment system

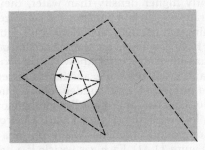

FIGURE 11.37

A diagrammatic illustration of the operation of a "light trap." The dashed line indicates the trajectory of a purple bacterium in a darkened field that contains a single illuminated area (white circle).

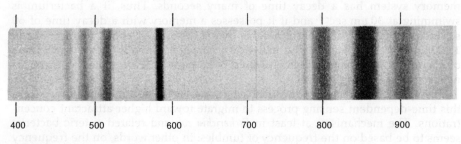

FIGURE 11.38

The pattern of phototactic accumulation of motile purple bacteria in a wet mount, which has been exposed to illumination in a spectrum. The cells accumulate massively at wavelengths corresponding to the absorption bands of their chlorophyll and carotenoids, which are both photosynthetically effective. The relatively weak accumulation around 500 nm corresponds to the positions of carotenoid absorption bands; the accumulations at 590, 800, 850, and 900 nm correspond to the positions of chlorophyll absorption bands. After J. Buder, "Zur Biologie des Bakteriopurpurins und der Purpurbakterien," *Jahrb. wiss. Botan.* **58**, 525 (1919).

(Figure 11.38). A careful quantitative study of the relative effectiveness of different wavelengths in mediating phototaxis has shown that the action spectrum for phototaxis by purple bacteria corresponds exactly to the action spectrum for the performance of photosynthesis.

SPECIAL PROCARYOTIC ORGANELLES

As already mentioned, most procaryotes do not form intracellular organelles bounded by unit membranes; the only possible exceptions are the thylacoids which house the photosynthetic apparatus of blue-green bacteria. However, three classes of procaryotic organelles are bounded by *nonunit membranes*, probably made up of protein in all cases. These are *gas vesicles*, *chlorobium vesicles*, and *carboxysomes* (*polyhedral bodies*).

• Gas vesicles and gas vacuoles

Since cells have a slightly higher density than water, they tend to sink in an aqueous medium at a rate which is a function of cell size. This effect can be counteracted by swimming against the gravitational pull. However, many aquatic procaryotes (Table 11.4) have developed another device to counteract gravitational pull: their cells contain gas-filled structures known as *gas vacuoles*. By light microscopy, gas vacuoles appear densely refractile, and have an irregular contour (Figure 11.39). If such cells are subjected to a sudden, sharp increase in hydrostatic pressure, the gas vacuoles collapse, and the cells simultaneously lose their buoyancy and become much less refractile (Figure 11.40). Electron microscopy shows that gas vacuoles are compound organelles made up of a variable number of individual gas vesicles (Figure 11.41). Each gas vesicle is a hollow cylinder, 75 nm in diameter with conical ends, and from 200 to 1000 nm in length. The vesicle

TABLE 11.4

Occurrence of gas vacuoles among procaryotes from aquatic habitats

A. Phototrophs
   Purple sulfur bacteria: *Lamprocystis, Thiodictyon,*
      Thiopedia, Amoebobacter
   Green bacteria: *Pelodictyon, Clathrochloris*
   Blue-green bacteria: *Microcystis, Spirulina* (some spp.),
      Oscillatoria (some spp.), *Pseudoanabaena, Anabaena*
         (some spp.), *Gloeotrichia* (some spp.), *Tolypothrix* (some spp.)

B. Chemotrophs
   Halobacterium
   Microcyclus
   Prosthecomicrobium
   Ancalomicrobium
   Pelonema
   Thiothrix

is bounded by a layer of protein 2 nm thick and is banded by regular rows of subunits ("ribs") running at right angles to its long axis (Figure 11.42).

Since the vesicles maintain a gas-filled space within the cell, it was originally supposed that the enclosing membrane is impermeable to gases. However, A. E. Walsby has shown that the membrane is freely permeable to all common gases (Figure 11.43). Therefore, the vesicles can neither store nor accumulate gas; the composition and pressure of the gas in the vesicle are functions of the dissolved

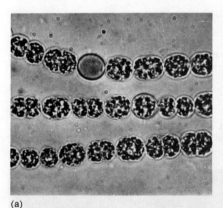
(a)

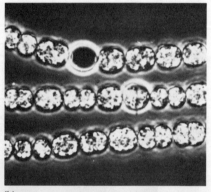
(b)

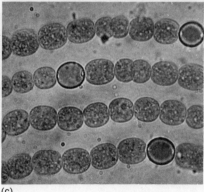
(c)

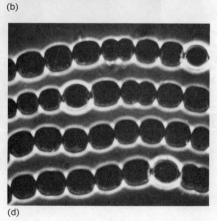
(d)

FIGURE 11.39

Filaments of a blue-green bacterium containing gas vacuoles, as visualized by bright field (a) and phase-contrast (b) illumination. Filaments from the same culture, after collapse by pressure of the gas vacuoles, are visualized by bright field (c) and phase contrast (d) illumination. The clear cells are heterocysts, which never contain gas vacuoles. From A. E. Walsby, "Structure and Function of Gas Vacuoles," *Bact. Revs.* **36**, 1 (1972).

(a)

(b)

FIGURE 11.40

Effect of sudden hydrostatic pressure on the turbidity and buoyancy of a suspension of gas vacuolate blue-green bacteria. Pressure applied to bottle at left; bottle at right untreated. (a) Appearance of suspensions immediately after pressure application; (b) appearance of suspensions several hours later. From A. E. Walsby, "Structure and Function of Gas Vacuoles," *Bact. Revs.* **36,** 1 (1972).

FIGURE 11.41 (left)

Electron micrograph of a thin section of *Oscillatoria,* showing the intracellular arrangement of the cylindrical gas vesicles which compose gas vacuoles (×25,800). Courtesy of Germaine Cohen-Bazire.

FIGURE 11.42 (right)

Electron micrograph of purified gas vesicles from *Oscillatoria,* negatively stained with uranyl acetate (×103,000). The vesicles are still inflated. Note the regular, banded fine structure of the vesicle wall. Courtesy of Germaine Cohen-Bazire.

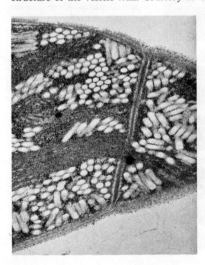

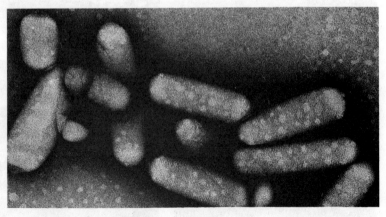

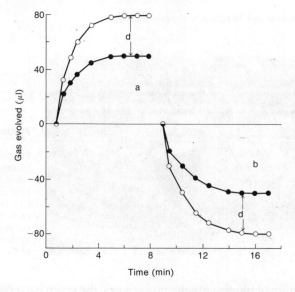

FIGURE 11.43

Top section, a, shows release of gas from two suspensions of a blue-green bacterium, following a sudden reduction of the pressure of the overlying gas phase. The two suspensions were identical, except that one (o) contained gas vacuoles and in the other (●) the gas vacuoles have been collapsed by application of pressure prior to the experiment shown. Lower section, b, shows uptake of gas by the same suspensions after a sudden return of the overlying gas phase to atmospheric pressure. The difference, d, corresponds to the volume of gas enclosed within the vacuoles. Redrawn from A. E. Walsby, "Structure and Function of Gas Vacuoles," *Bact. Revs.* **36**, 1 (1972).

gases in the surrounding medium. Water is excluded from the interior of the vesicles in the course of their formation and growth. This conclusion is confirmed by the observation that after pressure collapse, the vesicles do not recover; the cell can reacquire gas-filled vesicles only by *de novo* synthesis of these structures.

Two factors are probably important in maintaining a gas phase within vesicles. One is the structural rigidity of the enclosing protein membrane, which must be able to resist the various forms of pressure that normally act upon it (Figure 11.44). The other is the chemical character of the membrane protein. Its outer surface is clearly hydrophilic, being readily wettable, but the fact that liquid water cannot enter and accumulate within vesicles implies that the inner wall must be strongly hydrophobic.

As shown in Table 11.4, gas vacuoles occur in procaryotes that are markedly diverse in metabolic and physiological respects. The only common denominator of all these organisms is ecological: they occur in aquatic habitats. There can be little doubt, accordingly, that the function of gas vacuoles is to enable their possessors to regulate the buoyancy of the cell in order to occupy a position in the water column that is optimal for their metabolic activity with respect to light intensity, dissolved oxygen concentration, or the concentration of other nutrients.

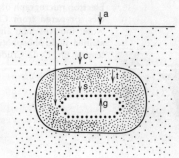

FIGURE 11.44

A diagrammatic representation of the sources of pressure which act upon the wall of a gas vesicle within the cell: a, atmospheric pressure; h, hydrostatic pressure; t, turgor pressure; c, surface tension pressure at cell wall; s, surface tension pressure at wall of vesicle; g, pressure of gas in vacuole.

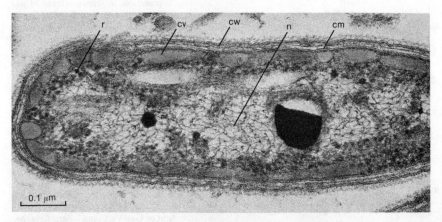

FIGURE 11.45

Electron micrograph of a thin section of the green bacterium *Pelodictyon,* showing the relationship of the chlorobium vesicles (cv) to other parts of the cell (×81,800): cw, cell wall; cm, cell membrane; r, ribosomes; n, nucleoplasm. Photo courtesy of Germaine Cohen-Bazire.

• **Chlorobium vesicles**

In one group of photosynthetic procaryotes, the green bacteria, the photosynthetic apparatus has a distinctive intracellular location: it is housed in a series of cigar-shaped vesicles, arranged in a cortical layer that immediately underlies, but it is physically distinct from, the cell membrane (Figure 11.45). These structures, detectable only by electron microscopy, are 50 nm wide and 100 to 150 nm long, being enclosed by a single-layered membrane 3 to 5 nm thick (Figure 11.46). The photosynthetic pigments are entirely contained within them. In isolated, broken vesicles, small subunits, about 10 nm wide with a central hole can be observed; these are probably the sites of the photosynthetic apparatus. The green bacteria are, accordingly, unique among photosynthetic organisms by virtue of the fact that the photosynthetic apparatus is not integrated into a unit membrane system.

• **Carboxysomes (polyhedral bodies)**

A number of photosynthetic bacteria (blue-green bacteria, certain purple bacteria) and chemolithotrophic bacteria (nitrifying bacteria, thiobacilli) contain structures termed *polyhedral bodies,* 50 to 500 nm wide, with polygonal profiles, which are surrounded by a monolayer membrane about 3.5 nm wide and which have a granular content (Figure 11.47). These structures have been recently isolated

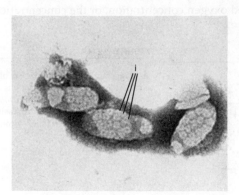

FIGURE 11.46

Electron micrograph of isolated chlorobium vesicles, prepared from *Chlorobium thiosulfatophilum.* Note intravesicular inclusions (i). Negatively stained with phosphotungstate, ×68,000. Courtesy of Dr. D. L. Cruden.

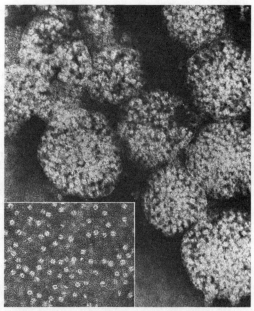

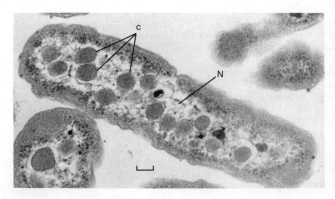

FIGURE 11.47 (left)

Electron micrograph of a section of *Thiobacillus*, containing numerous carboxysomes (c); N designates nucleoplasm. Bar indicates 0.1 μM. From J. M. Shiveley, F. L. Ball, and B. W. Kline, "Electron Microscopy of the Carboxysomes (Polyhedral Bodies) of *Thiobacillus neapolitanus*," *J. Bact.* **116**, 1405 (1973).

FIGURE 11.48 (right)

Electron micrographs of purified carboxysomes from *Thiobacillus*, and (inset) of the enzyme carboxydismutase, isolated from them. Negatively stained, × 108,000. From J. M. Shiveley, F. L. Ball, D. H. Brown, and R. E. Saunders, "Functional Organelles in Prokaryotes: Polyhedral Inclusions (Carboxysomes) of *Thiobacillus neapolitanus*," *Science* **182**, 584 (1973).

(Figure 11.48); they contain most of the cellular content of ribulose diphosphate carboxylase (carboxydismutase), the key enzyme in the fixation of CO_2 associated with the operation of the Calvin-Benson cycle. They have been termed *carboxysomes*, and evidently represent the site of CO_2 fixation in these autotrophic procaryotes.

THE PROCARYOTIC CELLULAR RESERVE MATERIALS

A variety of cellular reserve materials may occur in procaryotic organisms; they are frequently detectable as granular cytoplasmic inclusions.

• **Nonnitrogenous organic reserve materials**

Two chemically different kinds of nonnitrogenous organic reserve materials, each of which can provide an intracellular store of carbon or energy, are widespread among procaryotic organisms (Table 11.5). They are glucose-containing polysaccharides (α-1,4-glucans) such as starch and glycogen; and a polyester of β-

TABLE 11.5

The distribution of nonnitrogenous organic reserve
materials among procaryotes[a]

A. Glycogen
   Blue-green bacteria (most representatives)
   Enteric bacteria (most genera, except those listed under B)
   Sporeformers: many *Bacillus* and *Clostridium* species

B. Poly-β-hydroxybutyrate
   Enteric bacteria: genera *Beneckea* and *Photobacterium*
   Pseudomonas (many species)
   Azotobacter group (*Azotobacter, Beijerinckia, Derxia*)
   Rhizobium
   Moraxella (some species)
   Spirillum
   Sphaerotilus
   Bacillus (some species)

C. Both glycogen and poly-β-hydroxybutyrate
   Blue-green bacteria (a few species)
   Purple bacteria

D. No demonstrable reserve material
   Green bacteria
   Pseudomonas (many species)
   Acinetobacter

[a] This list is partial and includes only groups in which the nature of
reserve materials has been systematically investigated.

hydroxybutyric acid, poly-β-hydroxybutyric acid. The former class of substances
also occur as reserve materials in many eucaryotic organisms. Poly-β-hydroxy-
butyric acid is, however, uniquely found in procaryotic groups. Procaryotic organisms do not store neutral fats, which commonly occur as reserve materials in
eucaryotic organisms: poly-β-hydroxybutyric acid could, accordingly, be considered the procaryotic equivalent of this type of storage material.

As a general rule, only one kind of reserve material is formed by a given species.
Thus, many bacteria of the enteric group and anaerobic sporeformers (*Clostridium*)
synthesize only glycogen or starch as a reserve material, whereas many *Pseudomonas, Azotobacter, Spirillum,* and *Bacillus* species synthesize only poly-β-hydroxy-
butyrate. Certain bacteria can, however, synthesize both types of reserve material;
this is characteristic of the purple bacteria. Finally, it should be noted that a few
bacteria (e.g., the fluorescent species of the genus *Pseudomonas*) do not synthesize
any specific nonnitrogenous organic reserve material.

Bacterial polysaccharide reserves are usually deposited more or less evenly
throughout the cytoplasm, in areas detectable with the electron microscope but
not visible with the light microscope. The presence of large amounts of such
reserves in cells can be revealed by treatment with a solution of iodine in potassium iodide, which stains unbranched polyglucoses such as starch dark blue and
branched polyglucoses such as glycogen reddish brown. Deposits of poly-β-
hydroxybutyrate, however, are readily visible with the light microscope, occurring
as refractile granules of variable size scattered through the cell. They are specifically stainable with Sudan black, a property also shown, in other groups, by

353 neutral fat deposits. For this reason, bacterial poly-β-hydroxybutyrate granules have sometimes been incorrectly identified as fat reserves.

As a general rule, the cellular content of these reserve materials is relatively low in actively growing cells: they accumulate massively when cells are limited in nitrogen but still have a carbon energy source available. Under such circumstances, nucleic acid and protein synthesis are impeded and much of the assimilated carbon is converted to reserve materials, which may accumulate until they represent as much as 50 percent of the cellular dry weight. If such cells are then deprived of an external carbon source and furnished with an appropriate nitrogen source (e.g., NH_4Cl), the reserve materials can be used for the synthesis of nucleic acid and protein (Figure 11.49). Essentially, the synthesis of polyglucoses or poly-β-hydroxybutyrate represents a device for accumulating a carbon store in a form that is *osmotically inert*. In the case of poly-β-hydroxybutyrate, polymer synthesis also represents a method of *neutralizing an acidic metabolite*, since the free carboxyl group of β-hydroxybutyric acid is eliminated through the formation of the ester bonds between the subunits of the polymer. The cell can thus accommodate a very large store of such materials, whereas an equivalent intracellular accumulation of free glucose or β-hydroxybutyric acid could have catastrophic physiological consequences.

The accumulation and subsequent reutilization of carbon reserves is mediated by special enzymatic machinery, under close regulatory control. Poly-β-hydroxybutyrate is formed through a side path on the metabolic route of fatty acid synthesis (Figure 11.50). The native polymer granules into which it is incorporated have associated with them a complex system for degradation of the polymer. The polymer in native granules cannot be attacked until the granules have been "activated" by an enzyme that requires Ca^{2+} ions. This enzyme may be proteolytic, since its effect can be mimicked by trypsin. The activated granules become subject to the action of a depolymerase, which hydrolyzes the polymer to a dimeric ester;

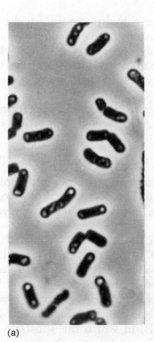 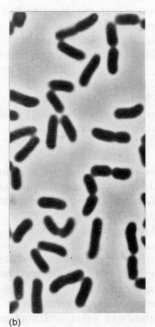

(a) (b)

FIGURE 11.49

The formation and utilization of poly-β-hydroxybutyric acid in *Bacillus megaterium* (phase contrast, ×2,200). (a) Cells grown with a high concentration of glucose and acetate. All cells contain one or more granules of poly-β-hydroxybutyric acid (light areas). (b) Cells from the same culture after incubation for 24 hours with a nitrogen source, in the absence of an external carbon source. Almost all the polymer granules have disappeared. Courtesy of J. F. Wilkinson.

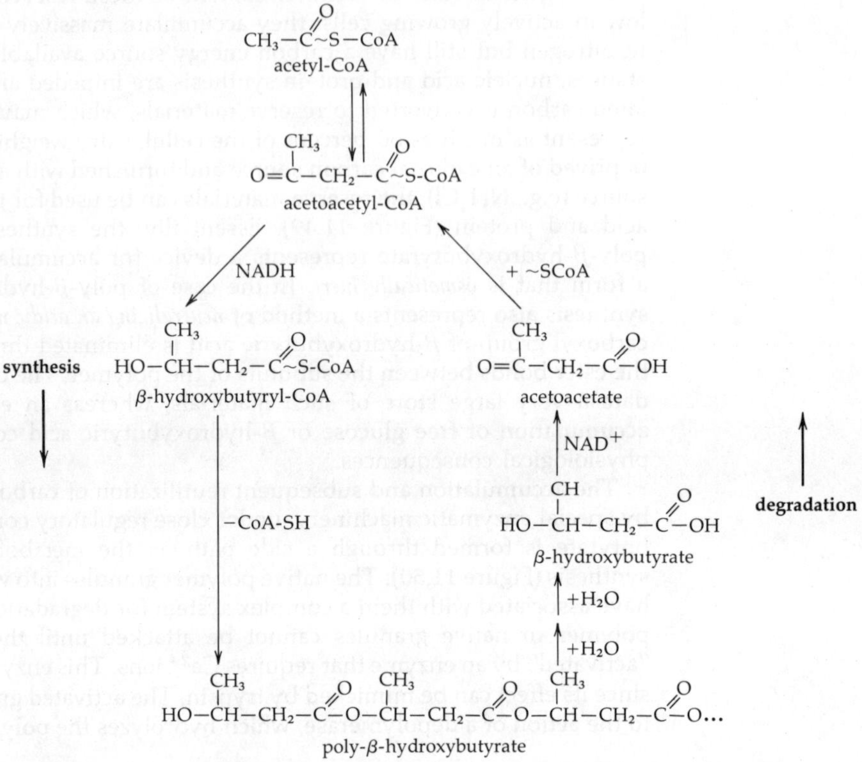

$$CH_3-\overset{\displaystyle O}{\overset{\displaystyle \|}{C}}\sim S-CoA$$

acetyl-CoA

$$O=\overset{\displaystyle CH_3}{\overset{\displaystyle |}{C}}-CH_2-\overset{\displaystyle O}{\overset{\displaystyle \|}{C}}\sim S-CoA$$

acetoacetyl-CoA

NADH   + ∼SCoA

synthesis

$$HO-\overset{\displaystyle CH_3}{\overset{\displaystyle |}{C}}H-CH_2-\overset{\displaystyle O}{\overset{\displaystyle \|}{C}}\sim S-CoA$$

β-hydroxybutyryl-CoA

$$O=\overset{\displaystyle CH_3}{\overset{\displaystyle |}{C}}-CH_2-\overset{\displaystyle O}{\overset{\displaystyle \|}{C}}-OH$$

acetoacetate

NAD⁺

$$HO-\overset{\displaystyle CH_3}{\overset{\displaystyle |}{C}}H-CH_2-\overset{\displaystyle O}{\overset{\displaystyle \|}{C}}-OH$$

β-hydroxybutyrate

degradation

−CoA-SH

+H₂O

+H₂O

FIGURE 11.50

The reactions involved in the synthesis and degradation of poly-β-hydroxybutyrate.

$$HO-\overset{\displaystyle CH_3}{\overset{\displaystyle |}{C}}H-CH_2-\overset{\displaystyle O}{\overset{\displaystyle \|}{C}}-O-\overset{\displaystyle CH_3}{\overset{\displaystyle |}{C}}H-CH_2-\overset{\displaystyle O}{\overset{\displaystyle \|}{C}}-O-\overset{\displaystyle CH_3}{\overset{\displaystyle |}{C}}H-CH_2-\overset{\displaystyle O}{\overset{\displaystyle \|}{C}}-O...$$

poly-β-hydroxybutyrate

the ester is then converted to free β-hydroxybutyric acid by a specific dimerase. A remarkable feature of this system is that *the depolymerase cannot hydrolyze the chemically purified polymer;* the only substrate that it can attack is the activated polymer granule. Even relatively mild treatments of the native polymer granules (e.g., freezing and thawing) may render them unutilizable by the intracellular polymer-degrading system.

The bacterial synthesis of glycogen is initiated by the formation of ADP-glucose from glucose-1-phosphate and ATP, through the action of the enzyme ADP-glucose pyrophosphorylase:

$$ATP + G\text{-}1\text{-}P \rightleftharpoons ADP\text{-glucose} + \textcircled{P} - \textcircled{P},$$

followed by the transfer of a glucosyl unit to an acceptor molecule of α-1,4-glucan, mediated by glycogen synthetase. The degradation of glycogen leads to the formation of glucose-1-phosphate, being mediated by glycogen phosphorylase (Figure 11.51).

The synthesis of glycogen in procaryotes is regulated at the level of ADP-glucose pyrophosphorylase, an allosteric enzyme that is inhibited by AMP, ADP, or inorganic phosphate, and activated by intermediates in carbohydrate dissimilation such as pyruvate, fructose-6-P, and fructose-1,6-diP. With respect alike to its mechanism and to its site of regulation, glycogen synthesis in procaryotes resembles starch synthesis in algae and higher plants. It differs from glycogen synthesis in yeast and mammals, in which the substrate for the synthetase is

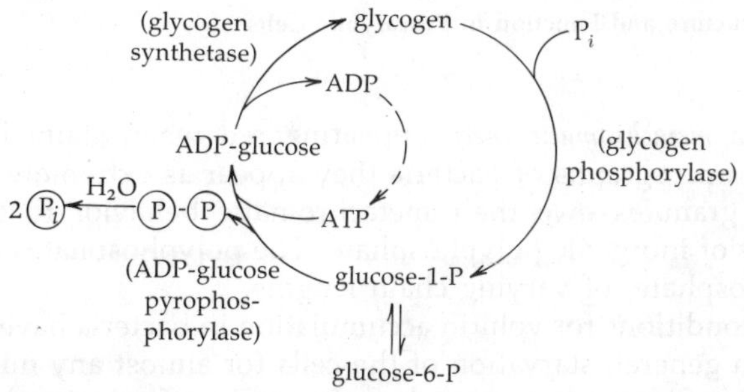

$$
\text{glucose-6-P}
$$

(diagram of the glycogen synthesis and degradation cycle: glycogen synthetase converts ADP-glucose to glycogen, producing ADP; glycogen + P$_i$ → glucose-1-P via glycogen phosphorylase; ADP-glucose pyrophosphorylase with 2 P$_i$, H$_2$O, P–P and ATP; glucose-1-P ⇌ glucose-6-P)

FIGURE 11.51

The reactions involved in the bacterial synthesis and degradation of glycogen.

UDP-glucose, and allosteric regulation occurs at the level of glycogen synthetase, not ADP-glucose pyrophosphorylase.

• **Nitrogenous reserve materials**

As a rule, procaryotes do not produce intracellular nitrogenous organic reserve materials. However, many of the blue-green bacteria accumulate a nitrogenous reserve material known as *cyanophycin* when cultures approach the stationary phase. Cyanophycin granules, which have a distinctive structured appearance in electron micrographs (Figure 11.52), have recently been isolated and characterized as a copolymer of arginine and aspartic acid. This material, which can represent as much as 8 percent of the cellular dry weight, is rapidly degraded when growth is reinitiated. The formation of cyanophycin does not occur through the normal mechanism of protein synthesis, since it accumulates in cells when protein synthesis has been arrested by treatment with chloramphenicol.

• **Volutin granules**

Many microorganisms, both procaryotic and eucaryotic, may accumulate *volutin granules*, which are stainable with basic dyes such as methylene blue (Figure 11.53). These bodies are also sometimes termed *metachromatic granules*, because they

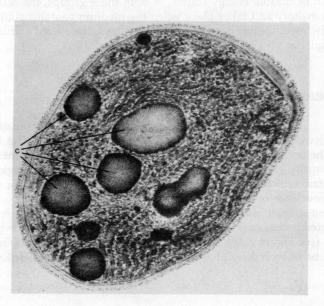

FIGURE 11.52

Electron micrograph of a section of a unicellular blue-green bacterium, containing cyanophycin granules (c), (28,800) Courtesy of Dr. Mary Mennes Allen.

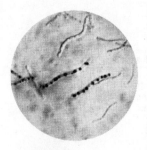

FIGURE 11.53

Volutin ("metachromatic granules") in the cells of a *Spirillum,* demonstrated by staining with methylene blue (×850). From George Giesberger, *Beitrage zur Kenntnis der Gattung Spirillum Ehbg* (1936), p. 46.

exhibit a *metachromatic effect,* appearing red when stained with a blue dye. In electron micrographs of bacteria they appear as extremely electrondense bodies. Volutin granules owe their metachromatic behavior to the presence of large amounts of inorganic polyphosphate. The polyphosphates are linear polymers of orthophosphate, of varying chain lengths.

The conditions for volutin accumulation in bacteria have been studied in some detail. In general, starvation of the cells for almost any nutrient leads to volutin formation. Sulfate starvation is particularly effective and leads to a rapid and massive accumulation of polyphosphate. When cells that have built up a polyphosphate store are again furnished with sulfate, the polyphosphate rapidly disappears, and tracer experiments with ^{32}P show that the phosphate is incorporated into nucleic acids. The volutin granules therefore appear to function primarily as an intracellular phosphate reserve, formed under a variety of conditions when nucleic acid synthesis is impeded. The formation of polyphosphate occurs by the sequential addition of phosphate residues to pyrophosphate, ATP serving as the donor:

$$\text{P—P} + \text{ATP} \longrightarrow \text{P—P—P} + \text{ADP}$$

$$(\text{—P—})_n + \text{ATP} \longrightarrow (\text{—P—})_{n+1} + \text{ADP}$$

The degradation of polyphosphate, if it occurs by the reversal of this reaction, might also provide a source of ATP for the cell, although this function has not so far been firmly established.

• **Sulfur inclusions**

Inclusions of inorganic sulfur may occur in two physiological groups: the purple sulfur bacteria, which use H_2S as a photosynthetic electron donor, and the filamentous, nonphotosynthetic organisms, such as *Beggiatoa* and *Thiothrix,* which use H_2S as an oxidizable energy source. In both these groups, the accumulation of sulfur is transitory and takes place when the medium contains sulfide; after the sulfide in the medium has been completely utilized, the stored sulfur is further oxidized to sulfate.

THE NUCLEUS

• **Recognition and cytological demonstration of bacterial nuclei**

Basic dyes, which selectively stain the chromatin of the eucaryotic nucleus, stain most bacterial cells densely and evenly. The basophilic property of the bacterial cell is caused by the abundance of ribosomes, which confers an unusually high nucleic acid content on the cytoplasmic region. Hence, in order to stain selectively the bacterial nucleus, the fixed cells must first be treated with ribonuclease or with dilute HCl, which hydrolyzes the ribosomal RNA. Subsequent staining with a basic dye then reveals the bacterial nuclei as dense, centrally located bodies of irregular outline; 2 to 4 are present in an exponentially growing cell (see Figure 3.18). The growth and division of bacterial nuclei in living cells can be observed by phase-contrast microscopy, provided that the cells are suspended in a medium (e.g., a concentrated protein solution), which enhances the slight difference in contrast between nucleoplasm and cytoplasm (see Figure 3.17).

The study of bacterial nuclear structure by electron microscopy was initially impeded by the difficulty of obtaining good fixation of the nuclear material. Once this problem had been solved, the nucleus was revealed in electron micrographs (Figure 11.54) as a region closely packed with fine fibrils of DNA. This region is not separated from the cytoplasm by a membrane, and contains no evident structures, apart from the DNA fibrils.

• **The bacterial chromosome**

By 1960 the cytological information about the structure of the bacterial nucleus had been complemented by genetic studies of *E. coli,* which suggested the presence of a single, circular linkage group. This in turn implied that each nucleus should contain a single, circular chromosome. If such were indeed the case, the fibrils of DNA revealed in the nuclear region by electron microscopy should represent sections of an extremely long circular molecule of DNA, highly folded to form a compact mass.

In 1963 J. Cairns succeeded in extracting DNA from *E. coli* under conditions that minimize its shearing. Cells were grown with ^{3}H-labeled thymine, so that only their DNA would be radioactive, and were placed in a chamber sealed at one end with glass and at the other end with a membrane filter. The cells were then gently lysed by dialysis against a solution of a detergent such as sodium lauryl sulfate or against a solution of lysozyme and ethylenediaminetetraacetic acid (EDTA). Further treatments were then carried out to dissociate any proteins from the DNA; all the treatments were done by allowing the reagents to *diffuse* into and out of the chamber through the membrane filter, so that the material was never subjected to mechanical agitation. Finally, the membrane was punctured, the chamber was drained, and the membrane was mounted on a slide for radioautography.

During the final draining of the chamber, the DNA of individual bacterial cells was spread out on the surface of the membrane filter. Examination of the developed radioautographs showed that the DNA was present as extremely long threads, the longest of which were slightly more than 1 mm in length. Furthermore, a few of the threads were *circular* (Figure 11.55). These threads are contained within cells which have an average length of approximately 2 μm.

The length of 1 mm for the DNA thread agrees well with the amount of DNA per nucleus as determined chemically, assuming that the radioactive structure in Cairn's pictures is an extended double helix. This amount of DNA represents approximately 5×10^6 base pairs, with a molecular weight of about 3×10^9.

DNA is a highly charged molecule, since adjacent bases are linked by phosphate groups, each with an ionized hydroxyl group. The resulting negative charges must therefore be balanced by an equivalent number of cationic groups. In eucaryotes this occurs through association of the chromosomal DNA with basic proteins (histones). Such proteins do not, however, occur in bacterial cells. Charge neutralization is effected by polyamines, such as spermine and spermidine (see Chapter 9), and by Mg^{2+}.

Cairn's autoradiographic demonstration of the circular structure of the bacterial chromosome also revealed its mode of replication (see Figure 11.55). Replication begins at one point on the circumference of the chromosome, and the replication fork then move along the chromosome until the entire structure has been dupli-

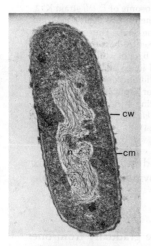

FIGURE 11.54

Thin section of a dividing cell of a unicellular procaryotic organism, *Bacillus subtilis* ($\times$ 20,000): n, nucleus; cm, cytoplasmic membrane; cw, cell wall. Courtesy of C. F. Robinow.

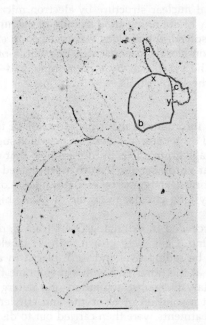

FIGURE 11.55

Autoradiograph of the chromosome of *E. coli* strain K12, labeled with tritiated thymidine for two generations and extracted as described in the text. The scale at the bottom represents 100 μm. Inset: A diagram of the same structure, showing regions (a, b, c) in which both strands contain tritium (double solid lines) and in which only one strand is labeled (one solid and one dashed line); x indicates the starting point for replication (the replicator); y indicates the replicating fork. From J. Cairns, "The Chromosome of *Escherichia coli*," *Cold Spring Harbor Symp. Quant. Biol.* **28,** 43 (1964).

cated. The process involves an initial breakage of one strand in the double helix at the initiation site, followed by rotation and progressive unwinding of the double helix, as shown schematically in Figure 11.56.

• **The replicon hypothesis**   During bacterial growth, dividing nuclei appear to separate gradually from one another, without undergoing a marked change in shape (see Figure 3.17). In eucaryotes, the separation of daughter chromosomes during mitosis is effected by their attachment to some microtubules of the mitotic apparatus, which draw

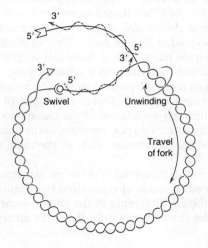

FIGURE 11.56

The replication of a circular double helix. One strand of the duplex is broken; the other then provides a swivel, permitting unwinding at the replication fork. Newly synthesized complementary strands are shown in dashed lines.

them toward the two poles of the spindle. However, no such apparatus exists in procaryotes. How, then, is separation of bacterial chromosomes achieved, following their replication?

On the basis of Cairns' observations, an explanatory hypothesis known as the *replicon hypothesis* was developed by F. Jacob, S. Brenner, and F. Cuzin in 1963 (see Figure 11.57). With minor modifications, this hypothesis still seems to provide the most satisfactory explanation of the events of bacterial chromosome replication and separation. Its essential points can be summarized as follows:

1. When chromosome replication begins, the chromosome is attached at a specific point on its circumference to a *replicator site* on the bacterial cell membrane. The enzymatic machinery responsible for DNA replication (see Chapter 7) is associated with this site.

2. The replication fork is located at the replicator site, and its passage along the chromosome involves movement of the chromosome past the attachment site.

3. Prior to the initiation of replication, a new replicator site adjacent to the old one is formed on the membrane, and the free end of the broken DNA strand is attached to it.

4. Separation of daughter chromosomes is effected by the localized synthesis of the membrane, in an annular region situated between the old and new attachment sites. Membrane growth consequently spreads the attachment sites farther and farther apart.

Several of the specific predictions of the replicon hypothesis have been subsequently confirmed. There is a considerable amount of electron microscopic evi-

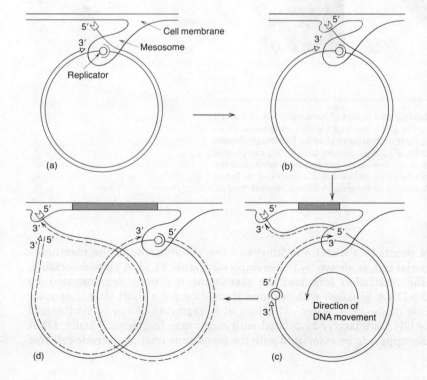

FIGURE 11.57

A schematic drawing illustrating the model of Jacob, Brenner, and Cuzin for the replication of the bacterial chromosome. (a) The chromosome is attached to the cell membrane at the replicator site, which serves as a swivel. One DNA strand is broken. (b) The 5' end of the broken strand attaches to a new site in the membrane. (c) The chromosome rotates counterclockwise past the attachment site, which carries the enzymatic machinery of replication. Newly synthesized DNA strands are shown as dashed lines. Separation of attachment sites by localized membrane synthesis (shaded area) has begun. (d) Replication has been completed.

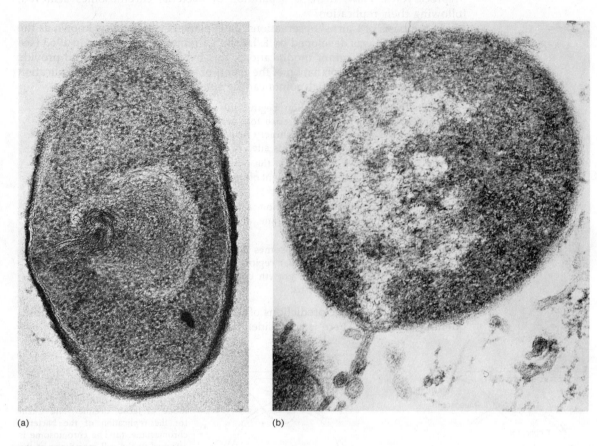

(a)   (b)

FIGURE 11.58

Electron micrographs of thin sections showing attachment of bacterial DNA to the cell membrane. (a) A *B. subtilis* cell, showing the DNA in contact with a mesosome. From A. Ryter and F. Jacob, "Membrane et Segregation Nucléaire chez les Bactéries," *Protides of the Biological Fluids* **15**, 267 (1967). Courtesy of Elsevier, Amsterdam. (b) A protoplast of *B. subtilis* prepared by lysozyme treatment. A mesosome has been extruded, and the DNA of the cell has been pulled up against the cell membrane at that site. From A. Ryter, "Association of the Nucleus and the Membrane of Bacteria: A Morphological Study," *Bacteriol. Rev.* **32**, 39 (1968).

dence that points to a direct attachment of the chromosome to the membrane in the bacterial cell, as shown by the examples in Figure 11.58. A close association between the replication fork and the membrane has been demonstrated by labeling the DNA of *E. coli* with tritiated thymidine for a short time, just prior to extraction of the cells. After extraction, the fragments of newly synthesized radioactive DNA are largely associated with membrane fragments. Finally, DNA polymerases appear to be associated with the membrane fraction in crude cell-free extracts.

361

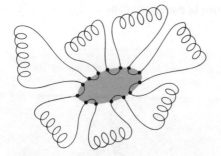

FIGURE 11.59

A schematic drawing illustrating the proposed structure of the folded chromosome of *E. coli*. The chromosome is shown as seven loops, each twisted into a superhelix (the actual number is much greater) and held together by a core of RNA (shaded area). After A. Worcel and E. Burgi, "On the Structure of the Folded Chromosome of *Escherichia coli*," *J. Mol. Biol.* **71**, 127 (1972).

The model of chromosome replication shown in Figure 11.57 does not take into account, however, the more recent finding that chromosome replication in *E. coli* is normally *bi-directional*, two replication forks moving simultaneously in opposite directions from a single initiation site (see Chapter 7). It is not yet clear how to represent this in a schematic drawing such as Figure 11.57, in relation to the fixed location of the replicating enzyme complex.

• **The isolation of bacterial nuclei**

If the bacterial chromosome consisted merely of a huge circular DNA molecule, folded in a random manner, its orderly replication and segregation could not possibly occur: some organization at a higher level must therefore be imposed on the chromosome. The recent isolation of structures that appear to be intact bacterial nuclei has provided a few clues to the nature of this organization. The structures in question have been obtained by gentle lysis of lysozyme-treated bacteria with nonionic detergents in 1.0 M NaCl. In addition to DNA, they contain a substantial amount of RNA and protein. They are rapidly sedimentable and are of low viscosity (in contrast to unfolded DNA of high molecular weight, which is extremely viscous).

Depending on the conditions of lysis, two structures of this type are obtainable. Lysis at 25°C releases structures with sedimentation coefficients of 1,300 to 2,200 S, which appear to be folded chromosomes, associated with a considerable amount of RNA and protein. The protein is largely RNA polymerase; and the RNA is newly transcribed single RNA strands. RNAse treatment causes a rapid increase in viscosity, which indicates that some of the associated RNA is responsible for holding the DNA into a compact form. The DNA in these structures is folded into a number (between 12 and 80) of supercoiled loops. In the light of these facts, the folded chromosome can be represented schematically as shown in Figure 11.59.

If lysis is conducted at a lower temperature (0 to 4°C), structures having considerably higher rates of sedimentation (3,000 to 4,000 S) are obtained. Electron microscopy (Figure 11.60) shows that they consist of folded chromosomes, attached to either one or two membrane fragments, from which they can be dissociated by gentle treatment. These observations show that binding of the chromosome to the membrane involves very weak forces. No folded chromosomes attached to membrane can be isolated from cells that have completed a round of DNA synthesis. Accordingly, it is possible that the resting chromosome may not be membrane-bound *in vivo*, membrane attachment taking place as a preliminary to the next round of replication.

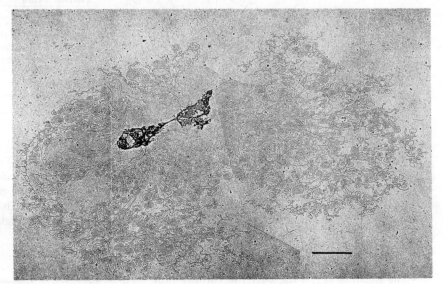

FIGURE 11.60

Electron micrograph of the isolated folded chromosome of *Escherichia coli*, attached to a fragment of the cell membrane (dark, irregular area in center of figure). The bar represents 2 μm. From H. Delius and A. Worcel, "Electron Microscopic Visualization of the Folded Chromosome of *Escherichia coli*," *J. Mol. Biol.* **82**, 107 (1974).

FURTHER READING

Books   LEIVE, L. (editor), *Bacterial Membranes and Walls.* New York: Marcel Dekker, 1973.

Reviews and original articles

ADLER, J., "Chemotaxis in Bacteria," *Science* **153**, 708 (1966).

CAIRNS, J., "The Bacterial Chromosome and Its Manner of Replication as Seen by Autoradiography," *J. Mol. Biol.* **6**, 208 (1963).

COHEN-BAZIRE, G., "The Photosynthetic Apparatus of Procaryotic Organisms," in *Biological Ultrastructure: The Origin of Cell Organelles,* P. Harris (editor). Corvallis: Oregon State University Press, 1971.

CRUDEN, D. L., and R. Y. STANIER, "The Characterization of Chlorobium Vesicles and Membranes from Green Bacteria," *Arch. Mikrobiol.* **72**, 115 (1970).

DAWES, E. A., and P. J. SENIOR, "The Role and Regulation of Energy Reserve Polymers in Microorganisms," *Advan. Microbiol. Physiol.* **10**, 136 (1973).

JACOB, F., S. BRENNER, and F. CUZIN, "On the Regulation of DNA Replication in Bacteria," *Cold Spring Harbor Symp. Quant. Biol.* **28**, 329 (1963).

KÉPÈS, A., and F. AUTISSIER, "Topology of Membrane Growth in Bacteria," *Biochem. Biophys. Acta* **265**, 443 (1972).

MACNAB, R. M., and D. E. KOSHLAND, "The Gradient-Sensing Mechanism in Bacterial Chemotaxis," *Proc. Natl. Acad. Sci. U.S.* **69**, 2509 (1972).

SCHLEIFER, K. H., and O. KANDLER, "Peptidoglycan Types of Bacterial Cell Walls and Their Taxonomic Implications," *Bact. Revs.* **36**, 407 (1972).

363 SHIVELEY, J. M., "Inclusion Bodies of Procaryotes," *Ann. Rev. Microbiol.* **28,** 167 (1974).

———, F. BALL, D. H. BROWN, and R. E. SAUNDERS, "Functional Organelles in Prokaryotes: Polyhedral Inclusions (Carboxysomes) of *Thiobacillus neapolitanus*," *Science* **182,** 584 (1973).

SIMON, R. D., "Cyanophycin Granules from the Blue-Green Alga *Anabaena cylindrica*," *Proc. Natl. Acad. Sci. U.S.* **68,** 265 (1971).

SUTHERLAND, I. W., "Bacterial Exopolysaccharides," *Advan. Microbiol. Physiol.* **8,** 143 (1972).

WALSBY, A. E., "Structure and Function of Gas Vacuoles," *Bact. Revs.* **36,** 1 (1972).

WORCEL. A., and E. BURGI, "On the Structure of the Folded Chromosome of *Escherichia coli*," *J. Mol. Biol.* **81,** 127 (1972).

12

THE VIRUSES

Long before the discovery of the microbial world, the term *virus* was used to denote any agent capable of producing disease. The word is a Latin one and originally meant "venom" or "poisonous fluid." Ideas concerning the causation of infectious disease were necessarily vague and abstract until the nineteenth century, when specific microbial agents of disease were first recognized. In the early days of microbiology, these microbial agents, whether bacteria, fungi, or protozoa, were often indiscriminately referred to as "viruses." The term is no longer used in this general sense.

THE DISCOVERY OF FILTERABLE VIRUSES

In 1892 D. J. Ivanowsky found that an infectious extract from tobacco plants with mosaic disease retained its infectivity after passage through a filter able to prevent the passage of bacteria. He assumed that the infectious agent was a small microorganism. During the following two or three decades, it was shown that many major diseases of both plants and animals are caused by similar infectious agents, so small that they cannot be seen with the light microscope. Since the basic criterion that served to differentiate these forms from the more familiar microbial agents of disease was their ability to pass through filters with pores sufficiently fine to retain even very small bacteria, they came to be known collectively as "filterable viruses." With the passage of time, the adjective "filterable" was gradually dropped, and the word *virus* became a *specific* group designation for these ultramicroscopic, filterpassing infectious agents.

Most of the scientists who studied viruses during the first decades of the twentieth century assumed that they were simply another class of microorganisms which differed in size but not in any really fundamental biological respect from the better-known kinds of microorganisms. Studies of their behavior in the laboratory led to the conclusion that they were *obligate intracellular parasites,* able to multiply only within the host cells. However, since obligate intracellular parasitism is also a character of some other microorganisms, this could not be considered a distinctive property of viruses.

The first indications that the viruses might be different *in nature* from cellular organisms had, however, been obtained not long after Ivanowski's discovery of the filterability of the tobacco mosaic virus. In the course of confirming Ivanowski's observations, M. W. Beijerinck discovered that the virus of tobacco mosaic disease could be precipitated from a suspension by alcohol without losing its infectious power and was capable of diffusing through an agar gel. These are properties never shown by a living organism, and Beijerinck accordingly concluded that the virus was not a living organism but rather a "fluid infectious principle." Nearly 40 years passed, however, before this brilliant intuition was confirmed. In 1935 W. M. Stanley showed that the infectious principle of the same virus could be crystallized and that the crystals consisted largely of protein. At first, this was taken to mean that the virus was a protein molecule. The first assumption proved oversimplified; a few years later, purified tobacco mosaic virus was found to contain, in addition to protein, a much smaller but constant amount of ribonucleic acid. The infectious principle of a virus is therefore not a protein molecule but a molecular complex, built up from two different kinds of macromolecule: protein and nucleic acid. The nucleic acid is specific to the virus and may be either DNA or RNA.

The first viruses to be described were agents of disease for higher plants and animals. About 1915 F. W. Twort and F. d'Herelle independently discovered that bacteria are susceptible to infection by ultramicroscopic, filter-passing agents, designated as *bacteriophages* (i.e., eaters of bacteria). This designation is often shortened to *phages.* Although d'Herelle very early stressed fundamental similarities between bacteriophages and plant and animal viruses, it was some time before the bacteriophages were universally recognized to be true viruses.

THE GENERAL PROPERTIES OF VIRUSES

A virus alternates in its life cycle between two phases, one extracellular and the other intracellular. In its *extracellular phase,* it exists as an inert, infectious particle, or *virion.* The virion consists of one or more molecules of nucleic acid, either DNA or RNA, contained within a protein coat, or *capsid;* this "nucleocapsid" may, as in the case of certain animal viruses, be enclosed within a membranous *envelope.*

In its *intracellular phase,* a virus exists in the form of replicating nucleic acid, either DNA or RNA. During the intracellular phase, the genetic material of the virus is not only replicated by the host cell but also serves as a genetic determinant for the synthesis by the cell of specific *viral proteins.* These proteins include the *subunits,* or *capsomers,* from which the capsid of the mature virus particle is assembled.

• **The nucleic acid of the mature virion** The mature virion contains a *single kind of nucleic acid,* although the type of nucleic acid present in the virion varies from one virus to another. Most viruses contain either double-stranded DNA or single-stranded RNA, but single-stranded DNA

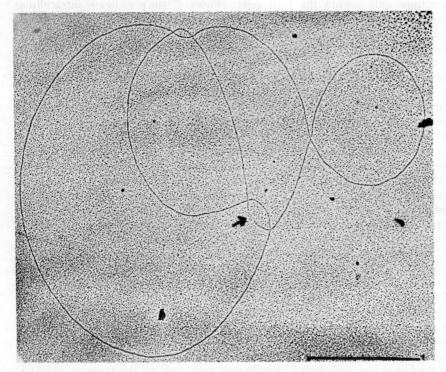

FIGURE 12-1

Lambda phage DNA. The scale marker represents 1 μm; the length of the DNA molecule is 16.3 μm. The arrow points to a region of discontinuity, believed to contain the "sticky ends" described in the text. From H. Ris and B. L. Chandler, "The Ultrastructure of Genetic Systems in Prokaryotes and Eukaryotes," *Cold Spring Harbor Symp. Quant. Biol.* **28**, 1 (1963).

is found in some viruses (e.g., coliphages fd and φX174, and minute virus of mice), and double-stranded RNA is found in the virions of reoviruses and of certain plant viruses.

When nucleic acid is extracted from virions under conditions that minimize shearing, it is generally found—with the exception of the large RNA viruses—that each virion contains a *single molecule of nucleic acid*. The length of the molecule is constant for a given virus but varies from a few thousand nucleotides (or nucleotide pairs) to as many as 250,000 nucleotides (or nucleotide pairs) in different viruses. If we take 1,000 nucleotides to be the size of an average gene (see Chapter 13), the smallest viruses contain fewer than 10 genes while the largest viruses contain several hundred.

The DNA of several bacterial and animal viruses is circular (Figure 12.1). In other DNA viruses, as in all RNA viruses, the nucleic acid of the virion is linear. In one case, that of bacteriophage λ, the DNA of the virion is linear but becomes circular immediately after it penetrates the host cell (see Figure 12.16). Circularity is also characteristic of bacterial chromosomes (e.g., *Escherichia coli*) and of plasmids: viruslike genetic elements found in bacteria (see Chapter 15).

The replication of the various types of viral nucleic acid will be discussed in a later section.

• The architecture of the nucleocapsid

Most virions fall into two classes with respect to the architecture of their nucleocapsids: they are either helical or polyhedral. These basic structures may be further

367 modified: in many bacteriophages a polyhedral head is joined to a helically constructed tail, while in many animal viruses the nucleocapsid is enclosed in a membranous envelope. One bacteriophage—a DNA phage whose host is a marine pseudomonad—has also been found to produce membrane-enveloped virions. The lipid-containing envelopes of the viruses which possess them are derived from the cell membranes of their hosts.

The most thoroughly studied helical virion is that of the tobacco mosaic virus (Figure 12.2). The single-stranded RNA molecule lies in a groove formed in the helically arranged capsid; the complete virion forms a rod-shaped structure containing about 2,000 identical capsomers. The virion of the tobacco mosaic virus is typical of many plant viruses; some of the animal viruses also form helical nucleocapsids, but these are always irregularly coiled within an envelope. Some examples are shown in Figure 12.3.

In the polyhedral virions the nucleic acid is packed in an unknown manner within a hollow, polyhedral head. In many of the polyhedral plant and animal virions the capsid is an icosahedron—a regular polyhedron with 12 corners, 20 triangular faces, and 30 edges.

The capsid of such a virion is composed of two types of capsomers: pentamers, which are situated at the corners, and hexamers, which fill the triangular faces (Figure 12.4). Each capsomer is a multimeric protein, the pentamer containing five identical subunits and the hexamer six. A subunit may consist of a single polypeptide chain; in one case, however, the subunit has been found to contain two different polypeptide chains.

The size of an icosahedral virus is determined by the number of capsomers it contains. This number follows the laws of crystallography: only certain numbers are possible in an icosahedral structure. For example, the smallest possible icosahedral capsid would have 12 pentamers and no hexamers, the next smallest would have 12 pentamers and 20 hexamers, and so on. The largest known icosahedral virion, that of an insect virus, contains 1,472 capsomers.

FIGURE 12.2

(a) A drawing of the structure of tobacco mosaic virus. For clarity, part of the ribonucleic acid chain is shown without its supporting framework of protein. From A. Klug, and D. C. D. Caspar, "The Structure of Small Viruses," *Adv. Virus Res.* **1**, 225 (1960). (b) Electron micrograph of tobacco mosaic virus particles in phosphotungstic acid. From S. Brenner and R. W. Horne, "A Negative Staining Method for High Resolution Electron Microscopy of Viruses," *Biochim. Biophys. Acta* **34**, 103 (1959). Courtesy of R. W. Horne.

0   100 A

(a)

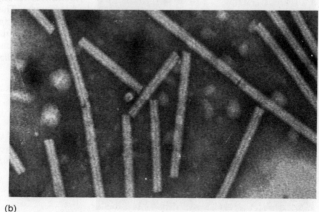

(b)

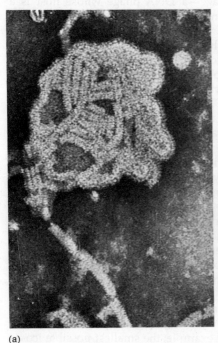

(a)

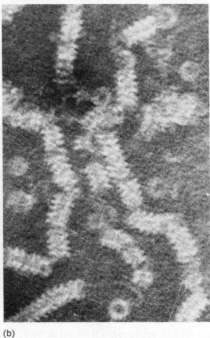

(b)

FIGURE 12.3

The nucleocapsid of enveloped viruses. (a) A partially disrupted particle of Newcastle disease virus, releasing the internal component (nucleocapsid). (b) High-magnification electron micrograph of the nucleocapsid released from particles of mumps virus. From R. W. Horne et al., "The Structure and Composition of Myxoviruses. I. Electron Microscope Studies of the Structure of Myxovirus Particles by Negative Staining Techniques," *Virology* **11**, 79 (1960).

• **The envelope of certain animal viruses**

As shown in Figure 12.5, some animal viruses are liberated from the host cell by an extrusion process which coats the nucleocapsid with a layer of cell membrane. Not all the materials of the envelope are derived from normal cell components, however. In many cases, the envelope contains proteins that are determined by viral genes. The finished envelope is often a complex, highly organized structure containing several layers (Figure 12.6).

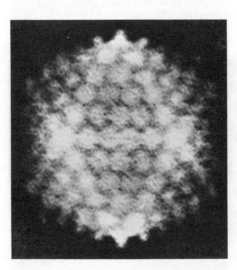

FIGURE 12.4

Adenovirus virion, showing the icosahedral array of capsomers. The capsomers at the vertices are surrounded by five nearest neighbors, all the others by six. From R. C. Valentine and H. G. Pereira, "Antigens and Structure of the Adenovirus," *J. Mol. Biol.* **13**, 13 (1965).

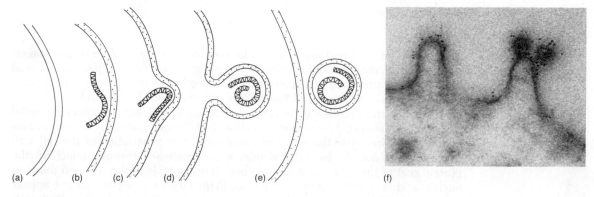

(a)   (b)   (c)   (d)   (e)   (f)

FIGURE 12.5

(a to e) Schematic representation of the liberation of an enveloped virus. The black dots represent new membrane proteins coded by viral genes. Modified from B. Davis et al., *Microbiology*. New York: Harper and Row, 1967. (f) Electron micrograph of a thin section through an animal cell that is liberating influenza virus particles. The material was treated with antibody against influenza virions; the antibody molecules were coupled to ferritin, which is extremely electron-dense due to bound iron atoms. Micrograph from C. Morgan et al., "The Application of Ferritin-Conjugated Antibody to Electron Microscopic Studies of Influenza Virus in Infected Cells," *J. Exptl. Med.* **114,** 825 (1961).

FIGURE 12.6

(a) A whole vaccinia particle, negatively stained with phosphotungstic acid. The ridges on the surface may be long rodlets or tubules. (b) A negatively stained vaccinia particle that has been centrifuged in a sucrose gradient. The particle has been partially disrupted, and it has lost its outer membrane. The remaining structure includes a biconcave inner core, containing the nucleic acid, two elliptical bodies, and a surrounding membrane. From S. Dales, "The Uptake and Development of Vaccinia Virus in Strain L Cells Followed with Labeled Viral Deoxyribonucleic Acid," *J. Cell Biol.* **18,** 51 (1963).

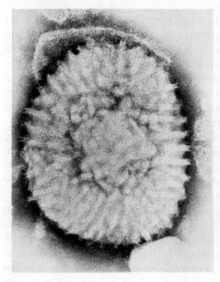
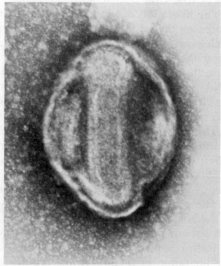

(a)   (b)

• **The reproduction of viruses**

The process of viral reproduction can be considered to take place in five stages: penetration of the host cell, synthesis of enzymes needed for viral nucleic acid replication, synthesis of viral constituents, assembly of the constituents to form mature virions, and release of the mature virions from the host cell.

The penetration process differs in bacterial, plant, and animal viruses. Bacterial and plant viruses must penetrate the cell wall of the host, whereas animal viruses may adsorb directly to the host cell membrane. The penetration of the cell wall by some, but not all, bacterial viruses is accomplished through injection: the protein coat of the virus remains adsorbed to the outside of the cell, and the viral nucleic acid is injected through the wall. Plant viruses, however, do not appear to be equipped with a specific device for penetration of the host cell wall: experimental infection requires mechanical injury of the plant cells. In nature, plant viruses are usually transmitted from one host plant to another by vectors such as insects, which inject the virus particles through the cell walls of the plant.

Animal viruses adsorb to the host cell membrane and may then be taken into the cell by phagocytosis. As a result of phagocytosis, the ingested virion is trapped within a membrane-bounded food vacuole (see Figure 29.4); it still must penetrate the cell membrane to reach the cytoplasm or nucleus, where it will be reproduced. In the case of a virus that possesses an envelope partially derived from the cell membrane of its previous host, penetration is facilitated by fusion of the viral envelope with the membrane of the new host cell (approximately the reverse of the process shown in Figure 12.5). In viruses lacking envelopes, including the phages, the manner in which the nucleocapsid penetrates the cell membrane is not known.

Only the viral nucleic acid of most bacterial viruses reaches the cytoplasm of the host cell. In plant and animal viruses, however, the entire nucleocapsid may reach the cytoplasm; consequently, the final step in the penetration process is the removal of the protein capsid, presumably by the action of proteolytic enzymes. Thus, penetration culminates with the appearance in the interior of the cell of free viral nucleic acid.

The liberation of viral nucleic acid within the host cell initiates two distinct processes: the synthesis of virus-specified proteins, and the replication of the viral nucleic acid itself. If the viral nucleic acid is RNA, it either serves directly as messenger RNA or is transcribed to form a messenger RNA strand with the opposite polarity. If the viral nucleic acid is DNA, it is first transcribed by a DNA-dependent RNA polymerase to form viral messenger RNA. In both cases, the viral messenger is translated by the host cell ribosomes to form the enzymes necessary for viral replication as well as the subunits of the viral capsid.

The viral nucleic acid serves as template for its own replication, complementary strands being synthesized by specific polymerases. The details of the mechanism differ, however, with the nature of the nucleic acid: DNA or RNA, single- or double-stranded, cyclic or linear. Some different modes of replication will be discussed in later sections of this chapter.

The processes of protein synthesis and replication lead to the accumulation in the host cell of numerous molecules of viral nucleic acid together with numerous subunits of the viral capsid. These combine with each other spontaneously to form complete nucleocapsids. The final stage in the reproductive process is the liberation of mature virions from the host cell. In animal cells this is often accomplished by extrusion through the cell membrane, whereas in bacteria it is usually accomplished by lysis of the host cell through the action of a viral enzyme.

• Infectious nucleic acid

We have seen that the host cell can synthesize complete virions from free viral nucleic acid liberated within the cytoplasm. Indeed, it is possible to infect host cells with free nucleic acid extracted from virions; the *efficiency* of this infection, however, is usually one-thousandth to one-millionth that of the intact virion. The capsid thus facilitates the penetration of the host cell by the virus by providing a mechanism for *adsorption* of the virion to the host cell surface and—in the case of some bacteriophages—by providing a mechanism for the injection of viral nucleic into the cell. The capsid may also serve to protect the viral nucleic acid from extracellular degradation.

The host-range specificity of a virus is determined largely by the specificity of its binding to receptor sites on the host cell. This binding involves specific adsorption sites on the surface of the viral capsid; consequently, the host range of free viral nucleic acid is much wider than that of the intact virion. In nature, however, the low efficiency of infection by free nucleic acid prevents the wider host range from being expressed.

• The differences between viruses and cellular organisms

From the foregoing account, it is clear the viruses represent a unique class of biological entities, different from cellular organisms. The differences between viruses and cells may be summarized as follows:

1. In its extracellular phase, the virion, a virus consists of a single kind of nucleic acid—DNA or RNA—combined with a protein capsid. The virion may include one or a few enzymes which play a role in the initiation of an infection, but the enzymatic complement is far from sufficient to reproduce another virion. In contrast, a cell always contains both DNA and RNA, along with an elaborate complement of enzymes capable of catalyzing all the reactions that are necessary for the reproduction of the cell.

2. A virus is reproduced by the assembly of nucleic acid replicas and capsid subunits, independently synthesized by the host cell. Growth of a cell, however, consists in the orderly increase in the amount of each of its constituent parts, during which the integrity of the whole is continuously maintained. Thus, the virion never arises directly from a preexisting virion, whereas the cell always arises directly from a preexisting cell. Cellular growth culminates in an increase in cell number by the process of cell division.

The nature of viral reproduction and the nature of cellular growth and division are thus completely different. At one time it was believed that certain infectious agents which were intermediate in size between typical cells and typical viruses also represented intermediate forms of life, but the criteria listed above now permit the unequivocal identification of these agents as cellular organisms in certain cases and as true viruses in others. Thus, the chlamydiae (the agents of psittacosis, lymphogranuloma venereum, and trachoma) are clearly cellular, whereas the agents of vaccinia and of smallpox—which are equally large—are clearly viral.

• The classification of viruses

The most widely used system for the classification of viruses is shown in Table 12.1. According to this system, introduced by A. Lwoff and his colleagues in 1962, viruses are grouped according to the properties of their virions: type of nucleic

TABLE 12-1

System for the classification of viruses

NUCLEIC ACID	CAPSID SYMMETRY	NAKED OR ENVELOPED	SIZE OF CAPSID, Å[a]	NUMBER OF CAPSOMERS	SPECIAL FEATURES	EXAMPLES BACTERIAL	ANIMAL	PLANT
RNA	Helical	Naked	175 × 3,000					Tobacco mosaic virus
		Enveloped	90				Myxoviruses	
			180				Paramyxoviruses	
	Polyhedral	Naked	200–250	32		Coliphage f2	Picornaviruses	Bushy stunt virus
			280	92	Double-stranded RNA		Reoviruses	
			700					
DNA	Helical	Naked	50 × 8,000		Single-stranded DNA	Coliphage fd		
	Polyhedral	Enveloped	90–100	12	Single-stranded DNA	Coliphage φX174	Poxviruses	
		Naked	220					
		Naked	450–550	72		Coliphages T2, T4, T6	Papovaviruses	
			600–900	252			Adenoviruses	
				162			Herpesviruses	
	Binal (polyhedral "heads," helical "tails")	Enveloped Naked	Head: 950 × 650 Tail: 170 × 1,150					

[a] Diameter, in case of polyhedral virion.

acid, capsid architecture, presence or absence of an envelope, and capsid size. Further subdivisions are based on other features of the virion, such as the number of nucleic acid strands (one or two); on features of viral development, such as the site of viral synthesis in the cell; and on host-virus interactions, as exemplified by host range.

This system is not intended to represent a natural, or phylogenetic, classification: that is, it does not attempt to express the evolutionary relationships between viruses, relationships that are completely obscure. Instead, it groups viruses according to common sets of chemical and structural features, which are constant properties that can be determined with precision.

It will be noted that host range plays only a minor role in this classification. The different methodologies required for the study of viruses in widely different hosts, however, have led to the general practice of grouping them according to whether their normal hosts are bacteria, animals, or plants.*

• Viroids A number of plant diseases are caused by transmissible agents that are much smaller than any of the known viruses. These agents, called *viroids*, have been shown to be extremely small RNA molecules, perhaps containing some double-stranded regions. Their molecular weights are estimated to fall in the range of 75,000 to 100,000 daltons; they do not possess capsids, and they appear to be transmitted "horizontally" from one plant to another by mechanical means, as well as "vertically" through the pollen or ovules. Whether they interfere with host functions directly, or through a translated polypeptide, is not yet known; nor is anything yet known about their mode of replication.

It is still an open question whether or not viroids occur in animal cells, and—if so—whether or not they are the etiologic agents of diseases which are assumed to be viral but for which the causative agent has not yet been found.

THE BACTERIAL VIRUSES

Nearly every species of bacterium investigated so far has been found to serve as host to one or more viruses (bacteriophages). Most of the research on bacteriophages, however, has been done with the phages that attack *Escherichia coli;* our discussion will therefore be mainly confined to results obtained with this group.

Both DNA and RNA phages are known. In most of the DNA phages the nucleic acid of the virion is double-stranded, although in a few cases it is single-stranded. The RNA phages discovered so far have single-stranded RNA in their virions. In all but one group of bacteriophages, the nucleic acid of the virion is contained within a polyhedral capsid; in many cases this is joined to a helical protein structure, or "tail," which serves as an adsorption organ. The exception to the polyhedral structure are the filamentous phages, of which the phage called fd is typical; fd is a rod-shaped structure containing single-stranded DNA. Some examples of phage virions are shown in Figures 12.7 and 12.11.

---

*For a long time, the only microorganisms known to serve as hosts to viruses were the bacteria. Recently, however, viruses have been found in some fungi, and particles resembling virions have been reported in several species of protozoa.

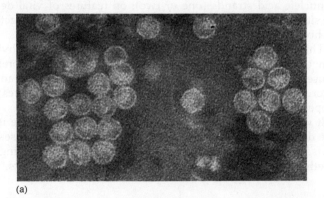

(a)

(b)

FIGURE 12.7

Bacteriophage virions: (a) icosahedral; (b) filamentous (helical symmetry). From D. E. Bradley, "The Structure of Some Bacteriophages Associated with Male Strains of *Escherichia coli*," *J. Gen. Microbiol.* **35**, 471 (1964).

• **The detection and enumeration of bacteriophage particles**

When a small number of virulent bacteriophage particles is added to a growing liquid culture of susceptible bacteria, some of the bacterial cells become infected. The infected cells show no obvious changes for a certain period, which commonly extends from 15 minutes to 1 hour or more, depending on the nature of the bacterium and phage in question and the conditions of cultivation. Then, quite suddenly, the infected cells undergo *lysis*. The lysis of the infected cell liberates a large number of new phage particles. These particles can in turn infect other cells in the population, with a repetition of the same cycle. Consequently, even if the number of infectious phage particles originally introduced is small relative to the number of bacterial cells, practically the entire bacterial population may be destroyed in a few hours.

It is very easy to determine the number of phage particles or infected bacterial cells in a suspension by spreading an appropriate dilution of this suspension over the surface of an agar plate that is evenly inoculated with a thin suspension of susceptible bacteria. After appropriate incubation, the surface of the plate shows a confluent layer of bacterial growth, except at those points where a phage particle or an infected cell has been deposited. Around such sites of infection, clear zones of lysis, or *plaques*, are formed as a result of the localized destruction of the film of bacteria by successive cycles of phage growth (Figure 12.8). By appropriate methods (e.g., differential centrifugation) one can separate infected cells from free phage particles in a mixture that contains both, and enumerate each separately by this *plaque-counting* method. If only the phage particles in a suspension are to be counted, the bacterial cells that are present can usually be killed by shaking an aliquot of the suspension with chloroform, to which bacteriophages are resistant.

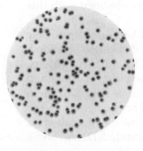

FIGURE 12.8

Plaques formed by bacteriophage T2. Courtesy of G. S. Stent.

The plaque method may be used to isolate from nature phages that are capable of attacking a particular species or strain of bacterium. A sample of material from the natural habitat of the bacterium is shaken with water; the supernatant is then sterilized by membrane filtration or by treatment with chloroform, and aliquots are mixed with suspensions of the bacterium and plated on agar. Any plaques that appear represent phage particles that were present in the natural material. The phage from a single plaque can be isolated by stabbing the plaque with a sterile inoculating needle and suspending the adhering material in a small volume of sterile diluent; such a suspension usually contains between 10^4 and 10^6 phage particles. Purification of the phage is then achieved by repeated serial isolations from single plaques.

To prepare large batches of phage for chemical or physical analysis, a culture of bacteria growing exponentially in liquid medium is inoculated with phage particles. The series of cycles of phage growth which ensues results in the lysis of most or all the cells in the culture; the culture is then freed of remaining cells and cellular debris by low-speed centrifugation and sterilized either by filtration or by treatment with chloroform. The resulting *sterile lysate* usually contains between 10^9 and 10^{12} bacteriophage particles per milliliter, together with soluble and particulate material liberated from the lysed cells.

The final purification of the bacteriophage is usually accomplished by ultracentrifugation: even the smallest phage particles will sediment at centrifugal forces exceeding $100,000 \times g$ (i.e., 100,000 times the force of gravity). As a preliminary step, it is often possible to precipitate the viral particles by treatment with ammonium sulfate or other agents.

THE DNA BACTERIOPHAGES:
THE LYTIC CYCLE OF INFECTION

The essential features of the lytic cycle of infection are shown by a fundamental quantitative experiment known as the *one-step growth experiment*. A culture of sensitive bacteria is mixed with a suspension of phage particles and incubated for a short time to permit the phage particles to become adsorbed on the bacterial cells. If the bacteria are in considerable excess, almost all the phage particles are adsorbed. The culture is then greatly diluted and allowed to grow. The evolution of the number of phage particles and infected bacteria is determined by periodic plaque counts. Figure 12.9 shows the typical result. After infection, the plaque counts remain constant for some time, since each infected bacterium

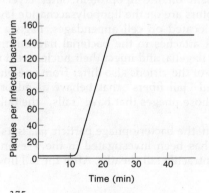

FIGURE 12.9

A typical one-step growth curve for a bacteriophage. The sudden increase in plaque count indicates that the host cells are lysing and liberating free phage.

serves as a center for the formation of a single plaque, no matter how many particles it may contain at the time that it is plated. This period is known as the *latent period*. It ends abruptly as the infected bacteria begin to lyse and liberate new phage particles. At this time, the plaque count rises very rapidly until all infected cells have lysed; this is known as the *burst period*. After lysis is complete, the plaque count remains more or less constant, even if there are uninfected cells left in the population, because the initial dilution of the culture largely prevents adsorption of the newly liberated phage particles by the remaining uninfected bacteria. The average number of new phage particles liberated by each infected cell is known as the *burst size*; in the actual experiment portrayed, it was about 150, but the value can vary greatly depending on the host-phage system studied and the conditions of the experiment.

Let us now consider in more detail the events that take place during the lytic cycle of infection. Most of the experiments on which this account is based have been done with a small group of phages that attack *Escherichia coli*. Certain phages, numbered T1 through T7 (T standing for "Type"), were arbitrarily chosen by the group of investigators that began the experimental attack on phage reproduction in 1939. Three of the phages, T2, T4, and T6, proved to be closely related and to have a number of features that make them especially suitable for experimental purposes. For example, the DNA of the "T-even" phages contains a unique base, 5-hydroxymethylcytosine, in place of cytosine; using a specific chemical assay for hydroxymethylcytosine, it is thus possible to follow the synthesis of viral DNA in infected cells in the presence of an excess of bacterial DNA.

For many years the T-even phages were assumed to be typical of phages in general. Electron microscopy, however, has revealed that the virions of the T-even phages are much more complex than those of other phages, and that their mechanisms of adsorption to and penetration of the host cell are quite specialized. The events which follow penetration, however, appear to be essentially the same in the T-even phages as in other DNA phages.

• **Adsorption and penetration**

The lytic cycle of infection begins when a bacteriophage particle undergoes a chance collision with a host cell. If the virion possesses an adsorption site that is chemically complementary to a specific *receptor site* on the bacterial cell surface, irreversible *adsorption* occurs. For some phages that attack Gram-negative bacteria, the receptors are present on the lipoprotein outer layer of the cell wall; for other phages, the receptors are on the lipopolysaccharide layer. For some phages, the receptor sites are located on cell appendages: flagella or pili. Figure 12.10 illustrates a phage that attaches to the bacterial flagellum; the attached phages slide to the base of the flagella, and inject their nucleic acid through the cell wall.

The adsorption sites of the virions also differ from one phage to another. Some phages have specialized "tail fibers" that behave as adsorption organs (Figure 12.11); in general, for those phages that have "tails," the tails serve as adsorption organs.

Following adsorption, the bacteriophage particle *injects* its DNA into the bacterial cell. This process has been investigated in the T-even phages, which have been shown to have contractile tail sheaths. After the tail fibers become adsorbed,

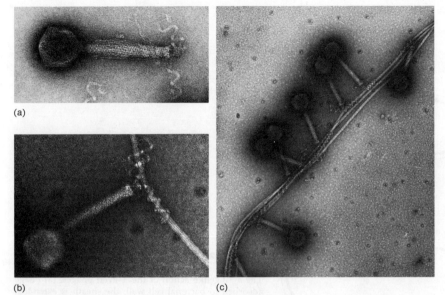

FIGURE 12.10

Electron micrographs of a bacteriophage of *Bacillus subtilis* which adsorbs on the bacterial flagellum. (a) A free phage particle with helical tail fibers (×120,000). (b) A phage particle that has adsorbed to a bacterial flagellum, around which the tail fibers are wrapped (×120,000). (c) A group of phage particles attached to several flagella (×61,000). From L. M. Raimondo, N. P. Lundh, and R. J. Martinez, "Primary Adsorption Site of Phage PBSI: the Flagellum of *Bacillus subtilis*," *J. Virol.* **2**, 256 (1968).

FIGURE 12.11

(a) An isolated particle of one of the T-even bacteriophages embedded in phosphotungstic acid. Note the filled head, contracted sheath, core, and tail fibers. From S. Brenner et al., "Structural Components of Bacteriophage," *J. Mol. Biol.* **1**, 281 (1959). (b) T-even phage components, with dimensions indicated in nm. From D. E. Bradley, "Ultrastructure of Bacteriophages and Bacteriocins," *Bacteriol. Revs.* **31**, 230 (1967).

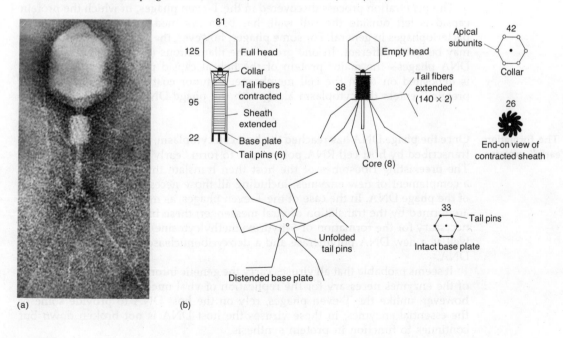

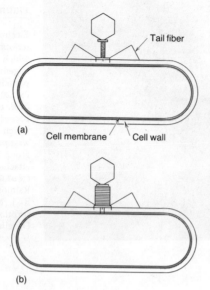

FIGURE 12.12

The syringelike action of the T-even phages. (a) Phage adsorbed to bacterial cell wall; the sheath is extended. (b) The sheath has contracted, driving the tail core through the cell wall.

the sheath contracts; the tail core is thus driven through the cell wall (Figure 12.12). When the tip of the core reaches the cell membrane, the DNA contents of the phage head are injected beneath the wall. The syringelike mechanism of the T-even phages is unique; it is not known how the other phages penetrate the cell walls of their hosts or how the DNA of any of the phages penetrates the cell membrane.

The penetration process discovered in the T-even phages, in which the protein capsid is left outside the cell wall, has been assumed to be characteristic of bacteriophages in general. For some phages, however, the penetration mechanism may be quite different. In one group—the filamentous (helical) single-stranded DNA phages—the major protein of the helical capsid penetrates the wall and is deposited on or in the cell membrane; a minor coat protein, called the "A protein," enters the cytoplasm along with the phage DNA.

• **The formation of "early" proteins**

Once the phage DNA has reached the host cell cytoplasm, part of it is immediately transcribed by host cell RNA polymerase to form "early" viral messenger RNA. The preexisting ribosomes of the host then translate this viral mRNA to form a complement of new enzymes, including all those necessary for the replication of the phage DNA. In the case of the T-even phages, as many as 11 new enzymes are formed by the translation of viral messenger; these include enzymes that are necessary for the formation of 5-hydroxymethylcytosine and its triphosphate, as well as a new DNA polymerase and a deoxyribonuclease that destroys the host DNA.

It seems probable that all viruses carry the genetic information for at least some of the enzymes necessary for the replication of viral nucleic acid. Many viruses, however, unlike the T-even phages, rely on the host DNA to provide some of the essential enzymes; in these viruses the host DNA is not broken down but continues to function in protein synthesis.

• The replication of bacteriophage DNA

The replication of bacteriophage DNA proceeds according to the general mechanism described in Chapter 7; the details of the mechanism differ according to whether the phage DNA replicates as a circular or as a linear molecule. The replication of a circular DNA molecule may occur in one of two modes. In one, called the *symmetric mode,* the two parental strands have equal roles; this is the type of replication described in Chapter 7. In the other, called the *asymmetric mode,* one strand remains unbroken and the other becomes a linear molecule. This type of replication, referred to as the *rolling-circle model,* is illustrated in Figure 15.10 for the circular sex factor of *Escherichia coli;* it is used by some phages as a mechanism for generating linear molecules from circular replicating intermediates.

In those phages in which the virion contains a molecule of single-stranded DNA, the injected DNA is rapidly converted to the double-stranded form by a bacterial DNA polymerase, following which replication proceeds in the usual semiconservative manner. Normal replication soon ceases, however, and is replaced by a new enzymatic process in which only one type of complementary strand is made from the double-stranded template. These new single strands are finally incorporated into progeny phage particles.

During replication, phage DNA molecules may undergo random base pair changes (*mutation*) and events of breakage and reunion. If a bacterial cell is simultaneously infected with two genetically different but related phages, the events of breakage and reunion produce recombinant phage DNA molecules. The processes of phage mutation and recombination are discussed in more detail in Chapters 13, 14, and 15.

• Maturation

Soon after the replication of phage DNA begins, that part not already transcribed is used as template for the synthesis of "late" viral messenger RNA. The translation of the late messenger results in the formation of a second set of viral-specified proteins, among them the subunits of the viral capsid. Simultaneously with the accumulation of capsid subunits, the viral DNA molecules undergo condensation: each molecule assumes a tightly packed, polyhedral shape as a result of its combination with a specific viral protein or "condensing principle" (Figure 12.13). The capsid subunits then combine with the condensates to form mature phage heads; in the T-even phages, maturation is completed by a process in which other subunits polymerize to form the tail, followed by assembly of tails and heads into complete virions.

The process of *self-assembly* is guided in an unknown manner by the products of certain viral genes. These "morphopoietic" gene products govern, for example, the positioning of special capsomers to form the corners of polyhedral phage heads. In phages containing mutant morphopoietic genes* the capsomers polymerize incorrectly to form bizarre structures such as "polyheads" or "polysheaths" (Figure 12.14).

• The liberation of mature virions

At the end of the latent period of lytic infection, another viral-specified "late" protein appears in the cell: phage lysozyme. This enzyme attacks the peptidoglycan layer of bacterial cell walls, hydrolyzing the linkages between the sugar residues backbone chains. The wall is thus progressively weakened until it is ruptured

---

*Such mutations are lethal and can thus be studied only as "conditionally expressed mutants," in which the mutation is expressed in one host but not in another (see Chapter 14).

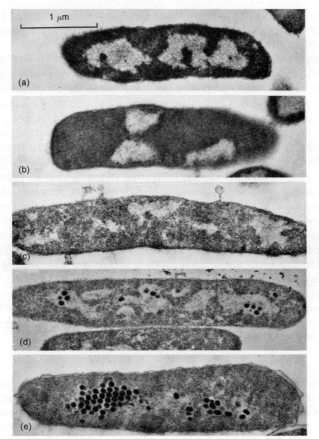

1 μm

(a)

(b)

(c)

(d)

(e)

FIGURE 12.13

The course of phage infection in the bacterium *E. coli*, illustrated by electron micrographs of successive thin sections of the cells (×25,700). (a) Uninfected cell. (b) Four minutes after infection. Note change in structure of the DNA-containing region (light areas of cell). (c) Ten minutes after infection. (d) Twelve minutes after infection. The first new phage bodies, or DNA condensates (dark spots), are developing within the DNA-containing regions of the cell. (e) Thirty minutes after infection (shortly before lysis). Many new phage bodies are evident in the infected cell. Courtesy of E. Kellenberger, E. Boy de la Tour, J. Sechaud, and A. Ryter.

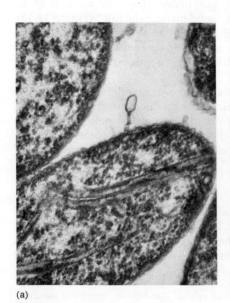

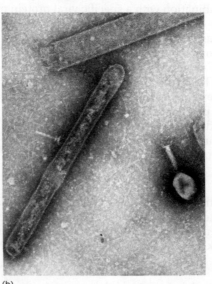

(a)

(b)

FIGURE 12.14

Polyheads of phage T4. (a) Intracellular polyheads in a partially lysed cell. Note the adsorbed phage particle. (b) Formaldehyde-fixed polyheads prepared in phosphotungstate. The preparation includes an intact phage particle for comparison. From R. Favre, E. Boy de la Tour, N. Segré, and E. Kellenberger, "Studies on the Morphopoiesis of the Head of Phage T-even. I. Morphological, Immunological, and Genetic Characterization of Polyheads," *J. Ultrastructure Res.* **13,** 318 (1965).

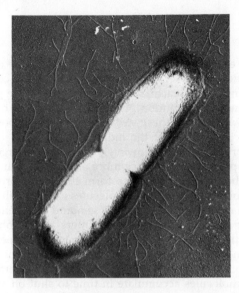

FIGURE 12.15

The liberation of filamentous phage by extrusion from living bacterial cells. The micrograph was made 30 minutes after the cells had been infected and washed free of unadsorbed phage particles. From P. H. Hofschneider and A. Preuss, "M13 Bacteriophage Liberation from Intact Bacteria as Revealed by Electron Microscopy," *J. Mol. Biol.* **7**, 450 (1963).

by the internal osmotic pressure of the cell, and the progeny phages are liberated into the environment along with the other contents of the cell.

An exception to the general rule that phages are liberated by lysis of the host cell is provided by the filamentous DNA phages, such as fd. As new phage coat proteins are synthesized, they are deposited at the cell membrane; maturation and liberation occur by the extrusion of phage DNA, which combines with coat protein during its passage through the membrane. The host cell remains viable and continues to grow throughout the liberation process (Figure 12.15).

LYSOGENY

In the lytic cycle of infection, described above for bacteriophages, every infected cell eventually lyses and liberates a crop of progeny virus particles. Many bacteriophages, however, are capable of an alternative interaction with their host: following penetration, the viral genome may reproduce in synchrony with that of the host, which survives and undergoes normal cell division to produce a clone of infected cells. In most of the progeny cells no viral structural proteins are formed, most viral genes having become *repressed*. In an occasional cell, however, derepression occurs spontaneously and the viral genome initiates a lytic cycle of development; the cell in which this occurs lyses and liberates mature virions.

This virus-host relationship is called *lysogeny*; infected cells that possess the latent capacity to produce mature phage particles are said to be *lysogenic*. Bacteriophages capable of entering this relationship are called *temperate*, and the viral genome present in the cells of a lysogenic culture is called *prophage*.

Thus, when a temperate phage particle infects a susceptible host, the entering phage genome has two alternative fates: it may commence rapid vegetative multiplication, culminating in the formation of mature virions and lysis of the host cell, or it may enter the prophage state, giving rise to a clone of lysogenic cells. When a suspension of phage particles is added to a culture of a susceptible host strain, so that every bacterium becomes infected, some of the cells undergo "productive" infection and lyse, while the remainder of the cells are lysogenized. The fraction that encounters each fate is determined by the genetic constitution of the host, the genetic constitution of the virus, and also by the environment; by experimenting with each of these variables, it has been possible to investigate the factors that decide the fate of the individual cell.

From such experiments, the following picture has emerged. In every newly infected cell, three processes take place more or less simultaneously: (1) the viral genome is transcribed and translated to form the "early proteins" of infection, including one or more types of *repressor* molecules which inhibit both viral replication and viral gene expression; (2) the viral genome commences replication; and (3) one or more of the replicated genomes enters the prophage state. The fate of the cell in which these three processes have begun then depends on the outcome of a "race" between repressor production and viral maturation: if mature virions and lysozyme are produced before repressor action can interfere, the cell will be lysed; if repressor molecules accumulate in time to shut off viral replication and gene expression before mature virions and phage lysozyme have been produced, the cell will be lysogenized. In the lysogenized cell, phage repressor molecules continue to be produced by the prophage and prevent both vegetative replication and maturation of the phage.

Repressor molecules do not interfere with the replication of prophage in synchrony with the cell. *Prophage replication* and *vegetative replication* are thus different processes; they will be discussed in the following sections.

• Lysogeny
of the λ type

Most of the research on the nature of prophage and the lysogenic state has been done with the DNA bacteriophage called *lambda* (λ), which was first discovered as a prophage present in strain K12 of *E. coli*.* When a culture of strain K12 is grown in the laboratory, about one cell in 10^4 lyses and liberates mature λ particles. As an even more rare event, an occasional daughter cell is formed that fails to receive a copy of the prophage at cell division; such a cell is said to have been "cured." When a cured (nonlysogenic) cell is isolated and cultivated, the λ-free cells can be reinfected and again lysogenized; if a genetic mutant of λ is used for this purpose, every lysogenic cell inherits the capacity to produce λ particles of the mutant type.

Such observations tell us that λ prophage consists of one or more copies of the λ genome. Where is the λ genome located in the lysogenic cell? What mechanism ensures its synchronous replication and the near-perfect segregation of its replicas to daughter cells?

The answers to these questions were provided by experiments in which lysogenic bacteria were allowed to recombine genetically with nonlysogenic (cured) bacteria or with lysogenic cells carrying prophages derived from different mutants of λ. As we shall discuss in Chapter 15, *E. coli* strain K12 is capable of recombination by conjugation, transduction, or transformation. In all three processes, seg-

*Strain K12 is one of a collection of *E. coli* strains that had been isolated from clinical specimens at the Stanford University Hospital. It was arbitrarily chosen for use in a pioneering project on bacterial genetics and has since become a standard strain for genetical research.

ments of the bacterial chromosome are transferred from one cell (the genetic donor) to another (the genetic recipient). The recipient thus becomes a partial diploid, or *merozygote*; recombination events (or "crossovers") in the merozygote produce recombinant chromosomes that are finally segregated to produce recombinant daughter cells.

As we will see in Chapter 15, recombination experiments have permitted the *mapping* of genes on the bacterial chromosome. If the donor and recipient differ in several genetic traits, the recombinants can be analyzed to determine which genes they have inherited from the donor parent and which from the recipient. The closer together that two genes lie on the donor chromosome, the higher the frequency with which they will be inherited together by the same recombinant cell; such genes are said to be closely *linked*. When nonlysogenic donor cells were recombined with recipient cells harboring λ prophage, *the prophage was found to be inherited as though it occupied a discrete site on the bacterial chromosome* between the *gal* and *bio* genes, which govern the enzymes of galactose utilization and biotin synthesis, respectively. Thus, the λ prophage is attached to the bacterial chromosome and replicates in synchrony with it. At cell division, every daughter cell receives a bacterial chromosome and with it the attached λ prophage.

The nature of the attachment between prophage and chromosome remained a mystery for almost 10 years after its existence was discovered. In 1962 A. Campbell suggested that attachment occurs by the process shown in Figure 12.16. According to the "Campbell model," the λ genome circularizes immediately after its penetration of the host cell. The λ genome possesses a region which pairs with a specific region on the bacterial chromosome, adjacent to the *gal* loci; *a single crossover event within the region of pairing inserts the λ genome into the continuity of the bacterial chromosome*. The inserted genome constitutes the λ prophage; it is henceforth replicated as a part of the bacterial chromosome, even after phage repressor has completely inhibited the replication of free λ DNA.

The Campbell model has been confirmed by a variety of experiments. The λ genome which is present in the mature virion has been shown to possess *cohesive ends*, which consist of single-stranded regions with complementary base sequences (Figure 12.17). Hydrogen bonding in these regions leads to circularization of the λ DNA; the circles are then covalently joined by the action of polynucleotide ligase.

The crossover event that integrates λ with the bacterial chromosome is brought about by a set of recombination enzymes which cut the DNA molecules and rejoin the broken ends in new combinations (see Chapter 15). The same enzymes, operating on the integrated structure, can bring about the *detachment* of the prophage, and such detachment has been shown to be the first step in the process that leads to the production of mature virus by an occasional lysogenic cell. This finding posed an apparent paradox: since *E. coli* strain K12 contains a full set of recombination enzymes, what prevents the detachment of λ in the great majority of the cells of a lysogenic culture? The paradox was solved by the discovery that the integration of λ with the bacterial chromosome requires the action of specific recombination enzymes, governed by λ genes: the integration of λ occurs before λ repressor has had time to appear; once it has appeared, however, it represses all λ genes, including those for the specific recombination enzymes. The detachment of λ is thus prevented, except in the rare cell in which the repressor is inactivated. The inactivation of the repressor, leading to the detachment of and its entry into the lytic cycle, will be discussed in a later section.

The events described above have been found to occur in the formation of

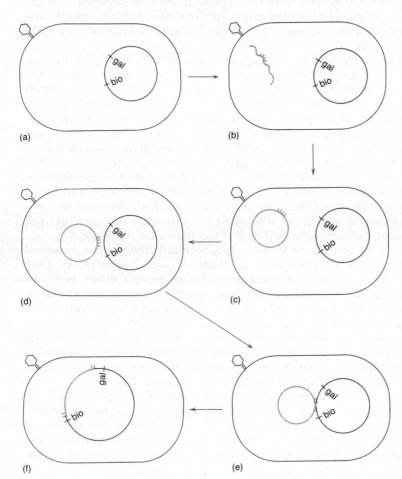

(a)

(b)

(c)

(d)

(e)

(f)

FIGURE 12.16

The formation of λ prophage. (a) Adsorption of the virus. (b) Injection of viral DNA. (c) Circularization of the viral genome. (d) Pairing of homologous regions on the viral and bacterial genomes. (e) A crossover event occurs within the region of pairing. (f) The two genomes have been integrated, forming a single circle. Note that the attachment site for λ is at a specific location, between the loci *gal* and *bio* (see Figure 15.14). The specific attachment site on λ is indicated by four short vertical lines on the DNA strand.

prophage by a number of other phages closely related to λ [e.g., phage φ80, which undergoes prophage insertion in the chromosome next to the *trp* loci (the genes governing the enzymes of tryptophan biosynthesis)]. The attachment sites of a number of prophages are shown on the genetic map of *E. coli* strain K12, which is reproduced in Figure 15.14.

• Lysogeny of the P1 type

A number of bacteriophages, of which phage P1 is the best studied, differ from the λ group of phages in two important respects. First, recombination experiments between lysogenic and nonlysogenic cells fail to reveal a discrete chromosomal location for P1 or the other phages of this type. The transfer of P1 prophage is occasionally observed, but no linkage to known bacterial genes is evident. Second, P1 and λ differ in the manner in which they form transducing particles, virions that contain fragments of host DNA. In Chapter 15 it will be seen that transducing particles of the λ type contain a major segment of phage genome, covalently linked

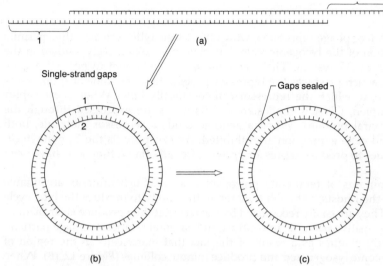

2

1

Single-strand gaps

1

2

(a)

Gaps sealed

(b)   (c)

FIGURE 12.17

The conversion of the λ genome from the linear form to the circular form. (a) The linear genome has single-stranded regions at each end that are complementary to each other. (b) The single-stranded regions are joined by hydrogen bonds between complementary bases. (c) The sugar-phosphate backbones are joined by polynucleotide ligase, forming a fully covalent circle.

to a segment of the host chromosome derived from the region adjacent to the prophage attachment site. As shown in Figure 15.20, a small error in the normal process of prophage detachment creates a transducing particle. Transducing particles of the P1 type, however, contain mainly host (bacterial) DNA; less than 10 percent is present as newly synthesized phage DNA. Furthermore, phages of the P1 type are capable of transducing any segment of the bacterial chromosome; in contrast, λ can transduce only those genes that are immediately adjacent to its attachment site.

Thus, both genetic mapping experiments and the analysis of transducing particles have failed to reveal a chromosomal location for prophage of the P1 type. The nonchromosomal location of P1 prophage was confirmed when P1 DNA was extracted from lysogenic cells and shown to consist of circular DNA structures of the same size as the DNA from P1 virions; it shows no association with bacterial DNA.

As an alternative to chromosomal attachment, it is very possible that prophages of the P1 type occupy attachment sites in the cell membrane, since (as will also be discussed in Chapter 15) such attachment sites appear to account for the orderly replication and segregation of the extrachromosomal genetic elements known as plasmids. For example, the cell membrane of *E. coli* strain K12 possesses attachment sites not only for the bacterial chromosome but also for F, a plasmid that behaves as a "fertility factor" in *E. coli*. As discussed in Chapter 15, both chromosome and F are replicated and segregated by a mechanism that involves their attachment to the cell membrane.

In summary, a phage of the P1 type can replicate in harmony with the cell as a prophage, presumably attached to the cell membrane, or it can undergo uncontrolled vegetative replication as part of a lytic cycle. The latter process is prevented in the majority of lysogenic cells by the action of a phage repressor, just as in the λ type of lysogeny. It is interesting that mutants of *E. coli* can be isolated in which λ prophage fails to integrate and instead establishes the P1 type of lysogenic relationship: i.e., it becomes an autonomously replicating element.

385

● **Repression and induction**

The mechanism of phage repression, which blocks the lytic cycle and thus permits the perpetuation of the lysogenic state, has been most extensively studied in the *E. coli* strain K12 (λ) system. The λ genome has been shown to possess a gene (the C_I gene) which produces the repressor protein; the λ genome also possesses a receptor site, to which the repressor can specifically bind. When the receptor site is unoccupied, DNA is replicated and its genes are expressed through the formation of viral proteins. When a genome binds a repressor molecule, both replication and gene expression are inhibited. An exception is the C_I gene itself, which continues to produce repressor protein after all other λ functions have been shut off.

Virulent mutants of temperate phage often occur. Such mutants are unable to lysogenize their hosts, although they retain the ability to produce the lytic cycle of infection. They are easily recognized by the fact that they produce clear plaques on a lawn of sensitive bacteria; in contrast, the parental temperate phage particles produce cloudy plaques as a result of the fact that many cells in the region of the plaque become lysogenized and produce minute colonies (Figure 12.18). When the virulent mutants are isolated, they prove to be of two different types: those in which the C_I gene has mutated so that the repressor is no longer produced and those in which the receptor is altered such that the repressor can no longer act.

Certain phages, including λ, are *inducible*: their prophages are caused to enter the lytic cycle of development by treatment of the host cell with any of a number of agents that produce lesions in DNA—ultraviolet irradiation, temporary thymine starvation, treatment with mitomycin C, or treatment with alkylating agents. The DNA lesions produced by these agents apparently bind λ repressor nonspecifically, making it unavailable for binding to the λ receptor site.

The massive induction of a λ-lysogenic culture can be brought about in two other ways. One way is to permit the cells to transfer their chromosomes (including the inserted λ prophage) to nonlysogenic recipients, by the process of conjugation (see Chapter 15). Since the cytoplasm of the recipient is free of repressor,

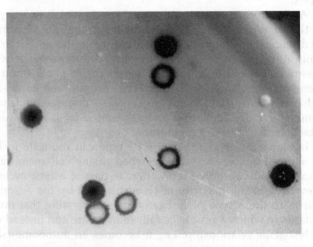

FIGURE 12.18

Plaques formed by the wild type and by a virulent mutant of the bacteriophage, lambda. The wild-type particles form cloudy plaques as a result of the growth of lysogenized cells; the mutant particles, which are unable to lysogenize the bacterial host, form clear plaques with sharp edges. Courtesy of C. Radding.

and little or no donor cytoplasm is transferred, the transferred prophages are immediately induced. This process is called *zygotic induction*. The other mode of induction occurs with a strain of λ in which the C_I gene has mutated so that the repressor is thermolabile. The mutant repressor functions normally at 37°C but is denatured at 43°C, unlike the wild-type repressor, which is stable at both temperatures. Thus, a lysogenic culture containing the mutant λ can be grown at 37°C and then massively induced by shifting the culture to 43°C.

As stated earlier, at each generation about one in 10^4 cells of a growing K12 culture lyses and liberates λ virions. This apparently reflects the spontaneous failure of repressor formation or action as a rare event in the cell population.

• **Immunity**

If the cells of a lysogenic culture are allowed to adsorb particles of a different phage, they are said to be *superinfected*. The DNA of the second phage will penetrate the lysogenic cells, but its ability to reproduce will depend on its sensitivity to the repressor produced by the resident prophage. If the superinfecting phage is resistant to the repressor, it will multiply and lyse the cells. If it is sensitive, the superinfection will be *abortive*: the DNA of the superinfecting phage will be neither replicated nor expressed and will be diluted out during subsequent growth of the cells. In that case, the lysogenic cells are said to be *immune* to the second phage.

Two phages that share sensitivity to the same repressor are termed *coimmune*. For example, if a cell harbors one strain of λ as prophage, it will be immune to superinfection with a second strain of λ. The exception, of course, will be the virulent mutant that owes its virulence to its resistance to its own repressor; such a virulent mutant will successfully superinfect any λ-lysogenic cell.

It is easy to demonstrate that immunity to a second phage is due to the same repressor that is responsible for the maintenance of lysogeny. For example, irradiating a culture that is lysogenic for an inducible phage will simultaneously induce the prophage and break the immunity barrier. If such a culture is super-infected immediately after irradiation, every cell will lyse and liberate a mixture of phage progeny arising from both the prophage and the superinfecting phage.

Immunity must be distinguished from the genetically controlled property of *phage resistance*. Phage resistance results from a mutation in the bacterial chromosome, resulting in the loss of a phage receptor on the cell surface. In contrast to an immune cell, a genetically resistant cell does not adsorb the phage.

• **Defective prophages**

From time to time, prophages undergo gene mutations or deletions (the loss of a segment of DNA spanning several genes) which render them incapable of normal development. For example, a mutation may cause the loss of ability to replicate in the vegetative state or to synthesize an essential viral protein. Such prophages are termed *defective*; their presence in the cell is revealed by the fact that the *cell retains its immunity to superinfection* by coimmune phages. Furthermore, if the defective prophage arose from an inducible phage, treatment of the cells with inducing agents may still derepress the functional genes and bring about lysis of the cell by phage lysozyme.

The existence of defective prophages in bacteria isolated from nature may thus be tested for by treatment of the cells with inducing agents. Lysis of the cells suggests the presence of an inducible, defective prophage; confirmation may be obtained by electron microscopic examination of the lysate, in which incomplete capsids may be observed. In some cases, it has been possible to demonstrate the presence of a defective prophage by superinfection of the cells with a related phage and recovery in the resulting lysate of *recombinant* phages that have inherited some genes from the defective prophage and others from the superinfecting phage. Alternatively, the lysate may contain a *mixture* of normal and defective phages, the latter having been enabled to reproduce by the superinfecting phage that provided the missing function.

THE RNA BACTERIOPHAGES

In a number of bacteriophages the genetic material in the virion consists of a molecule of single-stranded RNA.

When a molecule of viral RNA reaches the cytoplasm of the host cell, it is immediately recognized as messenger RNA by the ribosomes, which bind to it and initiate its translation into viral proteins. One such viral protein is a complex enzyme, *RNA replicase* (an RNA-dependent RNA polymerase). This enzyme brings about the replication of the viral RNA; it polymerizes the ribonucleoside triphosphates of adenine, guanine, cytosine, and uracil, using viral RNA as a template.

The first step in the process of RNA replication is the formation of a *double-stranded intermediate*, in which the entering viral RNA strand (called the "plus" strand) is used by the replicase as a template to form the complementary, or "minus," strand (Figure 12.19). The replicase now uses the double-stranded

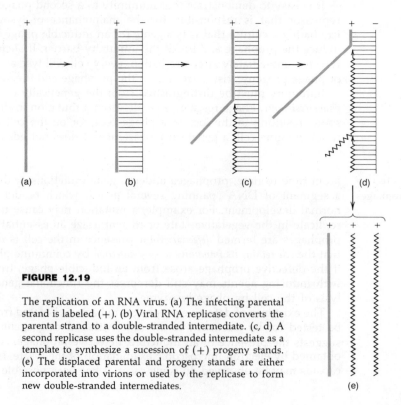

(a) (b) (c) (d)

(e)

FIGURE 12.19

The replication of an RNA virus. (a) The infecting parental strand is labeled (+). (b) Viral RNA replicase converts the parental strand to a double-stranded intermediate. (c, d) A second replicase uses the double-stranded intermediate as a template to synthesize a sucession of (+) progeny stands. (e) The displaced parental and progeny stands are either incorporated into virions or used by the replicase to form new double-stranded intermediates.

molecule as a template for the repeated synthesis of new plus strands, each new plus strand displacing the previous one from the replicative intermediate.

As the newly synthesized plus strands are released from the replicative intermediate, they are either used by the replicase to form a new double-stranded intermediate or are assembled into mature virions by the attachment of capsomers.

THE ANIMAL VIRUSES

• **Some important groups of animal viruses**

As the study of animal viruses has progressed, a number of different categories have been recognized, each of which groups together viruses possessing many characteristics in common. In some cases, the category is defined strictly on the basis of virion structure: the picornaviruses, for example, include all small RNA viruses ("picorna" is a composite of pico-, "small," and RNA). In other cases, the category is defined ecologically: the arboviruses, for example, include all arthropod-borne viruses. Some important groups of animal viruses are described in Table 12.2.

• **The detection and enumeration of animal virus particles**

Several methods for the enumeration of animal viruses are completely analogous to the plaque method described earlier for the enumeration of bacteriophages. In each of these methods, virus particles are placed on the surface of a layer of host cells; each particle initiates the infection of a single cell, which becomes a focus of infection for neighboring cells.

In the *pock method* the virus particles are placed on the surface of the chorioallantoic membrane of a chick embryo; the suspension of particles is introduced through a small hole drilled in the egg shell, which is then carefully sealed. The eggs are incubated until each virus has produced a visible lesion, or pock, on the membrane. The shell is then cut away and the pocks are counted.

In the *plaque method,* the virus particles are placed on the surface of a sheet of cells growing on the inner surface of a flat bottle. After allowing the virus particles to adsorb to the cells, the fluid medium is removed and the cells are covered with an overlay of soft agar. Virus particles released from the initially infected cells can spread to neighboring cells but are prevented from wider diffusion by the agar. After a period of incubation, each primary infected cell gives rise to a *plaque,* or zone, of infected cells. Methods for visualizing the plaques vary for different viruses. Some viruses produce a *cytopathic effect:* they kill their host cells. For these viruses, the plaques are revealed by flooding the cell layer with dyes that differentially stain live and dead cells (Figure 12.20). Other viruses do not kill their host cells, and their plaques are detected by flooding the cell layer with reagents that are specific for viral substances, such as fluorescent antibodies.

The virions of many animal viruses, including both enveloped and naked types, bind strongly to specific receptor sites on the surface of red blood cells. Since a virion has several adsorption sites, it can bind to two red blood cells simultaneously and form bridges between them. Thus, if a sufficient number of virions is mixed with a suspension of red blood cells, the latter are agglutinated. *Hemagglutination,* as this phenomenon is called, provides a rapid assay for those viruses which adsorb to red blood cells. In practice, serial dilutions of the virus preparation are tested for hemagglutinating activity, and the titer of the preparation is defined as the reciprocal of the highest dilution that has detectable activity. For example,

TABLE 12.2

Principal groups of animal viruses capable of causing infections in mammals

GROUP	PHYSICOCHEMICAL PROPERTIES	OTHER CHARACTERISTICS	EXAMPLES
Adenoviruses	Naked, icosahedral capsid containing double-stranded DNA	Ubiquitous agents of latent infections of lymphoid tissue; many produce tumors in experimental animals	Members of this group are designated as numbered "types"
Papovaviruses	Naked, icosahedral capsid containing double-stranded DNA	Production of tumors in animals	Polyoma, papilloma, SV-40
Herpesviruses	Enveloped, complex capsid containing double-stranded DNA	Tendency to cause latent infections	Herpes simplex, varicella (herpes zoster), pseudorabies
Poxviruses	Enveloped, helical (?) capsid containing double-stranded DNA	Very large; brick-shaped to ovoid; predilection for epidermal cells; two members (myxoma and fibroma viruses) cause tumors	Smallpox, vaccinia, cowpox
Picornaviruses	Naked, icosahedral capsid containing single-stranded RNA	Small size; cause enteric and/or respiratory infections	Poliovirus, Coxsackievirus, echovirus, rhinovirus
Reoviruses	Naked, icosahedral capsid containing double-stranded RNA	Large size; ubiquitous infectious agents but overt disease not produced	Members of this group are designated as numbered "types"
Arboviruses	Enveloped, probably icosahedral capsid containing single-stranded RNA	Arthropod-borne (mosquitos, ticks)	Yellow fever, equine encephalitis, dengue, Colorado tick fever, sindbis
Myxoviruses	Enveloped, helical capsid containing single-stranded RNA	Multiply in the nucleus; filamentous virions commonly formed	Influenza
Paramyxoviruses	Enveloped, helical capsid containing single-stranded RNA	Multiply in the cytoplasm; possess a hemolysin	Parainfluenza, mumps, measles, Newcastle disease, canine distemper
Leukoviruses	Enveloped capsid of uncertain architecture, containing single-stranded RNA	Production of tumors in animals	Murine and feline leukemia viruses, mouse mammary tumor virus, avian leucosis viruses, Rous sarcoma virus

if a preparation of virus shows hemagglutination at a dilution of 1:320 but not at 1:640, its titer is 320.

The hemagglutination assay described above is an example of the general method called *end-point titration*, in which one measures the highest dilution of a preparation to exhibit a given viral activity. Other activities that can be used for the end-point titration of viruses include killing of host animals and pathological effects on tissue cultures.

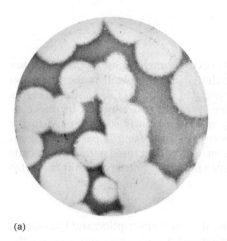
(a)

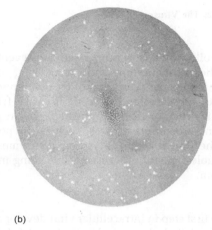
(b)

FIGURE 12.20

Plaques formed by encephalo-myocarditis virus on a layer of animal cells. Two genetically distinct plaque types are shown; the large plaques (a) are 10 to 12 mm diameter, and the small plaques (b) are 0.5 mm diameter. Courtesy of H. Liebhaber.

Any of the above methods can be used to assay the fractions produced at each step during the purification of a virus. When the preparation is sufficiently pure, the number of viral particles it contains can be directly counted in the electron microscope. When this is done, it is usually found that the number of *infectious units* is considerably lower than the number of morphologically recognizable particles. The ratio of infectious units to visible particles varies from 10^{-1} for some viruses to as low as 10^{-4} or 10^{-5} for others.

THE REPRODUCTION OF ANIMAL VIRUSES

• Adsorption and penetration

The adsorption of an animal virus to a host cell begins with the formation of noncovalent bonds between sites on the virion surface and receptor sites on the cell surface.

The adsorption sites on the virion vary in nature from one virus to another. In adenoviruses, for example, each pentamer at the corners of the icosahedral virion bears a small fiber that acts as an adsorption organ. In many of the enveloped viruses, the adsorption organs are the numerous spikes which stud the surface of the envelope (Figure 12.21). The receptor sites on the host cells are also varied. For the myxoviruses, for example, they consist of mucoproteins, while for poliovirus they consist of lipoproteins.

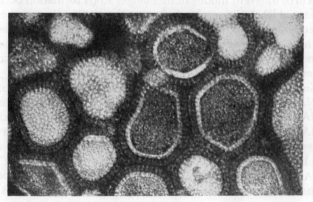

FIGURE 12.21

Particles of influenza virus, showing surface projections or "spikes." From R. W. Horne et al., "The Structure and Composition of the Myxoviruses. I. Electron Microscope Studies of the Structure of Myxovirus Particles by Negative Staining Techniques," *Virology* **11**, 79 (1960).

Following adsorption, a series of events occurs which includes the *penetration* of the cell membrane and the *uncoating* of the virion; the result is the liberation within the host cytoplasm of free viral nucleic acid. The exact sequence of events is still obscure, and it seems to vary from one virus to another. In one case (poliovirus), uncoating appears to begin while the virion is still attached to the membrane; in other cases, such as the poxviruses, uncoating is partially accomplished by the action of lysosomal enzymes on virions contained within phagocytic vacuoles; in still other cases, uncoating may take place entirely within the cytoplasm.

• **Intracellular development: DNA viruses**

The first step in intracellular viral development is the transcription and translation of "early" viral genes; the products of these genes are enzymes required for the replication of viral DNA. In most DNA viruses, early transcription takes place in the nucleus by means of a *host transcriptase* (DNA-dependent RNA polymerase). The exceptions are the poxviruses, whose virions contain their own transcriptase; as uncoating proceeds, the viral transcriptase is activated and "early" messenger RNA is produced in the cytoplasm of the host cell.

Translation of early viral messenger RNA by host ribosomes provides the enzymes for viral DNA replication, which—except for the poxviruses—occurs in the cell nucleus. The DNA is replicated semiconservatively and symmetrically; ultracentrifugation analyses of cell extracts show the viral DNA to be associated with lipid-containing particles, suggesting its attachment to a membrane during the replication process.

Sometime after the initiation of DNA replication, transcription and translation of "late" viral genes occur, leading to the production of capsid proteins. As in bacteriophage, maturation occurs by the self-assembly of capsid proteins and nucleic acid molecules to form nucleocapsids.

• **Intracellular development: RNA viruses**

The intracellular development of the RNA animal viruses parallels, in broad outline, that of the DNA viruses. It begins with the production of messenger RNA strands; these are translated to form both replicative enzymes and capsid proteins; the viral nucleic acid is replicated; and maturation occurs by the self-assembly of nucleocapsids.

Three different modes of messenger RNA formation occur among the different groups of RNA viruses. In one group, the picornaviruses, the virion RNA functions directly as a polygenic messenger; upon uncoating, it is immediately translated by host ribosomes to produce a single giant polypeptide, which is later cleaved into the several viral proteins. The same RNA molecule, released from the ribosomes, can then be replicated as described below.

In all other groups of RNA viruses, with the exception of the leukoviruses (RNA tumor viruses), the messenger RNA strand is the *complement* of the virion RNA strand, and is produced by the action of a virion-associated RNA-dependent RNA polymerase. The virion RNA is copied to form a series of monogenic messages, each of which is translated into a single species of viral protein by the host ribosomes; there is no distinction between "early" and "late" messages.

An entirely different mode of messenger RNA formation occurs in the leukoviruses. Here, the virion carries a polymerase which uses the virion RNA as a template for the synthesis of a *complementary DNA strand,* which is then replicated to double-stranded DNA. The virion-associated enzyme that catalyses these reactions is thus an *RNA-dependent DNA polymerase* and is called a *reverse transcriptase.* Both messenger RNA molecules and new virion RNA molecules are apparently synthesized on the double-stranded DNA template by normal DNA-dependent RNA polymerases. The significance of the DNA intermediate for oncogenesis (tumor formation) will be discussed in the section below on the tumor viruses.

In the nontumor RNA viruses, one of the viral proteins translated from viral messenger RNA is a viral *replicase* (an RNA-dependent RNA polymerase) which synthesizes new virion RNA strands (here designated +). This process, which is similar to that described earlier for the RNA phages, involves the initial formation of a double-stranded (+/−) RNA intermediate; the minus (−) complementary strand of this intermediate then serves as a template for the repeated synthesis of new virion (+) strands. Maturation again occurs by the self-assembly of nucleic acid molecules (in this case, RNA) and capsid proteins to form nucleocapsids.

It will be recalled that one class of viruses, the reoviruses, has double-stranded RNA in its virions. The intracellular development of these viruses depends on a combination of the processes described above: a virion-associated polymerase uses virion RNA to synthesize single-stranded messenger RNA molecules which are complementary to one of the parental strands; and a viral replicase, formed by translation of a viral message, converts messenger RNA to the double-stranded form which is incorporated into new virion particles.

• **Genetic recombination in animal viruses**

A number of animal viruses have been shown to undergo genetic recombination: mixed infection with two genetically marked viruses can yield progeny that inherit markers from both parents. Mutant characters employed as markers in these studies include altered plaque morphologies, resistance to inhibitory agents, and altered host range. Recombination has been shown to occur in several groups of DNA viruses, as well as in two RNA viruses, influenza (a myxovirus) and polio (a picornavirus). In the case of influenza virus, recombination reflects the random assortment of parental RNA molecules at the time that they are assembled into mature virions, each of which contains several RNA molecules. In the case of polio virus, a mechanism involving the pairing of RNA molecules, followed by breakage and reunion, has been inferred by analogy with the mechanism of DNA recombination.

• **The release of virions from the cell**

There are two general modes of virion release among the animal viruses. In some the nucleocapsid is extruded from the cell in such a manner that the virus particle acquires an outer envelope derived from the cell membrane (Figure 12.5). Prior to the extrusion, a number of viral proteins are incorporated into the cell membrane, so that the envelope contains both viral and host material. All the lipids of the envelope, however, are derived from the host membrane.

Other viruses are released through ruptures in the membranes; the free virions are naked nucleocapsids. This type of release usually follows death of the host cell and is characterized by the sudden appearance of a large number of free virions. In contrast, the release of virions by "budding off" of the membrane does not immediately kill the host cell, and the release process continues for many hours. In some cases, the host cells continue to multiply during the process and survive indefinitely. This situation leads to a *steady-state infection:* a clone of cell arises, all of which are continually liberating virus particles. Reinfection of the cells by extracellular virus is not necessary to maintain the steady state, since each daughter cell acquires intracytoplasmic virus particles at cell division.

THE TUMOR VIRUSES

• The discovery
of viruses as
etiological agents
of tumors

The characteristic tissues of animals are formed by the *regulated, limited growth* of their component cells. As a rare event, a cell may escape the normal regulatory processes and divide without restraint, forming an abnormal mass of tissue. Such masses are called *tumors* or *neoplasms;* the development of a tumor is called *neoplasia.*

Some tumors, such as most papillomas (warts) are *benign:* they remain localized and the animal is unharmed. Other tumors are malignant: their growth is invasive, so that the organ in which it occurs is damaged and the animal dies. Often, unregulated cells are released from the tumor and establish new neoplastic foci in other parts of the body. The development of malignant tumors is called *cancer.*

Tumors are usually named by appending the suffix *-oma* to the name of the tissue in which the tumor has arisen. Thus, a lymphoma is a cancer of the lymphoid tissue; a sarcoma is a cancer of fleshy, nonepithelial tissue; an adeno-carcinoma is a cancer of glandular tissue; and so on. An important exception to this system of nomenclature is *leukemia,* or cancer of the white blood cells.

The first evidence of a relationship between viruses and cancer was obtained as early as 1908, when V. Ellerman and O. Bang demonstrated that certain chicken leukemias could be transmitted to healthy birds by cell-free filtrates of diseased blood. Some years later, P. Rous was able similarly to transmit a chicken sarcoma. These discoveries received little attention at the time. Only in 1932, when R. E. Shope demonstrated a viral origin of rabbit papillomas, was importance attached to viruses as tumor agents.

An important step was taken when a *natural route of transmission* was found for a virus-induced cancer. In 1936 J. J. Bittner showed that mammary tumors of mice are caused by a virus that is transmitted from mother to offspring through the milk. Bittner's work led to an understanding of several important aspects of virus-induced cancer. First, an animal that is infected with a tumor virus at the time of birth may not develop a tumor until it has become an adult. Second, the ability of a virus to induce a tumor depends on certain environmental factors, such as the physiology of the host; mammary tumors of mice, for example, occur at high frequency only in animals undergoing hormonal stimulation characteristic of pregnancy. Even male mice will develop mammary tumors if injected with the hormone estradiol over a long period of time, provided that they are infected with the virus.

• The
transformation
of normal cells
into tumor cells

The induction of a tumor by a virus *in vivo* can be reproduced in tissue cultures. When a suspension of tumor virus particles is used to infect a susceptible tissue culture, some of the infected cells are transformed into a type that exhibits unregulated growth.

The process of *transformation,* as it is called,* causes a number of striking changes in the cell. Most animal cells in tissue culture exhibit the phenomenon of *contact inhibition:* the cells move about randomly by ameboid motion and divide repeatedly until they come into contact with one another; contact between cells inhibits both movement and cell division. The result is that normal cells in tissue culture form a monolayer on the surface of the glass vessel; transformed cells, however, do not exhibit contact inhibition, and they form tumorlike cell masses in tissue culture. Furthermore, cells transformed in tissue culture will initiate tumors when inoculated into host animals.

The efficiency of transformation varies greatly among tumor viruses. The Rous sarcoma virus, for example, transforms virtually every cell that it infects. Other tumor viruses may transform only one out of 10^3 to 10^5 infected cells. In order for transformation to be observed, it is necessary that the lethal effects of the virus be prevented. This is accomplished by using *nonpermissive host cells:* i.e., cells from a species that is incapable of supporting a lethal viral infection. For example, polyoma virus lethally infects mouse embryo or kidney cells, but not hamster or rat cells. The latter, which are thus nonpermissive, are transformed by polyoma virus at low efficiency.

• Types of tumor viruses

Oncogenic (tumor-producing) members are found within several groups of DNA viruses: adenoviruses, herpesviruses, poxviruses, and papovaviruses. Much of the experimental work on the mechanism of tumorigenesis has been done with members of this last group, which includes polyoma viruses (causing many types of tumors when injected into newborn mice), papilloma viruses (causing benign warts in man and other mammals), and SV-40, a simian virus isolated from cultures of rhesus monkey cells that were being used for the propagation of poliovirus in vaccine production.

Most attention has been focused on polyoma virus, because of its small size: it has only enough DNA to code for five proteins, and five genes have been identified by complementation tests between pairs of heat-sensitive mutants (such tests are described in Chapter 13). Two of the genes, which are transcribed late in infection, are involved in capsid formation, and two others, which are transcribed early, are involved in the transformation process. The product of one of these genes is probably involved in bringing about the integration of the viral DNA into the host genome (see below); the product of the other is apparently involved in the changes that free the host cell from its normal controls and confer on it the properties characteristic of the transformed state.

A limited group of RNA viruses are capable of oncogenesis: the leukemia, lymphoma, and sarcoma viruses of mice, cats, and chickens, and the mammary tumor virus of mice. They are similar in their physicochemical properties and are classified together as the *leukovirus group.* The ability of these viruses to cause tumors is apparently related to their unique mode of reproduction: as described earlier, these viruses possess a virion-associated reverse transcriptase, forming a DNA complement of virion RNA. This DNA complement is capable of entering the provirus state, presumably by integration into the host genome (see below).

---

* The term *transformation* is also used in biology to describe a totally different process: the formation of bacterial recombinants by the transfer of DNA extracted from donor cells (Chapter 15).

• The state of the viral genome in transformed cells

Following transformation by DNA viruses, cells of a virus-induced tumor or transformed cells of a tissue culture are completely free of detectable, infectious virus. Nevertheless, it can be demonstrated that part or all of the viral genome persists in the transformed cell and is reproduced at each cell generation. For example, when transformed cells are grown in mixture with susceptible, normal cells, under conditions that promote cell fusion, infectious virus particles may be liberated. In such cases, the fusion of a transformed cell with a normal cell allows the full expression of the latent viral genome.

The presence of at least part of the viral genome in a tumor cell can be demonstrated in two other ways. First, transformed cells can be shown to contain viral antigens—proteins that are coded for by viral genes. The nature of these antigens is not known; they are not capsid proteins, however. Second, transformed cells can be shown, by nucleic acid hybridization techniques, to contain viral DNA.

For several of the DNA tumor viruses, it has been shown that the viral DNA is *integrated into the DNA of the host cell*. Thus, when the DNA of transformed cells is fractionated and assayed for viral sequences by nucleic acid hybridization, these sequences are found to be covalently linked to host DNA sequences. Quantitative hybridization experiments indicate that more than one copy of the viral genome is present in each cell.

The evidence for the integration of DNA complementary to tumor virus RNA (formed by reverse transcriptase) is less direct. Such viral DNA sequences have been found in the nuclei of leukemic cells, for example, but a direct demonstration of their covalent linkage to host cell DNA has not yet been possible. The integration of viral DNA is suggested, however, by the presence in tumor virions of DNA nucleases and ligase—the enzymes that would be required for inserting a viral genome into host DNA by a series of breakage and rejoining events.

The ability of the tumor viruses to enter the *provirus state* means that they are capable of two modes of transmission: *horizontal transmission,* in which the virus passes from one cell to another by the release and adsorption of virions, and *vertical transmission,* in which the virus is transmitted from one cell generation to the next in form of a provirus. If the provirus enters a cell in the germ line of an animal, then it will also be vertically transmitted from one animal generation to the next. This possibility has formed the basis of a number of theories concerning the etiology of cancer, as discussed below.

• The relationship of viral transformation to lysogeny

The similarities between bacteriophage lysogeny, described earlier in this chapter, and animal virus transformation are striking. In both cases, the host cell acquires by infection a latent viral genome, or provirus, which replicates as part of—or in synchrony with—the host genome. Again, in both cases, one or a few viral genes remain actively expressed and confer new properties on the host cell. In particular, certain *surface properties* of the host cell are altered: lysogenized bacteria may acquire new surface antigens, and transformed mammalian cells lose the state of contact inhibition.

There appears to be one major difference between transformation and lysogeny, however. It will be recalled that the lysogenic state depends on the continued production of phage *repressor,* which shuts off the expression of all but a few phage genes. There is no evidence for the existence of repressors in transformed mammalian cells; instead, productive infection appears to be prevented by the *absence* from the nonpermissive host of factors required for viral development.*

• The role of viruses in human cancer: DNA viruses

Two groups of DNA viruses have been established as the etiological agents of *naturally occurring* cancers of animals: *papillomaviruses,* which cause malignant warts in rabbits (as well as benign warts in other animals, including man); and *herpesviruses,* which cause lymphomas in chickens (Marek's disease) and renal adeno-carcinomas in frogs.

Although a number of viruses found in human tissues can cause tumors when injected into animals (e.g., adenoviruses), few oncogenic viruses have ever been isolated from human cancers. The major exception to this rule is the herpesvirus called *Epstein-Barr* virus, or EB virus. EB virus was first isolated from a fatal form of human cancer called Burkitt lymphoma, and an apparently identical virus was later isolated from the cells of human nasopharyngeal carcinomas. The EB viruses are true tumor viruses, producing malignant lymphomas when injected into monkeys. They also have a unique effect on primate lymphocytes, which is the only type of cell that can be experimentally infected with EB virus: the infected cells are "transformed," acquiring the ability to grow continuously in culture and to produce malignant tumors in monkeys.

Recently, however, it has been found that a large fraction of the population of the United States carries EB viruses in their lymphocytes in the latent state, without producing overt disease. Liberated EB virions are found in the throats of individuals who are experiencing infectious mononucleosis (a nonmalignant disease), and these individuals exhibit a high titer of antibody against EB virus in their circulation. Thus, EB viruses are present in latent form in many healthy individuals, and they are released as mature virions in individuals suffering from Burkitt lymphoma and nasopharyngeal carcinoma, as well as in infectious mono-nucleosis. All of these EB viruses are indistinguishable from one another by existing laboratory methods: their DNAs hybridize with each other strongly, and they form identical antigens. They may, however, represent genetically different strains with differing pathogenicities.

It is not possible, despite the association of EB viruses with human diseases, to assign them an etiological role in either infectious mononucleosis or in human cancer. If they do play a role in these diseases, it is not clear why they affect so few of the individuals in which they occur. It is possible that other types of virus are involved in the induction of EB-associated tumors, and, indeed, an RNA virus has been found in Burkitt lymphoma cells.

The fact that other oncogenic DNA viruses have not been isolated from human tumors does not rule out a possible role for them in cancer. First of all, the lack of epidemiological evidence for contagion is inconclusive, since experience with tumor viruses in experimental animals shows that a very long time may elapse between infection and the appearance of a tumor. Second, the possibility exists that DNA tumor viruses are transmitted *vertically* from parent to child, causing cancers only when their transforming genes are activated by environmental factors such as chemical carcinogens or radiation.

---

*It should be remembered that a given bacterial host cell may undergo either lytic infection or lysogenization, depending on the competition between factors regulating repressor formation. In contrast, animal cells are either permissive or nonpermissive hosts: a given virus is always lethal in a permissive host, and transformation occurs only in a nonpermissive host.

• **The role of viruses in human cancer: RNA viruses**

A number of animal cancers have been shown to be caused by RNA tumor viruses: these are the leukemias, sarcomas, and lymphomas of mice, cats, and chickens, and the mammary tumors of mice. Each of these diseases has its analogy in a human cancer; using extremely sensitive detection techniques, S. Spiegelman and his associates have examined extracts of human cancer tissues for the presence of RNA viruses related to the etiological agents of the corresponding animal diseases.

Their procedure was to look for particles with the following properties: (1) a buoyant density falling within the normal range of the RNA tumor viruses, (2) the presence of 70S RNA, typical of RNA tumor viruses, (3) the presence of reverse transcriptase, (4) the ability of the reverse transcriptase to synthesize DNA complementary to the 70S RNA, and (5) sequence homology between the DNA so formed and the RNA of the tumor virus causing the analogous cancer in animals.

In each case they were successful: for example, particles with the above properties were found in human leukemic cells but not in normal cells, and the DNA synthesized endogenously by these particles hybridized to the RNA of mouse leukemia virus but not to other RNA tumor viruses. Similar homologies were found between the RNAs of particles from human mammary cancers, lymphomas, and sarcomas and the RNAs of the agents of the corresponding mouse cancers.

Spiegelman and his associates then went on to ask if similar particles could be isolated from human cancers for which no animal homology exists. Extracts were prepared from human brain, lung, and gastrointestinal tumors, and again particles were found with all of the above properties. (In this case, of course, the final test of homology with a known animal tumor virus RNA could not be made.)

Since the RNA-containing particles found in human cancers have not yet been shown to be infectious or oncogenic, final proof of an etiologic role is still lacking. Nevertheless, the specific homology of their base sequences with those of the RNA viruses causing analogous tumors in animals is fully in accord with such a relationship.

Even if RNA viruses were proved to carry the genes that govern the malignant state, the question of their origin and transmission would remain unanswered. With very few exceptions, believed to be laboratory artifacts, the RNA viruses found in animal tumors are poorly infective; and the infective RNA viruses that are capable of inducing tumors do so with very low efficiency. These facts, coupled with the ability of the tumors to enter the provirus state and undergo vertical transmission, have led to three different theories concerning the relationship of RNA viruses to cancer. They are the provirus theory, the oncogene theory, and the protovirus theory.

The *provirus theory* is the most simple and straightforward. According to this theory, the RNA tumor viruses are transmitted horizontally from individual to individual, entering the provirus state and causing cancers to appear at some later time as a result of the chance activation of their transforming genes. Occasionally they may be transmitted vertically. The provirus theory predicts that RNA tumor virus genomes should be found only in a fraction of the human population.

The *oncogene theory*, advanced by R. J. Huebner and G. J. Todaro, expresses a different view. According to their theory, the set of genes whose concerted activity leads to RNA virus production was acquired early in human evolution, either by infection and provirus integration or by endogenous genetic change. Included in this set of genes (which they refer to collectively as the *virogene*) is one or a set of genes whose activity brings about the malignant transformation: this gene, or genes, they call the *oncogene*.

The virogene and its included oncogene are thus proposed to be present in every cell of every human being, having persisted throughout human evolution as a result of serving some essential function. For example, the oncogene may play a role during embryogenesis, since antibodies against some tumor antigens cross-react with certain antigens of normal embryonic tissue. Some or all of the genes necessary for virus development and malignancy, however, are *repressed* under normal conditions, and may be derepressed under the influence of environmental agents (such as chemical or physical carcinogens). Derepression may lead to the production of virions, malignancy, or both.

The oncogene hypothesis gained credence when it was discovered that certain agents, such as bromodeoxyuridine, will induce the appearance of virions in many types of cells, both normal and malignant. These virions, which are detected by electron microscopy, are identical in appearance to the virions of the RNA tumor viruses. The *endogenous viruses*, as they are called, have not been shown to be oncogenic; indeed, many of them have not been capable of propagating further either in the host tissue of origin or in cultured cells from other species. Nevertheless, their appearance following induction in many types of cells suggests that all animal and human cells carry RNA viral genomes, presumably as DNA proviruses synthesized by reverse transcriptases.

The provirus and oncogene theories lead to different predictions concerning the DNAs of normal and cancer tissues. The provirus theory predicts that base sequences, complementary to the base sequences of the RNA tumor viruses, should be found in tumor tissue DNA but not in normal tissue DNA. The oncogene theory, however, predicts that such sequences should be present in the DNAs of all cells, the difference between cancer cells and normal cells being one of gene expression instead of gene presence.

Spiegelman's group tested these predictions by examining the DNAs of human leukemic cells and normal leukocytes for the presence of unique sequences. They found that leukemic cells, *but not normal leukocytes*, contain DNA sequences that are complementary to the sequences of the 70S RNA in the particles which we described earlier—particles that contain reverse transcriptase and whose RNA has sequence homology with the RNA of mouse luekemia virus. This finding supports the provirus hypothesis, at least with respect to human leukemia.

The *protovirus theory* was advanced by H. M. Temin, who—simultaneously with D. Baltimore—discovered the reverse transcriptases of the RNA tumor viruses. Temin proposes that reverse transcriptases, which have now been demonstrated in noninfected as well as in infected cells, play a role in normal embryogenesis. According to this theory, certain normal cell genes (the protoviruses) are duplicated by transcription into RNA and back into DNA; the duplicated segments are then reintegrated at new places in the genome (in the same cell or after transmission to neighboring cells), creating new gene sequences. These sequences result in new functions of the cell, as a normal developmental phenomenon; an occasional error in the process, however, may create an unusual set of genes whose expression results in the appearance of an RNA virus, malignancy, or both. According to the protovirus theory, then, RNA tumor viruses arise *de novo* as a perturbation of normal cellular differentiation; once formed, they can be transmitted horizontally and act as tumor viruses. Cancer, however, is visualized as arising mainly through aberrations in the normal developmental process.

The provirus theory is subject to direct experimental proof: if, for example, the particles found by Spiegelman and his colleagues to be uniquely present in cancer tissues can be shown to be oncogenic, their etiologic role in cancer will be confirmed. Since, on ethical grounds, experiments cannot be performed on humans, such tests will have to be done on laboratory animals, including primates. Thus, even if positive, the results cannot prove a role of the viruses in *human* cancers. Nevertheless, a suspected viral agent can be ruled out as the cause of a particular cancer if its genome can be shown to be absent from the DNA of the tumor cells.

The oncogene and protovirus hypotheses are more difficult to prove, because they rely on the demonstration that *all* normal cells either carry the genes for the production of RNA viruses and tumors or can generate them by gene duplications and rearrangements. Thus, while more and more evidence is appearing that implicates RNA viruses in the etiology of at least some human cancers, the problem of their exact roles remains unsolved.

FURTHER READING

Books   CASPAR, D. L. D., "Design Principles in Virus Particle Construction," in *Viral and Rickettsial Infections in Man*, F. Horsfall and I. Tamm (editors). Philadelphia: Lippincott, 1965.

FENNER, F., B. R. McAUSLAN, C. A. MIMMS, J. SAMBROOK, and D. O. WHITE, *The Biology of Animal Viruses*, 2nd ed. New York: Academic Press, 1974.

HERSHEY, A. D. (editor), *The Bacteriophage Lambda*. Cold Spring Harbor, N.Y.: Cold Spring Harbor Laboratory, 1971.

TOOZE, J. (editor), *The Molecular Biology of Tumour Viruses*. Cold Spring Harbor, N.Y.: Cold Spring Harbor Laboratory, 1973.

Reviews   BALTIMORE, D., "Expression of Animal Virus Genomes," *Bact. Rev.* **35** 235 (1971).

BARKSDALE, L., and S. ARDEN, "Persisting Bacteriophage Infections, Lysogeny and Phage Conversions," *Ann. Rev. Micriobiol.* **28**, 265 (1974).

CALENDAR, R., "The Regulation of Phage Development," *Ann. Rev. Microbiol.* **24**, 241 (1970).

DALES, S., "Early Events in Cell-Animal Virus Interactions," *Bact. Rev.* **37**, 103 (1973).

DIENER, T. O., "Viroids: The Smallest Known Agents of Infectious Disease," *Ann. Rev. Microbiol.* **28**, 23 (1974).

ECHOLS, H., "Developmental Pathways for the Temperate Phage: Lysis vs. Lysogeny," *Ann. Rev. Genetics* **6**, 157 (1972).

ECKHART, W., "Genetics of DNA Tumor Viruses," *Ann. Rev. Genetics* **8**, 301 (1974).

FENNER, F., "The Genetics of Animal Viruses," *Ann. Rev. Microbiol.* **24**, 297 (1970).

401 HERSKOWITZ, I., "Control of Gene Expression in Bacteriophage Lambda," *Ann. Rev. Genetics* **8**, 289 (1973).

JOKLIK, W. K., and H. J. ZWEERINK, "The Morphogenesis of Animal Viruses," *Ann. Rev. Genetics* **5**, 297 (1971).

LEMKE, P. A., and C. H. NASH, "Fungal Viruses," *Bact. Rev.* **3**, 29 (1974).

LWOFF, A., R. HORNE, and P. TOURNIER, "A System of Viruses," *Cold Spring Harbor Symp. Quant. Biol.* **27**, 51 (1962).

PADAN, E. and M. SHILO, "Cyanophages—Viruses Attacking Blue-Green Algae," *Bact. Rev.* **37**, 343 (1973).

SAMBROOK, J., "Transformation by Polyoma Virus and SV40," *Adv. Cancer Res.* **16**, 141 (1972).

TEMIN, H. M., "On the Origin of RNA Tumor Viruses," *Ann. Rev. Genetics* **8**, 155 (1974).

————, and D. BALTIMORE, "RNA-Directed DNA Synthesis and RNA Tumor Viruses," *Adv. Virus Res.* **17**, 129 (1972).

VALENTINE, R., R. WARD, and M. STRAND, "The Replication Cycle of RNA Bacteriophages," *Adv. Virus Res.* **15**, 1 (1969).

Original articles   BAXT, W. G., and S. SPIEGELMAN, "Nuclear DNA Sequences Present in Human Leukemic Cells and Absent in Normal Leukocytes," *Proc. Natl. Acad. Sci. U.S.* **69**, 3737 (1972).

MARVIN, D. A., and E. J. WACHTEL, "Structure and Assembly of Filamentous Bacterial Viruses," *Nature* **253**, 19 (1975).

MOLYNEUX, D. H., "Viruslike Particles in *Leishmania* Parasites," *Nature* **249**, 588 (1974).

13

MUTATION AND GENE FUNCTION
AT THE MOLECULAR LEVEL

The study of genetics was brought to the molecular level in one swift stroke when, in 1953, J. D. Watson and F. H. C. Crick published a proposal for the structure of DNA. The great significance of their model is that it provides an explanation in terms of molecular structure for both DNA replication and gene mutation. Since these processes underlie heredity and variation, respectively, it can safely be said that in the entire history of biological science only Darwin's recognition of the existence and fundamental mechanism of evolution has equaled their discovery in importance.

The Watson–Crick structure for DNA and its implications are now familiar to all students of biology; a brief summary of this topic was presented in Chapter 7. Given this background, it is now possible to define the gene, as well as mutation, in precise molecular terms. *A gene is a segment of DNA in which the sequence of bases determines, by transcription, the sequence of bases in an RNA molecule, and thus—by the process of translation—the sequence of amino acids in a polypeptide chain. Mutation can be defined as any permanent alteration in the sequence of bases of DNA*, even if this alteration does not have a detectable phenotypic effect.*

Although polypeptides vary considerably in chain length, the majority of those which have been accurately measured have a molecular weight of 30,000 to 40,000, corresponding to chains containing between 250 and 300 amino acid residues. If we take 300 amino acid residues to represent the size of an average polypeptide chain, then—given the triplet code—the average gene must contain about 1,000 nucleotide pairs.

---

*The term *gene* is also applied to certain segments of DNA that are transcribed but not translated (such as the genes coding for transfer RNAs and ribosomal RNAs) as well as to certain segments that are neither transcribed nor translated (such as the operator sites described on page 419). It should also be noted that, in the case of the RNA viruses, the genes are segments of RNA rather than of DNA.

A given gene can exist in a variety of different forms as a result of mutational changes in its nucleotide sequence; the different mutational forms of a gene are called *alleles*. The form in which a given gene exists in a microorganism as it is first isolated from nature (the wild-type organism) is defined as the *wild-type allele* of that gene; altered forms resulting from mutations are called *mutant alleles*.

THE CHEMICAL BASIS OF MUTATION

• **Types of mutation**   The sequence of nucleotides within a gene can be mutationally altered in any of several ways, the most frequent of which are base-pair substitutions, frame-shift mutations, and large deletions. In *base-pair substitution,* a base pair at a specific site in the wild-type allele, such as GC, is replaced in the mutant allele by a different base pair, such as AT or CG. In *frame-shift mutations,* so-named because they shift the "reading frame" of the translation process (see below), one or a few bases are inserted into, or deleted from, a specific site within the gene. In *large deletions,* a long sequence of bases, representing a major segment of the gene, is removed. Large deletions are irreversible; in contrast, base-pair substitutions and frame-shift mutations are reversible, by mechanisms which will be described later on in this chapter.

• **Mutagenesis**   Most of our understanding of the mechanisms of mutation derives from the results of experiments in which mutations are induced by chemical agents whose mode of action on DNA is known. Such mutagens include base analogues, which are incorporated into DNA in place of the natural bases; nitrous acid, which deaminates the purine and pyrimidine residues of DNA; acridines, which intercalate between the stacked bases of DNA, stretching the distance between them from the normal 3.4 to 6.8 A; and alkylating agents, which alkylate certain ring nitrogen atoms of the purine and pyrimidine bases.

Experiments on the chemical bases of mutation have been done almost exclusively with the bacterium *Escherichia coli* and with a number of phages for which *E. coli* is the normal host. In experiments with bacteria involving mutagens that are nontoxic, such as the base analogues, the mutagen is added to the growth medium for a number of generations, after which the cells are plated on a medium selective for a particular class of mutants. The class of mutants chosen is entirely a matter of technical convenience and experimental strategy, since *mutagens tend to raise the mutation rates of all genes more or less equally.* The mutation rate in the presence of the mutagen can be determined by various methods (several of which are described in Chapter 14) and compared with the rate in a control culture without mutagen.

For toxic mutagens, the experimental procedure that is generally followed is to treat a suspension of cells with the mutagen for a defined period of time. The cells are then washed free of mutagen and plated on a medium that selects for the growth of mutants in the surviving population. The results may be expressed quantitatively as the number of mutants induced per treated cell at a given level of survival.

Experiments on the mechanism of mutation are generally performed by inducing mutants with one mutagen and then testing their susceptibility to *reversion* by other mutagens. A common method of performing reversion tests is to spread a washed suspension of a mutant strain on agar that is selective for the revertant type; for example, auxotrophic cells may be spread on agar selective for prototrophic revertants. A drop of a solution of a mutagen is placed directly on the plate, so that the mutagen diffuses outward through the agar, forming a concentration gradient. If the mutant type is susceptible to reversion by that mutagen, a ring of revertant colonies will appear around the drop (Figure 13.1).

Mutagenesis experiments with bacteriophage are carried out in much the same way. Chemical agents that are capable of modifying existing DNA structures are added directly to suspensions of phage particles. The treated particles are then plated on lawns of appropriate bacterial strains to detect those particles with mutant properties, such as altered plaque type or host range. Agents that must be present during replication, such as base analogues, are added to host cells that have been infected with phage, and the liberated phage progeny are tested by plating on appropriate bacterial strains.

THE MOLECULAR BASIS OF MUTAGEN-INDUCED REVERSION   Simple reversion experiments of the types described above can only show that *the original phenotype is restored;* for example, if the primary mutation caused the wild-type strain to lose the activity of a particular biosynthetic enzyme, a revertant is a mutant that has regained that enzyme function.

Such a restoration of enzyme function can occur in different ways. First, it can occur by a true reverse mutation, which restores the original base-pair sequence. For example, if the primary mutation caused the substitution of an adenine-thymine (AT) base pair for a guanine-cytosine (GC) base pair, a true reverse mutation restores the GC pair at that position. Second, it can occur by substitution of a new base pair, different from that of the wild type or of the primary mutant, at the original site; e.g., the primary mutation may have been a GC $\longrightarrow$ AT change, and the reversion an AT $\longrightarrow$ CG (instead of an AT $\longrightarrow$ GC) change. In such a case, the revertant differs from wild type in one base pair, but enzyme activity may be normal—either because the code is degenerate and the revertant has the same amino acid as the wild type at that position, or because a new amino acid at that position is functionally equivalent to the amino acid of the wild type that it replaces.

Reversion may also occur as a result of a second mutation at a different site in the DNA, either within the same gene or in a different gene. Such second-site changes are called *suppressor mutations.* In *intragenic suppression* the second mutation produces a compensatory change in amino acid sequence which restores enzyme function; in *extragenic suppression* the second mutation alters a component of the translation system so as to correct the coding error caused by the first mutation. The mechanisms of suppression will be discussed later in this chapter; for the moment, it is sufficient to note that suppressor mutations are frequent.

In most experiments on mutational mechanisms the validity of the conclusions that are drawn rests on the certainty with which it can be inferred that *true reverse mutation* has occurred. In a true reverse mutation the phenotype of the revertant is indistinguishable from that of the wild type in all respects, and crosses between revertant and wild type fail to segregate the primary mutation; segregation should occur if the revertant carries both the primary mutation and a suppressor mutation.

Since both of these criteria are negative ones (*failure* to observe differences and *failure* to segregate a primary mutation), they cannot provide unequivocal

trpD	1	42	10	55

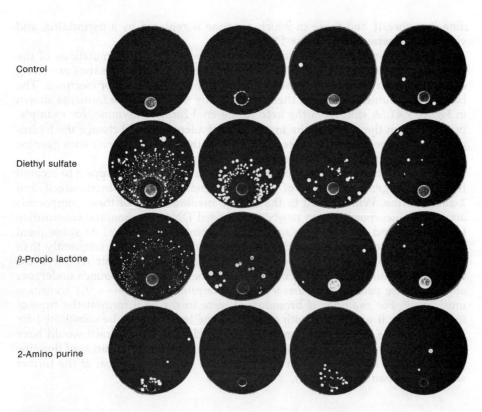

Control

Diethyl sulfate

β-Propio lactone

2-Amino purine

FIGURE 13.1

The induction of genetic reversions by chemical mutagens. Sixteen petri plates were spread with about 2×10^8 cells each of a tryptophan-dependent strain of *E. coli*. The agar contained sufficient tryptophan to allow about a tenfold increase in cell number, which is not enough to appear as visible growth to the naked eye. A drop of a mutagen solution (or sterile water as control) was placed on a filter-paper disc or in a hole made in the agar. The horizontal rows were treated with the mutagenic agents shown; the vertical rows represent different mutations in the trpD locus. Note that trpD1 is revertible by all three mutagens; trpD42 is revertible only by diethyl sulfate and β-propiolactone; trpD10 is revertible only by diethyl sulfate and 2-aminopurine; and trpD55 is revertible by none of the agents tested. From E. Balbinder, "The Fine Structure of the Loci tryC and tryD of *Salmonella typhimurium*. II. Studies of Reversion Patterns and the Behavior of Specific Alleles During Recombination, *Genetics* **47**, 545 (1962).

proof of true reverse mutation. Nevertheless, some of the mechanisms which have been deduced from experiments on reverse mutation have been directly confirmed by analysis of the amino acid sequence of enzymes isolated from wild type, mutant, and revertant strains.

MUTAGEN-INDUCED BASE-PAIR SUBSTITUTIONS   Mutations that represent the substitution of one base pair for another can be divided into two classes: those in which a purine is replaced by a different purine [or a pyrimidine by a different pyrimi-

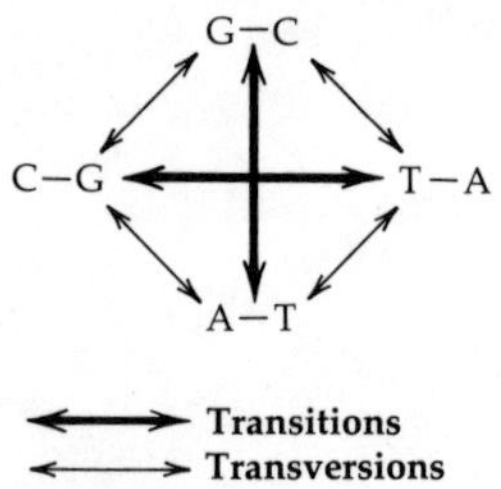

Transitions
Transversions

FIGURE 13.2

Base-pair substitutions. Those in which a purine is replaced by a different purine and a pyrimidine is replaced by a different pyrimidine are called *transitions*. Those in which a purine replaces a pyrimidine, or vice versa, are called *transversions*.

dine (*transitions*)], and those in which a purine is replaced by a pyrimidine, and vice versa (*transversions*) (Figure 13.2).

In their original paper, Watson and Crick pointed out that mutations of the type that we are now calling transitions should occur if, at the time they are acting as templates, any one of the bases underwent a *tautomeric shift* of electrons. The bases are presumed to exist in their energetically most probable forms, as shown in Figure 7.41. A shift from the keto to the enol form of thymine, for example, or a shift from the amino to the imino form of adenine would change the hydrogen-bonding specificities of these bases, so that thymine would pair with guanine and adenine would pair with cytosine (Figure 13.3).

The types of tautomeric shift postulated by Watson and Crick appear to account for the mutagenic activities of base analogues, notably 5-bromouracil and 2-aminopurine. When added to the culture medium, both of these compounds are readily incorporated into newly synthesized DNA, bromouracil substituting for thymine and aminopurine substituting mainly for adenine. At subsequent rounds of replication these analogues tautomerize much more frequently than do the natural bases, with the result that both compounds promote AT $\longrightarrow$ GC transitions. If, however, it happens that either one of these compounds undergoes abnormal base pairing *at the time it is being incorporated*, a GC $\longrightarrow$ AT transition must result. For example, if bromouracil were in the enol form at the time of incorporation, it would pair with guanine in the template and be substituted for cytosine. At the next replication cycle, by which time bromouracil would have resumed its more probable keto form, it would cause the incorporation of thymine in the opposite strand. Thus, an AT pair would eventually appear at the former site of a GC pair (Figure 13.4).

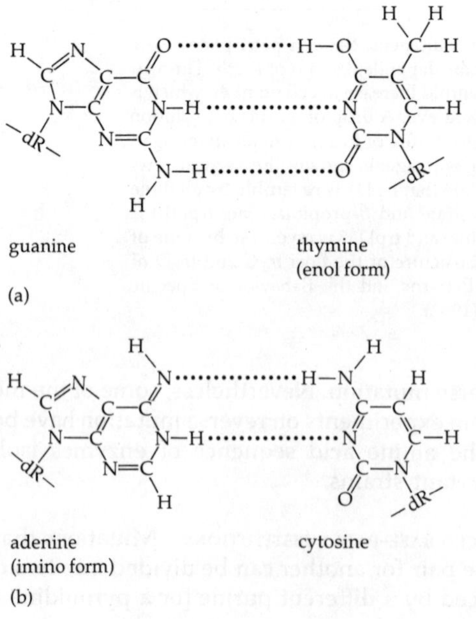

guanine

thymine
(enol form)

(a)

FIGURE 13.3

Changes in base pairing as the result of tautomeric shifts. In the enol form (a), thymine forms hydrogen bonds with guanine, instead of with adenine. In the imino form (b), adenine forms hydrogen bonds with cytosine, instead of with thymine. Similar shifts in guanine and cytosine will also cause changes in base pairing. (Compare the structures in this figure with those in Figure 7.41.)

adenine
(imino form)

cytosine

(b)

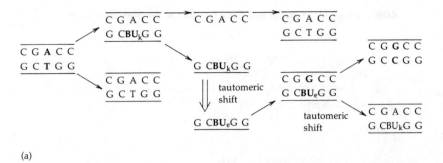

(a)

FIGURE 13.4

(a) When bromouracil (BU) is incorporated in its keto form (BU_k), a subsequent tautomeric shift during replication causes an AT → GC transition. (b) When BU is incorporated in its enol form (BU_e), a subsequent tautomeric shift during replication causes a GC → AT transition.

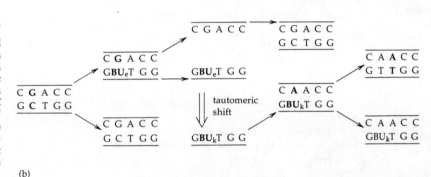

(b)

Nitrous acid is a third mutagen capable of inducing transitions in either direction. Nitrous acid deaminates the bases of DNA, a hydroxyl group replacing the amino group at the end of the reaction. As shown in Figure 13.5, deamination of the adenine of an AT pair will convert the adenine to hypoxanthine. At the next replication cycle, hypoxanthine will pair with cytosine, resulting in an AT ⟶ GC transition. Nitrous acid will also cause transitions in the opposite direction (GC ⟶ AT) by the deamination of cytosine to uracil.

Another transition inducer that has been studied is hydroxylamine. This compound reacts almost exclusively with cytosine, the net result of the reaction being the replacement of the amino group by a hydroxylamino group. The hydroxylamino derivative of cytosine undergoes frequent tautomerization, leading exclusively to GC ⟶ AT transitions.

The most powerful class of mutagens are the *alkylating agents*, some examples of which are mustard gas, ethylmethane sulfonate (EMS), and N-methyl-N′-nitro-N-nitrosoguanidine ("nitrosoguanidine" for short); the structures of these compounds are shown in Figure 13.6. They spontaneously transfer alkyl groups to ring nitrogen atoms of the bases in DNA; the most frequently observed reaction is the alkylation of the N-7 position of guanine, but a variety of other ring-nitrogen alkylations occur at lower frequencies.

Alkylation produces a number of chemical changes in DNA. The alkylation of guanine in the N-7 position causes ionization of the molecule, which could readily alter its base-pairing specificity; alkylation at certain other positions would block base pairing completely. Furthermore, the alkylation of the purines labilizes their bonds to deoxyribose, causing extensive depurination of the DNA. All of these changes probably contribute to the mutagenic action of the alkylating agents, which induce both transitions and transversions.

407

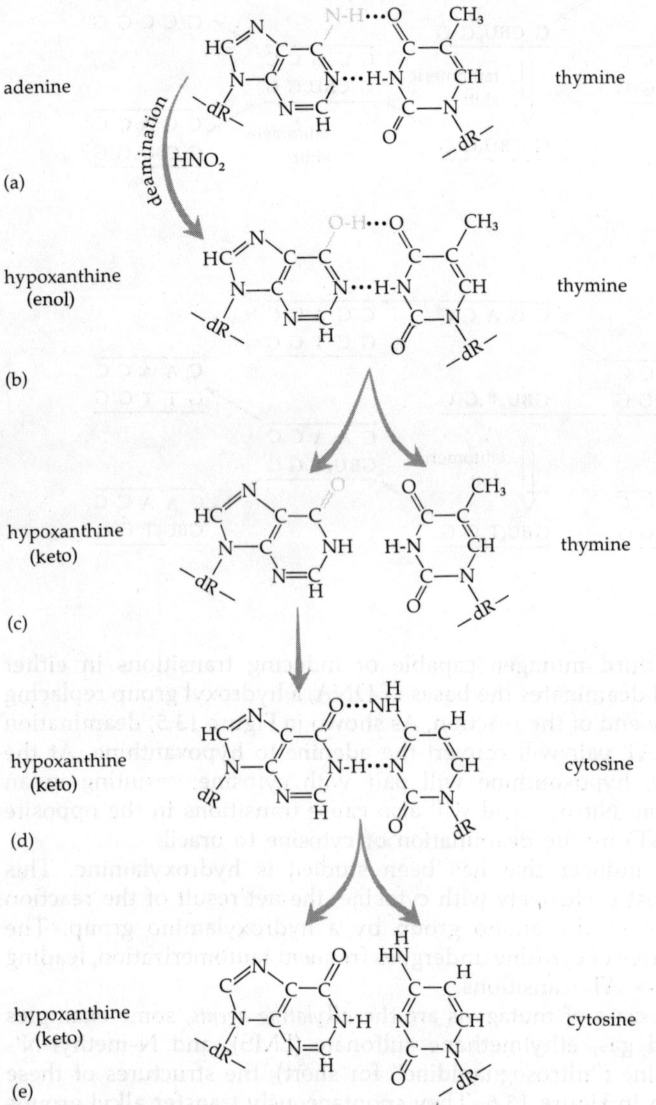

FIGURE 13.5

An example of nitrous acid-induced mutation. (a) An AT base pair in DNA. (b) Deamination of adenine to hypoxanthine (HX) produces an HXT base pair. (c) At strand separation, the hypoxanthine tautomerizes to the keto form. (d) At replication, a hypoxanthine-cytosine base pair is formed. (e, f) At the next round of replication, a CG base pair appears at a site formerly occupied by AT.

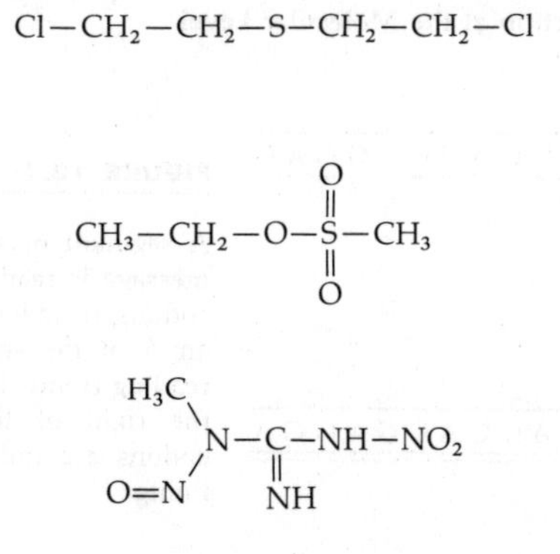

Cl—CH$_2$—CH$_2$—S—CH$_2$—CH$_2$—Cl   Sulfur mustard
[Di-(2-chloroethyl)-sulfide]

$$CH_3—CH_2—O—\overset{\overset{O}{\|}}{\underset{\underset{O}{\|}}{S}}—CH_3$$

Ethylmethane sulfonate
[EMS]

$$\underset{O=N}{\overset{H_3C}{\diagdown}}N—\overset{\overset{}{}}{\underset{\underset{NH}{\|}}{C}}—NH—NO_2$$

N-methyl-N'-nitro-N-nitrosoguanidine

FIGURE 13.6

The structures of some alkylating agents.

Nitrosoguanidine is the strongest chemical mutagen known; when used to treat *E. coli* under optimal conditions, it induces one or more mutations within each surviving cell. Typically, a cluster of several closely linked mutations is observed to occur within the region of the replicating fork of the DNA; the specificity for mutagenesis in this region is so high that it is possible to selectively mutagenize specific segments of the chromosome by adding the mutagen at precise times to populations of synchronously replicating cells.

MUTAGEN-INDUCED INSERTIONS AND DELETIONS   Acridine dyes, of which proflavin is the best-studied example, induce a unique class of mutations. Although they revert spontaneously and can be induced to revert at a higher rate by acridine dyes, such mutations are never reverted by base analogues, nitrous acid, or hydroxylamine. When reversion does occur, it usually proves to be the consequence of a second-site, or suppressor, mutation, within the same gene as the primary mutation. Furthermore, the suppressor mutation, when separated from the primary mutation by recombination, acts exactly like the primary mutation in every way: it inactivates the gene product and reverts only spontaneously or in response to acridine dyes. To put it another way, two acridine-induced mutations within the same locus may cancel each other out, so that the gene product remains functional.

Acridine-induced mutations exhibit an additional unique property: they are never "leaky" (i.e., they always produce a total loss of function of the gene product). Mutations of the transition type, however, are often leaky.

These special properties of acridine-induced mutants can be accounted for by assuming that *acridines cause the insertion or deletion of one or a few base pairs in DNA.* Since the message encoded in mRNA must be read as a series of triplets, insertion or deletion causes a shift in "reading frame," with the result that all codons beyond the site of mutation are changed (Figure 13.7).

Obviously, an insertion can be reversed by a deletion of the inserted base, and vice versa. It has been found, however, that a mutation caused by an insertion can be reversed by a deletion at a site many base pairs away; as shown in Figure 13.8, the revertant strain is normal except for a series of altered codons between the two mutational sites. The ability of such revertants to be selected means that many proteins can be functional even though extensive sequences of amino acids within their structures are altered.

The induction of insertions and deletions is related to the ability of acridines to intercalate between the stacked bases of DNA, as discussed earlier. It cannot

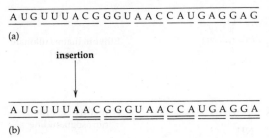

(a)

insertion

(b)

FIGURE 13.7

A segment of an mRNA molecule. The message is read as a series of triplets, or codons, from left to right. The insertion of an A at the seventh position shifts the reading frame, changing all the codons to the right of that point. The changed codons are indicated by double underlining.

be simply a matter of altering the DNA templates at the time of replication, however, since acridines will induce mutations in phages while they are replicating inside host bacteria, without having any mutagenic effect on the replicating bacterial DNA.

As an explanation of this curious phenomenon, it was suggested that acridines exert their mutagenic effect only in DNA that is undergoing *recombination*, since replicating phage DNA is known to recombine frequently, whereas bacterial DNA ordinarily does not. This proposal was supported by experiments with partial diploids of *E. coli*. Partially diploid zygotes are formed in bacterial conjugation as the result of incomplete transfer of the chromosome of one bacterial cell to another. When zygotes of *E. coli* are treated with an acridine, frame-shift mutations are induced in the diploid, but not in the haploid, region of the chromosome.

The process of recombination involves the formation of single-strand breaks in the paired DNA duplexes as an early event (see Chapter 15). This fact, together with the observation that acridines tend to act at sites characterized by short runs of identical base pairs, has led to the proposed mechanism of mutagenesis diagrammed in Figure 13.9. Acridines are suggested to stabilize the mispairing of strands which occurs when a single-strand break is formed adjacent to a repeat sequence.

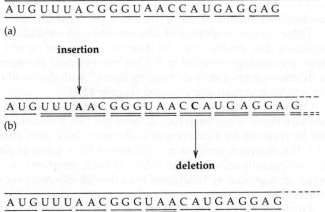

FIGURE 13.8

The reversion of a single insertion by a single deletion. Note that the revertant contains a series of altered codons (doubly underlined) between the mutant sites.

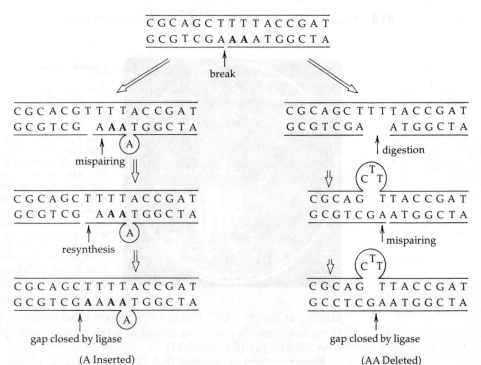

```
          C G C A G C T T T T A C C G A T
          G C G T C G A A A A T G G C T A
```

break

```
C G C A C G T T T T A C C G A T          C G C A G C T T T T A C C G A T
G C G T C G   A A A T G G C T A          G C G T C G A     A T G G C T A
```
          ↑ (A)   ↑ digestion
        mispairing ⇓⇓   ⇓⇓ (C T T)

```
C G C A G C T T T T A C C G A T          C G C A G   T T A C C G A T
G C G T C G   A A A T G G C T A          G C G T C G A A T G G C T A
```
          ↑ (A)   ↑ mispairing
        resynthesis ⇓   ⇓ (C T T)

```
C G C A G C T T T T A C C G A T          C G C A G   T T A C C G A T
G C G T C G A A A A T G G C T A          G C C T C G A A T G G C T A
```
          ↑ (A)   ↑
  gap closed by ligase   gap closed by ligase

      (A Inserted)   (AA Deleted)
```

**FIGURE 13.9**

A possible mechanism of frameshift mutation. A single-strand break occurs within a run of AT base pairs. (Left), mispairing followed by resynthesis and gap closure leads to the insertion of an A in one strand. (Right), digestion of two nucleotides, followed by mispairing and gap closure, leads to the deletion of two As from one strand.

MUTATIONS INDUCED BY ULTRAVIOLET LIGHT   DNA absorbs ultraviolet light strongly; the absorption maximum of DNA lies at a wavelength of 260 nm. Cells are rapidly killed by ultraviolet absorption, and a high rate of mutation occurs among the survivors.

When solutions of DNA are irradiated with ultraviolet light, two types of chemical changes take place. First, covalent bonds are formed between pyrimidine residues adjacent to each other in the same strand, forming *pyrimidine dimers*. These dimers distort the shape of the DNA molecule and interfere with normal base pairing. Second, pyrimidine residues are *hydrated* at the 4,5 double bond.

It is quite clear that much, if not all, of the mutagenic activity of ultraviolet light is attributable to the formation of pyrimidine dimers. The importance of pyrimidine dimers as a cause of ultraviolet light-induced mutation is shown by the fact that treatments leading to removal or cleavage of dimers reverse most of the mutagenic effect of ultraviolet light. If ultraviolet light-treated cells of some bacteria are immediately irradiated with visible light in the range 300 to 400 nm, for example, both mutation frequency and lethality are greatly reduced, a phenomenon known as *photoreactivation* (Figure 13.10). This has been shown to result from the activation, by light of the particular wavelengths in question, of an enzyme that hydrolyzes pyrimidine dimers.

All cells also possess an elaborate set of enzymes that effects the "dark repair" of ultraviolet light-damaged DNA. The repair process, which is apparently triggered by the distortion of the double helix in the region of a dimer, occurs in a series of steps. First, an endonuclease makes a cut in the sugar-phosphate backbone on either side of the dimer, *excising the dimer* as part of an oligonucleotide. Second, an exonuclease *widens the gap* by sequentially digesting the broken chain,
```

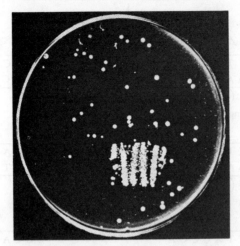

FIGURE 13.10

Photoreactivation. The surface of an agar plate was evenly and heavily inoculated with *E. coli,* and then was exposed to a dose of ultraviolet light sufficient to kill nearly all the cells. Thereafter, a portion of the plate was illuminated for a short time with visible light by focusing on its surface the image of a tungsten lamp filament. The plate was subsequently incubated in darkness until bacterial growth was complete. On most of the plate, only a few survivors of the ultraviolet irradiation gave rise to colonies. However, in the region illuminated with visible light, dense growth occurred. In this region the bacterial growth has the form of the tungsten lamp filament that was used for photoreactivation. From A. Kelner, "Ultraviolet Irradiated *E. coli,*" *J. Bacteriol.* **58**, 512 (1949).

starting at the 3' end.* Third, DNA polymerase *resynthesizes the missing segment,* using the opposite surviving strand as a template. Fourth, polynucleotide ligase *closes the final gap* (Figure 13.11).

Postirradiation conditions that favor such repair processes, while inhibiting DNA replication, cause many potential ultraviolet light-induced mutations to be lost. Conversely, conditions favoring replication but inhibiting repair increase the number of mutations in ultraviolet light-irradiated cells. The occurrence of a dimer-induced mutation therefore depends on the outcome of a race between the repair enzyme system and the replicating enzyme system. If the replicating enzymes reach the dimer first, a mutation may result; if the repair enzymes reach the dimer first, the dimer is excised and mutation is prevented.

The replication of DNA containing pyrimidine dimers ultimately leads to base-pair substitutions in the daughter duplexes. The mechanism of this process is complex, as illustrated in Figure 13.12. First, semiconservative replication produces two daughter duplexes with *single-strand gaps opposite pyrimidine dimers.* These duplexes are nonfunctional; recombination between sister duplexes, however, can reconstruct a fully normal duplex from the two damaged ones. *Most of the base-pair substitutions induced by ultraviolet light are formed by copying errors during the repair synthesis that constitutes one step in this recombination process.*† Mutants have been isolated in *E. coli* in which this repair process is much more accurate; in these mutants the mutagenic action of ultraviolet light is greatly diminished.

MUTATIONS INDUCED BY THE INSERTION OF A PROPHAGE   As described in Chapter 12, most prophages insert themselves into the bacterial chromosome at specific sites which have no other observable function. One phage, however, designated

---

*Endonucleases are enzymes that catalyze the hydrolysis of phosphodiester bonds within a polynucleotide chain. Exonucleases are enzymes that hydrolyze the terminal phosphodiester bond of a polynucleotide chain, splitting off the terminal nucleotide. Exonuclease action causes the *sequential digestion* of the chain.

†The mechanism of genetic recombination is discussed in Chapter 15, and illustrated in Figure 15.2.

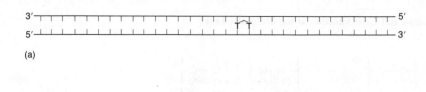

(a)

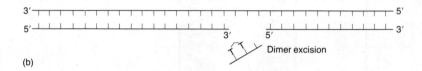

Dimer excision

(b)

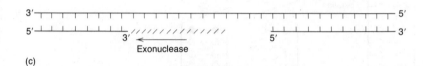

Exonuclease

(c)

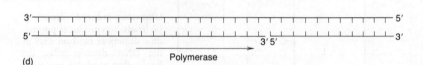

Polymerase

(d)

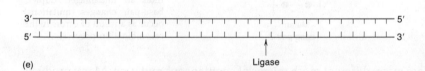

Ligase

(e)

FIGURE 13.11

Repair of DNA damaged by ultraviolet irradiation. (a) A thymine dimer is produced as a result of irradiation. (b) A specific endonuclease makes a cut on either side of the dimer, excising a small oligonucleotide. (c) The exposed 3' end is attacked by exonuclease, which widens the gap by sequentially digesting the strand. (d) DNA polymerase resynthesizes the missing segment. (e) Polynucleotide ligase seals the gap by joining the 3'-hydroxyl and 5'-phosphate.

mu, has the remarkable property of inserting into the bacterial chromosome at totally random sites within functional genes. Such an insertion constitutes a gross alteration in the base-pair sequence of the bacterial gene, and a normal gene product (e.g., enzyme) is not formed. Each lysogenized cell is thus an induced mutant for one or another gene; by appropriate selection following lysogenization with *mu*, one can isolate auxotrophic mutants or any other type characterized by the loss of a dispensable function.

• **Spontaneous mutations**

The term *spontaneous mutation* refers to those mutations which occur in the absence of known mutagenic treatment. The rate of spontaneous mutation is not a constant, however; many environmental conditions will affect it. In cells of *E. coli* grown at a very low rate in the chemostat, for example, the rate of spontaneous mutation depends on the nutrient that is limiting growth, and can vary widely. The mutation rate is lower in cells growing anaerobically than in cells growing aerobically.

413

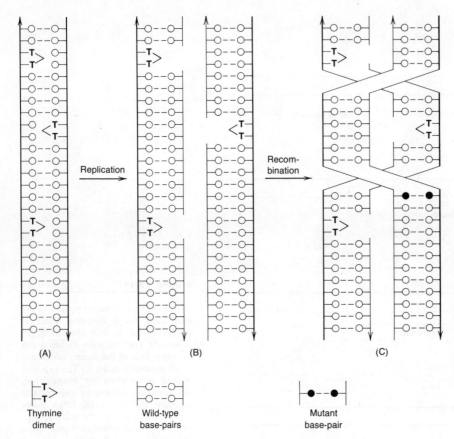

Thymine
dimer

Wild-type
base-pairs

Mutant
base-pair

FIGURE 13.12

Ultraviolet-induced mutation. (a) Irradiation produces pyrimidine dimers at random sites in the DNA double helix. (b) Replication produces two nonfunctional duplexes with gaps opposite dimers. (c) Sister-strand recombination reconstitutes an undamaged duplex; base-pair changes (mutations) occur as a result of errors in the recombination process.

There are many different mechanisms of spontaneous mutation. Many products or intermediates of cell metabolism are demonstrably mutagenic; these include peroxides, nitrous acid, formaldehyde, and purine analogues. Some spontaneous mutations may thus in reality be induced by endogenous mutagens. These would induce a mixture of types, including transitions, transversions, and possibly insertions and deletions. Reversion studies have been carried out that show that all these types do indeed occur among spontaneous mutations.

A mutation in one of the genes of *E. coli* and in a similar gene in *Salmonella typhimurium* has been shown to cause an increase in the spontaneous mutation rate for all loci by a factor of 100 to 1,000. The mutations caused by the new allele of the *mutator* locus are transversions. The product of the bacterial mutator gene has not been identified, but a similar mutator gene has been discovered in coliphage T4 and the product of this gene has been identified as DNA polymerase. The implications of this discovery are profound, since it demonstrates that the selection of nucleotides for the synthesis of a new strand of DNA depends not only on template-directed hydrogen bonding, but also on the fidelity of action of the polymerase. Since the mutationally altered polymerase causes a very high rate of "mispairings" leading to transversions, there is every reason to believe that the wildtype polymerases of phage and of bacteria are also responsible for spontaneous mispairings, although at a much lower rate than that of the altered polymerase.

Finally, a small fraction of spontaneous mutations has proved to reflect *deletions* of DNA segments. Their span varies from a large part of a single gene to several genes. For example, if one selects for mutants of *E. coli* strain B resistant to phage T1, a small fraction of the resistant mutants require tryptophan for growth. Genetic mapping experiments have shown that the genes for the T1 receptor and for the tryptophan biosynthetic enzymes are adjacent to each other, and in these *pleiotropic* mutants a spontaneous deletion has removed both. No known mutagen has been found to induce such large deletions; these are always spontaneous.

With the above mechanisms in mind, we can now understand two general ways in which environmental conditions could influence the rate of spontaneous mutation. First, the environment may be expected to influence appreciably the rate of formation of endogenous mutagens. Second, a number of the mechanisms described implicate the action of enzymes (DNA polymerase, recombination enzymes, repair enzymes), and both the synthesis and function of these enzymes may be subject to environmental influence.

THE EFFECTS OF MUTATION ON THE TRANSLATION PROCESS

• Nonsense
mutations

A codon is said to make "sense" if it is translated into the correct amino acid of the polypeptide for which the gene codes. A mutant codon that is translated into a different (incorrect) amino acid is said to contribute *missense* to the message, and the mutation which produced that codon is called a *missense mutation*.

Three mutant codons (UAG, UAA, and UGA) have been found to cause *premature chain termination*: when the ribosome reaches such a codon, the process of polypeptide chain elongation is terminated, and the incomplete polypeptide is released. Such codons are called *nonsense codons*, and the mutations producing them are called *nonsense mutations*.

The nature of the nonsense mutations was not understood at the time they were discovered. Their function being unknown, they were deliberately given the trivial names *amber* (mutations now known to produce the codon UAG) and *ochre* (mutations now known to produce the codon UAA). The codons so produced are commonly referred to as the *amber codon* (UAG) and the *ochre codon* (UAA). A corresponding name for the UGA codon has not come into common usage.

Following the discovery that nonsense codons cause premature chain termination, it was suggested that they might do so because there is no corresponding species of tRNA. This theory was confirmed by the studies of the binding of tRNA molecules to ribosomes in the presence of synthetic nucleotide triplets: of the 64 possible triplets, only UAG, UAA, and UGA do not stimulate the binding of any tRNA.

A glance at Table 7.8 will reveal the many possible base-pair substitutions that can change a "sense" codon to a nonsense codon: UGG, for example, which codes for tryptophan, changes to a nonsense codon (UGA) by a transition in the third letter.

The existence of three codons which can bring about chain termination raises the strong probability that one or more of them is the *natural chain terminator* in the translation process. As we shall see later, some groups of genes are transcribed into a single *polygenic messenger* RNA molecule; when a ribosome proceeds along a polygenic messenger, it must be given a signal to terminate one polypeptide chain and to start another. The termination signal is probably one of the nonsense codons.

• **Suppressor mutations**

The loss of activity of a protein resulting from a mutation can be at least partially restored by a second mutation at a different site. Mutations of the latter type are called *suppressor mutations,* and the genes in which they occur are called *suppressor genes.* The term *suppressor* is generally used to refer to the mutant allele of the suppressor gene.

Suppressor mutations can occur within the same gene as the primary mutation (*intragenic suppressors*) or within different genes (*extragenic suppressors*). Intragenic suppressors are of two types: one type causes an amino acid substitution that compensates for the primary missense mutation, restoring some function to the protein; the other type is an insertion or deletion that compensates for a primary frame-shift mutation.

The extragenic suppressors analyzed to date have been found to be mutations in genes coding for tRNAs (Figure 13.13). For example, a primary mutation in *E. coli* had changed a normal (sense) codon into the nonsense codon, UAG. A revertant of this mutant was isolated and was found to carry the original nonsense mutation together with an extragenic suppressor. The latter was shown to be a mutation in the gene coding for a *serine tRNA,* the revertant producing an altered tRNA with an anticodon that recognizes UAG. The revertant thus produces a normal polypeptide with serine at the position which—in the primary nonsense mutant—had been the site of premature chain termination.*

In a previous section of this chapter we showed that most suppressors of *frame-shift mutations* consist of compensating insertions or deletions within the same gene. Some frame-shift suppressors have been found to be *extragenic,* however, and these have proved to be mutations in tRNA genes. Thus, a frame shift resulting from the insertion of a single base can be restored to normal by a mutation that alters a tRNA anticodon so that it recognizes a sequence of four bases in messenger RNA instead of three.

Since the binding of tRNA to a codon in mRNA requires the presence of the ribosome, alterations in the ribosome might also affect the recognition process; in other words, an altered ribosome might occasionally permit a mutant codon to bind a different tRNA and thus to be suppressed. It has been found that streptomycin and chemically related antibiotics will bind to ribosomes derived from certain streptomycin-resistant mutants and will alter the reading of the genetic code in an *in vitro* system. The synthetic messenger polyuridylic acid, for example, normally promotes only the incorporation of phenylalanine into polypeptide, since it contains only UUU codons. In the presence of streptomycin, however, polyuridylic acid promotes the incorporation of isoleucine (codon: AUU), serine (codon: UCU), and other amino acids, as well as phenylalanine. This effect of streptomycin can lead to suppression of mutations *in vivo.* Some growth-factor-

---

*The revertant can still insert serine at its normal positions in the polypeptide, because the cell possesses several tRNAs for each amino acid, and the other serine tRNAs continue to function normally.

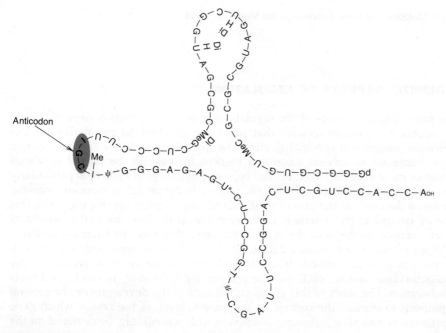

Anticodon

FIGURE 13.13

Schematic representation of alanine tRNA in one of its possible conformations. In this conformation there are four regions of hydrogen bonding between complementary bases. The anticodon, CGI, is complementary to the alanine codon, GCC (I has the same base-pairing specificity as G). Abbreviations: p, phosphate; A, adenosine-3′-phosphate; C, cytidine-3′-phosphate; A_{OH}, adenosine; DiHU, 5,6-dihydrouridine-3′-phosphate; DiMeG, N^2-dimethyl-guanosine-3′-phosphate I, inosine-3′-phosphate; MeG, 1-methyl-guanosine-3′-phosphate; MeI, 1-methylinosine-3′-phosphate; ψ, pseudouridine-3′-phosphate; T, ribothymidine-3′-phosphate; U, uridine-3′-phosphate; U*, a mixture of U and DiHU. From R. W. Holley et al., "Structure of a Ribonucleic Acid," *Science* **147**, 1462 (1965). Copyright 1965 by The American Association for the Advancement of Science.

dependent mutants, for example, will grow if furnished either with their required growth factor or with streptomycin; the cells grown on streptomycin are found to synthesize a small amount of the normal enzyme which had been lost as a consequence of the primary mutation.

Suppression by any of the above mechanisms has to be highly *inefficient* if the cell is to survive. A given suppression mechanism will alter the reading of the suppressible codon wherever it occurs in the DNA of the cell, and every codon must occur several times in almost every gene. Thus, if suppression is too efficient, it will lead to the inactivation of every protein in the cell. If suppression occurs at low efficiency, however, all the proteins in the cell will be made correctly most of the time. For example, when suppression operates with an efficiency of 5 percent, the mutant enzyme will be synthesized in its active form 5 percent of the time, whereas the other proteins in the cell will be made correctly 95 percent of the time. For many enzymes, 5 percent of wild-type activity is sufficient to permit growth of the cell. In fact, it is generally found that suppression restores the activity of the affected enzyme to a level that is ten percent or less of the level in the original wild type.

417

THE GENETIC ASPECTS OF REGULATION

The physiological aspects of the regulation of enzyme synthesis were discussed in Chapter 8. It will be recalled that substrates and cellular metabolites of low molecular weight can specifically affect the rate of synthesis of enzymes in two ways. Some act as *inducers,* causing a marked increase in the rate of synthesis of one or more specific enzymes; this type of physiological control is particularly frequent in the regulation of catabolic pathways. Some act as *repressors,* causing a marked decrease in the rate of synthesis of one or more specific enzymes; this type of control is characteristic of biosynthetic pathways, the end products of which repress synthesis of the specific enzymes that mediate their formation.

Much of our current knowledge about regulation of enzyme synthesis, at both the physiological and genetic levels, is derived from the intensive studies on the β-galactosidase system of *E. coli* carried out by J. Monod, F. Jacob, and their collaborators. The work of this group culminated in the development of a general hypothesis to explain the control, at the genetic level, of the rate at which gene expression occurs. The following discussion will, accordingly, be centered on the studies with the β-galactosidase system of *E. coli.*

The first clue to the mechanism by which this regulation is achieved was provided by the discovery of mutants of *E. coli* in which the synthesis of β-galactosidase had become constitutive. Genetic experiments showed that many of the mutations which make β-galactosidase synthesis constitutive map in one locus, which was named the *i locus* for "inducibility"; by present-day convention, this locus is now designated *lacI,* as one of a set of loci concerned with lactose utilization.

The product of the wild-type allele of the *lacI* locus ($lacI^+$) determines the inducible state; the product of the mutant allele ($lacI^-$) determines the constitutive state. To ascertain which of these two alleles is dominant, β-galactosidase formation was measured in *partial diploids* of *E. coli,* which were formed as transient zygotes in a mating between male and female cells.*

The zygotes, which were diploid and heterozygous for the *lacI* locus, were found to be inducible, establishing that $lacI^+$ is dominant over $lacI^-$. Evidently, the dominant $lacI^+$ allele produces a gene product that actively inhibits β-galactosidase formation; this product has been designated *repressor.*

The discovery of the *lac* repressor led to the general theory that inducible enzymes are ones for which the cell constantly synthesizes repressors; *the role of the inducer is to combine with and inactivate the repressor.* The theory was extended to explain end-product repression as well, by postulating the existence of genes that form repressors of biosynthetic enzymes. A repressor of this type would be nonfunctional as an inhibitor of enzyme formation, *unless activated by combination with the end product of the biosynthetic pathway.*

This extension of the repressor theory was soon confirmed by the discovery of genes which, when mutated, *derepress* the formation of biosynthetic enzymes, even in the presence of excess amounts of end products. The genes producing

*The nature and formation of zygotes in *E. coli* will be described at length in Chapter 15.

repressors have been given the general name *regulator genes* and are often symbolized by the letter R. The regulator gene for the arginine biosynthetic enzymes is thus *argR;* that for the tryptophan biosynthetic enzymes, *trpR;* and so on.

The general theory of genetic repression has been tested in cells that are diploid and heterozygous for regulator loci. Thus, an *argR⁺/argR⁻* diploid is repressed for the synthesis of the arginine biosynthetic enzymes, since the *argR⁺* allele of the diploid continues to synthesize a repressor that is activated by arginine. Regulator loci have now been identified for a number of catabolic and biosynthetic pathways.

The chemical nature of repressors remained obscure for some time. With the discovery that mutations in regulator genes can be suppressed by extragenic suppressors, however, it was necessary to conclude that they are proteins, since suppression acts only at the level of translation. The repressor protein produced by the *lacI* locus has indeed been isolated and identified by its ability to bind specific inducers; the mechanism by which repressors, as proteins, inhibit enzyme formation will be discussed in the following section.

• Operator genes   The existence of an inhibitor of enzyme synthesis, the repressor, implies the existence of a target in the cell for its action: there must be a *receptor site* that binds the repressor, shutting off synthesis of the specific enzyme.

From consideration of the general mechanism of protein synthesis, it is obvious that at least two possible sites exist: the repressor could bind either to a site on DNA, blocking transcription of the specific gene, or to a site on messenger RNA, blocking the translation of the specific mRNA molecule. Experiments with purified repressor and DNA corresponding to the *lac* region* have shown that the *lacI* repressor binds to DNA. Similarly, the repressors of the tryptophan (*trp*), arabinose (*ara*), and arginine (*arg*) genes, as well as the catabolite repressor (CAP) discussed in Chapter 8, have been shown to bind to DNA.

The segment of DNA that determines the binding site for repressor is called the *operator locus.* In the case of regulation of β-galactosidase synthesis, the repressor binds to the operator locus itself. The operator locus (*lacO*) can be identified by its capacity to mutate to a form (*lacO^c*) which cannot bind repressor, thus making enzyme synthesis constitutive.

An "operator constitutive" mutant can generally be distinguished from a "regulator constitutive" mutant of the *lacI⁻* type in partial diploids. Whereas a *lacI⁺/lacI⁻* diploid is inducible, by virtue of the production of repressor by the *lacI⁺* allele, a *lacO⁺/lacO^c* diploid† is constitutive, since the β-galactosidase gene adjacent to the *lacO^c* allele continues to function, even though repressor is present and able to bind to the *lacO⁺* allele on the other chromosome of the diploid.

To confirm that the operator locus controls the expression only of adjacent genes on the same chromosome, diploids constitutive for the operator locus and heterozygous for the β-galactosidase locus were prepared. The latter locus, which was designated Z (*lacZ* in the new terminology), was present in the diploids in two forms: *lacZ⁺*, making normal β-galactosidase, and *lacZ⁻*, making an inactive form of β-galactosidase which could be detected immunochemically. Diploids of

---

*Such DNA can be extracted from certain transducing phages which specifically incorporate a fragment of the bacterial chromosome that includes the *lac* genes (see Chapter 15).

†The *lacO* diploids were made with the use of F-*lac* episomes, which are described in Chapter 15 (*lacO*, wild-type allele; *lacO^c*, constitutive allele).

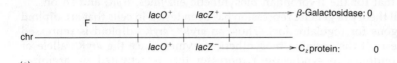

(a)

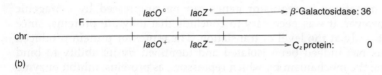

(b)

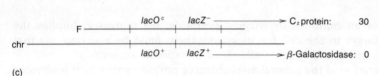

(c)

FIGURE 13.14

Test of the operator gene concept. (a) In a diploid in which both *lacO* alleles are wild type, neither β-galactosidase nor C$_Z$ protein is produced in uninduced cells. (b) When *lacZ*$^+$ is adjacent to the *lacO*^{c} allele, β-galactosidase is made constitutively. (c) When *lacZ*$^-$ is adjacent to the *lacO*^{c} allele, the C$_Z$ protein is made constitutively. The product of the gene adjacent to *lacO*$^+$ is not found in uninduced cells. Abbreviations: F, F factor; chr, chromosome. Numbers represent amounts of gene products in arbitrary units.

the types shown in Figure 13.14 were constructed, and both β-galactosidase and the inactive protein were measured in induced and uninduced cells. The results fully confirmed the theory: the *lacZ* allele adjacent to *lacO*$^+$ was always inducible, whereas the *lacZ* allele adjacent to *lacO*^{c} was always constitutive, although both were present in the same cell. This result ruled out the possibility that the operator produces a diffusible gene product which is able to influence the expression of a *lacZ* allele on the opposite chromosome of the pair.

• **Operons**

Early in the work on the genetic control of β-galactoside metabolism it was discovered that some mutants, unable to ferment lactose, contain normal amounts of β-galactosidase and are deficient for a specific permease. This permease actively transports β-galactosides into the cell; mutants lacking it can grow on lactose only if lactose is added to the medium at very high concentrations.

The mutations that affect the β-galactoside permease were all found to map in one locus, which was designated *Y* (*lacY*). Certain mutations were also observed to affect a third locus, *lacA*, governing a transacetylase of β-galactosides. Mapping experiments showed that the genes *lacO-lacZ-lacY-lacA* are tightly linked, forming a linear array in the order shown. When the experiments described above on the functions of the *lacI* and *lacO* genes were carried out, it was observed that the synthesis of β-galactosidase, permease, and transacetylase are always affected in an identical manner: all are inducible in *lacI*$^+$ cells and are constitutive in *lacI*$^-$ cells; all are inducible when their respective genes are adjacent to *lacO*$^+$, and all are constitutive when their respective genes are adjacent to *lacO*^{c}, even in diploid cells.

lacZ, *lacY*, and *lacA* thus behave as a unit of *coordinated expression*, and such a unit—together with its operator—is called an *operon*. Many other operons have since been discovered, functioning in both catabolic and biosynthetic pathways.

Since repressors are freely diffusible molecules, a regulator locus, in contrast to an operator locus, need not be immediately adjacent to the genes it regulates, and many regulator loci are indeed located on the chromosome some distance from the operons they control.

The operon provides an extremely efficient mechanism for the regulation of metabolic pathways. The enzymes of a metabolic pathway must function as a unit: if the pathway is operative, all its enzymes must function at roughly equivalent rates; if a pathway is not operative, none of its constituent enzymes is required for cell function. The clustering of the genes for the enzymes of a pathway provides a mechanism for coordinate control.

It only remains to ask how an operator, which has bound a molecule of repressor, can inhibit the transcription of a sequence of adjacent genes. This question has been answered by the discovery that all the genes of the operon are transcribed as a unit, forming a very long molecule of *polygenic messenger* RNA. The transcribing enzyme, RNA polymerase, binds to a site on the DNA called the *promoter*, and transcribes sequentially all the genes of the operon. The initiation of this process is sterically hindered when a repressor molecule is bound to the operator, which is immediately adjacent to the promoter. This mechanism is illustrated in Figure 13.15; in the case of the *lac* operon, the gene order is *promoter-lacO-lacZ-lacY-lacA*.

• **The translation of polygenic messenger RNA: polarity**

The transcription of an entire operon to form a single molecule of polygenic mRNA explains not only the ability of operator-bound repressor to inhibit the expression of all the genes in the operon simultaneously, but also the fact that nonsense mutations exert polarized effects on translation. The *polarity* resulting from a nonsense mutation is characterized by the greatly reduced synthesis of enzymes governed by all genes distal to the mutation (i.e., located on the opposite side of the mutation from the operator).

To illustrate the nature of a polarity effect, let us consider the case of a hypothetical operon containing five genes: operator-A-B-C-D. A nonsense mutation in A results in the total absence of enzyme A and reduced amounts of B, C, and D; a nonsense mutation in gene B results in the total absence of enzyme B and reduced amounts of enzymes C and D but normal synthesis of enzyme A. In some cases, polarity is absolute, the genes distal to the nonsense mutation being completely unexpressed.

Polarity effects can be interpreted as follows. The translation of a polygenic messenger must begin at the operator end, the only point to which ribosomes can attach. Once attached, the ribosome translates each gene in turn, an individual polypeptide being released when the ribosome reaches a chain-terminating codon. If a nonsense mutation produces a chain-terminating codon in the middle of a gene, however, the ribosome is discharged of its partial polypeptide, and such a *discharged ribosome has a high probability of becoming detached from the messenger* before it reaches the codon marking the beginning of the next gene. The messenger RNA, which is now unprotected in this region by a bound ribosome, is susceptible to nuclease attack and degradation.

Ribosomes which remain attached are able to reinitiate translation at the start of the next gene in the polygenic message, if the message has not been degraded. The closer the nonsense mutation to the end of the gene, the greater is the chance of this occurring before degradation can begin. Thus, the polarity effect is stronger for nonsense mutations which are located at the proximal (operator end) of the gene.

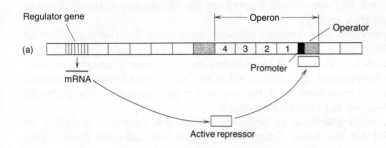

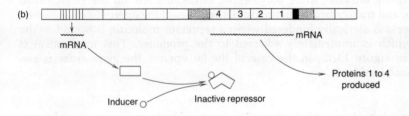

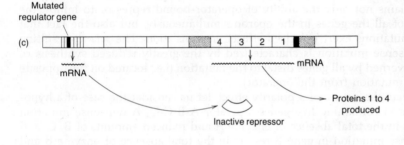

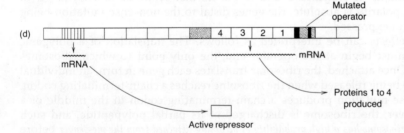

FIGURE 13.15

Regulation of an operon. The segmented horizontal bar represents a section of a chromosome, with each segment representing a gene. (a) Active repressor, produced by a regulator gene, binds to its specific operator and blocks transcription of the adjacent set of four genes. (b) An inducer molecule combines with repressor and inactivates it, preventing its binding to the operator. The operon is then expressed. (This situation is typical of an operon governing the enzymes of a catabolic pathway; in the case of biosynthetic pathways, the repressor is normally inactive unless it combines with a corepressor, such as the end product of the pathway.) (c) Inactive repressor is produced as a result of a mutation in the regulator gene; the operon is constitutively expressed. (d) A mutation in the operator prevents the binding of active repressor; the operon is constitutively expressed.

• **Diversity in the mechanisms of genetic regulation**   We have presented above only one mechanism of genetic regulation, that which is operative for the *lac* genes and their products. To recapitulate the essential features of this system, a series of genes with related functions are arranged in a sequence on the chromosome, adjacent to a special site called the *operator*. A *regulator gene* produces a product, the *lac repressor*, which binds to the operator and prevents the initiation of transcription of the *lac* genes. The *lac* genes remain unexpressed until an *inducer* enters the cell and combines with the repressor, inactivating it. The *lac* genes are then transcribed as a unit to form a polygenic messenger.

Many other operons have been detected, including sets of genes for the enzymes of biosynthetic pathways. In these cases, the repressor is normally inactive and is *activated* by combination with the end product of a specific biosynthetic pathway. This ability of a protein to be activated by a small molecule structurally unrelated to the substrate suggests that the protein is allosteric (see Chapter 8).

The genes for the enzymes of metabolic pathways are not always organized in operons. In *E. coli*, for example, many of the pathways are governed by genes that are scattered along the chromosome. In some of these systems, enzyme synthesis is coordinated by the product of a single regulator locus, indicating that one repressor can bind to a number of different receptors. In other cases of metabolic pathways governed by scattered genes, a common repressor is not involved. Instead, the different genes are separately regulated by the end products of the pathways, producing a much less precise coordination of their activities.

Some regulator gene products *activate*, rather than repress, the formation of enzymes. Several such cases have been discovered, including the regulation of alkaline phosphatase synthesis in *E. coli*. This enzyme allows the cell to obtain phosphate by the hydrolysis of organic phosphates in the medium. When the medium contains high levels of inorganic phosphate, the synthesis of alkaline phosphatase is repressed. This is brought about by the activation (by inorganic phosphate) of a repressor produced by a regulator gene. In the absence of inorganic phosphate, however, the repressor protein exists in a form in which it *activates* (rather than represses) the synthesis of alkaline phosphatase. Without this regulator gene product, no alkaline phosphatase is made, whether inorganic phosphate is present or absent. The effect of this arrangement is to render alkaline phosphatase synthesis doubly sensitive to the presence of inorganic phosphate, which converts an activator of enzyme synthesis to a repressor.

GENETIC COMPLEMENTATION

• Intergenic complementation: the *cis-trans* test of genetic function

At the beginning of this chapter we defined the gene as a segment of nucleic acid within which the sequence of bases determines the sequence of amino acids in a specific polypeptide chain. The gene is thus the *unit of genetic function*.

When two or more independently occurring mutations are observed to alter the same general property of a virus or cell, the question arises whether the different mutations have all occurred in the same gene or whether they have occurred in a number of different genes, all of which must function for the observed property to be normal. For example, wild-type cells of *E. coli* possess an enzyme, tryptophan synthetase, which catalyzes a complex reaction between indoleglycerol phosphate and serine to produce tryptophan plus glycerol phosphate (the final step in tryptophan biosynthesis: see Figure 7.22). Many mutants of *E. coli* have been isolated which lack tryptophan synthetase activity. Have all the mutations occurred in a single gene, or have they occurred in several different genes, all of which are involved in the formation of tryptophan synthetase? The latter would be true if tryptophan synthetase were composed of two or more *different* polypeptide chains.

Such a question can be tentatively answered by carrying out the *cis-trans test* of genetic function, devised by S. Benzer in the course of his studies on phage

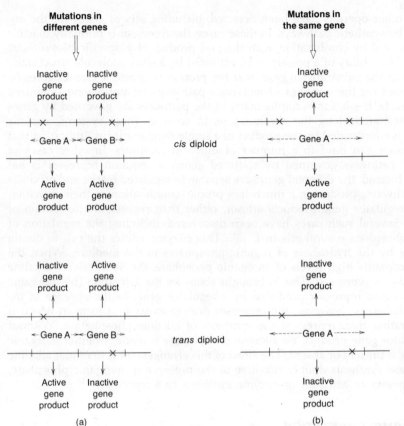

FIGURE 13.16

The *cis-trans* test of genetic function. (a) When the two mutations are in different genes, the *trans* diploid produces as much active gene product as does the *cis* diploid. (b) When the two mutations are in the same gene, only the *cis* diploid produces active gene product.

genetics. In this test two haploid genomes, each carrying an independent mutation affecting the same function, are brought together in the same diploid cell. The two mutations are said to be in the *trans* position. A second diploid, in which both mutations are present on the same genome, is also constructed; in this case the two mutations are said to be in the *cis* position. The two types of diploids are shown in Figure 13.16.

Inspection of Figure 13.16 will show that if the two mutations are in two different genes, both the *cis* and the *trans* diploids should produce the same amount of active gene product as a wild-type haploid cell. If, however, the two mutations lie in the *same* gene, the *cis* diploid will again produce the wild-type haploid amount of active gene product, but the *trans* diploid will produce a much lower amount or none at all.* The latter result therefore constitutes a "negative" *cis-trans* test.

Many pairs of tryptophan synthetase mutants were subjected to the *cis-trans* test. The results of the tests revealed that tryptophan synthetase is determined

---

* According to Figure 13.16, no functional gene product should appear in the *trans* diploid if both mutations lie in the same gene. Low amounts are often found, however, as a result of intragenic complementation. This phenomenon will be discussed in the next section.

by two genetic regions, designated A and B. *Trans* diploids carrying either two mutations in the A region or two mutations in the B region produce no tryptophan synthetase. When, however, any A-region mutation is put into a *trans* diploid with a B-region mutation, the wild-type level of tryptophan synthetase is formed.

The A and B regions thus constitute units of genetic function. Benzer coined the term *cistron* to denote a functional unit that has been identified by *cis-trans* tests; operationally, a cistron is a genetic region within which any two mutations produce a negative *cis-trans* test. Cistrons are generally presumed to correspond to genes; final confirmation, however, requires that the products of the cistrons be identified as separate polypeptides. This has been accomplished in the case of the A and B cistrons of tryptophan synthetase in *E. coli:* the two polypeptides, designated the A protein and B protein, combine to form an active molecule of enzyme.

The phenomenon described above, in which two genomes carrying mutations in different genes complement each other to restore the full wild-type phenotype, is called *intergenic complementation.* Complementation may also occur, however, between genomes carrying mutations at widely spaced sites within the *same* gene, as described in the following section.

• **Intragenic complementation** Many enzymes are polymeric proteins, containing two or more identical subunits. A single gene governs the structure of the subunit, several of which associate to form the active enzyme. When a mutation occurs in this gene, a defective subunit is formed and the enzyme is inactivated. When two different mutations in the same gene are brought together in a *trans* diploid, however, the two different types of defective subunits they produce may associate to produce an enzyme with partial activity (Figure 13.17). This phenomenon is called *intragenic complementation.*

For example, let us suppose that a haploid wild-type cell produces 100 units of a particular enzymatic activity and that 2 different haploid mutants each produce less than 1 unit. A *cis* diploid formed from the 2 mutants produces 100 units of activity. If a *trans* diploid formed from the 2 mutants is found to yield 30 units of activity, the results constitute a *positive complementation test*, indicating that the defective subunits produced by the 2 mutants are capable of complementing each other in the final polymer. These results, however, constitute a negative *cis-trans* test, since the *trans* diploid produces much less enzyme activity than does the *cis* diploid. The 2 mutations thus lie in the same cistron (see Figure 13.18).

MUTATIONS IN BACTERIOPHAGE

The nucleic acid of the mature virion is extremely stable; spontaneous mutations do not occur with a detectable frequency, even when suspensions of bacteriophage particles are stored for long periods of time.

Agents that react directly with DNA, however, can be used to induce mutations in extracellular phage particles. Thus, mutations can be induced by treating phage particles with such agents as nitrous acid, alkylating agents, or ultraviolet light. In each case, the nucleic acid is so altered that base-pair substitutions or other permanent changes in base sequence will result during the next replication cycle, when the treated particles infect new host cells.

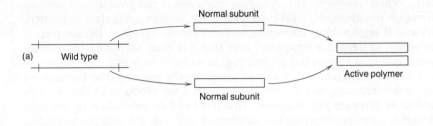

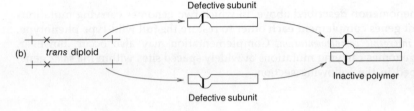

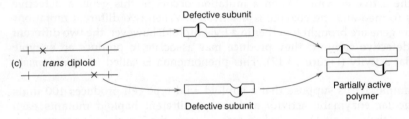

FIGURE 13.17

Intragenic complementation. (a) If, in a diploid cell, both alleles of the gene are wild type, the subunits that represent the gene products aggregate to produce a fully active polymeric enzyme. (b) If both alleles of the diploid carry the same mutation, the defective subunit produced will aggregate to form an inactive polymeric enzyme. (c) If the two alleles of the diploid carry two different mutations that are widely separated, the defective subunits produced may aggregate to form a polymer with partial activity.

Bacteriophage genomes that are replicating within their host cells, either as prophage or as vegetative phage, are subject to both spontaneous and induced mutation. All the mechanisms of mutation described at the beginning of this chapter are operative during phage replication; in fact, most of our current knowledge concerning these mechanisms has been derived from experiments on bacteriophage mutation.

Using bacteriophage T4, Benzer addressed himself to the following questions: Within a given gene, are all sites (e.g., base pairs) equally mutable? If not, do sites that are especially mutable with one mutagenic agent show similar mutability with other agents?

To answer these questions, Benzer carried out *fine-structure mapping* of mutations within the *rII* gene of phage T4. Phage particles carrying the wild-type allele of this gene form small, irregular plaques when plated on a lawn of sensitive cells of *E. coli* strain B. Phage particles carrying mutant alleles of the *rII* gene form large, sharply defined plaques: *rII* mutants are thus easily detected and may be isolated without difficulty.

Benzer discovered that wild-type (*rII⁺*) particles, but not *rII* mutant particles, can form plaques on a different host, *E. coli* strain K. In a mixed infection of strain

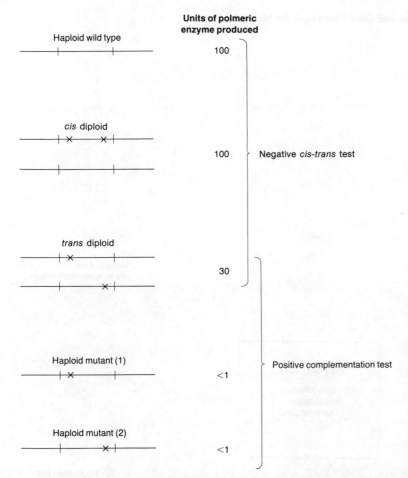

Haploid wild type

Units of polmeric
enzyme produced

100

cis diploid

100 — Negative *cis-trans* test

trans diploid

30

FIGURE 13.18

A comparison of the *cis-trans* test of
genetic function, and the complementation test. If the *trans* diploid produces
significantly less active enzyme than does
the *cis* diploid, the *cis-trans* test is negative
(indicating that the two mutations are in
the same gene). The ability of the different
defective subunits to complement each
other is indicated by the fact that the *trans*
diploid produces much more active
enzyme than does either haploid mutant
(positive complementation test).

Haploid mutant (1)

Positive complementation test

<1

Haploid mutant (2)

<1

B with two different *rII* mutants, most of the progeny will be of the parental types
and thus will fail to form plaques on strain K; if two *rII* mutant genomes *recombine*,
however, to form an *rII*⁺ recombinant, the resulting particle will be detected by
its ability to form a plaque on strain K.

Benzer isolated a large number of *rII* mutants and crossed them two at a time.
Each pair of mutants was used to mixedly infect *E. coli* strain B, and the progeny
were tested for the presence of recombinants able to form plaques on strain K.
The farther apart the two *rII* mutations, the higher the number of recombinants;
when the two *rII* mutations occupied the *same* site, however, no recombinants
were produced (Figure 13.19).

In this way Benzer constructed a map of the entire *rII* gene. Spontaneous
mutations were observed at over 300 distinct sites, each representing a different
base pair. The frequency of mutation at each site, however, was far from random:
two "hot spots" were observed, one that had mutated 52 times and the other
275 times. Furthermore, the distribution of mutational sites was different for
induced mutations, no 2 agents giving the same pattern of hot spots.

The hot spots cannot represent unusual base pairs, since only the four classical

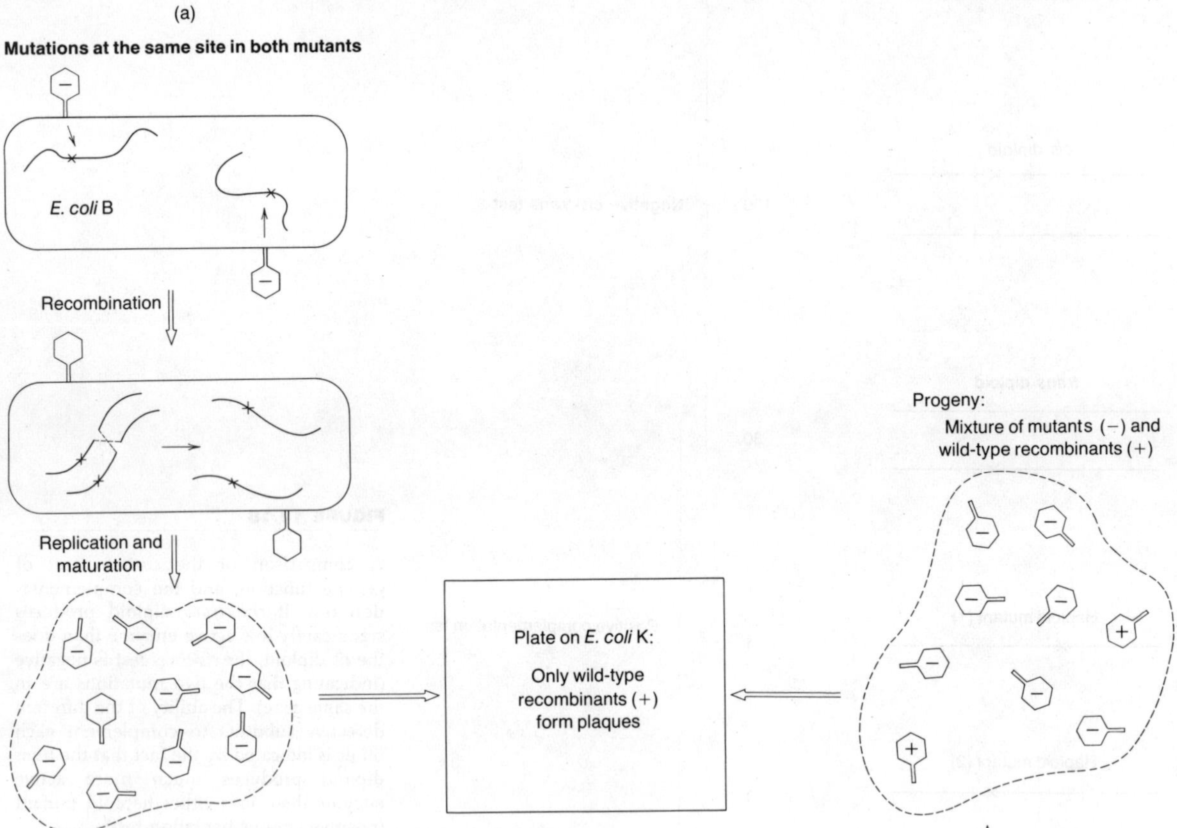

(a)

Mutations at the same site in both mutants

E. coli B

Recombination

Replication and maturation

Plate on *E. coli* K:

Only wild-type recombinants (+) form plaques

Progeny:
All mutant (−)

Progeny:
Mixture of mutants (−) and wild-type recombinants (+)

Replication and maturation

Double mutant

Wild type

Recombination

E. coli B

Mutation at a different site in each mutant

(b)

FIGURE 13.19

Fine-structure mapping of *rII* mutants. (a) Two *rII* mutants of phage T4 are allowed to infect jointly cells of *E. coli* B. Since both mutants carry mutations at the same site in the *rII* locus, recombination produces only the parental type of *rII* mutant, and no wild-type recombinants that can form plaques on *E. coli* K are produced. (b) The two *rII* mutants carry mutations at different sites. Crossing over between the sites produces a double mutant and a wild-type (+) recombinant. When the progeny are plated on *E. coli* K, the wild-type recombinants form plaques. The farther apart the two mutant sites, the higher the frequency of crossing over between them, and the larger the number of wild-type recombinants among the progeny of the cross.

base pairs are known to exist in T4 DNA. The unique susceptibility of the hot spots to mutation must then represent the influence of neighboring sequences of bases; the nature of this influence is not known.

FURTHER READING

Books BECKWITH, J. R., and D. ZIPSER (editors), *The Lactose Operon.* Cold Spring Harbor, N.Y.: Cold Spring Harbor Laboratory, 1970.

Chromosome Structure and Function. Cold Spring Harbor Symposia on Quantitative Biology, Vol. 38, 1973.

DRAKE, J. W., *The Molecular Basis of Mutation.* San Francisco: Holden-Day, 1970.

HARTMAN, P. E., and S. R. SUSKIND, *Gene Action,* 2nd ed. Englewood Cliffs, N.J.: Prentice-Hall, 1969.

HAYES, W., *The Genetics of Bacteria and Their Viruses,* 2nd ed. Oxford, England: Blackwell, 1968.

KING, R. C., *A Dictionary of Genetics,* 2nd ed. New York: Oxford University Press, 1972.

STENT, G. S., *Molecular Genetics.* San Francisco: Freeman, 1971.

WOESE, C. R., *The Genetic Code.* New York: Harper & Row, 1967.

Reviews AUERBACH, C., and B. J. Kilben, "Mutation in Eukaryotes," *Ann. Rev. Genetics* **5,** 163 (1971).

BECKWITH, J., and P. Rossow, "Analysis of Genetic Regulatory Mechanisms," *Ann. Rev. Genetics* **8,** 1 (1974).

BOYER, H. W., "DNA Restriction and Modification Mechanisms in Bacteria," *Ann. Rev. Microbiol.* **25,** 153 (1971).

CLARK, C. H. and D. M. SHANKEL, "Antimutagenesis in Microbial Systems," Bacteriol. Rev. **39,** 33 (1975).

ENGLESBERG, E., and G. WILCOX, "Regulation: Positive Control," *Ann. Rev. Genetics* **8,** 219 (1974).

FREESE, E., "Molecular Mechanism of Mutation," in *Chemical Mutagens: Principles and Methods for their Detection,* Vol. I, A. Hollaender (editor). New York: Plenum Press, 1971, pp. 1–56.

HARTMAN, P. E., and J. R. ROTH, "Mechanisms of Suppression," *Adv. in Genetics,* **16,** 1 (1973).

NEWCOMBE, H. B., "The Genetic Effects of Ionizing Radiations," *Ann. Rev. Genetics* **16,** 240 (1971).

REZNIKOFF, W. S., "The Operon Revisited," *Ann. Rev. Genetics* **6,** 133 (1972).

ROTH, J. R., "Frameshift Mutations," *Ann. Rev. Genetics* **8,** 319 (1974).

THE EXPRESSION OF MUTATION IN VIRUSES, CELLS, AND CELL POPULATIONS

14

All the properties of an organism are ultimately determined by its genes, including the genes of the organelles (the mitochondria and chloroplasts of eucaryotic cells) and plasmids, as well as the chromosomal genes in the nucleus. Each gene can exist in a variety of structural forms, or alleles, the nature of which was discussed in Chapter 13. The allelic states of all the genes in a cell constitute its *genotype*.

The structural and physiological properties of a cell constitute its *phenotype*. In the following sections we shall discuss the general ways in which the genotype of the cell determines its phenotype, and the manner in which changes in the genotype (mutations) bring about changes in the phenotype.

THE EFFECTS OF MUTATION ON PHENOTYPE

• The relation of phenotype to genotype

In Chapter 13 it was shown that each gene in the cell determines the structure of a single protein.* The structural and catalytic properties of the proteins determine in turn the anatomical and metabolic properties of the cell. The proteins interact with each other structurally to form the complex organelles of the cell, as well as functionally to form coordinated and regulated metabolic pathways. Thus, by altering the structure of a single protein, a mutation can bring about a profound change at the *secondary level* of cell structure and function.

* With the sole exception of operator genes, and genes that determine ribosomal RNA and tRNA.

• The effects of
mutation on
primary gene
products

For the purposes of this discussion, we shall consider the primary gene products of the cell to be proteins, since mutational effects on phenotype are usually the result of changes in protein structure. Such changes always consist of alterations in amino acid sequence resulting from mutational alterations in the base-pair sequence of DNA. The amino acid sequence may be altered by the substitution of one amino acid for another or by the deletion or insertion of a set of amino acids.

The alteration in amino acid sequence of a polypeptide may produce any one of the following alterations in the properties of the final protein product:

1. The protein may have an altered catalytic site; such alteration may cause the protein to be partially or completely inactivated as an enzyme.

2. The protein may become unusually sensitive to any of a number of agents: it may be inactivated by high or low temperatures, by an allosteric effector, by a metal ion, and so on. Thermolability as a consequence of mutation is a phenomenon of major interest, since it permits the study of mutations affecting indispensable cell functions (see below).

3. The polypeptide subunits of a polymeric protein may undergo abnormal association, with a consequent loss of normal catalytic or regulatory function.

4. There may be no detectable change in the physiochemical properties or catalytic activities of the protein. This may occur if one amino acid is substituted for another with very similar properties (e.g., the substitution of an acidic amino acid, glutamate, for another acidic amino acid, aspartate).

• Phenotypic
changes in
dispensable cell
functions

Depending on the environmental conditions, many cell functions are dispensable; mutations conferring the loss or alteration of these functions may thus be studied under conditions that permit the mutants to grow and divide. Such studies, often referred to collectively as *biochemical genetics*, have contributed a substantial portion of our knowledge concerning metabolic pathways, regulatory functions, and other aspects of cellular physiology.

The cell functions that are dispensable under some environmental conditions include the utilization of alternative forms of carbon, nitrogen, sulfur, or phosphate; the synthesis of the precursors of macromolecules and of coenzymes; the regulation of enzyme synthesis and activity; and the synthesis of various components of the cell surface: peptidoglycan, capsule, flagella, and pili. Some examples of the phenotypes that result from the loss of these functions will be briefly described in the following sections.

• Phenotypic
changes in
nutrition: sources
of elements

Many microorganisms are capable of utilizing *alternative* forms of carbon, nitrogen, sulfur, and phosphate. The loss of ability to use a particular form of an element (as a result of mutational inactivation of a protein involved in the pathway of utilization) is dispensable, provided that another utilizable form of that element is available. The activity lost by mutation may be either an enzymatic or a transport activity. The ability to use lactose as a source of carbon, for example, may result either from the loss of β-galactosidase activity or from the loss of β-galactoside transport (permease) activity.

Some of the catabolic pathways by means of which organic compounds are utilized were described in Chapter 6. Figure 6.13, for example, illustrates the pathways by means of which the pseudomonads utilize various aromatic compounds. Inspection of Figure 6.13 will reveal that the loss of any one enzyme of the convergent pathways will lead to the failure of the cell to utilize one or more aromatic compounds as sources of carbon and energy.

Many microorganisms are able to reduce sulfate and nitrate as sources of sulfhydryl and amino groups, respectively. In each case, a series of enzymatic reactions is required; the mutational loss of any of the enzymes involved will lead to the nutritional requirement for a more reduced form of sulfur or of nitrogen.

Most organisms can use organic phosphates as their source of phosphorus, hydrolyzing them to yield inorganic phosphate. In *E. coli*, for example, this is brought about by alkaline phosphatase, which is localized in the periplasmic space. The mutational loss of alkaline phosphatase activity confers on *E. coli* a nutritional requirement for inorganic phosphate.

• **Phenotypic changes in nutrition: growth factors**

When mutation causes the inactivation of an enzyme of a *dispensable biosynthetic pathway*, the cell becomes dependent on the environment for one or more growth factors. This phenotypic condition is termed *auxotrophy*; the original state of growth-factor independence is termed *prototrophy*. The dispensable biosynthetic pathways are restricted to those which produce small molecules capable of entering the cell and thus of serving as growth factors. The most common growth factors that may become required as a result of auxotrophic mutations are amino acids (precursors of proteins), purines and pyrimidines (precursors of nucleic acids), and vitamins (precursors of coenzymes).

The biosynthetic pathways for amino acids, purines, and pyrimidines, were outlined in Chapter 7. It will be noted that the purine and pyrimidine pathways involve phosphorylated intermediates and that the end products are nucleotides; neither the intermediates nor the end products can be supplied as growth factors to mutants that are blocked in these pathways, because the cell is impermeable to highly ionized (and thus to phosphorylated) compounds. Nevertheless, these pathways are dispensable if the medium contains the corresponding free purines or pyrimidines, since the cell is capable of taking up the free bases and converting them to nucleotides by the action of phosphoribosyl transferases (see Figure 7.15).

The biosynthetic pathways for the amino acids are so constituted that the loss of a single enzymatic activity may lead to one of several different *patterns of auxotrophy*. One pattern, which results from the loss of an enzyme prior to a branch in the pathway, is that of *multiple requirements*. For example, the loss of any enzymatic reactions leading to chorismic acid formation in the aromatic pathway (Figure 7.21) causes the cell to require phenylalanine, tyrosine, tryptophan, *para*-hydroxybenzoate, *para*-aminobenzoate, and 2,3-dihydroxybenzoate for growth. All six aromatic end products can be replaced, however, by shikimic acid, if the pathway is blocked prior to the formation of this compound. (Later intermediates, which are phosphorylated, cannot serve as growth factors.) Multiple requirements may also arise by the loss of an enzyme activity which is common to two different pathways; the inactivation of a single enzyme, for example, causes the cell to require both isoleucine and valine (Figure 7.25).

A second pattern produced by a single mutation is that of *alternative requirements*. This pattern arises as a consequence of the reversibility of certain reactions: the conversion of serine to glycine, for example, is freely reversible, so that a cell which is blocked in the biosynthesis of serine can use either serine or glycine as a growth factor (Figure 7.24).

• **Resistance and sensitivity to antimicrobial agents**

Mutational changes in protein structure can lead to increased resistance to antimicrobial agents, both chemical and physical. *Drug resistance** may reflect any of several different changes in primary gene products: the cell membrane may become impermeable to the drug; the cell may gain an enzymatic activity that confers on it the ability to degrade the drug or to inactivate it by chemical modification; or the component of the cell that is the primary target of drug action may acquire a reduced affinity for the drug.

Some structural analogues of normal metabolites are inhibitory to the growth of microorganisms. The reason for the inhibitory action of analogues varies; in some cases, the analogue is incorporated into a macromolecule in place of the normal metabolite, thus producing an inactive macromolecule (e.g., *p*-fluorophenylalanine is incorporated into protein in place of phenylalanine); in other cases, the analogue mimics the normal metabolite in producing end-product inhibition of the specific biosynthetic pathway, without replacing the metabolite for normal cell functions.

Resistance to analogues may thus arise in a number of different ways. Incorporation, for example, may be prevented by a mutation that alters the affinity of an amino acid-activating enzyme for the analogue, whereas resistance to false end-product inhibition may result from a change in the allosteric receptor of the sensitive enzyme.

Mutations can confer increased sensitivity, as well as increased resistance, to antimicrobial agents. The sensitivity of bacteria to ultraviolet light, for example, is greatly increased by the loss of one of the enzymes that effect the *repair* of DNA damaged by ultraviolet light. The enzymes of DNA repair were discussed in Chapter 13.

• **Phenotypic changes resulting from the loss of ability to synthesize components of the cell surface**

The structures of the cell which are exterior to the cell membrane are all dispensable under some environmental conditions. The cell wall, for example, is dispensable for bacterial cells growing in an isotonic medium; capsules, flagella, and pili are all dispensable under ordinary laboratory conditions. Mutations that deprive the cell of one or another of these structures are thus readily observed in the laboratory, and the changes can be traced to changes in the corresponding primary gene products.

Flagella and pili are each assembled from monomeric polypeptide subunits. Mutations in the genes for these polypeptides can prevent the synthesis of the polymerized appendages, and this in turn can produce secondary phenotypic changes. Flagellated cells, for example, may be sufficiently motile to produce spreading colonies on moist agar surfaces; nonflagellated mutants produce compact colonies on the same media.

The peptidoglycan layer of the bacterial cell wall contains diaminopimelic acid (DAP), which is formed as the last intermediate in the biosynthesis of lysine by bacteria (see Figure 7.18). If a mutation inactivates one of the enzymes that acts prior to the formation of DAP in the lysine biosynthetic pathway, the cell will

---

* Antimicrobial chemical agents used in chemotherapy are commonly called *drugs*.

require both DAP and lysine for growth. The cell will grow normally if provided with DAP, since DAP also furnishes a supply of lysine. If, however, the medium contains only lysine, the cell can synthesize proteins but lacks DAP for the synthesis of peptidoglycan, and its wall becomes osmotically fragile. Under these conditions, the cells will grow as L forms if placed in a medium of high osmotic strength; in ordinary media they will lyse. (L forms were discussed in Chapter 11.)

The production of a polysaccharide capsule by a bacterial cell involves a series of enzymatic reactions, forming a biosynthetic pathway. Cells that carry out the complete series of reactions are capsulated and form smooth, often mucoid, colonies; mutants, lacking one or another of the enzymes of the biosynthetic pathway, are noncapsulated and form "rough" colonies. This phenotypic change is accompanied, in pathogenic bacteria, by the loss of virulence, since the bacterial capsule affords the bacterium considerable resistance to phagocytosis in mammalian hosts.

• Pleiotropic mutations   A mutation that confers on the cell changes in two or more different phenotypic properties is said to be *pleiotropic*. Pleiotropy results when any gene product, either primary or secondary, has more than one function. We have already mentioned one example: a mutation that inactivates an enzyme in a biosynthetic pathway and thus deprives the cell of its capsule leads to the two secondary phenotypic changes: rough colony formation and the loss of virulence. Similarly, a mutation that inactivates a single component of the phosphotransferase system deprives the cell of the ability to utilize a large number of different, metabolically unrelated sugars.

THE SELECTION AND DETECTION OF MUTANTS

There are three general methods for selecting microbial mutants: selection based on relative growth; selection based on relative survival; and selection based on visual detection.

• Selection based on relative growth   In the first method the cells are plated on an agar-solidified medium, the composition of which permits only the desired type of mutant to form a visible colony; the wild-type, parental cells remain nondividing or are killed. For example, 10^6 or more *lac⁻* cells (unable to utilize lactose for growth) may be spread on agar containing lactose as the sole source of carbon; only the rare *lac⁺* mutants will grow and form colonies, the *lac⁻* cells being unable to divide. Similarly, 10^6 or more *his⁻* cells (unable to synthesize histidine) may be spread on agar lacking histidine; only the rare *his⁺* cells will grow and form colonies. The selective medium may, alternatively, be designed to actively inhibit or kill the parental cell population, rather than not to support its growth. For example, when a streptomycin-sensitive population of cells is spread on agar containing bactericidal concentrations of streptomycin, the parental cells are killed and the rare streptomycin-resistant mutants grow into visible colonies.

In the second method the selection is based on a condition in which growth of the desired mutant type is inhibited, and an agent is added which kills only the growing (parental) cell type. Thus, the desired mutant cells *survive* the lethal treatment, although they do not multiply during the selection process; they are transferred to a medium which supports their growth *after* the selection process has been completed. For example, penicillin kills only growing cells; when penicillin is added to a logarithmic phase culture of bacteria in liquid minimal medium,* the wild-type bacteria continue to grow until they are killed by the penicillin. If auxotrophic mutants are present, they fail to grow in the minimal medium and thus survive the penicillin treatment. At the end of the incubation period, the survivors are plated on a nutritionally complete medium and· the resulting colonies are picked and tested to confirm their auxotrophy and to determine their growth-factor requirements.

In the third method, plating conditions are designed so as to make colonies of the desired mutant type visually distinguishable from colonies of the wild type. For example, the colorless compound tetrazolium is reduced intracellularly to the brilliantly red, insoluble product, formazan, only within a narrow pH range. Thus, in a complete medium supplemented with a high concentration of a fermentable sugar, cells able to ferment that sugar lower the pH to the point where the dye is not reduced, and form white colonies. Mutant cells unable to ferment the provided sugar, however, reduce the tetrazolium intracellularly to formazan and produce bright red colonies. By this technique it is possible to detect a single fermentation-deficient mutant colony among 10^5 wild-type colonies on a petri dish.

In some cases, the only reagents which are able to stain mutant colonies differentially are also lethal. For example, one may wish to select mutants that form glycogen, which can only be detected by staining (and killing) the colonies with iodine. In such cases, the technique of *sib selection by replica plating is used:* a plate bearing thousands of colonies is replicated, as described below, and the replica plate is flooded with an iodine solution. If a glycogen-positive mutant colony is detected, an inoculum of *live* mutant cells can be recovered from the corresponding location on the original plate.

In replica plating, a piece of sterile velvet is stretched over a cylindrical block of wood or metal that is slightly smaller in diameter than a petri dish. The block is placed with the velvet surface facing upward; the petri dish with the lawn of bacterial colonies is inverted, and its surface is gently pressed against the velvet. The projecting fibers of the velvet, numbering thousands per square inch, act as inoculating needles, sampling every colony in the lawn. The petri dish is removed, and a fresh plate of agar is pressed against the velvet in order to receive an inoculum from each colony. The plates are identically oriented at each application of the velvet with respect to marks placed on their rims, so that the colonies that appear on the replica plate after incubation occupy positions congruent with those of their sibs on the original plate (Figure 14.1).

The inoculum on the velvet surface is usually large enough to permit a series of different agar plates to be sequentially "printed" from it. Thus, replica plating

*Minimal medium contains the minimal set of nutrients required for growth of the wild-type organism. For *E. coli*, this consists of mineral salts plus a carbon source such as glucose.

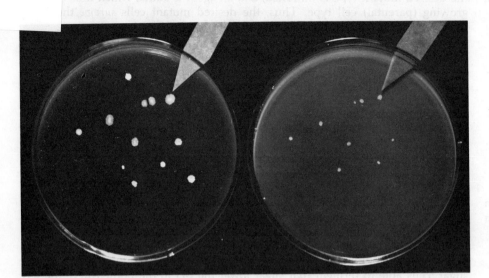

FIGURE 14.1

Replica plating. The master plate of nutrient agar (not shown) bore twelve colonies. One replica was prepared on nutrient agar (left) and one on a synthetic medium lacking growth factors (right). The two plates are similarly oriented, and the arrow points to sister replicas of one colony. Note that although twelve colonies developed on the complex medium, only nine were formed on the synthetic medium. The three colonies that failed to give replicas on the synthetic medium were made up of mutants that required growth factors for their development.

can also be used to test inocula from a very large number of colonies on a "master plate" for their ability to grow on as many as eight or ten different selective media. This technique has made practicable the multiple analyses which are basic to microbial and molecular genetics.

THE CONDITIONAL EXPRESSION OF GENE MUTATION

The mutations discussed in the preceding sections are expressed as functionally altered gene products, i.e., as enzymes that have lost their catalytic activity or as proteins that have lost their capacity to bind effectors or inhibitors. Many mutations, however, are *conditionally expressed*: under one set of conditions, called *permissive*, the mutation is not expressed, and the mutant cell forms a gene product that is functionally equivalent to that of the wild type; under a different set of conditions, called *nonpermissive*, the mutation is expressed and a functionally altered product is produced. Some mechanisms of conditional expression are described below.

The phenomenon of conditional expression has made possible the isolation and study of mutations which, when expressed, deprive the cell of *indispensable* *functions,* i.e., they are *lethal under all conditions of cultivation.* Mutations which inactivate DNA polymerase III, or RNA polymerase, or an amino-acyl tRNA synthetase, are examples of this class of mutations: the products of the reactions catalyzed by these enzymes are indispensable to the cell, and cannot be supplied from the environment. Such mutations, if expressed, are necessarily lethal.*

Many *nonlethal mutations* are also conditionally expressed. Thus, some mutants are auxotrophic under nonpermissive conditions, forming an inactive biosynthetic enzyme, and prototrophic under permissive conditions, forming an active biosynthetic enzyme. The expressed mutation is not lethal, since the end product of the biosynthetic pathway can be supplied from the medium.

The great utility of conditionally expressed mutations, particularly lethal mutations, is that the mutant can be cultured normally under permissive conditions. At a given moment, the mutant can be transferred to the nonpermissive environment and the events that follow the expression of the mutation can be followed by physical, chemical and microscopic analysis. For example, the effects on the cell of abruptly halting protein synthesis can be observed by switching an amino-acyl tRNA synthetase mutant from the permissive to the nonpermissive condition.

There are two different ways in which a mutation can be conditionally expressible. In one the *production of the altered protein* is conditional; in the other the altered protein is formed, but its *loss of normal function* is conditional.

• **The conditional production of an altered protein**

In Chapter 13, we saw one mechanism of conditional mutation expression, namely, *genetic suppression.* A bacteriophage carrying an amber (nonsense) mutation in a gene governing an indispensable function will develop normally in a permissive host (carrying an amber suppressor mutation), but it will develop abortively in a nonpermissive host (lacking the suppressor mutation). For example, if a nonpermissive host is infected with a phage T4 mutant carrying an amber mutation in the gene governing the head protein, the cells will produce only incomplete phages. Lysis will occur because of the production of phage lysozyme, but the lysate will contain only phage tail components.

Suppression (in the sense of translational correction) can also occur nongenetically. One type of *nongenetic suppression* was also mentioned in Chapter 13, namely, suppression by streptomycin. In some streptomycin-resistant mutants the binding of streptomycin to the ribosome alters the translation of the mutant codon, restoring the functional amino acid sequence. A number of conditionally expressed lethal mutations have been studied, using the presence of streptomycin as the permissive condition.

A second type of nongenetic suppression can be obtained by allowing the mutant cell to incorporate 5-fluorouracil (FU) into its messenger RNA. A mutant codon containing a U will, on replacement of uracil by FU, frequently be read as an A during translation, because of FU's hydrogen-bonding properties. This change may determine that the wild-type amino acid is inserted at that position, thus restoring the normal function of the protein.

* From now on we shall use the term *lethal* to mean *lethal under all conditions of cultivation.* Conditionally expressed mutations which, if expressed, are lethal have been widely referred to as *conditional lethal mutations.* This is confusing terminology, since it is not the lethality of the mutation which is conditional, but its expression. Even auxotrophy is conditionally lethal, i.e., in the absence of the required growth factor.

• **The conditional loss of normal protein function**

In many cases, a mutationally altered protein loses its normal function only under certain environmental conditions. Two classes of such mutations have been observed: *temperature-sensitive mutations* (including both heat- and cold-sensitive types) and *salt-remedial mutations*.

Heat-sensitive mutations are those that render the protein thermolabile at temperatures approaching the higher end of the physiological range of the organism. For example, many proteins of *E. coli* may be altered by mutation such that they are denatured at 40°C, while remaining intact at 30°C. Conversely, mutant proteins have been observed which cannot function at the lower end of the temperature range, remaining functional at higher temperatures: these are termed *cold-sensitive mutations*.

A number of mutants have been found which grow normally in media of high osmotic strength, but not in media of low osmotic strength. Called *salt remedial* mutants, they were found to have mutationally altered enzymes which are denatured unless stabilized by high salt concentrations.

The temperature-sensitive and salt-remedial mutations are thus cases in which the basis of the conditional expression is the retention or loss of function of an altered protein, rather than its formation or nonformation.

• **Conditionally expressed lethal mutations of bacteriophage**

Most viral gene products are indispensable for the production of normal virus particles: although some alterations in the coat proteins and viral-governed enzymes may be tolerated, most alterations prevent the normal synthesis and assembly of viral components.

The phenomenon of conditional expression, however, has permitted the isolation and study of mutations in virtually every type of phage gene. In phage T4, heat-sensitive and genetically suppressible nonsense mutations have been found for every phage function, and they have been mapped by the genetic recombination techniques discussed in Chapter 15. A simplified version of the genetic map of phage T4 is shown in Figure 14.2.

THE TIME COURSE OF PHENOTYPIC EXPRESSION OF MUTATION

When the primary effect of a mutation is the loss of a stable gene product, there may be a delay of several generations before the mutation is phenotypically expressed. This delay is called *phenotypic lag;* it may reflect either the time required for nuclear segregation or the time required for the dilution of the active gene product that is no longer synthesized.

• **Phenotypic lag: nuclear segregation**

Most bacteria contain an average of two to four nuclei per cell during the exponential phase of growth. These nuclei are genetically identical, since they are derived from a single nucleus that existed in the cell line one or two generations earlier. When a mutation occurs in one of the nuclei, the cell becomes a *heterocaryon*. If the mutation is one that causes the loss of a gene product, it will be *recessive* in the heterocaryon, since the other nuclei continue to make the gene

439

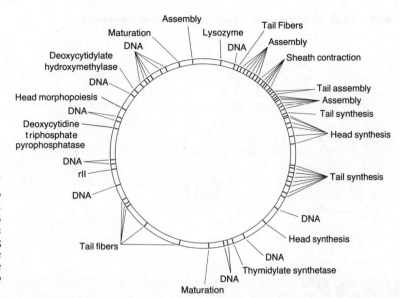

Assembly

Maturation   Lysozyme   Tail Fibers
DNA   DNA   Assembly

Deoxycytidylate   Sheath contraction
hydroxymethylase

DNA   Tail assembly
                                             Assembly
Head morphopoiesis   Tail synthesis

DNA   Head synthesis

Deoxycytidine
triphosphate
pyrophosphatase

DNA   Tail synthesis

rII

DNA

                                             DNA

                                       Head synthesis

Tail fibers

                                  DNA

                           Thymidylate synthetase

                  DNA

Maturation

FIGURE 14.2

A simplified genetic map of bacteriophage T4. The chromosome of T4 is represented by a ring; the dark lines in the ring represent genes. Note that many genes are clustered according to function.

product. Thus, one or two cell generations are required for nuclear segregation to produce a homocaryotic mutant cell, permitting a loss mutation to be phenotypically expressed. This situation is shown in Figure 14.3.

A bacterium is a *haploid* organism, because all its nuclei are identical. Even when genetic heterogeneity is produced by mutation or—as will be described in Chapter 15—by intercellular gene transfer, the haploid condition is restored within a few generations by the process of cell division and nuclear segregation.

• Phenotypic lag: dilution of active gene product

When a suspension of phage-sensitive bacteria is treated with a mutagenic agent (see Chapter 13) and the survivors are plated immediately on phage-coated agar, virtually no induced phage-resistant mutants develop. If, however, the survivors are permitted to undergo several generations of growth in nutrient broth before plating with phage, a large number of resistant mutants is obtained. The results of a typical experiment are shown in Figure 14.4, which shows that some of the induced mutations take as long as 14 generations to be expressed.

The basis of this delay in phenotypic expression became clear when the mechanism of phage resistance was elucidated. A sensitive bacterium adsorbs phage by means of specific receptors in the cell wall; resistant mutants lack these receptors. At the time that the sensitive cell undergoes the genetic change to resistance, its wall still possesses the preexisting receptor sites and the cell remains phenotypically sensitive. During subsequent growth, however, receptors are no longer synthesized and the old ones are diluted by the formation of new cell wall material. Phenotypic expression of the mutation to resistance is thus delayed until the mutated cells possess too few receptors to allow phage adsorption.

In general, a phenotypic lag occurs whenever the primary effect of a mutation is the loss of a stable gene product. Phenotypic lags will thus be observed in mutations from prototrophy to auxotrophy and in mutations that deprive the cell of the ability to use a particular form of carbon or nitrogen source. In each of these cases, the primary effect of the mutation is the loss of an enzyme activity;

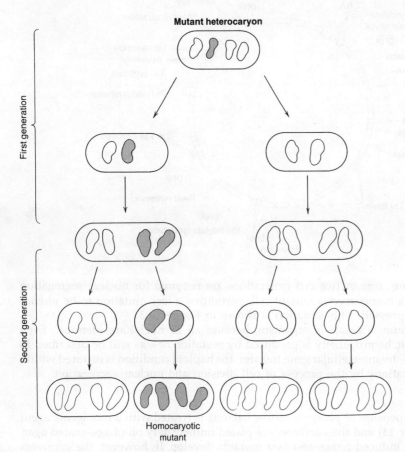

Mutant heterocaryon

First generation

Second generation

Homocaryotic
mutant

FIGURE 14.3

Bacterial cells with either two or four nuclei. If a mutation first occurs in a tetranucleate cell, two generations are required before a homocaryotic mutant cell appears. If the mutation is recessive, it cannot be expressed until the mutant nucleus has completely segregated from unmutated nuclei.

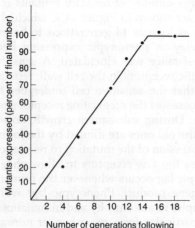

FIGURE 14.4

Delay in phenotypic expression of mutation. A suspension of phage-sensitive bacteria is treated with a mutagen and the survivors are plated on phage-coated agar. Only induced phage-resistant mutants appear. In this experiment the survivors were allowed to produce varying numbers of generations of growth before plating.

at the time of the mutation, the cell possesses a large number of enzyme molecules which must be diluted out by cell growth and division before the effect of the mutation is observed.

The practical significance of phenotypic lag is seen in the use of the "penicillin technique" for the selection of auxotrophic or other mutants with loss of enzymatic function. This technique exploits the fact that penicillin kills only those cells which are actively growing; if the survivors of mutagenic action are treated with penicillin immediately, the mutants are killed as effectively as the wild type. It is necessary to permit the survivors to undergo several generations of growth before treatment with penicillin, to allow the delayed phenotypic expression of "loss" mutations.

When a mutation results in the gain, rather than the loss, of a gene product, phenotypic expression is for all practical purposes immediate. If an auxotroph, for example, mutates to regain the ability to form a biosynthetic enzyme, active enzyme molecules begin to be synthesized immediately and to function in the biosynthetic pathway that had previously been blocked.

POPULATION DYNAMICS

A growing population of microbial cells is in a dynamic state with respect to the presence of mutant types. Two parameters are involved in this phenomenon: *mutation rate*, which can be assigned a rate constant, and *mutant frequency* (or mutant proportion), which is a variable parameter determined both by the rate of mutation and by the *rate of selection* of the mutant type.

• **The estimation of mutation rate**

For a population of microbial cells, the *mutation rate* can be defined as the *probability that any one cell will mutate during a defined interval of time*. The first measurements of bacterial mutation rates were performed by Luria and Delbrück in 1943, at a time when the nature of the genetic material and the mechanism of mutation were unknown. They chose as their time parameter the *division cycle*, or the bacterial generation; their formula for mutation rate is thus *the number of mutations per cell per generation*, averaged over many generations. As discussed in Chapter 13, most spontaneous mutations represent errors in template action which occur when the DNA double helix replicates. Since under normal conditions DNA replication and cell division are coordinated, the Luria–Delbrück formula is valid under normal physiological conditions of microbial growth.

In the Luria–Delbrück formula, mutation rate is the probability of a mutation occurring when one cell doubles in size and divides to form two cells. This series of events is called a *cell-generation*. The number of cell-generations can be simply determined for any culture, since each cell-generation increases the number of cells in the culture by one. Thus, the number of cell-generations equals the net increase in cells over the period of cultivation, and is given by the expression

$$n - n_0$$

where n is the final number of cells and n_0 the number of cells at time zero. A small correction has to be applied to this expression, since at the moment that the culture is sampled to measure n, the cells are in varying stages of completion of their next division cycle. In an exponentially growing, nonsynchronized culture the average progress toward the next generation is such that the true number of cell-generations accomplished by the culture is

$$\frac{n - n_0}{\ln 2} = \frac{n - n_0}{0.69}$$

where $\ln 2$ is the logarithm of 2 to the base e. The mutation rate is equal to the average number of mutations per cell-generation; we then have the equation:

$$a = \frac{m}{\text{cell-generations}} = \frac{m}{(n - n_0)/0.69} = (0.69)\frac{m}{n - n_0}$$

where a stands for the mutation rate and m for the average number of mutations occurring when n_0 cells increase in number to n cells.

To determine the mutation rate of a given culture, it is thus necessary to determine m. A simple way to achieve this is to allow the mutations to take place in a population of cells growing on a solid medium. Under such conditions, each mutation gives rise to a mutant clone that is fixed *in situ* and—with appropriate manipulations—can be detected as a single colony.

In practice, this means that a population of cells must be permitted to undergo a limited number of cell divisions on an agar plate, following which the conditions must be changed so that only the mutant clones can continue growing to form visible colonies. A variety of methods has been introduced to achieve these conditions; two examples will suffice.

1. Cells of an auxotrophic strain are spread on minimal agar containing a sufficient amount of the required growth factor to permit a limited number of divisions. Growth of the parental type then ceases; any prototrophic mutants, which no longer require the growth factor, are able to continue growth and to form visible colonies. The number of such mutants present in the inoculum must be subtracted; this number is determined by including one set of plates with no growth factor in the agar.

2. A population of streptomycin-sensitive cells is deposited on a membrane filter, and the membrane is placed on nutrient agar for a limited time. The membrane is then transferred to the surface of nutrient-streptomycin agar; the sensitive parents are killed, while resistant mutants that arose during growth on the nutrient agar form visible colonies. The number of resistant mutants present in the inoculum is determined by plating one set of membranes on nutrient-streptomycin agar at zero time.

In each of these examples the number of colonies per plate (corrected by subtraction of the number of mutants in the inoculum) equals the number of *mutations* per plate. It then remains only to determine the number of cell-generations per plate, which is accomplished by washing the cells off several plates in a known volume of liquid and performing a viable count. The number of cell-generations is, as stated above, equal to $(n - n_0)/0.69$; here n is the average number of cells per plate at the time that the parental population is killed or inhibited and n_0 the average number of cells per plate in the original inoculum.

As an example, 1.0×10^6 streptomycin-sensitive cells are deposited on each of a series of membrane filters. One set of filters is put on streptomycin agar at zero time, and the average number of resistant mutants in the inoculum is found to be 1.2 per filter. Another set of membranes is placed on nutrient agar and incubated for 6 hours. At the end of that time, half the membranes are placed on streptomycin agar, and half are used to determine the viable count. The viable count is found to be 1.0×10^9 cells per filter; the streptomycin plates, after incubation, show an average of 4.2 colonies per filter. The mutation rate, a, is calculated as follows:

$$a = (0.69) \frac{m - m_0}{n - n_0}$$

where m is the final number of mutant colonies and m_0 the number of mutants in the inoculum. Substituting the experimentally determined figures, we get:

$$a = (0.69) \frac{4.2 - 1.2}{(1.0 \times 10^9) - (1.0 \times 10^6)} = \frac{2.1}{1.0 \times 10^9}$$

The mutation rate to streptomycin resistance was thus 2.1×10^{-9} per cell-generation.

The number of mutations occurring in a *liquid* culture of microbial cells can be estimated by a statistical method, as shown by Luria and Delbrück in their original paper. A population of wildtype cells (e.g., streptomycin-sensitive) is used to inoculate a series of 20 or more cultures, each receiving an inoculum small enough to contain no streptomycin-mutants. The cultures are incubated until a high cell density has been reached, and then the entire contents of each tube are spread on a single plate of streptomycin agar. The plates are incubated until the colonies of streptomycin-resistant mutants are countable.

In most cases, the number of mutants found on the plate does not tell us the number of *mutations* which occurred in the corresponding tube; 16 mutants, for example, could have arisen from 16 mutations occurring during the last generation of the population, or from 1 mutation occurring 4 generations earlier. The exception is the plate with *zero mutants*: this represents a culture in which *zero mutations* took place.

Luria and Delbrück showed that the average number of mutations per culture is related to the fraction of cultures experiencing zero mutations by the zero term of the Poisson distribution:*

$$P_0 = e^{-\bar{m}}$$

where P_0 is the fraction of cultures with zero mutants (and hence zero mutations), and $\bar{m}$ is the average number of mutations per culture.

Solving for $\bar{m}$, we get:

$$\bar{m} = -\ln (P_0)$$

For example, if 30 percent of the cultures had no mutations, P_0 equals 0.3, and $\bar{m} = -\ln (0.3)$, or approximately 1.0. Suppose, in this example, that each tube grew from an inoculum of 1×10^4 cells to a final population size of 1×10^9 cells. The mutation rate is then calculated as follows:

$$a = (0.69) \frac{m}{n - n_0}$$

$$= \frac{(0.69)(1.0)}{(1 \times 10^9) - (1 \times 10^4)}$$

$$= \frac{0.69}{1 \times 10^9} = 0.69 \times 10^{-9}$$

---

*The Poisson distribution describes the proportion of subcultures of a population which will have experienced 0, 1, 2, 3, . . . , n mutations, when the subculture size is very large compared to the number of mutations.

In practice, a preliminary experiment is carried out to determine a suitable incubation period. If the cultures are allowed to grow too long, for example, every culture will have more than one mutant and the method cannot be applied.

• Mutational
equilibrium

There is a direct relationship between the mutation rate and the increase in the proportion of mutants in a culture at each generation, assuming that neither the mutant nor the parent type has a selective advantage over the other. Suppose, for example, that a culture is started from a small inoculum and that the first two mutations occur when there are 1×10^8 cells in the culture. The proportion of mutants in the culture is then 2×10^{-8}. The culture continues to grow, and at the next generation there are 2×10^8 parent cells. The mutants also divide, however, and there are now four mutants in 2×10^8 cells; the proportion thus remains 2×10^{-8}. If no further mutations were to occur, and the mutant cells divided as often as the parent cells, the proportion would remain constant.

Let us assume that mutations do continue to occur, however, with a probability of 2×10^{-8} per cell-generation. Then at each generation, for every 10^8 parent cells, two new mutants will be added to the culture, and the proportion of mutants to parents is increased by just that amount. This is shown in Table 14.1 and in Figure 14.5. In Figure 14.5 the slope of the line a/b directly represents the mutation rate; the units in which the slope is expressed are "mutants per 10^8 cells per generation."

What happens when such a population grows indefinitely? At first thought, one might expect the proportion of mutant cells to increase until it reaches 100 percent. This is prevented, however, by the phenomenon of *reverse mutation*. Many mutations are capable of mutating back to the original state, and this reverse mutation will have its own characteristic rate. When a population of bacteria has accumulated a high enough number of mutants, reverse mutations will become

TABLE 14.1

Hypothetical increase in proportion of mutants in a growing culture as the result of new mutations

| GENERATION | AVERAGE NUMBER OF PARENT CELLS DURING GENERATION | NUMBER OF MUTANT CELLS[a] | | | TOTAL | PROPORTION: MUTANTS/ PARENTS |
		NEW	OLD			
n	1×10^8	2			2	2×10^{-8}
$n+1$	2×10^8	4	4		8	4×10^{-8}
$n+2$	4×10^8	8	8	8	24	6×10^{-8}
$n+3$	8×10^8	16	16	16	16	64	8×10^{-8}
$n+4$	16×10^8	32	32	32	32	32	160	10×10^{-8}

[a] The numbers enclosed in the dotted line show how many mutants there would be if no further mutations took place after the first two. Note that the *proportion* of mutants then would have remained constant at 2×10^{-8}.

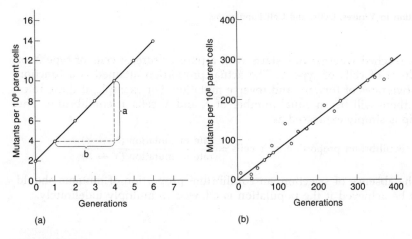

(a)   (b)

FIGURE 14.5

The increasing proportion of mutants in a culture as a result of spontaneous mutation. (a) The theoretical increase that results from exactly two new mutants per 10^8 cells appearing at each generation. The mutation rate (2.0×10^{-8} per generation) is expressed by the slope of the line, which is a/b, or $(6/10^8)/3$. (b) The results of an actual experiment in which the proportion of mutants in a culture has been determined by plating at successive times. The mutation rate is found to be 0.75×10^8 per generation since this is the slope of the plotted line.

significant; the proportion of mutants will ultimately level off at the point where the forward mutations and reverse mutations just balance each other. Assume, for example, that in a population of cells of type X, the mutation $X \rightarrow Y$ occurs at a certain rate. As the population grows, the proportion of Y mutants will increase. When there are sufficient Y cells, mutation $Y \rightarrow X$ will have a chance to occur, and eventually the population will reach a true equilibrium state in which the number of forward mutations ($X \rightarrow Y$) just equals the number of reverse mutations ($Y \rightarrow X$) at each generation. From then on, the proportion of mutants remains constant.

Figure 14.6 illustrates the fact that the same equilibrium proportion of Y

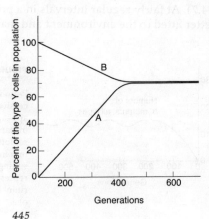

FIGURE 14.6

The attainment of an equilibrium proportion of mutants in a culture. In the case of curve A, the experiment was begun with a population having no cells of type Y. As a result of forward mutation, the proportion of type Y cells rose until they constituted about 70 percent of the population. At this point, back mutation and forward mutation just balanced each other. In the case of curve B, the experiment was begun with a pure culture of type Y cells. The proportion of Y cells decreased as the result of $Y \rightarrow X$ mutations, until again an equilibrium was reached at 70 percent.

mutants is reached whether one starts with a pure culture of cells of type X or a pure culture of cells of type Y. The actual proportion attained is a function of the relative rates of forward and reverse mutation. For example, if these rates are equal, there will be an equal number of X and Y cells at equilibrium. The relationship is simply expressed as:

$$\text{equilibrium proportion of Y cells} = \frac{\text{rate of mutation (X} \rightarrow \text{Y)}}{\text{rate of mutation (Y} \rightarrow \text{X)}}$$

Thus, in the absence of selection, an equilibrium proportion of mutants should eventually be achieved if the population is allowed to multiply indefinitely.

• Effects of selection on the proportions of mutant types

In the section above on mutational equilibrium we saw that the proportion of a given mutant type in a microbial population increases, in the absence of any selective advantage, in proportion to the mutation rate. Suppose, for example, that we have produced by mutagenesis a strain of *E. coli* requiring the amino acid histidine for growth. A pure culture of this strain, here designated h^-, is put on a slant. The fully grown slant culture will probably contain one h^+ mutant (able to synthesize histidine and hence not requiring it for growth) for every million or so h^- cells, and this proportion will increase with succeeding generations as the stock culture is transferred from slant to slant. The medium contains ample histidine, so there is no selective advantage for either type of cell. Assuming that about ten generations are accomplished on each slant and that the culture is transferred several times a year, one should expect that in a few years the culture would contain a greatly increased proportion of h^+ cells.

In practice, however, this rarely happens, even when calculations based on observed forward and reverse mutation rates predict that it will. Instead, we find that an apparent equilibrium is reached long before it should be, and always in favor of the genetic type with which the culture was started. The proportion of mutant cells in the culture increases only up to a very low value—perhaps 1×10^{-6}—and remains there indefinitely.

This puzzling observation has been found to result from the phenomenon of *periodic selection* (Figure 14.7). At fairly regular intervals in a population of bacteria, mutants arise that are better fitted to the environment and that eventually displace

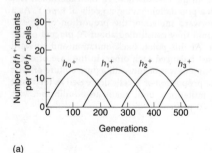

(a)

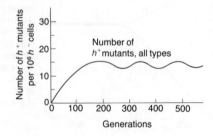

(b)

FIGURE 14.7

Periodic selection. (a) The successive appearance and disappearance of different h^+ mutants. (b) The same data are replotted in terms of total h^+ mutants of all types. The resulting slightly fluctuating curve represents the pseudoequilibrium level that the proportion of h^+ mutants reaches.

the parental type as a result of selection. We cannot always define the properties of the new mutant that give it this advantage. It might be an intrinsically faster growth rate, or it might be that the new type produces metabolic products that inhibit the parent type. In any case, the better adapted mutant overgrows the culture, only to be replaced in turn by a mutant that is still better adapted. The process of replacement may be repeated many times, for new mutants that can displace the predominant type from the population continue to arise. *This periodic change in the population has a direct effect on the equilibrium proportion of all other mutants.* Let us consider the specific case of the h^+ mutants mentioned earlier. Suppose that a better adapted mutant appears in the culture at the moment when the proportion of h^+ cells has risen to 1×10^{-6}. The better adapted mutant could theoretically arise from either an h^- cell or an h^+ cell, but since there are 10^6 h^- cells for every h^+ cell, the odds are a million to one that the new type will arise in the h^- population.

The better adapted mutant will thus have a selective advantage over all other cells in the population, which it will soon displace. Since the better adapted type is genetically h^-, all h^+ cells in the population should disappear as the result of selection.

The total disappearance of h^+ cells is prevented, however, by the occurrence in the better adapted h^- population of new mutations to h^+. Since the new h^+ cells are not at a selective disadvantage, they will increase in proportion until the cycle is started over again by the appearance of an even better adapted type.

This process can be expressed symbolically as follows. Let us call the original cells h_0^- and h_0^+ and the first better adapted mutant h_1^-. In the new h_1^- population, mutations to h_1^+ will occur. As time goes on and the proportion of h_0^+ cells drops, there is a corresponding increase in the number of h_1^+ cells. The cycle is repeated again and again. The h_1^- cells give rise to a still better adapted type, h_2^-, which displaces the h_1^- and h_1^+ cells. The loss of h_1^+ cells is compensated for by the appearance of h_2^+ mutants. The mutational pattern can be diagrammed as:

$$h_0^- \longrightarrow h_1^- \longrightarrow h_2^- \longrightarrow h_3^-$$
$$\downarrow \qquad\quad \downarrow \qquad\quad \downarrow \qquad\quad \downarrow$$
$$h_0^+ \qquad\quad h_1^+ \qquad\quad h_2^+ \qquad\quad h_3^+$$

The upper graph in Figure 14.7 shows the way that successive waves of h^+ mutants rise and fall in the population. The lower graph in Figure 14.7 shows the apparent stability of the population with respect to the characters h^+ and h^-, when h^+ mutants are considered as a single class.

The level that the proportion of mutants reaches can be called a *pseudoequilibrium*, because it is really the composite result of a series of discrete, nonequilibrium events. With ordinary mutation rates, which are very low, the occurrence of periodic selection results in the attainment of such pseudoequilibria. It is only when both the forward and back mutation rates are extremely high that true equilibria, as illustrated in Figure 14.6, can be attained. In such cases, the proportion of mutant cells rises so rapidly that better adapted mutants have an equal chance of appearing in the mutant or in the parent population.

Periodic selection is a subtle phenomenon, since the mutant type that is being experimentally observed (e.g., the h^+ mutant in the case described above) is not the one subject to selection. As the above example shows, the selection of one type of mutant in a population may prevent any other mutant type from increasing in proportion.

SELECTION AND ADAPTATION

• **The genetic variability of pure cultures**

As a general rule, any one gene has only one chance in about 100 million of mutating at each cell division. At first sight, therefore, mutation might appear too rare to be of much significance. Suppose, however, that we have a "pure culture" of a bacterium in the form of 10 ml of a broth culture that has grown to the stationary phase. Such a culture will contain about 10 billion cells; for any given gene, there may well be several thousand mutant cells present in the culture. Even during the growth a single bacterial colony, which may contain between 10^7 and 10^8 cells, a large number of mutants will arise (Figure 14.8).

Thus, a large population of bacteria is endowed with a high degree of potential variability, ready to come into play in direct response to changing environmental conditions. Because of their exceedingly short generation times and the consequent large sizes of their populations, these haploid organisms possess a store of latent variation despite the fact that they cannot accumulate recessive genes as can a population of diploid organisms. In practice, this means that no reasonably dense culture of bacteria is genetically pure; even a slight change in the medium may prove selective and bring about a complete change in the population within a few successive transfers. This explains, for example, why many "delicate" pathogenic bacteria, which prove difficult to cultivate when first isolated from their hosts, gradually become better and better adapted to the conditions of artificial media.

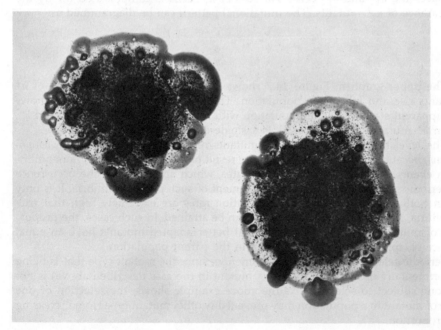

FIGURE 14.8

Two bacterial colonies showing papillae, which represent secondary growth of mutants that arose during the formation of the original colonies. From V. Bryson, in W. Braun, *Bacterial Genetics*. Philadelphia: Saunders, 1953.

So far we have considered only the selective forces that may operate in artificial cultures. In nature, however, selection acts in an even more stringent fashion. A microbe in the soil, for example, must be able not only to survive under a given set of physicochemical conditions, but also to survive in competition with the numerous other microbial forms that occupy the same niche. Any mutation that decreases, even to the slightest extent, the ability of the organism to compete, will be selected against and quickly eliminated. Nature tolerates little variation within microbial populations, for the laws of competition demand that each type retain the array of genes that confers maximum fitness.

As soon as an organism is isolated in pure culture, the selective pressures resulting from biological competition are removed. The isolated population becomes free to vary with respect to characters that are maintained stable in nature by selection. In adapting to existence in laboratory media, organisms may undergo genetic modifications that would lead to their speedy suppression in a competitive environment.

THE CONSEQUENCES OF MUTATION
IN CELLULAR ORGANELLES

Part of the genome of eucaryotic organisms is carried in the mitochondria and chloroplasts. Each kind of organelle contains and reproduces DNA that determines certain of its phenotypic properties. The properties of a chloroplast or a mitochondrion are thus controlled in part by nuclear and in part by organellar genes, both subject to change by mutation. Until recently, it has been difficult to select mutations that specifically affect organellar DNA. However, it has been found that mutations in yeast which confer resistance to certain antibiotics (those known to affect protein synthesis in bacteria) take place in the mitochondrial DNA. It has thus become possible to study experimentally the transmission of many different mitochondrial mutations. Each cell contains a population of mitochondria; hence, the relative growth rates of normal and mutated mitochondria determine the stability of such a mutation during vegetative growth. The situation is entirely comparable to that of a growing bacterial population which contains two genetically different kinds of cells, and the outcome can also be determined by environmental factors.

Organellar mutations have also been observed in the chloroplast of the unicellular alga, *Chlamydomonas*. R. Sager and her colleagues have induced mutations in chloroplast DNA affecting the ability of the cell to photosynthesize, as well as to resist the action of certain antibiotics.

MUTANT TYPES OF BACTERIOPHAGES

In addition to the conditionally expressed lethal mutations described earlier, phages can undergo mutations that produce nonlethal changes in phenotype. The most readily observed nonlethal changes are those which produce alterations in *plaque morphology and alterations in host range*.

When a wildtype phage, such as T4, is plated on a sensitive host, such as *E. coli* strain B, under carefully standardized conditions, the plaques that appear are homogeneous and characteristic in appearance: they are small, with irregular fuzzy edges. When a large number of plaques is examined, however, a few aberrant types are always observed; when particles from such plaques are picked and replated, the aberrant plaque type is found to breed true and thus to reflect a genetic mutation. Several mutant phenotypes are listed in Table 14.2.

Earlier in this chapter we described the occurrence of phage-resistant mutants in populations of phage-sensitive bacteria. These mutants owe their resistance to the production of altered surface receptors, such that they no longer adsorb wildtype phage particles; *E. coli* strain B, for example, can mutate to the state designated B/2, which does not adsorb phage T2. If 10^6 or more particles of T2 are plated on a lawn of B/2 cells, however, a few plaques appear; when particles from these plaques are isolated and purified, they are found to be *host-range mutants*, which can now adsorb to cells of B/2 as well as to cells of *E. coli* strain B. The mutation in this case consists of a base-pair change in the gene governing the structure of the tail-fiber proteins, which are the adsorption organs of phage T2. The mutant phage is designated T2*h*.

By plating cells of B/2 with the mutant phage, one can select a new class of mutant bacteria that is resistant to phage T2*h*. The entire cycle can now be repeated: a second-step host-range mutant of the phage can be selected, which can adsorb to the new resistant bacterium. Apparently, any altered configuration of the bacterial surface receptor can be matched by an alteration in the adsorption organ of the phage. In nature the mutational capacities of cell and virus permit both to exist: at any given moment there are both susceptible hosts available to the virus as well as cells that can resist viral attack.

TABLE 14.2

Some mutant types of T-even bacteriophages[a]

TYPE	PHENOTYPE	PRIMARY EFFECT OF MUTATION
Rapid lysis	Large plaques with sharp edges	Unknown
Minute	Very small plaques	Slow synthesis of phage, or precocious lysis of host cell
Host range	Adsorbs to bacteria that are resistant to wild-type phage	Altered polypeptides of the tail fibers
Cofactor-requiring	Requires a cofactor such as tryptophan for adsorption to host	Abnormal tail fibers bind to sheath, require cofactor to be released
Acriflavin-resistance	Forms plaques on agar containing concentrations of acriflavin that are lethal for wild-type phage	Causes host cell membrane to have reduced permeability for acriflavin
Osmotic shock	Survives rapid dilution from 3.0 M NaCl into distilled water	Alteration in head protein increases permeability of head
Lysozyme	Does not produce halo around plaque	Abnormal lysozyme synthesis

[a] Modified from G. Stent, *Molecular Biology of Bacterial Viruses.* San Francisco: Freeman, 1963.

FURTHER READING

Books   Braun, W., *Bacterial Genetics*, 2nd ed. Philadelphia: Saunders, 1965.

Goodenough, U., and R. P. Levine, *Genetics*. New York: Holt, Rinehart & Winston, 1974.

Hayes, W., *The Genetics of Bacteria and Their Viruses*, 2nd ed. Oxford, England: Blackwell, 1968.

Review article   Gots, J. S., and C. E. Benson, "Biochemical Genetics of Bacteria," *Ann. Rev. Genetics* **8,** 77 (1974).

Original articles   Levinthal, M., "Bacterial Genetics Excluding *E. coli*," *Ann. Rev. Microbiol.* **28,** 219 (1974).

Luria, S., and M. Delbrück, "Mutations of Bacteria from Virus Sensitivity to Virus Resistance," *Genetics* **28,** 491 (1943).

GENETIC RECOMBINATION

In evolution, natural selection operates not so much on single gene mutations as on the *new combinations of genes* which arise when, in a single cell, mutant genes from two different cells are brought together. This process is called *genetic recombination;* in this chapter we shall survey the mechanisms by which genetic recombination is brought about in bacteria, bacterial viruses, and protists.

RECOMBINATION IN BACTERIA

In molecular terms, recombination is the process by which a recombinant chromosome is formed from DNA derived from two different parental cells. Three processes that lead to the formation of recombinant chromosomes are known to occur in bacteria. In order of their discovery, these are *transformation, conjugation,* and *transduction.* They differ from the sexual process of eucaryotes in that a true fusion cell is not formed: instead, a part of the genetic material of a donor cell is transferred to a recipient cell. The recipient cell thus becomes diploid for only part of its genetic complement; such partial zygotes are called *merozygotes.*

The original genome of the recipient is termed the *endogenote,* and the fragment of DNA introduced into the recipient cell is termed the *exogenote.* Both the nature and the size of the exogenote differ in the three processes. In transformation, short pieces of double-stranded DNA, released into the medium from donor cells, are adsorbed to the surface of recipient cells and are taken into the cell by a process that results in the degradation of one strand. In transduction, a small, double-stranded fragment of DNA is brought from the donor cell into the recipient cell by a bacteriophage particle, in some cases attached to a piece of phage DNA. In conjugation, a single strand of DNA is transferred between cells which are in direct contact, and may represent a major fraction of the donor genome.

Before we discuss these different processes, some features common to all three will be described.

• The fate of the exogenote

If the exogenote has a sequence of base pairs which is homologous with a segment of the endogenote, pairing occurs rapidly and a recombinant chromosome is immediately formed by the integration of part or all of the exogenote with the endogenote.

If pairing and integration are prevented for any reason, the exogenote may undergo one of several alternative fates. If the exogenote carries the genetic elements necessary for its own replication, it may persist and replicate, so that the merozygote gives rise to a clone of partially diploid cells. This has been observed in special cases of both transduction and conjugation, as will be discussed later. If the exogenote lacks such elements, it may persist but not replicate, so that in the clone of cells that arise from the merozygote only one cell at any time is a partial diploid. This phenomenon, which has been observed only in transduction, is called *abortive transduction*.

Finally, the exogenote may be enzymatically degraded; this phenomenon is called *host restriction*.

• Restriction and modification of foreign DNA

The degradation of foreign DNA that has penetrated a bacterial cell was first discovered in bacteriophage infections. It will be profitable to describe this phenomenon before going on to consider how restriction operates in bacterial recombination.

The early literature on bacteriophage contains a number of reports of what were called "host-induced modifications of bacterial viruses." In each case, it was observed that the passage of a phage through one bacterial host greatly reduced its *efficiency of plating* on a second bacterial host. An example of such an experiment is given in Table 15.1: the table shows that particles of phage λ, produced during replication in *Escherichia coli* strain K12, plate with an efficiency* of 1.0 on strain K12 itself but with an efficiency of 1×10^{-4} on *E. coli* strain B. If, however, the phage particles released from strain B are tested, their properties are reversed: their efficiency of plating is now 1.0 on strain B but only 4×10^{-4} on strain K12. In other words, the phage particle can successfully infect only the type of host cell in which its DNA has been produced. If the DNA of the phage is "foreign" to the host cell, it fails to establish a productive infection in about 99.99 percent of the cells.

The failure of "foreign" phage to infect productively a bacterial host results from the existence in the host cell of a specific endonuclease that initiates the degradation of foreign DNA; if the phage is labeled with ^{32}P and allowed to inject its DNA into the foreign host, the radioactive DNA is rapidly released in the form of small fragments.

This phenomenon is called *restriction,* and the endonuclease is called a *restricting enzyme.* The corollary of this observation is that phage DNA which has been produced in a given type of host cell is no longer a substrate for the restricting enzyme; in other words, it has been enzymatically *modified* by the host, so that it is henceforth protected against restriction.

---

*A plating efficiency of 1.0 means that every particle causes a productive infection and hence a plaque. A plating efficiency of 1×10^{-4} means that only one particle in 10^4 can cause a productive infection of the host cell.

TABLE 15.1

Efficiency of plating of bacteriophages grown in different hosts[a]

HOSTS IN WHICH PHAGE IS GROWN	EFFICIENCY OF PLATING ON:	
	E. coli K12	*E. coli* B
E. coli K12	1.0	1×10^{-4}
E. coli B	4×10^{-4}	1.0

[a] Modified from W. Arber and D. Dussoix, "Host Specificity of DNA Produced by *Escherichia coli*. I. Host Controlled Modification of Bacteriophage λ," *J. Mol. Biol.* **5**, 18 (1962).

Going back to Table 15-1, we can now reconsider the results in terms of restriction and modification. The phage grown in strain K12 (we will call it λK) was specifically modified so that the restricting enzyme of strain K12 could not attack it. It was still a good substrate for the restricting enzyme of strain B, however, so that the DNA of λK was degraded in strain B. Nevertheless, in one out of 10^4 cells the phage succeeded in growing, having been modified by the *E. coli* strain B cell before degradation could begin. The DNA of the particles released from this cell carried the B type of modification; these particles, which we can call λB, became immune to attack by the restricting enzyme of *E. coli* strain B but were quickly degraded by the restricting enzyme of strain K12.

The modification process consists of the methylation of bases at a very few, highly specific sites on the DNA, which are also the sites of attack by the restricting enzyme. The DNA of the bacterial chromosome is, of course, modified in the same way, since otherwise the chromosome would be degraded by the restricting enzyme system. Restriction and modification also come into play when bacterial DNA is transferred from one cell to another by one of the recombination processes. For example, *E. coli* strain B will conjugate with *E. coli* strain K12, but the number of recombinants produced is one-thousandth of that produced in a similar cross between two strains derived from *E. coli* strain B or between two derived from *E. coli* strain K12.

The genes responsible for the production of the restricting and modifying enzymes are tightly linked on the bacterial chromosome. By recombination, it has been possible to produce a strain of K12 which carries the restriction and modification alleles of *E. coli* strain B; this K12 strain now recombines with *E. coli* strain B with high efficiency.

The existence and specificities of restricting and modifying enzymes produce a complex pattern of compatibilities between bacterial strains, with respect to the ability of the DNA of one to escape degradation in another. The pattern is further complicated by the fact that some phage genomes, as well as some plasmids,* also possess genes for restriction and modification. Thus, a derivative of *E. coli* strain K12 that is lysogenic for phage P1 is a poor recipient in recombination with K12 donors that are nonlysogenic.

---

*Plasmids will be discussed later on in this chapter.

• The integration of	Experiments with density-labeled DNA have established that recombination
exogenote and	between exogenote and endogenote occurs by the process of *breakage and reunion*
endogenote	of the parental DNA molecules.

• **The integration of exogenote and endogenote**

Experiments with density-labeled DNA have established that recombination between exogenote and endogenote occurs by the process of *breakage and reunion* of the parental DNA molecules. The most remarkable feature of the recombination process, and one that is essential to its success, is *the conservation of base-pair sequence*. With the exception of mutational differences, the recombinant DNA molecule must have the same base-pair sequence as that of the parents. The mechanism that ensures this precise alignment of the parental molecules is believed to involve the complementary pairing of bases between single-stranded regions of the parental double helices. For example, recombination between DNA fragments might occur as shown in Figure 15.1. Two fragments with overlapping base sequences undergo base pairing between complementary strands, so that the opposite strand in each case is displaced from the duplex and is present in the unpaired state. The displaced single strands are then digested by a cellular exonuclease, the digestion process continuing beyond the region of displacement. At some moment, DNA polymerase begins the resynthesis of the missing segment, and the process is completed by the rejoining of the strands through the action of polynucleotide ligase.

The resemblance of this process to that of DNA repair (Figure 13.11) is striking. Indeed, it is likely that the enzymes of digestion, resynthesis, and reunion, which effect DNA repair, are also involved in recombination.

When recombination occurs in the middle instead of at the ends of two DNA duplexes, the process must be more complicated. Again, it is believed that a single-stranded region of one parental duplex displaces the homologous region of a strand in the other duplex, but the exact sequence of events is still unknown. One possible sequence of events is shown in Figure 15.2. Whatever its exact mechanism, this must be the kind of recombination that occurs in eucaryotic chromosomal crossing over, as well as in the integration of two different replicons* in bacteria.

In transformation, as well as in the transfer of short pieces of DNA by conjugation, the situation is somewhat different. Here, isotope-labeling experiments show that only one strand of the exogenote is integrated into the double-stranded endogenote. A possible mechanism that could produce this result is shown in Figure 15.3. In the transfer of long pieces of chromosomal DNA by conjugation, and in the phage-mediated process of transduction, it is not yet known whether integration involves single- or double-stranded segments of the exogenote.

• **The segregation of the recombinant cell**

Bacteria are multinucleate, but the process of recombination involves only one nucleus in the recipient cell. Following integration of the exogenote, the multinucleate cell is thus a heterocaryon; the formation of a homocaryotic recombinant cell requires the same process of *nuclear segregation* as that described earlier for the formation of a homocaryotic mutant (Figure 14.3).

BACTERIAL TRANSFORMATION

• **The discovery of transformation**

Transformation was discovered in the pneumococcus (*Streptococcus pneumoniae*). The pneumococci in the sputum or tissues of a victim of pneumonia are always surrounded by large capsules consisting of polysaccharide; on agar plates the

*The *replicon* is the unit of replication. The integration of two replicons occurs when a plasmid such as the F factor, or the temperate phage λ, integrates with the bacterial chromosome.

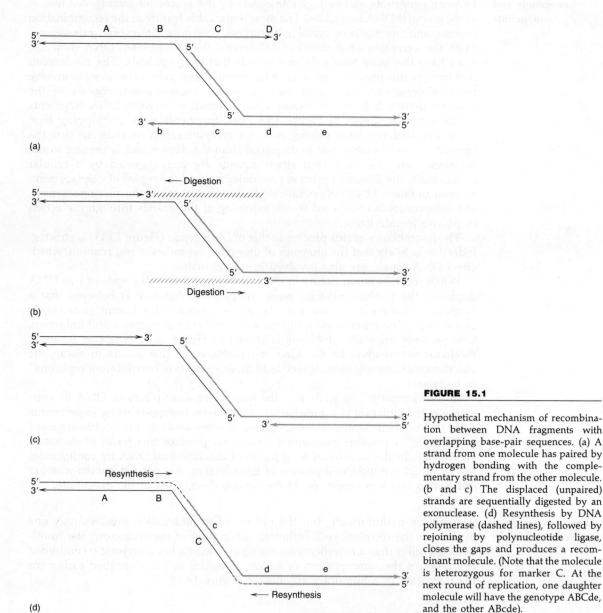

FIGURE 15.1

Hypothetical mechanism of recombination between DNA fragments with overlapping base-pair sequences. (a) A strand from one molecule has paired by hydrogen bonding with the complementary strand from the other molecule. (b and c) The displaced (unpaired) strands are sequentially digested by an exonuclease. (d) Resynthesis by DNA polymerase (dashed lines), followed by rejoining by polynucleotide ligase, closes the gaps and produces a recombinant molecule. (Note that the molecule is heterozygous for marker C. At the next round of replication, one daughter molecule will have the genotype ABCde, and the other ABcde.)

encapsulated cells form smooth (S) colonies. Pneumococci can be separated into a great many types on the basis of chemical differences between their capsular polysaccharides. The different types (designated I, II, III, and so on) can be distinguished immunochemically.

When an S strain (forming smooth colonies) is serially subcultured, R cells (forming rough colonies) appear in the population. R cells have no capsules and

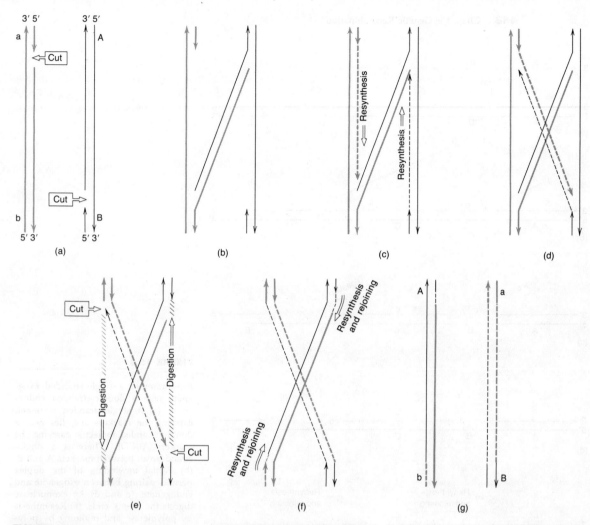

FIGURE 15.2

Hypothetical mechanism of recombination between DNA molecules. (a) One parental
molecule carries the markers AB, the other ab. Each parental molecule has a single strand
cut at a different position, between markers A and B. (b) The parental duplexes have
partially unwound, and a strand from one molecule has paired with a strand from the
other molecule. (c) Resynthesis occurs along the single-stranded region of each parental
molecule (dashed lines). (d) The newly synthesized strands pair with each other. (e) An
endonuclease makes a single-strand cut in each parental duplex, and the cut 3′ ends are
sequentially digested by an exonuclease. (f and g) Resynthesis and rejoining produce two
recombinant molecules, Ab and aB. After H. Whitehouse.

are avirulent. In 1928 F. Griffith observed that when a very large inoculum of
R cells, derived from what had originally been a type I smooth culture, was
inoculated under the skin of a mouse together with heat-killed S cells of type
II, the mouse died within a few days. Blood from such an animal yielded only
type II smooth cells. The dead type II cells thus had liberated something which

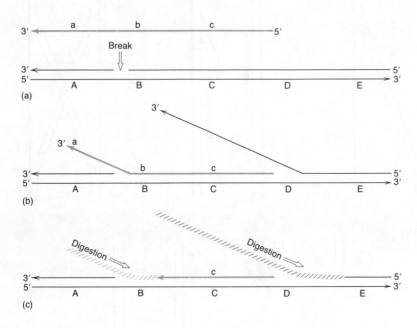

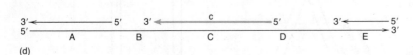

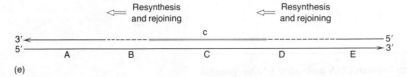

FIGURE 15.3

Integration of a single-stranded exogenote with a double-stranded endogenote. (a) A single-stranded fragment, carrying the markers abc, lies near a double-stranded molecule carrying the markers ABCDE. There is a single-strand break between markers A and B. (b) Partial unwinding of the duplex permits pairing between exogenote and endogenote. (c and d) An exonuclease digests the free 3′ ends. (e) Resynthesis by polymerase and rejoining by polynucleotide ligase produce a recombinant molecule, in which one of the strands carries the markers ABcDE.

conferred on the R cells the ability to make a new type of capsular polysaccharide. The transferred property proved to be heritable.

A few years afterward, other groups of investigators succeeded in carrying out such *type transformations* by mixing R cells and heat-killed S cells *in vitro*. It was later found that type transformations could be brought about with cell-free extracts of S cells. In other words, when a chemical substance extracted from S cells of type II is added to a culture of R cells derived from type I, some of the cells are genetically changed (transformed) to type II. The unidentified chemical substance responsible was called *transforming principle*. Transforming principle has two properties that are generally associated only with genes: (1) it is self-duplicating, for a culture of transformed cells can be extracted and shown to contain much more transforming principle than the amount used originally to

459 transform them; (2) it directs a specific function of the cell—the production of one type of polysaccharide.

• **The nature of transforming principle**

In 1944 O. T. Avery, C. M. MacLeod, and M. McCarty succeeded in purifying pneumococcal transforming principle and identified it as DNA. Until that time it was generally believed that the specificity of the gene is determined by the protein moiety of nucleoprotein; the chemical characterization of transforming principle provided the first direct evidence that DNA is the carrier of genetic information.

Since 1944 similar transformations have been effected in other genera of bacteria, notably in *Hemophilus, Neisseria,* and *Bacillus.* All attempts to carry out transformation in many other species have failed. The transformation of *E. coli* K12 strains was accomplished only after many unsuccessful attempts. The ability of this species to undergo transformation was found to depend on high concentrations of calcium ion in the medium, and is greatly increased by the use of mutant strains lacking certain deoxyribonucleases.

• **The transformation of genetic markers**

All mutant loci (or "markers") of a recipient cell are capable of being transformed. The genetic fragment that is transferred is usually very small, however; although it may carry a number of genes, it rarely carries more than one marker. This is an artifact that results from the method by which transforming DNA is usually prepared. In this method, which involves stirring the preparation with phenol to remove proteins and repeated precipitations with ethanol, the DNA is usually sheared into fragments which, even in the most carefully handled preparations, rarely have a molecular weight that exceeds 1×10^7. This represents about 0.3 percent of the bacterial chromosome, corresponding to about 15 genes.* However, only a few dozen mutant loci have been used in transformation experiments with any one species, so that the chance of finding two markers on the same DNA fragment is low. Nevertheless, such "linked transformations" have occasionally been observed.

When crude preparations of transforming DNA are used, in which shearing and enzymatic degradation have been minimized, it is possible to obtain the linked transformation of many markers. As much as a third of the chromosome can be transferred with such preparations.

Recombination usually occurs with a relatively low frequency, so that selective markers must be used; that is, the donor DNA must contain markers which permit the recombinants to be selected in the presence of a large excess of viable parental recipient cells. Two convenient types of selective markers are drug resistance and nutritional independence. For example, if the recipient culture is streptomycin sensitive and the donor culture is streptomycin resistant, transformation of even a few cells of the former can be detected by plating the DNA-treated culture on streptomycin-containing agar. In this case, the mutant locus determining resistance is a *selective marker*. Similarly, if the recipient culture is auxotrophic (e.g., requiring arginine for growth) and the donor culture is arginine independent,

---

*The average gene size is estimated as 1,000 base pairs, coding for a polypeptide of about 330 amino acid residues. The bacterial chromosome contains about 5×10^6 base pairs of DNA, which is the equivalent of 5,000 genes of average size.

transformation of the former can be detected by plating the DNA-treated population on agar lacking arginine.

Streptococcus, Hemophilus, and *Neisseria* require complex media for growth, so it is difficult to work with nutritional markers in these organisms. *Bacillus subtilis,* however, grows on a simple mineral medium containing a suitable carbon source, and many different auxotrophic mutants have been produced. Transformation of each auxotrophic type is readily detected using DNA taken from the wild-type strain and plating the treated auxotrophic recipient cells on *minimal medium* (medium containing the minimal number of nutrients capable of supporting growth of the wild type).

• The
transformation
process

The limiting factor in the yield of transformants is usually the *competence* of the recipient cell population to take up transforming DNA.

Competence is a physiological state that fluctuates greatly during the cell cycle. Very little is known about competence, except that competent cells produce a protein that can be isolated from the medium and used to confer competence on other cells. This protein could conceivably be a component of the membrane which catalyzes the uptake of DNA; it is equally possible that it is an enzyme which degrades some component of the cell surface, unmasking the receptor for DNA. The change from noncompetence to competence appears to reflect the *synthesis* of this protein, since competence does not develop in the presence of agents, such as chloramphenicol, which block protein synthesis.

The uptake of DNA has been extensively studied in the Gram-positive pneumococci. Three stages in the process have been recognized. In the first stage, double-stranded DNA fragments bind to sites that are present (or accessible) only during competence. The continuous incorporation of choline, a constituent of membrane phospholipids, is necessary for DNA binding. Choline incorporation takes place at the equatorial region of the cell, which is the site of nascent cross-wall formation, and it is likely that DNA is taken up only at this part of the cell surface. Pneumococcal cells are not specific with respect to the type of DNA which they will take up: they will take up calf thymus DNA, for example, as efficiently as homologous DNA. Recombinants are not formed, however, unless the exogenote and endogenote have sufficient homology in their base-pair sequences to permit pairing.

In the second stage the externally bound DNA is enzymatically cleaved at random sites to form fragments with a median molecular weight of 4 to 5×10^6. In the final stage, which is energy-dependent, the DNA fragments are taken into the cell. The penetration step is accompanied by the degradation of one strand of the DNA duplex, producing an intracellular single-stranded intermediate. Fragments with molecular weights less than about 5×10^5 are not taken up.

In a mutant strain of pneumococcus lacking the two major deoxyribonucleases, a residual activity has been found representing a third, minor class of nuclease. Mutants defective in the ability to be transformed have been isolated in this line; some of them have lost the residual nuclease; they are able to bind DNA normally, but are unable to degrade it or to take it into the cell. Thus, the minor class of deoxyribonuclease appears to be essential for transformation and to drive the uptake of DNA by the degradation of one strand.

The uptake process has also been studied in the Gram-negative bacterium, *Hemophilus influenzae*. Although the process in this organism resembles in general that described above for the Gram-positive pneumococcus, some important differences have been observed. One difference is in the specificity of the process: *Hemophilus* will take up only homologous DNA. Another difference is in the penetration step: the conversion to the single-stranded stage appears to occur simultaneously with the integration step, for no free single-stranded intermediate appears to be formed.

Hemophilus cells undergo cyclic changes in competence similar to those observed in the pneumococcus. Cyclic AMP has recently been found to play a role in the development of competence: the addition of this nucleotide to the medium can increase the level of competence in the cell population as much as ten thousand-fold.

We have already discussed the integration of transforming DNA in the endogenote. Integration takes place very rapidly; if transforming DNA carrying the linked markers A^+B^- is taken up by cells that are genetically A^-B^+, fragments of A^+B^+ DNA can be extracted from the cell shortly afterward.* This can be shown by testing extracts of the recipients for the presence of DNA which can "cotransform" an A^-B^- recipient to A^+B^+; A^+B^+ DNA is formed at a linear rate starting with almost no lag, reaching the half-maximum level after 6 minutes. This process takes place in the absence of significant DNA replication.

• **The occurrence of transformation in nature**

Transformation, as discovered in the laboratory, requires the artificial extraction of donor DNA, but it was recognized very early that recombination might take place by transformation in nature. To test this possibility, mixed cultures of genetically marked pneumococci were prepared under conditions in which many of the cells were lysing. As predicted, recombinants were produced as a result of the release of DNA from some cells and its uptake by others. Since recombination vastly increases the number of gene arrays upon which natural selection can act, recombination in nature—even at low frequency—must play a major role in the evolution of bacterial species.

BACTERIAL CONJUGATION

The discovery of transformation revealed for the first time the existence of recombination in bacteria. Once the existence of transformation in bacteria had been established, a search was undertaken for processes of genetic recombination that might resemble eucaryotic sexual reproduction more closely. In 1946 J. Lederberg and E. L. Tatum carried out an experiment with *E. coli* that was designed to reveal the occurrence of recombination by conjugation.

E. coli requires no growth factors. By mutagenesis, Lederberg and Tatum produced two auxotrophic strains of *E. coli* strain K12, differing from each other with respect to four genes governing biosynthetic enzymes. Each mutation conferred a different growth-factor requirement: one strain required the compounds biotin and methionine, and the other required the compounds threonine and leucine. The loci in which the mutations occurred have now been designated *bio*,

---

*The designations *A* and *B* represent two linked markers; + represents the wild-type allele and − the mutant allele in each case.

met, thr, and *leu,* respectively. The two parental genotypes can thus be partially described as follows:

parental type I: bio^- met^- thr^+ leu^+
parental type II: bio^+ met^+ thr^- leu^-

The + sign in the genotype indicates that the gene is functional, or wild-type. The − sign indicates that the gene is present as a mutant allele, producing an inactive enzyme (in these cases blocking biosynthetic pathways).

About 10^8 cells of each type were mixed together and plated on minimal medium containing none of the four growth factors. Although neither auxotroph could grow on this medium, a few hundred colonies developed. On isolation, these proved to have the genotype $bio^+met^+thr^+leu^+$ (i.e., they consisted of cells having the heritable capacity to synthesize all four growth factors). The first problem was to determine whether or not some sort of transforming principle was involved. Exhaustive attempts were made to find a diffusible chemical substance that could pass from one cell to another and bring about the observed genetic changes, but such attempts were uniformly negative. It was finally established by microscopic observation that recombination in *E. coli* requires direct cell-to-cell contact, or *conjugation.*

• **The role of plasmids in bacterial conjugation**

It was initially assumed that bacterial conjugation involves cell fusion with the formation of true zygotes. Ultimately, however, it was discovered that there are two mating types in *E. coli* and that during conjugation *one partner acts only as genetic donor, or male, and the other only as genetic recipient, or female.* A male is recognized by the fact that it can be killed with streptomycin or other agents and still retain its fertility, while a female is recognized by the fact that its fertility is destroyed by lethal agents. In other words, since the only function of the male is to transfer some of its DNA, it need not remain viable, whereas the female cell must remain viable in order for the zygote to develop. A pair of conjugating cells is shown in Figure 15.4.

When the recombinants issuing from a variety of crosses were analyzed for mating type, it was discovered that *maleness in bacteria is determined by a transmissible genetic element:* when male and female bacteria conjugate, every female cell is converted to a male. The genetic element governing the inherited property of maleness is called the *F factor* (for "fertility"); it is transmitted only by direct cell-to-cell contact.

Although every conjugating cell transfers the F factor, the transfer of chromosomal markers is a relatively rare, and random, event. Thus, F is an autonomous element, separate from the bacterial chromosome. In 1952 Lederberg coined the term *plasmid* as a generic name for all extrachromosomal hereditary determinants, of which F is an example. We now know that bacterial plasmids are small, circular molecules of DNA which carry the genes for their own replication. In many cases, they also carry genes which confer new properties on the host cell, such as resistance to drugs or the production of toxins. Finally, many plasmids carry genes which govern the process of conjugation. These include genes which determine

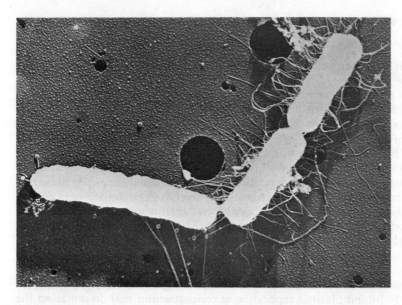

FIGURE 15.4

Mating cells of *E. coli* (×25,600). This electron micrograph was taken shortly after mixing together donor (Hfr) cells and recipient (F⁻) cells. Before mixing, the Hfr cells were "marked" by causing them to absorb inactive particles of bacteriophage; the F⁻ cells are easily recognized by the fact that in this strain they are covered heavily with pili. The micrograph clearly shows the conjugation bridge that has formed between the Hfr cell and one of the F⁻ cells. Note the bacteriophage particles adsorbed by their tails onto the Hfr cell. From T. F. Anderson, E. L. Wollman, and F. Jacob, "Sur les Processus de Conjugaison et de Recombination chez *E. coli*. III. Aspects Morphologiques en Microscopie Électronique," *Ann. Inst. Pasteur* **93**, 450 (1957).

certain *new surface structures* of the cell, such that it will form a conjugation bridge when it comes into contact with another bacterial cell. They also include genes whose products bring about the *transfer of a molecule of plasmid DNA* into the recipient.

Conjugation is thus a mechanism imposed on the bacterial cell by a plasmid, the normal result of which is the transfer of plasmid DNA. The transfer of the bacterial chromosome, as we shall see later, is a secondary consequence of plasmid transfer: in most, if not all, cases, it depends on the integration of chromosome and plasmid in the donor cell.

• **Types of plasmids**   Following the discovery of the nature of F, a large number of bacterial host properties have been found to be determined by plasmid-carried genes. Most plasmids have been classified on the basis of the host properties which initially led to their detection. Thus, there are the *R factors* (for "resistance") and the *Col factors* (for "colicinogeny") of the Gram-negative bacteria, the *penicillinase plasmids* of *Staphylococcus aureus*, the "*degradative*" plasmids of *Pseudomonas*, the *cryptic plasmids*, and so on.

Those plasmids which confer on their hosts the ability to transfer chromosomal markers, but no other readily detectable properties, have been called *F factors*; this group includes the original F factor discovered in *E. coli* K12, which has been designated as F1. Many plasmids, however, including some R factors and Col factors, can mobilize the chromosome for transfer, and some authors have used the term *sex factor* to describe any plasmid with this potential. The term *sex factor* has also been used in two other ways: (1) as a generic name for all plasmids which determine host conjugation and thus their own transfer, regardless of whether or not chromosomal markers are also transferred, and (2) to describe that set of genes on the plasmid whose products mediate the conjugational process.

• **The recognition of plasmids**

Plasmids can be recognized by both genetic and physical techniques. Genetically, the presence of a plasmid is revealed if a gene (or set of genes) governing one or more host properties can be shown to be unlinked to the bacterial chromosome, i.e., to replicate autonomously. Such autonomy has usually been inferred from the *independent transfer* of the plasmid and chromosome during conjugation—as in the case of F1—or from the *irreversible elimination* of the plasmid, either spontaneously or by such agents as heat, acridine dyes, or ultraviolet light. Under ordinary circumstances, the irreversible loss of a gene function might just as easily be interpreted as a nonrevertible mutation in a chromosomal gene. The induction of such loss by the above-mentioned agents, however, is characteristic of plasmids, such as F, whose nature has been well established by transfer experiments. Thus, in cases in which transfer does not occur, induced elimination at high frequency is considered strong evidence for the plasmid state. When a given set of genes, rather than a single gene, is regularly eliminated *en bloc*, the evidence is stronger still.

The elimination of plasmids by dyes and other agents reflects the ability of such agents to inhibit plasmid replication at concentrations that do not affect the chromosome. Thus, plasmid-free segregants arise at the time of cell division, if plasmid replication is prevented.

The criterion of elimination implies that plasmids carry only nonessential genes, i.e., they are defined as *dispensable, autonomous elements*. Indeed, if a plasmid were to acquire an essential gene, such as one required for protein synthesis by the cell, it would differ from the bacterial chromosome only in size. Such plasmids have been produced in the laboratory by genetic manipulation; the cells that carry them behave in every respect as though their genome consisted of two chromosomes of unequal length.

Plasmids can also be recognized by physical techniques. As we will discuss later, their small size and circularity make it possible to detect and even to isolate them as a unique fraction of DNA. Thus, the presence of a small circular DNA molecule that disappears when a marker is eliminated constitutes physical evidence for the presence of that marker on a plasmid. In some cases, cells can be transformed for a particular marker with plasmid DNA, proving the association directly.

PROPERTIES OF PLASMIDS

• **Molecular structure**

All plasmids for which the structure is known consist of circular, double-stranded DNA molecules. Of those for which size measurements have been made, several (F1 and certain R factors) have molecular weights in the range of 5×10^7 to 7×10^7. One (an R factor) has a molecular weight of only 1×10^7, and some of the cryptic plasmids are even smaller. Since the amount of DNA required to code for an average polypeptide with a molecular weight of 40,000 is about 6×10^5, F1 and other plasmids of similar size may contain as many as 100 genes.

Plasmids can also be characterized in terms of mean base composition [i.e., the percent of G + C (guanine + cytosine) and A + T (adenine + thymine) base pairs]. For example, the DNA of F1 has been found to contain two regions of markedly different base composition: one, comprising 90 percent of the molecule, contains DNA of 50 percent G + C, while the remaining 10 percent contains 44 percent G + C.

Differences in mean base composition permit the different plasmids to be separated from each other or from chromosomal DNA by ultracentrifugation in cesium chloride density gradients, since the density of DNA is proportional to its content of G-C base pairs. For example, when F1 is present in *Proteus mirabilis*, a bacterium containing chromosomal DNA of 40 percent G + C, centrifugation of total extracted DNA in cesium chloride produces a main band of chromosomal DNA and a smaller "satellite" band representing the plasmid (Figure 15.5). In *E. coli*, however, which has chromosomal DNA of 50 percent G + C, the densities of many plasmids are too close to that of the host to permit their detection by this method.

A newer method permits the detection of plasmids even when their density is similar to that of the host chromosome. When DNA is extracted from cells, the long molecules are inevitably broken into smaller fragments by the shear forces generated during the process of breaking the cells, stirring the extracts, etc. It has been discovered, however, that if the DNA is in the form of very small circular molecules, a portion of these will escape fragmentation. The DNA extracted from a cell containing plasmids thus contains both linear fragments and circular (plasmid) molecules. The latter can be separated from the former by ultracentrifugation, and can be highly purified. They can then be examined for

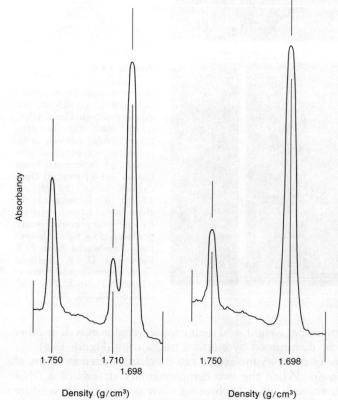

Absorbancy

1.750   1.710   1.750   1.698
         1.698

Density (g/cm³)   Density (g/cm³)

FIGURE 15.5

Effects of plasmid infection on the DNA of *Proteus mirabilis*, strain PM-1. Microdensitometer tracings of ultraviolet absorption photographs taken after equilibrium was reached in cesium chloride density gradient centrifugation: (left) DNA extracted from strain PM-1 infected with the plasmid F13; (right) DNA extracted from PM-1 before infection. The band at density 1.750 g/cm³ is a density standard that was added to both tubes. The band at density 1.698 g/cm³ is *Proteus* DNA; the small "shoulder" at density 1.710 is F13 DNA. From S. Falkow et al., "Transfer of Episomic Elements to *Proteus*. I. Transfer of F-Linked Chromosomal Determinants," *J. Bacteriol.* **87,** 209 (1964).

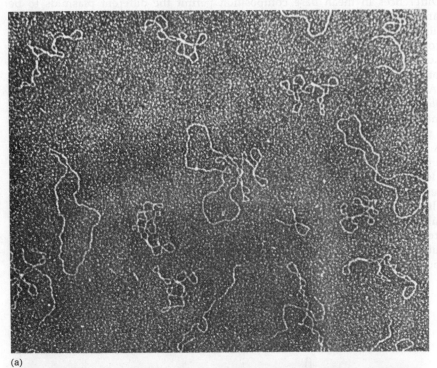

(a)

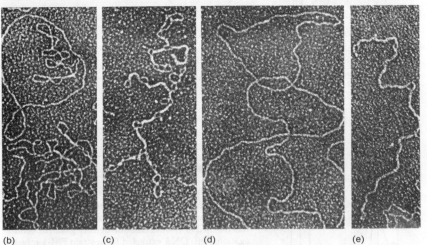

(b)   (c)   (d)   (e)

FIGURE 15.6

Electron micrographs of purified Col E_1 DNA. (a) Open and supercoiled circular DNA, 2.3 μm class. (b) An open 2.3 μm circular molecule juxtaposed to its supercoiled allomorph. (c) Supercoiled forms (2.3 μm and 4.7 μm). (d) Open circular forms (2.3 μm and 4.7 μm). (e) Supercoiled circular form, 4.7 μm, extracted by a different procedure which produces more tightly twisted coils. Reproduced from T. F. Roth and D. R. Helinski, "Evidence for Circular Forms of a Bacterial Plasmid," *Proc. Natl. Acad. Sci.* **58**, 650, 1967.

base composition and molecular weight, and their contour lengths as well as other structural features can be determined by electron microscopy (Figure 15.6).

Finally, different species of DNA molecules can be characterized in terms of their *base sequence homologies*. When the two complementary strands of a DNA duplex are separated from each other by heating, slow cooling of the mixture permits the complementary single strands to reform hydrogen-bonded base pairs and thus to reanneal. The extent of annealing between any two single strands depends on the extent of their base sequence homology: the longer the runs of bases which are complementary to each other, the more extensively the two strands will anneal. Furthermore, the more closely two DNAs are related to each other, in terms of evolutionary origin, the more nearly identical their base sequences; hence, the ability of heat-denatured DNA molecules of different origin to anneal with each other serves as a precise measure of their genetic relatedness.

The double-stranded DNA molecules that are formed when partially complementary single strands are annealed together are called *heteroduplexes*. The genetic relationships of a large number of plasmids have been determined by experiments on heteroduplex formation, as we shall discuss later.

• Replication

The general features of plasmid replication are assumed to be similar to those of chromosomal replication. In both cases, the circular DNA is assumed to be attached to the bacterial membrane at the replicating fork and to move past the attachment site as replication proceeds.

Although chromosome replication and plasmid replication are similar in their general outlines, some differences have been observed. For example, each type of plasmid seems to operate under its own genetically determined system of *replication control*; the rate of plasmid replication may differ widely from that of the chromosome under certain conditions. In a constant environment, however, a steady state is achieved in which the number of copies per cell of a given plasmid remains constant. When F⁺ cells of E. coli are growing exponentially in rich medium at 37°C, for example, there are two F plasmids per chromosome. Under such conditions, both chromosomes and plasmids are doubling during each cell generation; this fact implies a single mode of regulation for both types of replicons.

As we have mentioned earlier, many plasmids are uniquely sensitive to inhibition of their replication by certain agents, including acridine dyes. Acridine orange, for example, will prevent F replication at a concentration that has little effect on cell growth. Since plasmids represent much smaller targets than the chromosome for the nonspecific binding of acridine to DNA, this type of binding cannot account for their unique sensitivity; its true cause is still unknown. Empirical knowledge about this sensitivity is very useful, however, since it often permits cells to be "cured" of their plasmids.

• Integration and recombination

In Chapter 12 the integration of λ prophage with the bacterial chromosome was shown to occur by a single crossover event. A crossover is a recombination event mediated by specific enzyme systems. Extensive studies of recombination in λ phage have revealed three distinct recombination enzyme systems in λ-infected E. coli cells: (1) a system, determined by host genes, which catalyzes crossing over between two homologous regions of chromosomal DNA or two homologous regions of λ DNA; this system is inhibited in λ-infected cells. (2) A similar system, determined by phage genes; and (3) a phage-determined integration enzyme system, which promotes crossing over between a specific attachment site on λ and a specific attachment site on the E. coli chromosome. There is some evidence that the integration enzyme system owes its specificity entirely to its recognition of discrete base sequences in the two attachment sites, without these sites being homologous with each other.

A number of different plasmids are able to integrate with the host chromosome; a plasmid having this ability is called an *episome*. In most cases, it is not yet known whether such integrations are mediated by integrase-like enzymes, which recognize different base sequences in the plasmid and chromosome, or whether they are mediated by recombinases, which operate only on molecules that have undergone reciprocal base pairing in regions of base sequence homology. In either case, however, the host chromosome must possess an appropriate site if plasmid integration is to occur; thus, the ability of a plasmid to behave as an episome is *host specific*. For example, F1 behaves as an episome in *E. coli,* but not in *Proteus mirabilis*. From studies of annealing, it is known that F1 DNA has some homology with the *E. coli* chromosome but none with the *P. mirabilis* chromosome, and the failure of F1 to integrate in the latter host shows that there are also no specific attachment sites recognizable by an integrase.

Once F has integrated with its host chromosome, a second crossover event nearby will bring about its detachment, with a consequent exchange of the DNA segments lying between the crossover sites. In fact, any odd number of crossovers will serve to integrate two circular DNA structures, and any even number of crossovers will result in an exchange of segments (Figure 15.7). F-*lac,* for example, is a recombinant replicon which arose by two successive crossover events: one, which integrated F1 at a chromosomal site adjacent to the *lac* operon of *E. coli* K12, and a second, later event which caused F to detach bearing the *lac* genes.

Some col factors and R factors are also able to integrate with the chromosomes of certain hosts and, like F1, are able to detach again by a second crossover. Such processes, occuring again and again over a long period of evolutionary time, could well explain the heterogeneity of the DNA in certain plasmids, and the presence in most plasmids of genes that govern host cell functions; the formation of such plasmids as F-*lac* serves as a model for this hypothesis. Indeed, several plasmids carrying the *lac* gene exist naturally in rare *lac*$^+$ strains of *Proteus;* these species are normally *lac*$^-$.

Crossing over often occurs between two different plasmids, so that the integration of two or more plasmids and genetic exchanges between similar plasmids are common events. Thus, in certain hosts and under certain conditions many R factors have been found capable of separating into two or more independently replicating circles. Genetic exchanges, involving even numbers of crossovers, occur between homologous plasmids at a rate sufficient to permit the construction of genetic maps of the plasmids.

Recombinant plasmids can also be created *in vitro* by the enzymatic treatment of isolated plasmid DNA, and can be reintroduced into host bacteria by the process of transformation. The first step in this process is to cleave the isolated plasmid at one specific site, by means of a purified restriction endonuclease of the type described at the beginning of this chapter. The enzyme converts the circular DNA molecule into a linear one, by cutting the two strands at sites which are about five or six base pairs apart; the resulting molecule thus has short single-stranded ends which are complementary to each other.*

---

*This DNA structure is exactly analogous to the virion form of phage λ DNA, as shown in Figure 12.17.

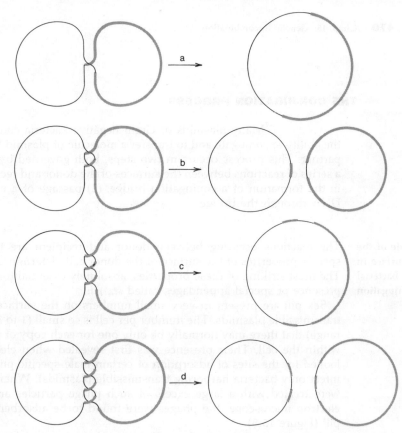

FIGURE 15.7

(a and c) Odd numbers of crossovers within a pairing region bring about integration. (b and d) Even numbers of crossovers bring about the exchange of segments.

Using the same enzyme, another plasmid is similarly cleaved; when the two linear plasmids are mixed together, their single-stranded ends can pair with each other to form a single circle of DNA, half coming from one plasmid and half from the other. Treatment with a second enzyme, polynucleotide ligase, is then used to join the two plasmids covalently.

This procedure has been used not only to join together two different plasmids, but also to insert a length of DNA from a higher animal into a plasmid. For example, the DNA that codes for ribosomal RNA in the frog, *Xenopus*, can be isolated, highly purified and cleaved with the restriction enzyme. When the fragments of rDNA, as it is called, are mixed with plasmid DNA that has been cleaved with the same enzyme and then treated with ligase, some circular DNA molecules are formed consisting of the plasmid with an inserted length of frog rDNA. When the recombinant plasmid is introduced into a host bacterium, the rDNA is replicated and expressed as part of the established plasmid, forming frog ribosomal RNA (although such RNA is not incorporated into the bacterial ribosomes). The enzymatic synthesis and establishment of artificial recombinant plasmids thus provide a means of amplifying indefinitely any type of nonplasmid DNA; it has considerable potential for the large-scale preparation and analysis of human genes such as those involved in the inherited diseases of man.

THE CONJUGATION PROCESS

Many plasmids of Gram-negative bacteria confer on the host cell the ability to conjugate and to transfer a molecule of plasmid DNA to its conjugal partner. This process occurs in two steps, both governed by plasmid genes: (1) a series of reactions between the surfaces of the donor and recipient cell, resulting in the formation of a conjugation bridge; (2) passage of a molecule of plasmid DNA through the bridge.

• **The role of the cell surface in bacterial conjugation**

The reactions occuring between donor and recipient are the consequence of specific properties of the surface of the donor cell, determined by plasmid genes. The most striking of these properties, absolutely essential for conjugation, is the presence of special appendages called *sex pili*.

Sex pili are present in very small numbers on the surface of cells that carry transmissible plasmids. The number per cell is so small (1 to 10 being the normal range) that there may normally be only one for each copy of the plasmid present within the cell. Their presence was first revealed when electron microscopists looked for the sites of adsorption of certain male-specific phages (phages which infect only bacteria harboring transmissible plasmids). When male (donor) cells were treated with a large excess of such phage particles and examined in the electron microscope, the phages were found to be adsorbed exclusively to sex pili (Figure 15.8).

A specific receptor site for conjugation also appears to be present on recipient

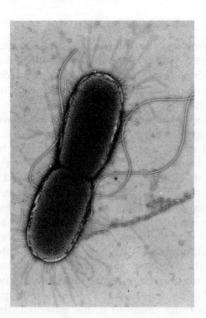

FIGURE 15.8

A dividing cell of an Hfr strain of *E. coli*, showing three types of appendages. The long, curved, thick appendages are flagella. The short, thin, straight appendages are ordinary pili. The long, thin appendage (lower right) coated with phage particles is an F pilus. Courtesy of Judith Carnahan and Charles Brinton.

cells. This site is also the specific adsorption site of a single-stranded DNA phage. Mutant female strains which are resistant to this phage fail to adsorb it, and they are also defective in conjugation. The defect is in the entry process, since the mutants undergo normal recombination by phage transduction.

The first step in the conjugation process is the attachment of the recipient cell to the tip of a sex pilus [Figure 15.9(a)]. Within a few minutes the two cells move into a position of direct contact [Figure 15.9(b)], probably by retraction of the pilus into the male cell.

Cells connected by a sex pilus have been isolated by micromanipulation and subjected to genetic analysis. Some of the cells so isolated produce recombinant clones, indicating that donor DNA can pass through or along the sex pilus. Most genetic transfer, however, occurs only after the cells come into direct contact through pilus retraction. The intercellular passage of DNA is a highly specific process, since little or no cellular material other than DNA is transferred.

• **The transfer of plasmid DNA**

Through experiments in which radioactive or density-labeled precursors were incorporated into plasmid DNA either before or during transfer, the following picture of the transfer process has emerged.

Prior to conjugation, the plasmid exists in the donor cell as a double-stranded, circular DNA molecule. Contact between sex pilus and recipient cell wall triggers the process diagrammed in Figure 15.10(a): one strand of the plasmid DNA is broken at the replication origin and the duplex unwinds, the broken strand entering the recipient cell beginning with its 5′ end. Complementary strands are synthesized by DNA polymerase in both donor and recipient; the process is thus analogous to normal plasmid replication, with the difference that, at completion, one replica is located in the recipient while the other remains in the donor cell. Circularization of the plasmid in the recipient is accomplished by the action of

(a)
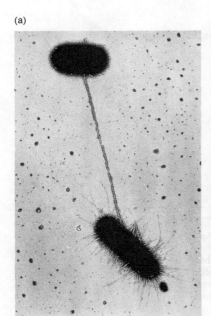

(b)
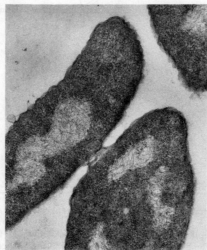

FIGURE 15.9

(a) A male and a female cell joined by an F pilus. One F pilus has been "stained" with male-specific RNA phage particles. The male cell also possesses ordinary pili, which do not absorb male-specific phages and which are not involved in conjugation. Courtesy of Judith Carnahan and Charles Brinton. (b) Electron micrograph of a thin section of a mating pair that has come into direct contact through retraction of the F pilus. From J. D. Gross and L. G. Caro, "DNA Transfer in Bacterial Conjugation," *J. Mol. Biol.* **16**, 289 (1966).

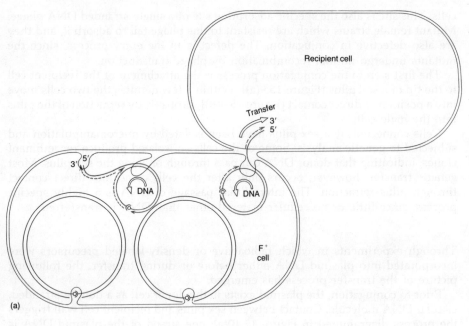

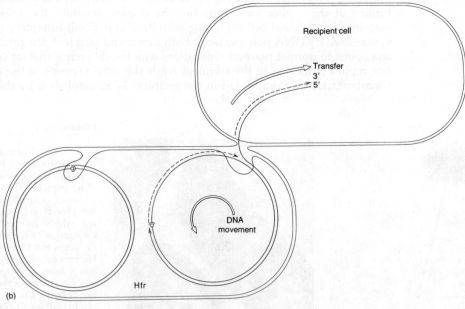

FIGURE 15.10

The hypothetical mechanism of DNA transfer as a consequence of F replication. (a) An F+ cell, containing two autonomous F replicons and two chromosomes, is shown conjugating with a recipient cell. Replication of F is proceeding according to the mechanism outlined in Figure 11.57. The F at the left is replicating within the host cell; the F at the right is being driven into the recipient by replication. (b) The same process takes place when an Hfr cell conjugates. Now, however, chromosomal DNA is also transferred, as a result of its integration with F. (Compare with Figure 15.11.)

DNA ligase. Although the transfer of one strand of the plasmid DNA duplex is usually accompanied by replication, such replication is not essential for the transfer process.

If a bacterium harbors two plasmids, one of which can promote conjugation and the other not, the former may bring about the simultaneous transfer of the latter, i.e., the latter is *mobilized*. Mobilization occurs if the nonconjugational plasmid lacks one or more gene functions (e.g., pilus formation) that the conjugational plasmid can provide. Mobilization also occurs when the two plasmids are either permanently or transiently integrated by a genetic crossover. The bacterial chromosome may similarly be mobilized by integration with a conjugational plasmid [Figure 15.10(b)].

F-MEDIATED CHROMOSOME TRANSFER IN *E. COLI*

• The F+ and
Hfr states

Cells that harbor an F factor are called F+; cells that do not are called F−. In a population of F+ cells the integration of F and chromosome by crossing over occurs about once per 10^5 cells at each generation; the cells in which this occurs, and the clones which arise from them, are called Hfr, for "high frequency of recombination."

Integration does not always occur at the same site on the bacterial chromosome. There are eight or ten preferred sites at which F1 has been observed to integrate; these are presumed to represent base-pair regions homologous with regions on F1. In addition, there are some other sites at which F1 integration occurs with extreme rarity.

The integration process, as we have discussed earlier, is reversible. In a population of Hfr cells, detachment by a second crossover occurs at roughly the same rate as integration in an F+ population. Thus, every F+ population contains a few Hfr cells, and every Hfr population contains a few F+ cells.

• DNA transfer by
Hfr donor cells

When a suspension of Hfr cells is mixed with an excess of F− cells, every Hfr cell will attach to an F− cell and initiate DNA transfer. Since F and chromosome have integrated to form a single replicon, chromosomal DNA, as well as F DNA, passes into the recipient.

The order in which chromosomal markers move into the recipient depends on the chromosomal site at which F has integrated, as well as on the polarity of the integration event. In Figure 15.11, for example, the orientation of F at the time of integration is such that, in the resulting Hfr cell, chromosomal markers will be transferred in the order *thr, leu, pro, lac, . . . , met*. Integration may occur at other sites, however, and with the opposite polarity, producing different orders of marker transfer as shown in Figure 15.12. The polarity of integration is presumably determined by the homologous base sequences in the attachment sites on F and chromosome.

The linear transfer of chromosomal markers by Hfr donors was discovered by E. Wollman and F. Jacob in 1956, in the course of experiments on the kinetics of recombinant formation. Such kinetic experiments were performed as follows. Samples from a mating mixture of Hfr and F− cells were withdrawn at intervals and agitated in a blender for several minutes to shear apart mating pairs. Each sample was plated on several different selective media to determine the numbers of different types of recombinants that had been formed by the time of sampling.

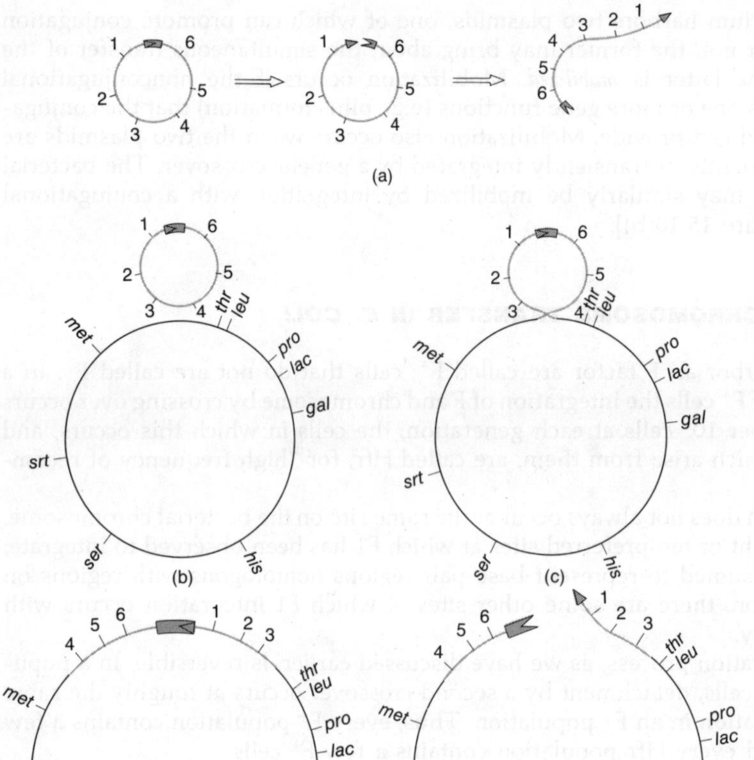

FIGURE 15.11

The breakage and transfer of the Hfr chromosome as a consequence of its integration with F. (a) F is shown as a circle, which has a special site of breakage between markers 1 and 6. Breakage is followed by transfer (during conjugation), such that F markers penetrate the recipient in the order 1-2-3-4-5-6. (b) F has paired with a chromosomal site between *met* and *thr*. (c) A crossover in the region of pairing integrates the two circles. (d) Same as (c), but redrawn as a single circle. (e) Breakage at the special F site leads to transfer causing F markers 1-2-3 to enter the recipient first, followed by chromosomal markers *thr, leu, pro, lac,* and so on; F markers 4-5-6 enter last. Recombinants will be males if they receive all six F markers; thus the terminal end of the chromosome must be transferred to produce an Hfr recombinant.

The results of such an experiment are shown in Figure 15.13; the markers studied are designated simply as A, B, C, and D. As this figure shows, the longer the mating couples are allowed to conjugate before being separated, the more genetic material is transferred from the Hfr cells to the F⁻ cells. For example, if the mating pairs are separated after 10 minutes, only gene A will have been transferred. If conjugation is allowed to proceed for 15 minutes, genes A and B are transferred; after 20 minutes genes A, B, and C are transferred; and so on.

These experiments showed that the donor cell slowly injects a chromosomal strand into the receptor cell, so that the chromosomal genes enter the receptor cell in sequence. If conjugation is not interrupted, it proceeds until the entire chromosome has been injected, a process that in *E. coli* requires about 90 minutes at 37°C.

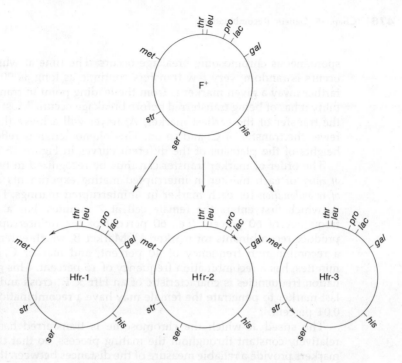

FIGURE 15.12

Diagram showing the origin of three different Hfr males from a common F⁺ parent. Arrowheads on chromosomes indicate the leading point and direction of chromosome transfer in each case.

The slopes of the lines in Figure 15.13 show that the donor cells are not synchronized with respect to the initiation of transfer. Some cells begin to transfer almost immediately, but in others there is a variable delay. Within 25 minutes, however, all donor cells begin to transfer. The intercept of each curve with the abscissa indicates the time at which a given gene arrives in the zygote in the first mating couples to initiate transfer.

If a mating is allowed to proceed without artificial interruption, considerable

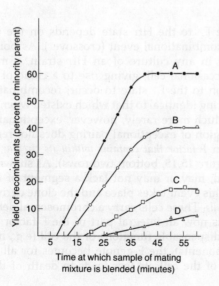

Time at which sample of mating mixture is blended (minutes)

FIGURE 15.13

The kinetics of recombinant formation. See text for explanation. After E. Wollman and F. Jacob.

spontaneous chromosome breakage occurs. The time at which any given break occurs is random; very few transfers continue as long as 90 minutes. Thus, the farther away a given marker is from the leading point in transfer, the less probability it has of being transferred before breakage occurs. Most couples will achieve the transfer of the earliest marker, A; fewer will achieve the transfer of B; still fewer the transfer of C; and so on. This phenomenon is reflected in the relative heights of the plateaus of the different curves in Figure 15.13.

The order of marker transfer can thus be recognized in two ways: by the *time of entry* of each marker in interrupted mating experiments and by the *frequency of recombination* for each marker in uninterrupted matings. For example, marker A, which first enters the female cell at 7 minutes, has a final recombination frequency of 60 percent (i.e., 60 percent of the uninterrupted mating couples produce recombinants for marker A). Marker B, which enters at 12 minutes, has a recombination frequency of 40 percent; and marker C, which enters at 17 minutes, has a recombination frequency of 15 percent. This gradient of recombination frequencies is characteristic of an Hfr $\times$ F$^-$ cross and is so steep that the last marker to penetrate the female may have a recombination frequency of only 0.01 percent.

The speed at which the chromosome is transferred has been found to be relatively constant throughout the mating process, so that the times of entry of markers provide a reliable measure of the distances between them. Since the entire chromosome, representing 5×10^6 base pairs, is transferred in 90 minutes, each minute on the time scale in Figure 15.13 is equivalent to approximately 5×10^4 base pairs. A genetic map of the chromosome of *E. coli* K12 is shown in Figure 15.14; the positions of markers that are half a minute or more apart have been determined by measurements of entry times in interrupted mating experiments; those markers which lie too close together to be resolved by such procedures have been mapped by linkage analyses in phage-mediated transduction, a process that will be described later in this chapter.

• **The origin and behavior of F′ strains**

The change from the F$^+$ to the Hfr state depends on the integration of F and chromosome by a recombinational event (crossover). As pointed out earlier, this process is reversible: in any culture of an Hfr strain, F may detach from the chromosome in an occasional cell, giving rise to a clone of F$^+$ cells.

For a true reversion to the F$^+$ state to occur, recombination must take place within a region of pairing identical to that which existed when integration occurred (Figure 15.15, top). Much more rarely, however, exceptional pairing takes place. A crossover in the region of exceptional pairing does not regenerate an ordinary F factor but rather an F *factor that contains within its circular structure a segment of chromosomal DNA* (Figure 15.15, bottom two rows). As shown in Figure 15.15, the F-*genote*, as it is called, may or may not lack a segment of F DNA.

The cell in which this event takes place and the clone derived from it are called *primary F′* (F-prime) *cells*. These cells carry a chromosomal deletion, corresponding to the segment that is now an integral part of the F factor. When this segment includes genes for indispensable functions of the cell (e.g., nucleic acid polymerases, ribosomal components), the F-genote becomes for all practical purposes a second chromosome of the cell; its loss leads to death of the cell.

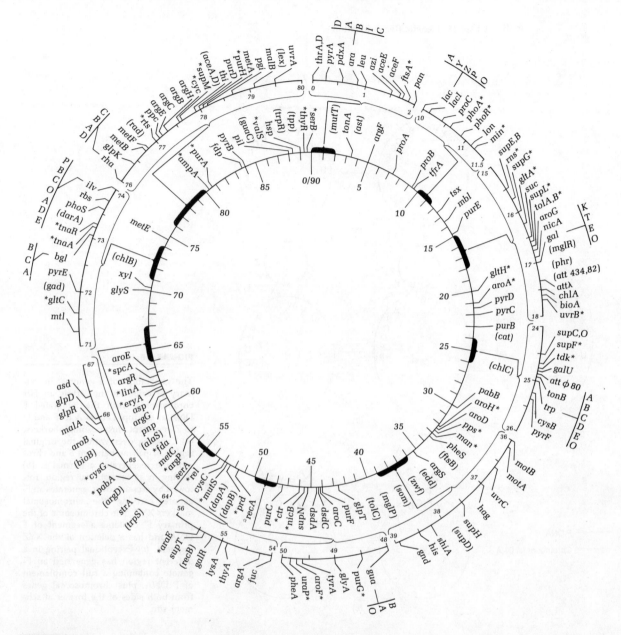

FIGURE 15.14

The genetic map of *E. coli* K12. The units of distance along the inner circle represent the relative times of entry of the markers during conjugal transfer at 37°C. The zero point has arbitrarily been chosen as the *thrA* locus. Certain parts of the map (e.g., the region from 9 to 10 minutes) are displayed on arcs of the outer circle to provide an expanded time scale for crowded regions. Markers in parentheses are only approximately mapped; those marked with an asterisk are more precisely mapped than markers in parentheses, but their positions relative to adjacent markers are not exactly known. Arrows show the direction of messenger RNA transcription for the loci concerned. From A. L. Taylor and C. D. Trotter, "Linkage Map of *Escherichia coli* Strain K12," *Bacteriol. Rev.* **36**, 504 (1972).

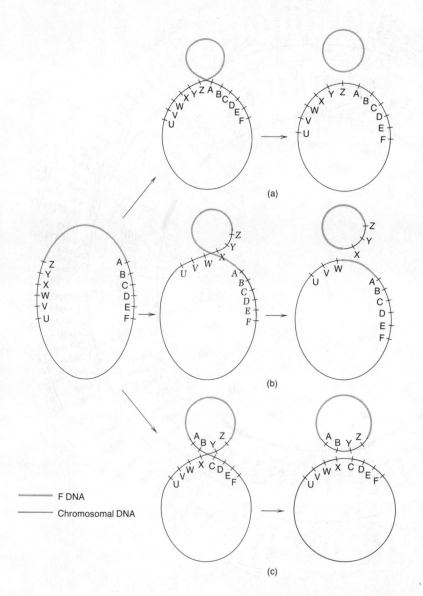

F DNA

Chromosomal DNA

FIGURE 15.15

The generation of F-genotes, in primary F′ cells. At the left is an Hfr chromosome, with the integrated F DNA at the top. Letters A to F and U to Z represent chromosomal markers. (a) Crossing over within the original region of pairing between F and chromosome regenerates a normal F. (b) Pairing in an exceptional region, followed by crossing over, generates an F-genote carrying the chromosomal markers XYZ. The chromosome of the primary F′ contains a segment of F DNA and has a deletion of the XYZ segment. (c) Exceptional pairing in a different region has generated an F-genote containing a full complement of F DNA, plus chromosomal genes from both sides of the former attachment site.

When primary F′ cells are mated with cells of an F⁻ strain, F is transferred (together with the integrated chromosomal segment) with high efficiency. Chromosomal transfer, however, occurs with an efficiency of 10^{-5} per cell or less, the primary F′ cells having the same low probability of F integration with chromosome as do ordinary F⁺ cells.

An F-genote that is transferred to an F⁻ cell reestablishes its circular form and becomes an autonomous replicon, producing a *secondary F′ cell*. The secondary F′ cell differs from the primary F′ cell in that the chromosomal segment of the F-genote is also present in the host chromosome; the secondary F′ cell is thus a *partial diploid* (Figure 15.16), and the F-genote is a dispensable element.

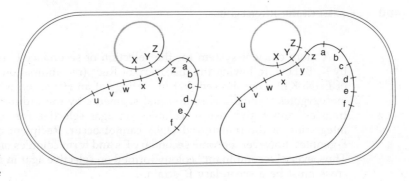

FIGURE 15.16

A secondary F′ cell. The cell shown is a heterozygous diploid for genes X, Y, and Z.

————— F DNA

————— Chromosomal DNA

The region of diploidy in a secondary F′ cell makes it a moderately efficient donor of chromosomal genes, since pairing and recombination occur readily in such a region. In fact, a culture of a secondary F′ strain is always a mixture of F′ and Hfr cells, the two types being in dynamic equilibrium, as shown in Figure 15.17. When such a culture is mated with an F⁻ strain, those cells which at the moment of mating contain a detached F-genote transfer an F-genote alone; other cells, in which the F-genote and chromosome are at that moment integrated, transfer the entire integrated structure.

The first F′ strain was discovered accidentally during the isolation of high-frequency donor cells from an Hfr culture that had reverted to the F⁺ state; one of the high-frequency cell types that was isolated proved to be a primary F′. A method was then devised for the selection of secondary F′ strains. Designating the markers transferred by an Hfr strain as A⁺ through Z⁺, it will be recalled that Z⁺, the last marker transferred, does not normally appear in recombinants before 90 minutes of mating. If the Hfr culture contains a rare F′ cell, however, in which marker Z⁺ is attached to the F factor, that cell will transfer F-Z⁺ within the first 10 or 20 minutes. The procedure is thus to interrupt mating after about 20 minutes and to select for the rare recipient cell that has received the Z⁺ marker. These are usually found to be secondary F′ strains, carrying F-Z⁺. It was by this procedure that the F-*lac* diploids described in Chapter 13 were isolated.

FIGURE 15.17

The dynamic equilibrium between the integrated and detached states of the F-genote in a secondary F′ cell.

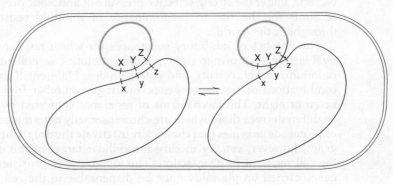

————— F DNA

————— Chromosomal DNA

A more efficient system for the selection of secondary F' strains was devised by K. B. Low, following the discovery of Rec⁻ (recombination-deficient) mutants of *E. coli* K12. Low showed that in matings of an Hfr strain with a Rec⁻F⁻ strain, merozygotes receiving chromosomal segments by the ordinary process of DNA transfer cannot give rise to colonies on agar selective for recombinants, since integration of the transferred DNA cannot occur. Recipient cells which receive F-genotes, however, become secondary F's and form colonies on the selective agar. Thus, every "recombinant" colony formed on selective agar in an Hfr × F⁻ (Rec⁻) cross must be a secondary F' strain.

The dynamic equilibrium between the integrated and attached states of the F-genote in a secondary F' cell, and the dynamic equilibrium in an F⁺ or Hfr culture, differ only in the respective rates of attachment and detachment. In the case of the secondary F', both integration and detachment occur about once every 10 cell-generations; in the F⁺ ⇌ Hfr system, both integration and detachment occur with a frequency of about 10^{-5} per cell-generation.

THE MAJOR GROUPS OF PLASMIDS

• R factors

During an outbreak of bacterial dysentery in Japan in 1955, a strain of *Shigella dysenteriae* was isolated that proved to be simultaneously resistant to four drugs: sulfanilamide, streptomycin, chloramphenicol, and tetracycline. Although such multiple resistance had not been encountered previously, strains of this type became more and more prevalent, so that by 1964 over 50 percent of all *Shigella* hospital isolates in Japan were of the multiple-resistance type.

In 1962 a multiply resistant strain of *Salmonella* was isolated during an outbreak of gastroenteritis at a London hospital, and by 1965 the incidence of multiply resistant *Salmonella* in England had risen to 61 percent. Since then, multiply resistant enteric bacteria have been isolated wherever they have been sought.

K. Ochiai, T. Akiba, and their coworkers in Japan discovered that the genetic determinants for drug resistance are transferable *en bloc* by conjugation, and further studies, particularly in the laboratory of T. Watanabe, established that the resistance genes are linked in various combinations as parts of transmissible plasmids, now called *R factors*. The rapid spread of R factors among populations of enteric bacteria, under the strong selective pressure of antibiotic prophylaxis and therapy, accounts for the rise to predominance of multiply resistant enteric bacteria throughout the world.

The number of inhibitory substances for which resistance may be mediated by R factors has grown to eight or more antibiotics, several heavy metals (mercury, cadmium, nickel, cobalt), and sulfonamides. Different R factors have different combinations of resistance genes, ranging in number from one to as many as seven or eight. The mechanisms of resistance conferred by these genes tend to be different from those which are chromosomally determined. The plasmid genes often encode enzymes that chemically inactivate the drug. Chromosomal resistance genes, however, usually modify the cellular target of the drug so as to render the cell resistant to drug action. This difference is consistent with the fact that genes carried on plasmids must be dispensable to the cell under at least some conditions.

An R factor comprises two distinct segments of DNA: one, called *RTF* (*resistance transfer factor*), carries the genes for replication and transmission of the plasmid; another consists of one or more sequentially linked *r-determinants* (*resistance determinants*). The RTFs of several R factors have been isolated and found to have molecular weights of about 6×10^7, while their sets of r-determinants have molecular weights of about 1×10^7. The RTF and r-determinant segments of an R factor also have different G-C contents, and thus different buoyant densities, suggesting different evolutionary origins.

Indeed, several R factors are capable of dissociating into independently replicating RTFs and r-determinants; the latter thus appear to be self-sufficient replicons, but they are incapable of undergoing transfer unless mobilized by an RTF. The stability of the association varies greatly with the R factor and the host. Some R factors have been observed never to dissociate in one bacterial host, but to do so readily in another. The reversible association presumably reflects the integration of the RTF and r-determinants.

E. S. Anderson and his coworkers have found that some *Salmonella* strains in nature carry only RTFs, which were detected in the following way. A tester strain was produced (by the early interruption of mating with an R factor donor) which contained an r-determinant but no RTF. The tester was first mated with a potential RTF donor, and then with a drug-sensitive recipient. The transfer of an RTF from the donor was detected by the acquired ability of the tester to transfer its drug resistance to the final recipient (Figure 15.18).

Thus, some (if not all) r-determinants are capable of autonomous replication but not of self-transfer. The genes controlling the conjugation process are located exclusively on RTFs, whose function in drug-sensitive cells is entirely unknown.

The many R factors that have been isolated from enteric bacteria fall into a number of discrete *compatibility groups*. A compatibility group is defined as a group of plasmids, no two of which can be stably established in the same host cell. Two or more plasmids representing *different* compatibility groups, however, can replicate side by side in the same host. As we will discuss later, in the section on penicillinase plasmids, the phenomenon of incompatibility probably reflects the competition between all members of a given group for a common membrane

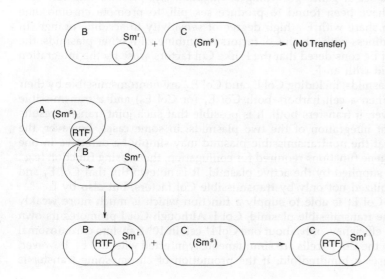

FIGURE 15.18

Testing for the presence of an RTF. Strain B, the tester strain, carries an r-determinant for streptomycin resistance which is not self-transferable; C is a streptomycin-sensitive, plasmid-free strain. When cells of types B and C are mixed (upper row), no transfer occurs. If strain B is first mated with strain A, however, and then mixed with strain C, its streptomycin-resistance determinant is transferred, revealing the presence in strain A of a mobilizing RTF. (As a further control, a mixture of strains A and C alone is shown not to produce streptomycin-resistant recombinants.)

attachment site. The members of a compatibility group appear to be related to each other by two other criteria. First, they form a sex pilus of the same chemical type, as shown by the adsorption patterns of male-specific phages. Second, their DNAs show base-pair sequence homologies in experiments on heteroduplex formation.

All R factors carry repressor genes that regulate pilus production. When a population of cells is freshly infected with an R factor, most of the cells form sex pili and temporarily behave as donors. Repressor action sets in rapidly, however, and pili are both diluted out and lost as the population grows. Thus, in an established culture of R^+ cells, only a few cells form R pili and are active donors.

Some R factors are able to promote the transfer of the *E. coli* chromosome. Stable Hfrs have been isolated from R^+ populations, and R factors which behave like F have been described. These observations show that some R factors, like F1, can reversibly integrate with the *E. coli* chromosome.

• **Col factors**   Many bacterial strains liberate proteinaceous toxins called *bacteriocins*, which are active only against closely related strains. Toxins of this type which are liberated by strains of *E. coli* are called *colicins*. Most are simple proteins; several different types have been isolated which kill sensitive cells by different mechanisms. *E. coli* strains are generally immune to the colicins which they produce themselves.

The genetic determinants for a number of different colicins have been shown by P. Fredericq to be carried on plasmids and never to be integrated into the chromosome. These plasmids are called *colicinogeny (Col) factors*. The principal ones that have been studied are Col B, Col E_1, Col E_2, Col I, and Col V; the colicins which they produce are correspondingly lettered.

Some Col factors, including Col B, Col I, and Col V, carry "fertility determinants," i.e., a set of genes governing conjugation and plasmid transfer. Two Col V factors have been found to produce sex pili, to promote chromosome transfer, and to share with F a high degree of sensitivity to acridine orange. In view of the readiness with which Col factors recombine with other plasmids, the possibility must be considered that the F-like Col factors arose by the integration of a Col plasmid with an F.

Other Col plasmids, including Col E_1 and Col E_2, are nontransmissible by their own means. When a cell harbors both Col E_1 (or Col E_2) and a transmissible plasmid, however, it transfers both. It is possible that such joint transfer results from a transient integration of the two plasmids; in some cases, however, the data suggest that the nontransmissible plasmid may simply be defective in one or more of the gene functions required for conjugation, the missing function (e.g., enzyme) being supplied by the active plasmid. It is noteworthy that Col E_1 and Col E_2 are mobilized not only by transmissible Col factors, but also by F.

Conversely, Col E_1 is able to supply a function which is much more weakly expressed in the transmissible plasmid, Col I. Although Col I promotes its own transfer at high efficiency, only about one Col I^+ cell in 10^8 transfers chromosomal markers. When the Col I^+ cells are simultaneously infected with Col E_1, however, this rate increases a hundredfold. If the promotion of chromosome transfer is

483   due to its integration with Col I, perhaps Col E produces a more active integrating enzyme.

• The "degradative" plasmids of *Pseudomonas*

The pseudomonads are characterized by their ability to catabolize a great variety of unusual organic compounds, by special metabolic pathways. I. C. Gunsalus, A. M. Chakrabarty, and their collaborators have discovered that the enzymes for a number of these special pathways are determined by plasmid genes. In each case, a plasmid determines the enzymes of a single catabolic pathway or a portion thereof. For example, the enzymes of camphor degradation are determined by the "camphor plasmid" of *P. putida*, the enzymes of octane degradation by the "octane plasmid," and the enzymes of salicylate degradation by the "salicylate plasmid." These plasmids are transmissible within and between pseudomonad species by conjugation.

• The penicillinase plasmids of *Staphylococcus aureus*

Staphylococcus aureus is a Gram-positive bacterium. It is an important pathogen of humans that causes skin and wound infections. These were, at first, readily treated with penicillin, but penicillin-resistant staphylococci appeared very soon, and by 1950 most strains isolated in hospitals the world over were highly resistant.

This high level of resistance is due, in every case, to the production of a potent enzyme, *penicillinase*, which destroys the antibiotic by hydrolyzing the β-lactam ring within its structure. The gene for this enzyme (p^+), along with others involved in its regulation, has been shown by R. Novick to be located (in most strains) on a plasmid. The evidence includes the following:

1. Penicillin-resistant cells lose their resistance abruptly and irreversibly, at a rate of approximately one per thousand cell divisions (Figure 15.19).
2. A number of other markers, including resistance to mercury ions (Hg^r) and to erythromycin (ero^r) are closely linked to p^+ during phage transduction. When the transducing phage particles are irradiated with ultraviolet light, all three markers are simultaneously inactivated (in terms of their ability to produce recombinant transductants among the phage recipients) with identical single-hit kinetics.
3. Strains which carry p^+-Hg^r-ero^r lose all three markers *en bloc*, and they regain all three in a single transduction event.

Although not all staphylococcal plasmids carry penicillin-resistance determi-

FIGURE 15.19

Staphylococcus colonies that have been flooded with N-phenyl-1-naphthylamineazo-O-carboxybenzene plus penicillin. The dark sectors contain penicillinase-positive cells; the light sectors contain penicillinase-negative cells. Courtesy of R. Novick.

nants, the "penicillinase plasmids" have been studied most extensively. They have been assigned to a variety of types, designated α, β, γ, etc., based on the markers which they carry as well as on the chemically different penicillinases which they produce.

The different types show *compatibility relationships* analogous to those described earlier for certain R factors. When two incompatible, genetically marked plasmids are introduced into the same cell, 95 percent of the cells formed during the first cell division segregate one or the other type, and segregation is completed within a few more cell divisions. In such segregants, recombinant plasmids are very frequent, and the frequencies with which different markers recombine independently form the basis for constructing a genetic map of the plasmids. In contrast, two compatible plasmids rarely recombine.

These observations have led to the following hypothesis. Every plasmid carries a "maintenance compatibility" locus (*mc*), by means of which the plasmid attaches to a site in the cell (presumably the cell membrane). Such attachment is necessary for plasmid replication. Compatible plasmids carry different *mc* loci, permitting them to occupy different attachment sites in the cell. Incompatible plasmids, however, carry the same *mc* locus and must compete for a single attachment site. When two incompatible plasmids are placed in the same cell, they are capable of integrating with each other (by a single crossover), but segregate rapidly as a result of further crossovers; if the crossovers are far enough apart, recombinants for the known markers are detected.

This hypothesis is supported by a variety of observations: plasmid mutants in which the *mc* region is deleted lose their ability to replicate autonomously, and both chromosomal and plasmid mutations have been observed markedly to lower plasmid stability. The chromosomal gene in which the former class of mutations occurs is presumably involved in the determination of the cellular attachment site, since it specifically affects the stability of one or the other (but not both) of the compatibility groups. The plasmid gene in which the latter class of mutations occurs is probably the *mc* locus itself; the plasmid mutation for instability does not segregate from the *mc* allele with which it was first associated.

Interactions between penicillinase plasmids and the host chromosome have also been observed. In one type of interaction, a plasmid marker introduced by phage transduction becomes stably integrated into the chromosome. A plasmid bearing the same marker may become integrated into the bacterial chromosome at this site, by crossing over in the region of homology. The reverse process—the "extraction" of a chromosomal marker by a plasmid—may also occur.

Although the staphylococcal plasmids do not promote conjugation, they can be transferred from cell to cell by phage transduction. This process appears to occur at a rate sufficient to balance the rate of spontaneous loss, and thus maintain the plasmids in natural populations of staphylococci. Prior to the widespread use of penicillin in medical practice, the equilibrium level of penicillinase plasmids in the population was probably very low; the selective action of penicillin, however, has brought the plasmid-containing strains to predominance in many localities.

• **Cryptic plasmids**   When it was discovered that plasmid DNA could be isolated in the ultracentrifuge as small, covalently closed circular molecules, a search for similar molecules was made among a variety of bacterial species. In almost every species examined, such DNA molecules were found to exist; they are called *cryptic plasmids* because they do not carry any genes which can as yet be detected phenotypically.

It thus seems that the presence of plasmids is the general rule, rather than the exception, in bacterial cells. Some hypotheses concerning their origins are presented below.

• Phylogenetic relationships between plasmids, viruses, and chromosomes

As described earlier in this chapter, plasmids exhibit the following properties:

1. They are circular molecules of double-stranded DNA.
2. They are capable of autonomous replication in the host cell.
3. Some are capable of integrating with each other and with host chromosomes; some are also capable of exchanging genetic segments with each other and with host chromosomes.
4. Many plasmids carry genes that affect the phenotype of the host cell.
5. Some, but not all, plasmids promote their own transfer by means of host cell conjugation.
6. Transmissible plasmids may bring about the transfer of other plasmids residing in the same cell, or of the host chromosome itself.
7. Plasmids are dispensable to the host cell, at least under some conditions.

Of these properties, all but the last three are also properties of the chromosome. Thus, a nontransmissible plasmid differs from a bacterial chromosome only in its dispensability. Even this criterion can be compromised, since F' cells can be isolated in which genes essential for bacterial growth are present on the F-genote but not on the chromosome. In such cases, the F-genote is no longer dispensable, and must, by our definition, be considered a second chromosome instead of a plasmid.

The relationship between plasmids and DNA phages is equally close. DNA phages share with plasmids all of the properties listed above except that of promoting conjugation.* Phages differ from plasmids, however, in their ability to be liberated from the host cell as protein-coated, infectious particles. It has been proposed that transmissible plasmids are closely related to some of the filamentous DNA phages, the sex pilus being an analogue of the phage coat. Since it appears that plasmid DNA may be transferred through the pilus during conjugation, the analogy becomes very strong indeed.

All three genetic elements are capable of recombining with each other, as indicated in the following diagram:

$$\text{chromosome}$$
$$\text{plasmid} \longleftrightarrow \text{phage}$$

Thus, many phages and plasmids can integrate with the chromosome, and recombination has also been observed between a plasmid (one of the R factors) and a phage, producing a phage which carries a resistance gene.

There are many situations in which the distinctions between the three types of genetic elements break down completely. In the prophage state, for example, phage P1 replicates as an autonomous, dispensable genetic element; in this state

---

* Even this statement has its exception, since a phage that infects a species of *Pseudomonas* promotes the transfer of chromosomal markers by conjugation. Since some pseudomonads harbor F factors, this unique phage may have acquired its fertility determinants by recombination with such a plasmid.

it is indistinguishable from a plasmid, just as (in the special case we mentioned earlier) a plasmid may be indistinguishable from a chromosome. What distinguishes the three categories, therefore, is the *potentialities* of their members: plasmids *may* be dispensable, and they *may* promote conjugation. Phages *may* be liberated as infectious particles.

To put it another way, the three classes are to some extent *interconvertible*. A phage may become, by loss of function, a plasmid. A plasmid may become, by acquiring essential genes, a chromosome. Our problem, in trying to explain the evolutionary origin of these elements, is thus to deduce the time sequence in which their differences arose. At what point did replicating DNA molecules acquire the ability to form protein coats and thus become infectious particles? Did plasmids arise from phages, or phages from plasmids? Or did both arise by separate evolutionary pathways from bacterial chromosomes?

It is not difficult to imagine a nontransmissible plasmid arising from the chromosome, since it is known that chromosomal fragments arising by deletion (i.e., F-genotes) can circularize. It would only be necessary for the deleted segment to include the chromosomal replicator site in order for it to become a nontransmissible plasmid. Such an event would not be stable, however, since the newly arisen plasmid would be incompatible with the chromosome; it would be excluded from the chromosomal attachment site. Thus, the evolution of a stable, nontransmissible plasmid would require, as a simultaneous event, a mutation creating a new attachment specificity of the plasmid.

Much more difficult to imagine are the evolutionary events giving rise to the transmissibility of DNA, either in conjugating systems or in the liberation of phage particles. These questions will probably never be answered unequivocally, but they emphasize the close relationships that exist at present between plasmids, chromosomes, and bacterial viruses.

THE OCCURRENCE OF CONJUGATION IN DIFFERENT GROUPS OF BACTERIA

• **Gram-negative bacteria**
In general, a Gram-negative bacterium carrying a conjugational plasmid can mate with a wide variety of other Gram-negative bacteria and transfer plasmid DNA. The efficiency of interspecific and intergenic mating varies widely, however: Table 15.2 shows the range that is observed when F-*lac* is transferred from cells of one genus of Gram-negative enteric bacteria to cells of another. F+ and F' strains have been created in many bacteria of the enteric group by the transfer of suitable plasmids from *E. coli* K12. In some cases, these plasmids have undergone integration with the chromosome of the recipient, forming Hfr donor cells; such Hfrs have been produced in *Salmonella, Yersinia pseudotuberculosis,* and *Erwinia amylovora.*

When Hfr cells are used as donors in intergeneric matings, chromosomal DNA is transferred with frequencies comparable to those shown for F-*lac* in Table 15.2. Recombinants are not found in such matings, however, unless there is sufficient base-pair homology between donor and recipient chromosomes to permit pairing and crossing over. Thus, chromosomal recombinants are formed in matings among *Escherichia, Salmonella,* and *Shigella,* among which homology is high, but never in matings between *Escherichia* and *Proteus,* in which no homology is detectable.

TABLE 15.2

Efficiency of conjugation between different genera of
Gram-negative bacteria[a]

F-*lac* DONOR	RECIPIENT	FREQUENCY OF F-*lac* TRANSFER[b]
Escherichia coli	*Escherichia coli*	10^{-1}–10^{-3}
Escherichia coli	*Salmonella typhosa*	10^{-4}–10^{-5}
Escherichia coli	*Proteus mirabilis*	10^{-5}–10^{-6}
Salmonella typhosa	*Escherichia coli*	10^{-4}–10^{-5}
Salmonella typhosa	*Proteus mirabilis*	10^{-4}–10^{-5}
Salmonella typhosa	*Serratia marcescens*	10^{-7}–10^{-8}
Salmonella typhosa	*Vibrio comma*	10^{-5}–10^{-6}
Proteus mirabilis	*Escherichia coli*	10^{-5}–10^{-6}
Proteus mirabilis	*Proteus mirabilis*	$<10^{-10}$

[a] Data supplied by L. S. Baron.
[b] These frequencies do not reflect the fact that higher-frequency recipients can be obtained from wild-type populations.

Plasmid-mediated conjugation occurs not only in the enteric Gram-negative bacteria, but also in the pseudomonads. The degradative plasmids described earlier mediate conjugation between different species of *Pseudomonas,* and an F-like sex factor has been discovered that promotes chromosomal transfer at low frequency between strains of *Pseudomonas aeruginosa.*

• **Gram-positive bacteria**

The genetics of the actinomycete, *Streptomyces coelicolor,* has been explored in detail by D. Hopwood, G. Sermonti, and their associates. Recombinants arise when spores of two genetically marked strains are plated together in selective media. Recombination requires direct contact between hyphae, and is mediated by a plasmid, SCP1. The original strain, "IF," carries SCP1 in the autonomous state, and is weakly self-fertile. From this have been derived high-frequency donor "NF" strains, carrying SCP1 integrated at a specific chromosomal site, and "UF" strains, which show high fertility as recipients, and which have lost SCP1.

There are certain parallels between the SCP1 system of *S. coelicolor* and the F1 system of *E. coli* K12. In IF × UF crosses (as in F⁺ × F⁻ crosses), every donor cell transfers the conjugational plasmid, but the frequency of chromosomal transfer is several orders of magnitude less. In NF × UF crosses (as in Hfr × F⁻ crosses), every mating event leads to transfer of long chromosomal fragments. However, UF × UF crosses yield recombinants at low frequencies, but matings between two F⁻ strains do not. Furthermore, in NF × UF crosses, the fragment of chromosome that is transferred extends symmetrically in both directions from the SCP1 attachment site, and every recipient receives the integrated plasmid, becoming an NF donor. In contrast, Hfr × F⁻ crosses always show a polarity of transfer, and the integrated plasmid is acquired by the recipient cell only in the very rare event of a complete chromosome transfer.

The SCP1 plasmid has been observed to "pick up" a fragment of the chromosome bearing the gene *cysB;* SCP1-*cysB* is analogous to F-*lac,* and cells which carry it behave like the F' donors described earlier for *E. coli* K12.

By measuring the frequencies with which different donor genes were inherited together in multifactorial crosses, Hopwood was able to construct a genetic map of *S. coelicolor*. All of the markers analyzed were found to lie on a single linkage group, which is circular. A recombination map has been constructed by similar means for another actinomycete, *Nocardia mediterranei*. Homologous markers exhibit the same relative positions on the two maps, indicating a close phylogenetic relationship between the two species.

TRANSDUCTION BY BACTERIOPHAGE

• **The discovery of transduction**   In 1952 N. Zinder and J. Lederberg began experiments to determine whether conjugation occurred between two strains of *Salmonella*. When they mixed strains carrying appropriate genetic markers and plated them on media selective for recombinants, colonies appeared at low frequency.

In a series of further experiments, Zinder and Lederberg discovered that this genetic exchange had not occurred by conjugation, but rather that one of the parent strains had released particles of a temperate bacteriophage that had infected cells of the other parent strain. Many of the receptor cells survived the infection, and some of the survivors proved to have received fragments of genetic material derived from the cells in which the phage had originally grown.

In *transduction*, as this phenomenon was named, a small piece of bacterial chromosome is incorporated into a maturing phage particle. When this particle infects a new host cell, it injects the genetic material from the former host. The recipient thus becomes a partial zygote.

The phage discovered by Zinder and Lederberg, called P22, transduces all *Salmonella* markers with roughly equal efficiency. When other temperate phages of *Salmonella* and *E. coli* were examined, however, only a few appeared to have this capability. In the meantime, the discovery had been made that certain temperate phages of *E. coli* occupy fixed positions on the bacterial chromosome when in the prophage state. For example, λ always attaches between the *bio* and *gal* loci, as shown in Figure 15.14. When λ was tested for the ability to transduce markers between strains of *E. coli*, it was found that it could transduce the *gal* loci or the *bio* loci, but not markers located farther from the attachment site.

Transduction of the type mediated by P22, in which any chromosomal marker has a roughly equal chance of being incorporated into a phage coat, is called *generalized transduction*. Transduction of the type mediated by λ, in which only those loci immediately adjacent to the prophage attachment site are incorporated, is called *specialized transduction*. Specialized and generalized transduction differ in the mechanism by which the genetic material of the transducing phage particles is formed.

Transduction can take place between various members of the enteric group of bacteria and has also been observed in *Pseudomonas*, *Staphylococcus*, and *Bacillus*. Since most species of bacteria harbor temperate phages, transduction may well be the most common mechanism of recombination in bacteria.

In a culture of *E. coli* lysogenic for λ, every cell carries a prophage inserted into the chromosome at a site between the *gal* and *bio* loci. When such a culture is induced, the prophage in each cell detaches by a recombinational event within the normal region of pairing between a chromosomal site and a site on the prophage. As a rare event, however, pairing occurs between two different sites, and the recombinational event generates a circle of DNA consisting of a major fraction of λ, together with a segment of chromosomal DNA from one or the other side of the attachment site (Figure 15.20). If the excised DNA includes the *gal* genes, the resulting phage particle is called λ*dg*, standing for "λ defective, carrying the *gal* loci." If the excised DNA includes the *bio* genes, the resulting phage particle is called λ*dbio*. The total amount of DNA that can be packaged in a phage particle is fixed; thus, if *gal* or *bio* genes are included, a corresponding amount of λ DNA must be missing. This explains why the transducing particle is always defective for some λ functions, and it also explains why *gal* and *bio* genes are never found in the same transducing particle.

The induction of a λ-lysogenic *gal*+ culture produces a lysate containing about

The formation of λ*dg*. (a) Normal pairing followed by a single crossover event regenerates the wild-type λ genome. (b) Abnormal pairing followed by a single crossover event generates the λ*dg* genetic element.

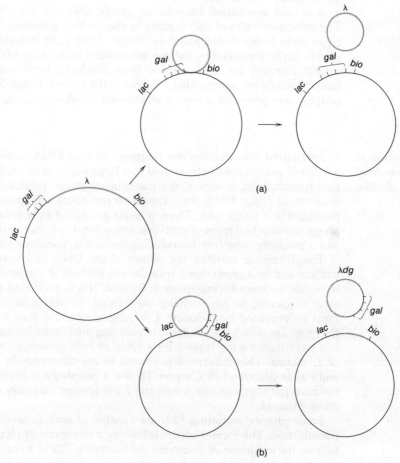

489

one λ*dg* for every 10^5 normal phage particles. When this lysate is used to infect a culture of nonlysogenic *gal⁻* bacteria, some of the cells receive λ*dg* DNA, which becomes integrated into the chromosome as a prophage. Each of these cells, however, still retains its *gal⁻* chromosomal locus and thus becomes a *gal⁺/gal⁻* partial diploid. *Specialized transduction thus involves the addition to the chromosome of a defective prophage carrying a segment of DNA taken from its former attachment site.*

If the infection of the *gal⁻* culture is carried out at a high phage multiplicity, any cell receiving a λ*dg* particle also receives a particle of normal λ. Such a cell becomes a double lysogen, carrying both a normal and a λ*dg* prophage. When a culture derived from the double lysogen is induced, the normal λ prophage provides the phage functions for which λ*dg* is defective. The result is that both types of phage mature, and the lysate that is produced contains equal numbers of λ and λ*dg* phages. When this lysate is now used to transduce a fresh *gal⁻* culture, roughly half the phage particles transduce the *gal⁺* loci. This phenomenon is called *high-frequency transduction.*

A second specialized transducing phage, φ80, has been discovered in *E. coli*. The attachment site of φ80 is close to the *trp* loci, governing the enzymes of the tryptophan biosynthetic pathway. Phage φ80*dt* (φ80 defective, carrying the *trp* genes) can be produced by the procedure used to produce λ*dg*. Another derivative of φ80, carrying the *lac* genes, has been produced from a strain in which F-*lac* had previously been integrated adjacent to the *trp* loci. These defective transducing phages have provided a source of purified DNA containing specific genes.

• The mechanism of generalized transduction

In specialized transduction the fragment of host DNA carried by a transducing particle is always covalently linked to a large segment of phage DNA. In generalized transduction, however, the transducing phage particles contain only small amounts of phage DNA; they appear to constitute a fragment of bacterial DNA packaged in a phage coat. Thus, a lysate produced by a generalized transducing phage contains two types of particles: a majority type, containing only phage DNA, and a minority type (the transducing particles), containing mainly host DNA.

The difference between the nature of the DNA in a specialized transducing particle and in a generalized transducing particle is consistent with the manner in which the transducing lysate is formed. If λ is produced in its host by a lytic cycle of growth, no λ*dg* particles are formed. To obtain such particles, the lysate must be prepared by inducing a λ-lysogenic culture (i.e., λ must be present in the prophage state). Generalized transducing phages, including phage P1, however, can yield transducing lysates by a cycle of lytic growth as well as by induction of a lysogen. This difference is related to the difference in prophage state of λ and P1: as described in Chapter 12, the λ prophage is normally integrated into the bacterial chromosome, while the P1 prophage normally exists as an autonomous plasmid.

Some phages, including P22, are capable of both generalized and specialized transduction. The former occurs following a lytic cycle of phage growth; the latter follows the induction of lysogenic cells carrying P22 as a chromosomally attached prophage.

• The fate of the exogenote formed by transduction

In either type of transduction the exogenote frequently pairs with the endogenote, and donor genes are incorporated into the recombinant chromosome by crossing over within the region of pairing. The partial zygote then segregates a haploid recombinant cell. In restricted transduction the exogenote may also persist and replicate as part of a prophage, giving rise to a heterozygous, partially diploid clone.

The exogenote formed in specialized transduction may persist and function but not replicate, so that in the clone arising from the partial zygote only one cell at any given moment contains an exogenote. This relatively common phenomenon is called *abortive transduction*. If, for example, a phage particle injects a *his*⁺ gene into a *his*⁻ bacterium and the transduction is abortive, the following events occur. The *his*⁺ exogenote functions in the partial zygote, producing enzyme for the synthesis of histidine. If plated on agar lacking histidine, the zygote grows and divides, but the exogenote does not replicate. Thus, one daughter cell does not receive the *his*⁺ gene and can continue to divide only until the initial supply of enzyme has been diluted out by growth. The other daughter cell, however, is a heterogenote like its parent and continues to form enzyme. At its next division, it again segregates one *his*⁻ cell containing a limited supply of enzyme and one *his*⁺/*his*⁻ heterogenote.

The result is the production of a *slow-growing minute colony*, which becomes barely visible after a few days of growth. When a minute colony of this type is picked and restreaked, it gives rise to one minute colony again, since only one of the cells that were picked contains the *his*⁺ gene. The events in abortive transduction are shown in Figure 15.21.

In some cases, a given suspension of phage will produce 10 times as many abortive transductions as normal ones. An abortive transduction results when recombination fails to occur between the exogenote and endogenote. This failure can be overcome by ultraviolet irradiation of the transduced cells. The ability of ultraviolet light to convert abortive transductions to normal transductions is consistent with its known ability to stimulate recombination in many diploid systems.

• Phage conversion

Certain properties of bacterial cells are controlled solely by phage genes. Such properties are manifested only by lysogenic or phage-infected strains and never by phage-free strains. For example, *Corynebacterium diphtheriae* produces toxin only if it is infected with a particular strain of phage. Every phage-infected cell produces toxin, but phage-free cells do not and are incapable of mutating to toxigenicity. Similarly, certain antigenic components of the cell wall of *Salmonella* are produced in every cell infected with a particular phage and are always absent if the phage is absent. The acquisition of a new property solely as a result of infection by a phage is called *phage conversion*.

Conversion thus differs from specialized transduction, which involves the transfer by a phage of a gene that is normally found in the bacterial chromosome. There is a second difference between conversion and specialized transduction: converting phage particles are completely normal and possess all necessary phage functions, whereas specialized transducing phages are usually *defective* for some phage functions, having lost some phage genes by exchange. With these exceptions, however, phage conversion and high-frequency restricted transduction are very similar.

It is not difficult, in the light of current knowledge, to visualize a process of

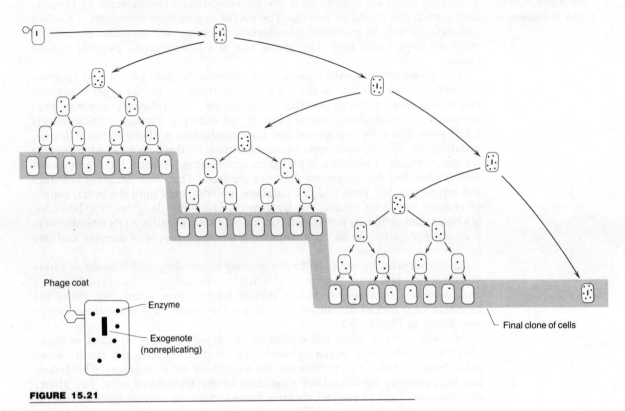

Phage coat

Enzyme

Exogenote
(nonreplicating)

Final clone of cells

FIGURE 15.21

Diagrammatic representation of the events that take place in abortive transduction. After
H. Ozeki.

evolution in the course of which both bacterial and phage chromosomes evolved
from a common ancestor. The ability of bacterial genes to replace phage genes,
and vice versa, points to a close phylogenetic relationship between bacterial
viruses and their hosts. Thus, a gene controlling a given host property may be
found only in the bacterium, only in the phage (conversion), or in both (transduction).

RECOMBINATION IN BACTERIAL VIRUSES

A bacterial cell may be simultaneously infected with two closely
related phages which differ from each other in a number of genetic markers. When
the host cell lyses, the progeny phage particles are found to include not only the
two parental types, but also a variety of genetic recombinants.

The first phage recombination studies, carried out by A. Hershey, involved
two of the mutant types described in Chapter 14: rapid lysis (r) and host range
(h). A double mutant of phage T2 was obtained in two steps: an r mutant was
isolated from a mutant plaque and was plated on *E. coli* strain B/2 to obtain

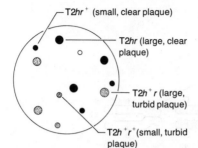

— T2hr⁺ (small, clear plaque)

— T2hr (large, clear plaque)

— T2h⁺r (large, turbid plaque)

— T2h⁺r⁺(small, turbid plaque)

FIGURE 15.22

Diagrammatic representation of the appearance of the plaques of four types of phages.

the double mutant, T2rh. (The parental T2, carrying the wild-type alleles r^+ and h^+, can be designated T2+ +.)

Hershey infected *E. coli* strain B with T2rh and T2+ + simultaneously and observed that each infected cell yielded a mixed burst containing four types: the two parental types and the recombinants T2+h and T2r+. All four types could be scored simultaneously by plating the progeny on a "mixed indicator": in this case, a lawn of cells containing equal numbers of *E. coli* strain B and *E. coli* strain B/2. The four plaque types that result are shown in Figure 15.22.

In the course of a mixed infection, both parental genomes as well as any recombinant genomes continue to replicate until lysis of the host cell occurs. Recombination can occur between any two genomes at any time during replication. The rate of such recombination is very high, averaging more than one such event per particle per replication cycle.

RECOMBINATION IN EUCARYOTIC PROTISTS

• **Meiotic recombination**

In eucaryotic organisms that have sexual life cycles, chromosomal recombination occurs during the process of meiosis. The stages in meiosis are illustrated in Figure 3.10. This process is presented again in Figure 15.23 to show how genetic markers contributed by the two parental types recombine during meiosis. Note that new combinations arise in two ways: (1) by the reassortment of parental chromosomes and (2) by the exchange of segments between paired homologous chromatids through the process of crossing over.

If one of the parents contributes two markers which are on different chromosomes, such as markers A and E in Figure 15.23, these markers have a 50 percent chance of being transmitted to the same haploid segregant. Two markers which enter the zygote on the same parental chromosome, however, such as markers A and B in Figure 15.23, will remain together ("linked") unless separated by a crossover. Since the probability of a crossover occurring between two markers is proportional to the distance between them, the relative distances between any two linked parental markers can be determined by measuring the relative frequencies of recombination between them during meiosis. By carrying out such measurements for many pairs of genes, genetic maps have been constructed for several fungi, including *Neurospora crassa*, *Aspergillus nidulans*, and *Saccharomyces cerevisiae*. A genetic map based on meiotic recombination has also been constructed for the unicellular green alga, *Chlamydomonas reinhardi*.

The DNA of the eucaryotic chromosome can be inferred from several lines of evidence to be a single, continuous molecule of double-stranded DNA, coiled

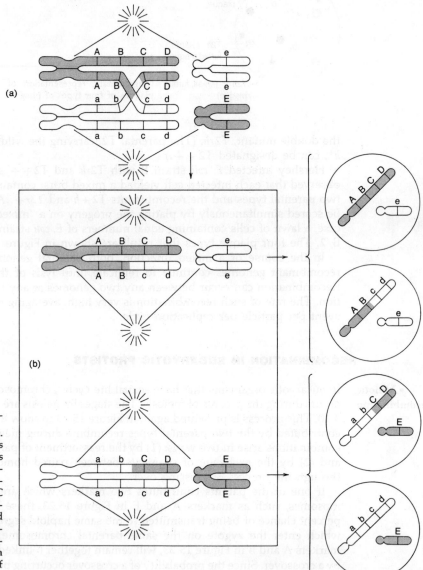

FIGURE 15.23

Meiotic recombination. (a) Two heterozygous pairs of homologous chromosomes. A crossover has occurred between markers B and C of the larger chromosome. (b) The second meiotic division produces four haploid products, each possessing a different combination of the five markers.

and folded in a manner that is not yet understood. When the chromosome replicates it forms two *chromatids*, each with a double-stranded daughter molecule of DNA. The mechanism by which crossing over occurs between two adjacent chromatids is believed to be the same as that in viruses and procaryotic organisms; a molecular model of crossing over has already been presented in the section on bacterial recombination (Figure 15.2).

• Gene conversion   At the beginning of meiosis, each chromosome is represented by a tetrad of chromatids, two from each parent. At the end of meiosis, the four chromatids have been segregated among the four haploid progeny (Figure 15.23). A heterozygous marker (such as marker A) is represented by four alleles, two of which are A and two of which are a, and must appear in the same numbers among the four haploid progeny, i.e., it must show a 2:2 segregation. Such segregations are called *Mendelian*, having first been observed and interpreted by Mendel.

In many protists, notably in the ascomycete group of fungi, the four products of meiosis are readily isolated and scored for their inheritance of parental markers. Such *tetrad analyses*, as they are called, reveal the expected Mendelian segregation of heterozygous markers in most, but not all, cases. In exceptional cases, segregations of 3:1 are observed; for example, the four alleles of marker A in Figure 15.23, having entered meiosis as A, A, a, and a, may be found among the four progeny as A, A, A, and a. Thus, one of the a alleles may be converted to an A allele during the meiotic process. This phenomenon, which often takes place in the region of a crossover, is called *gene conversion*.

Although the detailed molecular mechanism of gene conversion is still obscure, it can be explained in general terms by the following sequence of events: (1) At random sites along the four DNA duplexes, present in each meiotic tetrad, single-stranded breaks occur. (2) At each break, a single-strand gap is formed by exonuclease attack. (3) The gap is repaired by the action of DNA polymerase, *using one of the opposite parental strands as template*. (4) The gap is closed by polynucleotide ligase.

Since similar events occur during crossing over (see Figure 15.2), it is evident that gene conversion may occur within the crossover region. It may also occur, however, in the absence of any detectable crossover: thus, when gene B is converted to b during the meiosis of an ABC/abc diploid, markers A and C may or may not be found to have recombined by crossing over.

• Mitotic recombination   Many protists undergo repetitive cell division in the diploid state. Such diploid cells may eventually enter meiosis, as in the case of sporulating yeasts, or they may remain diploid indefinitely. During mitotic divisions of these diploid cells, homologous chromosomes may undergo pairing and crossing over just as in meiosis, although at a much lower frequency. This process is diagrammed in Figure 15.24; note that the consequence of a mitotic crossover between the centromere and a given heterozygous marker is that, in 50 percent of such events, *that marker and all others distal to it become homozygous*.

The daughter cells which receive exchange chromatids possess the same total set of genes as the parent, but new gene combinations are now present in the form of recombinant chromosomes. The daughter cell may thus be changed *phenotypically*, since homozygosity will permit the expression of a recessive phenotype. In Figure 15.24, for example, let us assume that marker D governs pigment formation in fungal spores, the D (dominant) allele producing the wild-type, green pigment and the d (recessive) allele producing a mutant yellow pigment. Spores that carry D/d will be green; spores that carry d/d, however, as a result of a mitotic crossover, will be yellow.

The phenomenon of mitotic crossing over can serve as the basis for genetic mapping. In Figure 15.24, for example, genes A, B, C, and D can be inferred to be linked together in that order because a single crossover between centromere and A causes all four genes to become homozygous simultaneously, while a

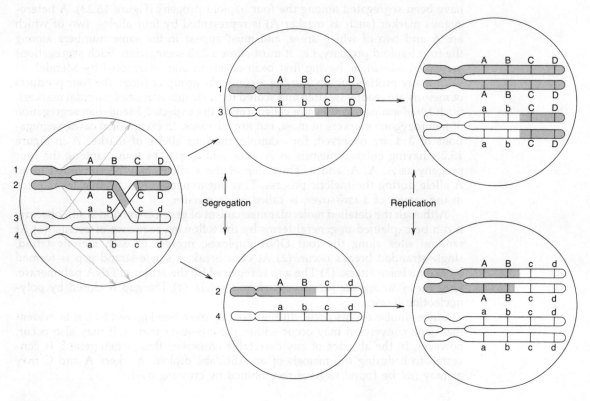

FIGURE 15.24

Mitotic recombination. Only the larger chromosome of Figure 15.23 is shown here, again with a crossover between markers B and C. At mitosis, one chromatid from each parental chromosome migrates to the same pole; in this instance, chromatids 1 and 3 have gone to one pole, and chromatids 2 and 4 to the other. As a result, one daughter cell has become homozygous for the dominant C and D alleles, and the other daughter has become homozygous for the recessive c and d alleles.

crossover between A and B causes only genes B, C, and D to become homozygous, and so on. As in meiotic mapping, the relative distances between pairs of markers are estimated on the basis of the relative frequencies of crossing over between them. The frequency of mitotic recombination can be greatly increased by a variety of inducing agents, most of which are also mutagenic.

• Heterocaryosis and parasexuality in mycelial fungi

Among mycelial fungi, hyphal fusions are not restricted to heterothallic individuals of opposite mating types; they can occur between any two strains belonging to the same species. If the two strains belong to the same mating type, so that eventual completion of the sexual cycle cannot follow hyphal fusion, the result is the formation of a single organism containing nuclei derived from both parental strains; such an organism is a heterocaryon. The nuclei derived from each parental mycelium migrate into the cytoplasm of the other; because the mycelia are coenocytic, migration, coupled with nuclear division, eventually leads to a complete mingling of the two types of nuclei.

In most cases, the two types of nuclei do not fuse but multiply by mitosis independently of one another. The phenotypic properties of a heterocaryon are determined by both types of nuclei, so that heterocaryons have the properties of hybrids between the two parental strains, even though all their nuclei may be haploid. The two types of nuclei may become dissociated if uninucleate conidia are produced. However, since many fungi produce multinucleate conidia, heterocaryosis may be perpetuated through the asexual life cycle.

On rare occasions other than when fruiting bodies are being formed, nuclear fusion may occur within a heterocaryotic mycelium, to form a true diploid nucleus derived from the two types of haploid nuclei. This was first detected in *Aspergillus* through the use of mutant strains which produce conidia with altered pigmentation. Mutants forming yellow and white conidia, respectively, instead of the normal green conidia of the wild type, were used to prepare a heterocaryon. Since the conidia of *Aspergillus* are uninucleate, each spore head of this heterocaryon normally produces chains of both yellow and white spores. The presence of diploid nuclei is revealed by the occasional appearance of a chain of diploid spores, easily detected as a result of their green color, characteristic of the wild type. Such spores can be isolated and used to propagate diploid mycelia. Haploidy is never reestablished by meiosis, but can be attained by the successive losses of one member of each chromosome pair. The resulting haploids are recombinants between the two parental forms, since the individual chromosome may be derived from either parent and crossovers between chromosome pairs occur during mitosis in the diploid state. This phenomenon is termed *parasexuality*: it has the genetic consequences of a sexual process, but does not involve the usual sequence of gametic fusion and meiosis. It provides a mechanism for genetic recombination in imperfect fungi, but it is probably of little importance in fungi with a sexual stage.

• **Recombination of mitochondrial genes in yeast**

The mitochondria of yeast, as of other eucaryotes, resemble bacteria in their sensitivity to certain inhibitors of protein synthesis which have no effect on the ribosomes of the eucaryotic cytoplasm. These inhibitors include chloramphenicol, erythromycin, and spiramycin.

A large series of mutants of *Saccharomyces cerevisiae* have been isolated which are resistant to one or another of these antibiotics. In each case, when a resistant haploid strain is crossed with a sensitive haploid strain and the resulting diploid allowed to undergo meiotic sporulation, the resistance character is found to give 4:0 segregations, i.e., it is inherited as a "cytoplasmic" factor. Thus, the ribosomal alteration that is responsible for resistance to the antibiotic is not only present in mitochondria, but is also determined by a mitochondrial gene.*

An extensive series of crosses carried out by P. Slonimski and his associates has established that recombination occurs between mitochondrial genes contributed by the two parental yeast cells. From their data it has been possible to construct a partial linkage map of the mitochondrial genome.

*Evidence that the antibiotic resistance genes of yeast are located in the mitochondria and not elsewhere in the cytoplasm came from a demonstration that they may be lost in certain respiratory mutants which can be shown to have lost segments of mitochondrial DNA.

• **Recombination of chloroplast genes in *Chlamydomonas***

Chlamydomonas reinhardi is a unicellular green alga which possesses one nucleus, one chloroplast, and several mitochondria (Figure 15.25). It is facultatively photosynthetic, and it can grow in the dark with acetate as carbon and energy source. It has a sexual life-cycle controlled by two mating type alleles of a single gene, called *mt*; the mating types, and their allele determinants, are called *mt+* and *mt−*, respectively (Figure 15.26).

R. Sager and her colleagues have isolated a large number of mutants which exhibit cytoplasmic inheritance. These include photosynthesis-deficient mutants, which require acetate for growth in the light; temperature-sensitive mutants; and mutants resistant to a number of antibiotics that inhibit protein synthesis in chloroplasts, but not in the cytoplasm. Each of these mutations exhibits *uniparental ("maternal") inheritance* in sexual crosses: in every case, a 4 : 0 segregation is observed, all four haploid progeny inheriting the allele contributed by the *mt+* parent. Exceptions to this rule (presence of alleles from both parents among the haploid progeny), occur in less than 1 percent of the zygotes.

Several lines of evidence point to the chloroplast as the location of the maternally inherited genes, most of which control known chloroplast functions, e.g., photosynthesis and protein synthesis by chloroplast ribosomes. The uniparental mode of inheritance was later shown by Sager to reflect the *total degradation of the zygote chloroplast DNA originating from the mt− parent.* In the exceptional zygotes

FIGURE 15.25

Diagrammatic cross section of *Chlamydomonas.* (CW), cell wall; (CV), contractile vacuole; (M), mitochondrion; (E), eyespot; (PY), pyrenoid body; (S), starch grain; (N), nucleus; (Nu), nucleolus. There is a single chloroplast. From R. Sager, *Cytoplasmic Genes and Organelles.* New York: Academic Press, 1972.

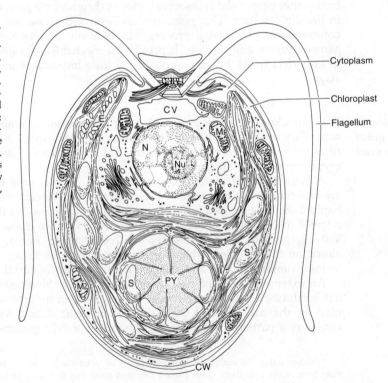

Cytoplasm

Chloroplast

Flagellum

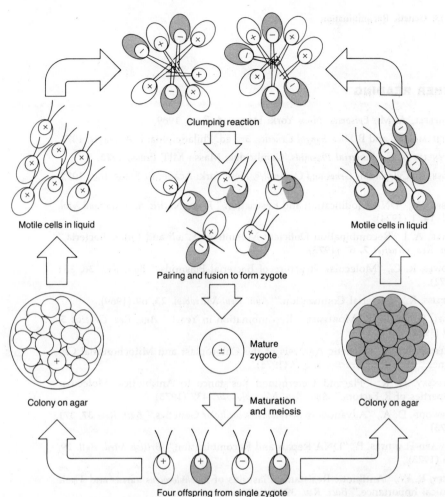

Clumping reaction

Motile cells in liquid

Motile cells in liquid

Pairing and fusion to form zygote

Colony on agar

Colony on agar

Mature zygote

±

Maturation and meiosis

Four offspring from single zygote

FIGURE 15.26

The life cycle of *Chylamy-domonas reinhardi*, showing the segregation of mating type, denoted by (+) and (−), and of an unlinked nuclear gene pair denoted by light and dark shading. The haploid motile cells undergo a clumping reaction prior to pairing and fusion. From R. Sager, "Inheritance in the Green Alga, *Chlamydomonas reinhardi*," *Genetics* **40**, 476 (1955).

such degradation does not occur. Thus, the meiotic products which arise from the exceptional zygotes contain heterozygous chloroplast DNA, and they transmit this heterozygous DNA through subsequent mitotic divisions. At each mitotic division, chloroplast segregants appear which exhibit one or the other parental phenotype; in multifactorial crosses the recombinants appear in patterns that suggest linkage between various chloroplast genes.

In order to obtain sufficient data for a full genetic analysis of the segregation process, Sager looked for ways to increase the frequency of zygotes exhibiting biparental inheritance of chloroplast markers. She found that a majority of the zygotes would do so if the *mt⁺ parent was irradiated with UV light prior to mating;* under these conditions, the DNA of the *mt⁻* parent does not break down, and over 50 percent of the zygotes behave as chloroplast heterozygotes. Using this procedure, Sager was able to obtain segregants and recombinants in high enough numbers to permit the construction of a detailed genetic map of the chloroplast chromosome.

499

FURTHER READING

Books   CAMPBELL, A. M., *Episomes*. New York: Harper & Row, 1969.

FINCHAM, J. R., and P. DAY, *Fungal Genetics*, 3rd ed. Philadelphia: F. A. Davis, 1971.

MEYNELL, G. G., *Bacterial Plasmids*. Cambridge, Mass.: MIT Press, 1973.

SAGER, R., *Cytoplasmic Genes and Organelles*. New York: Academic Press, Inc., 1972.

Reviews   ARBER, W., "DNA Modification and Restriction," *Progr. Nucleic Acid Res. and Mol. Biol.* **14**, 1 (1974).

CLARK, A. J. "Recombination Deficient Mutants of *E. coli* and Other Bacteria," *Ann. Rev. Genetics* **7**, 67 (1973).

CLOWES, R. C., "Molecular Structure of Bacterial Plasmids," *Bact. Rev.* **36**, 361 (1972).

CURTISS, R., "Bacterial Conjugation," *Ann. Rev. Microbiol.* **23**, 69 (1969).

FOGEL, S., and R. K. MORTIMER, "Recombination in Yeast," *Ann. Rev. Genetics* **5**, 219 (1971).

GILLHAM, N. W., "Genetic Analysis of the Chloroplast and Mitochondrial Genomes," *Ann. Rev. Genetics* **8**, 347 (1974).

HELINSKY, D. R., "Plasmid Determined Resistance to Antibiotics: Molecular Properties of R Factors," *Ann. Rev. Microbiol.* **27**, 437 (1973).

HOPWOOD, D. A., "Advances in *Streptomyces coelicolor* Genetics," *Bact. Rev.* **37**, 371 (1973).

HOWARD-FLANDERS, P., "DNA Repair and Recombination," *British Med. Bull.* **29**, 226 (1973).

LACEY, R. W., "Antibiotic Resistance Plasmids of *Staphylococcus aureus* and Their Clinical Importance," *Bact. Rev.* **39**, 1 (1975).

LOW, K. B., "*Escherichia coli* K-12 F-prime Factors, Old and New," *Bact. Rev.* **36**, 587 (1972).

NOTANI, K., and J. K. SETLOW, "Mechanism of Bacterial Transformation and Transfection," *Progr. Nucleic Acid Res. and Mol. Biol.* **14**, 39 (1974).

NOVICK, R. P., "Bacterial Plasmids," in *C.R.C. Handbook IV* (A. I. Laskin and H. A. Lechevalier, editors), p. 537. Cleveland. C.R.C. Press, 1974.

PONTECORVO, G., "The Parasexual Cycle in Fungi," *Ann. Rev. Microbiol.* **10**, 393 (1956).

RADDING, C. M., "Molecular Mechanisms in Genetic Recombination," *Ann. Rev. Genetics* **7**, 87 (1973).

SANDERSON, K. E., "Linkage Map of *Salmonella typhimurium*, Edition IV," *Bact. Rev.* **36**, 558 (1972).

————, H. ROSS, L. ZIEGLER, and P. H. MÄKELÄ, "F+, Hfr and F' Strains of *Salmonella typhimurium* and *Salmonella abony*," *Bact. Rev.* **36**, 608 (1972).

TAYLOR, A. L., and C. D. TROTTER, "Linkage Map of *Escherichia coli* Strain K-12," *Bact. Rev.* **36**, 504 (1972).

501   WILLETTS, N., "The Genetics of Transmissible Plasmids," *Ann. Rev. Genetics* **6**, 257 (1972).

Original articles   COSLOY, S. D., and M. OISHI, "The Nature of the Transformation Process in *E. coli* K12," *Mol. Genetics* **124**, 1 (1973).

DAVIDOFF-ABELSON, R., and D. DUBNAU, "Kinetic Analysis of the Products of Donor DNA in Transformed Cells of *B. subtilis*," *J. Bacteriol.* **116**, 154 (1973).

KIRBY, R., L. F. WRIGHT, and D. A. HOPWOOD, "Plasmid-Determined Antibiotic Synthesis and Resistance in *Streptomyces coelicolor*," *Nature* **254**, 265 (1975).

LACKS, S., B. GREENBERG, and M. NEUBERGER, "Role of a DNase in the Genetic Transformation of *D. pneumoniae*," *Proc. Natl. Acad. Sci. U.S.* **71**, 2305 (1974).

MESELSON, M. S., and C. M. RADDING, "A General Model for Genetic Recombination," *Proc. Natl. Acad. Sci. U.S.* **72**, 358 (1975).

MORROW, J. F., S. N. COHEN, A. C. Y. CHANG, H. W. BOYER, H. GOODMAN, and R. B. HELLING, "Replication and Transorption of Eukaryotic DNA in *E. coli*," *Proc. Natl. Acad. Sci. U.S.* **71**, 1743 (1974).

NOVOTNY, C. P., and P. FIVES-TAYLOR, "Retraction of *F. pili*," *J. Bacteriol.* **117**, 1306 (1974).

REINER, A. M., "*E. coli* Females Defective in Conjugation and in Adsorption of a Single-Stranded DNA Phage," *J. Bacteriol.* **119**, 183 (1974).

SAGER, R., and Z. RAMANIS, "The Mechanism of Maternal Inheritance in *Chlamydomonas*: Biochemical and Genetic Studies," *Theor. and Appl. Genetics* **43**, 101 (1973).

SCHUPP, T., R. HUTTER, and D. A. HOPWOOD, "Genetic Recombination in *Nocardia mediterranei*," *J. Bacteriol.* **121**, 128 (1975).

VAPNEK, D., and W. D. RUPP, "Asymmetric Segregation of the Complementary Sex-factor DNA Strands During Conjugation in *Escherichia coli*," *J. Mol. Biol.* **53**, 287 (1970).

The art of biological classification is known as *taxonomy*. It has two functions: the first is to identify and describe as completely as possible the basic taxonomic units, or *species;* the second, to devise an appropriate way of arranging and cataloging these units.

SPECIES: THE UNITS OF CLASSIFICATION

The notion of a species is complex. Speaking broadly, a species consists of an assemblage of individuals (or, in microorganisms, of clonal populations) that share a high degree of phenotypic similarity, coupled with an appreciable dissimilarity from other assemblages of the same general kind. The recognition of species would not be possible if natural variation were continuous, so that an intergrading series spanned the gap between two assemblages of markedly different phenotype. However, it became evident early in the development of biology that, among most groups of plants and animals, reasonably sharp discontinuities do separate the members of a group into distinguishable assemblages. Hence, the notion of the species as the base of taxonomic operation proved workable.

Every assemblage of individuals shows some degree of internal phenotypic diversity, because genetic variation is always at work. Hence, it becomes a matter of scientific tact to decide what *degree* of phenotypic dissimilarity justifies the breaking up of an assemblage into two or more species; or, to put the matter another way, how much internal diversity is permissible in a species. Opinions on this question vary. Taxonomists themselves can be broadly divided into two groups: "lumpers," who set wide limits to a species, and "splitters," who differentiate species on more slender grounds.

For plants and animals that reproduce sexually, a species can be defined in genetic and evolutionary terms. As long as a sexually reproducing population is free to interbreed at random, its total gene pool undergoes continuous redistribution, and new mutations, the source of phenotypic variation, are dispersed throughout the population. Such an interbreeding population may evolve in response to environmental changes, but it will evolve with reasonable uniformity. *Divergent evolution,* eventually leading to the emergence of new species, can occur only if a segment of the population becomes reproductively isolated in an environment that is different from that occupied by the rest of the population. Reproductive isolation is probably always geographic in the first instance; a physical barrier of some sort (for example, a mountain range or a body of water) is interposed between two parts of the initially continuous population. Within each of these subpopulations, a common gene pool is maintained by interbreeding, but through chance mutation and selection, the two subpopulations are now free to evolve along different lines. They will continue to diverge, as long as the geographical barrier persists. Eventually, the cumulative differences become so great that *physiological* isolation is superimposed on geographic isolation; members of the two populations are no longer capable of interbreeding if they are brought together. Hence, even if the two populations subsequently commingle once more, their gene pools remain permanently separated; a point of no return has been reached. These evolutionary considerations lead to a dynamic definition of the species, as a stage in evolution at which actually or potentially interbreeding arrays have become separated into two or more arrays physiologically incapable of interbreeding. This definition is, in fact, an *explanation* of the origin of specific discontinuities in nature. At the same time, it provides an experimental criterion for the recognition of species differences: inability to interbreed.

Since most microorganisms are haploid, and reproduce predominantly by asexual means, the concept of the species which has emerged from work with plants and animals is evidently inapplicable to them. A microbial species cannot be considered an interbreeding population: the two offspring produced by the division of a bacterial cell are reproductively isolated from one another, and, in principle, they are free to evolve in a divergent manner. Genetic isolation is to some degree reduced by sexual or parasexual recombination in eucaryotic microorganisms and by the special mechanisms of recombination distinctive of procaryotes. However, it is very difficult to assess the evolutionary effect of these recombinational processes, since the frequencies with which they occur in nature are unknown. In procaryotes the problem is further complicated by episomic transfer, which is relatively nonspecific, and permits exchanges of genetic material among bacteria of markedly different genetic constitution.

Since the dynamics of microbial evolution are so unlike the dynamics of evolution of plants and animals, there is no theoretical basis for the assumption that microbial evolution has led to phenotypic discontinuities that would justify the recognition of species. However, the experience of microbial taxonomists has shown that when many strains of a given microbial group are thoroughly analyzed, they can usually be divided into a series of discontinuous clusters: it is such *clusters of strains* that the microbial taxonomist recognizes empirically as species. Further insights into the dynamics of microbial evolution may eventually permit a formal definition of the microbial species; if so, this will most likely be different from the species definition applicable to plants and animals.

In bacterial populations, genetic change can occur so rapidly by mutation that it would be unwise to distinguish species on the basis of differences in a small number of characters, governed by single genes. Accordingly, the best working definition of a bacterial species is the following: a group of strains that show a high degree of overall phenotypic similarity and that differ from related strain groups with respect to many independent characters.

• The characterization of species

Ideally, species should be characterized by complete descriptions of their phenotypes or—even better—of their genotypes. Taxonomic practice falls far short of these ideals; in most biological groups, even the phenotypes are only fragmentarily described, and genotypic characterizations are extremely rare.

As a general rule, the phenotypic characters that can be most easily determined are structural or anatomical ones that can be directly observed. For this reason, biological classification is still based, at most levels, almost entirely on structural properties. Virtually the only exception is the classification of bacteria. The extreme structural simplicity of bacteria offers the taxonomist too small a range of characters upon which to base adequate characterizations. Hence, the bacterial taxonomist has always been forced to seek other kinds of characters—biochemical, physiological, ecological—with which to supplement structural data. The classification of bacteria is based, to a far greater extent than that of any other biological groups, on *functional* attributes. Most bacteria can be identified only by finding out what they can do, not simply how they look.

This confronts the bacterial taxonomist with an additional problem. To find out what a bacterium can do, he has to perform experiments with it. The number of possible experiments that can be performed is extremely large, and although all will reveal facts, the facts so revealed will not necessarily be taxonomically significant ones, in the sense of contributing to a differentiation of the organism under study from related assemblages. Consequently, the bacterial taxonomist can never be sure that he has performed the right experiments for taxonomic purposes; he may well have failed to perform certain experiments that would have shown him significant clustering in a collection of strains, and therefore erroneously conclude that he is dealing with a continuous series. There is no obvious way to get around this difficulty, except to make phenotypic characterizations as exhaustive as possible.

• The naming of species

According to a convention known as the *binomial system of nomenclature*, every biological species bears a latinized name that consists of two words. The first word indicates the taxonomic group of immediately higher order, or *genus* (plural, *genera*) to which the species belongs, and the second word identifies it as a particular species of that genus. The first letter of the generic (but not of the specific) name is capitalized, and the whole phrase is italicized: *Escherichia* (generic name) *coli* (specific name). In contexts in which no confusion is possible, the generic name is often abbreviated to its initial letter: *E. coli*.

A rigid and complex set of rules governs biological nomenclature; the rules are designed to keep nomenclature as stable as possible. The specific name given to a newly recognized species cannot be changed, unless it can be shown that the organism has previously been described under another specific name, in which case the older name has priority. Unfortunately, the same stability does not govern the generic half of the name, since the arrangement of related species into genera is an operation that can be carried out in different ways and that often changes in the course of time as new information becomes available. For example, *E. coli* has in the past been placed in the genus *Bacterium*, as *Bacterium coli* and in the genus *Bacillus*, as *Bacillus coli*. These three names are synonyms, since they all refer to one and the same species. This consequence of the binomial system can be very confusing, and taxonomic descriptions usually list all such synonyms in order to minimize the confusion. Binomial nomenclature is used for all biological groups except viruses. The virologists are currently divided over the best way to designate members of this group; some wish to extend the binomial system to the viruses, whereas others would prefer another system, which gives in coded form information about the properties of the organism.

In bacterial taxonomy, when a new species is named, a particular strain is designated as the *type strain*. Type strains are preserved in culture collections; if one is lost, a *neotype strain*, which resembles as closely as possible the description of the type strain, is chosen. The type strain is important for nomenclatural purposes, since the specific name is attached to it. If other strains, originally included in the same species, prove on subsequent study to deserve recognition as separate species, they must receive new names, the old specific name resting with the type strain and related strains.

In the taxonomic treatment of a biological group, the individual species are usually grouped in a series of categories of successively higher order: genus, family, order, class, and division (or phylum). Such an arrangement is known as a *hierarchical* one, because each category in the ascending series unites a progressively larger number of taxonomic units in terms of a progressively smaller number of shared properties. It should be noted that the genus has a position of special importance, since according to the rules of nomenclature a species cannot be named unless it is assigned to a genus. The allocation of a species to a taxonomic category higher than the genus does not carry any essential *nomenclatural* information; it is merely indicative of the position of an organism, relative to other organisms, in the system of arrangement adopted.

THE PROBLEMS OF TAXONOMIC ARRANGEMENT

In dealing with a large number of different objects, some system of orderly arrangement is essential for purposes of data storage and retrieval. It does not matter what criteria for making the arrangement are adopted, provided that they are unambiguous and convenient. Books can be arranged in different ways: for example, by subject, by author, or by title. Different individuals tend to adopt different systems, depending on their particular needs and tastes. Such a system of classification, based on arbitrarily chosen criteria, is termed an *artificial* one.

The earliest systems of biological classification were largely artificial in design. However, as knowledge about the anatomy of plants and animals increased, it became evident that these organisms conform to a number of *major patterns* or *types*, each of which shares many common properties, including ones that are not necessarily obvious upon superficial examination. Examples of such types are the mammalian, avian, and reptilian types among vertebrate animals. The first system of biological classification that attempted to group organisms in terms of such typological resemblances and differences was developed in the middle of the eighteenth century by Linnaeus. The Linnaean arrangement was more useful than previous artificial arrangements, since the taxonomic position of an organism furnished a large body of information about its properties: to say that an animal belongs to the vertebrate class Mammalia immediately tells one that it possesses all those properties which distinguish mammals collectively from other vertebrates. Because Linnaean classification expressed the *biological nature* of the objects that it classified, it became known as a *natural system* of classification, in contrast to preceding artificial systems.

• **The phylogenetic approach to taxonomy**

When the fact of biological evolution was recognized, another dimension was immediately added to the concept of a natural classification. For biologists of the eighteenth century, the typological groupings merely expressed *resemblances*; but for post-Darwinian biologists, they revealed *relationships*. In the nineteenth century the concept of a "natural" system accordingly changed: it became one that grouped organisms in terms of their *evolutionary affinities*. The taxonomic hierarchy became in a certain sense the reflection of a family tree. The analogy between a family tree and a hierarchy cannot be pushed very far, however, since the dimension of *time*, implicit in a family tree, is completely absent from a taxonomic hierarchy of extant organisms. This point was not very clearly grasped by many evolutionary biologists, for whom taxonomy suddenly acquired a new goal: the restructuring of hierarchies to mirror evolutionary relationships. Such a taxonomic system is known as a *phylogenetic system*.

Reflection and experience have shown, however, that the goal of a phylogenetic classification can seldom be realized. The course that evolution has actually followed can be ascertained only from direct historical evidence, contained in the fossil record. This record is at best fragmentary and becomes almost completely illegible in Precambrian rocks more than 400 million years old. By the beginning of the Cambrian period, most of the major biological groups that exist had already made their appearance; vertebrates and vascular plants are the principal evolutionary newcomers in postcambrian time. For these two groups, the fossil record is, accordingly, reasonably complete, and the main lines of vascular plant and vertebrate evolution can be retraced with some assurance. For all other major biological groups, the general course of evolution will probably never be known, and there is not enough objective evidence to base their classification on phylogenetic grounds.

• **Numerical taxonomy**

For these and other reasons, most modern taxonomists have explicitly abandoned the phylogenetic approach, in favor of a more empirical one: the attempt to base taxonomic arrangement upon quantification of the similarities and differences among organisms. This was first suggested by Michel Adanson, a contemporary of Linnaeus, and is known as *Adansonian* (or *numerical*) *taxonomy*. The underlying assumption is that, provided each phenotypic character is given equal weighting, it should be possible to express numerically the taxonomic distances between organisms, in terms of the number of characters they share, relative to the total number of characters examined. The significance of the numerical relationships so determined is greatly influenced by the number of characters examined; these

TABLE 16.1

The determination of similarity coefficient and matching coefficient for two bacterial strains, both characterized with respect to many $(a + b + c + d)$ different characters

---

Number of characters positive in both strains: a

Number of characters positive in strain 1 and negative in strain 2: b

Number of characters negative in strain 1 and positive in strain 2: c

Number of characters negative in both strains: d

Similarity coefficient $(S_J) = \dfrac{a}{a + b + c}$

Matching coefficient $(S_S) = \dfrac{a + d}{a + b + c + d}$

---

should be as numerous and as varied as possible, to obtain a representative sampling of phenotype.

Until recently, the Adansonian approach appeared impractical because of the magnitude of the numerical operations involved. This difficulty has been obviated by the advent of computers, which can be programmed to compare data for a large number of characters and organisms and to compute the degrees of similarity. For any pair of organisms, the calculation of similarity can be made in two slightly different ways (Table 16.1). The similarity coefficient S_J does not take into account characters negative for both organisms, being based only on positive matches; the matching coefficient S_S includes both positive and negative matches in the calculation.

After similarity (or matching) coefficients have been calculated pairwise for all the organisms under study, the data are arranged in a similarity matrix, an example of which is shown in Figure 16.1(a) for a series of 10 bacterial strains. By inspection, such a matrix can be reordered so as to bring into juxtaposition similar strains [Figure 16.1(b)]. The data can then be transposed into a dendrogram (Figure 16.2), as a basis for determining taxonomic arrangement in terms of numerical relationships. The two dotted vertical lines in Figure 16.2 indicate similarity levels which might be considered appropriate for recognizing two different taxonomic ranks (e.g., genus and species).

Numerical taxonomy does not have the evolutionary connotations of phylo-

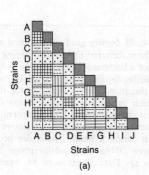

Strains

A B C D E F G H I J
Strains

(a)

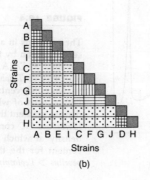

Strains

A B E I C F G J D H
Strains

(b)

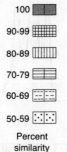

100

90-99

80-89

70-79

60-69

50-59

Percent
similarity

FIGURE 16.1

Similarity matrices for a collection of 10 bacterial strains, designated as A through J: (a) similarity matrix before rearrangement; (b) similarity matrix after rearrangement. The strains have been so ordered as to bring into juxtaposition strains that most closely resemble one another in overall phenotype. After P. H. A. Sneath, "The Construction of Taxonomic Groups," in *Microbial Classification*, G. C. Ainsworth and P. H. A. Sneath (editors). New York: Cambridge University Press, 1962.

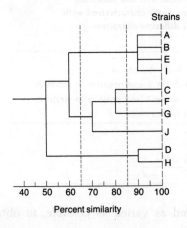

Strains

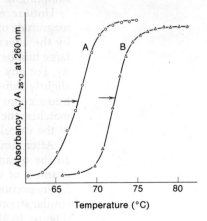

FIGURE 16.2

A dendrogram showing similarity relationships among the 10 bacterial strains, using the data from Figure 16.1. The two dotted vertical lines indicate possible similarity levels at which successive ranks (e.g., genus and species) in the taxonomic hierarchy might be established. After P. H. A. Sneath, "The Construction of Taxonomic Groups," in *Microbial Classification*, G. C. Ainsworth and P. H. A. Sneath, (editors). New York: Cambridge University Press, 1962.

FIGURE 16.3

Melting curves determined optically for two samples of bacterial DNA. Curve A: DNA of *Lactobacillus acidophilus* (T_m 67.7°C). Curve B: DNA of *Leptospira* sp. (T_m 72.1°C). The ordinate expresses the absorbancy at 260 nm of the DNA sample at each temperature on the abscissa, relative to its absorbance at 25°C. The midpoint of the absorbancy increase (arrows) is the temperature (T_m) at which approximately half of the hydrogen bonds holding together the DNA double helices have been broken. This temperature is directly related to the GC content of the DNA sample; the higher the GC content, the higher the T_m. Data courtesy of M. Mandel.

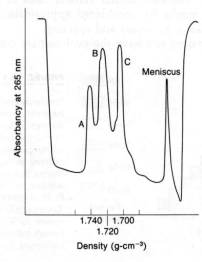

FIGURE 16.4

The positions in a CsCl density gradient assumed after centrifugation by three different DNAs of differing GC content. (A) DNA of a bacteriophage of *Bacillus subtilis*. (B) DNA of *Thiobacillus novellus*. (C) DNA of *Leptospira* sp. Note that centrifugation sharply separates the three DNAs, each of which bands to a position in the CsCl density gradient that corresponds to its GC content: the lower the GC content of a given DNA, the lower the density at which it forms a band. The order of GC content for the three DNAs is *B. subtilis* phage > *T. novellus* > *Leptospira* sp. Data courtesy of M. Mandel.

genetic taxonomy, but it provides a more objective and stable basis for the construction of taxonomic groupings. Perhaps its greatest advantage is that it cannot be applied at all until a relatively large number of characters have been determined, so that its use encourages a thorough examination of phenotypes. Furthermore, the analyses are open to continuous revision and refinement as more characters in a given group are determined.

NEW APPROACHES TO BACTERIAL TAXONOMY

The growth of molecular biology has opened up a number of new approaches to the characterization of organisms, which have had a profound impact on the taxonomy of bacteria. Of particular value are certain techniques that give insights into genotypic properties and thus complement the hitherto exclusively phenotypic characterizations of these organisms. Two kinds of analysis performed upon isolated nucleic acids furnish information about genotype: the analysis of the base composition of DNA and the study of chemical hybridization between nucleic acids isolated from different organisms.

• The base composition of DNA: its determination and significance

DNA contains four bases: adenine (A), thymine (T), guanine (G), and cytosine (C). For double-stranded DNA, the base-pairing rules (see Chapter 7) require that A = T and G = C. However, there is no chemical restriction on the molar ratio (G + C):(A + T). Early in the chemical study of DNA, analyses showed that this ratio in fact varies over a rather wide range in DNA preparations from different organisms, and subsequent work has revealed that the base composition of DNA is a character of profound taxonomic importance, particularly among microorganisms.

Although DNA base composition may be determined chemically, after hydrolysis of a DNA sample and separation of the free bases, it can be determined more easily by physical methods, and these are now the ones principally used. The "melting temperature" of DNA (i.e., the temperature at which it becomes denatured, by breakage of the hydrogen bonds that hold together the two strands) is directly related to G + C content, because hydrogen bonding between GC pairs is stronger than that between AT pairs. Strand separation is accompanied by a marked increase in absorbance at 260 nm, the absorption maximum of DNA, and this can be easily measured in a spectrophotometer. When a DNA sample is gradually heated, the absorbance increases as the hydrogen bonds are broken and reaches a plateau at a temperature at which the DNA has all become single-stranded (Figure 16.3). The midpoint of this rise, the melting temperature (T_m), is a measure of the G + C content. The G + C content of DNA may also be determined by subjecting a DNA sample to centrifugation in a CsCl gradient, and determining optically the position at which the DNA bands in the gradient, which affords a precise measure of its density (Figure 16.4). This method can be used because the density of DNA is also a function of the (G + C):(A + T) ratio.

Physical methods of analysis also provide an indication of the *molecular heterogeneity* of a DNA sample. If every molecule of DNA had the same G + C content, both the thermal transition in a melting curve and the band position in a CsCl gradient would be extremely sharp. The steepness of the curve for thermal transition and the narrowness of the band in a gradient are therefore directly related to the homogeneity of G + C content in a population of DNA molecules. Even when DNA has been considerably fragmented by shearing, preparations from most organisms remain relatively homogeneous by these criteria, which indicates that the mean G + C content varies little in different parts of the genome. The only major exceptions are preparations from organisms that contain two genetic elements of different G + C content. Thus, DNA of mitochondrial or chloroplast origin may differ appreciably in G + C content from the nuclear DNA, in preparations from certain eucaryotic organisms; and there is sometimes a marked molecular heterogeneity in the DNA of a bacterium that harbors a plasmid. In such cases, the minor constituent may form a distinct *satellite band* in a CsCl gradient; this phenomenon provided one of the clues that led to the discovery of DNA in mitochondria and chloroplasts.

Since no DNA preparation shows *absolute* molecular homogeneity, the G + C content is always a *mean* value and represents the peak in a normal distribution curve.

• **The taxonomic implications of DNA base composition**

The mean DNA base compositions characteristic of the nuclear DNA in major groups of organisms are shown in Figure 16.5. In both plants and animals the ranges are relatively narrow and quite similar, centering about a value of 35 to 40 mole percent G + C. Among the protists the ranges are much wider. The widest range of all occurs among the procaryotes, in which the range extends from about 30 to 75 mole percent G + C. If, however, one examines the mean G + C content of many different strains that belong to a *single microbial species*, the values are closely similar or identical, as shown by the data for several *Pseudomonas* species assembled in Table 16.2. Each bacterial species, accordingly, has DNA with a characteristic mean G + C content; this can be considered one of its important specific characters.

Why do organisms differ so widely in the mean base composition of their DNA? This cannot reasonably be ascribed to coding differences, since there is considerable evidence that the genetic code is universal. It could reflect differences between organisms in the amino acid composition of the cellular proteins, some having a preponderance of amino acids coded for by G + C-rich triplets, others a preponderance coded for by A + T-rich triplets. There is a certain amount of evidence to suggest that this factor does play a role. In addition, as a result of the degeneracy of the genetic code, substantial differences in mean G + C content could occur, even between two organisms of identical amino acid composition, if there were systematic selection for the use of specific triplets. Thus, one organism might systematically employ a preponderance of G + C-rich triplets to code for amino acids represented by several triplets, and another might systematically employ a preponderance of A + T-rich triplets to code for the same amino acids. It is possible that both factors have played a role in producing the existing biological divergences with respect to mean G + C content.

One point is clear: *a substantial divergence between two organisms with respect to mean DNA base composition reflects a large number of individual differences between the specific base sequences of their respective DNAs.* It is *prima facie* evidence for a major genetic divergence and hence for a wide evolutionary separation. The very broad span of values characteristic of the procaryotes, accordingly, reveals the great evolu-

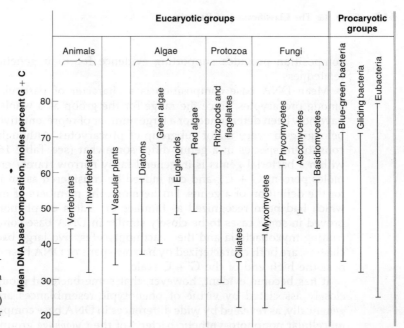

FIGURE 16.5

The ranges of mean DNA base composition (moles percent G + C) characteristic of major biological groups.

tionary diversity of this particular biological group, and it also suggests its evolutionary antiquity.

However, two organisms with identical mean DNA base compositions may differ greatly in genetic constitution. This is evident from the very similar base ratio values for DNA from all plants and animals. Hence, major evolutionary divergence is not *necessarily* expressed by a divergence of mean base composition. When two organisms are closely similar in their DNA base composition, this fact can be construed as indicative of genetic and evolutionary relatedness only if the organisms *also* share a large number of phenotypic properties in common or are known to resemble one another in genetic constitution (e.g., different strains that belong to a single bacterial species). In such a case, near-identity of DNA base

TABLE 16.2

Constancy of G + C content in strains of bacteria belonging to a given species[a]

Pseudomonas SPP.	NUMBER OF STRAINS EXAMINED	G + C CONTENT OF DNA, MOLE PERCENT (MEAN VALUE ± STANDARD DEVIATION)
P. aeruginosa	11	67.2 ± 1.1
P. acidovorans	15	66.8 ± 1.0
P. testosteroni	9	61.8 ± 1.0
P. cepacia	12	67.6 ± 0.8
P. pseudomallei	6	69.5 ± 0.7
P. putida	6	62.5 ± 0.9

[a] From M. Mandel, *J. Gen. Microbiol.* **43**, 273 (1966).

composition provides supporting evidence for their genetic and evolutionary relatedness.

Mean DNA base composition is a character of particular taxonomic value among procaryotes, since the range for the group as a whole is so wide. Values have now been determined for a large number of representative strains and species belonging to every major subgroup of procaryotes. Although the values for the constituent species in a genus differ somewhat (see Table 16.2), the total range within a bacterial genus is in general fairly narrow (rarely greater than 10 to 15 moles percent G + C), and can indeed be considered as an important character for the definition of a genus. Furthermore, the members of multigeneric clusters which had been recognized as families or orders on phenotypic grounds have proved in some cases to be closely similar in DNA base composition. Thus, the fruiting myxobacteria and the euactinomycetes, two large bacterial multigeneric clusters, are both characterized by narrow spans of DNA base composition that lie near the high end of the G + C scale.

It has become evident, however, that some bacterial groups which had been closely associated by virtue of phenotypic resemblances are in fact unrelated genetically, as revealed by wide differences in DNA base composition. The gliding, unicellular nonphotosynthetic bacteria of the *Cytophaga* group, similar to fruiting myxobacteria with respect to vegetative cell structure, lie almost at the opposite end of the G + C scale; and the same situation occurs among the Gram-positive cocci of the *Staphylococcus* and *Micrococcus* groups (Figure 16.6).

• Nucleic acid hybridization   Upon rapid cooling of a solution of thermally denatured DNA, the single strands remain separated. However, if the solution is held at a temperature from 10 to 30°C below the T_m value, specific reassociation ("annealing") of complementary strands to form double-stranded molecules occurs. There is always some random pairing, but since it is the complementary regions that form the most stable duplexes, their reassociation is favored.

Shortly after this phenomenon had been discovered, it was shown that when DNA preparations from two related strains of bacteria are mixed and treated in this manner, *hybrid DNA molecules are formed*. One bacterial strain was grown in a medium containing D_2O, so that its DNA was "heavy" as a result of deuterium incorporation. After the two DNA samples had been mixed, denatured, and annealed, hybrid molecules could be detected by centrifugation in a CsCl gradient,

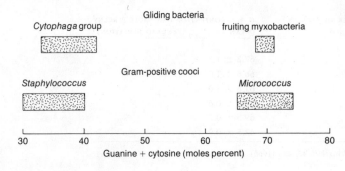

FIGURE 16.6

Ranges of mean DNA base composition of certain representatives in two bacterial assemblages defined by phenotypic properties: gliding bacteria and Gram-positive cocci. The wide separations with respect to DNA base composition between the *Cytophaga* group and the fruiting myxobacteria and between the genera *Staphylococcus* and *Micrococcus* reveal that they are genetically unrelated.

where they formed a band intermediate in position between those of the "light" and "heavy" duplexes (Figure 16.7). When similar experiments were conducted with DNA preparations from two unrelated bacteria, no hybridization could be detected; upon annealing, duplexes were formed only by specific pairing between single strands originally derived from the same DNA.

The discovery of the reassociation of single-stranded DNA molecules from different biological sources to form hybrid duplexes laid the foundations of an entirely new approach to the study of genetic relatedness in bacteria. *In vitro* experiments on DNA-DNA reassociation permit an assessment of the overall degree of genetic homology between two bacteria. Furthermore, since duplexes can also be formed between single-stranded DNA and complementary RNA strands, analogous DNA-RNA reassociations can be performed. If the RNA preparations consist of either tRNAs or rRNAs, such experiments permit an assessment of the genetic homology between two bacteria *with respect to specific, relatively small segments of the chromosome:* those that code the base sequences either of the transfer RNAs or of the ribosomal RNAs. As will be discussed later, DNA-rRNA reassociations are of particular taxonomic interest.

• **The techniques and interpretations of reassociation experiments**

The density gradient method (Figure 16.7) is too cumbersome for routine use, and has the further disadvantage of detecting only reassociations between complementary strands of very high homology. A variety of simpler methods for measuring nucleic acid reassociation have been developed. All are based on the same general principle: formation of duplexes between two denatured DNA samples, one of which is labeled with a radioisotope; separation of the duplexes from residual single-stranded nucleic acid; and measurement of their radioactivity. A point of reference is always required: it is provided by the DNA of a *reference*

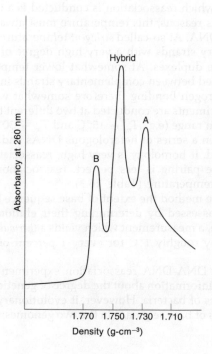

FIGURE 16.7

An experiment demonstrating the formation of hybrid DNA molecules through reassociation of single-stranded (denatured) DNA molecules to form double helical DNA. DNA was prepared from two different strains of *Pseudomonas aeruginosa;* strain A was grown in a normal medium and strain B was grown in a medium containing "heavy water" (D_2O) and $N^{15}H_4Cl$. Consequently, the DNA of strain B, although identical in base composition to the DNA of strain A, had a higher density as a result of its content of heavy atoms. The two DNAs were isolated, denatured by heating, mixed, and then annealed, after which residual single-stranded molecules were eliminated by treatment with a specific nuclease. The preparation was then centrifuged in a CsCl gradient. Three peaks of double-stranded DNA are apparent in the density gradient. Peak A corresponds to "light" double-helical DNA, formed by reassociation of single-stranded DNA from strain A; peak B corresponds to "heavy" double-stranded DNA, formed by reassociation of single-stranded DNA from strain B. Between these two peaks, a third peak of intermediate density occurred; this corresponds to hybrid double-stranded DNA, formed by specific reassociation of single strands, one of which was derived from strain A, and one from strain B. Data courtesy of M. Mandel.

strain, which is prepared both labeled and unlabeled. The amount of reassociation between these two homologous DNAs is determined, and is assigned an arbitrary value of 100. The amount of reassociation between the reference DNA and DNAs from heterologous strains can then be measured and expressed as a percentage of the normalized value for the homologous DNA-DNA reassociation. Depending on the particular technique used, either the reference DNA or the heterologous DNA may be labeled. The same principles apply to DNA-RNA reassociation experiments.

The simplest technique currently employed for studies of nucleic acid reassociation is the *hydroxyapatite method.* Hydroxyapatite is a calcium phosphate gel which, under defined conditions, specifically absorbs nucleic acid duplexes, but not single strands. Nucleic acid reassociation between labeled and unlabeled DNAs is allowed to occur in solution, after which the duplexes are absorbed on hydroxyapatite, which is washed to eliminate single-stranded material. The absorbed duplexes are then eluted, either by increasing the ionic strength of the solvent or by raising the temperature; and the radioactivity of the eluate is determined. In this method it is essential that the unlabeled DNA be present in very large (several thousandfold) excess, in order to prevent reassociation between complementary radioactive single strands.

In every type of reassociation experiment the temperature of reassociation, the ionic strength of the solvent, and the mean length of the DNA fragments must be standardized, since these factors all affect the rate of duplex formation. When they are held constant, reassociation is determined solely by the *DNA concentration* and the time of incubation. Provided that these two parameters are correctly chosen, complete reassociation of complementary strands will occur.

The temperature at which reassociation is conducted is a factor of paramount importance. For obvious reasons, this temperature must always be below the T_m value of the reference DNA. At so-called *stringent* temperatures, 10 to 15°C below T_m, only complementary strands with a very high degree of base sequence homology can form stable duplexes. At somewhat lower temperatures, stable duplexes can also be formed between complementary strands in which base pairing is less perfect and hydrogen bonding therefore somewhat weaker.

If reassociation experiments are conducted at two different temperatures within the specific reassociation range (e.g., $T_m - 15°C$ and $T_m - 30°C$), the *degree of base sequence homology* between a series of heterologous DNAs and the reference DNA can be roughly assessed. If homology is very high, reassociation is little affected by temperature; if base pairing is less perfect, reassociation will be markedly reduced at the higher temperature (Table 16.3).

In the hydroxyapatite method the extent of base sequence homology of hybrid duplexes can also be assessed by determining their elution from the gel as a function of temperature, a measurement which yields a *thermal elution profile* (Figure 16.8). The $T_{m\ (e)}$ falls by roughly 1°C for every 1 percent of mismatched bases in the hybrid.

Carefully controlled DNA-DNA reassociation experiments can thus provide much semiquantitative information about the degree of genetic homology between related strains or species of bacteria. However, if evolutionary divergence has led to numerous differences of base sequence in the two genomes, specific DNA-DNA

TABLE 16.3

The effect of incubation temperature on DNA duplex formation between radioactive DNA prepared from *Escherichia coli* and unlabeled DNAs from other bacteria belonging to the enteric group[a]

	RELATED DUPLEX FORMATION AT	
	60°C ($T_m - 30$°C)	75°C ($T_m - 15$°C)
E. coli/E. coli[b]	100	100
E. coli/Shigella boydii	89	85
E. coli/Salmonella typhimurium	45	11
E. coli/Enterobacter hafniae	21	4

[a] Data from D. J. Brenner and S. Falkow, *Adv. Genetics* **16**, 81 (1971).
[b] Control, using labeled and unlabeled DNA from the reference strain of *E. coli*: values normalized to 100.

reassociation becomes too weak to be measured. The range of organisms among which genetic homology is detectable can be greatly extended by parallel studies on DNA-rRNA reassociation, because the relatively small portion of the bacterial genome that codes for ribosomal RNAs has a much more highly conserved base sequence than the bulk of the chromosomal DNA. As a result, it is frequently possible to detect by DNA-rRNA reassociation relatively high homology between

FIGURE 16.8

Thermal elution profiles from hydroxyapatite of hybrid DNA-DNA duplexes formed between labeled DNA of *Escherichia coli* and unlabeled DNA of *E. coli*, *Shigella dysenteriae* and *Citrobacter freundii*. The relatively low $T_{m(e)}$ of the *E. coli/C. freundii* hybrid duplexes is an indication of considerable base mismatching. The $T_{m(e)}$ of the *E. coli/S. dysenteriae* hybrid duplexes, almost identical with that of the *E. coli/E. coli* duplexes, indicates a much higher degree of base pairing. (a) Elution of duplexes formed at $T_m - 30$°C; (b) elution of duplexes formed at $T_m - 15$°C. Note that the *E. coli/C. freundii* duplexes formed at the higher annealing temperature are much better paired than those formed at the lower annealing temperature. After D. Brenner and S. Falkow, *Adv. Genetics* **16**, 81 (1971).

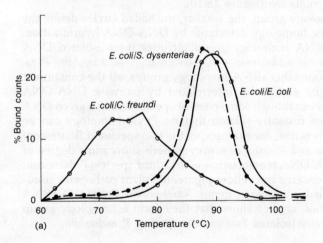

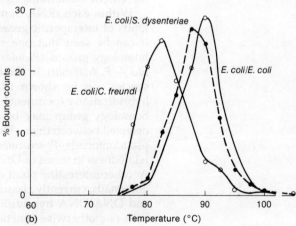

TABLE 16.4

Nucleic acid homologies between *Pseudomonas acidovorans* and
four other *Pseudomonas* species, as revealed by parallel
DNA-DNA and DNA-rRNA hybridization experiments[a]

	RELATIVE DUPLEX FORMATION	
	DNA-DNA	DNA-rRNA
P. acidovorans/P. acidovorans	100	100
P. acidovorans/P. testosteroni	33	92
P. acidovorans/P. delafieldii	0	89
P. acidovorans/P. facilis	0	87
P. acidovorans/P. saccharophila	0	79

[a] Data from N. J. Palleroni, R. Kunisawa, R. Contopoulou, and
M. Doudoroff, *Int. J. Syst. Bacteriol.* **23**, 333 (1973).

the genomes of two bacteria which show no significant homology by DNA-DNA
reassociation (Table 16.4).

In any given bacterial group the value of nucleic acid reassociation studies is
directly related to the number of strains and species that have been compared.
Really extensive comparative data are now available for only two major bacterial
groups: the enteric bacteria and the aerobic pseudomonads. The insights which
have been obtained into the genetic relationships among about 30 species of
aerobic pseudomonads, currently classified in the genera *Pseudomonas* and *Xanthomonas*, are shown schematically in Figure 16.9. Each large, shaded circle in this
diagram embraces a single rRNA homology group. The species included in each
group all show a much higher degree of relatedness with one another than with
any species belonging to other rRNA homology groups. In fact, the *intergroup* level
of rRNA homology among aerobic pseudomonads are no greater than those with
the enteric bacterium, *Escherichia coli* (Figure 16.10).

Within each rRNA homology group, the smaller, unshaded circles define the
limits of interspecific genetic homology detectable by DNA-DNA hybridization.
It can be seen that one rRNA homology group includes three isolated DNA
homology groups (*Pseudomonas acidovorans* + *P. testosteroni*; *P. saccharophila*; *P. facilis* + *P. delafieldii*). In the four other rRNA homology groups, all the constituent
species can be shown to be genetically interrelated by pairwise DNA-DNA
hybridization experiments, even though some pairs of species within a given DNA
homology group may be so distantly related that no DNA homology can be
detected between them. This is true, for example, of the two species of flourescent
pseudomonads, *P. aeruginosa* and *P. syringae*; however, both show some degree of
relatedness in terms of DNA-DNA reassociation with a third species, *P. fluorescens*.
Another interesting point concerns the yellow-pigmented plant pathogenic pseudomonads, currently classified in a separate genus, *Xanthomonas*. Both DNA-rRNA
and DNA-DNA hybridization studies show that they form a homology group
with one otherwise genetically isolated *Pseudomonas* species, *P. maltophilia*.

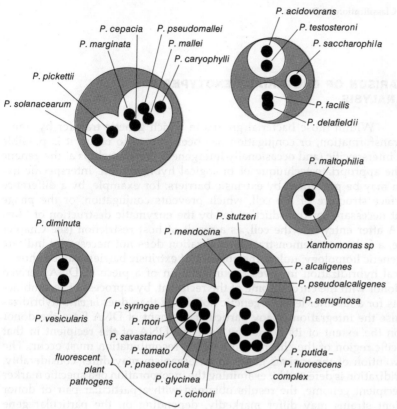

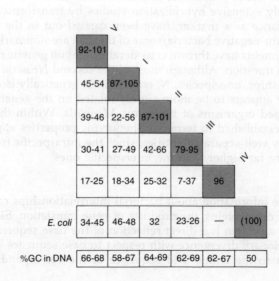

FIGURE 16.9

A schematic diagram of genetic relationships among aerobic pseudomonads, as revealed by nucleic acid reassociation *in vitro*. Each large, shaded circle represents an rRNA homology group, determined by DNA-rRNA hybridizations. The distances between the five groups are arbitrary, since their interrelations are not known. The smaller white circles define DNA homology groups. The black dots within each white circle represent species (or biotypes) among which relatedness can be shown by DNA-DNA hybridization: the distance between any two black dots is a rough measure of the degree of shared genetic homology. After N. J. Palleroni et al., *Int. J. Syst. Bacteriol.* **23**, 333 (1973).

FIGURE 16.10

A similarity matrix of rRNA homologies (determined by DNA-rRNA reassociation) among 30 species of aerobic pseudomonads. These data permit a clear-cut division of these bacteria into five homology groups (I–V). Intragroup homologies (shaded squares) are never less than 79 percent; intergroup homologies lie in the range of 7 to 66 percent, and are not much higher than those obtainable in DNA-rRNA reassociation experiments between aerobic pseudomonads and the enteric bacterium, *Escherichia coli*, for which data are also shown. The mean DNA base compositional ranges for the five homology groups (bottom line) are similar, illustration of the fact that this property is not in itself a good indication of genetic relatedness. From N. J. Palleroni et al., *Int. J. System. Bacteriol.* **23**, 333 (1973).

THE COMPARISON OF BACTERIAL GENOTYPES BY GENETIC ANALYSIS

Within those bacterial groups in which genetic transfer by transduction, transformation, or conjugation has been shown to occur, it is possible to explore interspecific (and occasionally intergeneric) relationships at the genetic level by the appropriate technique of biological hybridization. Interspecific hybridization may be prevented by extrinsic barriers: for example, by a difference in the surface structure of the cell, which prevents conjugation, or the phage attachment necessary for transduction, or by the enzymatic destruction of "foreign" DNA after entry into the cell, as a result of host restriction (see Chapter 15). Hence, a failure to demonstrate hybridization does not necessarily indicate a lack of genetic homology, unless the absence of extrinsic barriers can be shown.

Biological hybridization involves the integration of a piece of DNA derived from the donor with the chromosome of the recipient, by a process of recombination. It tests for a far finer level of genetic homology than does *in vitro* hybridization, because the integration of any particular segment of DNA from the donor depends on the extent of its homology with the DNA of the recipient in that small, specific region of the chromosome where recombination must occur. The rates of evolution of the various genes in an organism may differ considerably. Since hybridization is detected by examining the incorporation of a specific marker into the recipient genome, the results obtained with a particular pair of donor and recipient strains may differ markedly, depending on the particular gene selected as a marker. This has been shown very clearly in experiments on interstrain hybridization in *Bacillus,* where the integration of a donor marker for streptomycin resistance consistently occurred with much higher frequencies than the integration of donor nutritional markers (Table 16.5).

Particularly extensive hybridization studies by transformation, using streptomycin resistance as a marker, have been carried out in the *Neisseria–Moraxella* group of Gram-negative bacteria; some of the data are summarized in Figure 16.11. These experiments have thrown considerable light on genetic relationships within the group in question. Although most of the coccoid *Neisseria* show close genetic interrelationships, one species, *N. catarrhalis,* is genetically isolated from the rest and actually appears to be more closely related on the genetic level of some of the rod-shaped organisms of the genus *Moraxella*. Within the *Moraxella* group, the species established in terms of phenotypic properties appear to correspond to genetically well-separated entities, since the intraspecific frequencies of transformation are far higher than the interspecific ones.

• **The study of relatedness at the level of gene translation**

Considerable information about bacterial interrelationships can be derived from the study of cell proteins, the products of gene translation. Since the amino acid sequence of a protein is a direct reflection of the base sequence of the encoding gene, evolutionary divergence with respect to base sequence leads to divergence with respect to the amino acid sequence of the corresponding protein. These

TABLE 16.5

Effect of the choice of genetic marker on the frequencies of genetic recombination observed in interstrain and interspecies hybridization by transformation[a]

| DONOR STRAIN | RELATIVE FREQUENCY OF TRANSFORMATION (NORMALIZED TO VALUES OBTAINED WITH HOMOLOGOUS DONOR) | | |
	STREPTOMYCIN	TRYPTOPHAN	LEUCINE
B. subtilis 168-2 (homologous)	1.0	1.0	1.0
B. subtilis 6633	0.6	$<5 \times 10^{-5}$	$<5 \times 10^{-5}$
B. subtilis var. niger	0.54	$<5 \times 10^{-5}$	1.4×10^{-3}
B. pumilus	2.3×10^{-2}	$<5 \times 10^{-5}$	$<5 \times 10^{-5}$
B. licheniformis	4×10^{-3}	$<5 \times 10^{-5}$	$<5 \times 10^{-5}$
B. megaterium	$<2 \times 10^{-5}$	$<5 \times 10^{-5}$	$<5 \times 10^{-5}$
B. polymyxa	$<2 \times 10^{-5}$	$<5 \times 10^{-5}$	$<5 \times 10^{-5}$

[a] A strain of *B. subtilis* 168-2, which was streptomycin-sensitive and required tryptophan and leucine as growth factors, was transformed for streptomycin-resistance or nutritional independence with DNA from several different donors (all streptomycin-resistant and prototrophic for the two amino acids). Data of D. Dubnau, I. Smith, P. Morell, and J. Marmur, *Proc. Natl. Acad. Sci. U.S.* **54**, 491 (1965).

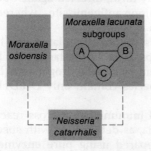

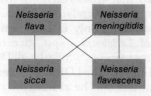

FIGURE 16.11

A simplified diagram illustrating genetic relationships among the *Moraxella* and *Neisseria* groups, as inferred from studies on transformation frequencies for a particular genetic marker, streptomycin resistance. Solid lines connecting boxes or circles indicate intergroup transformation frequencies ranging from 10^{-2} to 10^{-3} times those observed within each group; dashed lines, transformation frequencies from 10^{-3} to 10^{-5} times those observed within each group. The absence of connecting lines indicates that intergroup transformations have not been detected. Note that in terms of this criterion, *"Neisseria" catarrhalis* appears to be more closely related to *Moraxella* spp. than to other *Neisseria* spp., with which it has been classified on the basis of morphological resemblances (coccoid cell form).

changes are frequently reflected by changes in the immunological properties of proteins, which can be readily detected by rather simple techniques.

A protein bears numerous distinct antigenic determinants, each reflecting a specific feature of its molecular structure (for example, a short sequence of amino acids). An antiserum prepared against a pure protein (for example, a bacterial enzyme) contains antibodies directed against each of these antigenic determinants. An antiserum is tested by allowing it to diffuse through an agar gel toward a source of crude extract of the bacterium that contains the enzyme; a single, sharp line of precipitate forms in the zone of antigen-antibody interaction (Figure 16.12). Only one line is produced, because each of the various antigenic determinants is carried by single enzyme molecules, which diffuse through the gel at a uniform rate.

When the antigen against which the antiserum was produced, known as the *homologous* antigen (H), is compared in a double diffusion experiment with a *heterologous* antigen, h (i.e., a protein of the same nature, but prepared from another organism), two precipitation patterns are possible (assuming that H and h share some common antigenic determinants). Complete fusion of the precipitation lines (Figure 16.13) shows that all antigenic determinants on H are also carried by h. Partial fusion, associated with a spur directed toward the well containing h (Figure 16.14) shows that H and h share some common antigenic sites, but that H possesses one or more sites not present on h.

When two heterologous antigens, h_1 and h_2 are similarly compared with anti-H antiserum, three precipitation patterns are possible. Complete fusion of the two precipitation lines shows that h_1 and h_2 possesses antigenic sites in common with H and are identical. A spur directed toward h_1 indicates that h_2 has at least one additional site in common with H that is absent from h_1. By pairwise comparisons of heterologous antigens that form spurs against one another, it is thus possible to establish their order of decreasing antigenic similarity to H; $H > h_2 > h_1$, etc.

Occasionally, double spurs may form, directed against both h_1 and h_2. In this case, H, h_1, and h_2 all possess common antigenic sites, whereas others are common only to H and h_1, or to H and h_2.

This approach to the analysis of protein relatedness has been extensively applied in certain bacterial groups, notably the lactic acid bacteria. These organisms all produce NAD-dependent lactic dehydrogenases (LDHs), specific either for D- or for L-lactate. Some species contain only D-LDH, some only L-LDH, some both.

Table 16.6 shows the extent of immunological cross-reactivity of the D-LDHs and L-LDHs of lactic acid bacteria, as determined with specific antisera directed against D-LDH and L-LDH, prepared using pure enzymes isolated from two *Lactobacillus* species. It can be seen that cross-reactions with these two antisera by crude extracts of *Lactobacillus* spp. are correlated with the enzymatic activities present: for example, extracts containing only D-LDH activity cross react only with anti-D-LDH. All D-LDHs of *Lactobacillus* spp. give immunological cross-reactions as do the D-LDHs of *Leuconostoc* spp. Similarly, all D-LDHs of *Lactobacillus* spp. are cross-reacting proteins. However, the LDHs of bacteria belonging to other genera do not react with either antiserum. The absence of a reaction could mean either that their LDHs are proteins completely unrelated to those of the lactobacilli or that they have diverged so far, by successive amino acid substitutions, that they no longer possess any shared antigenic determinants.

 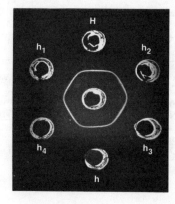 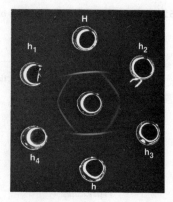

FIGURE 16.12 (left)

A double diffusion experiment in an agar gel, to illustrate the antigen-antibody reaction between an antiserum, prepared against a pure bacterial protein, and the homologous antigen. The well labeled anti-pc contained an antiserum prepared against pure phycocyanin, a chromoprotein isolated from the blue-green bacterium *Aphanocapsa sp.* One antigen well (pc) contained pure phycocyanin. The other (ext) contained a crude, cell-free extract of *Aphanocapsa sp.* A single, sharp precipitin line (composed of the specific antigen-antibody complex) is produced between each antigen well and the antibody well, and the two lines fuse together completely. The antiserum is specific for phycocyanin, and does not contain antibodies directed against any of the other proteins present in the crude cell-free extract.

FIGURE 16.13 (center)

A double diffusion experiment to compare the reactions of an antiserum with the homologous antigen and with a series of heterologous antigens. The antiserum (center well) was prepared against the purified D-lactic dehydrogenase of *Lactobacillus leichmannii*, and was tested against crude cell-free extracts of this bacterium (wells labeled H) and of four other bacterial strains (wells labelled h_1, h_2, h_3, and h_4), belonging to the species *L. bulgaricus*, *L. lactis*, and *L. delbrueckii*. There is complete fusion of the precipitin lines between H, h_1, h_2, h_3, and h_4; this shows that all the antigenic determinants carried by the D-lactic dehydrogenase of *L. leichmannii* are present on the D-lactic dehydrogenases of the other three species examined. From F. and C. Gasser, "Immunological Relationships among Lactic Dehydrogenases in the Genera *Lactobacillus* and *Leuconostoc*," *J. Bact.* **106**, 113 (1971).

FIGURE 16.14 (right)

A double diffusion experiment to compare the reactions of an antiserum with the homologous antigen and with a series of heterologous antigens. The antiserum (center well) directed against the D-lactic dehydrogenase of *Lactobacillus leichmannii*, was tested against crude, cell-free extracts of this bacterium (wells labeled H) and of four different strains of *Lactobacillus fermenti* (wells h_1, h_2, h_3, h_4). Spurs are formed between H and each of the heterologous antigens, h_1–h_4, the spurs being directed towards the wells that contain the heterologous antigens. This shows that the D-lactic dehydrogenase of *L-fermenti* carries some, but not all, of the antigenic determinants associated with the enzyme of *L. leichmannii*. Courtesy of F. Gasser.

Two antisera directed against the D-LDHs of other species of *Lactobacillus* have also been prepared and used for comparative immunological studies. The combined results obtained with all three antisera permit the construction of a map showing the antigenic relatedness among the D-LDHs from many species of *Lactobacillus* and *Leuconostoc* (Figure 16.15). If one assumes that changes in the amino

TABLE 16.6

Immunological cross-reactions obtained with cell-free extracts of various bacteria[a]

	PRESENCE OF NAD-DEPENDENT LACTIC DEHYDROGENASES		CROSS-REACTIONS WITH ANTISERAL PREPARED AGAINST	
	D-	L-	D-LACTIC DEHYDROGENASE OF *L. leichmannii*	L-LACTIC DEHYDROGENASE OF *L. acidophilus*
LACTIC ACID BACTERIA				
Lactobacillus:				
leichmanii	+	−	+	−
delbrueckii	+	−	+	−
lactis	+	−	+	−
bulgaricus	+	−	+	−
jensenii	+	−	+	−
acidophilus	+	+	+	+
jugurti	+	+	+	+
salivarius	−	+	−	+
casei	−	+	−	+
Leuconostoc spp.	+	−	+	−
Streptococcus spp.	−	+	−	−
Pediococcus spp.	+	−	−	−
OTHER BACTERIA				
Bifidobacterium spp.	−	+	−	−
Butyribacterium rettgeri	+	−	−	−

(Cross-reacting lactic dehydrogenases — applies to the first block; Noncross-reacting lactic dehydrogenases — applies to the second block)

[a] Tested with antisera directed against a D-lactic dehydrogenase and an L-lactic dehydrogenase purified from two different lactobacilli.

acid sequence of LDHs occur at a relatively uniform rate, such a map provides a rough indication of the evolutionary distances among the species in question. Similar maps of antigenic relatedness among lactic acid bacteria have been prepared by comparative immunological studies of two other enzymes common to many species in this group (aldolase, glucose-6-P dehydrogenase). With minor exceptions, these maps are concordant with the anti-D-LDH map. It therefore appears probable that the divergence of most enzymes within a bacterial group occur at similar rates.

• **Regulatory mechanisms as markers of relatedness**

Recent comparative studies on the patterns of regulation of biosynthetic and dissimilatory pathways common to many bacterial groups suggest that the specific mode of regulation of a given pathway is a conservative and group-specific property. A striking illustration is provided by the regulation of two branched biosynthetic pathways, those which lead to the synthesis of aromatic amino acids and amino acids of the aspartate family, respectively. In each the primary site of regulation, both by repression and by feedback inhibition, is the first enzyme

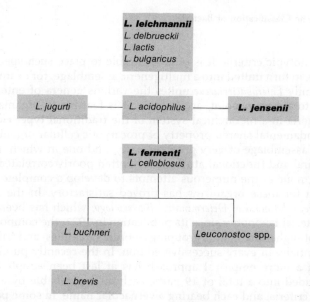

FIGURE 16.15

A map showing the antigenic relationships among the D-lactic dehydrogenases of *Lactobacillus* and *Leuconostoc* spp., as determined from the cross-reactions with three anti-D-lactic dehydrogenases, prepared against the enzymes of the three species shown in heavy type. The *Lactobacillus* species enclosed in each box are similar (or closely similar) in antigenic properties. After F. and C. Gasser, "Immunological Relationships among Lactic Dehydrogenases in the Genera *Lactobacillus* and *Leuconostoc*," *J. Bacteriol.* **106**, 113 (1971).

of the pathway: deoxy-arabino-heptulosonic acid phosphate synthetase (DAHP synthetase) in the aromatic pathway and aspartokinase (AK) in the aspartate pathway.

Escherichia coli and related enteric bacteria have three isofunctional DAHP synthetases and three isofunctional AKs, each of the three enzymes being subject to repression or feedback inhibition by a specific end product of the pathway. In most other bacteria so far examined the first reaction in each pathway is catalyzed by a single enzyme, subject to different and, in general, more complex modes of regulatory control by the end products. Mediation of these two reactions by isofunctional enzymes subject to independent regulatory control therefore appears to be a distinctive character of the bacteria of the enteric group. The only other bacteria in which it is known to exist are facultatively anaerobic Gram-negative bacteria with polar flagella (genera *Vibrio*, *Aeromonas*, *Beneckea*, and *Photobacterium*). Apart from the difference from the classical enteric bacteria with respect to the site of flagellar insertion, these bacteria have many properties suggestive of a relationship to the enteric group (see Chapter 20). The shared mechanisms for the regulation of the aromatic and aspartate pathways are evidence of relatedness.

THE MAIN OUTLINES OF BACTERIAL CLASSIFICATION

The recently developed molecular approaches to the analysis of bacterial relationships have provided a very valuable supplement to earlier purely phenotypic characterizations. It is now possible to recognize among the bacteria a considerable number of subgroups that appear to be natural ones, in the sense that their members are all of common evolutionary origin. Many of these groups (e.g., the enteric bacteria, some homology groups of aerobic pseudomonads) contain a large number of species, readily distinguishable by both phenotypic and genotypic criteria. It is often desirable to place such species in two or more genera, in turn united into a multigeneric assemblage, for example, a family. Thus, the family *Enterobacteriaceae* unites the various genera of enteric bacteria.

Up to a certain point, accordingly, it is feasible to organize the classification of bacteria in a hierarchical system of the traditional type. However, apart from the fundamental shared property of procaryotic cellular organization, the bacteria are an assemblage of very great diversity; and one in which, furthermore, major structural and functional attributes are often poorly correlated with one another. Hence, none of the numerous attempts to develop a complete hierarchical classification for these organisms has proved satisfactory. In the successive editions of *Bergey's Manual of Determinative Bacteriology*, which has been the major treatise on bacterial taxonomy since its publication in 1923, the composition and arrangement of higher taxonomic groupings—orders, families, and tribes—have changed substantially in every successive edition. In the recently published (1974) eighth edition a more empirical approach has at last been adopted. The bacteria are subdivided into a total of 19 parts, each distinguishable by a few readily determined criteria, and each bearing a vernacular name. In some parts the constituent genera are simply described in arbitrary sequence; in other parts they are grouped into families and sometimes into orders. The arrangement is in effect a partial, or *fragmented*, hierarchy. It is symptomatic of the difficulties involved in establishing hierarchies among bacteria that several parts in which genera are grouped into families also contain appendices headed *Genera of Uncertain Affiliation*, made up of genera that cannot even be fitted into the limited hierarchy employed.

Insofar as the goal of *Bergey's Manual* is to facilitate the identification of bacteria, it matters very little how the constituent genera are arranged, provided that an effective system of keys for determining the generic position of an unknown organism is available. A detailed multiple key for this purpose is included in the eighth edition, and is one of its most valuable features.

Simply to indicate the general organization adopted in the latest edition of *Bergey's Manual*, a summary outline of the 19 parts is presented below.

KINGDOM PROCARYOTAE

Division I. Cyanobacteria (not further treated)

Division II. Bacteria

Part 1. Phototrophic Bacteria.
   One order, 3 families, 18 genera

Part 2. Gliding Bacteria
   Two orders, 8 families, 21 genera; also 6 genera of uncertain affiliation

Part 3. Sheathed Bacteria
   7 genera

Part 4. Budding and/or Appendaged Bacteria
   17 genera

Part 5. Spirochaetes
   One order, 1 family, 5 genera

525

Part 6. Spiral and Curved Bacteria
One family, 2 genera; also 4 genera of uncertain affiliation

Part 7. Gram-Negative Aerobic Rods and Cocci
Five families, 14 genera; also 6 genera of uncertain affiliation

Part 8. Gram-Negative Facultatively Anaerobic Rods
Two families, 17 genera; also 9 genera of uncertain affiliation

Part 9. Gram-Negative Anaerobic Bacteria
One family, 3 genera; also 6 genera of uncertain affiliation

Part 10. Gram-Negative Cocci and Coccobacilli
One family, 4 genera; also 2 genera of uncertain affiliation

Part 11. Gram-Negative Anaerobic Cocci
One family, 3 genera

Part 12. Gram-Negative Chemolithotrophic Bacteria
Two families, 17 genera

Part 13. Methane-Producing Bacteria
One family, 3 genera

Part 14. Gram-Positive Cocci
Three families, 12 genera

Part 15. Endospore-Forming Rods and Cocci
One family, 5 genera; also 1 genus of uncertain affiliation

Part 16. Gram-Positive Asporogenous Rod-Shaped Bacteria
One family, 1 genus; also 3 genera of uncertain affiliation

Part 17. Acinomycetes and Related Organisms.
Four genera not assigned to a family; 1 family with 2 genera; 1 order with 8 families and 31 genera

Part 18. Rickettsias
Two orders, 4 families, 18 genera

Part 19. Mycoplasmas
One class, 1 order, 2 families, 2 genera; also 2 genera of uncertain affiliation.

The following chapters (17 through 24) provide surveys of the properties of some of the major bacterial subgroups. The organization adopted follows to some extent the structure of the latest edition of *Bergey's Manual*, but it is considerably more simplified and sometimes brings together groups which are treated in separate parts of the *Manual*.

FURTHER READING

Books AINSWORTH, G. C., and P. H. A. SNEATH (editors), *Microbial Classification*, Twelfth Symposium of the Society for General Microbiology. Cambridge, England: University Press, 1962.

BUCHANAN, R. E., and N. E. GIBBONS, (editors), *Bergey's Manual of Determinative Bacteriology*, 8th ed. Baltimore: Williams & Wilkins. 1974.

LOCKHART, W. R., and J. LISTON, (editors), *Methods for Numerical Taxonomy*. Bethesda, Md.: American Society for Microbiology, 1970.

MARTIN, S. M., and V. B. D. SKERMAN (editors), *World Directory of Collections of Cultures of Microorganisms.* New York: Wiley Interscience, 1972.

SKERMAN, V. B. D., *A Guide of the Identification of the Genera of Bacteria,* 2nd ed. Baltimore: Williams and Wilkins Co., 1967.

SNEATH, P. H. A., and R. R. SOKAL, *The Principles and Practices of Numerical Classification,* Freeman, 1973.

Reviews and original articles

BRENNER, R. J., and S. FALKOW, "Molecular Relationships Among Members of the Enterobacterraceae," *Adv. Genetics* **16,** 81 (1971).

GASSER, J., and C. GASSER, "Immunological Relationships Among Lactic Dehydrogenases in the Genera *Lactobacillus* and *Leuconostoc,*" *J. Bact.* **106,** 113 (1971).

JONES, D., and P. H. A. SNEATH, "Genetic Transfer and Bacterial Taxonomy," *Bact. Revs.* **34,** 40 (1970).

LONDON, J., and K. KLINE, "Aldolase of Lactic Acid Bacteria: A Case History in the Use of an Enzyme as an Evolutionary Marker," *Bact. Revs.* **37,** 453 (1973).

MANDEL, M., "New Approaches to Bacterial Taxonomy," *Ann. Rev. Microbiol.* **23,** 239 (1969).

PALLERONI, N. J., R. KUNISAWA, R. CONTOPOULOU, and M. DOUDOROFF, "Nucleic Acid Homologies in the Genus *Pseudomonas,*" *Int. J. System. Bact.* **23,** 333 (1973).

SNEATH, P. H. A., "Computer Taxonomy," in J. R. Norris and D. W. Ribbons (editors), *Methods in Microbiology,* **7A,** 29. New York: Academic Press, 1972.

17

THE PHOTOSYNTHETIC PROCARYOTES

The Gram-negative procaryotes include three distinct and well-defined photo-synthetic groups. The *blue-green bacteria* perform oxygenic photosynthesis and possess a pigment system similar in basic respects to that of photosynthetic eucaryotes. This group has long been treated by botanists as one of the major classes or divisions of the algae. However, the typically procaryotic cell structure of these organisms, which was clearly established about 1960, identifies them unambigously as bacteria. As discussed in Chapter 5 (pp. 119–153), the group is large and structurally diverse, including many different types of filamentous and unicellular organisms. Movement, when it occurs, is by gliding.

Purple bacteria and *green bacteria* perform anoxygenic photosynthesis and possess unique pigment systems that confer on them spectral properties unlike those of all other phototrophs. With one exception, they are all unicellular. Most green bacteria are small, immotile, rod-shaped organisms; the purple bacteria are rods, cocci, or spirilla, frequently motile by means of flagella. Because of the evident structural resemblances of these organisms to nonphotosynthetic bacteria, their taxonomic position has never been questioned; they have been included among the bacteria since their discovery in the mid-nineteenth century. However, the nature of their metabolism remained controversial for many decades.

About 1885 W. Engelmann suggested that the purple bacteria might be photo-synthetic organisms, as a result of his discovery that they are phototactic, and that their growth is favored by light. Repeated attempts to demonstrate oxygen production by these organisms in the light failed; the absence of this property appeared at the time to be strong evidence against Engelmann's hypothesis. Moreover, S. Winogradsky had shown that some purple bacteria can oxidize H_2S to sulfate with transient intracellular accumulation of elemental sulfur, an unusual property also possessed by certain chemoautotrophic bacteria (see Chapter 18). About 1905 W. Molisch observed that other purple bacteria can grow, either in

527

the light or in the dark, in complex organic media, and do not oxidize H_2S. The seemingly irreconcilable reports of Engelmann, Winogradsky, and Molisch remained without a coherent explanation until 1930, when C. B. van Niel first recognized and defined the various metabolic versions of anoxygenic photosynthesis and demonstrated that it is the characteristic mode of energy-yielding metabolism in both purple and green bacteria.

The history of the photosynthetic procaryotes has thus been complex, marked by a series of misunderstandings either about their biological position or about their metabolic properties. Another chapter was added in 1971 by W. Stoeckenius, who discovered a unique process of light-dependent ATP synthesis in bacteria which had been long known and had previously been regarded as chemoheterotrophs. These organisms, the extreme halophiles of the genus *Halobacterium*, synthesize a special carotenoid that can undergo photochemical transformations coupled with net ATP synthesis. The relations of *Halobacterium* to light will be described in the last section of this chapter.

THE FUNCTIONAL PROPERTIES OF PHOTOSYNTHETIC PROCARYOTES

Tables 17.1 and 17.2 summarize and compare some of the major properties associated with photosynthetic function in purple bacteria, green bacteria, and blue-green bacteria. Comparative data for the chloroplasts of two major algal groups are also included.

The two tabulations make clear how closely the photosynthetic machinery of the blue-green bacteria resembles that of eucaryotes, and more particularly that of one algal group, the red algae. Blue-green bacteria and red algae share the following unique functional properties: chlorophyll a is their only chlorophyllous

TABLE 17.1

The structure of the photosynthetic apparatus and the mechanisms of photosynthesis in procaryotes and in chloroplasts

| | PROCARYOTES | | | CHLOROPLASTS OF | |
	PURPLE BACTERIA	GREEN BACTERIA	BLUE-GREEN BACTERIA	RED ALGAE	GREEN ALGAE AND PLANTS
Cell structure which contains photosynthetic apparatus	Cell membrane	Chlorobium vesicles	Thylacoids and phycobilisomes	Thylacoids and phycobilisomes	Thylacoids
Photosystems					
I	+	+	+	+	+
II	−	−	+	+	+
Reductants used for CO_2 assimilation	H_2S, H_2, or organic compounds	H_2S, H_2	H_2O	H_2O	H_2O
Principal photosynthetic carbon source	CO_2 or organic compounds	CO_2 or organic compounds	CO_2	CO_2	CO_2

TABLE 17.2

Pigments and lipids of the photosynthetic apparatus in procaryotes and chloroplasts

	PROCARYOTES			CHLOROPLASTS OF	
	PURPLE BACTERIA	GREEN BACTERIA	BLUE–GREEN BACTERIA	RED ALGAE	GREEN ALGAE AND PLANTS
1. Pigments Chlorophylls	Bacterio a or Bacterio b	Minor Bacterio a Major Bacterio c d or e	Chlorophyll a	Chlorophyll a	Chlorophylls a and b
Predominant carotenoids Alicyclic (β-carotene and related oxycarotenoids)	−	−	+	+	+
Aryl carotenoids	+ or −	+	−	−	−
Aliphatic, bearing methoxyl groups	− or +	−	−	−	−
Phycobiliproteins					
Allophycocyanin	−	−	+	+	−
Phycocyanin	−	−	+	+	−
Phycoerythrin	−	−	+ or −	+ or −	−
2. Lipids					
Monogalactosyl-diglyceride	−	+	+	+	+
Digalactosyl-diglyceride	−	−	+	+	+
Polyunsaturated fatty acids	−	−	+ or −	+	+

pigment; the chromproteins known as *phycobiliproteins* are their major light-harvesting pigments; and these chromoproteins are not integrated into the thylacoids, as are both chlorophylls and carotenoids, but are instead localized in special structures known as *phycobilisomes*, which are attached to the outer surfaces of the thylacoids.

The purple and green bacteria differ from other phototrophs with respect to nearly all the characters listed. The photosynthetic apparatus has a unique intracellular location in each group: it is incorporated into a topologically complex cell membrane in purple bacteria, and it is housed in special organelles (chlorobium vesicles) in green bacteria (see Chapter 11 for further details). Lacking photosystem II, these organisms cannot use H_2O as a reductant for CO_2 assimilation; this function is assumed by other reduced inorganic compounds. They can use organic compounds either as sole photosynthetic carbon sources or in conjunction with CO_2; under these conditions, an inorganic reductant is not required. Finally, the pigments and lipids of the photosynthetic apparatus are largely group-specific in both purple and green bacteria.

• **The chlorophylls of procaryotes**

The chlorophylls of purple and green bacteria have the same basic structure and are synthesized through the same biosynthetic pathway as plant chlorophylls. However, because they are confined to these two bacterial groups, they are termed *bacteriochlorophylls* (bchls). Figure 17.1 and Table 17.3 show the chemical differences between the five known bchls and chlorophyll a. The bchls fall into two chemical subclasses: bchls a and b; bchls c, d, and e. This is correlated with their biological distributions. The cells of purple bacteria contain only one form of bchl: either bchl a or bchl b, depending on the species. The pigment is in part responsible for light harvesting and in part associated with the photochemical reaction centers.

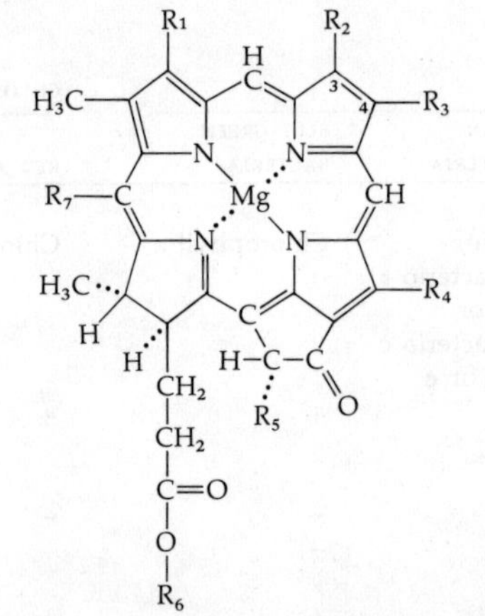

FIGURE 17.1

The chemical structure common to chlorophylls. The nature of variable substituents (R_1 through R_7) in chlorophyll a and in the various bacteriochlorophylls is shown in Table 17.3.

TABLE 17.3

Chemical differences between chlorophyll a and bacteriochlorophylls

CHLOROPHYLL	R_1	R_2	R_3	R_4	R_5	R_6	R_7
a	$-CH=CH_2$	$-CH_3$	$-C_2H_5$	$-CH_3$	$-\overset{\text{O}}{\underset{\|}{C}}-OCH_3$	Phytyl	$-H$
Bacterio a	$-\overset{\|}{\underset{\text{O}}{C}}-CH_3$	$-CH_3{}^a$	$-C_2H_5{}^a$	$-CH_3$	$-\overset{\text{O}}{\underset{\|}{C}}-O-CH_3$	Phytyl or geranyl-geranyl	$-H$
Bacterio b	$-\overset{\|}{\underset{\text{O}}{C}}-CH_3$	$-CH_3{}^b$	$=CH-CH_3{}^b$	$-CH_3$	$-\overset{\text{O}}{\underset{\|}{C}}-O-CH_3$	Phytyl	$-H$
Bacterio c	$-\overset{\text{H}}{\underset{\text{OH}}{C}}-CH_3$	$-CH_3$	$-C_2H_5$	$-C_2H_5$	$-H$	Farnesyl	$-CH_3$
Bacterio d	$-\overset{\text{H}}{\underset{\text{OH}}{C}}-CH_3$	$-CH_3$	$-C_2H_5$	$-C_2H_5$	$-H$	Farnesyl	$-H$
Bacterio e	$-\overset{\text{H}}{\underset{\text{OH}}{C}}-CH_3$	$-\overset{\text{H}}{\underset{\|}{\underset{\text{O}}{C}}}=O$	$-C_2H_5$	$-C_2H_5$	$-H$	Farnesyl	$-CH_3$

[a] No double bond between C—3 and C—4; additional —H atoms at C—3 and C—4.
[b] No double bond between C—3 and C—4; additional —H atom at C—3.

531   The cells of green bacteria always contain a major and minor form of bchl. Depending on the species, the major form is either bchl c, bchl d, or bchl e; it has a light-harvesting role. The minor pigment in all green bacteria is bchl a; and it is the form of chlorophyll in the photochemical reaction centers.

The three light-harvesting chlorophylls which occur among green bacteria (bchls c, d, and e) all resemble chlorophyll a rather closely in spectral properties (Table 17.4). However, the two forms of bchl (a and b) which are the sole chlorophylls of purple bacteria share a seemingly minor special chemical feature (saturation of one double bond in the ring system, between C-3 and C-4) which has a profound effect on their spectral properties, changing the positions of all the major absorption peaks relative to those of other chlorophylls. The long wavelength peak of bchls a and b lies over 100 nm farther toward the red end of the spectrum than that of other bacterial and plant chlorophylls, being situated very close to the infrared region of the spectrum.

• The carotenoids of photosynthetic procaryotes

Carotenoids are always associated with the photosynthetic apparatus, and have two functions in photosynthesis. They often serve as light-harvesting pigments, absorbing light in the blue-green region of the spectrum, between 400 and 550 nm; their relative contribution to this function is major in some photosynthetic organisms, minor in others. In addition, they play an indispensable role as quenchers of chlorophyll-catalyzed photooxidation, thus protecting the photosynthetic apparatus from photooxidative damage (see Chapter 10, p. 311).

Among phototrophs, carotenoid composition tends to be both complex and group specific. However, the eucaryotic phototrophs all contain as major carotenoids the bicyclic carotenes and related oxygen-containing carotenoids. Such pigments are likewise characteristic of blue-green bacteria; many of these organisms also contain carotenoid glycosides, which are group specific (Figure 17.2).

In purple bacteria, carotenoid composition is extraordinarily diverse; over 30 pigments of this type occur in the different members of the group, and none is common to all. However, the carotenoids of purple bacteria are group-specific compounds, in the sense that they do not occur in other photosynthetic organisms. Most of them are aliphatic (open-chain) compounds, which often bear methoxyl groups; a few are aryl carotenoids, bearing an aromatic ring at one end of the chain (Figure 17.3). Aryl carotenoids are characteristic of nearly all green bacteria;

TABLE 17.4

Spectroscopic properties of purified chlorophylls in ether solution

	ABSORPTION MAXIMA (WAVELENGTHS IN nm); RELATIVE PEAK HEIGHTS IN PARENTHESES					
Chlorophyll a			430 (100)	615 (13)	662 (77)	
Bacterio c			428 (100)	622 (29)	660 (63)	
Bacterio d			424 (100)	608 (17)	654 (61)	
Bacterio e			458 (100)	593 (12)	647 (32)	
Bacterio a	358 (100)	391 (68)	577 (29)			773 (126)
Bacterio b	368 (100)	407 (87)	582 (30)			795 (96)

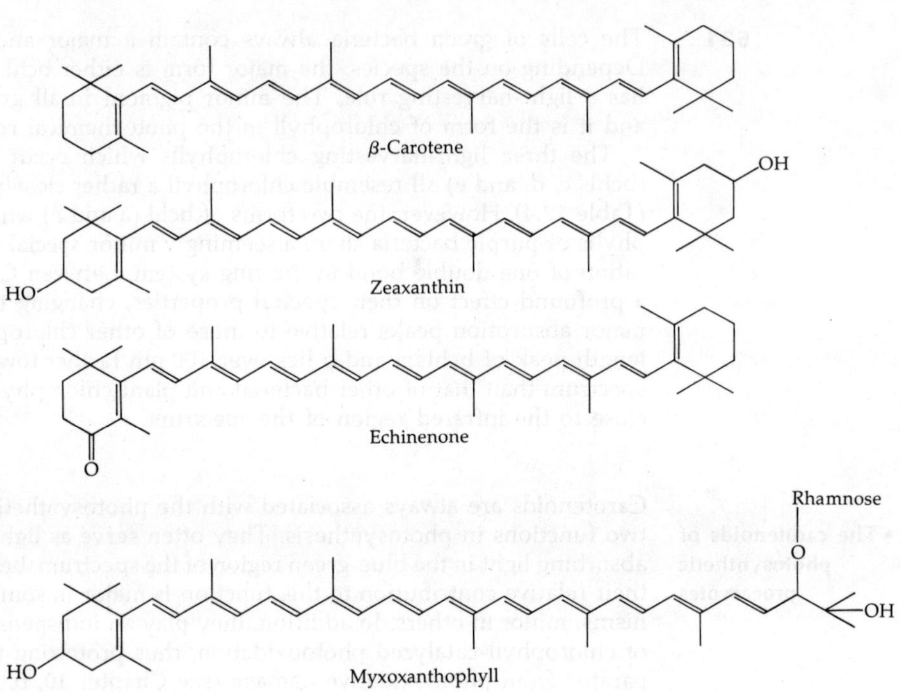

FIGURE 17.2

Some carotenoids of blue-green bacteria. The bicyclic hydrocarbon, β-carotene, is a major pigment in all members of the groups, and its oxyderivatives, zeaxanthin and echinone, occur in many. The two former carotenoids are common chloroplast pigments in eucaryotes. Myxoxanthophyll, a carotenoid glycoside, is an example of the group-specific carotenoids synthesized by certain blue-green bacteria.

β-Carotene

Zeaxanthin

Echinenone

Myxoxanthophyll

Rhamnose

FIGURE 17.3

Some representative examples of the types of carotenoids characteristic of the purple bacteria. Most of these compounds are aliphatic (open-chain) carotenoids, often bearing tertiary hydroxyl or methoxyl (—OCH₃) groups. Aryl carotenoids, bearing an aromatic ring at one end of the chain (exemplified by okenone) occur only in a few purple sulfur bacteria.

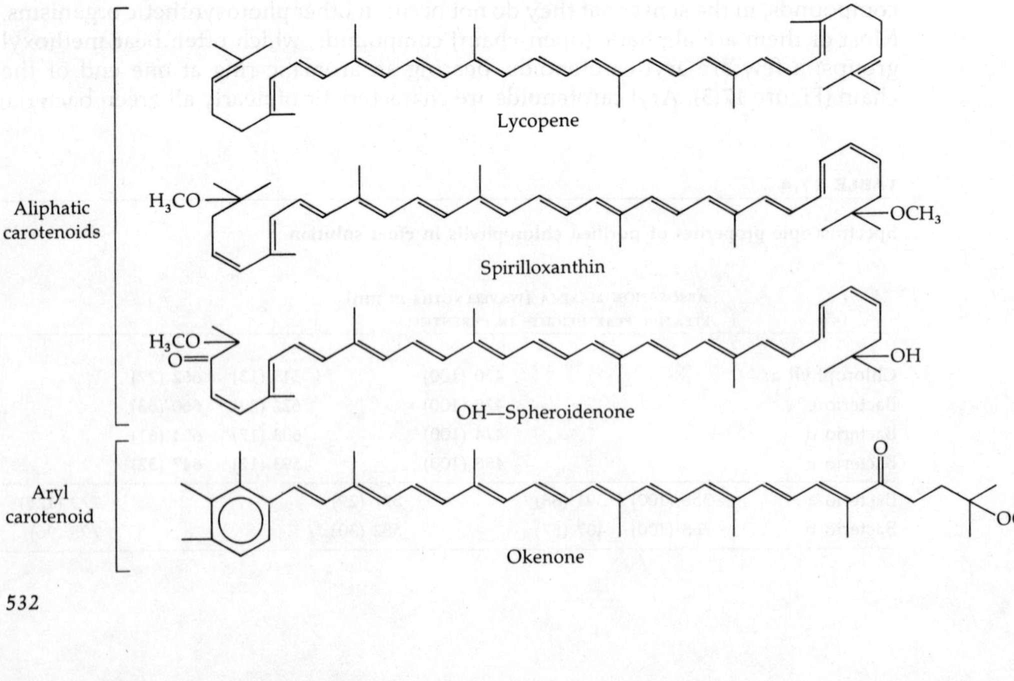

Aliphatic carotenoids

Lycopene

H₃CO — Spirilloxanthin —OCH₃

H₃CO — OH—Spheroidenone —OH

Aryl carotenoid

Okenone —OCH₃

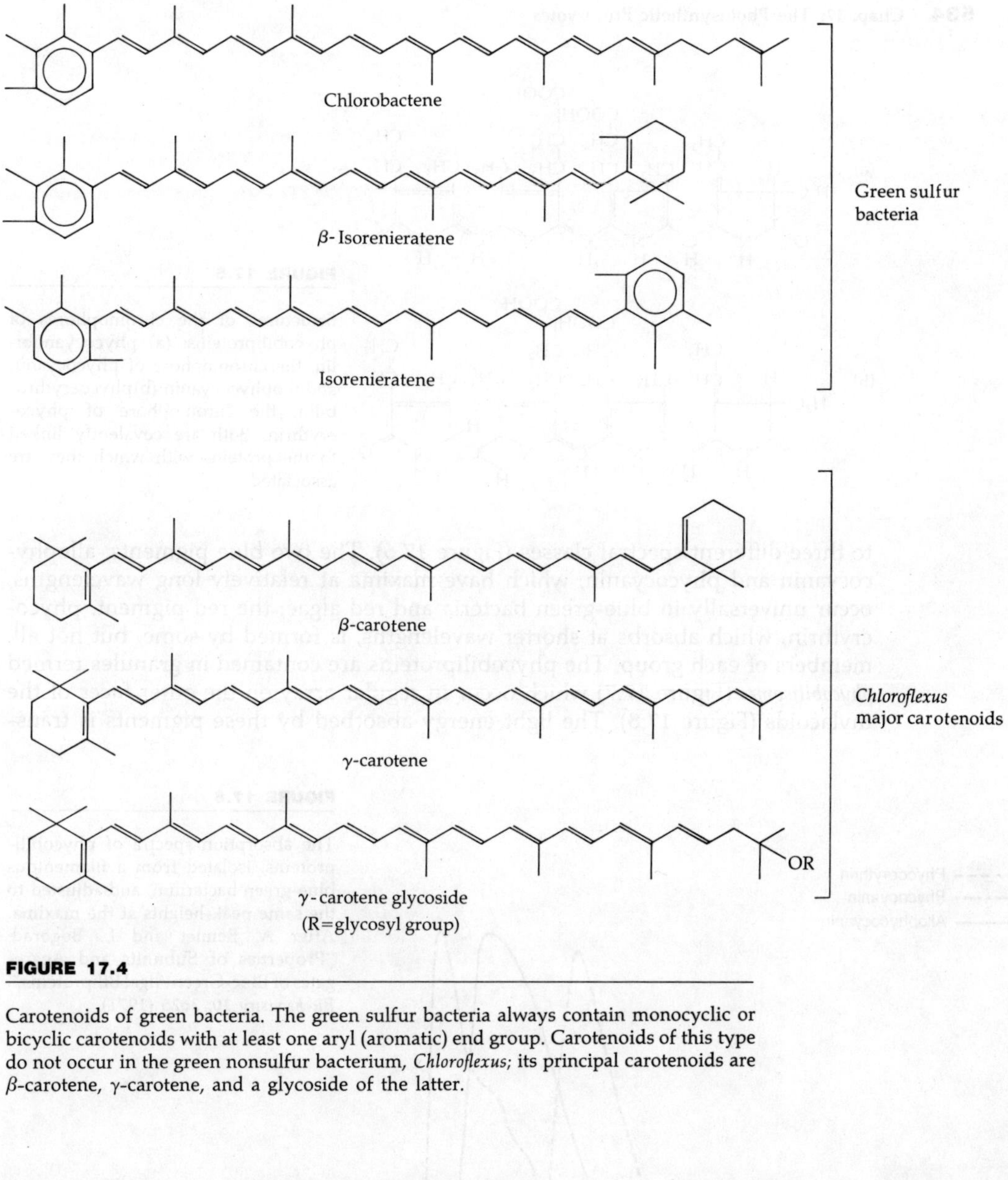

FIGURE 17.4

Carotenoids of green bacteria. The green sulfur bacteria always contain monocyclic or bicyclic carotenoids with at least one aryl (aromatic) end group. Carotenoids of this type do not occur in the green nonsulfur bacterium, *Chloroflexus*; its principal carotenoids are β-carotene, γ-carotene, and a glycoside of the latter.

however, *Chloroflexus* contains β- and γ-carotene, as well as glycosides of the latter pigment (Figure 17.4).

• **Phycobiliproteins**   The phycobiliproteins, which are the major light-harvesting pigments of both blue-green bacteria and red algae, are water-soluble proteins that contain covalently bound linear tetrapyrroles (bilins) as chromophores (Figure 17.5). They absorb light in a broad region near the middle of the visible spectrum, and belong

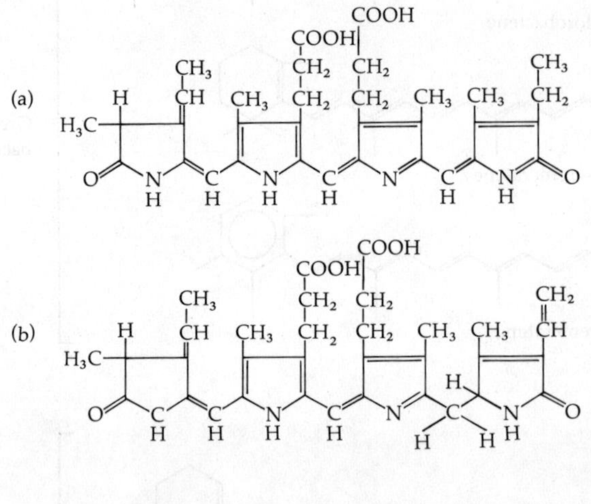

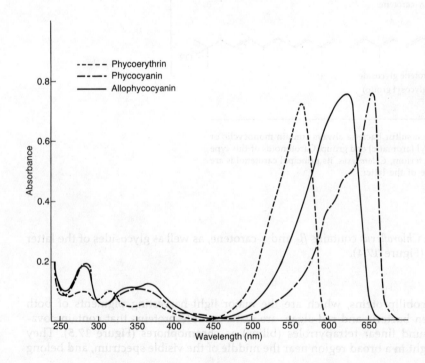

(a)

(b)

FIGURE 17.5

Structures of the chromophores of phycobiliproteins: (a) phycocyanobilin, the chromophore of phycocyanin and allophycocyanin; (b) phycoerythrobilin, the chromophore of phycoerythrin. Both are covalently linked to the proteins with which they are associated.

to three different spectral classes (Figure 17.6). The two blue pigments, allophycocyanin and phycocyanin, which have maxima at relatively long wavelengths, occur universally in blue-green bacteria and red algae; the red pigment, phycoerythrin, which absorbs at shorter wavelengths, is formed by some, but not all, members of each group. The phycobiliproteins are contained in granules termed *phycobilisomes* (Figure 17.7) which occur in regular array on the outer faces of the thylacoids (Figure 17.8). The light energy absorbed by these pigments is trans-

FIGURE 17.6

The absorption spectra of phycobiliproteins, isolated from a filamentous blue-green bacterium, and adjusted to the same peak heights at the maxima. After A. Bennet and L. Bogorad, "Properties of Subunits and Aggregates of Blue-Green Algal Biliproteins," *Biochemistry* **10**, 3625 (1971).

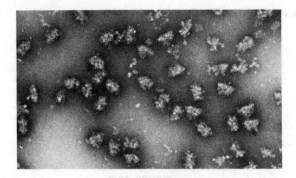

FIGURE 17.7

Isolated phycobilisomes, prepared from the red alga, *Porphyridium*. Electron micrograph of a negatively stained preparation. Courtesy of Dr. E. Gantt.

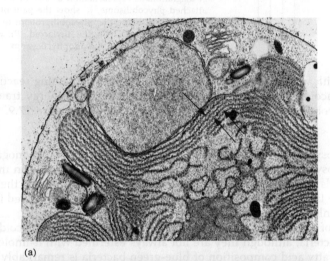

(a)

(b)

FIGURE 17.8

Electron micrographs of cellular thin sections, showing the arrangement of phycobilisomes (arrows) on the surface of the thylacoids in a chloroplast of the red alga *Porphyridium* (a) and in the cell of a blue-green bacterium *Aphanocapsa* (b). (a) Courtesy of Dr. E. Gantt; (b) courtesy of Dr. G. Cohen-Bazire.

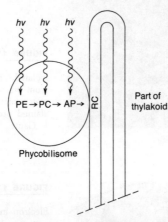

FIGURE 17.9

A diagrammatic representation of part of a thylacoid with an attached phycobilisome, to show the path of energy transfer from the light-harvesting phycobiliproteins to a photochemical reaction center (RC) in the thylacoid. PE = phycoerythrin; PC = phycocyanin; AP = allophycocyanin.

ferred with very high efficiency to the chlorophyll-containing reaction centers in the thylacoids; the three spectral classes constitute an energy transfer chain in the phycobilisomes, as indicated diagrammatically in Figure 17.9.

• Lipids of the photosynthetic apparatus

In photosynthetic eucaryotes, two classes of glycolipids, monogalactosyl and digalactosyl diglycerides (Figure 17.10) are specifically located in the chloroplast thylacoids and constitute about 80 percent of the total lipid of these structures. They are largely esterified with α-linolenic acid, a polyunsaturated fatty acid that occurs only in the chloroplast.

Both of these galactolipids are similarly present in the thylacoids of all blue-green bacteria, although they are not always esterified with α-linolenic acid. The cellular fatty acid composition of blue-green bacteria is remarkably varied; some have a high content of α-linolenic acid; some contain other polyunsaturated fatty acids; and some have the fatty acids characteristic of bacteria in general, i.e., exclusively saturated and monounsaturated compounds.

The green bacteria contain only one of these galactolipids, monagalactosyl-diglyceride; purple bacteria contain neither. Polyunsaturated fatty acids do not occur in the lipids of either group.

• Ferredoxins

In all photosynthetic organisms the nonheme iron proteins known as *ferredoxins* play an important role as electron carriers in the photosynthetic transport chain.

FIGURE 17.10

The structures of monogalactosyl - diglyceride (left) and digalactosyl-diglyceride (right). R_1 and R_2 denote esterifying fatty acids.

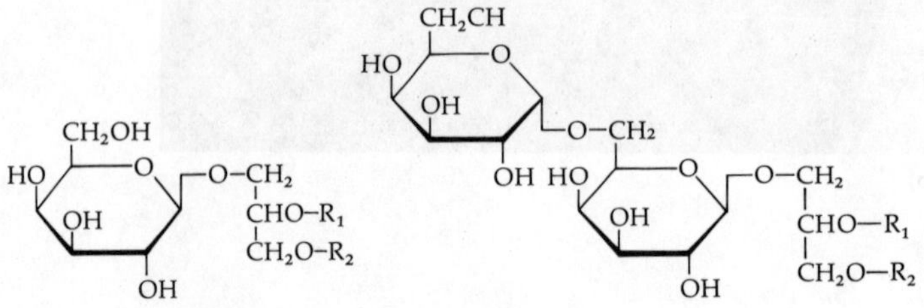

TABLE 17.5

Chemical and spectroscopic properties of some ferredoxins

	BIOLOGICAL SOURCE				
	Clostridium SPP. (ANAEROBIC CHEMOHETERO- TROPHS)	*Chlorobium* (GREEN BACTERIUM)	*Chromatium* (PURPLE BACTERIUM)	*Anabaena* (BLUE-GREEN BACTERIUM)	SPINACH CHLOROPLAST
Molecular weight	6,000	6,000	10,000	12,000	12,000
Fe, atoms/mole	8	8	8	2	2
Labile sulfide, atoms/mole	8	8	8	2	2
Positions of absorption maxima (nm) in reduced form		280, 300, 390		276, 330, 420, 463	

These proteins, which are of low redox potential, also function as electron carriers in strictly anaerobic, nonphotosynthetic bacteria. As shown in Table 17.5, the ferredoxin of blue-green bacteria closely resembles chloroplast ferredoxin in chemical properties, but the ferredoxins of purple and green bacteria do not, being more similar to those of nonphotosynthetic bacteria.

• The cellular absorption spectra of photosynthetic procaryotes

Although not all pigments in the photosynthetic apparatus are equally effective in harvesting light (chlorophyll a is far less effective than phyobiliproteins in blue-green bacteria), the cellular absorption spectra of photosynthetic organisms provide a rough indication of the spectral regions that are utilized for the performance of photosynthesis. Figure 17.11 compares the cellular absorption spectra of several different photosynthetic procaryotes; in each case, the specific contributions to light absorption made by chlorophylls, carotenoids, and phycobiliproteins are indicated. It is evident that the cellular absorption spectra characteristic of each group of photosynthetic procaryotes are distinctive, and to a considerable extent complementary. In blue-green bacteria, light is absorbed largely between 550 and 700 nm (by phycobiliproteins and by chlorophyll a). In green bacteria the major absorption band lies considerably farther toward the red region, between 700 and 800 nm; it is attributable to the light-harvesting bchls (c, d, or e) characteristic of this group. In purple bacteria the major absorption bands lie largely in the infrared region, being represented by one or more peaks, attributable either to bchl a or bchl b. The position of this band in purple bacteria that contain bchl b is situated beyond 1,000 nm, very close to the spectral limit beyond which light can no longer mediate photochemical reactions, as a result of the low energy content of the light quanta.*

---

*The energy content of the light quantum is an inverse function of wavelength.

FIGURE 17.11

Cellular absorption spectra of five representative photosynthetic procaryotes, to show the characteristic differences in the positions of the major absorption bands. The approximate contributions to cellular light absorption by the major classes of photosynthetic pigments, and the types of chlorophyll present in each organism are indicated on the figure. The double peak of phycobiliprotein light absorption in the spectrum of the blue-green bacterium illustrated reflects the presence of both phycoerythrin (maximum: 675 nm) and phycocyanin (maximum: 625 nm). Allophycocyanin absorption (maximum: 650 nm) is masked by the phycocyanin peak.

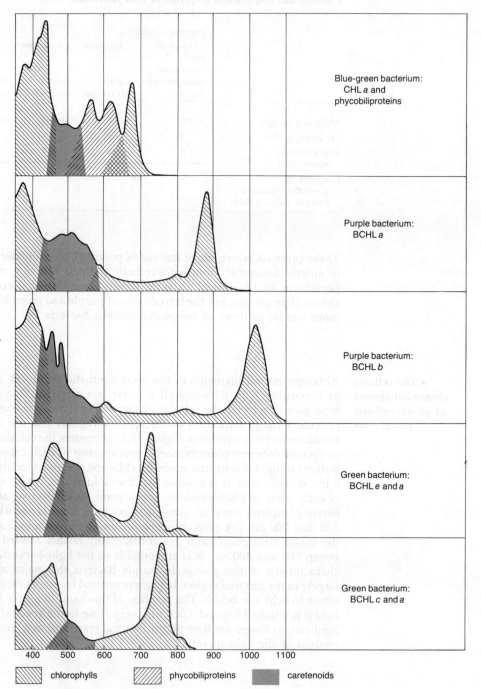

Blue-green bacterium:
CHL *a* and
phycobiliproteins

Purple bacterium:
BCHL *a*

Purple bacterium:
BCHL *b*

Green bacterium:
BCHL *e* and *a*

Green bacterium:
BCHL *c* and *a*

chlorophylls   phycobiliproteins   caretenoids

TABLE 17.6

The positions of the long wavelength peaks of the chlorophylls
of procaryotes in organic solvents (ether or acetone)
and in the intact cells

	POSITION OF PEAKS (nm) IN		
BIOLOGICAL GROUP	ORGANIC SOLVENTS	CELLS	MAGNITUDE OF *in vivo* SHIFT (nm)
Blue-green bacteria			
chl a	662	680–685	18–23
Purple bacteria			
bchl a	773	850–910	78–137
bchl b	795	1,020–1,035	225–240
Green bacteria			
bchl c	660	750–755	90–95
bchl d	654	725–735	71–79
bchl e	647	715–725	68–78
bchl a[a]	773	805–810	32–37

[a] This bchl, the only one common to all green bacteria, is always a minor pigment in this group, and is represented in the cellular absorption spectrum by a peak that is very small relative to that of the major, group-specific bchl (c, d, or e).

Broadly speaking, the cellular absorption spectra of photosynthetic eucaryotes resemble those of blue-green bacteria, though only red algae show major peaks in the region between 550 and 630 nm, where phycobiliproteins absorb. The differences in the light-absorbing properties of the various groups of photosynthetic organisms are of profound ecological significance, as will be discussed later (see p. 558).

As previously mentioned, the bchls of purple bacteria have a unique spectral character, with a major absorption band at a much longer wavelength than that of other chlorophylls. However, the light-harvesting bchls of the green bacteria closely resemble chlorophyll a in intrinsic spectral character, their long wavelength absorption bands being located at slightly shorter wavelengths than that of chlorophyll a (see Table 17.6). Hence, the positions of the bchl peaks in the cellular absorption spectra of green bacteria might have been expected to lie close to that of chlorophyll a in the cells of blue-green bacteria. In fact, they occur at considerably longer wavelengths (Figure 17.11). This shift is caused by a modification of the intrinsic spectral properties of chlorophylls *in vivo*, which results from the way they are linked to the proteins of the photosynthetic apparatus. In all phototrophs the chlorophyll peaks *in vivo* occur at longer wavelengths than the peaks of the extracted pigments; however, the magnitude of this *in vivo* wavelength shift is not constant, and it is far greater in purple and green bacteria than in blue-green bacteria and eucaryotes (Table 17.6). Thus, the intrinsic spectral properties of the bacteriochlorophylls only in part account for the ability of purple and green bacteria to perform photosynthesis with light of very long wavelengths; this is also determined to a considerable extent by the nature of the chlorophyll-protein complexes present in the photosynthetic apparatus.

• **The colors of photosynthetic procaryotes**

The common names of the three groups of photosynthetic procaryotes are not always well correlated with the color of their cells, as judged visually. Since the major chlorophyll absorption bands of purple bacteria lie in the infrared, to which the eye is blind, the visible color of these organisms is determined largely by their carotenoid complement. The green bacteria appear yellow-green, orange, or brown, again depending on their carotenoid composition.

The phycobiliproteins of blue-green bacteria contribute largely to light absorption in the visible region, and the visible color of the cells is therefore much influenced by the phycobiliprotein complement. If phycoerythrin is absent, the cells appear blue-green; if it is present, they may appear red, violet, brown, or almost black.

THE BLUE-GREEN BACTERIA

The structural diversity of blue-green bacteria (see p. 137) is paralleled by a considerable genetic diversity, as revealed by their mean DNA base composition (Figure 17.12). The total span is almost as wide as that for procaryotes as a whole, and several distinct compositional groups occur among the unicellular representatives.

With respect to nutritional and metabolic properties, however, the group is relatively uniform. All are photoautotrophs; and growth factors are rarely required, though some marine blue-green bacteria have a specific requirement for vitamin B_{12}. Assimilation of CO_2 occurs through the Calvin cycle, with the formation and deposition of glycogen as an intracellular reserve material.

Many blue-green bacteria are *obligate phototrophs*, being wholly incapable of dark growth at the expense of organic sources of carbon and energy. In members of the group that can grow in the dark, the growth rate is very low relative to that in the light; it occurs only at the expense of glucose and a few other sugars which are dissimilated by aerobic respiration. The limited range of utilizable organic carbon and energy sources reflects the universal absence of a functional TCA cycle; blue-green bacteria lack a key enzyme of this cycle, α-ketoglutarate dehydrogenase, a metabolic peculiarity which they share with many obligate anaerobic chemoautotrophs and methylotrophs (see Chapter 18). Respiratory metabolism occurs exclusively through the oxidative pentose phosphate cycle, the reactions of which are in large part common to those of the Calvin cycle (Figure 17.13). The obligate photoautotrophy characteristic of many blue-green bacteria appears to be caused by the absence of the specific permeases necessary for the uptake of exogenous sugars by the cell, since the enzymatic machinery of the oxidative pentose phosphate cycle is present in all blue-green bacteria. This pathway permits the generation of ATP in the dark, through endogenous respiration of the cellular glycogen store.

• **Nitrogen fixation**

Blue-green bacteria are the only organisms able to perform oxygenic photosynthesis that can also fix nitrogen; many (though not all) are vigorous nitrogen fixers. Such organisms have the simplest known nutritional requirements, since they can grow in the light in a mineral medium, using CO_2 as a carbon source and N_2 as a nitrogen source.

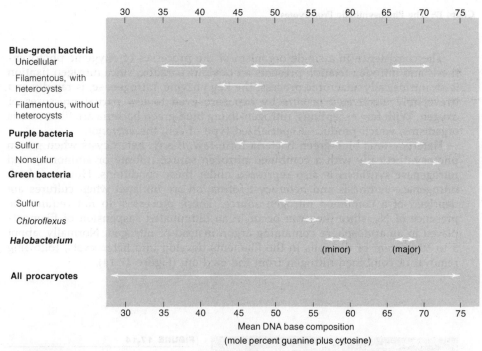

FIGURE 17.12

The ranges of mean DNA base composition among the photosynthetic procaryotes and *Halobacterium*. The range among unicellular blue-green bacteria is particularly wide; they include several subgroups of markedly different DNA base composition. Among purple sulfur bacteria two different base compositional subgroups are also evident. The DNA from *Halobacterium* is heterogeneous, the cells containing a major and a minor DNA component of markedly different compositions.

FIGURE 17.13

A simplified diagram to show the interrelationships between the primary pathways of carbon metabolism in blue-green bacteria. Substrates and end products are enclosed in square boxes. Reactions specific to the Calvin cycle are shown as white arrows; reactions specific to dark respiratory metabolism as heavy black arrows. RUDP: ribulose-1,5-diphosphate; PGA; phosphoglycerate; GAP, glyceraldehyde phosphate; F-6-P, fructose-6-phosphate; G-6-P, glucose-6-phosphate; 6-PG, 6-phosphogluconate; Ru-5-P, ribulose-5-phosphate; G-1-P, glucose-1-phosphate; ADPG, ADP-glucose.

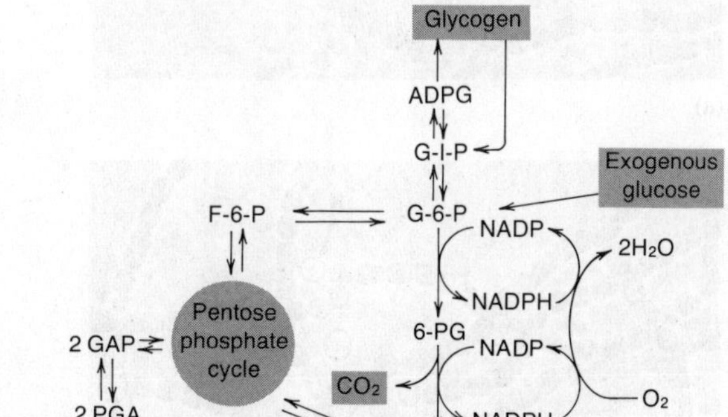

541

The coexistence in a single organism of the processes of oxygenic photosynthesis and nitrogen fixation presents an obvious paradox, since nitrogen fixation is an intrinsically anaerobic process; the key enzyme, nitrogenase, is rapidly and irreversibly inactivated *in vitro* by exposure even to low partial pressures of oxygen. With few exceptions, nitrogen-fixing blue-green bacteria are filamentous organisms, which produce a specialized type of cell, the *heterocyst*.

Heterocystous blue-green bacteria form few, if any, heterocysts when grown photosynthetically with a combined nitrogen source (nitrate or ammonia); and nitrogenase synthesis is also repressed under these conditions. However, both nitrogenase synthesis and heterocyst formation are initiated when cultures are deprived of a combined nitrogen source. These processes do not require the presence of N_2, since they can occur in an illuminated suspension of filaments placed in an atmosphere containing argon instead of nitrogen. Normally, about 5 to 10 percent of the cells in the filaments develop into heterocysts, following removal of combined nitrogen from the medium (Figure 17.14).

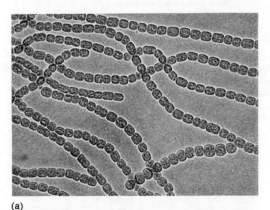

(a)

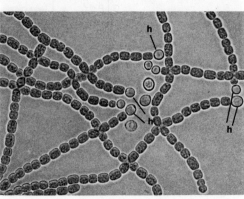

(b)

FIGURE 17.14

The effect of the nitrogen source on heterocyst formation in a blue-green bacterium, *Anabaena* sp. (a) Filaments from a culture grown with ammonia as a nitrogen source; (b) filaments from a culture grown with N_2 as a nitrogen source; h, heterocyst ($\times$616). Courtesy of R. Rippka.

543 The differentiation of a heterocyst from a vegetative cell is accompanied by the synthesis of a new, thick outer wall layer; extensive reorganization of the thylacoids, which become concentrated near the two poles of the cell; and the formation of constricted and specialized polar connections at the points where the heterocyst is attached to adjacent vegetative cells (Figure 17.15). Mature heterocysts no longer contain nuclei, and hence can neither divide nor revert to vegetative cells. They have an almost normal content of chlorophyll a, but are devoid of phycobiliproteins (Figure 17.16). Although heterocysts retain photosystem I activity, they have completely lost photosystem II, and can therefore neither fix CO_2 nor produce O_2 in the light.

The conditions that favor heterocyst formation, coupled with the special properties of these cells, suggest that heterocysts are the specific cellular sites of nitrogen fixation under aerobic conditions in the light. The retention of photosystem I enables the heterocyst to generate ATP by cyclic photophosphorylation. The loss of photosystem II provides an intracellular environment in which O_2 is not produced, and which is thus favorable for the maintenance of nitrogenase function. However, this loss also prevents CO_2 assimilation, so that the heterocyst is dependent on adjacent vegetative cells for a supply of intermediary metabolites which can furnish the reducing power necessary to reduce N_2 to ammonia. The nongrowing heterocysts in turn furnish the vegetative cells of the filament with the fixed nitrogen compounds necessary for their growth. In most heterocystous blue-green bacteria, heterocysts are spaced at regular intervals along the filament. As the filament elongates, by growth and division of the component vegetative cells, new heterocysts are differentiated midway between the preexisting ones.

Although there is strong circumstantial evidence in favor of the hypothesis that nitrogen fixation under aerobic conditions by heterocystous blue-green bacteria occurs exclusively in heterocysts, this hypothesis has not yet been fully substantiated. However, indirect support for it has been obtained from the study

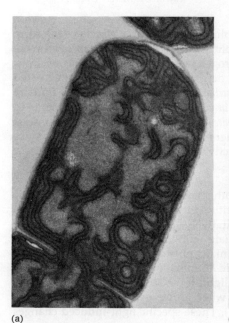

(a)

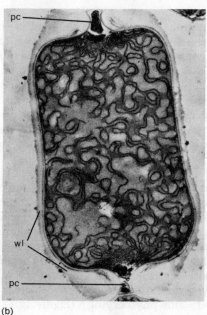

(b)

FIGURE 17.15

Electron micrographs of thin sections of *Anabaena cylindrica*, showing the principal differences in ultrastructure between the vegetative cell (a) and the heterocyst (b). Note the extra wall layers (wl) that surround the heterocyst, and the narrowed and specialized polar connections (pc) to adjacent vegetative cells in the filament ($\times 17,000$). (a) from S. A. Koolasooriya, N. J. Lang, and P. Fay, "The Heterocysts of Blue-Green Algae III. Differentiation and Nitrogenase Activity," *Proc. Roy. Soc. London B* **181**, 199 (1972); (b) from P. Fay and N. J. Lang, "The Heterocysts of Blue-Green Algae I. Ultrastructural Integrity after Isolation," *Proc. Roy. Soc. London B* **178**, 185 (1971).

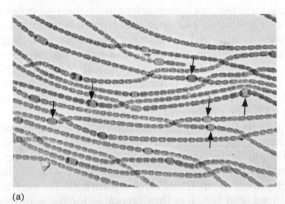

(a)

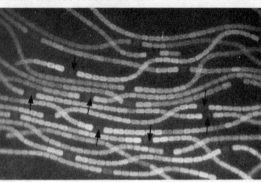

(b)

FIGURE 17.16

Photomicrographs of heterocyst-containing filaments of *Anabaena cylindrica,* to show the distribution of chlorophyll and phycocyanin. (a) Transmission image taken with blue light, preferentially absorbed by chlorophyll. The densities of vegetative cells and heterocysts (arrows) are similar, showing that they do not differ significantly in chlorophyll content. (b) Fluorescence image, taken under conditions which specifically reveal the fluorescence of phycocyanin. Vegetative cells are brilliantly fluorescent, whereas heterocysts (arrows) are barely visible, showing that they contain little (if any) phycocyanin. Courtesy of Dr. Marcel Donze.

of nitrogenase synthesis by filamentous blue-green bacteria which never form heterocysts (*Oscillatoria* and *Plectonema* strains). These organisms can synthesize nitrogenase in the absence of combined nitrogen in the light, *provided that conditions are initially anaerobic;* no nitrogenase can be detected in control cultures incubated in air. The nitrogenase synthesized by *Oscillatoria* and *Plectonema* under initially anaerobic conditions is rapidly destroyed *in vivo* by exposure of the filaments to air. Such blue-green bacteria are clearly *potential* nitrogen fixers; but in contrast to heterocyst formers, they cannot grow aerobically at the expense of N_2, since they are unable to maintain nitrogenase activity under these growth conditions.

• Regulation of
pigment synthesis

Synthesis of photosynthetic pigments and other components of the photosynthetic apparatus by blue-green bacteria is constitutive. Even after many transfers on sugar-containing media in the dark, facultatively heterotrophic strains retain a normal pigment complement and can initiate immediate growth when returned to a mineral medium in the light. However, many phycoerythrin-producing strains exhibit an interesting response to chromatic illumination, known as *complementary chromatic adaptation.* When grown in green light, these strains contain a high ratio of phycocyanin to phycoerythrin, whereas when grown in red light, they synthesize very little phycoerythrin (Figure 17.17). These specific light-induced changes

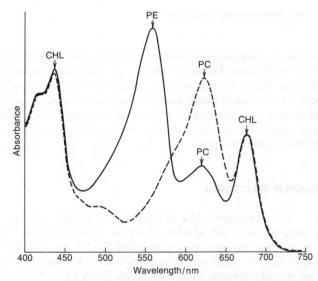

FIGURE 17.17

The effect of chromatic illumination on phycobiliprotein synthesis by a blue-green bacterium, *Pseudoanabaena* sp. Absorption spectra of cells grown in green light (solid line) and in red light (dashed line). The positions of the absorption maxima of chlorophyll *a* (CHL), phycocyanin (PC), and phycoerythrin (PE) are indicated by arrows. Cells grown in red light contain virtually no phycoerythrin; those grown in green light have a high content of phycoerythrin, and a much reduced phycocyanin in content. Courtesy of Mrs. N. Tandeau de Marsac.

of phycobiliprotein content enable the cells to absorb most effectively the wavelengths of light that are available. The mechanism of chromatic adaptation is not known, but there are indications that it is mediated by a regulatory light-sensitive pigment, similar to but not identical with phytochrome, a biliprotein which is an important photoregulator of plant growth and differentiations.

• **Ecology**
The blue-green bacteria occupy a far wider range of habitats than do other photosynthetic procaryotes, occurring in all environments that support the growth of algae: the sea, fresh water, and soil. Furthermore, they develop in certain habitats from which photosynthetic eucaryotes are largely or completely excluded. Nitrogen-fixing representatives are conspicuous in environments where combined nitrogen is a limiting nutrient, notably in tropical soils, which are often nitrogen-poor. Certain thermophilic blue-green bacteria grow abundantly in neutral or alkaline hot springs, where they are the predominant members of the photosynthetic population. The temperature ranges of thermophilic blue-green bacteria vary, but some unicellular forms can grow at temperatures in excess of 70°C. They are excluded by their relatively high pH range from acid hot springs, of which the characteristic photosynthetic inhabitant is a red alga, *Cyanidium caldarum*, which has a low pH optimum and is the only truly thermophilic photosynthetic eucaryote. However, its temperature maximum (approximately 56°C) is considerably below that of many thermophilic blue-green bacteria.

Deserts are an extreme environment in which the microbial photosynthetic population consists almost entirely of unicellular blue-green bacteria. These organisms grow in microfissures just below the surface of rocks, where small amounts of moisture are trapped and where sufficient light penetrates to permit photosynthesis. The ability to tolerate extreme fluctuations of temperature, which often become very high during the day, is important to their survival in the desert habitat.

In lakes that have undergone eutrophication (artificial enrichment with mineral nutrients, notably phosphate and nitrate), a massive development of unicellular and filamentous blue-green bacteria characteristically occurs during the warmer months of the year. These are largely gas vacuolate forms; in calm weather, the population floats to the surface, accumulating there to produce a so-called "bloom." Subsequent death and decomposition of the bloom promotes a massive development of chemoheterotrophic microorganisms, which may have catastrophic effects on the animal population of the lake because they deplete the dissolved oxygen supply.

THE PURPLE BACTERIA

Taxonomically speaking, the purple bacteria are a small group, consisting of about 30 species. They are unicellular and reproduce by binary fission or, in a few species, by budding. Most are motile by flagella; a few are immotile. Gas vacuoles are formed by some. Despite the small size of this group, it is genetically diverse, since the mean DNA base composition ranges from 46 to 73 moles percent GC (Figure 17.12).

All purple bacteria are, at least potentially, photoautotrophs, capable of growing anaerobically in the light with CO_2 as the principal carbon source and reduced inorganic compounds as the electron donor. Under these conditions, the Calvin cycle is the principal pathway of carbon assimilation. However, the purple bacteria can also develop photoheterotrophically under anaerobic conditions in the light at the expense of organic compounds, of which acetate is the most widely utilized. Under these circumstances, cell material is derived largely from the organic substrate, although CO_2 may also be assimilated.

Concomitant CO_2 uptake becomes important if the organic substrate is more reduced than cell material, since the reductive assimilation of CO_2 provides a means of absorbing excess electrons derived from the organic substrate, and thus maintaining redox balance. This point may be illustrated by considering the paths of photometabolism by purple bacteria of two fatty acids, acetate and butyrate. Both compounds are rapidly assimilated by many purple bacteria under anaerobic conditions in the light, and they are initially converted in large part to a reserve material formed by all purple bacteria, poly-β-hydroxybutyric acid, through the sequences of reactions shown in Figure 17.18. The photosynthetic assimilation of acetate occurs in the absence of CO_2, but butyrate assimilation requires the presence of CO_2.

The conversion of acetate to poly-β-hydroxybutyrate is a reductive process:

$$2n\text{CH}_3\text{COOH} + 2n\text{H} \longrightarrow (\text{C}_4\text{H}_6\text{O}_2)_n + 2n\text{H}_2\text{O}$$

All purple bacteria possess the machinery of the TCA cycle and can thus generate reductant by the anaerobic oxidation of acetate through this cycle:

$$\text{acetate} + 2\text{H}_2\text{O} \longrightarrow 2\text{CO}_2 + 8\text{H}$$

This permits a reductive conversion of acetate to poly-β-hydroxybutyrate, the balanced equation for the overall reaction being:

$$9n\text{CH}_3\text{COOH} \longrightarrow 4(\text{C}_4\text{H}_6\text{O}_2)_n + 2n\text{CO}_2 + 6n\text{H}_2\text{O}$$

As this equation shows, carbon assimilation is remarkably efficient, some 90 percent of the organic substrate being converted to cellular reserve material. This

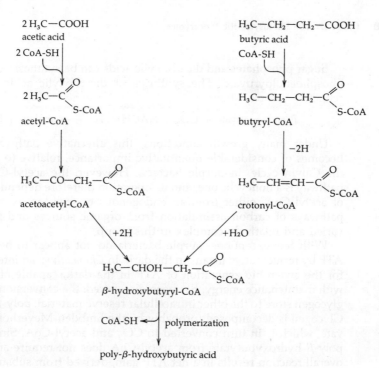

FIGURE 17.18

The conversions of acetic and butyric acids to poly-β-hydroxybutyric acid by nonsulfur bacteria.

high efficiency is possible only because the photochemical reactions (cyclic photophosphorylation) can furnish potentially unlimited quantities of ATP, necessary for the initial activation of acetate (i.e., formation of acetyl-CoA).

The formation of poly-β-hydroxybutyrate from butyrate is an oxidative process:

$$nCH_3CH_2CH_2COOH \longrightarrow (C_4H_6O_2)_n + 2nH$$

It can therefore proceed anaerobically *only if a hydrogen acceptor is available*. The role of acceptor is played by CO_2, which is assimilated via the Calvin cycle and converted to glycogen, the other storage material formed by all purple bacteria. Symbolizing glycogen as $(CH_2O)_n$, the coupled photoassimilation can be represented as:

$$2nC_4H_8O_2 + nCO_2 \longrightarrow 2(C_4H_6O_2)_n + (CH_2O)_n + nH_2O$$
$$\text{butyrate} \qquad\qquad \text{poly-}\beta\text{-} \qquad \text{glycogen}$$
$$\text{hydroxybutyrate}$$

The anaerobic photoassimilation of butyrate is thus obligatorily coupled with CO_2 assimilation, both processes being driven by ATP derived from cyclic photophosphorylation; and operation of the TCA cycle, essential for the anaerobic assimilation of acetate, plays no role in the process.

The synthesis of poly-β-hydroxybutyrate does not, in itself, represent a *de novo* synthesis of cell material. In order for this reserve product to be used as a general source of cell constituents, the constituent acetyl units must be converted to pyruvate. Like many anaerobic chemoheterotrophic bacteria, the purple bacteria can synthesize pyruvate from acetyl units by the ferredoxin (FD)-mediated reaction:

$$acetyl\text{-}CoA + CO_2 + FDH_2 \longrightarrow CH_3COCOOH + CoA + FD$$

Sugar phosphates and dicarboxylic acids can be synthesized from pyruvate via phosphoenolpyruvate. The synthesis of dicarboxylic acids involves a second reductive carboxylation:

$$\text{P-enol pyruvate} + CO_2 + NADH + H^+ \longrightarrow \text{malate} + NAD^+ + \textcircled{P}$$

Under many growth conditions, this alternative pathway of CO_2 fixation becomes of considerable quantitative importance, relative to fixation of CO_2 via the Calvin cycle, in purple bacteria. However, the acetyl-CoA-malate fixation pathway is a noncyclic one, and its operation therefore depends on the availability of acetyl-CoA, either from an endogenous or from an exogenous source. The pathways of carbon assimilation from organic sources and from CO_2 are thus varied and relatively complex in this group.

With few exceptions, purple bacteria do not appear to be able to synthesize ATP by fermentative means in the dark. In *Chromatium,* an interesting mechanism for the anaerobic generation of ATP in the dark, capable of providing the cell with maintenance energy, has been discovered: the conversion of the intracellular glycogen store to the other intracellular reserve material, poly-β-hydroxybutyrate. Glycogen is decomposed (probably by the Embden-Meyerhof pathway) to pyruvate, which is in turn converted to CO_2 and acetyl-CoA. Since the synthesis of poly-β-hydroxybutyrate from acetyl-CoA does not require an input of ATP, the overall reaction results in a net ATP gain, derived from substrate-level phosphorylation during the conversion of glycogen to pyruvate. The overall reaction can be represented as:

$$(C_6H_{10}O_5)_n + nH_2O \longrightarrow (C_4H_6O_2)_n + 2nCO_2 + 6nH$$

It depends on the availability of a suitable hydrogen acceptor; in *Chromatium* this role is assumed by the intracellular deposits of elemental sulfur, which are reduced to H_2S:

$$3nS + 6nH \longrightarrow 3nH_2S$$

The formation of poly-β-hydroxybutyrate from glycogen is thus not a fermentation, but an *endogenous anaerobic respiration,* using S as an electron acceptor.

• **Physiological subdivisions among the purple bacteria**

It is customary to recognize two subgroups among the purple bacteria; the distinctions between them are both physiological and ecological (Table 17.7). *Purple sulfur bacteria* have a predominantly photoautotrophic mode of metabolism, based on the use of H_2S as an electron donor, and are strict anaerobes. *Purple nonsulfur bacteria* have a predominantly photoheterotrophic mode of metabolism. They are sensitive to H_2S, their growth being inhibited by low concentrations of sulfide, even though some can oxidize sulfide anaerobically in the light if the concentration is kept very low.

Whereas the purple sulfur bacteria are obligate phototrophs, many purple nonsulfur bacteria can grow well aerobically in the dark. Such strains possess an aerobic electron transport chain, and are thus endowed with respiratory capacity. A few of them can also grow (though very slowly) anaerobically in the dark, through the fermentation of pyruvate or sugars.

The purple nonsulfur bacteria typically occur in fresh water lakes or ponds,

TABLE 17.7

Characters that distinguish the two subgroups of purple bacteria

	PURPLE SULFUR BACTERIA	PURPLE NONSULFUR BACTERIA
Principal mode of photosynthesis	Photoautotrophic	Photoheterotrophic
Range of photoassimilable organic substrates	Narrow	Broad
Aerobic dark growth	−	+ or −
Ability to oxidize H_2S	+	+ or −
Accumulation of S^0 as intermediate in H_2S oxidation SO_4^{2-}	+	−
H_2S toxicity	Usually low	Usually high
Use of SO_4^{2-} as sulfur source	+ or −	+
Growth factors required	+ or −	+ or −
Nature of growth factors, if required	Vitamin B_{12}	Thiamine and/or biotin and/or niacin

where organic matter is present but sulfide is either absent or present at low concentrations. The typical habitats of the purple sulfur bacteria are sulfide-rich waters, where sulfide is generated by the activity of sulfate-reducing bacteria. The borderline between these two subgroups is somewhat hazy, since all purple nonsulfur bacteria can grow photoautotrophically with H_2, and sometimes also with reduced inorganic sulfur compounds.

• **Purple sulfur bacteria** The characters that distinguish genera of purple sulfur bacteria are shown in Table 17.8, and some typical representatives are illustrated in Figure 17.19. The characteristic photometabolism of these organisms involves assimilation of CO_2, largely

TABLE 17.8

The genera of purple sulfur bacteria[a]

GENUS	CELL SHAPE AND ARRANGEMENT	MOTILITY	GAS VACUOLES	SITE OF SULFUR DEPOSITION
Thiospirillum	Single, helical	+	−	Intracellular
Ectothiorhodospira	Single, vibroid	+	−	Extracellular
Chromatium	Single, cylindrical	+	−	Intracellular
Thiocystis	Single, spherical	+	−	Intracellular
Thiosarcina	Cubical packets, spherical	+	−	Intracellular
Thiocapsa	Single, spherical	−	−	Intracellular
Lamprocystis	Single, spherical	+	+	Intracellular
Thiodictyon	Single, cylindrical	−	+	Intracellular
Thiopedia	Flat rectangular plates, spherical	−	+	Intracellular
Amoebobacter	Single, spherical	−	+	Intracellular

[a] Characters common to all: division by binary fission; if motile, by polar flagella.

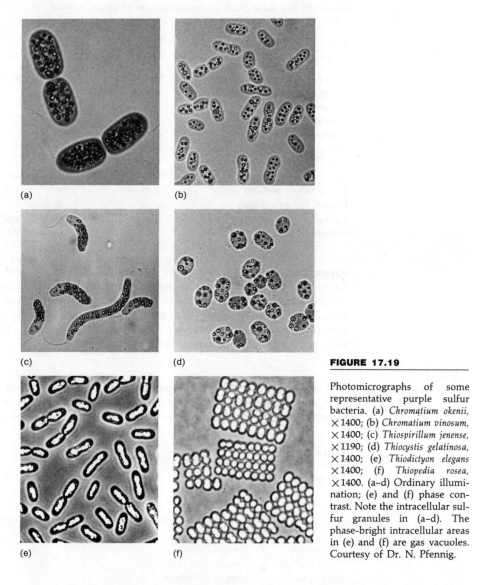

FIGURE 17.19

Photomicrographs of some representative purple sulfur bacteria. (a) *Chromatium okenii,* ×1400; (b) *Chromatium vinosum,* ×1400; (c) *Thiospirillum jenense,* ×1190; (d) *Thiocystis gelatinosa,* ×1400; (e) *Thiodictyon elegans* ×1400; (f) *Thiopedia rosea,* ×1400. (a–d) Ordinary illumination; (e) and (f) phase contrast. Note the intracellular sulfur granules in (a–d). The phase-bright intracellular areas in (e) and (f) are gas vacuoles. Courtesy of Dr. N. Pfennig.

through the Calvin cycle, ATP being provided by cyclic photophosphorylation; reducing power is provided by H_2S which is oxidized anaerobically, via elemental sulfur, to sulfate. The overall reaction can be represented schematically as:

$$2CO_2 + H_2S + 2H_2O \longrightarrow 2(CH_2O) + H_2SO_4$$

Some (but not all) purple sulfur bacteria can use other reduced inorganic sulfur compounds (S, thiosulfate, sulfite) in place of H_2S as exogenous reductants. The biochemistry of the oxidation of these reduced sulfur compounds by purple sulfur bacteria is complex and not well established. It probably involves enzymatic mechanisms similar to those operative in the oxidation of reduced inorganic sulfur compounds by aerobic chemoautotrophs. The terminal step, oxidation of SO_3^{2-} to SO_4^{2-}, has been well characterized in enzymatic terms. Both in thiobacilli and in purple sulfur bacteria it involves the formation of an adenylic acid derivative, adenylphosphosulfate (APS):

$$SO_3^{2-} + AMP \rightleftharpoons APS + 2e$$

This reaction is mediated by an enzyme of complex structure (adenylyl sulfate reductase); which contains FAD as well as several nonheme iron groups. In purple sulfur bacteria this enzyme also bears two heme groups, absent from the *Thiobacillus* enzyme.

The oxidation of H_2S by the purple sulfur bacteria always leads to a massive but transient accumulation of elemental sulfur, since this first step is much more rapid than the ensuing oxidation of S^0 to SO_4^{2-}. In most of these organisms the elemental sulfur is deposited within the cell, as refractile globules. However, *Ectothiorhodospira* spp. excrete sulfur into the medium, and subsequently reabsorb it prior to further oxidation. In addition to reduced inorganic sulfur compounds, nearly all species can use H_2 as a reductant for CO_2 assimilation.

However, the photometabolism of purple sulfur bacteria is never obligatorily photoautotrophic, since all these organisms can photoassimilate some organic compounds, acetate being a universal substrate. Some of them require a small amount of H_2S for photoheterotrophic growth, using it as a cellular sulfur source, since they are unable to perform an assimilatory sulfate reduction. The only organic growth factor required is vitamin B_{12}, an essential nutrient for a few species.

• Purple nonsulfur bacteria

The distinguishing properties of the three genera of purple nonsulfur bacteria are listed in Table 17.9, and photomicrographs of some typical representatives are shown in Figure 17.20. The only purple bacteria which reproduce by budding rather than by binary fission are members of this subgroup; they include *Rhodomicrobium* and some *Rhodopseudomonas* spp.

The range of organic compounds which can be photoassimilated by purple nonsulfur bacteria is quite wide: it includes fatty acids, other organic acids, primary and secondary alcohols, carbohydrates, and even aromatic compounds. Species capable of respiratory metabolism can grow aerobically in the dark, by the oxidation of the same range of organic substrates that they photoassimilate

TABLE 17.9

The genera of purple nonsulfur bacteria[a]

GENUS	CELL FORM	FLAGELLAR INSERTION	MODE OF CELL DIVISION	PROSTHECAE
Rhodospirillum	Helical	Polar	Binary fission	−
Rhodopseudomonas	Cylindrical or ovoid	Polar	Binary fission in some spp., Budding in some spp.	−
Rhodomicrobium	Ovoid	Peritrichous	Budding	+

[a] Common characters: always motile; never produce gas vacuoles; never accumulate sulfur within the cell.

anaerobically in the light. This does not necessarily involve the operation of the same metabolic pathways, however. As previously discussed, the photoassimilation of butyrate is obligatorily coupled with CO_2 assimilation and does not involve the operation of the TCA cycle. When it serves as a substrate for dark respiratory metabolism, much of the substrate is oxidized to acetyl-CoA, which is then converted to CO_2 through the TCA cycle, in order to provide ATP by oxidative phosphorylation.

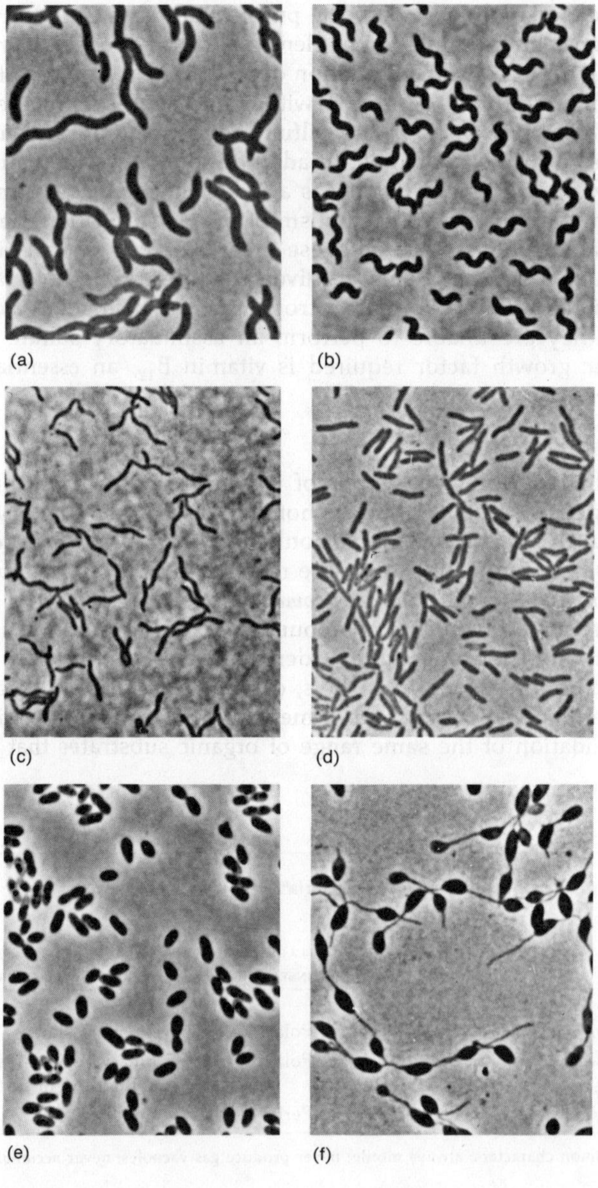

(a)

(b)

(c)

(d)

(e)

(f)

FIGURE 17.20

Photomicrographs of some representative purple nonsulfur bacteria, all ×1400. (a) *Rhodospirillum rubrum;* (b) *Rhodospirillum fulvum;* (c) *Rhodospirillum tenue;* (d) *Rhodopseudomonas gelatinosa;* (e) *Rhodopseudomonas sphaeroides* (f) *Rhodomicrobium vannielii.* (a–d) ordinary illumination; (e) and (f) phase contrast. Courtesy of Dr. N. Pfennig.

The rule that photoassimilable organic substrates can also be respired by purple bacteria has one interesting exception. Some of these organisms can photoassimilate benzoate anaerobically in the light, but are completely unable to use it as a respiratory substrate. The photometabolism of benzoate occurs through a unique *reductive* pathway, the initial steps of which convert benzoate to a saturated dicarboxylic acid, pimelate:

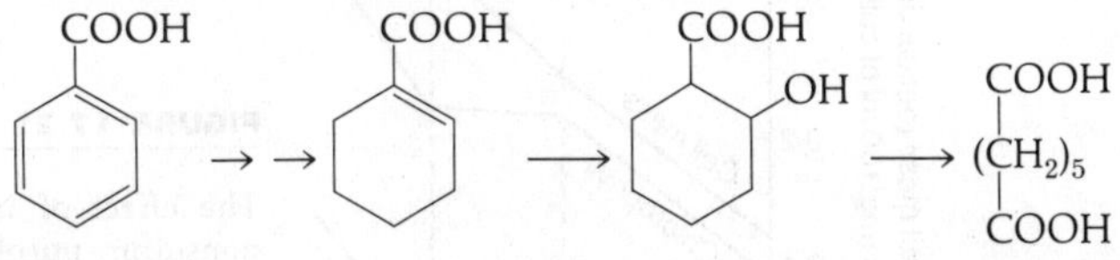

The enzymes that catalyze these reactions are exceedingly oxygen-sensitive, the photoassimilation of benzoate being immediately arrested if cells are exposed even to traces of O_2. The absence from purple bacteria of the enzymes that catalyze an oxygenative pathway of benzoate dissimilation, characteristic of benzoate-utilizing aerobic chemoheterotrophs, accounts for their inability to use this compound as a respiratory substrate.

As previously mentioned, purple nonsulfur bacteria are not necessarily incapable of photoautotrophic growth with reduced inorganic sulfur compounds. One species, *Rhodopseudomonas palustris,* has long been known to utilize thiosulfate; and several species have recently been found to be capable of oxidizing H_2S, provided that its concentration is kept low. However, their sulfide metabolism is not like that of purple sulfur bacteria. Some species oxidize H_2S only to elemental sulfur, which is excreted into the medium; others oxidize it to sulfate, but without the intermediate accumulation of S^0.

Most of the purple nonsulfur bacteria require vitamins, and their growth rate is frequently improved by the provision of amino acids. Various combinations of biotin, thiamin, and niacin are the typical vitamin requirements; a requirement for vitamin B_{12}, characteristic of some purple sulfur bacteria, does not occur.

• **Effects of O_2 on growth and pigment synthesis in purple nonsulfur bacteria**

None of the purple nonsulfur bacteria is killed by exposure to air; however, some of these organisms cannot use O_2 as a terminal electron acceptor, and therefore they cannot grow aerobically in the dark. Others grow at least as rapidly under aerobic conditions in the dark as they do under anaerobic conditions in the light. However, aerobic growth leads rather rapidly to an almost complete loss of the photosynthetic pigment system. This is a consequence of the fact that, even at relatively low partial pressures, O_2 is a potent repressor of pigment synthesis by purple bacteria, exerting this effect even in the presence of light. Light itself is not required for pigment synthesis, as shown by the fact that species possessing a fermentative metabolism maintain a high pigment content through many generations of anaerobic growth in the dark.

The aerobic growth of purple bacteria consequently leads, either in the dark or in the light, to a progressive dilution of the cellular pigment content. This is a purely physiological phenomenon, immediately reversed when cells are returned to anaerobic growth conditions (Figure 17.21). Consequently, the photosynthetic development of all purple bacteria, anaerobes and facultative aerobes alike, is possible only in an O_2-free environment. Under anaerobic conditions in the light, both the growth rate and the differential rate of bacteriochlorophyll synthesis are governed by light intensity. As light intensity is increased, the growth rate increases and the cellular bacteriochlorophyll content declines (Figure 17.22).

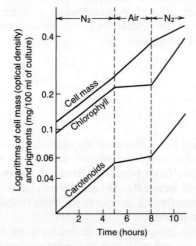

FIGURE 17.21

The effect of introducing oxygen into a culture of the nonsulfur purple bacterium, *Rhodopseudomonas spheroides*, growing exponentially under anaerobic conditions (N_2 atmosphere) in the light. Note that the change from N_2 to air does not diminish the growth rate but almost instantly abolishes chlorophyll and carotenoid synthesis. When anaerobic conditions are reestablished about 3 hours later, photosynthetic pigment synthesis resumes at a rapid rate. After G. Cohen-Bazire, W. R. Sistrom, and R. Y. Stanier, "Kinetic Studies of Pigment Synthesis by Non-Sulfur Purple Bacteria," *J. Cellular Comp. Physiol.* **49**, 25 (1957).

• **Nitrogen fixation and hydrogen production**

Many of the purple bacteria have been shown to fix nitrogen. However, effective nitrogenase synthesis occurs only under anaerobic conditions, probably because the enzyme is rapidly inactivated in the cells by exposure to oxygen. Purple bacteria in which nitrogenase activity has been induced will produce H_2 vigorously, if furnished with a suitable organic or inorganic electron donor and placed in the light in a N_2-free atmosphere (e.g., in helium or argon). This curious reaction is mediated by nitrogenase, an enzyme of relatively low substrate specificity which can reduce hydrogen ions:

$$2H^+ + 2e \xrightarrow{\text{(ATP)}} H_2$$

The reaction is strictly light-dependent, since photophosphorylation must provide the substantial quantities of ATP required for nitrogenase activity. Under these

FIGURE 17.22

The effect of light intensity on the growth rate and specific cellular content of bacteriochlorophyll of *Rhodospirillum rubrum*, growing photoheterotrophically in the absence of O_2.

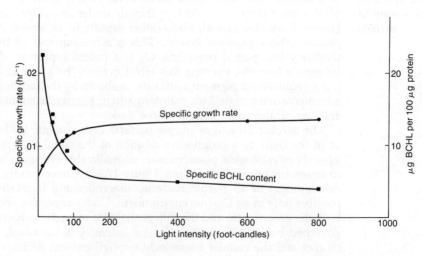

special conditions, purple bacteria can effect a complete anaerobic oxidation of organic substrates, converting them to CO_2 and H_2, as exemplified by the oxidation of acetate:

$$CH_3COOH + 2H_2O \longrightarrow 2CO_2 + 4H_2$$

Photohydrogen production is specifically inhibited by N_2; the inhibition is competitive, since this gas is the normal substrate for nitrogenase.

THE GREEN BACTERIA

The green bacteria comprise an even smaller taxonomic group than the purple bacteria; there are only nine recognized species placed in five genera (Table 17.10). Photomicrographs of typical representatives are shown in Figure 17.23. The span of DNA base composition is comparatively narrow: 48 to 58 moles percent GC (Figure 17.12). With respect to its physiological and nutritional properties, the group shows interesting parallelisms to the purple bacteria. Most members—the green sulfur bacteria—are counterparts of the purple sulfur bacteria. However, the recently discovered thermophilic green bacterium, *Chloroflexus*, closely resembles the purple nonsulfur bacteria in its metabolic and nutritional properties.

• The green sulfur bacteria   These organisms are small, permanently immotile, rod-shaped bacteria; four genera are recognized on the basis of structural characters. The members of this subgroup are strictly anaerobic photoautotrophs, which use H_2S, other reduced inorganic sulfur compounds, or H_2 as electron donors. Elemental sulfur arising from H_2S oxidation is deposited extracellularly (as in *Ectothiorhodospira*), prior to oxidation to sulfate. Since green sulfur bacteria cannot use sulfate as a sulfur source, they require sulfide to meet biosynthetic needs when growing with H_2 as an electron donor; some also require vitamin B_{12}. Nitrogen fixation is of common occurrence. In all these respects, the analogies to purple sulfur bacteria

TABLE 17.10

Genera of green bacteria

	CELL FORM AND ARRANGEMENT	GLIDING MOTILITY	GAS VACUOLES	PROSTHECAE
Green sulfur bacteria				
Chlorobium	Straight or curved rods, single or short chains	−	−	−
Prosthecochloris	Ovoid, single, or short chains	−	−	+
Pelodictyon	Chains of rods, forming nets	−	+	−
Clathrochloris	Chains of rods in loose, trellislike aggregates	−	+	−
Green nonsulfur bacteria				
Chloroflexus	Long filaments composed of rods	+	−	−

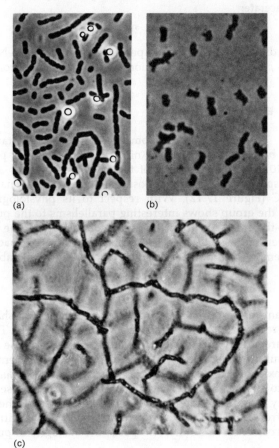

(a)

(b)

(c)

FIGURE 17.23

Photomicrographs (phase contrast) of green sulfur bacteria. (a) *Chlorobium limicola*, ×1500; note extracellular sulfur granules. (b) *Prosthecochloris aestuarii*, ×2300; the prosthecae are just detectable by light microscopy, conferring an irregular outline on the profile of the cells. (c) *Pelodictyon clathratiforme*, ×1500, showing the characteristic net formation; the phase-bright areas in some of the cells are gas vacuoles. Courtesy of Dr. N. Pfennig.

are evident. Indeed, purple and green sulfur bacteria commonly coexist in illuminated, sulfide-rich anaerobic aquatic environments and have essentially overlapping natural distributions.

There is a marked difference, however, with respect to carbon nutrition in the two groups. None of the green sulfur bacteria can grow photoheterotrophically, using organic compounds as their sole or principal carbon source in the absence of an inorganic reductant. They can photoassimilate acetate, but only if H_2S and CO_2 are simultaneously provided. These organisms do not synthesize poly-β-hydroxybutyrate as a reserve material, and acetate is assimilated exclusively through the reductive synthesis of pyruvate, from acetyl-CoA and CO_2, thus serving directly as a precursor of cell material.

There has been a good deal of uncertainty about the major route of CO_2 assimilation in the green sulfur bacteria. It has been suggested that the primary reaction is always a reductive synthesis of pyruvate, followed by other CO_2-fixation reactions which in turn permit a regeneration of acetyl-CoA, thus constituting a cyclic fixation pathway. However, green sulfur bacteria contain high levels of carboxydismutase, and there is now increasing evidence that—at least in the absence of exogenous acetate—CO_2 is very largely assimilated through the Calvin cycle. Accordingly, the green sulfur bacteria probably do not constitute an exception to the general rule that the Calvin cycle is the principal route of CO_2 fixation in all autotrophs, photosynthetic and chemosynthetic.

• Green
nonsulfur bacteria:
the *Chloroflexus*
group

In 1971 B. Pierson and K. Castenholz discovered a new category of green bacteria, the *Chloroflexus* group. Although these organisms differ from green sulfur bacteria in their structure, nutrition, metabolism, and ecology, they possess two properties that clearly identify them as green bacteria: the presence in the cells of chlorobium vesicles and of bacteriochlorophylls c and a as the major and minor chloroplyllous pigments, respectively.

The *Chloroflexus* group consists of filamentous, gliding organisms, the filaments sometimes attaining a length of 300 μm. They are thermophils and develop abundantly in neutral or alkaline hot springs at temperatures in the range of 45°C to 70°C. The masses of intertwined filaments form orange to dull green mats several millimeters thick, often closely associated with unicellular thermophilic blue-green bacteria of the *Synechococcus* type. Their natural habitat is often partly aerobic; since the synthesis of bacteriochlorophylls by *Chloroflexus* is repressed by oxygen, the bacteriochlorophyll content of the mats is often low, being largely masked by the orange carotenoids which are abundantly formed by *Chloroflexus* under all growth conditions, and by the chlorophyll a in the accompanying cells of *Synechococcus*. Hence, the filaments of which the mats are composed had long been interpreted as gliding, nonphotosynthetic bacteria. However, the isolation of pure cultures of *Chloroflexus* has permitted a conclusive demonstration of their ability to perform anoxygenic photosynthesis. Growth is most rapid in complex media, incubated anaerobically in the light; and under these conditions the filaments have a high content of bacteriochlorophylls c and a. No growth occurs anaerobically in the dark. Growth is good in complex media incubated aerobically either in the light or in the dark, though under these conditions the bacterio-chlorophyll content of the cells is very low. The nutritional requirements of *Chloroflexus* are complex and not yet precisely determined.

Chloroflexus is evidently a typical photoheterotroph and facultative chemoheterotroph. However, it grows in hot springs which have a low content of organic matter. The organism appears to derive organic nutrients from the blue-green bacteria with which it is naturally associated. The two organisms can be successfully maintained in the laboratory as two-membered cultures, grown in the light on a mineral medium.

In its mode of movement and filamentous structure, *Chloroflexus* resembles some filamentous, nonheterocyst-forming blue-green bacteria; and its carotenoids include β-carotene, which is similarly a major carotenoid in blue-green bacteria. However, its other chemical properties, its cellular fine structure, and its anoxygenic mode of photosynthesis show that its affinities lie rather with the green sulfur bacteria. Since the latter organisms are all immotile, their possible relations to motile groups of procaryotes were previously obscure. However, the gliding motility characteristic of *Chloroflexus* suggests that the green bacteria as a whole should probably be placed among the gliding procaryotes. In this respect, they differ from the purple bacteria which, if motile, bear flagella.

• Ecological
restrictions imposed
by anoxygenic
photosynthesis

For the performance of photosynthesis, anoxygenic phototrophs require anaerobic conditions and either organic compounds or reduced inorganic compounds other than water. These limitations do not apply to blue-green bacteria and photosynthetic eucaryotes. The purple and green bacteria are confined to a limited range of special habitats, and their quantitative contribution to photosynthetic productivity in the biosphere is negligible. They are exclusively aquatic and grow in bodies of water which provide the indispensable combination of anaerobiosis, light and the nutrients specific for these organisms. These conditions occur principally in two types of aquatic environments, similar in chemical respects but differing markedly in the quality of light available. One consists of shallow ponds, relatively rich in organic matter, CO_2, H_2, and often H_2S, produced by anaerobic bacteria in the underlying sediment. Except near the air-water interface, occupied by blue-green bacteria and algae, the water is essentially oxygen-free. Hence, purple and green bacteria can grow close to the water surface, where light intensity is high, but are usually covered by a growth of oxygenic phototrophs. It is in this environment that the ability of purple green bacteria to absorb light of very long wavelengths, transmitted by overlying oxygenic phototrophs, becomes of critical importance for their survival. The light used for photosynthesis is almost entirely absorbed by bacteriochlorophylls, in the far red and infrared regions.

The second environment in which purple and green bacteria abound occurs at a considerable depth in lakes, particularly so-called *meromictic* lakes which are characterized by a permanent stratification of the water. The warmer, aerobic upper layer is underlain at depths of 10 to 30 meters by a stagnant layer that is cold and oxygen-free. The anoxygenic phototrophs occur in a narrow horizontal band, situated just within the anaerobic layer (Figure 17.24). Water samples from this depth are often brightly colored as a result of their content of purple and green bacteria, the population density being far greater than that of oxygenic phototrophs in the upper, aerobic layers. However, at the depth where the purple and green bacteria find the anaerobic conditions necessary for development, *the overlying water column itself becomes an effective light filter*, transmitting only green and blue-green light, of wavelengths between 450 and 550 nm. The role of light-harvesting pigments is largely assumed by carotenoids, not by bacteriochlorophylls; and anoxygenic phototrophs from this environment typically have a very high carotenoid content.

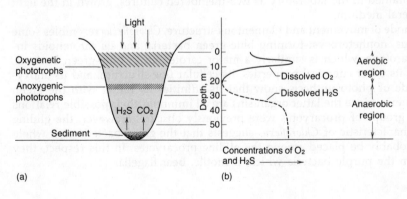

FIGURE 17.24

Diagram of the structure of a meromictic lake, showing (a) the vertical distributions of oxygenetic phototrophs and anoxygenetic phototrophs and (b) the relative concentrations in the water profile of dissolved oxygen and H_2S.

THE EVOLUTION OF PHOTOSYNTHESIS

During much of its early history the surface of the earth remained oxygen-free; the earliest forms of life of necessity obtained energy by anaerobic metabolic processes. It is now believed that the emergence of living organisms occurred at the expense of accumulated organic matter, synthesized in a prebiotic phase of planetary evolution; this organic matter also furnished an energy supply for the first forms of life. However, biological evolution would have come to an early end, had not some members of the community acquired the ability to use light as an energy source by a process analogous to—but no doubt much simpler than—contemporary anoxygenic photosynthesis. Until about 3 billion years ago, anoxygenic photosynthesis remained the dominant form of phototrophic metabolism. About that time, however, molecular oxygen appeared in the biosphere, almost certainly as a consequence of the development, in some members of the primitive photosynthetic community, of oxygenic photosynthesis. This crucial event required a profound modification of the photosynthetic apparatus: the evolution of a second type of photosynthetic reaction center, capable of oxidizing water, which permitted the use of this universally available inorganic compound as a source of reducing power. As the oxygen concentration of the biosphere increased, phototrophs which possessed the older, anoxygenic machinery of photosynthesis were progressively displaced, surviving in anaerobic niches largely as a result of their special light-harvesting pigment systems, which enabled them to avoid direct competition with oxygenic phototrophs for light. Two such isolated groups, the purple and green bacteria, have managed to persist until the present time.

The properties of the contemporary blue-green bacteria provide conclusive evidence that oxygenic photosynthesis emerged within the framework of the procaryotic cell. Indeed, fossil microorganisms identifiable as blue-green bacteria have been discovered in the oldest known fossil-bearing sedimentary formations (flints), which date from an era not very long after the change from an anaerobic to an aerobic biosphere (~1.5 billion years).

At a later stage of biological evolution, the capacity for the performance of oxygenic photosynthesis became implanted in certain lines of eucaryotes. This probably occurred initally by the introduction into the host cell of photosynthetic procaryotic endosymbionts, which subsequently lost their genetic autonomy and became reduced to organelles (chloroplasts). In the context of this hypothesis, the close structural and functional resemblances between the photosynthetic machinery of blue-green bacteria and the chloroplasts of red algae suggest the likelihood of a direct origin of the red algal chloroplast from blue-green bacterial endosymbionts. The possible procaryotic origins of other types of chloroplasts are more obscure.

HALOBACTERIUM AND ITS RELATION TO LIGHT

Highly saline environments (salt lakes, brines) harbor large populations of a small and distinctive group of Gram-negative bacteria: immotile cocci (*Halococcus*) and polarly flagellated rods (*Halobacterium*). Despite the differences of cell form, these organisms share a number of common properties, many of which are clearly adaptations to the high salinity and high light intensity of their natural habitat. The *minimum* NaCl concentration that permits growth is 2 to 2.5 M, the optimum 4–5 M; the Mg^{2+} requirement is also very high (optimum 0.1–0.5 M), whereas the K^+ requirement is lower (about 0.025 M). Peptidoglycans are absent from the cell walls of both groups. The intracellular salt concentration is at least as high as the extracellular one, though the intracellular ionic composition is different, composed largely of Na^+, K^+, and Cl^-. The biochemical machinery of the cell (enzymes, ribosomes) is not only salt-tolerant, but salt-requiring, and functions effectively only at near-saturated salt concentrations. Both groups have unique cellular lipids, not esterified with fatty acids, but containing a long-chain alcohol (dihydrophytol) in ether linkage to glycerol.

A characteristic feature of the extreme halophils is their content of red carotenoids, which are incorporated into the cell membrane. The carotenoids of *Halobacterium* have been shown to protect the cells from photochemical damage by the high light intensities characteristic of the natural environment (see Chapter 10, p. 311 for a more detailed account of this carotenoid function).

Extreme halophiles are aerobic organisms with complex nutritional requirements and are commonly cultivated in peptone-containing media; amino acids are the preferred carbon and energy sources. They frequently develop as colored patches on salted dried fish and hides as a result of treatment with salt that contains these bacteria.

Until very recently it was assumed that these extreme halophiles are aerobic chemoheterotrophs, with an exclusively respiratory metabolism. However, this interpretation had to be modified in 1971 with respect to *Halobacterium*, as a result of the discoveries of W. Stoeckenius and his collaborators. These discoveries developed from an unexpected observation. When *Halobacterium* is grown in liquid cultures subject to O_2 limitation, the cells synthesize a chemically modified cell membrane. In aerobically grown cells the cell membrane is red, as a result of its high carotenoid content. Oxygen limitation induces synthesis of a new purple membrane component (Figure 17.25). This component is laid down as a series of discrete patches, embedded in the red membrane, and can account for as much as half its total area. The purple areas are readily distinguishable in electron micrographs of freeze-etched cells, since they differ from red membrane in surface structure (Figure 17.26).

If cells of *Halobacterium* are lysed by dilution of the suspending medium, the red part of the membrane disaggregates, while the purple patches remain intact and can be isolated by differential centrifugation. In addition to lipid (25 percent of the dry weight), they contain only one species of protein, a chromoprotein called *bacteriorhodopsin*, because of its similarities to the visual pigment of the vertebrate retina, rhodopsin. Both these colored proteins contain the same chromophore, the C_{20} carotenoid retinal (Figure 17.27). Both are rapidly bleached by light, undergoing a complex series of photochemical transformations. The bleaching of rhodopsin is accompanied by the separation of retinene from the protein, whereas the bleaching of bacteriorhodopsin is not. When isolated purple membrane fragments are exposed to light, the bleaching of bacteriorhodopsin (Figure 17.28) is accompanied by a release of protons (Figure 17.29). Both effects are reversed in the dark. When intact cells of *Halobacterium* containing purple membrane are exposed to light, the bleaching of rhodopsin is accompanied by a release

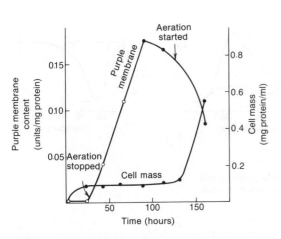

FIGURE 17.25 (left)

The effect of O_2 on the growth and synthesis of purple membrane in a liquid culture of *Halobacterium*. When the culture is no longer aerated, growth ceases, and the purple membrane content of the cells starts to rise; when aeration is resumed, growth once again starts and the purple membrane content of the cells declines. After D. Oesterhelt and W. Stoeckenius, "Functions of a New Photoreceptor Membrane," *Proc. Natl. Acad. Sci. U.S.* **70**, 2853 (1973).

FIGURE 17.26 (right)

Electron micrograph of a freeze-etched preparation of *Halobacterium* cells in 4.3 M NaCl, showing surface structure of the cell membrane. The areas composed of purple membrane (pm) are recognizable by their smooth surfaces. ×35,000. Courtesy of Dr. W. Stoeckenius.

of protons from the cells. In effect, a *proton gradient across the cell membrane is established*, and is maintained as long as illumination is continued. The formation of the proton gradient permits net ATP synthesis by the illuminated cells, a fact established by the following ingenious series of experiments.

Placed under anaerobic conditions in the dark, *Halobacterium* has no available means of synthesizing ATP, and the intracellular ATP pool falls to a low level. *A normal ATP pool level can be restored either by the introduction of air or by illumination of the cells.* The relative effectiveness of different wavelengths of light in restoring the intracellular ATP pool under anaerobic conditions clearly identifies bacteriorhodopsin as the cellular light receptor. Cells which lack purple membrane do not show this response to light. Accordingly, the bacteriorhodopsin of the purple membrane acts as a *light-driven proton pump*, the operation of which can be coupled with ATP synthesis. *Halobacterium* can thus synthesize ATP by a mechanism of photophosphorylation, which does not involve the participation of chlorophyll-containing photochemical reaction centers, characteristic of all photosynthetic organisms hitherto known.

The adaptive value of this device in the strongly illuminated natural environment of *Halobacterium* is obvious. The solubility of O_2 in a saturated salt solution is much lower than in pure water; hence, *Halobacterium*, which has a strictly aerobic dark metabolism, is often exposed to low concentrations of dissolved oxygen.

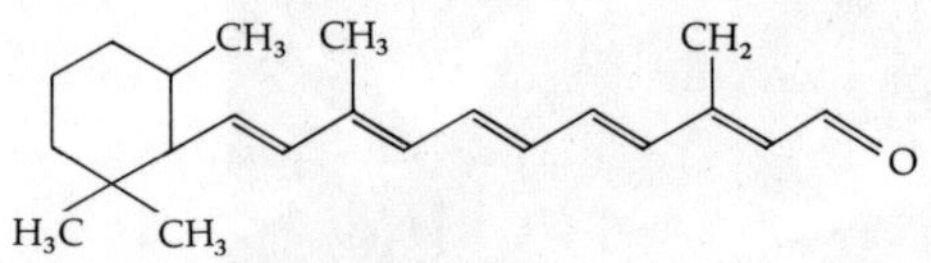

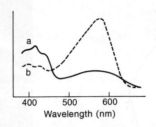

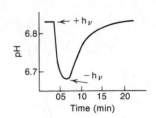

FIGURE 17.27 (left)

The structure of retinal, the C_{20} carotenoid which is the light-sensitive chromophore of the vertebrate visual chromoprotein, rhodopsin, and the purple membrane chromoprotein, bacteriorhodopsin, of *Halobacterium*.

FIGURE 17.28 (center)

The effect of light on the absorption spectrum of isolated purple membrane, containing bacteriorhodopsin. Spectra measured at the temperature of liquid nitrogen ($-196°C$). (a) Illuminated preparation; (b) preparation kept in the dark. After D. Oesterhelt and W. Stoeckenius, "Functions of a New Photoreceptor Membrane," *Proc. Natl. Acad. Sci. U.S.* **70**, 2853, (1973).

FIGURE 17.29 (right)

Changes of pH resulting from the illumination of an aqueous suspension of isolated purple membrane. Illumination ($+hv$) causes a drop in pH (release of protons), which is reversed when illumination ceases ($-hv$). After D. Oesterhelt and W. Stoeckenius, "Functions of a New Photoreceptor Membrane," *Proc. Natl. Acad. Sci. U.S.* **70**, 2853 (1973).

This derepresses purple membrane synthesis, just as oxygen deprivation derepresses bacteriochlorophyll synthesis in purple nonsulfur bacteria. *Halobacterium* thereby acquires the ability to make ATP by a light-mediated mechanism that can operate in the absence of oxygen.

It is of some interest to note that the extreme halophiles also include one purple sulfur bacterium, *Ectothiospira halophila*. This organism is a strict anaerobe and develops in salt-rich habitats that contain sulfide, which makes them unfavorable for *Halobacterium*. The ecological ranges of the two organisms do not therefore overlap.

FURTHER READING

Books   CARR, N. G., and B. A. WHITTON (editors), *The Biology of Blue-Green Algae*. Oxford, England: Blackwell, 1973.

FOGG, G. E., W. D. P. STEWART, P. FAY, and A. E. WALSBY, *The Blue-Green Algae*. New York: Academic Press, 1973.

KONDRAT'EVA, E. N., *Photosynthetic Bacteria*. London: Oldbourne Press, 1965.

Reviews and original articles   DANON, A., and W. STOECKENIUS, "Photophosphorylation in *Halobacterium halobium*," *Proc. Natl. Acad. Sci. U.S.* **71**, 1234 (1974).

563   GEST, H., "Energy Conversion and Generation of Reducing Power in Bacterial Photosynthesis," *Adv. Microb. Physiol.* **7**, 243 (1972).

HANSEN, T. A., and H. VAN GEMERDEN, "Sulfide Utilization by Purple Nonsulfur Bacteria," *Arch. Microbiol.* **86**, 49 (1972).

LARSEN, H., "The Halobacteria's Confusion to Biology," *Antonie van Leeuwenhoek* **39**, 383 (1973).

OESTERHELT, D., and W. STOECKENIUS, "Functions of a New Photoreceptor Membrane," *Proc. Natl. Acad. Sci. U.S.* **70**, 2853 (1973).

PFENNIG, N., and H. G. TRÜPER, "The Phototrophic Bacteria" in *Bergey's Manual of Determinative Bacteriology*, 8th ed. (R. E. Buchanan and N. E. Gibbons, editors). Baltimore: Williams and Wilkins, 1974.

PFENNIG, N., "Photosynthetic Bacteria," *Ann. Rev. Microbiol.* **21**, 286 (1967).

PIERSON, B. K., and R. W. CASTENHOLZ, "A Phototrophic Gliding Filamentous Bacterium of Hot Springs, *Chloroflexus aurantiacus*, gen. and sp. *nov.*," *Arch. Microbiol.* **100**, 5 (1974).

STEWART, W. D. P., "Nitrogen Fixation by Photosynthetic Microorganisms," *Ann. Rev. Microbiol.* **27**, 283 (1973).

STANIER, R. Y., "The Origins of Photosynthesis in Eucaryotes" in *Evolution in the Microbial World*, Symposium 24, Soc. Gen. Microbiol. London: Cambridge University Press, 1974.

TRÜPER, H. G., and L. A. ROGERS, "Purification and Properties of Adenyl Reductase from the Phototrophic Sulfur Bacterium, *Thiocapsa roseopersicina*," *J. Bact.* **108**, 1112 (1971).

VAN GEMERDEN, H., "Utilization of Reducing Power in Growing Cultures of *Chromatium*," *Arch. Microbiol.* **64**, 111 (1968).

GRAM-NEGATIVE BACTERIA: THE CHEMOAUTOTROPHS AND METHYLOTROPHS

18

Two highly specialized physiological groups of aerobic Gram-negative bacteria will be discussed in this chapter. They are the *chemoautotrophs* (alternative name: *chemolithotrophs*), which can derive the energy required for growth from the oxidation of inorganic compounds, and the *methylotrophs*, which can derive both the energy and the carbon required for growth from methane and other one-carbon organic compounds. Although readily distinguishable from one another by their different energy sources, the two groups show some interesting and suggestive analogies.

Both the chemoautotrophs and the methylotrophs are in turn made up of several physiological subgroups, distinguished by their specific energy sources, and also to some degree by the degree of their specialization for the autotrophic (or methylotrophic) way of life. They are all of wide natural distribution in soil and water, and play important roles in the cycles of elements in the biosphere (see Chapter 25). The total number of species represented is small, but both groups are diverse in structural as well as physiological respects.

THE CHEMOAUTOTROPHS

By definition, a chemoautotroph can grow in a strictly mineral medium in the dark, deriving its carbon from CO_2 and its ATP and reducing power from the respiration of an inorganic substrate. This mode of life, which exists only among procaryotes, was discovered between 1880 and 1890 by S. Winogradsky: his pioneering studies on several of the principal subgroups provided a solid foundation for all later work on chemoautotrophy. Winogradsky showed that two other remarkable properties are characteristic of the chemoautotrophs:

1. High specificity with respect to the inorganic energy source.
2. Frequent inability to use organic compounds as energy and carbon sources; indeed, their growth is sometimes adversely affected by organic compounds.

• **Utilizable substrates**

The inorganic materials capable of supporting chemoautotrophic growth include H_2S and other reduced forms of sulfur; ammonia and nitrite; molecular hydrogen; and ferrous iron (Fe^{2+}). In the biosphere they are in part produced through the metabolic activities of other organisms and in part of geochemical origin. It should be noted that certain of them are chemically unstable under aerobic conditions: H_2S is readily oxidized to elemental sulfur in contact with air; ferrous iron also undergoes autooxidation in neutral or alkaline solutions, although it is stable under acid conditions. The chemical instability of H_2S and Fe^{2+} has seriously impeded the isolation and study of some organisms that use these substrates.

The substrate specificities of the chemoautotrophs permit the recognition of four major subgroups (Table 18.1). *Nitrifying bacteria* use reduced inorganic nitrogen compounds as energy sources. The substrate specificity within this subgroup is very high; its members either oxidize ammonia to nitrite, or nitrite to nitrate; none can oxidize both these reduced nitrogen compounds. *Sulfur-oxidizing bacteria* use H_2S, elemental sulfur, or its partially reduced oxides, as energy sources; all these substances are converted to sulfate. One member of this group can in addition use ferrous iron as an energy source. *Iron bacteria* can oxidize reduced iron and manganese, but not reduced sulfur compounds; however, their status as true chemoautotrophs remains in some doubt. The *hydrogen bacteria* use molecular hydrogen as an energy source.

• **The nitrifying bacteria**

In the middle of the nineteenth century circumstantial evidence indicated that the oxidation of ammonia to nitrate in natural environments is a microbial process.

TABLE 18.1

Physiological groups of aerobic chemoautotrophs

GROUP		OXIDIZABLE SUBSTRATE	OXIDIZED PRODUCT	TERMINAL ELECTRON ACCEPTOR	TAXONOMIC STRUCTURE
Nitrifying bacteria	Ammonia oxidizers	NH_3	NO_2^-	O_2	4 genera; 5 species
	Nitrite oxidizers	NO_2^-	NO_3^-	O_2	3 genera; 3 species
Sulfur oxidizers[a]		H_2S, S, $S_2O_3^{2-}$	SO_4^{2-}	O_2; sometimes NO_3^-	3 genera; 10 species[b]
Iron bacteria[c]		Fe^{2+}	Fe^{3+}	O_2	Several genera
Hydrogen bacteria		H_2	H_2O	O_2; sometimes NO_3^-	Classified with chemoheterotrophs in several genera; about 10 species

[a] One species can also use Fe^{2+} as an energy source.
[b] Includes only those members of the group that have been isolated in pure culture; many other sulfur oxidizers have been described from nature, but never isolated (see Table 18.3).
[c] None so far isolated and shown to grow as chemoautotrophs in pure culture.

However, many attempts to isolate the causal agents by the use of conventional culture media failed completely. This problem was solved in 1890 by S. Winogradsky, who succeeded in isolating pure cultures of nitrifying bacteria by the use of strictly inorganic media. The causal agents proved to be small, Gram-negative, rod-shaped bacteria: the ammonia oxidizer, *Nitrosomonas,* and the nitrite oxidizer, *Nitrobacter.* They develop best under neutral or alkaline conditions; since the oxidation of ammonia to nitrite results in considerable acid formation, the growth medium for *Nitrosomonas* must be well buffered (for example, by the addition of insoluble carbonates). Growth of both these organisms is slow, minimal generation times approximating 24 hours; and the growth yields, expressed in terms of the quantity of the inorganic substrate oxidized, are low. Winogradsky showed that these bacteria are obligate autotrophs, incapable of development in the absence of the specific inorganic energy source.

In recent years a few other ammonia and nitrite oxidizers have been discovered. They resemble the two classical prototypes physiologically, but they are remarkably diverse in structural respects (Table 18.2). The group is a small one, consisting of eight species, assigned to seven genera, distinguished by structural properties and by the specific oxidizable substrate. Both the ammonia oxidizers and the nitrite oxidizers have narrow ranges of DNA base composition, the values for the latter being significantly higher than for the former.

The diversity of the nitrifying bacteria in gross cell structure is paralleled by a curious diversity in fine structure. In some genera the cell membrane is devoid of intrusions; in others there are extensive intrusions, which may be vesicular, lamellar, or tubular (Figures 18.1 and 18.2).

Obligate autotrophy is the rule in the nitrifying bacteria, with the exception of some *Nitrobacter* strains, which have been shown to use acetate as a carbon and energy source; however, these strains grow much more slowly with acetate than with nitrite.

TABLE 18.2

The genera of nitrifying bacteria

ENERGY-YIELDING REACTION	CELL FORM	FLAGELLA	MEMBRANE INTRUSIONS	DNA BASE COMPOSITION MOLES (%) G + C	OBLIGATE AUTOTROPHY	GENUS
$NH_3 \longrightarrow NO_2^-$	Rod	Subpolar[b]	Lamellar	50–51	+	*Nitrosomonas*
	Tight spiral	Peritrichous[b]	None	54	+	*Nitrosospira*
	Sphere	Peritrichous[b]	Lamellar	50–51	+	*Nitrosococcus*
	Irregular, lobed	Peritrichous[b]	Vesicular	54–55	+	*Nitrosolobus*
$NO_2^- \longrightarrow NO_3^-$	Rod, often pear-shaped[a]	—	Lamellar	60–62	+ or −	*Nitrobacter*
	Long, slender rod	—	None	58	+	*Nitrospina*
	Sphere	polar	Tubular	61	+	*Nitrococcus*

[a] Reproduction by budding; all other nitrifiers reproduce by binary fission.
[b] Some strains are nonmotile.

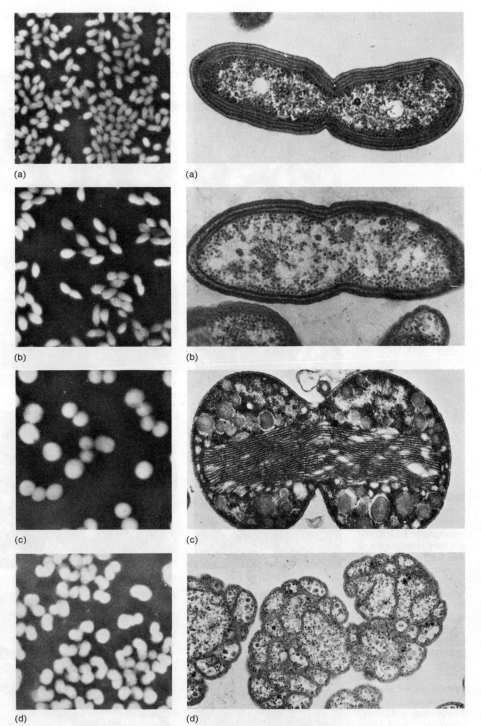

(a)

(a)

(b)

(b)

FIGURE 18.1

Phase contrast photomicrographs (at left) and electron micrographs of thin sections (at right) of nitrifying bacteria. (a) *Nitrosomonas europea*, ×2,200 and ×32,500. (b) *Nitrosomonas* sp., ×2,200 and ×39,600. (c) *Nitrosocystis oceanus*, ×2,200 and ×23,800. (d) *Nitrosolobus multiformis*, ×2,200 and ×22,000. From S. W. Watson and M. Mandel, "Comparison of the Morphology and Deoxyribonucleic Acid Composition of 27 Strains of Nitrifying Bacteria," *J. Bact.* **107,** 563 (1971).

(c)

(c)

(d)

(d)

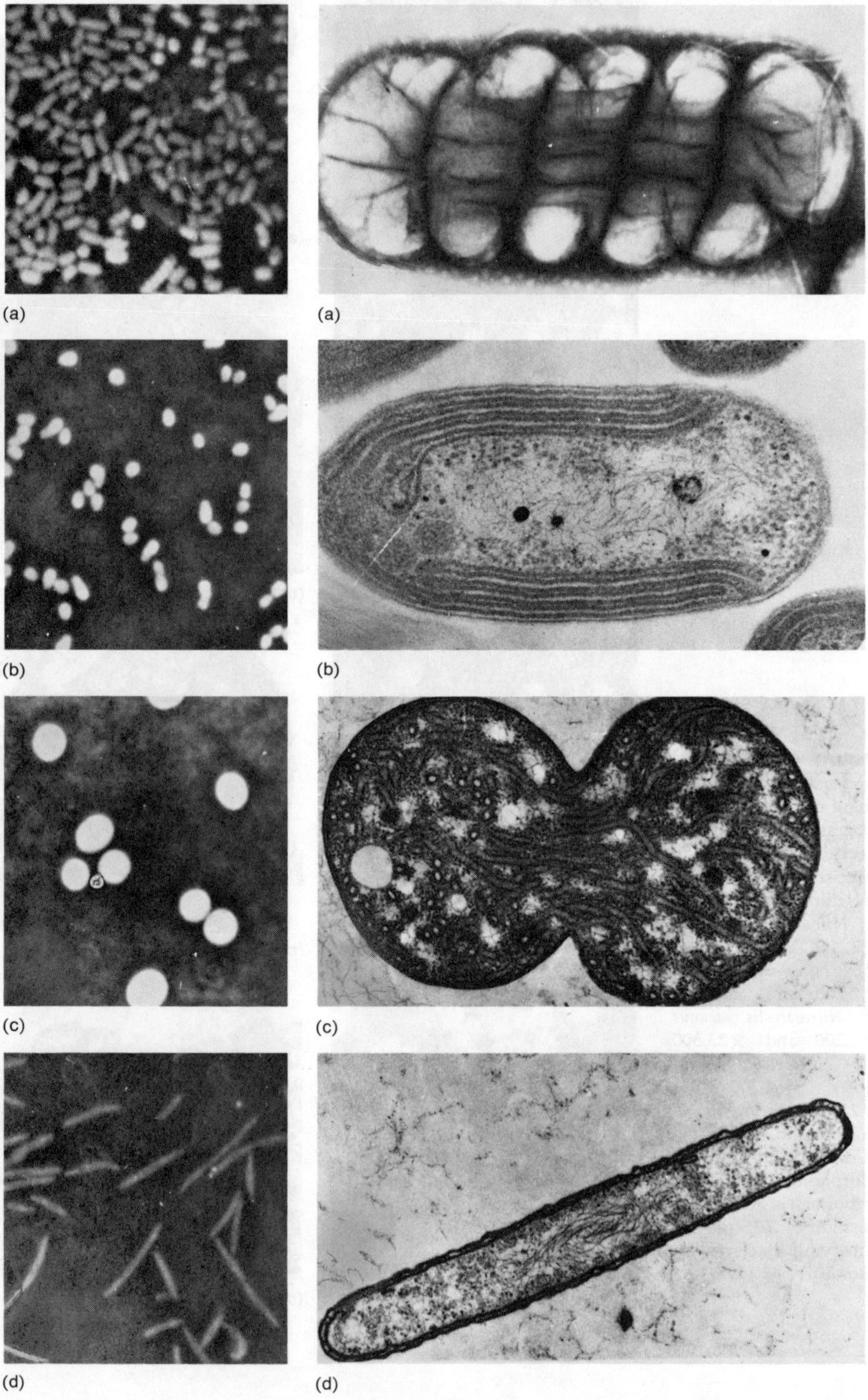

FIGURE 18.2

Phase contrast photomicrographs (at left) and electron micrographs of thin sections (at right) of nitrifying bacteria. (a) *Nitrosospira briensis,* ×2,200 and ×35,300. (b) *Nitrobacter winogradskyi,* ×2,200 and ×63,200. (c) *Nitrococcus mobilis,* ×2,200 and ×21,000. (d) *Nitrospina gracilis,* ×2,200 and ×37,500. From S. W. Watson and M. Mandel. "Comparison of the Morphology and Deoxyribonucleic Acid Composition of 27 Strains of Nitrifying Bacteria," *J. Bact.* **107,** 563 (1971).

TABLE 18.3

Bacteria which oxidize H₂S with formation of intracellular sulfur deposits but have not been isolated and grown in pure culture

1. Filamentous gliding organisms:
   Beggiatoa, Thiothrix, Thioploca
2. Very large unicellular gliding organisms:
   Achromatium
3. Unicellular rod-shaped or spiral organisms, immotile or motile by flagella:
   Cells rod-shaped, immotile: *Thiobacterium*
   Cells rod-shaped, polar flagella: *Macromonas*
   Cells round or ovoid, peritrichous flagella: *Thiovulum*
   Cells spiral, polar flagella: *Thiospira*

• Sulfur oxidizers

In the course of his pioneering studies on chemoautotrophy, Winogradsky examined the properties of the chemotrophic filamentous gliding bacteria of the *Beggiatoa-Thiothrix* group, which occur characteristically in certain sulfide-rich environments and often contain massive inclusions of elemental sulfur. He showed that these organisms can oxidize H₂S, initially to elemental sulfur which accumulates in the cells; the stored sulfur is subsequently further oxidized to sulfate. A variety of other aerobic bacteria (Figure 18.3) have been subsequently shown to oxidize H₂S in a similar manner, with transient intracellular sulfur deposition (Table 18.3). However, none of the organisms listed in Table 18.3 has been isolated in pure culture because of the technical difficulty of growing them aerobically at the expense of H₂S. Current knowledge of this physiological group of chemoautotrophs is based almost entirely on work with unicellular sulfur oxidizers which have small cells and do not accumulate sulfur within the cell (Figure 18.4). These members of the group can be purified and grown without difficulty (Figure 18.5), as a result of their ability to oxidize chemically stable reduced forms of sulfur, notably thiosulfate and elemental sulfur. Most of them are small, polarly flagellated rods, placed in the genus *Thiobacillus* and occur widely in both marine and terrestrial environments. The spiral organism, *Microthiospira* occurs in marine mud; *Sulfolobus*, an organism of irregular cell form, is confined to thermal habitats (Table 18.4). The properties of some *Thiobacillus* species are shown in Table 18.5. It may

TABLE 18.4

Sulfur-oxidizing chemoautotrophs which have been isolated in pure culture

	Thiobacillus	*Thiomicrospira*	*Sulfolobus*
	(8 spp.)	(1 sp.)	(1 sp.)
Cell form	Rods	Spirals	Lobed spheres
Flagella	Polar	Polar	Immotile
DNA base composition (moles % G + C)	34–70	48	60–68
pH range	Variable	5.0–8.5	0.5–6.0
Autotrophy	Variable	Obligate	Facultative

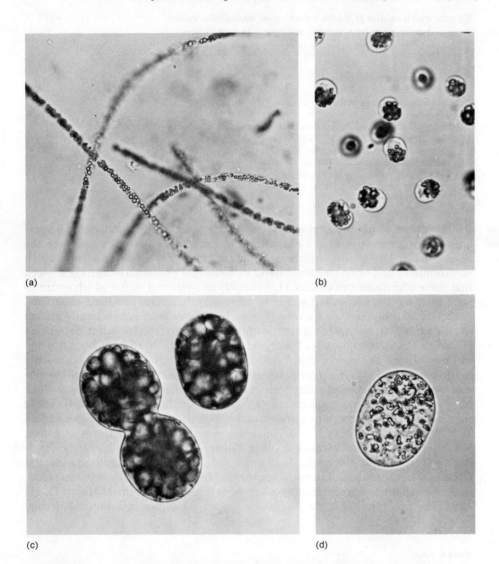

(a)

(b)

(c)

(d)

FIGURE 18.3

Some large, colorless, sulfur-oxidizing bacteria which accumulate sulfur internally. (a) Filaments of *Beggiatoa,* ×900. (b) *Thiovulum,* ×701. (c) and (d) *Achromatium,* ×700. The cells of this very large bacterium contain numerous calcium carbonate inclusions, shown in (c); the cell in (d) has been treated with dilute acetic acid, which has dissolved the inclusions of calcium carbonate, revealing the sulfur granules which are also present. Courtesy of Dr. J. W. M. La Rivière. (a) From J. W. M. La Rivière, "The Microbial Sulfur Cycle and Some of Its Implications for the Geochemistry of Sulfur Isotopes," *Geologischer Rundschau* **55,** 568 (1966). (c) and (d) from W. E. de Boer, J. W. M. La Rivière, and K. Schmidt, "Some Properties of *Achromatium oxaliferum,*" *Antonie van Leeuwenhoek* **37,** 553 (1971).

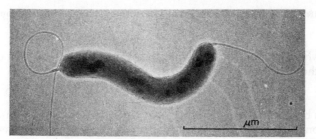

FIGURE 18.4

The colorless sulfur-oxidizing bacterium *Thiomicrospira*. Electron micrograph, showing the polar flagella. From J. G. Kuenen, and H. Veldkamp, *"Thiomicrospira pelophila, gen. n., sp. n., a New Obligately Chemolithotrophic Colorless Sulfur Bacterium," Antonie van Leeuwenhoek* **38**, 241 (1972).

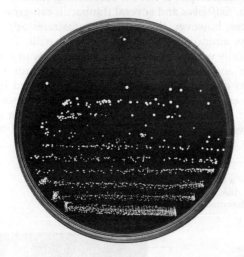

FIGURE 18.5

Thiobacillus thioparus colonies growing on a mineral agar plate with thiosulfate as the energy source. The colonies are pale yellow and very refractile, as a consequence of the deposition of elemental sulfur among the cells.

be noted that the range of DNA base composition in the group is rather wide, particularly among the thiobacilli. None of the small-celled sulfur oxidizers contain the extensive membranous intrusions that are characteristic of many nitrifying bacteria.

The chemoautotrophic growth of these organisms is rapid, some having generation times as short as 2 hours when growing at the expense of thiosulfate. A common and striking feature of the group is their extreme acid tolerance; some species can grow at a pH as low as 1 to 2 and fail to grow at pH values above 6. These organisms are often found in special environments in which the pH is maintained at a low level by their metabolic activities, since the oxidation of reduced sulfur compounds to sulfate results in considerable acid formation. The thermophile *Sulfolobus* is an inhabitant of acid, sulfur-rich hot springs, where sulfide of geochemical origin is immediately converted to elemental sulfur at the high ambient temperature of the water (70 to 85°C) in contact with air. *Sulfolobus* grows attached to the sulfur particles (Figure 18.6) which serve as its oxidizable substrate.

A specialized, man-made environment in which thiobacilli are abundant is the acid drainage water discharge from mines that contain metal sulfide minerals, notably iron pyrite (FeS_2). The predominant species in this habitat are the strongly acidophilic species *Thiobacillus thiooxidans*, which rapidly oxidizes elemental sulfur, and *T. ferrooxidans*, which can derive energy from the oxidation both of reduced sulfur compounds and Fe^{2+}. Since this organism can grow at pH values of 2 to

4, where ferrous iron is chemically stable, its ability to grow chemoautotrophically at the expense of the reaction:

$$4Fe^{2+} + 4H^+ + O_2 \longrightarrow 4Fe^{3+} + 2H_2O$$

can be rigorously established, but so far this has not been achieved for classical "iron bacteria," which have higher pH ranges (see next section).

Obligate chemoautotrophy is not the rule among sulfur oxidizers as it is among nitrifying bacteria. *Sulfolobus* and several thiobacilli can grow with organic carbon and energy sources; however, the utilizable substrates appear to be confined to glucose and a few amino acids. One of the thiobacilli, *T. intermedius*, can grow chemoautotrophically with thiosulfate, but fails to grow in a mineral medium with glucose as carbon and energy source, unless the medium is supplemented with yeast extract. This curious paradox was recently resolved by the discovery that *T. intermedius* has an absolute requirement for a reduced source of sulfur, which can be met either by thiosulfate or by a sulfur-containing amino acid.

The rapid growth and relative nutritional versatility of the thiobacilli makes

FIGURE 18.6

Sulfolobus acidocaldarius. (a) Electron micrograph of a thin section, ×83,000. From T. D. Brock, K. M. Brock, R. T. Belly, and R. L. Weiss, *Arch. Mikrobiol.* 84, 64, 1972. (b) Fluorescent photomicrograph of cells stained with acridine orange attached to elemental sulfur crystal, ×750.

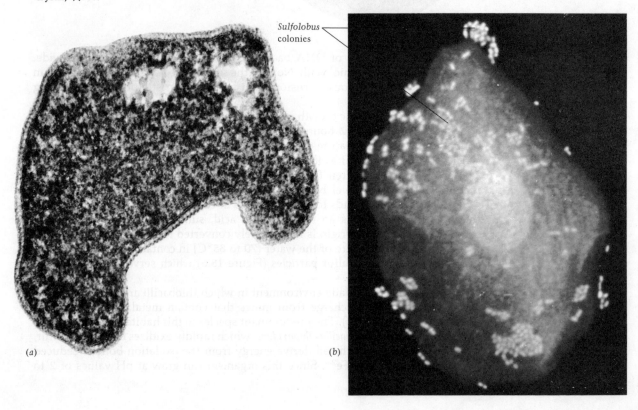

Sulfolobus colonies

(a)   (b)

TABLE 18.5

Characteristics of some *Thiobacillus* species

	T. thiooxidans	T. thioparus	T. ferrooxidans	T. novelins	T. intermedius
Obligate autotrophy	+	+	Variable	−	−
Utilizable energy sources:					
H_2S, S, $S_2O_3^{2-}$	+	+	+	+	+
Fe^{2+}	−	−	+	−	−
Glutamate	−	−	−	+	+
Glucose	−	−	+ (some strains)	−	+
pH range	0.5–6.0	4.0–7.5	1.4–6.0	5.0–9.2	2.0–7.0

them the most favorable objects for experimental analysis of the physiological problems associated with the chemoautotrophic modes of life (see p. 578).

• **The iron bacteria**   Certain freshwater ponds and springs have a high content of reduced iron salts. It has long been known that a distinctive bacterial flora is associated with such habitats. These *iron bacteria* form natural colonies that are heavily encrusted with ferric oxide. However, since most iron springs are neutral or alkaline, ferrous iron undergoes rapid spontaneous oxidation in contact with the atmosphere, so it has proved very difficult to ascertain the role which it plays in the metabolism of these bacteria. The most conspicuous iron bacteria are filamentous, ensheathed bacteria of the *Sphaerotilus* group, in many of which the sheaths are encrusted with iron oxide. They can be readily grown as chemoheterotrophs and so isolated in pure culture. Although it is possible to show that such pure cultures will accumulate iron oxide on the sheaths, there is no evidence that such deposition is a physiologically significant process or that these bacteria can develop as chemoautotrophs. Chemoautotrophic growth seems more probable in the case of another structurally distinctive iron bacterium, *Gallionella*. All attempts to obtain cultures of *Gallionella* in organic media have failed, but it has been grown artificially (although not in the pure state) in a mineral medium containing a deposit of ferrous sulfide as a source of reduced iron. The use of this virtually insoluble ferrous salt minimizes the problem of spontaneous oxidation of iron under neutral conditions. In these cultures, *Gallionella* will form cottony colonies attached to the wall of the vessel. Much of the "colony" is inorganic: the small, bean-shaped bacterial cells are located at the branched tips of the excreted stalks, which are heavily impregnated with ferric hydroxide (Figure 18.7).

• **Hydrogen bacteria**   Many species of aerobic bacteria possess the ability to grow chemoautotrophically with molecular hydrogen. In contrast to other groups of chemoautotrophs, the hydrogen bacteria are all nutritionally versatile organisms that can use a wide range of organic compounds as carbon and energy sources. Most of these bacteria were formerly placed in a special genus, *Hydrogenomonas*. However, their facultative chemoautotrophy does not appear to justify a generic separation from similar organisms that are chemoheterotrophs, and they are now classified in a series

(a)

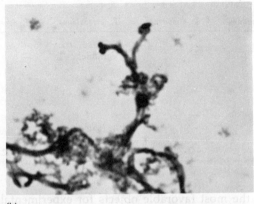

(b)

(c)

FIGURE 18.7

The iron bacterium, *Gallionella*. (a) Flocculent colonies (consisting largely of ferric hydroxide) growing attached to the glass in a liquid culture. (b) Light micrograph of the edge of a colony, showing cells attached to the tips of a branched stalk, ×2,430. (c) Electron micrograph of a single cell, attached to the tip of the stalk, which is impregnated with ferric hydroxide. Courtesy of R. S. Wolfe. Photographs (a) and (b) reproduced from S. Kucera and R. S. Wolfe, "A Selective Enrichment Method for *Gallionella Ferruginea*," *J. Bacteriol.* **74**, 347 (1957). Micrograph (c) from *Principles and Applications in Aquatic Microbiology*, Chap. 5. New York: Wiley, 1964.

TABLE 18.6

The generic assignments of facultatively chemoautotrophic hydrogen bacteria

CELL FORM	GRAM REACTION	FLAGELLAR INSERTION	DNA BASE COMPOSITION, MOLES % G + C	GENUS	H₂-OXIDIZING SPECIES
Rods	−	Polar	61–72	*Pseudomonas*	*P. saccharophila, P. facilis, P. ruhlandii*
Rods	−	Peritrichous	67–68	*Alcaligenes*	*A. eutrophus, A. paradoxus*
Cocci	−	Immotile	66	*Paracoccus*	*P. denitrificans*
Branched rods	+	Immotile	65–70	*Nocardia*	*N. opaca*

of genera that contain phenotypically similar nonautotrophic bacteria (Table 18.6). A few hydrogen bacteria are evidently related to nonautotrophic species; for example, the hydrogen bacterium *Pseudomonas facilis* closely resembles the nonautotrophic *P. delafieldii* in general phenotype, and shows a high level of genetic homology with this species, as evidenced by DNA-DNA hybridization *in vitro*.

• **The metabolic basis of chemoautotrophy**

There is good evidence in all physiological groups of chemoautotrophs that the assimilation of CO_2 occurs through the reactions of the Calvin cycle (see Chapter 7, p. 190). When grown chemoautotrophically, cells contain high levels of the two enzymes specific to this pathway, carboxydismutase and phosphoribulokinase. However, in facultatively autotrophic thiobacilli and in hydrogen bacteria, synthesis of these enzymes is often largely or partly repressed when cells grow with organic substrates. Many obligately autotrophic thiobacilli and nitrifying bacteria possess carboxysomes, the specialized procaryotic organelles which contain carboxydismutase (see Chapter 11, p. 350).

In order to drive the reactions of primary carbon assimilation, the chemoautotrophs must obtain both ATP and reducing power (reduced pyridine nucleotide) through the oxidative dissimilation of the inorganic substrate. The relative effectiveness with which they do so is indicated by some comparative measurements of growth yields (Table 18.7).

• **The oxidation of molecular hydrogen**

In principle, the generation of both ATP and reduced pyridine nucleotide presents no unusual problems, provided that the oxidation of the inorganic substrate is coupled, like the oxidation of organic substrates, to the reduction of NAD^+; part of the NADH so formed can be reoxidized *via* the respiratory electron transport chain, with an accompanying synthesis of ATP by oxidative phosphorylation, and part used to drive the reductive steps of carbon assimilation. This appears to be the situation in hydrogen bacteria, several of which have been shown to contain a soluble *hydrogen dehydrogenase*, catalyzing the reaction:

$$H_2 + NAD^+ \longrightarrow NADH + H^+$$

This enzyme has been purified from *Nocardia opaca* and has two interesting properties. It is oxygen-sensitive, being inactivated when exposed to air in a neutral medium, and it requires Ni^{2+} (partly replaceable by Mg^{2+}) for activity.

These properties of hydrogen dehydrogenase clarify some puzzling earlier observations in the nutritional behavior of hydrogen bacteria. Although hydro-

TABLE 18.7

Growth yields of some chemoautotrophic bacteria[a]

ORGANISM	GROWTH YIELD
Pseudomonas facilis	12 g/g-mol. H_2
Thiobacillus neapolitanus	4 g/g-mol. $S_2O_3^{2-}$
Thiobacillus ferrooxidans	0.35 g/g-atom Fe^{2+}

[a] Expressed as grams (dry weight) of cell material synthesized per gram-molecule or gram-atom of substrate oxidized.

gen-oxidizing pseudomonads show normal aerobic behavior when grown with organic substrates, some of them become microaerophilic under chemoautotrophic growth conditions, developing at the expense of H_2 only with partial pressures of oxygen considerably lower than atmospheric. The hydrogen bacterium *Alcaligenes eutrophus* has a specific requirement for Ni^{2+} (not previously known to be an essential element for bacteria) when grown chemoautotrophically, but not when grown with organic substrates.

• **Oxidative phosphorylation and reduced pyridine nucleotide synthesis in other chemoautotrophs**

The mechanisms by which substrate oxidation fulfills the biosynthetic requirements of chemoautotrophs other than hydrogen bacteria appear to be considerably more complex. This problem can be illustrated by the specific example of the nitrite oxidizers. In this group, oxidation of the substrate involves the removal of only one pair of electrons:

$$NO_2^- + H_2O \longrightarrow NO_3^- + 2H^+ + 2e^-$$

However, the potential of the NO_2^-/NO_3^- couple ($E_0' + 0.42$) is far too high to drive the reaction:

$$NAD^+ + H^+ + 2e^- \longrightarrow NADH$$

of which the E_0' is -0.32.

Particulate fractions of cell-free extracts prepared from *Nitrobacter*, which contain the electron transport system of the cell, can mediate an oxidation of nitrite by O_2. Analysis of this system shows that the reaction involves a transfer of electrons from nitrite *via* cytochromes a_1 and a_3 to O_2, accompanied by the synthesis of one mole of ATP per mole of substrate oxidized (Figure 18.8). Accordingly, the generation of ATP in this mode of chemoautotrophy is satisfactorily accounted for. However, the formation of reduced pyridine nucleotide is not. Recent studies have shown that NAD^+ can be reduced in cell-free preparations by nitrite at the expense of added ATP, which mediates a reverse electron transfer from reduced cytochrome a_1 (the initial acceptor of electrons in the transport chain); several moles of ATP are required to drive the successive reductive steps leading to NAD^+ reduction, as shown schematically in Figure 18.9. Thus, much of the ATP derived from the oxidation of nitrite by molecular oxygen (mediated by the terminal components of the transport chain) must be expended to push electrons from nitrite up an unfavorable thermodynamic gradient in order to reduce pyridine nucleotide.

Current evidence suggests that analogous processes of reverse electron trans-

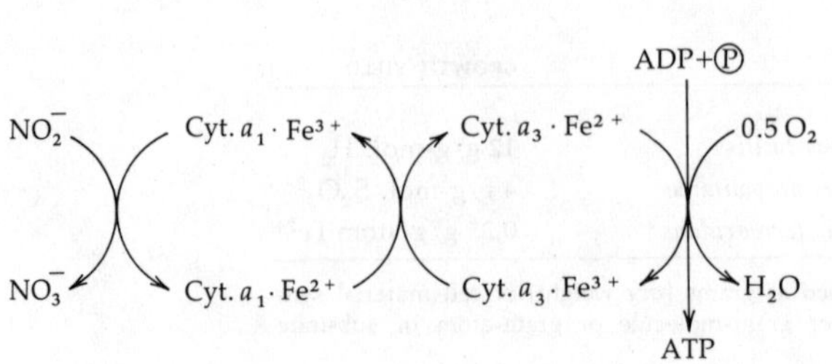

FIGURE 18.8

The coupling of ATP synthesis with nitrite oxidation in *Nibrobacter*.

$$NO_2^- \qquad Cyt.\ a_1 \cdot Fe^{3+} \qquad Cyt.\ a_3 \cdot Fe^{2+} \qquad 0.5\ O_2$$

$$ADP + \textcircled{P}$$

$$NO_3^- \qquad Cyt.\ a_1 \cdot Fe^{2+} \qquad Cyt.\ a_3 \cdot Fe^{3+} \qquad H_2O$$

$$ATP$$

FIGURE 18.9

Probable pathway of energy-linked reverse electron transfer from nitrite to NAD$^+$ in *Nitrobacter*.

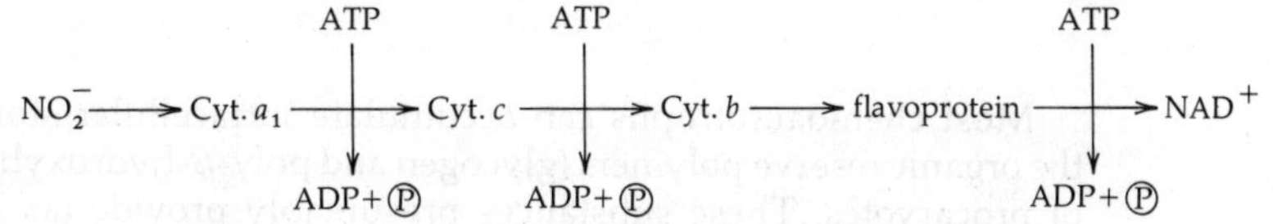

$$NO_2^- \longrightarrow Cyt.\ a_1 \xrightarrow[\substack{ATP \\ ADP+\textcircled{P}}]{} Cyt.\ c \xrightarrow[\substack{ATP \\ ADP+\textcircled{P}}]{} Cyt.\ b \longrightarrow flavoprotein \xrightarrow[\substack{ATP \\ ADP+\textcircled{P}}]{} NAD^+$$

port are required to reduce pyridine nucleotides at the expense of the other oxidizable inorganic substrates used by chemoautotrophs: Fe^{2+}, NH_3, and reduced sulfur compounds. However, none of these systems has been studied in such detail as has nitrite oxidation. Furthermore, considerable uncertainties remain concerning the step reactions and biochemical mechanisms involved in the oxidation of ammonia to nitrite and of reduced sulfur compounds to sulfate. The oxidation of ammonia (valence of N: -3) to nitrite (valence of N: $+3$) involves removal of six electrons, and therefore involves at least two intermediates in which the nitrogen atom has intermediate valences. There is considerable evidence that the first step is mediated by an oxygenase, which attaches an oxygen atom derived from O_2 to ammonia, with formation of hydroxylamine (NH_2OH)(valence of N: -1). Later intermediates have not been identified. A possible sequence might be:

$$NH_4^+ \xrightarrow{(O_2)} NH_2OH \longrightarrow [N_2O \longrightarrow H_2N_2O_2] \longrightarrow NO_2^-$$

The oxidation of H_2S to sulfate involves removal of eight electrons. The rapid intracellular formation of elemental sulfur by many H_2S-oxidizing chemolithotrophs identifies this substance as the first stable intermediate, although the enzymatic mechanism of its formation is not known. The penultimate intermediate is sulfite. The nature of the reactions that lead to its formation are controversial; several possible mechanisms have been postulated. Both the mechanism and the enzymology of the oxidation of sulfite to sulfate are well established. This reaction is the only one associated with the oxidation of an inorganic compound that permits energy conservation through a substrate level phosphorylation: it proceeds *via* an adenylated intermediate, adenosyl-phosphosulfate (APS):

$$SO_3^{2-} + AMP \longrightarrow APS + 2e^-$$

$$APS + \textcircled{P} \longrightarrow SO_4^{2-} + ADP$$

Since ATP can be formed from ADP by the adenylate kinase reaction:

$$2ADP \longrightarrow ATP + AMP$$

ATP is formed as a consequence of the substrate level of phosphorylation of AMP.

• **The phenomenon of obligate autotrophy**

Primarily on the basis of his experience with nitrifying bacteria, Winogradsky proposed that an inability to grow at the expense of organic substrates is an intrinsic property of chemoautotrophs. Current knowledge shows that this is not correct: obligate autotrophy, although well-nigh universal among the nitrifiers, is a variable character among the sulfur oxidizers, and is completely absent from hydrogen bacteria. Many attempts have been made to find an explanation for this phenomenon, and for its nonrandom distribution among the major physiological groups of chemoautotrophs.

Most chemoautotrophs can accumulate intracellular stores of one or both of the organic reserve polymers (glycogen and poly-β-hydroxybutyrate) characteristic of procaryotes. These substances presumably provide (as in other bacteria) endogenous carbon and energy reserves. Hence, there are good reasons to believe that chemoautotrophs possess the biochemical machinery for the metabolism of these reserve materials.

One possible explanation for the failure of these organisms to use exogenous organic compounds as carbon and energy sources might be a generalized impermeability to organic compounds, caused by the absence of specific permeases. However, experiments with several obligate sulfur oxidizers and nitrifiers show that this hypothesis is not tenable. The incorporation of radioactivity into cell material can be readily demonstrated when ^{14}C-labeled organic substrates are added to cultures of obligate autotrophs growing at the expense of their inorganic energy source; such uptake ceases in the absence of the oxidizable inorganic substrate. A number of amino acids and acetate are assimilated at relatively high rates. However, little if any $^{14}CO_2$ is produced from either carbon atom of ^{14}C-acetate; the metabolism of this compound is strictly assimilatory in obligate autotrophs. Furthermore, acetate provides only a small fraction (10 to 15 percent) of newly synthesized cell carbon. In comparable experiments with facultative autotrophs (e.g., *Thiobacillus intermedius*), acetate carbon makes a much larger fractional contribution to newly synthesized cell carbon. Biochemical studies have shown that the failure of many obligate autotrophs to utilize acetate more effectively is attributable to the absence of a functional TCA cycle: these organisms lack a key enzyme, α-ketoglutarate dehydrogenase, and have unusually low levels of both succinic and malic dehydrogenase. In the absence of α-ketoglutarate dehydrogenase, the other enzymes associated with the cycle cannot mediate an oxidation of acetyl-CoA; they function as two separate pathways, which have purely biosynthetic roles (Figure 18.10). The dicarboxylic acid branch, (fed by carboxylation of phosphoenolpyruvate) operates in the reverse of its customary direction, to provide precursors of the amino acids of the aspartate family and of porphyrins; the citrate branch (fed by oxaloacetate and acetyl-CoA) operates in its customary direction to produce α-ketoglutarate, the precursor of amino acids of the glutamate family. The relatively small contribution of exogenous acetate to newly synthesized cell carbon, accordingly, reflects the fact that it enters only those biosynthetic pathways for which acetyl-CoA is a specific precursor: in addition to the pathway shown in Figure 18.10, they include the pathways of lipid synthesis and of leucine synthesis (see Chapter 7 for further details).

For an organism with a strictly respiratory mode of metabolism, the absence of α-ketoglutarate dehydrogenase prevents dissimilation of most organic substrates, with the possible exception of glucose and a few other sugars, which can be oxidized to CO_2 through the pentose phosphate pathway; this mode of sugar dissimilation occurs in many blue-green bacteria, a group which similarly lacks a functional TCA cycle (see p. 540).

A second possible explanation for the inability of many chemoautotrophs to derive energy from the oxidation of organic compounds may lie in the special character of their electron transport chains. With the exception of hydrogen oxidation, the oxidation of inorganic substrates by chemoautotrophs involves the entry of electrons into the transport chain at the level of the cytochromes, and

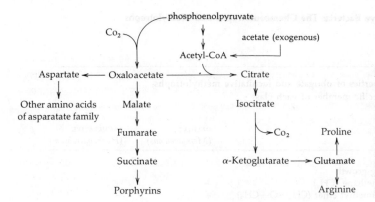

phosphoenolpyruvate

FIGURE 18.10

The biosynthetic roles of reactions normally associated with the operation of the tricarboxylic acid
in organisms that cannot convert α-ketoglutarate
to succinate. Note that carbon from exogenous
acetate can enter the amino acids of the glutamate
family via citrate and α-ketoglutarate, but cannot
enter those of the aspartate family via succinate
and oxaloacetate as it does in organisms with a
functional TCA cycle.

consequently it entails synthesis of reduced pyridine nucleotides by energy-
mediated reverse electron transport. It is therefore possible that the transfer of
electrons to the respiratory chain via NADH (the major site of entry from organic
substrates) is in some way impeded. The adaptive value of a mechanism to prevent
reoxidation of NADH via the transport chain is evident, in view of the high
energetic cost of NADH synthesis. An explanation of obligate autotrophy in terms
of such a mechanism can also account for the absence of obligately autotrophic
hydrogen bacteria, since H_2 is the only oxidizable inorganic substrate oxidized
via NADH. However, evidence to support or disprove this hypothesis is lacking,
primarily because the structure and function of electron transport chains in
chemoautotrophs are still largely unexplored.

• **Growth inhibition
by organic
compounds**

Winogradsky concluded from his physiological studies on the nitrifying bacteria
that organic compounds are not simply unutilizable, but are actually toxic for
these organisms. However, later work has not confirmed this contention; organic
substances are not in general growth inhibitory for chemoautotrophs when they
are added to cultures at concentrations tolerated by chemoheterotrophs. However,
some examples of *specific* inhibition, particularly by amino acids, have been described. Growth of some thiobacilli is arrested by low concentrations (1 to 10 mm)
of single amino acids, the specific patterns of inhibition varying somewhat from
strain to strain. In every such case that has been carefully analyzed, the growth
inhibition is attributable to the end-product inhibition of an enzyme which
mediates an early step in a biosynthetic pathway; growth can be restored by
adding other end products of the pathway affected. Thus, phenylalanine inhibition
is relieved by tyrosine and tryptophan. Entirely comparable cases of specific
inhibition have been observed in chemoheterotrophic bacteria; the phenomenon
is, accordingly, not specific to chemoautotrophs.

THE METHYLOTROPHS

The distinguishing property of these organisms is the ability to
derive both carbon and energy from the metabolism of one-carbon compounds
containing a methyl (CH_3—) group or of compounds containing two or more
methyl groups that are not directly linked to one another (e.g., dimethyl ether,

TABLE 18.8

The nutritional properties of obligate and facultative methylotrophs, exemplified by a specific member of each class

	OBLIGATE (*Methylomonas*)	FACULTATIVE (*Hyphomicrobium*)
Substrates supporting growth		
C$_1$ compounds — methane (CH$_4$)	+	−
C$_1$ compounds — dimethyl ether (CH$_3$—O—CH$_3$)	+	−
C$_1$ compounds — methanol (CH$_3$OH)	+	+
C$_1$ compounds — formate (HCOO$^-$)	−	+
C$_2$ compounds — ethanol	−	+
C$_2$ compounds — acetate	−	+
C$_4$ compound β-OH-butyrate	−	+

CH$_3$—O—CH$_3$). By far the most abundant compound of this class in nature is the gas methane (CH$_4$), which occurs in coal and oil deposits and is continuously produced on a large scale in anaerobic environments by the methanogenic bacteria (see Chapter 24, p. 704). Methanol is formed as a breakdown product of pectins and other naturally occurring substances that contain methyl esters or ethers. Methylated amines [(CH$_3$NH$_2$, (CH$_3$)$_2$NH, (CH$_3$)$_3$N)], and their oxides occur naturally in plant and animal tissues.

The dissimilation of these highly reduced methyl compounds is almost always respiratory, being mediated by strict aerobes. The only known exceptions to this rule are the anaerobic utilizations of methanol by methanogenic bacteria and purple bacteria. Only procaryotes are known to oxidize methane; however, methanol can serve as a growth substrate for certain yeasts.

The procaryotic methylotrophs fall into two primary physiological subgroups: obligate and facultative methylotrophs (Table 18.8). The obligate methylotrophs are all able to grow with methane; the only other substrates that support growth are dimethyl ether and methanol. The facultative methylotrophs can grow with methanol and/or methylamines, but not with methane. Other growth substrates often include formate, as well as a few simple C$_2$ and C$_4$ compounds; however, the range is always small, as shown by the specific example included in Table 18.8.

• **The obligate methylotrophs**

The growth of knowledge about this group has been slow. Its first representative, *Methylomonas methanica*, a Gram-negative, polarly flagellated rod, was described almost 70 years ago, but for many decades it remained the only known methane oxidizer. The development of improved methods for the enrichment and purification of methane oxidizers has recently led to the discovery of many new types, all basically similar in nutritional respects, but remarkably varied in cell structure (Figure 18.11). Electron microscopy of thin sections (Figure 18.12) shows that all possess complex intruded membrane systems, which are of two types: stacks of flattened discs (type I membranes), and pairs of membranes, usually running

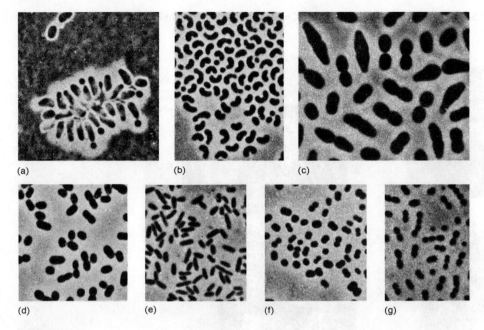

(a)

(b)

(c)

(d)

(e)

(f)

(g)

FIGURE 18.11

Phase-contrast photomicrographs (all ×1,600) of some obligate methylotrophs. (a) *Methylosinus;* (b) *Methylocystis;* (c) *Methylobacter;* (d) and (e) two species of *Methylomonas;* (f) and (g), two species of *Methylococcus.* From R. Whittenbury, K. C. Phillips, and J. F. Wilkinson, "Enrichment, Isolation and Some Properties of Methane-Utilizing Bacteria," *J. Gen. Microbiol.* **61,** 205 (1970).

parallel to the cytoplasmic membrane in the peripheral cytoplasm (type II membranes). In their topology and complexity, these membrane systems resemble those found in certain of the nitrifying bacteria. In structural terms, obligate methylotrophs can be assigned to five different groups (Table 18.9). Some form resting stages, resistant to desiccation. These structures are of two types: cysts, resembling the cysts of *Azotobacter*, and so-called *exospores*, which are small, spherical cells budded off from one pole of the mother cell (Figure 18.13).

TABLE 18.9

Subgroups of obligate methylotrophs

CELL STRUCTURE	MEMBRANE TYPE	SUBGROUP
Polarly flagellated rods, often pear-shaped, often rosettes; forming exospores	II	*Methylosinus*
Immotile curved rods, forming rosettes; no exospores	II	*Methylocystis*
Polarly flagellated rods	I	*Methylomonas*
Polarly flagellated rods and cocci; form cysts	I	*Methylobacter*
Immotile cocci	I	*Methylococcus*

581

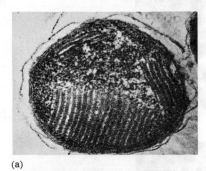

(a)

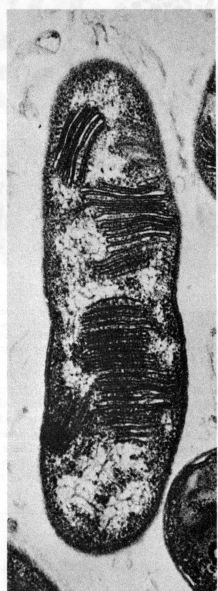

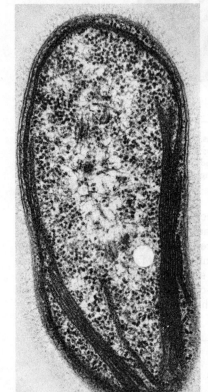

(c)

(b)

FIGURE 18.12

Electron micrographs of thin sections of three obligate methylotrophs, showing the two types of membrane systems characteristic of these organisms. (a) *Methylococcus* (type I membrane system), ×45,900. (b) *Methylomonas* (type I membrane system) ×22,900. (c) *Methylosinus* (type II membrane system), ×45,900. From S. L. Davies, and R. Whittenbury, "Fine Structure of Methane and Other Hydrocarbon-Utilizing Bacteria," *J. Gen. Microbiol.* **61**, 227 (1970).

Methane is the best growth substrate; methanol is toxic for many strains, and hence must be provided at very low concentrations. The growth rates under optimal conditions are low (doubling times of 4 to 6 hours). A few organic substrates incapable of supporting growth can be oxidized: formate is oxidized to CO_2, ethylene and ethanol to acetaldehyde. Both nitrate and ammonia can be

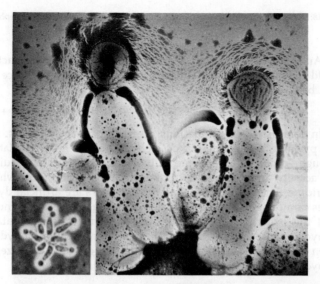

FIGURE 18.13

The budding off of exospores from one cell pole in *Methylosinus*. Electron micrograph of a negatively stained preparation of whole cells, showing the wrinkled appearance of the surface of the exospore, and the fine fibers which extend out from it, ×12,000. Insert: a similar group of cells forming exospores, negatively stained with India ink and observed by phase-contrast microscopy, ×1,400. From R. Whittenbury, S. L. Davies, and S. L. Davey, "Exospores and Cysts Formed by Methane-Utilizing Bacteria," *J. Gen. Microbiol.* **61**, 219 (1970).

used as nitrogen sources; however concentrations of ammonia greater than 0.05 percent reduce the rate of growth, since ammonia is an inhibitor of methane oxidation. In media containing ammonia, trace amounts of nitrite are formed; the methane oxidizers appear to be mini-nitrifiers, although there is no evidence that they can derive energy from this quantitatively insignificant oxidation.

• **Facultative methylotrophs**

Although the obligate methylotrophs can grow at the expense of methanol, enrichments with this substrate invariably yield other types of organisms, which are called *facultative methylotrophs*. Aerobic enrichments with either methanol or methylamines lead to the isolation of Gram-negative, polarly flagellated rods, none of which has been well characterized. The best-known facultative methylotroph is a budding organism, *Hyphomicrobium*. It is a powerful denitrifier and can be specifically enriched by the use of a medium containing methanol and nitrate, incubated anaerobically.

• **The metabolism of methyl compounds**

The chemical pathway for the complete oxidation of methane and methanol is straightforward:

$$CH_4 \longrightarrow CH_3OH \longrightarrow HCHO \longrightarrow HCOOH \longrightarrow CO_2$$

However, its enzymology still presents many puzzles. Tracer experiments with $^{18}O_2$ have shown that the oxygen atom in methanol, the first intermediate, is derived from O_2, not water. Hence, the initial attack on methane (as on ammonia) involves an oxygenation.

It is conceivable that methanol is the direct product; however, the recent discovery that methyl ether ($CH_3—O—CH_3$) is a substrate for all methane oxidizers suggests that it may be the primary product of the initial oxygenation. Certain Gram-positive bacteria can use ethyl ether ($C_2H_5—O—C_2H_5$) as a growth substrate; this compound is decomposed by oxygenative attack, with formation of ethanol and acetaldehyde:

$$C_2H_5—O—C_2H_5 + O_2 + NADH + H^+$$
$$\longrightarrow CH_3CH_2OH + CH_3CHO + NAD^+ + H_2O$$

An analogous oxidation of dimethyl ether by methane bacteria might therefore yield methanol and formaldehyde, although the occurrence of this reaction has not been demonstrated:

$$CH_3{-}O{-}CH_3 + O_2 + NADH \longrightarrow CH_3OH + HCHO + NAD^+ + H_2O$$

In contrast to the oxidations of other primary alcohols, which are mediated by pyridine nucleotide-linked dehydrogenases, the oxidation of methanol is brought about by an enzyme of unusual properties. It requires ammonium ions for activity and contains an unknown prosthetic group which appears to be a pteridine. The only known electron acceptors are artificial dyes; pyridine nucleotides cannot be reduced by this dehydrogenase. There is some evidence to suggest that the oxidation of formaldehyde to formate is also catalyzed by methanol dehydrogenase, although NAD-linked formaldehyde dehydrogenases have also been detected in methylotrophs. Formate oxidation is catalyzed by an NAD-linked dehydrogenase.

• Carbon assimilation by methylotrophs

The methylotrophs are able to derive all cell carbon from C_1 compounds, i.e., the organic substrate or CO_2. Tracer experiments in fact show that a large fraction of cell carbon is derived from the oxidizable substrate, not from CO_2. The source of cell carbon is in fact the metabolic intermediate, formaldehyde: two unique and distinct cyclic pathways for its conversion to cell material have been discovered among methylotrophs.

The *ribose phosphate pathway* is in many ways analogous to the Calvin pathway for the assimilation of CO_2. The key reactions are the addition of formaldehyde to ribose-5-P with formation of the phosphorylated ketohexose, allulose-6-P, which is then epimerized to fructose-6-P. Simplified representations of the ribose phosphate and Calvin cycles are shown in Figures 18.14 and 18.15; the main differences between them are summarized in Table 18.10.

The *serine pathway* (Figure 18.16) involves a completely different set of cyclic reactions. A C_1 unit derived from formaldehyde is transferred, following its addition to tetrahydrofolic acid (THF), to the amino acid glycine, with the forma-

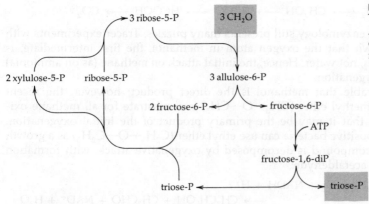

FIGURE 18.14

A simplified diagram of the ribose phosphate cycle for assimilation of formaldehyde in some methylotrophs.

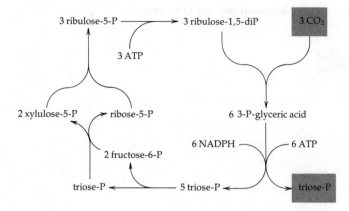

FIGURE 18.15

A simplified diagram of the Calvin cycle to show its analogy with the ribose phosphate cycle.

tion of serine:

$$C_1\text{-THF} + CH_2NH_2COOH \longrightarrow CH_2OHCHNH_2COOH + THF$$

Serine is converted through a series of reactions to phosphoglyceric acid, part of which is assimilated through conversion to triose phosphate, and part of which serves to regenerate glycine, the primary C_1 acceptor. The reactions of glycine regeneration are complex. They involve the conversion of phosphoglyceric acid via phosphoenolpyruvic acid to malate, with an accompanying fixation of CO_2. Malate is cleaved to glyoxylate and acetyl-CoA; the latter intermediate gives rise to a second molecule of glyoxylate, after passage through the glyoxylate cycle.

The distribution of these two cyclic pathways among methylotrophs has been examined. In facultative methylotrophs only the serine pathway occurs. Among obligate methylotrophs the serine pathway operates in organisms with a type II membrane system (*Methylosinus-Methylocystis*), and the pentose phosphate pathway operates in organisms with a type I membrane system (*Methylomonas-Methyl-obacter-Methylococcus*)(see Table 18.11).

The mechanisms of carbon assimilation characteristic of these organisms probably explain why obligate methylotrophs cannot grow at the expense of formate, even though it is the terminal intermediate in methane oxidation. If formate cannot be reduced to formaldehyde, it cannot provide a reduced carbon

TABLE 18.10

Principal differences between the ribose-phosphate cycle for the fixation of formaldehyde and the Calvin cycle for the fixation of CO_2

	CALVIN CYCLE	RIBOSE-PHOSPHATE CYCLE
C_1 acceptor	Ribulose-1,5-diP	Ribose-5-P
Product of fixation step	C_3 compound (phosphoglyceric acid)	C_6 compound (allulose phosphate)
Reductive step for regeneration of sugar phosphates	Reduction of phosphoglyceric acid to triose phosphate	Not required

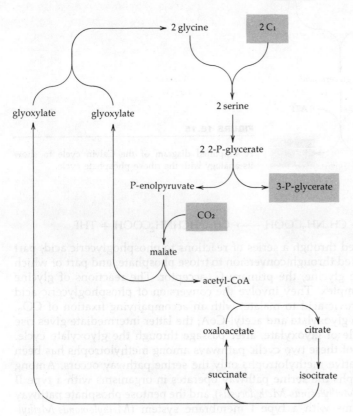

FIGURE 18.16

The serine pathway coupled with the glyoxylate cycle, used for the assimilation of formaldehyde by some methylotrophs.

TABLE 18.11

Distribution of some major intermediary metabolic pathways among methylotrophs

	OBLIGATE METHYLOTROPHS		
	TYPE I MEMBRANE SYSTEM	TYPE II MEMBRANE SYSTEM	FACULTATIVE METHYLOTROPHS
Calvin cycle	−	−	$+^a$ or −
Ribose phosphate cycle	+	−	−
Serine pathway, glyoxylate cycle	−	+	+
Tricarboxylic acid cycle	−	+	+

^{a} Operative in formate utilizers, during growth at the expense of this substrate.

source, and carbon assimilation must necessarily occur at the expense of CO_2. Some facultative methylotrophs can use formate as a growth substrate; and in one member of this group, carbon assimilation during growth with formate has been shown to occur via the Calvin cycle. The ability to synthesize the two key enzymes of this cycle, phosphoribulokinase and carboxydismutase, is therefore a precondition for growth at the expense of formate.

• Occurrence and role of the tricarboxylic acid cycle in methylotrophs

In all facultative methylotrophs the machinery of the glyoxylate cycle is necessarily present, since it plays a key role in carbon assimilation from C_1 substrates. However, growth of these organisms at the expense of substrates containing more than one carbon atom (e.g., acetate) implies that they also possess the complete enzymatic machinery necessary to operate the tricarboxylic acid cycle, an inference which has been confirmed for *Hyphomicrobium*.

There is an interesting dichotomy among the obligate methylotrophs: those organisms which assimilate formaldehyde via the pentose phosphate pathway lack α-ketoglutarate dehydrogenase, and therefore lack a functional tricarboxylic acid cycle. This key enzyme is, however, present in obligate methylotrophs that assimilate formaldehyde through the serine pathway.

In terms of enzymatic constitution, accordingly, there is a sharp dichotomy among obligate methylotrophs. Those with type II membrane systems closely resemble facultative methylotrophs with respect to their enzymatic machinery; those with type I membrane systems have an enzymatic constitution analogous to (but not identical with) that of many obligate chemoautotrophs.

ORIGINS OF CHEMOAUTOTROPHS AND METHYLOTROPHS

In view of the simplicity of their nutritional requirements, chemoautotrophs and methylotrophs were once regarded as "primitive" organisms, possibly representative of the earliest forms of life on earth. This notion of their place in evolution is now untenable, for two reasons. First, their biochemical machinery (and even their cellular fine structure) is at least as complex as that of most chemoheterotrophic bacteria. Second, there is now good evidence to support the view that the earliest living organisms arose on an anaerobic earth, where the oceans contained an abundance of preformed organic matter. The shift to an oxygen-rich biosphere occurred much later (probably about 2 billion years ago); this major geochemical change can be plausibly explained only as a consequence of the evolution of oxygenic photosynthesis. In this evolutionary scenario the appearance of aerobic chemoautotrophs and methylotrophs on earth was dependent on the development of oxygenic photosynthesis. It is therefore conceivable that the chemoautotrophs and methylotrophs arose from procaryotic ancestors which performed either oxygenic or anoxygenic photosynthesis, by loss of the photosynthetic apparatus and adaptation to a new function of the photosynthetic electron transport chain. Unusual properties shared by some members of these two major procaryotic assemblages, one photosynthetic and other nonphotosynthetic, include several elaborate and distinctive types of internal membrane systems; absence of a functional tricarboxylic acid cycle; presence of the Calvin cycle or its analogue, the pentose phosphate cycle; and location of a key enzyme of the Calvin cycle (carboxydismutase) in carboxysomes.

FURTHER READING

Reviews and original articles

ALEEM, M. I. H., "Oxidation of Inorganic Nitrogen Compounds," *Ann. Rev. Plant Physiol.* **21**, 67 (1970).

KELLY, D. P., "Autotrophy: Concepts of Lithotrophic Bacteria and Their Organic Metabolism," *Ann. Rev. Microbiol.* **25**, 177 (1970).

PECK, H. D., "Energy-Coupling Mechanisms in Chemolithotrophic Bacteria," *Ann. Rev. Microbiol.* **22**, 489 (1968).

QUAYLE, J. R., "The Metabolism of One-Carbon Compounds by Microorganisms." *Adv. Microb. Physiol.* **7**, 119 (1972).

RITTENBERG, S. C., "The Obligate Autotroph—the Demise of a Concept," *Antonie v. Leeuwenhoek* **38**, 457 (1972).

WATSON, S. W., "Taxonomic Considerations of the Family *Nitrobacteriaceae* Buchanan," *Int. J. Systematic Bacteriol.* **21**, 254 (1971).

WHITTENBURY, R., K. C. PHILLIPS, and J. F. WILKINSON, "Enrichment, Isolation and Some Properties of Methane-Utilizing Bacteria," *J. Gen. Microbiol.* **61**, 205 (1970).

19

GRAM-NEGATIVE BACTERIA: AEROBIC CHEMOHETEROTROPHS

The groups of Gram-negative bacteria which are photosynthetic were discussed in Chapter 17; and those which synthesize ATP by the oxidation of inorganic compounds or reduced C_1 compounds were discussed in Chapter 18. Chapters 19 through 21 will describe the major groups of Gram-negative bacteria which are chemoheterotrophs and which use organic compounds containing more than one carbon atom as energy and carbon sources. These organic substrates may be dissimilated either by respiration or by fermentation. Chapter 19 presents a survey of groups which possess a purely respiratory metabolism and which, if motile, bear flagella. Most gliding Gram-negative bacteria are also organisms with a purely respiratory metabolism; they will be described separately in Chapter 21.

The characters that distinguish the major groups to be reviewed in this chapter are shown in Table 19.1. Primary taxonomic differentiation is based on structural properties, particularly cell shape and the mode of flagellar insertion (polar or peritrichous). However, certain groups (and many constituent genera within groups) are distinguished by additional properties of a physiological or ecological nature.

The great majority of the bacteria belonging to the groups listed in Table 19.1 are dependent on O_2 as a terminal electron acceptor, and hence are *strict aerobes*, for which molecular oxygen is always an essential nutrient. Denitrifying bacteria provide the only exception to this rule; since they can use nitrate in place of O_2 as a terminal electron acceptor, these organisms can also grow anaerobically. Anaerobic growth through denitrification can be readily distinguished from fermentative growth by its strict dependence on the provision of nitrate in quantities sufficient to meet respiratory needs. Denitrification results in vigorous gas formation, since N_2 is the major endproduct of nitrate reduction.

The respiratory transport chains of Gram-negative chemoheterotrophs differ widely in composition, particularly with respect to cytochrome components. The

TABLE 19.1

Major subgroups among aerobic Gram-negative chemoheterotrophs

CELL SHAPE	FLAGELLAR INSERTION	MOLES % G + C IN DNA	OTHER DISTINCTIVE PROPERTIES	GROUP	GENERA INCLUDED
Rods	Polar	58–70	None	Aerobic pseudomonads	Pseudomonas Xanthomonas
Rods	Peritrichous or subpolar	58–70	Some form nodules or galls on plants		Alcaligenes Rhizobium Agrobacterium
Rods	Polar or subpolar	59–65	Prosthecate; special division cycle	Caulobacter group	Caulobacter Asticcacaulis
Rods	Peritrichous, rarely polar	57–70	Free-living, aerobic nitrogen fibers	Azotobacter group	Azotobacter Azomonas Beijerinckia
Rods	Polar or peritrichous	55–64	Oxidize organic substrates incompletely	Acetic acid bacteria	Gluconobacter Acetobacter
Rods	Polar or subpolar	69–70	Form sheaths	Sphaerotilus group	Sphaerotilus Leptothrix
Helical	Polar	30–65	None	Spirillum group	Spirillum Aquaspirillum Oceanospirillum Campylobacter Bdellovibrio
Cocci or short rods	Nonflagellate	40–52	None	Moraxella-Neisseria group	Neisseria Branhamella Moraxella Acinetobacter

oxidase test, frequently employed in the identification of these organisms, reveals differences with respect to the nature of the cytochromes in the transport chain. Oxidase-positive aerobes immediately form colored products when their cells are mixed with a solution of *p*-phenylenediamine or a related oxidizable amine; the reaction is associated with the presence of a cytochrome of the c type in the transport chain. Oxidase-negative aerobes, which do not give this reaction, have transport chains without a cytochrome c component.

The respiratory dissimilation of most inorganic substrates, whether linked to the reduction of O_2 or of nitrate, leads to the formation of CO_2 as the principal or sole oxidized product, and involves the operation of the tricarboxylic acid cycle as the mechanism of terminal oxidation. However, terminal oxidation cannot occur until the primary substrate has been converted to acetyl-CoA or to other metabolites which are intermediates of the tricarboxylic acid cycle. The ability to use a wide and varied range of organic compounds as sole sources of carbon and energy, which is characteristic of many of the aerobic bacteria discussed in this chapter, therefore depends on their possession of numerous special metabolic pathways for substrate oxidation, all of which converge on the tricarboxylic acid cycle. Some of these pathways occur in many of the groups listed in Table 19.1.

TABLE 19.2

Groups of Gram-negative bacteria in which hexose dissimilation occurs through the Entner-Doudoroff pathway

---

A. Aerobes with a purely respiratory metabolism

    Aerobic pseudomonads
    Caulobacter group
    Azotobacter group
    Agrobacterium
    Rhizobium
    Spirillum

B. Facultative anaerobes able to ferment sugars

    Zymomonas

For example, the respiratory dissimilation of sugars almost always takes place in these groups through the Entner-Doudoroff pathway (see Figure 6.7), which yields two moles of acetyl-CoA per mole of hexose oxidized. Indeed, as shown in Table 19.2, the Entner-Doudoroff pathway is typical for Gram-negative bacteria with a purely respiratory metabolism, and rarely operates in the fermentation of hexoses.

Another major metabolic route for the respiratory conversion of primary substrates to intermediates of the tricarboxylic acid cycle is the β-ketoadipate pathway, present in many groups of Gram-negative chemoheterotrophs (Table 19.3). Its two convergent branches serve principally for the dissimilation of aromatic compounds, the six ring carbon atoms in the various primary substrates being eventually converted to acetyl-CoA and succinate. Other points of entry to the tricarboxylic acid cycle are fumarate (formed in the oxidation of pyridine derivatives such as nicotinic acid) and α-ketoglutarate (formed in the oxidation of pentoses by certain Gram-negative aerobes). The metabolic interrelations between some of these primary pathways and the tricarboxylic acid cycle are shown schematically in Figure 19.1.

Most of the organic substrates that can be dissimilated aerobically by denitrifying bacteria can also be dissimilated by these organisms under anaerobic conditions, with nitrate as a terminal electron acceptor. However, this rule does not hold for aromatic substrates and for certain other substrates, such as aliphatic

TABLE 19.3

Gram-negative bacteria in which the dissimilation of aromatic compounds occurs through the β-ketoadipate pathway

---

Aerobic pseudomonads (genus *Pseudomonas*)
Alcaligenes
Azotobacter
Moraxella group (genus *Acinetobacter*)

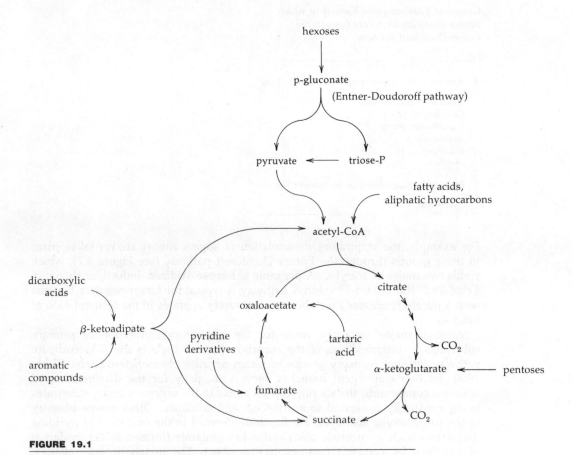

FIGURE 19.1

Convergence of various pathways for the aerobic dissimilation of organic substrates with the reactions of the tricarboxylic acid cycle.

hydrocarbons, because one or more early steps in substrate breakdown are catalyzed by oxygenases, for which molecular oxygen is an obligatory cosubstrate. As a result, these pathways can function only under aerobic conditions.

One of the most striking collective properties of the aerobic chemoheterotrophs is their nutritional range; there are probably very few naturally occurring organic compounds that cannot be used by some of these bacteria as principal sources of carbon and energy. This can be very simply shown by enrichment culture experiments. When a synthetic medium containing almost any organic compound as sole source of carbon and energy is inoculated with soil or water and incubated aerobically, the predominant bacterial population that develops is largely composed of Gram-negative chemoheterotrophs. In many natural environments these bacteria appear to be the principal agents for the aerobic mineralization of organic material. The members of the Azotobacter group have an additional natural role of great importance, as agents of nitrogen fixation: they are virtually the only chemoheterotrophs which can fix nitrogen aerobically.

The acetic acid bacteria provide a striking exception to the general rule that the respiratory dissimilation of organic substrates yields CO_2 as the principal end product. These organisms oxidize many organic compounds with the accumulation of large amounts of organic products, the formation of acetic acid from ethanol being the most characteristic of these partial substrate oxidations.

The Gram-negative aerobic chemoheterotrophs include a considerable number of bacteria which are either pathogenic for, or live in close association with, plants or animals. Some of these associations will be discussed in the following pages.

• The aerobic pseudomonads

Among the many groups of Gram-negative flagellated rods that contain DNA with a base composition in the range of 58 to 70 moles percent G + C (Table 19.1), the organisms known as *aerobic pseudomonads* have received the most extensive study. It must be emphasized that the present limits of this group are somewhat arbitrary, its distinction from such organisms as *Alcaligenes, Agrobacterium, Rhizobium,* and the acetic acid bacteria being based on characters the taxonomic significance of which has not yet been carefully evaluated. The primary criteria for assigning a Gram-negative, aerobic chemoheterotroph to this particular group is the mode of flagellar insertion, which is polar (Figure 19.2). However, some rods with polar flagella are excluded (e.g., the Caulobacter group, some acetic acid bacteria, and some members of the Azotobacter group) on the basis of special characters. About 30 species are now recognized among the aerobic pseudomonads; with the exception of yellow-pigmented plant pathogens, assigned to the genus *Xanthomonas,* they are placed in the genus *Pseudomonas.* Nucleic acid hybridization has revealed that the aerobic pseudomonads are a group of considerable internal heterogeneity, the constituent species being assignable to a total of five major and isolated genetic homology groups (see p. 517). We shall describe here some of the species representative of four of these homology groups: fluorescent pseudomonads; the pseudomallei group; the acidovorans group; and the Xanthomonas group (Table 19.4).

A somewhat variable but distinctive property of fluorescent pseudomonads (Table 19.5) is the production of a yellow-green, water-soluble pigment, which diffuses into the medium and is fluorescent under ultraviolet light. These bacteria do not require growth factors, and they do not synthesize poly-β-hydroxybutyrate

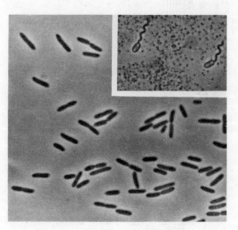

FIGURE 19.2

Phase contrast photomicrograph of cells of *Pseudomonas aeruginosa* ($\times 1,100$). Inset (upper right): flagella stain of *Pseudomonas stutzeri,* showing the polar monotrichous flagellation characteristic of many *Pseudomonas* species ($\times 1,290$). Courtesy of N. J. Palleroni.

TABLE 19.4

Differential properties of four subgroups among the aerobic pseudomonads

	DNA BASE COMPOSITION, MOLES % G + C	POLY-β-HYDROXY-BUTYRATE AS RESERVE MATERIAL	PIGMENTATION SOLUBLE, FLUORESCENT	PIGMENTATION CELLULAR, YELLOW	GROWTH FACTORS REQUIRED	USE AS PRINCIPAL CARBON AND ENERGY SOURCES GLUCOSE	L-THREONINE	NORLEUCINE
Fluorescent group	60–67	–	+	–	–	+	–	–
Pseudomallei group	67–69	+	–	–	–	+	+	–
Acidovorans group	61–69	+	–	–	–	–	–	+
Xanthomonas group	64–69	–	–	+	–	+	–	–

TABLE 19.5

Distinctions between the principal species of fluorescent pseudomonads

	MOLES % G + C IN DNA	NUMBER OF FLAGELLA PER CELL	PRODUCTION OF PYOCYANIN	GROWTH AT 41°C	HYDROLYSIS OF GELATIN	OXIDASE REACTION	DENTRIFI-CATION
P. aeruginosa	67	1	+	+	+	+	+
P. putida	62	>1	−	−	−	+	−
P. fluorescens	60–63	>1	−	−	+	+	V[a]
P. syringae	58–60	>1	−	−	+	−	−

[a] V; variable within species.

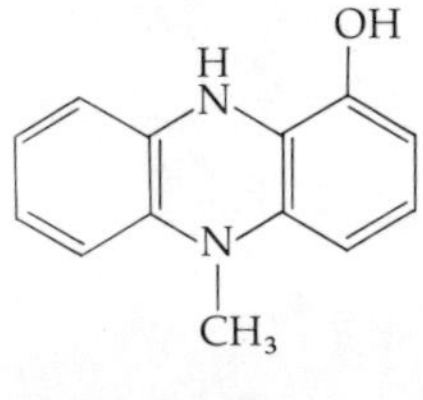

FIGURE 19.3

The structure of pyocyanin, the blue pigment of *Pseudomonas aeruginosa*.

as a cellular reserve material. In addition to the yellow fluorescent pigments, pyocyanin, a blue phenazine pigment (Figure 19.3), is characteristic of the species *P. aeruginosa*. This species, *P. fluorescens*, and *P. putida* are common members of the microflora of soil and water, and are all nutritionally highly versatile, being able to use 60 to 80 different organic compounds as sole sources of carbon and energy. For this reason, they have been much studied by microbial biochemists as biological material for the elucidation of the special metabolic pathways involved in the dissimilation of different classes of organic compounds. They have also recently become accessible to genetic study, following the discovery of conjugational and transductional systems of genetic transfer within the group. One outcome of this work has been the discovery that the genetic determinants governing certain of the special pathways of substrate dissimilation (e.g., the pathway for camphor dissimilation) are carried on plasmids, transmissible from strain to strain (see Chapter 15).

P. aeruginosa, which has a considerably higher temperature maximum than *P. fluorescens* and *P. putida*, is sometimes pathogenic for man. It belongs to the category of opportunistic pathogens, which do not normally exist in animal hosts, but which can establish infections in individuals whose natural resistance has been reduced. Thus, *P. aeruginosa* typically causes infections, not infrequently fatal, in victims of severe burns and in cancer patients who have been treated with immunosuppressive drugs.

The fluorescent pseudomonads also include organisms that are pathogenic for plants; the many varieties, which differ in host range, are assigned to one species, *P. syringae*. These plant pathogens are true parasites, readily distinguishable from the free-living soil and water species by their physiological and biochemical properties. They are less versatile nutritionally; and their growth rates, both in synthetic and in complex media, are much lower. They are also the only oxidase negative members of the fluorescent group.

The pseudomallei group (Table 19.6), like the fluorescent group, are nutritionally versatile organisms which do not require growth factors. Although usually pigmented, they never produce a yellow-green diffusible fluorescent pigment; and they all synthesize poly-β-hydroxybutyrate as a reserve material. The prototype of this group, *P. pseudomallei*, was originally discovered as the agent of melioidosis, a highly fatal tropical disease of man and other mammals. Even in the tropical areas where melioidosis is endemic, it is a relatively rare disease, typically contracted through the contamination of wounds with soil or mud. In fact, *P. pseudomallei* appears to be, like *P. aeruginosa*, an opportunistic pathogen which is a normal

TABLE 19.6

Distinguishing characters of the three principal species of the pseudomallei group

| | NUMBER OF FLAGELLA | MOLES % G + C IN DNA | DENITRIFICATION | EXTRACELLULAR HYDROLYSIS OF | | CARBON SOURCES FOR GROWTH | | | |
				STARCH	POLY-β-HYDROXYBUTYRATE	D-XYLOSE	D-RIBOSE	ERYTHRITOL	m-HYDROXYBENZOATE
P. pseudomallei	>1	69	+	+	+	+	+	+	−
P. mallei	0	69	V[a]	V	V	+	+	−	−
P. cepacia	>1	67	−	−	−	+	+	−	+

[a] V: variable within species.

member of the microflora of soil and water in the tropics. However, the closely related species *P. mallei* is a true parasite, causing a disease of horses known as glanders. *P. mallei* is unable to survive in nature in the absence of its specific animal host. It is the only aerobic pseudomonad that is permanently immotile: its inclusion in the group is based on its close genetic relationship and its phenotypic similarity to *P. pseudomallei.*

The pseudomallei group also contains several species that occur in soil and are occasionally pathogenic for plants; they are exemplified by *P. cepacia,* notable for its extreme nutritional versatility; it can use over 100 different organic compounds as carbon and energy sources. These compounds are listed in Table 2.3

The soil bacteria of the acidovorans group, comprising the species *P. acidovorans* and *P. testosteroni* (Table 19.7) are nonpigmented; they accumulate poly-β-hydroxybutyrate and do not require growth factors. The nutritional spectrum of these bacteria is a distinctive one. Many of the organic substrates commonly utilized by the fluorescent and pseudomallei groups cannot support their growth; these include glucose and other aldose sugars, polyamines such as putrescine and spermine, and several amino acids, notably arginine and lysine. Nevertheless, they can grow at the expense of several organic acids and amino acids rarely, if ever, utilized by members of other groups: these include glycollate, muconate, and norleucine.

Another interesting property of the acidovorans group is the possession of metabolic pathways for the dissimilation of widely utilized substrates that are unlike those found in other aerobic pseudomonads. For example, *p*-hydroxybenzoate, dissimilated through the β-ketoadipate pathway by all other aerobic pseudomonads capable of utilizing this substrate, is dissimilated through a special (*meta* cleavage) pathway by the acidovorans group (Figure 19.4).

Certain plant pathogenic pseudomonads have long been placed in a special genus, *Xanthomonas,* distinguished by the production of yellow cellular pigments. Nucleic acid homology studies have shown that these xanthomonads are a genetically isolated subgroup, although distantly related to one other aerobic pseudomonad, *Pseudomonas maltophilia.* In contrast to the groups so far discussed, both xanthomonads and *P. maltophilia* require organic growth factors, including methionine and (for xanthomonads) a few of the B vitamins.

It was long assumed that the yellow cellular pigments of xanthomonads are carotenoids, but chemical studies have shown that these pigments are of unique chemical composition, being brominated derivatives of aryl octanes, of the general structure shown in Figure 19.5. Pigments of this type are not known to occur in any other bacteria; thus, the chemical nature of the cellular pigments is a highly distinctive property of the xanthomonas group.

• Gram-negative rods of the genera *Alcaligenes, Agrobacterium,* and *Rhizobium*

The Gram-negative aerobic chemoheterotrophs with rod-shaped cells include many representatives in which flagellar insertion is not polar, and which are hence excluded by definition from the aerobic pseudomonads. These bacteria normally bear few (1 to 4) flagella and the flagellar insertion, which is not easy to determine unambiguously, is often described as "subpolar" or as "degenerately peritrichous." Apart from the practical difficulty of determining this character, there are considerable grounds for doubting whether or not it really possesses the taxonomic importance that has traditionally been ascribed to it. An extension of the study of genetic interrelationships by nucleic acid hybridization, which has been so successful in revealing the internal relationships of the aerobic pseudomonads,

TABLE 19.7

Distinguishing characters of the two principal species of the acidovorans group

	MOLES % G + C IN DNA	CARBON SOURCES FOR GROWTH			
		FRUCTOSE	TESTOSTERONE	ETHANOL	L-TRYPTOPHAN
P. acidovorans	69	+	−	+	+
P. testosteroni	62	−	+	−	−

to these nonpolarly flagellated rods will probably aid considerably in classifying their taxonomic status, which is now most unclear.

At the present time, the members of the genera *Rhizobium* and *Agrobacterium* are distinguished by their special relationships (described below) to plants. The genus *Alcaligenes* is used as a repository for other Gram-negative, rod-shaped aerobes in which flagellar insertion is nonpolar.

Of these groups, the genus *Rhizobium* has been studied in greatest detail, because of its major agricultural importance. The rhizobia can invade the root hairs of leguminous plants and initiate the formation of root nodules, within which they develop as intracellular symbionts and fix nitrogen (see Chapter 27 for further discussion). However, pure cultures of rhizobia cannot grow at the expense of N_2. This curious fact has been recently explained. Although rhizobia are strict aerobes, their nitrogenase is extremely susceptible to oxygen inactivation within the cell. Hence the enzyme is active only in cultures exposed to very low O_2 concentrations, which severely limit growth. The root nodule must therefore provide a special environment, which favors the maintenance of high rhizobial nitrogenase activity. Six species of *Rhizobium* are recognized, primarily on the basis of their host ranges among leguminous plants, these ranges being relatively narrow and specific. The genus is almost certainly a heterogeneous one, being divisible into two groups of species that differ in mode of flagellar insertion and in growth rate on complex media (Table 19.8). Although the rhizobia persist for some time in soil, they are probably unable to grow in competition with members of the free-living soil microflora. Hence, when a leguminous plant is introduced into an area where it has not been previously cultivated, inoculation of the seeds with the specific rhizobial symbiont is essential to ensure satisfactory nodulation and nitrogen fixation.

The organisms of the genus *Agrobacterium* are responsible for the formation of galls or tumors on the roots or stems of many different plant families. The principal species is *A. tumefaciens;* the nature of its interaction with the plant host presents a fascinating and still unsolved problem. Although bacterial infection is essential to initiate tumor formation, the bacteria soon disappear from the tumor, which continues to develop in their absence. This suggests that bacterial infection leads to a transmissible genetic modification of the infected plant cells, but the nature and mechanism of the genetic change so induced remain unknown.

● **The Caulobacter group** The members of the Caulobacter group are primarily aquatic organisms, which occur in both fresh water and the sea. These prosthecate bacteria are distinguished

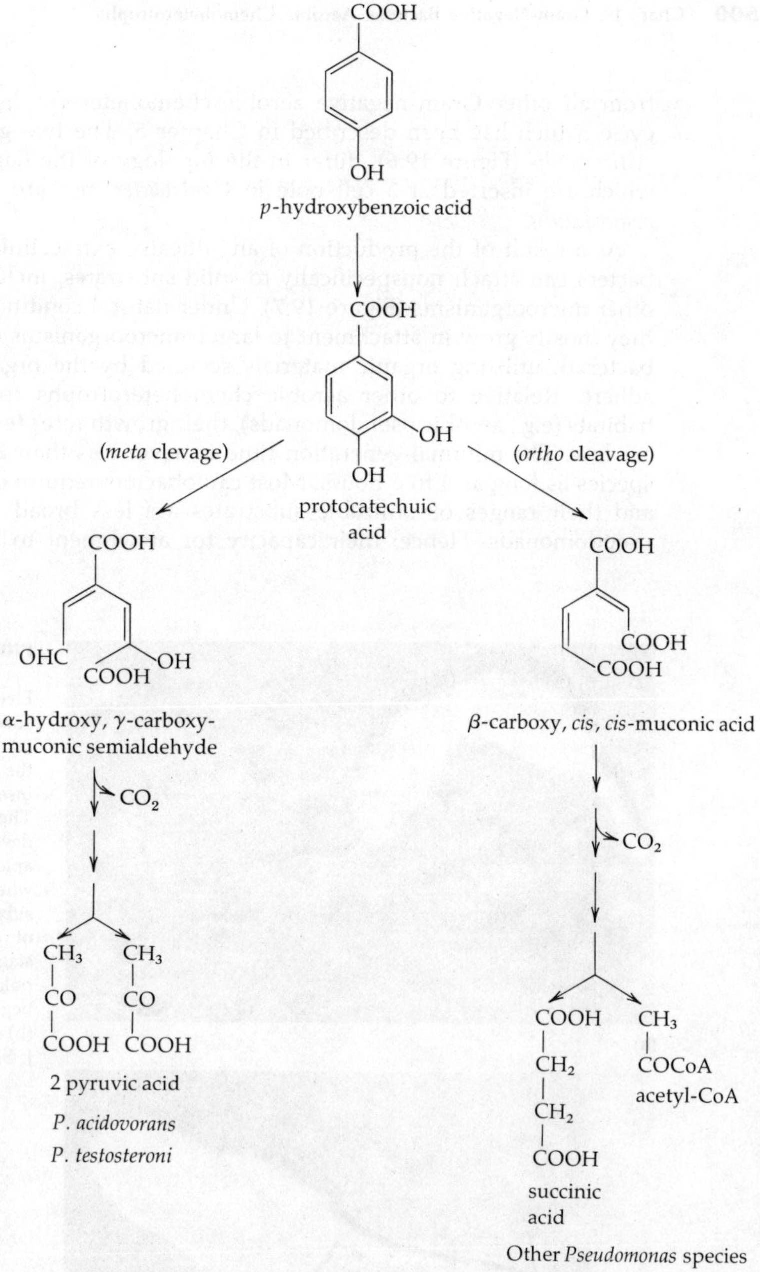

COOH

OH

p-hydroxybenzoic acid

COOH

OH

OH

protocatechuic
acid

(*meta* clevage)

(*ortho* cleavage)

COOH

OHC

COOH

OH

α-hydroxy, γ-carboxy-
muconic semialdehyde

CO_2

CH₃

CO

COOH

CH₃

CO

COOH

2 pyruvic acid

P. acidovorans

P. testosteroni

COOH

COOH

COOH

β-carboxy, *cis*, *cis*-muconic acid

CO_2

COOH

CH₂

CH₂

COOH

succinic
acid

CH₃

COCoA

acetyl-CoA

Other *Pseudomonas* species

FIGURE 19.4

The divergent pathways
for the oxidation of *p*-hy-
droxybenzoate among
aerobic pseudomonads.

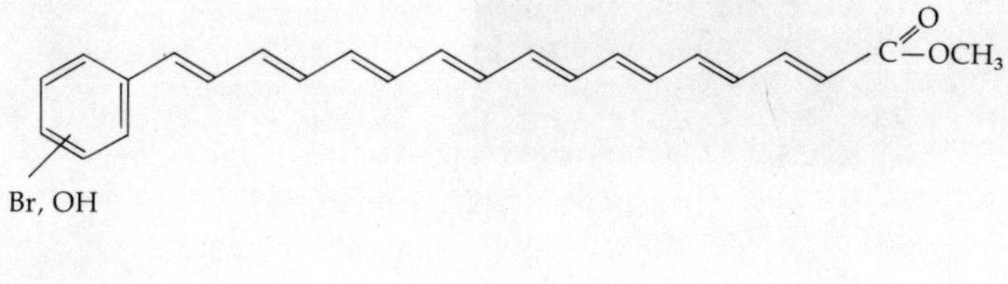

$C{=}O$
OCH_3

Br, OH

FIGURE 19.5

The structure of one of the brominated aryl
octanes which are the group-specific pigments of
xanthomonads. The bromine atom is attached to
the ring, but the point of attachment is uncertain.

from all other Gram-negative aerobic chemoheterotrophs by their unique life cycle, which has been described in Chapter 5. The two genera, *Caulobacter* and *Asticcacaulis* (Figure 19.6), differ in the topology of the flagellum and prostheca, which are inserted at a cell pole in *Caulobacter*, but are subpolar or lateral in *Asticcacaulis*.

As a result of the production of an adhesive extracellular holdfast, the caulo-bacters can attach nonspecifically to solid substrates, including the cell walls of other microorganisms (Figure 19.7). Under natural conditions, it is probable that they mostly grow in attachment to larger microorganisms (algae, protozoa, other bacteria), utilizing organic materials secreted by the organisms to which they adhere. Relative to other aerobic chemoheterotrophs from the same aquatic habitats (e.g., aerobic pseudomonads), their growth rates (even in complex media) are low. The minimal generation time is never less than 2 hours, and for many species as long as 4 to 6 hours. Most caulobacters require organic growth factors, and their ranges of utilizable substrates are less broad than those of aerobic pseudomonads. Hence, their capacity for attachment to other microorganisms

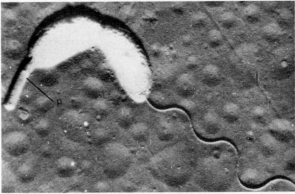

(a)

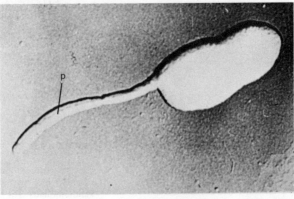

(b)

FIGURE 19.6

Electron micrographs of metal-shadowed cells of a *Caulobacter* (a) and *Asticcacaulis* (b), to show the difference in the site of insertion of the prostheca (p). The dividing *Caulobacter* cell has developed a flagellum at the apical pole, inserted at the site where the new prostheca will subsequently develop. The cell of *Asticcacaulis* is at an early stage of division, and the sub-polar flagellum has not yet begun to develop. (a) ×20,000; (b) ×34,000. Courtesy of Dr. J. S. Poindexter.

TABLE 19.8

The two subgroups of the genus *Rhizobium*

	MOLES % G + C OF DNA	FLAGELLA INSERTION	FLAGELLA NUMBER	GROWTH ON COMPLEX MEDIA	PREFERRED CARBON SOURCES	VITAMIN REQUIREMENTS	REPRESENTATIVE SPECIES	PLANT HOST RANGE
Subgroup I	59–63	Peritrichous	2–6	Rapid	Glucose, fructose	Biotin (some)	*R. leguminosarum*	Pea, vetch, lentil
Subgroup II	62–66	Subpolar	1	Slow	Pentoses	None	*R. lupini*	Lupins

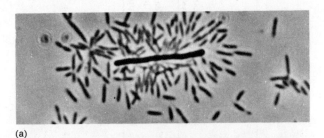

(a)

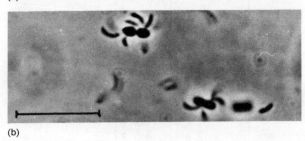

(b)

FIGURE 19.7

The attachment of *Caulobacter* to (a) *Bacillus* and (b) *Azotobacter* (scale marker, 1 μm). From J. L. S. Poindexter, "Biological Properties and Classification of the *Caulobacter* group," *Bact. Rev.* **28**, 231 (1962).

appears to be an important factor for successful competition with other aerobic chemoheterotrophs in nature.

• **The Azotobacter group**   These rod-shaped organisms possess a property that does not occur in any other group of Gram-negative chemoheterotrophs: the ability to fix nitrogen under aerobic growth conditions. In view of the extreme sensitivity of nitrogenase to oxygen inactivation, the existence of this ability in bacteria which are strict aerobes appears paradoxical. It is evident that the Azotobacter group must possess special mechanisms for the protection of nitrogenase, since facultatively anaerobic nitrogen-fixing chemoheterotrophs (e.g., *Bacillus polymyxa*) can maintain nitrogenase activity only when growing in the absence of oxygen. The azotobacters have extraordinarily high respiratory rates, far in excess of those of all other aerobic bacteria, and this may prevent O_2 penetration to the intracellular sites of nitrogenase activity. It has also been suggested that nitrogenase exists in the *Azotobacter* cell in a special conformational state which renders it oxygen-resistant.

The members of the *Azotobacter* group have oval to rod-shaped cells, which are large (as much as 2 μm wide) in most species. They form poly-β-hydroxybutyrate as a reserve material. Cultures on solid media have a characteristic mucoid appearance (Figure 19.8), since these organisms produce large amounts of extracellular polysaccharide. The members of the genera *Azotobacter* and *Azomonas* (Table 19.9) are common in temperate regions in neutral or alkaline soils and waters. In tropical regions the far more acid-tolerant members of the genus *Beijerinckia* are the prevalent members of the aerobic, nitrogen-fixing soil microflora; they can grow at pH values as low as 3, and thus are well adapted to the relatively acid soils characteristic of this climatic zone.

The members of the genus *Azotobacter* (but not of the other two genera) produce distinctive resting cells known as *cysts* (Figure 19.9). These structures, which arise

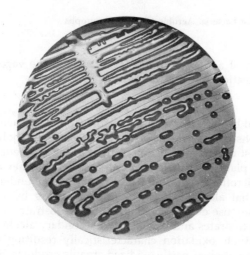

FIGURE 19.8

A streaked plate of *Azotobacter vinelandii*, showing the smooth, glistening colonies typical of *Azotobacter*. Courtesy of O. Wyss; reproduced from his *Elementary Microbiology*. New York: Wiley, 1963.

TABLE 19.9

Properties of the principal genera of the *Azotobacter* group

	MODE OF FLAGELLAR INSERTION	MOLES % G + C IN DNA	FORMATION OF CYSTS	MINIMAL pH FOR GROWTH	HABITAT
Azotobacter	Peritrichous	63–66	+	5.5	Neutral to alkaline soils
Azomonas	Peritrichous or polar	53–59	–	4.5	Water[a]
Beijerinckia	Peritrichous or absent	55–60	–	3.0	Acid soils in tropical regions

[a] One species is a soil microorganism.

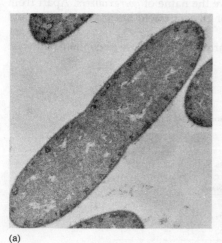

(a)

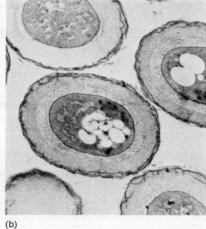

(b)

FIGURE 19.9

Electron micrographs of thin sections of vegetative cells and cysts of *Azotobacter vinelandii*. (a) Vegetative cells ($\times 17,800$). (b) Cysts ($\times 10,700$). Both from O. Wyss, M. G. Neumann, and M. D. Socolofsky, "Development and Germination of the *Azotobacter* Cyst," *J. Biophys. Biochem. Cytol.* **10**, 555 (1961).

by the deposition of additional outer layers around the vegetative cell wall, are resistant to desiccation but not to heat.

• **The acetic acid bacteria**   These rod-shaped bacteria are distinguishable from other aerobic Gram-negative chemoheterotrophs by a series of physiological and metabolic characters, which seem to be at least in part a reflection of their ecology. Acetic acid bacteria occur on the surface of plants, particularly flowers and fruits. They develop abundantly as a secondary microflora in decomposing plant material under aerobic conditions, following an initial alcoholic fermentation of sugars by yeasts. Under these circumstances, they use ethanol as an oxidizable substrate, converting it to acetic acid. Many carbohydrates and primary and secondary alcohols can also serve as energy sources, their oxidation characteristically resulting in the transient or permanent accumulation of partly oxidized organic products. Since most members of the group have relatively complex growth factor requirements, they are usually grown in media with a complex base (e.g., yeast extract), supplemented with an oxidizable substrate. These bacteria are markedly acidophilic, growing at pH values as low as 4, with an optimum between pH 5 and 6.

By virtue of their capacity to convert many organic compounds almost stoichiometrically to partly oxidized organic end products, the acetic acid bacteria are a group of considerable industrial importance. Their major industrial use is in the manufacture of vinegar, by the acetification of ethanol-containing materials (e.g., wine, cider).

Two genera (Table 19.10) are distinguished by differences in flagellation and in oxidative capacities. The *Gluconobacter* group do not have a functional tricarboxylic acid cycle, and cannot oxidize acetate; hence, they oxidize ethanol (and other substrates convertible to acetate) with the stoichiometric accumulation of acetic acid; for this reason they are known in the vinegar industry as *underoxidizers*. The *Acetobacter* group possess a functional tricarboxylic acid cycle, and hence they can oxidize acetate to CO_2. When growing at the expense of ethanol, they convert this substrate rapidly to acetic acid, which is then oxidized more slowly to completion, a property to which they owe the name of *overoxidizers*. Apart from this difference, the oxidative metabolism of all acetic acid bacteria is similar, and it has a number of unusual features.

Sugars are oxidized to CO_2 exclusively through the pentose phosphate pathway:

TABLE 19.10

The distinguishing characters of the genera *Gluconobacter* and *Acetobacter*

GENUS	FLAGELLAR INSERTION	MOLES % G + C IN DNA	PRESENCE OF TRICARBOXYLIC ACID CYCLE
Gluconobacter	Polar or absent	60–64	−
Acetobacter	Peritrichous or absent	55–65	+

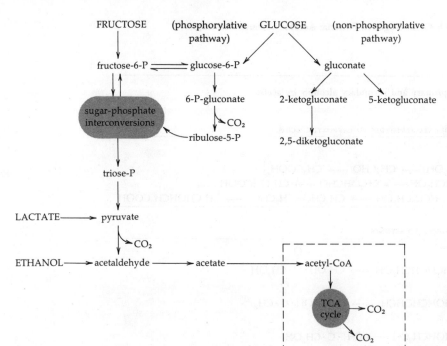

FRUCTOSE (phosphorylative GLUCOSE (non-phosphorylative
pathway) pathway)

fructose-6-P $\rightleftharpoons$ glucose-6-P gluconate

6-P-gluconate 2-ketogluconate 5-ketogluconate

sugar-phosphate
interconversions CO_2

ribulose-5-P 2,5-diketogluconate

triose-P

LACTATE $\longrightarrow$ pyruvate

CO_2

ETHANOL $\longrightarrow$ acetaldehyde $\longrightarrow$ acetate $\longrightarrow$ acetyl-CoA

TCA
cycle CO_2

CO_2

FIGURE 19.10

Major pathways of oxidative metabolism in acetic acid bacteria. The terminal reactions of acetate metabolism (square box) do not occur in *Gluconobacter*. Primary substrates are capitalized.

the Entner-Doudoroff pathway, otherwise common in aerobic chemoheterotrophs, does not operate in this group. The metabolism of pyruvate is also unusual. Whereas in most aerobes, pyruvate is oxidized to acetyl-CoA and CO_2, the acetic acid bacteria decarboxylate it nonoxidatively to acetaldehyde. Since acetaldehyde is also the first intermediate in ethanol dissimilation, it lies at the point of metabolic convergence between the oxidation of substrates metabolized through pyruvate and the oxidation of ethanol (Figure 19.10).

In addition to the oxidative pentose phosphate pathway, the acetic acid bacteria oxidize glucose through a second, nonphosphorylative pathway which results in the accumulation of partly oxidized products—gluconate and ketoacids derived from it (Figure 19.10). Growth at the expense of glucose is therefore always accompanied by the conversion of a large part of the substrate to these acidic derivatives.

In addition to the oxidation of primary alcohols with accumulation of the corresponding carboxylic acids, exemplified by the oxidation of ethanol to acetic acid, the acetic acid bacteria can oxidize many secondary alcohols to ketones. Only polyalcohols can be attacked in this manner, since the secondary alcoholic group which undergoes oxidation must be adjacent to another alcoholic group, primary or secondary. Some of the oxidations of primary and secondary alcohols carried out by acetic acid bacteria are listed in Table 19.11.

The species *Acetobacter xylinum* is one of the few procaryotes able to synthesize the polysaccharide, cellulose. This substance is formed as a result of growth at the expense of glucose and certain other sugars, and is deposited outside the cells in the form of a loose, fibrillar mesh (Figure 19.11). In liquid cultures this organism forms a tough, cellulosic pellicle which encloses the cells and can attain a thickness of several centimeters.

FIGURE 19.11

Extracellular formation of cellulose by *Acetobacter xylinum* (×860). The bacterial cells are entangled in a mesh of cellulose fibrils. From J. Frateur, "Essai sur la systématique des acétobacters," *La Cellule* 53, 3 (1950).

TABLE 19.11

Examples of the oxidations of primary and secondary alcohols by acetic
acid bacteria

1 OXIDATIONS OF PRIMARY ALCOHOLS VIA ALDEHYDES TO CARBOXYLIC ACIDS

SUBSTRATE	REACTION
Ethanol	$CH_3CH_2OH \longrightarrow CH_3CHO \longrightarrow CH_3COOH$
Ethylene glycol	$CH_2OHCH_2OH \longrightarrow CH_2OHCHO \longrightarrow CH_2OHCOOH$
1,3-butanediol	$CH_3CHOHCH_2CH_2OH \longrightarrow CH_3CHOHCH_2CHO \longrightarrow CH_3CHOHCH_2COOH$

2. OXIDATIONS OF SECONDARY ALCOHOLS TO KETONES

SUBSTRATE	REACTION
Glycerol	$CH_2OHCHOHCH_2OH \longrightarrow CH_2OH \cdot \overset{O}{\overset{\|}{C}} \cdot CH_2OH$
2,3-butanediol	$CH_3CHOHCHOHCH_3 \longrightarrow CH_3CHOH \cdot \overset{O}{\overset{\|}{C}} \cdot CH_3$
1,2-propanediol	$CH_3CHOHCH_2OH \longrightarrow CH_3 \cdot \overset{O}{\overset{\|}{C}} \cdot CH_2OH$
Sugar alcohols (oxidation to keto-sugars)	mannitol $\longrightarrow$ fructose sorbitol $\longrightarrow$ sorbose erythritol $\longrightarrow$ erythrulose

● **The Sphaerotilus
group**

The *Sphaerotilus* group are relatively large rod-shaped organisms which grow
as chains of cells enclosed in tubular sheaths (Figure 19.12), usually attached to
solid substrates by basal holdfasts. Reproduction occurs by the liberation of motile
cells, bearing polar or subpolar flagella, from the open apex of the sheath. These
bacteria are all aquatic, the two principal genera being distinguished by their
habitats and by the structure of their sheaths. *Sphaerotilus* forms thin sheaths,
normally without incrustations, and is found in slowly running streams contaminated with sewage or other organic matter, where it grows as long, slimy, attached
tassels. It also develops in aerobic sewage treatment ponds. The members of the
genus *Leptothrix* are common in uncontaminated fresh water containing iron salts,
and their sheaths are heavily encrusted with hydrated ferric or manganic oxides
(Figure 19.13(b)).

The sheaths are chemically complex, made up of proteins, polysaccharides,
and lipids. Labeling experiments show that new sheath material is laid down only
at the apical tip of the filament, presumably by the apical cell in the growing
chain.

The respiratory substrates are alcohols, organic acids, and sugars. Vitamin B_{12}
is required, and although inorganic nitrogen sources can be utilized, these bacteria
grow best when provided with amino acids as nitrogen sources.

The ability of *Leptothrix* spp. to deposit ferric hydroxide in the sheath led
Winogradsky to suggest that these bacteria might be iron-oxidizing chemoautotrophs. However, subsequent work has failed to substantiate this hypothesis.

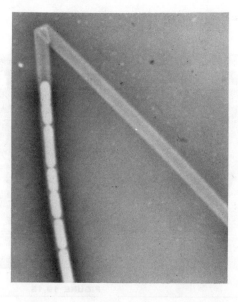

FIGURE 19.12

Sphaerotilus, showing a chain of cells enclosed within the sheath. From J. L. Stokes, "Studies on the Filamentous Sheathed Iron Bacterium *Sphaerotilus natans,"* J. Bacteriol. **67,** 281 (1954).

Although it is clear that the oxidation of ferrous and manganous salts is mediated by *Leptothrix* spp., the process appears to be the result of a nonspecific reaction, catalyzed by sheath proteins, and consequently leading to a massive deposit of the metal oxides on or in the sheath. Such a process evidently cannot provide the enclosed cells with ATP; and all current evidence suggests that *Leptothrix* spp., like *Sphaerotilus,* are obligate chemoheterotrophs.

• **The Spirillum group**

The spirilla are aerobic chemoheterotrophs characterized by the possession of helical cells, bearing bipolar flagellar tufts (Figure 19.14). They are all aquatic bacteria, common in both fresh water and marine environments. The group is composed of three genera (Table 19.12).

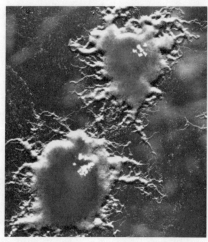

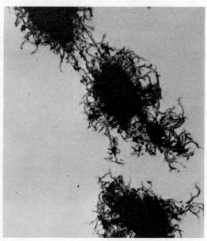

FIGURE 19.13

Colonies of two *Sphaerotilus* species. (a) *Sphaerotilus natans* (×22). (b) *Sphaerotilus discophorus,* growing on agar containing MnCO$_3$ (×146); the dark color of the colonies is caused by heavy incrustation of the sheaths with manganic oxide. Courtesy of J. L. Stokes, and M. A. Rouf.

(a)

(b)

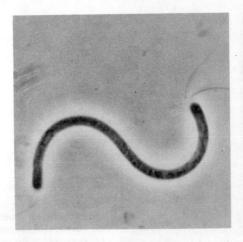

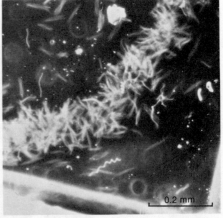

FIGURE 19.14

A single cell of the largest spirillum, *S. volutans* (phase contrast). Note the flagellar tufts at one pole of the cell. From N. R. Krieg, "Cultivation of *Spirillum volutans* in a Bacteria-free Environment," *J. Bacteriol.* **90**, 817 (1965).

FIGURE 19.15

Typical aerotactic pattern of *Spirillum* in a wet mount, showing the accumation of the cells in a very narrow band some distance from the air-water interface. From N. R. Krieg, "Cultivation of *Spirillum volutans* in a Bacteria-free Environment," *J. Bacteriol.* **90**, 817 (1965).

The substrates most commonly used as carbon and energy sources include a limited range of amino acids and organic acids; sugars are not utilized, except by one species which can grow with fructose. A physiological property, shared to some degree by all spirilla, which is uncommon in strictly aerobic bacteria is a *preference for low oxygen tensions*. The microaerophilic tendencies of these bacteria are shown by their behavior in wet mounts; the highly motile cells accumulate in a dense, narrow band, located at some distance from the edge of the cover slip (Figure 19.15). Despite this, most spirilla can form colonies on the surface of agar media. However, *Spirillum volutans*, a species with very large cells, cannot be isolated on streaked plates exposed to air. It is an *obligate microaerophile*, which can initiate growth only in an oxygen-depleted environment containing 3 to 9 percent O_2, instead of the normal atmospheric concentration of 20 percent.

TABLE 19.12

Differential properties of the *Spirillum* group

GENUS	CELL DIAMETER, μm	MOLES % G + C IN DNA	OBLIGATELY MICRO- AEROPHILIC	GROWTH IN PRESENCE OF NaCl			HABITAT
				NONE	3%	9%	
Spirillum	1.4–1.7	38	+	+	−	−	Fresh water
Aquaspirillum	0.2–1.4	50–65	−	+	−	−	Fresh water
Oceanospirillum	0.3–1.2	42–48	−	−	+	+	Marine

TABLE 19.13

The *Campylobacter-Bdellovibrio* group

GENUS	CELL STRUCTURE	MOLES % G + C IN DNA	ECOLOGICAL PROPERTIES
Campylobacter	Curved or helical rods, 0.3–0.8 by 0.5–5 μm, single polar flagellum	30–35	Parasites of mammals
Bdellovibrio	Curved or helical rods, 0.3–0.4 by 0.8–1.2 μm, single polar flagellum	43–50	Parasites of other Gram-negative bacteria

• *Campylobacter* and *Bdellovibrio*

Possibly related to the spirilla are two genera of parasitic bacteria with very small curved or helical cells, bearing single polar flagella (Table 19.13). The members of the genus *Campylobacter* are animal parasites or pathogens with complex and as yet undefined nutritional requirements. They are pronouncedly microaerophilic and, like most spirilla, utilize amino acids and organic acids as energy sources, being unable to grow at the expense of carbohydrates. The members of the genus *Bdellovibrio* are parasites of other Gram-negative bacteria, attaching to, penetrating, and killing the host cells. Many have not yet been grown in the absence of their bacterial hosts; some, which are facultative parasites, can be grown on complex media. The properties of *Bdellovibrio* are further discussed in Chapter 28.

• The Moraxella and Neisseria group

The members of this group are nonflagellated short rods or cocci (Figure 19.16). The DNA base compositions range from approximately 40 to 50 moles percent G + C.

Generic distinctions are shown in Table 19.14. An unusual property shared by the parasitic members of the group is their marked penicillin sensitivity; except in strains that have acquired penicillin resistance, growth is inhibited by penicillin at levels of 1 to 10 units/ml, whereas the inhibitory concentration for Gram-negative bacteria usually lies between 100 and 1,000 units/ml. The coccoid organisms of the genera *Neisseria* and *Branhamella* are parasites found in the mucous membranes of mammals; the genus *Neisseria* includes two human pathogens, the agents of gonorrhea and meningitis. As shown in Table 19.14, the two

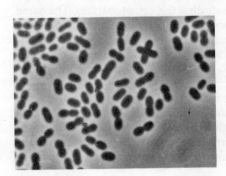

(a)

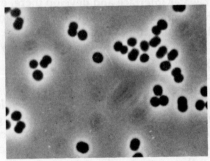

(b)

FIGURE 19.16

The *Moraxella-Neisseria* group (phase contrast, ×2,200). (a) *Moraxella osloensis*. (b) *Neisseria catarrhalis*.

TABLE 19.14

The distinguishing properties of the genera in the *Moraxella-Neisseria* group

GENUS	CELL SHAPE	PLANES OF DIVISION	MOLES % G + C IN DNA	RESISTANCE TO PENICILLIN, 10 UNITS/ML	GROWTH FACTORS REQUIRED	OXIDASE REACTION
Neisseria	Cocci	2	47–52	−	+	+
Branhamella	Cocci	2	40–45	−	+	+
Moraxella	Rods	1	40–46	−	V[a]	+
Acinetobacter	Rods[b]	1	39–47	+	−	−

[a] V: variable within group.
[b] Rod in exponential phase; becomes spherical in stationary phase.

genera are very similar in phenotypic respects, but they differ slightly in their ranges of DNA base composition. Their separation is based primarily on genetic considerations: studies by DNA transformation have shown no genetic affinities between the members of the two genera, whereas *Branhamella* is genetically related to the rod-shaped organisms of the genus *Moraxella*. The rod-shaped organisms of the genera *Moraxella* and *Acinetobacter* (Figure 19.16), although similar in many phenotypic respects and in DNA base composition, give no evidence of genetic relationships, either by transformation or by nucleic acid hybridization *in vitro*. *Moraxella*, like the *Neisseria-Branhamella* group, consists of parasites on the mucous membranes of vertebrates, whereas the *Acinetobacter* group are free-living bacteria with an exceptionally wide natural distribution. The two genera are most easily distinguished by the oxidase test and penicillin sensitivity, as shown in Table 19.14.

The members of the *Acinetobacter* group are nutritionally versatile chemoheterotrophs. The range of substrates used as sole carbon and energy sources parallels that of the aerobic pseudomonads; and like these organisms, the *Acinetobacter* group are common in soil and water, from which they can be selected by similar enrichment procedures. The outcome of such enrichment cultures appears to be largely determined by a secondary environmental factor, aeration. In unshaken liquid enrichment cultures the aerobic pseudomonads often predominate, since as a result of their motility and aerotactic behavior, these organisms can occupy the air-water interface and develop there, excluding the acinetobacters. The same enrichment medium, receiving the same inoculum, will often yield predominantly acinetobacters, if mechanically agitated to ensure full aeration; under these conditions, the selective advantage possessed by aerobic pseudomonads through their ability to occupy the well-aerated surface layer of an unshaken medium disappears.

The acinetobacters cannot use hexoses as carbon and energy sources, a characteristic which they share with the *Branhamella-Moraxella* group. Nevertheless, many *Acinetobacter* strains can produce acid from glucose and other sugars, when grown in a complex medium that contains these carbohydrates. Acid production results from the oxidation of aldose sugars to the corresponding sugar acids, as exemplified by the reaction:

$$\text{glucose} + O_2 \longrightarrow \text{gluconic acid}$$

611   It is mediated by a single, highly nonspecific, aldose dehydrogenase, which can oxidize at least ten different sugars, including pentoses, hexoses, and disaccharides.

FURTHER READING

Books   ASAI. T., *Acetic Acid Bacteria, Classification and Biochemical Activities.* Baltimore: University Park Press, 1973.

CLARKE, P. H., and M. H. RICHMOND (editors), *Genetics and Biochemistry of Pseudomonas.* New York: John Wiley, 1975.

Reviews and original articles   ANDREWES, A. G., S. HERTZBERG, S. LIAAEN-JENSEN, and M. P. STARR, "Xanthomonas Pigments. 2. The *Xanthomonas* 'Carotenoids'—Non-Carotenoid Brominated Arylpolyene Esters," *Acta Chem. Scandinavica,* **27,** 2383 (1973).

BAUMANN, P., M. DOUDOROFF, and R. Y. STANIER, "A Study of the *Moraxella* Group. I. Genus *Moraxella* and the *Neisseria catarrhalis* Group," *J. Bact.* **95,** 58 (1968).

———, "A Study of the *Moraxella* Group. II. Oxidase-Negative Species (Genus *Acinetobacter*)," *J. Bact.* **95,** 1520 (1968).

DeLEY, J., "Comparative Biochemistry and Enzymology in Bacterial Classification," in *Microbial Classification,* G. C. Ainsworth and P. H. A. Sneath (editors). Cambridge, England: Cambridge University Press, 1962, p. 164.

DIXON, R. O. D., "Rhizobia (with Particular Reference to Relations to Host Plants)," *Ann. Rev. Microbiol.* **23,** 137 (1969).

HENRIKSEN, S. D., "*Moraxella, Acinetobacter* and the *Mimae,*" *Bact. Revs.* **37,** 522 (1973).

HYLEMON, P. B., J. S. WELLS, N. R. KRIEG, and H. W. JANNASCH, "The Genus *Spirillum:* A Taxonomic Study," *Int. J. System. Bact.* **23,** 340 (1973).

JENSEN, H. L., "The Azotobacteriaceae," *Bact. Revs.* **18,** 195 (1954).

MULDER, E. G., "Iron Bacteria, Particularly Those of the *Leptothrix-Sphaerotilus* Group," *J. Appl. Bact.* **27,** 151 (1964).

PALLERONI, N. J., and M. DOUDOROFF, "Some Properties and Taxonomic Subdivision of the Genus *Pseudomonas,*" *Ann. Rev. Phytopathol.* **10,** 73 (1972).

PALLERONI, N. J., R. KUNISAWA, R. CONTOPOULOU, and M. DOUDOROFF, "Nucleic Acid Homologies in the Genus *Pseudomonas,*" *Int. J. System. Bact.* **23,** 333 (1973).

POINDEXTER, J. S., "Biological Properties and Classification of the *Caulobacter* Group, *Bact. Revs.* **28,** 231 (1964).

SCHMIDT, J. L., "Prosthecate Bacteria," *Ann. Rev. Microbiol.* **25,** 93 (1971).

STANIER, R. Y., N. J. PALLERONI, and M. DOUDOROFF, "The Aerobic Pseudomonads: A Taxonomic Study," *J. Gen. Microbiol.* **43,** 159 (1966).

STARR, M. P., and R. J. SEIDLER, "The Bdellovibrios," *Ann. Rev. Microbiol.* **25,** 649 (1971).

VÉRON, M., and R. CHATELAIN, "Taxonomic Study of the Genus *Campylobacter* Sebald and Véron," *Int. J. System. Bact.* **23,** 122 (1973).

WHITE, L. O., "The Taxonomy of the Crown-Gall Organism *Agrobacterium tumefaciens* and Its Relationship to Rhizobia and Other Agrobacteria," *J. Gen. Microbiol.* **72,** 565 (1972).

20

THE ENTERIC GROUP AND
RELATED ORGANISMS

The organisms treated in this chapter constitute one of the largest well-defined groups among the Gram-negative, nonphotosynthetic eubacteria. They have small, rod-shaped cells, either straight or curved, not exceeding 0.5 μm in width. Some are permanently immotile; motile representatives include organisms with peritrichous flagella, with polar flagella, and with "mixed" (polar-peritrichous) flagellation. They can be distinguished from all other Gram-negative eubacteria of similar structure by the property of *facultative anaerobiosis*. Under anaerobic conditions, energy is provided by fermentation of carbohydrates; under aerobic conditions, a wide range of organic compounds can serve as substrates for respiration.

The classical representative is *Escherichia coli*, one of the most characteristic members of the normal intestinal flora of mammals. Closely related to this organism are the coliform bacteria of the genera *Salmonella* and *Shigella*. They are pathogens, responsible for such intestinal infections as bacterial dysentery, typhoid fever, and bacterial food poisoning.

Clearly related to these coliform bacteria, but of different ecology, are the genera *Enterobacter*, *Serratia*, and *Proteus*, which occur primarily in soil and water, and the plant pathogens of the genus *Erwinia*. Together with the coliform bacteria, these genera have long been placed in a single family, the *Enterobacteriaceae*. They constitute the enteric group, as classically defined.

In recent years it has been recognized that certain bacteria pathogenic for animals which were formerly placed in the ill-defined genus *Pasteurella* are in fact members of the enteric group. These organisms are now classified in the genus *Yersinia*. They include *Yersinia pestis*, the agent of bubonic plague, an infection markedly different in both mode of transmission and symptomatology from the enteric infections.

The bacteria so far discussed are either immotile or peritrichously flagellated. The primary importance which was for so long accorded to the mode of flagellar insertion as a taxonomic character impeded the recognition that some polarly flagellated bacteria are also allied to the enteric group, and most appropriately treated as members of it. These are all aquatic bacteria, which occur either in freshwater (*Vibrio, Aeromonas*) or in the marine environment (*Beneckea, Photobacterium*). Many of the marine forms show "mixed" (polar-peritrichous) flagellation. Some are animal pathogens; these include two species which cause intestinal diseases (*Vibrio cholerae, Beneckea parahemolytica*).

As yet, there is no generally accepted collective name for this entire assemblage. It will be termed here the *enteric group,* but it must be emphasized that this designation includes genera (the polarly flagellated group) that fall outside the traditional confines of the family *Enterobacteriaceae.*

COMMON PROPERTIES OF THE ENTERIC GROUP

Most members of the enteric group can use a considerable number of simple organic compounds as substrates for respiratory metabolism: organic acids, amino acids, and carbohydrates are universally utilized. Under aerobic conditions, all these bacteria grow well in conventional complex bacteriological media, the nitrogenous constituents of which (amino acids and peptides) provide oxidizable substrates. Under anaerobic conditions, however, growth becomes strictly dependent on the provision of a fermentable carbohydrate. Some monosaccharides, disaccharides, and polyalcohols are fermented by all members of the enteric group. The utilization of polysaccharides is less common; however, pectin is attacked by many of the plant pathogens (*Erwinia*), and chitin and alginic acid by many of the marine *Beneckea* species.

Although it is customary to grow enteric bacteria on complex media, the minimal nutritional requirements of these organisms are usually simple. In many genera no growth factors are required (e.g., *Escherichia, Enterobacter, Serratia,* most *Salmonella* species). Auxotrophic representatives usually have very simple growth factor requirements. A requirement for nicotinic acid is particularly frequent, occurring in many species of the genera *Proteus, Erwinia,* and *Shigella.* A specific requirement for tryptophan exists in *Salmonella typhi,* and for methionine in some *Photobacterium* species.

The regulation of amino acid biosynthesis has been studied in many members of the enteric group, and this work has revealed distinctive regulatory patterns that appear to set these organisms apart from other bacteria. For example, the initial step in the biosynthesis of amino acids of the aspartate family, conversion of aspartic acid to aspartylphosphate, is always mediated in the enteric group by three isofunctional aspartokinases: aspartokinase I, which is both inhibited and repressed by threonine; aspartokinase II, repressed (but not inhibited) by methionine; and aspartokinase III, both inhibited and repressed by lysine. This particular mode of regulation of the aspartate pathway has not been demonstrated in any bacteria outside the enteric group.

The mean DNA base composition for members of the enteric group is rather wide, extending from 37 to 63 moles percent (Table 20.1). With the exception of *Proteus,* in which one species differs widely from the rest in DNA base composition, the ranges within each genus are narrow. The values for the closely related organisms of the three genera *Escherichia, Salmonella,* and *Shigella* are not significantly different. The total span of base composition for the "classical" enteric

TABLE 20.1

The ranges of mean DNA base composition characteristic of the members of the enteric group

		MOLES % GUANINE + CYTOSINE IN DNA				
		40	45	50	55	60
PERITRICHOUS FLAGELLATION OR IMMOTILE	Escherichia Salmonella Shigella			←——→		
	Yersinia		←→			
	Proteus	←——→		←→		
	Ewinia				←————→	
	Enterobacter				←————→	
	Serratia				←——→	
POLAR OR "MIXED" FLAGELLATION	Vibrio		←——→			
	Aeromonas					←——→
	Beneckea			←——→		
	Photobacterium	←——→				

bacteria of the family *Enterobacteriaceae* (37 to 59 moles percent) corresponds rather closely to that for the polarly flagellated members (39 to 63 moles percent).

• Fermentative metabolism   Sugar fermentation by enteric bacteria occurs through the Embden-Meyerhof pathway. The products vary, both qualitatively and quantitatively. These fermentations have one characteristic biochemical feature, however, which is not encountered in any other bacterial fermentations. This is a special mode of cleavage of the intermediate, pyruvic acid, to yield formic acid:

$$CH_3COCOOH + CoASH \longrightarrow CH_3COSCoA + HCOOH$$

Formic acid is, therefore, frequently a major fermentative end product. It does not always accumulate, however, since some of these bacteria possess the enzyme formic hydrogenlyase, which splits formic acid to CO_2 and H_2:

$$HCOOH \longrightarrow CO_2 + H_2$$

In such organisms, formic acid is largely replaced as a fermentative end product by equimolar quantities of H_2 and CO_2.*

The most frequent mode of fermentative sugar breakdown in the enteric group

*Formation of molecular hydrogen as an end product of sugar fermentation is also characteristic of many sporeformers of the genera *Clostridium* and *Bacillus* (see Chapter 22). The biochemical mechanism responsible for its production is, however, different. In sporeformers, hydrogen is formed as a *direct* product of pyruvic acid cleavage:

$$CH_3COCOOH + CoASH \longrightarrow CH_3COSCoA + H_2 + CO_2$$

is the *mixed-acid fermentation*, which yields lactic, acetic, and succinic acids; formic acid (or CO$_2$ and H$_2$); and ethanol. This fermentation is characteristic of the genera *Escherichia*, *Salmonella*, *Shigella*, *Proteus*, *Yersinia*, and *Vibrio*, and it occurs in some *Aeromonas*, *Beneckea*, and *Photobacterium* species. The ratios of the end products may vary considerably, both from strain to strain and in a single strain grown under different environmental conditions (e.g., at different pH values). This variability reflects the fact that the end products arise from pyruvic acid through three independent pathways (Figure 20.1).

In some enteric bacteria, sugar fermentation gives rise to an additional major end product, 2,3-butanediol, which is formed from pyruvic acid by a fourth independent pathway (Figure 20.2). This *butanediol fermentation* is characteristic of *Enterobacter* and *Serratia*, most species of *Erwinia*, and some *Aeromonas*, *Beneckea*, and *Photobacterium* species. The formation of butanediol is accompanied by increased formation of the reduced end product ethanol, since the formation of butanediol from glucose results in a net generation of reducing power:

$$C_6H_{12}O_6 \longrightarrow CH_3CHOHCHOHCH_3 + 2CO_2 + 2H$$

Hence, neutral end products (butanediol and, to a lesser extent, ethanol) predominate over the acidic end products; as a result, the total amount of acid formed per mole of glucose fermented is considerably less in a butanediol fermentation than in a simple mixed-acid fermentation. This is shown by the fermentation balances for a number of different species assembled in Table 20.2.

The formation of gas as a result of sugar fermentation is a character of considerable differential value in the enteric group, since it distinguishes the gasformers of the genus *Escherichia* from the pathogens of the *Shigella* group and *Salmonella typhi* which ferment sugar without gas production. In a simple mixed-acid fermentation, gas can be formed only by the cleavage of formate; gas production therefore reflects the possession of formic hydrogenlyase. This enzyme is not, of course, essential for fermentative metabolism and can be lost by mutation without effect on fermentative capacity. In fact, experience has shown that "anaerogenic" (i.e., nongas-producing) strains of such a typical gas producer as

FIGURE 20.1

Pathways of formation from pyruvic acid of the characteristic end products (boldface) of a mixed-acid fermentation.

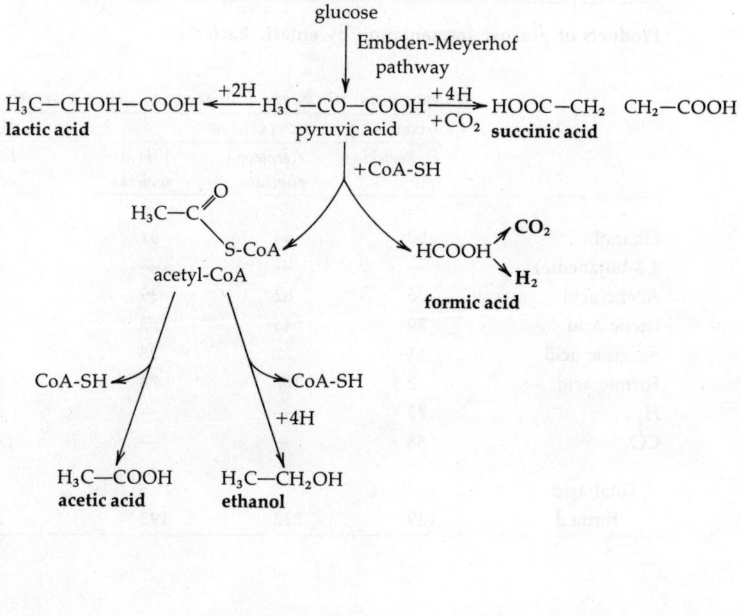

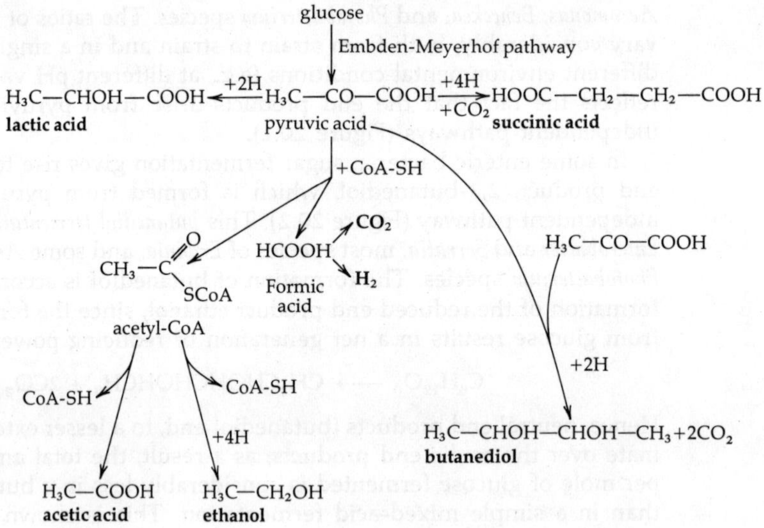

glucose

Embden-Meyerhof pathway

$$H_3C—CHOH—COOH \xleftarrow{+2H} H_3C—CO—COOH \xrightarrow[+CO_2]{+4H} HOOC—CH_2—CH_2—COOH$$

lactic acid   pyruvic acid   **succinic acid**

+CoA-SH

$CH_3—C\overset{O}{\big|}$ HCOOH $\xrightarrow{} CO_2$ $H_3C—CO—COOH$

SCoA   Formic H_2

acetyl-CoA   acid

FIGURE 20.2

Pathways of formation from pyruvic acid of the characteristic end products (boldface) of a butanediol fermentation.

CoA-SH   CoA-SH

+4H

$H_3C—COOH$ $H_3C—CH_2OH$
acetic acid   **ethanol**

+2H

$H_3C—CHOH—CHOH—CH_3 + 2CO_2$
butanediol

Escherichia coli exist in nature. Hence, although gas production is a useful differential character in the enteric group, it is by no means an infallible one.

The bacteria that perform a butanediol fermentation also differ with respect to the possession of formic hydrogenlyase. Members of the genus *Enterobacter* almost always contain formic hydrogenlyase and are vigorous gas producers; members of the genus *Serratia* do not contain the enzyme and produce little or no visible gas, as judged by the customary criterion (formation of a bubble in an inverted vial placed in the fermentation tube). This may appear paradoxical,

TABLE 20.2

Products of glucose fermentation by enteric bacteria

	PRODUCTS, MOLES PER 100 MOLES OF GLUCOSE FERMENTED				
	MIXED–ACID FERMENTATIONS			BUTANEDIOL FERMENTATIONS	
	Escherichia coli	*Aeromonas punctata*	*Vibrio cholerae*	*Enterobacter aerogenes*	*Serratia marcescens*
Ethanol	50	64	51	70	46
2,3-butanediol	—	—	—	66	64
Acetic acid	36	62	39	0.5	4
Lactic acid	79	43	53	3	10
Succinic acid	11	22	28	—	8
Formic acid	2.5	105	73	17	48
H_2	75	—	—	35	—
CO_2	88	—	—	172	116
Total acid formed	129	232	193	20	70

since the formation of butanediol from sugars is accompanied by a considerable net production of CO_2. However, this gas is very soluble in water, so that most (or all) of the CO_2 produced tends to remain dissolved in the medium. When CO_2 is the sole gaseous product of a bacterial fermentation, special cultural methods may be required to demonstrate its formation, a point discussed in connection with the lactic acid bacteria (Chapter 23).

Another character of considerable diagnostic importance in the enteric group is the ability to ferment the disaccharide, lactose, which depends on possession of β-galactosidase. Effective lactose utilization also depends on possession of a specific galactoside permease which facilitates entry of lactose into the cell. Strains lacking permease but containing β-galactosidase cannot take up lactose at a sufficient rate to produce a prompt and vigorous fermentation and will normally be classified as nonfermenters of this sugar. Lactose fermentation is characteristic of *Escherichia* and *Enterobacter* and is absent from *Shigella, Salmonella,* and *Proteus*. It should be noted that some *Shigella* strains produce β-galactosidase but cannot ferment lactose because they lack permease.

The fermentative characteristics of the various genera have been summarized in Table 20.3.

• **Some physiological characters of differential value**   A few physiological and biochemical characters are of considerable use in distinguishing major subgroups within the enteric group. One is the oxidase reaction, the mechanism of which was discussed in Chapter 19 (see p. 590). The "classical" enteric group (*Escherichia-Salmonella-Shigella-Proteus-Erwinia-Enterobacter-Serratia*) and

---

TABLE 20.3

Summary of fermentative patterns in the enteric group and related organisms

I. MIXED-ACID FERMENTATION
   A. Produce CO_2 + H_2 (contain formic hydrogenlyase):
      Escherichia
      Proteus
      Salmonella (most spp.)
      Photobacterium (some spp.)
   B. No gas produced (formic hydrogenlyase absent):
      Shigella
      Salmonella typhi
      Yersinia
      Vibrio
      Aeromonas (some spp.)
      Photobacterium (some spp.)
      Beneckea (most spp.)

II. BUTANEDIOL MODIFICATION OF MIXED-ACID FERMENTATION
   A. Produce CO_2 + H_2 (contain formic hydrogenlyase):
      Enterobacter
      Aeromonas hydrophila
      Photobacterium phosphoreum
   B. Produce only CO_2 (formic hydrogenlyase absent); visible gas formation slight or undetectable:
      Serratia
      Erwinia herbicola and *E. carotovora*
      Beneckea alginolytica

the genus *Yersinia* are all oxidase negative. Most polar flagellates, however, are oxidase positive; this is true of *Vibrio, Aeromonas,* and most species of *Beneckea* and *Photobacterium.*

Nearly all groups of enteric bacteria synthesize glycogen as the sole organic cellular reserve material. The formation of poly-β-hydroxybutyrate as a cellular reserve material is a property confined to the marine bacteria of the genera *Beneckea* and *Photobacterium,* although it does not occur in all species.

The members of the genera *Beneckea* and *Photobacterium* can be readily distinguished from all other members of the enteric group by their *salt requirements.* Indigenous marine bacteria have an absolute requirement for sodium ions, no growth occurring if sodium salts are omitted from the medium. Furthermore, the magnitude of this requirement is considerable, concentrations of Na$^+$ ranging from about 100 to 300 mM being necessary to assure growth at maximal rate. A specific Na$^+$ requirement cannot be demonstrated for most nonmarine bacteria (apart from extreme halophiles). Furthermore, both Mg^{2+} and Ca^{2+} are required for marine bacteria at much higher concentrations than those which satisfy these essential mineral requirements for other bacteria. In all these respects, the members of *Beneckea* and *Photobacterium* are typical marine bacteria; they cannot grow in media that contain low concentrations of Na$^+$, Mg^{2+}, and Ca^{2+}, which are adequate to support growth at maximal rates of other enteric bacteria.

GENETIC RELATIONS AMONG ENTERIC BACTERIA

The discovery of conjugational and transductional gene transfer in the enteric group has permitted rather detailed studies of genetic relatedness among certain of its members. Chromosomal hybrids can be produced between *E. coli* and members of both the genera *Salmonella* and *Shigella,* which indicates that all these bacteria share a very high degree of genetic homology. This is confirmed by the chromosomal maps of the two species that have been subjected to the most detailed genetic analysis, *Escherichia coli* and *Salmonella typhimurium.* As shown by partial maps of these two species (Figures 20.3 and 20.4), many

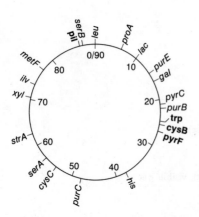

FIGURE 20.3

Partial genetic map of *E. coli* strain K12. Compare with the partial genetic map of *Salmonella typhimurium* shown in Figure 20.4 and with the more complete genetic map of *E. coli* shown in Figure 15.14. The numbers along the inside of the circle are units of distance and equal the number of minutes required to transfer that length of the chromosome during conjugal transfer at 37°C (see Chapter 15). Different genetic loci which have been identified and mapped are indicated by three-letter or four-letter symbols. Over 200 loci have been mapped in *E. coli* strain K12 and almost as many in *S. typhimurium,* of which approximately 100 have the same functions in the two species. Of these 100, only a few have different map locations in the two species. The locus for pilus formation (pil) for example, is at minute 88 in *E. coli* but at minute 23 in *S. typhimurium; E. coli* contains the genes for lactose utilization (*lac*) at minute 10, whereas the *lac* genes are totally absent from *S. typhimurium.* Finally, the chromosomal segment *trp–cysB–pyrF,* located at minute 25 on the *E. coli* chromosome, is inverted in *S. typhimurium.* Modified from A. Taylor and C. Trotter, "Linkage Map of *Escherichia coli* Strain K-12" *Bact. Rev.* **36,** 504 (1972).

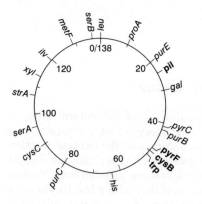

FIGURE 20.4

Partial genetic map of *S. typhimurium.* Compare with the partial genetic map of *E. coli* strain K12 shown in Figure 20.3; the similarities and differences are described in the legend to Figure 20.3. In both cases, the maps have been assigned units of distance corresponding to the number of minutes required to transfer corresponding map segments during conjugal transfer at 37°C. The difference in total "length" (90 minutes vs. 134 minutes) probably does not represent a difference in physical length but rather a difference in speed of transfer or in measurement techniques. Modified from K. Sanderson, "Linkage Map of *Salmonella typhimurium* Edition IV," *Bact. Rev.* **36**, 558 (1972).

markers are located at corresponding positions on the two chromosomes. The very close genetic relatedness of the *Escherichia, Salmonella,* and *Shigella* groups is also indicated by the high levels of intergeneric DNA-DNA hybridization obtained *in vitro.* However, the formation of chromosomal hybrids between this subgroup and representatives of other enteric genera (*Proteus, Enterobacter, Serratia*) is rare, if not totally absent. Moreover, DNA-DNA hybridization experiments indicate that the degree of genetic homology between the *Escherichia-Salmonella-Shigella* subgroup and the other groups of enteric bacteria is rather low. Bacteria belonging to many other genera in the enteric group can acquire plasmids by conjugational transfer from donor strains of *E. coli* (although with widely differing efficiencies of conjugation; see Table 15.2) and subsequently maintain them as extrachromosomal elements. Both substituted F factors (e.g., F-*lac*) and R factors conferring various patterns of drug resistance can be widely disseminated in this manner among the members of the enteric group (Table 20.4).

TABLE 20.4

Genetic evidence concerning relationships among the enteric bacteria and related groups[a]

I. ORGANISMS BETWEEN WHICH HYBRIDIZATION OF CHROMOSOMAL GENES CAN OCCUR

Genetic donors:	Genetic recipients:
Escherichia coli	Many *E. coli* strains *Shigella flexneri* and other *Shigella* spp. *Salmonella typhimurium, S. typhi*
Salmonella typhimurium	*Salmonella typhi,* other *Salmonella* spp.
Shigella flexneri	*Escherichia coli, Salmonella typhi*

II. ORGANISMS BETWEEN WHICH EPISOMES CAN BE TRANSFERRED

A. Substituted F factors of *E. coli* (e.g., F-*lac*):
   Salmonella spp.; *Shigella* spp.; *Serratia marcescens; Proteus* spp.;
   Enterobacter-Klebsiella group; *Vibrio cholerae; Yersinia pestis*

B. Drug resistance (R) factors:
   Escherichia coli; Shigella spp.; *Salmonella* spp.; *Serratia marcescens;*
   Enterobacter–Klebsiella group; *Proteus* spp.; *Vibrio cholerae; Yersinia pestis;*
   Aeromonas hydrophila

---

[a] Data courtesy of D. J. Brenner, R. V. Citarella, and Stanley Falkow.

TAXONOMIC SUBDIVISION OF THE ENTERIC GROUP

A simplified scheme for the subdivision of the enteric group is shown in Tables 20.5 and 20.6. A reasonably satisfactory primary separation can be made on the basis of the mode of flagellar insertion and the oxidase reaction. The oxidase negative groups which, if motile, possess only peritrichous flagella, include the classical representatives of the *Enterobacteriaceae*, together with *Yersinia*. As shown in Table 20.5, DNA base composition, together with a few biochemical and physiological characters, permits the recognition of four major subgroups (I–IV). It must be noted that the specialists who have been concerned with the taxonomy of the *Enterobacteriaceae* have created a very large number of genera among these organisms, for the most part distinguished by phenotypic differences so minor that other bacterial taxonomists would employ them (at best) to distinguish species. A few examples will illustrate this problem. The maintenance of the three genera *Escherichia*, *Salmonella*, and *Shigella* cannot really be justified in view of the close genetic relationships now known to exist among them. The generic distinction is based on pathogenic properties alone. Indeed, additional genera (*Arizona, Citrobacter, Edwardsiella*) have been proposed for certain members of this complex. The differences between *Enterobacter* and *Serratia* similarly do not justify a generic separation; and two additional genera (*Klebsiella* and *Hafnia*) have been proposed for organisms very similar to *Enterobacter*. Finally, the genus *Erwinia* is united solely by the dubious character of plant pathogenicity. It is internally heterogeneous, and some (but not all) of the species included resemble the *Enterobacter-Serratia* group. *Erwinia amylovora*, however, clearly does not. The second major group of enteric bacteria consists of organisms that bear polar flagella or display mixed polar-peritrichous flagellation and, with some exceptions, are oxidase positive (Table 20.6). A primary separation within this group can be made on the basis of ionic requirements, which distinguish *Beneckea-Photobacterium* from *Vibrio-Aeromonas*. One character of considerable utility in this group is flagellar structure: the polar flagella of *Vibrio* and *Beneckea* spp. are relatively thick, being enclosed by a sheath that is made up of an extension of the cell membrane. The polar flagella of *Aeromonas* and *Photobacterium* are not sheathed.

• Group I:
Escherichia-
Salmonella-Shigella

The members of this group are all inhabitants of the intestinal tract of man and other vertebrates. The principal generic distinctions are shown in Table 20.7. It should be noted that certain strains of intestinal origin—the so-called *paracolon group*—have characters intermediate between those shown for the genera in Table 20.7, so that distinctions are not always as clear-cut as the table suggests. The additional genera, *Arizona* and *Citrobacter*, were created for the intermediate forms of the paracolon group, but this generic hypertrophy really does nothing to help the problem of differentiation.

Highly detailed, intraspecific subdivisions among the species of group I have been made on the basis of the immunological analysis of the surface structures of the cell. The extreme specificity of antigen-antibody reactions makes it possible to recognize differences in these respects between strains of a bacterial species

TABLE 20.5

Taxonomic subdivision of the peritrichously flagellated enteric bacteria and related immotile forms[a]

MAJOR SUBGROUP	DNA BASE COMPOSITION, MOLE % G + C	MOTILITY	PRODUCTION OF BUTANEDIOL	PRODUCTION OF $H_2 + CO_2$	UREASE	TRYPTOPHAN DEAMINASE	CONSTITUENT GENERA	OTHER GENERIC NAMES FREQUENTLY APPLIED TO SOME MEMBERS OF GROUP
I	50–53	V[b]	–	V	–	–	*Escherichia, Salmonella, Shigella*	*Arizona* and *Citrobacter* for "intermediate" types
II	50–59	V	+	V	–	–	*Enterobacter, Serratia, Erwinia*	*Klebsiella* and *Hafnia*
III	37–50	+	–	+	+	+	*Proteus*	*Providencia*
IV	46–47	V	–	–	+	–	*Yersinia*	Formerly placed in *Pasteurella*

[a] Straight rods, immotile or motile by peritrichous flagella; oxidase negative.
[b] V denotes variable within group.

TABLE 20.6

Taxonomic subdivision of polarly flagellated enteric bacteria[a]

GENUS	DNA BASE COMPOSITION, MOLES % G + C	SHEATHED POLAR FLAGELLA	POLY-β-HYDROXY-BUTYRIC ACID FORMED	BIOLUMIN-ESCENCE	Na^+ REQUIREMENT	FERMENTATIVE CHARACTERS: BUTANEDIOL PRODUCED	FERMENTATIVE CHARACTERS: PRODUCTION OF $H_2 + CO_2$
Aeromonas	57–63	–	–	–	–	V	V
Vibrio	45–49	+	–	–	–	–	–
Beneckea	45–47	+	V[b]	V	+	V	–
Photobacterium	39–43	–	+	+	+	V	V

[a] Straight or curved rods; motile by polar flagella, some showing "mixed" polar-peritrichous flagellation; mainly oxidase positive.
[b] V denotes variable within group.

TABLE 20.7

Internal differentiation of the major genera of group I

CHARACTERISTICS	Escherichia	Salmonella	Shigella
Pathogenicity for man or animals	—	V^a	+
Motility	V	+	—
Gas ($CO_2 + H_2$) from fermentation of glucose	+	$+^b$	—
Fermentation of lactose	+	—	—
β-galactosidase	+	—	V
Utilization of citrate as carbon source	—	+	—
Production of indole from tryptophan	+	—	V

[a] V denotes a variable reaction within the group.
[b] Except for *S. typhosa*.

that are indistinguishable on the basis of other phenotypic criteria. Three classes of surface antigens have been extensively explored among the enteric bacteria of group I: the O antigens, which are the polysaccharide components of the lipopolysaccharides in the outer layer of the cell wall; the K antigens, which are capsular polysaccharides; and the H antigens, which are the flagellar proteins. Many of these organisms possess two sets of genetic determinants for different flagellar antigens that are subject to alternate phenotypic expression, a phenomenon known as *phase variation*. On any given cell the flagella are of one antigenic type, but as the cell multiplies, variants of the alternate type arise with a certain probability. Cultures of such biphasic strains thus contain two specific sets of H antigens.

In the genus *Salmonella* the detailed analysis of O and H antigenic structure (including phase variation of the H antigens) has made it possible to distinguish many hundreds of different *serotypes*; comparable analyses of *Escherichia* and *Shigella* are less extensive. The principal utility of these systems of antigenic classification is not taxonomic, but *epidemiological*. The serotype of a pathogenic *Salmonella* strain is a marker that permits its recognition (and hence allows one to follow its transmission) where other phenotypic characters do not.

Escherichia coli and some members of the paracolon group are components of the normal intestinal flora and give rise to disease only under exceptional conditions. The genera *Salmonella* and *Shigella* comprise pathogens that cause a wide variety of enteric diseases in man and other animals. In all cases, entry occurs through the mouth; and the small intestine is the primary locus of infection, although some of these pathogens may subsequently invade other body tissues and cause more generalized damage in the infected host. The members of the genus *Shigella* are the agents of a specifically human enteric disease, bacterial dysentery. In the genus *Salmonella* both the host range and the variety of diseases produced are much broader. *Salmonella typhi* and *S. paratyphi*, the agents of typhoid fever, are specific pathogens of man, whereas certain other species (e.g., *S. typhi-murium*) are specific pathogens of other mammals or of birds. The great majority of the *Salmonella* group, however, have a low host specificity. They exist, often without causing disease symptoms, in the intestine and in certain tissues of animals or birds. If these forms gain access to and develop in foods, their subsequent ingestion by man can give rise to food poisoning. This can be defined as a gastrointestinal infection which is usually acute but transient, although in some cases it may assume a graver form. Outbreaks of food poisoning often have an epidemic character, because food preparation on a large scale provides many favorable opportunities for the growth of these organisms.

• Group II:
Enterobacter-Serratia-Erwinia

Enterobacter aerogenes, the prototype of this group, is common in soil and water, and sometimes also occurs in the intestinal tract. Similar bacteria, distinguished from *E. aerogenes* by permanent immotility and the presence of capsules, occur in the respiratory tract. Although they are customarily classified in a separate genus, *Klebsiella*, it is questionable whether or not a specific distinction, let alone a generic one, is justified. A biochemical property that distinguishes some (though not all) *Enterobacter* strains from other enteric bacteria is the ability to fix nitrogen. This property can be expressed only under anaerobic growth conditions, since the nitrogenase of these bacteria is rapidly denatured in the cells in the presence of oxygen. Although other enteric bacteria do not possess nitrogenase naturally, the *nif* genes which govern its synthesis have recently been transferred from *E. aerogenes* to *Escherichia coli* in which they can be phenotypically expressed. This piece of genetic engineering was accomplished by the use of a transducing phage, P1.

Serratia marcescens, also a common soil and water organism, differs from *Enterobacter* principally by its failure to produce formic hydrogenlyase (little or no visible gas formed during sugar fermentation) and by its inability or weak ability to ferment lactose (Table 20.8). Many (but by no means all) strains of *Serratia* produce

TABLE 20.8

Internal differentiation of the major genera of group II

CHARACTERISTICS	Enterobacter	Serratia	Erwinia
Motility	V[a]	+	+
CO$_2$ + H$_2$ formed by glucose fermentation	+	−	−
Lactose fermentation	+	−(+)[b]	V
β-galactosidase	+	+	+
Gelatin liquefied	−	+	+(−)[c]
Pectinolytic enzymes produced	−	−	V
Yellow cellular pigments	−	−	V
Red cellular pigments	−	+(−)	−
Plant pathogens or parasites	−	−	+

[a] V denotes a variable reaction within the group.
[b] −(+) denotes predominantly negative, with rare positive strains.
[c] +(−) denotes predominantly positive, with rare negative strains.

a characteristic red cellular pigment, *prodigiosin*, a tripyrrole derivative (Figure 20.5).

Relative to the enteric bacteria so far discussed, the representatives of the genus *Erwinia* constitute a very heterogeneous group, in which three principal subgroups are now recognized, exemplified by the species *Erwinia amylovora, E. carotovora,* and *E. herbicola.*

E. amylovora is the agent of fire blight, a necrotic disease of pears and related plants. This species is notable for its limited range of utilizable sugars and its requirement for organic growth factors, characters absent from other erwinias. *E. carotovora* causes soft rots of the storage tissues of many plants, an action attributable in part to its ability to produce pectolytic enzymes, which destroy the pectic substances that serve as intracellular cementing materials in plant tissues. *E. herbicola,* which produces yellow cellular pigments, occurs commonly on the leaf surfaces of healthy plants; some strains are plant pathogens. Similar pigmented enteric bacteria have occasionally been isolated from human sources, although their pathogenicity for man remains uncertain.

• **Group III:** *Proteus*   The members of the *Proteus* group are probably soil inhabitants, although they are found in particular abundance in decomposing animal materials. The relatively low G + C content of their DNA distinguishes most species from the groups so far discussed, from which they are also distinguishable by certain physiological properties. These include strong proteolytic activity (gelatin is rapidly liquefied) and ability to hydrolyze urea. Most members of the *Proteus* group are very actively motile and can spread rapidly over the surface of a moist agar plate, a phenomenon known as *swarming.* A curious feature of the swarming phenomenon is its periodicity: it occurs in successive waves, separated by periods of quiescence and growth. This produces a characteristic zonate pattern of development on an agar plate (Figure 20.6).

• **Group IV:** *Yersinia*   The genus *Yersinia* contains two or three species which are agents of disease in rodents. *Yersinia pestis* can be transmitted by fleas from its normal rodent hosts to man; it is the cause of human bubonic plague, a disease that has been responsible throughout human history for massive epidemics with a very high mortality. In man the disease can also be transmitted through the respiratory route. Both in their mode of transmission and in their symptoms, the diseases caused by *Yersinia* species are entirely different from the major enteric diseases, such as dysentery and typhoid fever.

The members of the genus *Yersinia* carry out a mixed–acid fermentation without production of H_2 and CO_2; in this respect, they resemble the *Shigella* group. They produce β-galactosidase and have a powerful urease. Motility is a variable char-

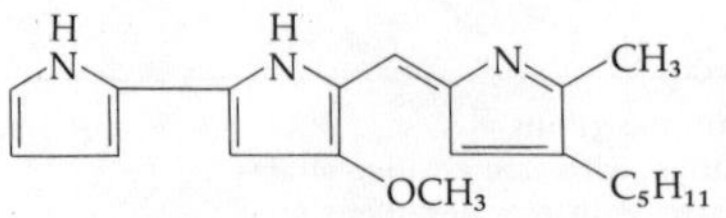

FIGURE 20.5

The structure of prodigiosin, the red pigment formed by *Serratia*.

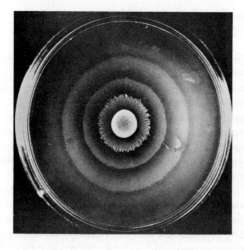

FIGURE 20.6

Swarming of *Proteus* on the surface of nutrient agar plate. The plate was inoculated in the center with a drop of a bacterial suspension and was photographed after incubation at 37°C for 20 hours. From H. E. Jones and R. W. A. Park, "The Influence of Medium Composition on the Growth and Swarming of *Proteus*," *J. Gen. Microbiol.* **47**, 369 (1967).

acter; *Y. pestis* is permanently immotile. The G + C content of the DNA is significantly lower than that of the *Escherichia-Salmonella-Shigella* group. A cultural character that distinguishes them from the Enterobacteriaceae is their relatively slow growth on complex media.

• **The polar flagellates:**
Aeromonas-Vibrio-Photobacterium-Beneckea

The polarly flagellated facultative anaerobes have been studied far less intensively and systematically than the Enterobacteriaceae, and their classification is still uncertain. Provisionally, four groups may be recognized (Table 20.6).

The genus *Aeromonas* contains organisms which differ with respect to the nature of sugar fermentation. *Aeromonas hydrophila* performs a butanediol fermentation, accompanied by H_2 and CO_2 production, similar to that of *Aerobacter*. *Aeromonas punctata* and *A. shigelloides* perform a mixed-acid fermentation without gas production, similar to that of *Shigella*. *Aeromonas shigelloides* also closely resembles the *Shigella* group in DNA base composition and has been shown to share certain somatic antigens with this enteric group. Some authors place it in a separate genus, *Plesiomonas*.

Aeromonas hydrophila and *A. punctata* are widespread in fresh water; the former is capable of causing disease in both frogs and fish. *A. shigelloides* appears to be an inhabitant of the intestinal tract, and it has been implicated in human gastroenteritis.

The limits of the genus *Vibrio* are somewhat controversial. Several of the marine organisms now assigned to *Beneckea* and *Photobacterium* have been included in it, largely on the basis of the curvature of their cells. However, the significance of this structural character seems increasingly doubtful; and if marine enteric bacteria are excluded, the genus *Vibrio* becomes essentially reduced to one species: *V. cholerae*, a water-transmitted pathogen that causes the gastrointestinal disease known as *cholera*.

The bacteria of the genera *Beneckea* and *Photobacterium* are among the most abundant chemoheterotrophs in marine environments. They occur in sea water, in the intestinal tract and on the body surfaces of marine animals. Many of these bacteria can decompose chitin and alginic acid. *Beneckea parahemolytica* is a frequent cause of human gastroenteritis in Japan, where raw fish is commonly consumed.

The property of bioluminescence (Figure 20.7), common to all members of the genus *Photobacterium*, also occurs in one *Beneckea* species.

The light emitted by luminous bacteria is blue-green in color, being confined to a rather narrow spectral band, with a maximum near 490 nm. Cells emit light continuously, provided that oxygen is present. As shown in Figure 20.7, an unaerated cell suspension rapidly becomes dark, as a result of the depletion of the dissolved oxygen supply by bacterial respiration. Indeed, the production of light by a suspension of luminous bacteria is one of the most sensitive methods known for the detection of traces of oxygen.

Light emission can be obtained in cell-free extracts of luminous bacteria, and biochemical studies have revealed the nature of the reaction involved. It is mediated by the enzyme luciferase, which is a mixed function oxidase. The substrates for the light-emitting reaction are three in number: reduced flavin mononucleotide ($FMNH_2$); molecular oxygen; and a long-chain saturated aldehyde, containing more than 8 carbon atoms (R—CHO). The overall reaction catalyzed by luciferase can be formulated as follows:

$$FMNH_2 + O_2 + R\text{—}CHO \longrightarrow FMN + H_2O + R\text{—}COOH + hv$$

The study of chemical model systems, in which simple organic molecules can be treated in a manner that causes them to emit light (the phenomenon of chemiluminescence) shows that the light-emitting molecule, A, must be brought into an electronically excited state A^*. Part of the excitation energy is then emitted as a photon (hv), with return of the molecule to its ground state:

FIGURE 20.7

Luminous bacteria photographed by their own light: left, a streaked plate of *Photobacterium phosphoreum*; right, two flasks containing a suspension of the same organism in a sugar medium. A stream of air was passed continuously through the flask on the right during the photographic exposure. The bacteria in the unaerated flask on the left had exhausted the dissolved oxygen and had ceased to luminesce except at the surface, where organisms were exposed to the air.

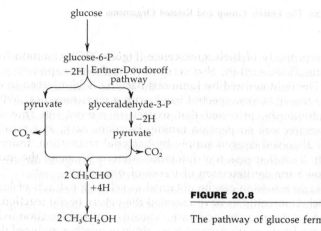

glucose

↓

glucose-6-P

−2H | Entner-Doudoroff
pathway

pyruvate   glyceraldehyde-3-P

↓ −2H

CO_2 ←   pyruvate

↓ CO_2

2 CH_3CHO

+4H

2 CH_3CH_2OH

FIGURE 20.8

The pathway of glucose fermentation by *Zymomonas*.

$$A \xrightarrow{\text{excitation}} A^*$$
$$A^* \longrightarrow A + h\nu$$

A comparable event is presumed to occur during the reaction mediated by luciferase, and it has been postulated that the immediate product of $FMNH_2$ oxidation is electronically excited flavin mononucleotide, FMN*. Light emission would then result from the return of FMN* to the ground state:

$$FMN^* \longrightarrow FMN + h\nu$$

Luminous bacteria are widely distributed in the oceans. They often live in highly specialized symbiotic associations with certain fishes and cephalopods. The animal hosts of luminous bacteria produce light organs, which consist of saclike invaginations on the body surface, each organ containing a dense and brightly luminescent population of bacteria. One of these associations is further discussed in Chapter 28.

• *Zymomonas*   The bacteria of the genus *Zymomonas* are polarly flagellated, Gram-negative rods that occur in fermenting plant materials. Like enteric bacteria, they are facultative anaerobes that have both respiratory and fermentative capacity. However, the sugar fermentation characteristic of *Zymomonas* is a unique one that sharply distinguishes these bacteria from the enteric group. Only glucose, fructose, and sucrose can be fermented; they are converted to equimolar quantities of ethanol and CO_2 (Figure 20.8). As in yeast, pyruvate is decarboxylated nonoxidatively, with the formation of acetaldehyde, subsequently reduced to ethanol. However, the conversion of glucose to pyruvate occurs through the Enter-Doudoroff pathway, not the Embden-Meyerhof pathway.

COLIFORM BACTERIA IN SANITARY ANALYSIS

The enteric diseases caused by the coliform bacteria are transmitted almost exclusively by the fecal contamination of water and food materials. Transmission through contaminated water supplies is by far the most serious source of infection and was responsible for the massive epidemics of the more serious enteric diseases (particularly typhoid fever and cholera) that periodically scourged all countries until the beginning of the present century. Today these diseases are almost unknown in most parts of the Western world, although cholera has recently reappeared in the countries bordering the Mediterranean. Their eradication was achieved primarily by appropriate sanitary controls. An essential part of this operation was *the development of bacteriological methods for ascertaining the occurrence of fecal contamination in water and foodstuffs.*

It is seldom possible to isolate enteric pathogens directly from contaminated water because they are usually present in small numbers, unless contamination from an infected individual has been recent and massive. To demonstrate the fact of fecal contamination, it is sufficient to show that the sample under examination contains bacteria known to be specific inhabitants of the intestinal tract, even though they may not be agents of disease. The bacteria that have principally served as indices of such contamination are the fecal streptococci (discussed in Chapter 23) and *E. coli.* The methods of sanitary analysis developed by bacteriologists differ somewhat from country to country.

One method for detecting *E. coli* is to inoculate dilutions of the sample under test into tubes of lactose broth, which are then incubated at 37°C, and examined after 1 and 2 days for acid and gas production. Cultures showing acid and gas formation are then streaked on a special medium, with a composition that facilitates recognition of *E. coli* colonies. One of the media most commonly used is a lactose-peptone agar containing two dyes, eosin and methylene blue (EMB agar). On this medium, *E. coli* produces blue-black colonies with a metallic sheen, whereas the other principal member of the group capable of fermenting lactose with acid and gas production, *Enterobacter aerogenes* (not necessarily indicative of fecal contamination) produces pale pink mucoid colonies without a sheen (Figure 20.9). For a final distinction between these two organisms, a series of physiological tests, known as the *IMViC tests,* can be performed on material from an isolated colony. The typical results obtained with the two species are shown in Table 20.9. Of these four tests, the Methyl Red and Voges-Proskauer tests are the most significant, since they indirectly reveal the mode of fermentative sugar breakdown. Both are performed on cultures grown in a glucose-peptone medium. The Methyl Red test affords a measure of the final pH: this indicator is yellow at a pH of 4.5 or higher and red at lower pH values. A positive test (red color) is therefore indicative of substantial acid production, characteristic of a mixed-acid fermentation. The Voges-Proskauer test is a color test for acetoin, an intermediate in

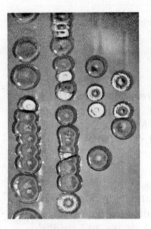

FIGURE 20.9

A plate of EMB agar streaked with a mixture of *Escherichia coli* and *Enterobacter aerogenes*. The colonies of *E. coli* are relatively small and appear light as a result of their metallic sheen. Courtesy of N. J. Palleroni.

TABLE 20.9

IMViC tests for the differentiation between *Escherichia coli* and *Enterobacter aerogenes*

	TYPICAL REACTIONS			
	INDOLE	METHYL RED	VOGES-PROSKAUER	CITRATE
Escherichia coli	+	+	−	−
Enterobacter aerogenes	−	−	+	+

the formation of butanediol from pyruvic acid; a positive reaction is therefore indicative of a butanediol fermentation. The test for indole production from tryptophan, performed on a culture grown in a peptone medium rich in tryptophan, is a test for the presence of the enzyme tryptophanase, which splits tryptophan to indole, pyruvate, and ammonia. This enzyme is present in many bacteria of the enteric group and related forms (including *E. coli*) but is not found in *E. aerogenes*. The citrate utilization test determines ability to grow in a synthetic medium containing citrate as the sole carbon source. This ability is lacking in most strains of *E. coli*, as a result of the absence of a citrate permease.

The first step of the analytical procedure described above is relatively nonspecific, since many bacteria, not even necessarily members of the Enterobacteriaceae, can grow at 37°C in lactose broth with acid and gas production. A much more specific primary enrichment of *E. coli* can be achieved by the Eijkman method: use of a glucose broth, incubated at 44°C. This slight elevation of incubation temperature elimates *Enterobacter aerogenes* and most other organisms that ferment lactose with gas production but permits growth of *E. coli*.

FURTHER READING

Surprising as it may seem, there are no books or comprehensive reviews that give good general accounts of the properties of the enteric group. However, a considerable amount of information about these organisms is contained in the section devoted to their description (Part 8, Gram-negative facultatively anaerobic rods) in the eighth edition of *Bergey's Manual of Determinative Bacteriology* (Baltimore: Williams and Wilkins, 1974). The following references provide specific information concerning some of the constituent subgroups.

Reviews and original articles

ANDERSON, E. S., "The Ecology of Transferable Drug Resistance in the Enterobacteria," *Ann. Rev. Microbiol.* **22,** 131 (1968).

BAUMANN, P., L. BAUMANN, and M. MANDEL, "Taxonomy of Marine Bacteria: The Genus *Beneckea*," *J. Bact.* **107,** 268 (1971).

BRENNER, D. J., "Deoxyribonucleic Acid Reassociation in the Taxonomy of Enteric Bacteria," *Int. J. Syst. Bact.* **23,** 298 (1972).

REICHELT, J. L., and P. BAUMANN, "Taxonomy of the Marine, Luminous Bacteria," *Arch. Mikrobiol.* **94,** 283 (1973).

STARR, M. P., and A. K. CHATTERJEE, "The Genus *Erwinia*: Enterobacteria Pathogenic to Plants and Animals," *Ann. Rev. Microbiol.* **26,** 389 (1972).

The Gram-negative chemoheterotrophs include several groups of unicellular or filamentous gliding bacteria. Gliding movement requires contact of the cell with a solid substrate. It is considerably slower than flagellar movement, and no locomotor organelles have so far been demonstrated. In unicellular gliding bacteria, translative movement is often accompanied by rapid flexing of the cells, which suggests that the cell wall may be less rigid than that of other Gram-negative bacteria. However, the cell walls of gliding organisms do not appear to differ significantly, either in ultrastructure or in chemical composition, from those of other Gram-negative bacteria; they contain a thin inner layer of peptidoglycan, chemically similar to the peptidoglycans of other Gram-negative bacteria, and an outer layer with the profile of a unit membrane.

THE MYXOBACTERIA

The myxobacteria are differentiated from other gliding bacteria by two properties: their special developmental cycle and the composition of their DNA, which is of very high G + C content (67 to 71 moles percent).

Myxobacterial vegetative cells are small rods, which multiply by binary transverse fission. On solid substrates these organisms form flat, spreading colonies with irregular borders, made up of small groups of advancing cells (Figure 21.1). The migrating cells produce a tough slime layer that underlies and gives coherence to the colony. Under favorable conditions, the vegetative cells aggregate at a number of points in the inner area of the colony, and fruiting bodies then differentiate from these cellular aggregates. Each fruiting body is made up of slime and bacterial cells; when mature, it acquires a definite size, form, and color. The pigments of these bacteria are carotenoids, associated with the cells. As the fruiting

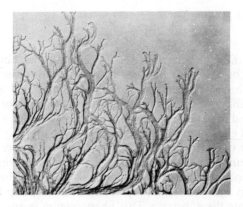

FIGURE 21.1

Edge of the growth of a *Sorangium* species on agar (×19). Courtesy of M. Dworkin and H. Reichenbach.

bodies mature, the cells within them become converted to resting cells, known as *myxospores*. Upon subsequent germination, each myxospore gives rise to a vegetative rod.

The myxobacteria are soil organisms and are usually detected in nature through the development of their fruiting bodies on solid substrates: the bark of trees, decomposing plant material, and in particular the dung of animals. They can be isolated from soil by a number of special enrichment methods. The most effective method is to place sterilized dung particles on the surface of a large sample (50 to 200 g) of moistened soil. The myxobacteria develop on the particles, forming fruiting bodies on the dung surface after 1 to 3 weeks of incubation.

Myxobacteria are strict aerobes which grow poorly (if at all) on conventional complex media. They fall into two nutritional subgroups: bacteriolytic and cellulolytic organisms. The majority of species are bacteriolytic, growing at the expense of bacteria and other microorganisms, living or dead. Living host cells are killed by antibiotics, secreted by the myxobacteria; these substances have not yet been characterized chemically. The host cells are then lysed through the action of extracellular enzymes, which include proteases, nucleases, and lipases. Myxobacterial growth occurs at the expense of the soluble hydrolytic products. The exceptionally favorable nature of dung as a substrate is largely attributable to its high bacterial content. The bacteriolytic myxobacteria are most readily cultivated on the surface of mineral agar, over which a heavy suspension of host cells has been spread. Growth in liquid media is often poor. The minimal nutritional requirements of these organisms are not well known, since many species will not grow readily except at the expense of microbial cells.

The most easily cultivable representatives (*Myxococcus* spp.) can grow in a defined medium, containing a mixture of amino acids, which provide a source of carbon, nitrogen, and energy; carbohydrates are not utilized.

A few species, all of the genus *Polyangium*, have quite different nutrient requirements. They are active cellulose-decomposers, and grow in a medium with a mineral base, supplemented either with cellulose or with its hydrolytic products (the soluble sugars cellobiose and glucose).

Most myxobacteria (exceptions: some *Myxococcus* spp., the cellulose-decomposing *Polyangium* spp.) do not readily form fruiting bodies on media that support good vegetative growth. Fructification is generally favored by cultivation on media that are poor in nutrients, but the specific factors that control the process are not understood. Since the classification of these bacteria is very largely based on the structure of their fruiting bodies, the difficulty of obtaining reproducible fructification can be a serious obstacle to their identification.

The distinguishing properties of some of the principal genera are shown in Table 21.1. A primary division can be made on the basis of the structure of the vegetative cells and myxospores. In the genera listed under group I the vegetative cells have tapered ends (Figure 21.2a), and the myxospores are considerably more refractile than the vegetative cells. In the genera listed under group II the vegetative cells are of uniform diameter, with rounded ends (Figure 21.2b); and the myxospores do not differ from vegetative cells, either in form or in refractility. When growing on agar plates, the representatives of group II form colonies that erode and penetrate the substrate. This phenomenon is apparently physical, and does not involve an enzymatic attack on agar. Among the representatives of group I, erosion of agar by the colonies does not usually occur.

The formation of myxospores in the genus *Myxococcus* involves a striking morphogenetic change: the long, slender vegetative cells shorten and round up to produce spherical myxospores, which are much more refractile than the vegetative cells. The fruiting bodies of most species are glistening, colored droplets, uniformly filled with myxospores.

In *Cystobacter, Melittangium,* and *Stigmatella* the myxospores are shortened refractile rods which develop within cysts, each of which encloses a large number of individual myxospores. The cysts of *Cystobacter* lie on the surface of the substrate, embedded in a mass of slime; those of *Melittangium* and *Stigmatella* develop at the apices of a colorless stalk, which consists largely of hardened slime. The stalk is unbranched and bears a single cyst in *Melittangium;* the stalk is branched and bears multiple cysts in *Stigmatella.*

Fructification of *Polyangium* and *Chondromyces* leads to the formation of fruiting bodies analogous in form to those of *Cystobacter* and *Stigmatella,* respectively. However, the myxospores within the cysts are nonrefractile and cannot be distinguished microscopically from vegetative cells.

TABLE 21.1

Properties of some genera of fruiting myxobacteria

GENUS	FORM OF MYXOSPORE	STRUCTURE OF FRUITING BODY		COLONIES ERODE AGAR SURFACE	NUTRITIONAL CHARACTER	
		MYXOSPORES IN CYSTS	CYSTS BORNE ON STALKS		BACTERIOLYTIC	CELLULOLYTIC
Group I: Vegetative cells tapered, myxospores refractile						
Myxococcus	Sphere	−		Variable	+	−
Cystobacter	Rod	+	−	−	+	−
Melittangium	Rod	+	+, Simple	−	+	−
Stigmatella	Rod	+	+, Branched	−	+	−
Group II: Vegetative cells of uniform width, myxospores not refractile						
Polyangium	Rod	+	−	+	+ (some)	+ (some)
Chondromyces	Rod	+	+, Branched	+	+	−

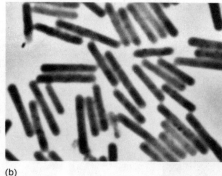

FIGURE 21.2

Vegetative cells of myxobacteria. (a) *Myxococcus xanthus;* (b) *Chondromyces crocatus.* Both ×2,050, dark phase contrast. Courtesy of Dr. H. McCurdy.

(a)

(b)

In *Stigmatella* and *Chondromyces* the mature fruiting body has a treelike form, numerous cysts being formed at the tips of the much-branched stalk. Successive stages in the differentiation of such a complex fruiting body are shown in Figure 21.3. The range of fruiting body structure in the genera described is illustrated in Figure 21.4.

The properties of myxospores have been examined primarily in the genus *Myxococcus.* Although they are more heat-resistant than vegetative cells, the difference in this respect is not great. However, they are much more resistant to desiccation, and can survive for months or years in the dry state. The formation of these resting structures in fruiting bodies that are raised off the surface of the substrate no doubt facilitates their physical dispersion. In the genus *Myxococcus* the myxospore is both the resting structure and the unit of dispersion. However, in cyst-forming genera the unit of dispersion is not the individual myxospore, but the cyst which contains many myxospores. This fact is most evident if one compares the respective modes of germination. The fruiting bodies of most *Myxococcus* spp. are deliquescent, and each of the myxospores liberated by their breakdown can germinate under favorable conditions, giving rise through subsequent growth to a vegetative colony. In cyst-forming genera, germination of myxospores is accompanied by a rupture of the wall of the enclosing cyst, with the release of hundreds of vegetative cells (Figure 21.5). The cells remain in association and give rise to a single vegetative colony. Cyst germination thus permits the rapid buildup of a vegetative population, prior to the initiation of cell division.

In *Myxococcus* spp., myxospore formation can be induced experimentally in

FIGURE 21.3

Successive stages (a–d) in the development of the apical region of a fruiting body of *Chondromyces apiculatus*, taken over a period of 4 hours at 27° C. Frames from film E 779, *Publ. wiss. Filmen, Sekt. biologie*, **7**, 245 (1974), prepared by the Institut für den wissenschaftlichen Film, Göttingen. Courtesy of Dr. Hans Reichenbach.

(a)

(b)

(c)

(d)

(a)   (b)   (c)

(d)   (e)   (f)

FIGURE 21.4

Fruiting bodies of myxobacteria. (a) *Melittangium lichenicolum* (×232). (b) *Stigmatella aurantiaca* (×318). (c) *Cystobacter fuscus* (×106). (d) *Polyangium* sp. (×573). (e) *Chondromyces pediculatus* (×94). (f) *Myxococcus xanthus* (×32). (a–d) Courtesy of M. Dworkin and H. Reichenbach; (e–f) courtesy of H. McCurdy.

(a)

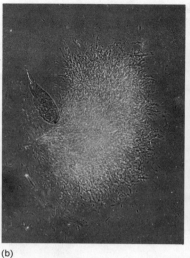
(b)

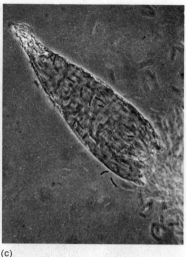
(c)

FIGURE 21.5

Cyst germination in *Chondromyces apiculatus*. (a) Mature fruiting body, bearing cysts, ×104. (b) Germination of a detached cyst on the surface of an agar plate, showing the emergence of a large population of vegetative cells, derived from the enclosed myxospores; phase contrast, ×120. (c) The empty wall of a germinated cyst; phase contrast, ×485. Courtesy of Dr. Hans Reichenbach.

vegetative cells, without the usual preliminary events of aggregation and fructification. The addition of 0.5 m glycerol to a suspension of vegetative cells causes a massive conversion to myxospores within 2 hours; starvation for specific amino acids (methionine, or phenylalanine and tyrosine) has the same effect.

• **Algicidal nonfruiting myxobacteria**

The death of natural populations of green algae and cyanobacteria is often caused by myxobacteria. Many of these algicidal myxobacteria have not been observed to form fruiting bodies, so that their taxonomic position is uncertain; however, the DNA base composition of these organisms is similar to that of fruiting myxobacteria, lying in the range of 69 to 71 moles percent G + C. Consequently, they are referred to by the noncommittal name of *myxobacters*. In contrast to bacteriolytic fruiting myxobacteria, the myxobacters have simple nutrient requirements; they grow well in liquid media of defined composition, and can use glucose or other carbohydrates as sole sources of carbon and nitrogen.

Some of the myxobacters which kill cyanobacteria produce extracellular enzymes that destroy the peptidoglycan layer of the host cell. The lytic enzymes include a protease of very low molecular weight (8,000 daltons) and broad substrate specificity, which hydrolyzes peptide bonds of peptidoglycans. Other myxobacters kill cyanobacteria only by cell-to-cell contact (Figure 21.6); the enzymatic mechanism of the attack is not known.

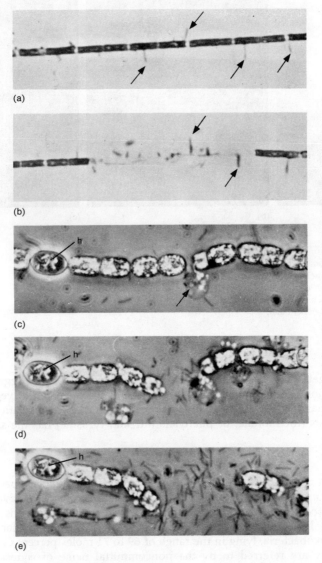

(a)

(b)

(c)

(d)

(e)

FIGURE 21.6

The attack by a myxobacterium on filaments of two blue-green bacteria. (a) Filament of *Oscillatoria redekei*, showing polar attachment of myxobacterial cells (arrows) at various points along the filament. (b) Filament of *O. redekei* after lysis of some of the component cells; the myxobacterium is still attached to the lysed cells (arrows). (c, d, e) Time lapse sequence, showing lysis by the myxobacterium of a filament of *Anabaena flosaquae*. This blue-green bacterium contains gas vacuoles, which accounts for the phase-bright appearance of the vegetative cells. The filament contains a heterocyst (h), which is not susceptible to attack. In (c) the arrow points to a lysed vegetative cell, the destruction of which has caused a break in the filament. All ×1,260. From M. J. Daft, and W. D. P. Stewart, "Light and Electron Microscope Observations on Algal Lysis by Bacterium CP-1," *New Phytol.* **72,** 799 (1973).

THE CYTOPHAGA GROUP

The gliding bacteria of the Cytophaga group do not form fruiting bodies, and they differ markedly from the myxobacteria (including the nonfruiting myxobacters) in their DNA base composition, which lies in the range of 30 to 50 moles percent G + C. Some of these bacteria produce chains of cells, 100 μm or more in length, a character that does not occur in myxobacteria. However, other cytophagas grow as single, slender rods, and cannot be distinguished by vegetative cell structure from myxobacteria. Resting cells are formed only in the genus *Sporocytophaga*, which produces spherical, refractile microcysts, similar in structure and development to the myxospores of *Myxococcus*. The base composition of the DNA is therefore a character of primary importance to distinguish the members of the Cytophaga group from the myxobacteria, and in particular from the nonfruiting myxobacters.

The principal genera of the Cytophaga group (Table 21.2) are distinguished primarily by their nutritional properties. *Cytophaga* and *Sporocytophaga* spp. can hydrolyze and grow at the expense of complex polysaccharides, whereas *Flexibacter* spp. are not polysaccharide decomposers. Most of these bacteria are strict aerobes; a few *Cytophaga* and *Flexibacter* spp. are facultative anaerobes and can ferment carbohydrates. The major fermentation products are fatty acids and succinic acid; carbon dioxide is required in substrate amounts for fermentative growth.

The most active aerobic cellulose-decomposing bacteria in soil are certain species of *Cytophaga* and *Sporocytophaga*. They can be readily enriched from soil in a medium with a mineral base containing filter paper as the sole source of carbon and energy. Upon initial isolation, these bacteria are unable to grow at the expense of any other organic substrate; by subsequent selection, mutants able to grow with the soluble sugars, glucose and cellobiose, can be obtained. The cellulolytic ability of these organisms is remarkable; when streaked on a sheet of filter paper placed on the surface of a mineral agar plate, they completely destroy its fiber structure, the attacked areas being converted into slimy colored patches filled with bacterial cells (Figure 21.7). Another distinctive property of the cellulose-decomposing soil cytophagas is the necessity for direct contact with the cellulosic substrate; this behavior suggests that the primary attack on cellulose is mediated by an exoenzyme which is nondiffusible, remaining bound to the cell surface. As a consequence, the cells in a culture containing cellulose adhere closely to the cellulose fibers, often in a very regular alignment, the rod-shaped cells being oriented parallel to the polysaccharide fibrils (Figure 21.8).

Other species of the genus *Cytophaga* grow at the expense of the polysaccharides chitin and agar (Figure 21.9). Agar is the commercial name of a complex mixture of polygalactans which serve as structural polymers in certain marine algae and which do not occur in terrestrial environments. The agarolytic cytophagas are, in consequence, all organisms of marine origin, whereas chitinolytic members of the group occur both in soil and in sea water.

The hydrolysis of chitin and agar is mediated by inducible, extracellular enzymes, and hence does not require direct contact with the substrate. The chitinolytic and agarolytic cytophagas are also much less specialized nutritionally than the cellulose-decomposing species. They can all use a wide range of soluble

TABLE 21.2

Principal genera of the *Cytophaga* group

	VEGETATIVE STRUCTURE	FORMATION OF MYCROCYSTS	MOLES % G + C IN DNA	GROWTH AT EXPENSE OF CELLULOSE, CHITIN OR AGAR
Cytophaga	Single rods	−	33–42	+
Sporocytophaga	Single rods	+	36	+
Flexibacter	Rods, single or in chains	−	31–43	−

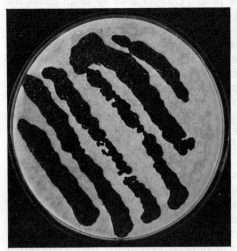

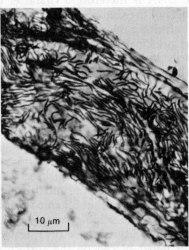

FIGURE 21.7 (left)

Agar plate covered with a layer of filter paper and streaked with a culture of *Cytophaga*. Note that the filter paper has been completely dissolved where growth has occurred.

FIGURE 21.8 (right)

Cellulose fiber heavily attacked by *Cytophaga* (stained preparation). Note the characteristic regular arrangement of the cells. From S. Winogradsky, *Microbiologie du Sol*. Paris: Masson, 1949. Reprinted with permission of M. Manigault and the publisher.

10 µm

sugars as carbon and energy sources, and most can grow in complex nitrogenous media (peptone or extract) in the absence of carbohydrates.

The members of the genus *Flexibacter* are widely distributed in soil, freshwater, and marine environments. They include several species specifically pathogenic for fish. Infections caused by *Flexibacter columnaris* often assume epidemic proportions with massive mortalities in fish hatcheries during the warmer months of the year. The pathogens of the *Flexibacter* group are perhaps the most important bacterial agents of disease in fish, both in the ocean and in freshwater.

FILAMENTOUS, GLIDING CHEMOHETEROTROPHS

In addition to the Cytophaga group, the gliding bacteria with DNA of low G + C content include a number of filamentous chemoheterotrophs. The principal genera are listed in Table 21.3.

Saprospira and *Vitreoscilla* are organisms which develop as flexible, gliding filaments as much as 500 µm long, made up of cells 2 to 5 µm in length. The filaments of *Saprospira* are helical (Figure 21.10), those of *Vitreoscilla* straight (Figure 21.11). Both groups occur largely in aquatic environments. The genus *Simonsiella* is distinguished by the formation of flattened, ribbon-shaped filaments (Figure 21.12), which are motile only when the broad surface of the filament is in contact with the substrate. These bacteria are aerobic members of the microflora of the oral cavity of man and other animals. Their nutritional requirements are not precisely known; they develop best in complex media, supplemented with serum or blood.

The most highly differentiated of the gliding chemoheterotrophs is *Leucothrix*, a marine organism which grows as an epiphyte on seaweeds, and is also found in decomposing algal material. The very long filaments are immotile and are attached to substrates by an inconspicuous basal holdfast. Reproduction occurs not by random fragmentation of the filament, as in *Saprospira*, *Vitreoscilla*, and

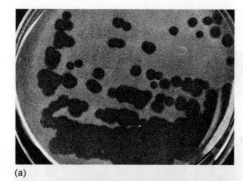

(a)

(b)

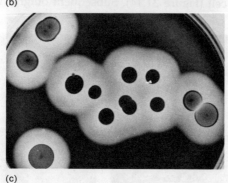

(c)

FIGURE 21.9

The decomposition of chitin and agar by members of the cytophaga group. (a) Plate culture of *Cytophaga johnsonae*, growing on an agar medium containing a suspension of the insoluble polysaccharide, chitin. The chitin has been decomposed by extracellular enzymes in the clear areas beneath the spreading, translucent colonies. (b and c) Two photographs of a plate culture of *Cytophaga fermentans*, growing on a complex medium containing 1 percent agar. (b) is photographed by reflected light, to reveal the depressions resulting from agar decomposition around the colonies. In (c) the plate has been flooded with an I–KI solution, which is decolorized in the areas of agar decomposition, and thus reveals the extent of the diffusion zones of agarase around the colonies. (a) Courtesy of Dr. H. Veldkamp. (b and c) from H. Veldkamp, "A Study of Two Marine, Agar-Decomposing, Facultatively Anaerobic Myxobacteria," *J. Gen. Microbiol.* **26**, 331 (1961).

TABLE 21.3

Major genera of filamentous, gliding chemoheterotrophs

	SHAPE OF FILAMENT	MOTILITY OF FILAMENT	PRESENCE OF HOLDFAST	MODE OF REPRODUCTION	MOLES % G + C IN DNA
Saprospira	Helical cylinder	+	−	Fragmentation of filament	35–48
Vitreoscilla	Straight cylinder	+	−	Fragmentation of filament	44
Simonsiella	Flattened ribbon of cells	+	−	Fragmentation of filament	Not determined
Leucothrix	Straight cylinder	−	+	Release of single gliding cells from filament apex	46–51

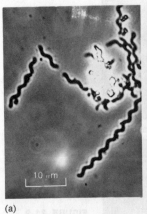
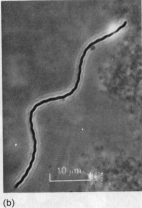

(a)   (b)

FIGURE 21.10

Phase contrast photomicrographs of *Saprospira* ($\times$451): (a) *S. albida;* (b) *S. grandis.* Courtesy of Ralph Lewin.

Simonsiella, but by the breaking off of ovoid cells, single or in short chains, from the apical end of the filament (Figure 21.13). These reproductive cells (sometimes termed gonidia) are capable of gliding movement; when released in large numbers, they aggregate to form rosettes, held together by the holdfasts that are located at one pole of the cell (Figure 21.14). Subsequent outgrowth of the cells in such

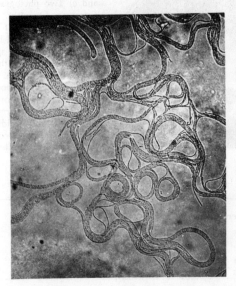

FIGURE 21.11

Vitreoscilla filaments growing on the surface of an agar plate. Courtesy of E. G. Pringsheim.

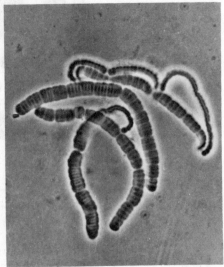

FIGURE 21.12

Simonsiella, a gliding bacterium which forms ribbon-shaped filaments of flattened cells (phase contrast, $\times$440). Some of the filaments are viewed on edge, and they appear much thinner than the filaments that lie flat. Courtesy of Mrs. P. D. M. Glaister.

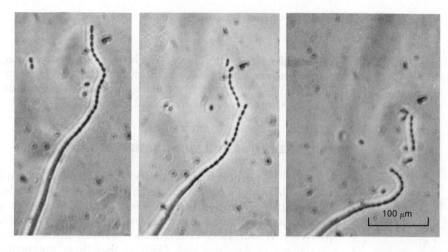

FIGURE 21.13

Successive pictures of a *Leucothrix* filament showing liberation of gonidia (phase contrast, ×309). From Ruth Harold and R. Y. Stanier, "The Genera *Leucothrix* and *Thiothrix*," *Bact. Rev.* **19**, 49 (1955).

FIGURE 21.14

Successive stages in the aggregation of *Leucothrix* gonidia to form rosettes (phase contrast photographs of a single field, taken over a period of 55 minutes, ×279). From Ruth Harold and R. Y. Stanier, "The Genera *Leucothrix* and *Thiotrix*," *Bact. Rev.* **19**, 49 (1955).

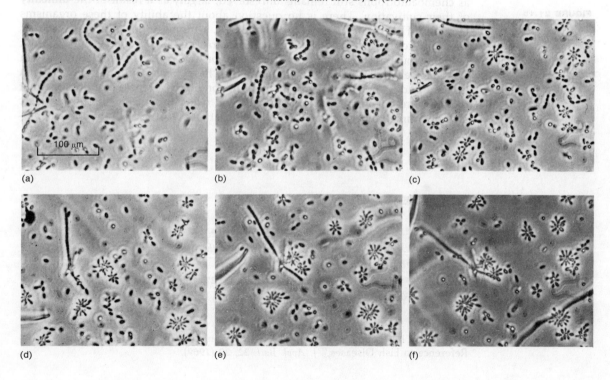

rosettes gives rise to colonies made up of numerous radiating filaments, all attached to a central mass of holdfast material. *Leucothrix* can grow in a seawater medium containing a variety of simple organic compounds (principally organic acids) as sole sources of carbon and energy.

THE FILAMENTOUS SULFUR-OXIDIZING BACTERIA

Aerobic marine or freshwater habitats where H_2S is present often contain filamentous gliding bacteria, packed with refractile inclusions of sulfur (Figure 21.15). The two principal representatives of this ecological group are *Beggiatoa* and *Thiothrix*, structural counterparts of *Vitreoscilla* and *Leucothrix*, respectively. About 1885 Winogradsky showed through experiments on natural populations of *Beggiatoa* and *Thiothrix* that these organisms oxidize H_2S under aerobic conditions. The oxidation leads initially to the intracellular deposition of sulfur. If the filaments are maintained in the absence of H_2S the sulfur inclusions gradually disappear, as a result of their further oxidation to H_2SO_4. Addition of H_2S to suspensions of the sulfur-depleted filaments results in a rapid reappearance of intracellular sulfur inclusions. Winogradsky also observed that the maintenance of crude cultures of *Beggiatoa* is dependent on the periodic provision of H_2S. It was largely this work that led to his initial formulation of the concept of chemoautotrophy.

In view of this history, it is ironic that the status of *Beggiatoa* and *Thiothrix* as chemoautotrophs still remains unresolved, largely as a result of the difficulty of their cultivation. There is no question about the ability of these organisms to oxidize H_2S to S and subsequently H_2SO_4, a property not possessed by their structural counterparts, *Vitreoscilla* and *Leucothrix*; and it is probable, although not unambiguously established, that this oxidation can provide the organisms with ATP. However, the strains of *Beggiatoa* that have been purified have all been isolated under chemoheterotrophic growth conditions. Hence, it is possible that they require an organic carbon source, even when provided with an inorganic energy source, and thus may not conform strictly to the definition of chemoautotrophs.

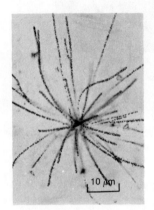

FIGURE 21.15

A characteristic rosette of *Thiothrix*, the filaments of which are filled with sulfur droplets (×475). Courtesy of E. J. Ordal.

FURTHER READING

Reviews   DWORKIN, M., "Biology of the Myxobacteria," *Ann. Rev. Microbiol.* **20,** 75 (1966)

HAROLD, R., and R. Y. STANIER, "The Genera *Leucothrix* and *Thiothrix*," *Bact. Rev.* **19,** 49 (1955).

SORIANO, S., "Flexibacteria," *Ann. Rev. Microbiol.* **27,** 155 (1973).

STANIER, R. Y., "The Cytophaga Group," *Bact. Rev.* **6,** 143 (1942).

Original articles   ANDERSON, J. I. W., and D. A. CONROY, "The Pathogenic Myxobacteria with Special Reference to Fish Diseases," *J. Appl. Bact.* **32,** 30 (1969).

643   McCurdy, H. D., "Studies on the Taxonomy of the *Myxobacterales*. I. Record of Canadian Isolates and Survey of Methods," *Can. J. Microbiol.* **15**, 1453 (1969).

——, "Studies on the Taxonomy of the *Myxobacterales*. II. *Polyangium* and the Demise of the *Synangiaceae*," *Int. J. Syst. Bact.* **20**, 283 (1970).

——, "Studies on the Taxonomy of the *Myxobacterales*. III. *Chondromyces* and *Stigmatella*," *Int. J. Syst. Bact.* **21**, 40 (1971).

Pringsheim, E. G., "The *Vitreoscillaceae*: A Family of Colourless, Gliding, Filamentous Organisms," *J. Gen. Microbiol.* **5**, 124 (1951).

Steed, P. D. M., "Simonsiellaceae *fam. nov.* with Characterization of *Simonsiella crassa* and *Alysiella filiformis*," *J. Gen. Microbiol.* **29**, 615 (1963).

Thaxter, R., "On the Myxobacteriaceae, a New Order of Schizomycetes," *Botan. Gaz.* **17**, 389 (1892).

Voelz, H., "The Physical Organization of the Cytoplasm in *Myxococcus xanthus* and the Fine Structure of Its Components," *Arch. Mikrobiol.* **57**, 181 (1967).

GRAM-POSITIVE BACTERIA: UNICELLULAR ENDOSPOREFORMERS

22

Many unicellular Gram-positive bacteria share the ability to form a distinctive type of dormant cell known as an *endospore*. Endospores (Figure 22.1) can be readily recognized microscopically by their intracellular site of formation, their extreme refractility, and their resistance to staining by basic aniline dyes that readily stain vegetative cells. They are not normally formed during active growth and division; their differentiation begins when a population of vegetative cells passes out of the exponential growth phase as a consequence of nutrient limitation. Typically, one endospore is formed in each vegetative cell. The mature spore is liberated by lysis of the vegetative cell in which it has developed. Free endospores have no detectable metabolism, but for many years (often decades) they retain the potential capacity to germinate and develop into vegetative cells. This state of total dormancy is known as *cryptobiosis*. Endospores are in addition highly resistant to heat, ultraviolet and ionizing radiations, and many toxic chemicals. Their heat resistance is frequently taken advantage of in the isolation of spore-forming bacteria; these organisms can be selected by subjecting suspensions of source materials to a thermal pretreatment sufficient to kill all vegetative cells.

The unicellular endosporeformers all reproduce by binary transverse fission, and with one exception they are rod-shaped. If determined on exponentially growing cells, the Gram-reaction is positive; however, many sporeformers rapidly become Gram-negative upon entering the stationary phase. Motility is widespread, but not universal, and is effected by means of peritrichous flagella.

The sporeformers are chemoheterotrophs; dissimilation of organic substrates occurs by aerobic respiration, anaerobic respiration, or fermentation. Growth factors are required by some, but not all.

The typical habitat of spore-forming bacteria is soil. A few species are pathogenic for either insects or vertebrates. Most pathogenic sporeformers cause disease by toxin production; and few of them are able to invade animal tissues.

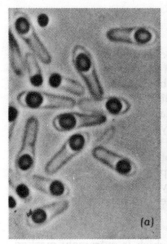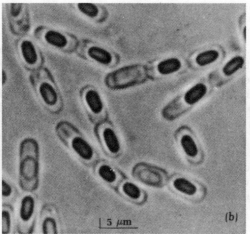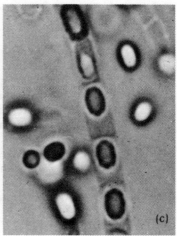

| 5 μm |

(a)　　　　　(b)　　　　　(c)

FIGURE 22.1

Sporulating cells of *Bacillus* species: (a) unidentified bacillus from soil; (b) *B. cereus*; (c) *B. megaterium*. From C. F. Robinow, in The Bacteria (I. C. Gunsalus and R. Y. Stanier, editors), Vol. 1, p. 208. New York: Academic Press, 1960.

• **Classification of the endosporeformers**

The primary taxonomic subdivision of the sporeformers (Table 22.1) is based on their *oxygen relations*. The genera *Clostridium* and *Desulfotomaculum* are strict anaerobes, although *Clostridium* spp. show some variation with respect to oxygen sensitivity. As a rule, vegetative cells are rapidly killed by exposure to air, although spores are not. The distinction between *Clostridium* and *Desulfotomaculum* is metabolic. *Clostridium* spp. synthesize ATP by fermentative means, the modes of fermentation being remarkably diverse with respect both to substrates and products. *Desulfotomaculum* spp. perform anaerobic respiration with sulfate as a terminal electron acceptor, a mechanism of ATP synthesis also characteristic of nonsporeforming Gram-negative strict anaerobes of the genus *Desulfovibrio* (see Chapter

TABLE 22.1

Properties of the genera of unicellular endosporeformers

GENUS	VEGETATIVE CELL SHAPE	RELATIONS TO OXYGEN	DISSIMILATORY METABOLISM	RANGE OF DNA BASE COMPOSITION, MOLES % G + C
Bacillus	Rods	Aerobes; some facultative anaerobes	Aerobic respiration, also fermentation or denitrification in some	33–51
Sporosarcina	Cocci	Strict aerobes	Aerobic respiration	40–43
Clostridium	Rods	Strict anaerobes	Fermentation	22–28
Desulfotomaculum	Rods	Strict anaerobes	Anaerobic respiration with SO_4^{2-} as terminal electron acceptor	42–46

645

24). The sulfate-reducers contain heme proteins associated with their anaerobic electron transport chain, whereas the clostridia do not synthesize heme proteins.

The genera *Bacillus* and *Sporosarcina* consist of aerobes, either strict or facultative, which possess an aerobic electron transport chain.

Aerobic sporeformers with rod-shaped cells are placed in one genus, *Bacillus*, although they constitute a heterogeneous group in physiological and biochemical respects. A species with spherical cells is placed in the genus *Sporosarcina*; physiologically it closely resembles one of the subgroups of *Bacillus*.

As shown in Table 22.1, the DNA of most endospore-forming bacteria is of low GC content. This chemical feature is particularly striking in the clostridia, which have DNA of lower GC content than any other procaryotes.

• **Peptidoglycan structure**

Variability of the cell wall peptidoglycan structure, which is extremely marked in Gram-positive bacteria of the actinomycete line (see Chapter 23), occurs to a limited extent in sporeformers (Figure 22.2 and Table 22.2). Most bacilli and clostridia synthesize a peptidoglycan in the vegetative cell wall of the type that is well-nigh universal in Gram-negative procaryotes, containing *meso*-diaminopimelic acid as a diamino acid, directly cross-linked to D-alanine; the distribution of other forms of wall peptidoglycans is shown in Table 22.2. It should be noted, however, that the integument of the endospore contains a peptidoglycan of unique structure, probably similar in all sporeformers and different in numerous respects from the peptidoglycans associated with the vegetative cell wall; this question is further discussed on p. 663.

• **The possible production of endospores by other types of bacteria**

The gut contents of herbivorous animals often harbor very large bacteria, unicellular or filamentous, which produce intracellular structures that may be endospores. One example is *Metabacterium*, each cell of which contains several sporelike bodies (Figure 22.3). Since none of these bacteria has so far been cultivated in artificial media, nothing is known about their growth cycles or the properties of the spores that they form.

Recent work has, however, shown beyond question that true endospores, having the structural, chemical and physiological properties of the endospores of unicellular bacteria, are produced by the members of one small group of Actinomycetes, *Thermoactinomyces*. These organisms and their spores will be discussed in Chapter 23 (p. 701).

• **The aerobic sporeformers**

Most aerobic sporeformers are versatile chemoheterotrophs capable of utilizing a considerable range of simple organic compounds (sugars, amino acids, organic acids) as respiratory substrates, and in some cases also of fermenting carbohydrates. A few species require no organic growth factors; others may require amino acids, B vitamins, or both. The majority are mesophiles, with temperature optima in the range of 30 to 45°C; however, the genus also contains a number of thermophilic representatives that grow at temperatures as high as 65°C.

The mesophilic species of the genus *Bacillus* can be divided into three subgroups, distinguished by the structure and intracellular location of the endospore (Table 22.3).

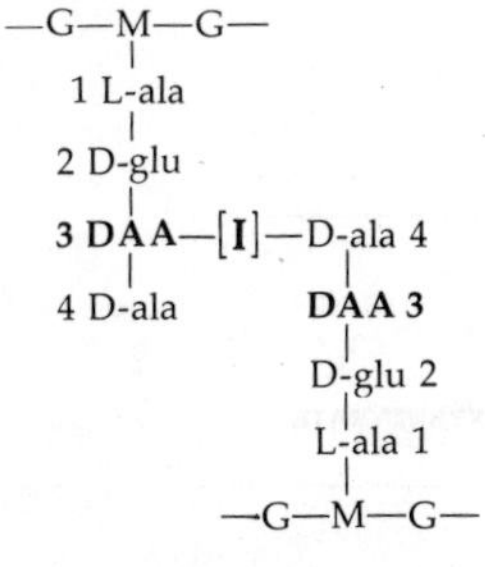

—G—M—G—
|
1 L-ala
|
2 D-glu
|
3 DAA—[I]—D-ala 4
| |
4 D-ala   DAA 3
                    |
                D-glu 2
                    |
                L-ala 1
                    |
              —G—M—G—

FIGURE 22.2

Peptidoglycan structure in spore-forming bacteria. The variable structural features (boldface) are the diamino acid (DAA) in position 3 of the peptide chain and the presence or absence of additional amino acids, comprising an interpeptide bridge (I) between cross-linked peptide chains. M, G, ala, and glu represent respectively, *N*-acetylglucosamine, *N*-acetylmuramic acid, and glutamic acid.

TABLE 22.2

Distribution of peptidoglycan types in spore-forming bacteria

DIAMINO ACID IN POSITION 3	INTERPEPTIDE BRIDGE	
Meso-diaminopimelic acid	Absent	*Bacillus* (most spp.) *Clostridium* (most spp.)
L-lysine	Present	*Sporosarcina* *Bacillus sphaericus* *B. pasteurii*
L-lysine	Absent	*Clostridium septicum* *C. tertium* *C. paraputrificum*
L, L-diaminopimelic acid	Present	*Clostridium pectinovorum* *C. fallax* *C. perfringens*

Within group I, a further distinction can be made between the *B. subtilis* and *B. cereus* groups on the basis of cell size and on the presence or absence of poly-β-hydroxybutyrate as a cellular reserve material (Table 22.4).

Most species of group I can grow anaerobically at the expense of sugars. They carry out a distinctive fermentation, in which 2,3-butanediol, glycerol, and CO_2 are the major end products, accompanied by small amounts of lactate and ethanol. The fermentation can be approximately represented as:

$$3 \text{ glucose} \longrightarrow 2 \text{ 2,3-butanediol} + 2 \text{ glycerol} + 4CO_2$$

Glucose is initially dissimilated through the Embden-Meyerhof pathway, to the level of triose phosphate, at which point a metabolic divergence occurs. Part of the triose phosphate is converted to pyruvate, from which butanediol and CO_2 are produced, as shown in Figure 22.4. However, the formation of butanediol from pyruvate results in the reoxidation of only part of the NADH formed in the conversion of triose phosphate to pyruvate:

$$2 \text{ triose-P} + 2NAD^+ + 4ADP + 2 \text{℗} \longrightarrow 2 \text{ pyruvate} + 4ATP + 2NADH + 2H^+$$

$$2 \text{ pyruvate} + NADH + H^+ \longrightarrow 2CO_2 + 2\text{,3-butanediol} + NAD^+$$

TABLE 22.3

Subgroups of mesophilic *Bacillus* species

	STRUCTURE OF SPORULATING CELL				
GROUP	ENDOSPORE SHAPE	ENDOSPORE POSITION IN VEGETATIVE CELL	VEGETATIVE CELL DISTENDED BY SPORE	MAJOR REPRESENTATIVES	
I	Oval	Central	−	*B. subtilis* group *B. cereus* group *B. fastidiosus*	
II	Oval; spore coat thick and ridged	Central	+	*B. polymyxa* group	
III	Spherical	Terminal	+	*B. pasteurii* group	

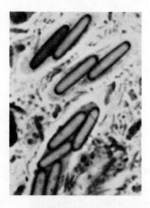

FIGURE 22.3

Metabacterium polyspora: smear from intestinal tract of guinea pig, showing three sporulating cells, each containing two or more rod-shaped endospores (×1,700). Courtesy of C. F. Robinow.

TABLE 22.4

Distinguishing properties of the *B. subtilis* and *B. cereus* groups

B. subtilis GROUP: VEGETATIVE CELLS <0.8 μ WIDE; DO NOT FORM POLY-β-HYDROXYBUTYRATE AS RESERVE MATERIAL

	ANAEROBIC GROWTH WITH:	
	SUGARS	NITRATE (DENITRIFICATION)
B. subtilis	−	−
B. licheniformis	+	+

B. cereus GROUP: VEGETATIVE CELLS >1.0 μ WIDE; FORM POLY-β-HYDROXYBUTYRATE AS RESERVE MATERIAL

	MOTILITY	ANAEROBIC GROWTH WITH SUGARS	REQUIRE GROWTH FACTORS	PATHOGENICITY
B. cereus	+	+	+	(none)
B. anthracis	−	+	+	Pathogenic for mammals
B. thuringensis	+	+	+	Pathogenic for insects
B. megaterium	+	−	−	(none)

Redox balance is reestablished by a concomitant reduction of triose phosphate to glycerol:

$$\text{triose-P} + \text{NADH} + \text{H}^+ \longrightarrow \text{glycerol} + \textcircled{P} + \text{NAD}^+$$

B. subtilis, unlike most other species of group I, cannot grow anaerobically at the expense of glucose, probably because it cannot reduce triose phosphate to glycerol; in the presence of air this species carries out a fermentation of glucose with the formation of large amounts of 2,3-butanediol.

B. licheniformis can also grow anaerobically at the expense of nonfermentable organic substrates when furnished with nitrate, since it is a vigorous denitrifier; it is the only *Bacillus* species with this capacity.

A distinctive species cluster in group I consists of *B. cereus*, one of the most abundant aerobic sporeformers in soil, and two related pathogens, *B. anthracis* and *B. thuringensis*. In all three species the endospores are enclosed by a loose outer coat known as the *exosporium* (Figure 22.5), which is not formed by other bacilli. Certain strains of *B. cereus* produce distinctive loosely spreading colonies, superficially resembling those of fungi (Figure 22.6). *B. anthracis* is the agent of anthrax, a disease of cattle and sheep that is also transmissible to man. It is one of the few spore-forming bacteria that is a true parasite, in the sense that it is able to develop massively within the body of the animal host. Apart from its pathogenic properties, the biochemical mechanisms of which have been extensively studied (see Chapter 29), it differs from *B. cereus* by its permanent immotility.

B. thuringensis is the causal agent of paralytic disease of the caterpillars of many

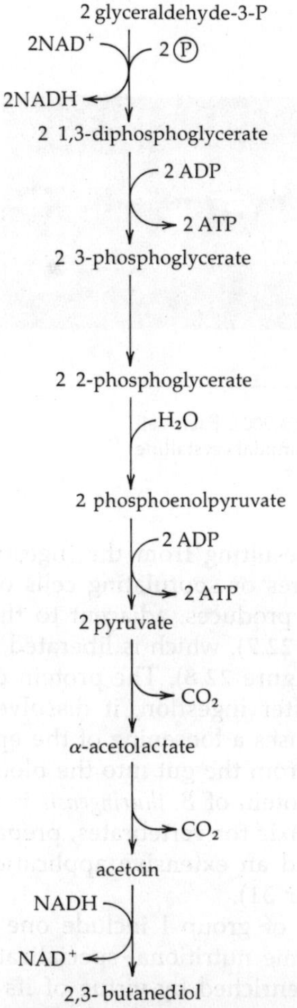

649

2 glyceraldehyde-3-P

2NAD⁺ ⟍ 2 (P)

2NADH ⟵

2 1,3-diphosphoglycerate

2 ADP

→ 2 ATP

2 3-phosphoglycerate

2 2-phosphoglycerate

H₂O

2 phosphoenolpyruvate

2 ADP

→ 2 ATP

2 pyruvate

CO₂

α-acetolactate

CO₂

acetoin

NADH

NAD⁺ ⟵

2,3- butanediol

FIGURE 22.4

Conversion of triose phosphate to 2,3-butanediol.

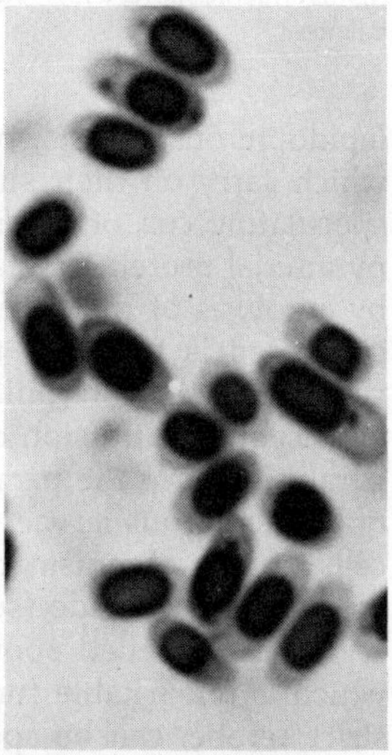

FIGURE 22.5

Mature endospores of *Bacillus cereus* (×3,600). Each stained spore is surrounded by a less deeply stained exosporium. Courtesy of C. F. Robinow.

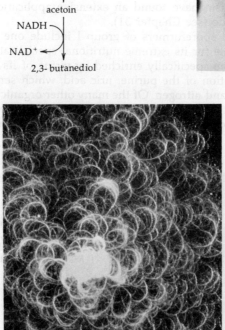

FIGURE 22.6

A colony of *Bacillus cereus var. mycoides* (×1.29). Courtesy of David Cornelius and C. F. Robinow.

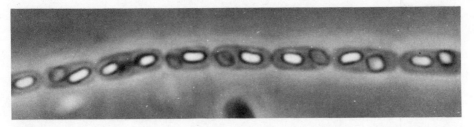

FIGURE 22.7

A chain of sporulating cells of *Bacillus thuringensis* (phase contrast, ×3,900). Each cell contains, in addition to the bright, refractile spore, a less refractile bipyramidal crystalline inclusion. Courtesy of P. FitzJames.

lepidopterous insects, paralysis resulting from the ingestion of plant materials which carry on their surface spores or sporulating cells of the bacterium. Each sporulating cell of *B. thuringensis* produces, adjacent to the spore, a regular bipyramidal protein crystal (Figure 22.7), which is liberated, along with the spore, by autolysis of the parent cell (Figure 22.8). The protein of which the crystal is composed is toxic for insects; after ingestion, it dissolves in the alkaline gut contents of the caterpillar and causes a loosening of the epithelial gut wall, with a consequent diffusion of liquid from the gut into the blood. This leads to rapid paralysis. Since the parasporal protein of *B. thuringensis* is toxic for a wide range of lepidopterous larvae, but nontoxic for vertebrates, preparations of sporulating cells of this organism have found an extensive application in agriculture as a biological insecticide (see Chapter 31).

The large-celled sporeformers of group I include one species, *B. fastidiosus*, which is remarkable for its extreme nutritional specialization. This organism, a strict aerobe, can be specifically enriched by virtue of its ability to perform an oxidative dissimilation of the purine, uric acid, which serves as its sole source of carbon, energy, and nitrogen. Of the many other organic compounds that have been tested, only two, allantoic and glyoxylic acids, can be used as carbon and energy sources. Both are intermediates in the pathway of uric acid dissimilation (Figure 22.9).

FIGURE 22.8

Electron micrograph of free crystals and a free spore surrounded by an exosporium of a crystal-forming *Bacillus* (metal-shadowed preparation, ×7,500). From C. L. Hannay and P. FitzJames, "The protein crystals of *B. thuringensis*." *Can. J. Microbiol.* **1**, 694 (1955).

uric acid

FIGURE 22.9

allantoic acid   glyoxylic acid

The structures of uric acid and of two metabolites derived from it; these three compounds are the only substrates utilizable by *Bacillus fastidiosus* as sources of carbon and energy. The mode of attack on uric acid and allantoin is indicated by dashed lines, and the carbon atoms which are converted to glyoxylate are shown in heavy type. Urea and CO_2 are the other products of ring oxidation.

FIGURE 22.10

Electron micrograph of a transverse section of a spore of *Bacillus polymyxa*, showing the elaborate stellate contour of the spore coat ($\times 32{,}400$). Courtesy of R. G. E. Murray.

The two principal *Bacillus* species of group II, *B. polymyxa* and *B. macerans*, form spores with a distinctive, thick outer coat which bears a series of raised ridges, giving a star-shaped profile to the spore (Figure 22.10). Both are fermentative organisms, dissimilating starch and pectins as well as monosaccharides; and good growth occurs only in the presence of a utilizable carbohydrate. *B. polymyxa* carries out a butanediol fermentation, chemically distinct from that of the group I *Bacillus* species; in addition to 2,3-butanediol, the main products are ethanol, CO_2, and H_2. The products of sugar fermentation of *B. macerans* are ethanol, acetone, acetate, formate, CO_2, and H_2. The production of H_2 as a major end product and the failure to produce glycerol distinguish the sugar fermentations of these species from those of group I. Another distinctive property of both *B. polymyxa* and *B. macerans* is the ability to fix N_2 when grown under anaerobic conditions; they are the only *Bacillus* species known to possess this property.

The *Bacillus* species of group III (*B. sphaericus* and *B. pasteurii*), which are distinguished by the formation of spherical spores located terminally in the sporulating cell, form a distinct subclass of bacilli in several physiological and metabolic respects. As shown in Table 22.2, their wall peptidoglycan is chemically distinct from that of other aerobic sporeformers. They lack fermentative capacity and are unable to utilize carbohydrates effectively as respiratory energy sources. The principal substrates of respiratory metabolism are amino acids and organic acids. Many (though not all) produce large amounts of the enzyme urease, which catalyzes a hydrolysis of urea:

$$CO(NH_2)_2 + H_2O \longrightarrow CO_2 + 2NH_3$$

Although this reaction does not result in ATP formation, it plays an important physiological role for the ureolytic bacilli. The principal ureolytic species, *B. pasteurii*, cannot grow on conventional complex media (e.g., nutrient broth) at neutral pH, unless such media are supplied with urea. Although urea is a neutral substance, its hydrolysis is accompanied by a considerable production of alkali, since two moles of ammonia are formed per mole of urea decomposed. Hence, a urea-containing medium inoculated with *B. pasteurii* rapidly becomes highly alkaline, as a result of ammonia formation. The specific urea requirement of this bacterium can be replaced by ammonia, in conjunction with a high initial pH (8.5). No other monovalent cation will replace ammonia for *B. pasteurii*; and it is required for respiration as well as growth. Therefore, in reality, the urea requirement of *B. pasteurii* reflects a specific requirement for ammonia and for a high pH. The only species of the genus *Sporosarcina*, *S. ureae*, closely resembles

B. pasteurii in metabolic and physiological respects as well as in cell wall composition. Indeed, these two species (which have similar DNA base compositions) are genetically related, as shown by nucleic acid hybridization experiments.

The ecological advantage conferred on these sporeformers by their strong ureolytic ability can be readily demonstrated by a simple enrichment experiment. If a peptone medium containing 2 to 5 percent of urea is inoculated with unheated soil, the population which develops consists exclusively of ureolytic sporeformers, despite the fact that a wide variety of chemoheterotrophic aerobic bacteria can grow in pure culture on this medium. In an enrichment culture all competing bacteria are killed by the high concentration of free ammonia produced from urea by the ureolytic sporeformers.

• **Thermophilic bacilli** A special physiological group among the aerobic sporeformers consists of the extreme thermophiles, which are capable of growing at temperatures as high as 65°C and usually fail to grow at temperatures below 45°C. There are probably a number of species with this attribute, but their taxonomy has not been studied in detail. The characteristic environment for these organisms is decomposing plant material, in which the heat generated by microbial metabolic activity cannot be readily dissipated. The classical example is a moist haystack, in which the rise of temperature is so marked that it sometimes leads to spontaneous combustion. As the temperature rises, in the first place through the metabolic activities of mesophilic microorganisms, the primary microbial population is displaced by extreme thermophiles, principally bacilli and actinomycetes.

THE ANAEROBIC SPOREFORMERS: GENUS *CLOSTRIDIUM*

Anaerobic sporeformers with a fermentative mode of metabolism (genus *Clostridium*) were discovered by Pasteur in the middle of the nineteenth century when he demonstrated that some of these organisms carry out a fermentation of sugars accompanied by the formation of butyric acid. Shortly afterward it was recognized that clostridia are also the principal agents of the anaerobic decomposition of proteins ("putrefaction"). Toward the end of the nineteenth century it became evident that some clostridia are agents of human or animal disease. Like other members of the group, the pathogenic clostridia are normal soil inhabitants, with little or no invasive power; the diseases they produce result from the production of a variety of highly toxic proteins (exotoxins). Indeed, botulism (cause by *C. botulinum*) and less serious types of clostridial food poisoning (caused by *C. perfringens*) are pure intoxications, resulting from the ingestion of foods in which these organisms have previously developed and formed exotoxins. The other principal clostridial diseases, tetanus (cause by *C. tetani*) and gas gangrene (caused by several other species) are the results of wound infections; tissue damage leads to the development of an anaerobic environment which permits localized growth and toxin formation by these organisms. Some clostridial toxins (those responsible for botulism and tetanus) are potent inhibitors for nerve function. Others (those responsible for gas gangrene) are enzymes that cause tissue destruction; they include lecithinases, hemolysins, and a variety of proteases (see Chapter 29).

Although over 60 *Clostridium* species have been described, the conventional taxonomic treatment of this group is not satisfactory. The highly diverse mechanisms of dissimilatory metabolism which occur in this genus probably provide the soundest basis for its taxonomic subdivision, and the group will be described here primarily in terms of these properties.

• The butyric
acid bacteria

Many clostridia perform a fermentation of soluble carbohydrates, starch or pectin, with the formation of acetic and butyric acids, CO_2 and H_2. These *butyric acid bacteria* grow poorly (if at all) in complex media devoid of a fermentable carbohydrate. Two other characters are distinctive of this subgroup: they synthesize as cellular reserve material a starchlike polysaccharide, detectable microscopically by its deep purple color in cells treated with an iodine solution; and many fix nitrogen very actively, a property otherwise absent from the clostridia.

The butyric fermentation is initiated by a conversion of sugars to pyruvate through the Embden-Meyerhof pathway. The pathways for the formation of end products from pyruvate are shown in Figure 22.11. Pyruvate undergoes a thiolytic cleavage to acetyl-CoA, CO_2, and H_2; acetate is derived from acetyl-CoA via acetyl phosphate, with an accompanying synthesis of ATP. Butyrate synthesis occurs through an initial condensation of two moles of acetyl-CoA, to form acetoacetyl-

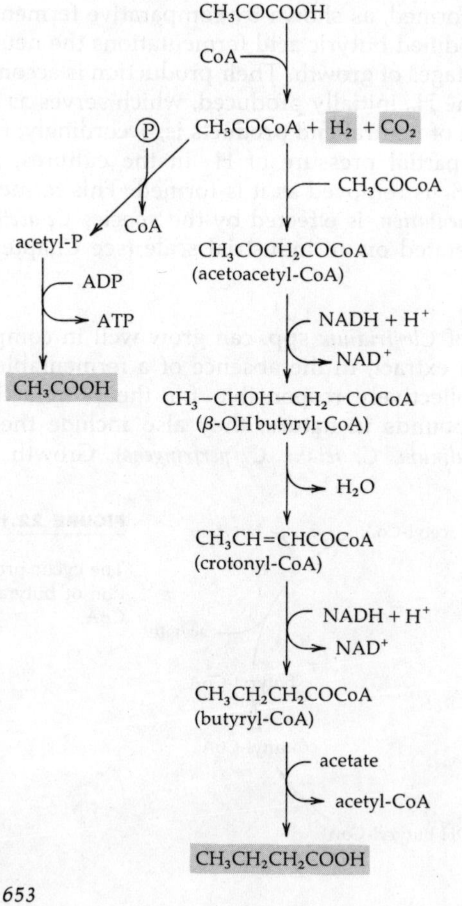

FIGURE 22.11

Pathways for the formation of the end products of the butyric acid fermentation (tinted boxes) from pyruvate.

653

CoA, which is then reduced to butyryl-CoA. Butyrate is then produced by CoA transfer to acetate:

$$\text{butyryl-CoA} + \text{acetate} \longrightarrow \text{acetyl-CoA} + \text{butyrate}$$

The pathway that leads to butyrate formation is therefore cyclic, involving a resynthesis of acetyl-CoA from free acetate, as shown in Figure 22.12: in effect, the precursors of butyrate are one mole of acetyl-CoA and one mole of free acetate. The overall reaction for the butyric acid fermentation of glucose can be roughly represented by the equation:

$$4\ \text{glucose} \longrightarrow 2\ \text{acetate} + 3\ \text{butyrate} + 8CO_2 + 8H_2$$

Since a net synthesis of ATP accompanies the formation of acetate from acetyl-CoA, the ATP yield per mole of glucose fermented is approximately 2.5 moles.

Some of the butyric acid bacteria form additional neutral compounds (butanol, acetone, isopropanol, small amounts of ethanol) from sugars. With the exception of ethanol (formed by the reduction of acetyl-CoA), these neutral products arise by divergences from the normal pathway of butyrate formation, as shown in Figure 22.13; and their formation is accompanied by a reduction in the amounts of butyrate and H_2 formed, as shown by comparative fermentation balances (Table 22.5). In these modified butyric acid fermentations the neutral products typically arise in the later stages of growth. Their production is accompanied by a reutilization of some of the H_2 initially produced, which serves as a reductant of NAD^+. The accumulation of neutral end products is, accordingly, favored by the maintenance of a high partial pressure of H_2 in the cultures, and it can be largely prevented if the H_2 is removed as it is formed. This fermentation, known as the *acetone-butanol fermentation*, is effected by the species *C. acetobutylicum* (Table 22.5) and has been operated on an industrial scale (see Chapter 31).

• **The anaerobic dissimilation of amino acids by clostridia**

A large number of *Clostridium* spp. can grow well in complex media containing peptones or yeast extract, in the absence of a fermentable carbohydrate. These organisms are collectively responsible for the putrefactive decomposition of nitrogenous compounds in nature; they also include the principal pathogenic clostridia (*C. botulinum*, *C. tetani*, *C. perfringens*). Growth in complex media is

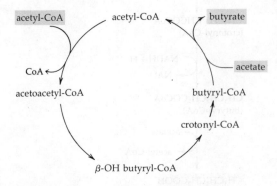

FIGURE 22.12

The cyclic process which leads to formation of butyrate from acetate and acetyl-CoA.

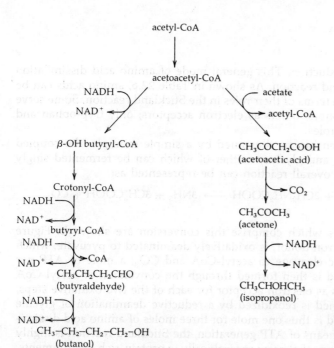

acetyl-CoA

↓

acetoacetyl-CoA

NADH
NAD⁺

β-OH butyryl-CoA

↓

Crotonyl-CoA

NADH
NAD⁺

butyryl-CoA

NADH
NAD⁺ → CoA

CH₃CH₂CH₂CHO
(butyraldehyde)

NADH
NAD⁺

CH₃–CH₂–CH₂–CH₂–OH
(butanol)

acetate
→ acetyl-CoA

CH₃COCH₂COOH
(acetoacetic acid)

→ CO₂

CH₃COCH₃
(acetone)

NADH
NAD⁺

CH₃CHOHCH₃
(isopropanol)

FIGURE 22.13

Pathways for the formation of butanol, acetone, and iso-propanol from acetyl-CoA.

accompanied by the formation of ammonia, CO_2, H_2S, fatty acids, and a variety of other volatile substances, often having unpleasant odors. The nature of the specific fermentable substrates and the pathways involved in their dissimilation long remained unknown; they have been elucidated during the past 40 years, largely through the work of L. H. Stickland and the work of H. A. Barker and his associates.

Probably the most widespread mechanism of amino acid dissimilation among these organisms is the fermentation of pairs of amino acids, one of which acts as an electron donor, undergoing oxidation, while the other acts as an electron

TABLE 22.5

The products of fermentation of glucose by three species of butyric acid bacteria[a]

	C. lactoacetophilum	C. acetobutylicum	C. butylicum
Butyric acid	73	4	17
Acetic acid	28	14	17
CO_2	190	221	204
H_2	182	135	78
Ethanol	—	7	—
Butanol	—	56	59
Acetone	—	22	—
Isopropanol	—	—	12

[a] moles/100 moles of glucose fermented

acceptor, undergoing reduction. This general mode of amino acid dissimilation is known as the Stickland reaction. As shown in Table 22.6, amino acids can be assigned to two series in terms of their roles in the Stickland reaction. Some serve uniquely as electron donors, others as electron acceptors; only tryptophan and tyrosine can play both roles.

The Stickland reaction can be illustrated by a simple example: the coupled fermentation of alanine and glycine, neither of which can be fermented singly by most clostridia. The overall reaction can be represented as:

$$CH_3CHNH_2COOH + 2CH_2NH_2COOH \longrightarrow 3NH_3 + 3CH_3COOH + CO_2$$
$$\text{(alanine)} \qquad\qquad \text{(glycine)}$$

The various reactions which comprise this conversion are shown in Figure 22.14. Alanine, the electron donor, is oxidatively deaminated to pyruvate, which then undergoes thiolytic cleavage to acetyl-CoA and CO_2; a mole of ATP per mole of alanine oxidized is then formed through the conversion of acetyl-CoA to acetate. NAD^+ serves as electron acceptor for each of the two oxidative steps, and the NADH so formed is reoxidized by a reductive deamination of glycine to acetate. The ATP yield is thus one mole for three moles of amino acid dissimilated. Considered as a means of ATP generation, the Stickland reaction is highly advantageous for organisms that grow anaerobically in protein-rich environments, since virtually all the constituent amino acid of proteins can be utilized as energy sources.

Many clostridia are able to ferment specific, single amino acids, a mode of dissimilatory metabolism that may or not be accompanied by the ability to perform the Stickland reaction (Table 22.7). The pathway for the dissimilation

TABLE 22.6

Classification of amino acids in terms of their roles as electron donors and electron acceptors in the Stickland reaction[a]

ELECTRON DONORS		ELECTRON ACCEPTORS	
AMINO ACID	RELATIVE RATE OF OXIDATION	AMINO ACID	RELATIVE RATE OF REDUCTION
Alanine	100	Glycine	100
Leucine	100	Proline	100
Isoleucine	100	Hydroxyproline	100
Norleucine	100	Ornithine	100
Valine	76	Arginine	80
Histidine	37	Tryptophan	67
Phenylalanine	28	Tyrosine	25
Tryptophan	17	Cysteine	22
Tyrosine	16	Methionine	15
Serine	16		
Asparagine	13		

[a] The relative rates of oxidation (or reduction) shown were determined for the species *C. sporogenes*.

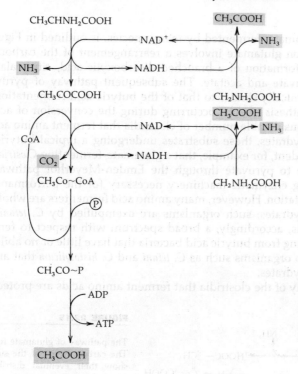

Alanine oxidation

Glycine reduction

CH_3CHNH_2COOH

NAD^+

NH_3 $NADH$

$CH_3COCOOH$

NAD

CoA

$NADH$

CO_2

$CH_3Co{\sim}CoA$

P

$CH_3CO{\sim}P$

ADP

ATP

CH_3COOH

CH_3COOH

NH_3

CH_2NH_2COOH

CH_3COOH NH_3

CH_2NH_2COOH

FIGURE 22.14

Mechanism of the Stickland reaction, with alanine as electron donor and glycine as electron acceptor.

TABLE 22.7

The amino acids fermented by some species of *Clostridium*

| | SINGLE AMINO ACIDS THAT CAN BE FERMENTED | | | | | | | | | | | | | |
SPECIES	Alanine	Arginine	Aspartate	Cysteine	Glutamate	Histidine	Leucine	Lysine	Methionine	Phenylalanine	Serine	Threonine	Tyrosine	ABILITY TO PERFORM STICKLAND REACTION
C. botulinum											+			+
C. cochlearium					+									−
C. perfringens					+						+	+		
C. propionicum	+			+							+	+		−
C. sporogenes		+		+			+		+	+	+	+		+
C. sticklandii		+						+	+	+				+
C. tetani			+		+	+				+				−
C. tetanomorphum			+	+	+	+				+			+	−

of glutamate, fermented by many species, is outlined in Figure 22.15. The initial attack on glutamate involves a rearrangement of the carbon skeleton and leads to the formation of a branched dicarboxylic acid, citramalate, which is cleaved to pyruvate and acetate. The subsequent pathway of pyruvate dissimilation is biochemically similar to that of the butyric acid fermentation (see Figure 22.11), the synthesis of ATP occurring during the conversion of acetyl-CoA to acetate.

A considerable number of clostridia that ferment amino acids can also ferment carbohydrates, these substrates undergoing a typical butyric acid fermentation. It is evident, for example, that a glutamate-fermenting *Clostridium* that can convert glucose to pyruvate through the Emden-Meyerhof pathway possesses the remaining enzymatic machinery necessary for the performance of a butyric acid fermentation. However, many amino acid fermenters are wholly unable to ferment carbohydrates; such organisms are exemplified by *C. tetani* and *C. histolyticum*. There is, accordingly, a broad spectrum with respect to fermentable substrates extending from butyric acid bacteria that have little or no ability to ferment amino acids to organisms such as *C. tetani* and *C. histolyticum* that are unable to ferment carbohydrates.

Many of the clostridia that ferment amino acids are proteolytic organisms and

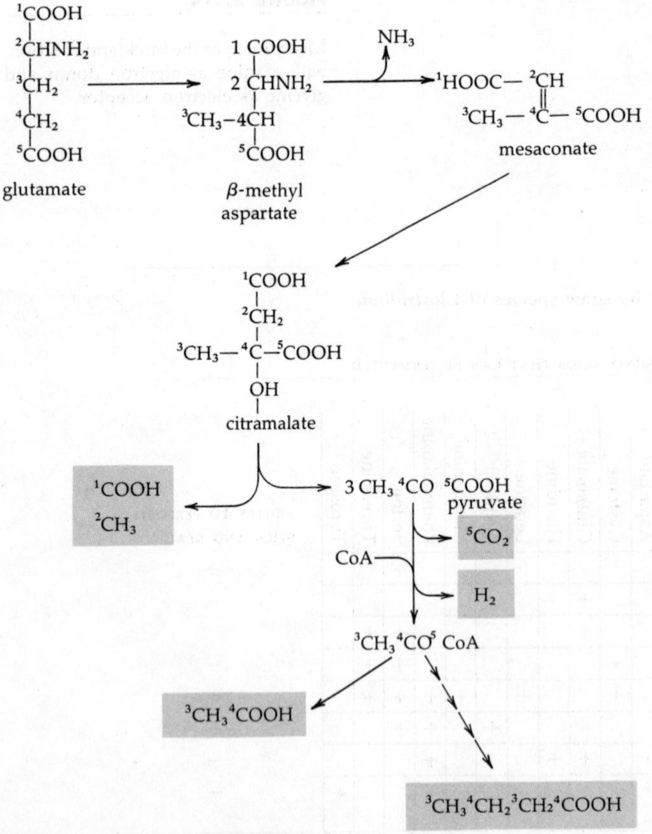

FIGURE 22.15

The pathway of glutamate fermentation by clostridia. The carbon atoms of the substrate are numbered to show their eventual distribution in end products (shaded boxes).

produce a wide diversity of proteases; several hydrolytic enzymes of this type with different substrate specificities are often produced by a single organism. Proteolysis is of course a necessary preliminary step in the production of fermentable substrates from proteins by members of this group which obtain energy through amino acid fermentations. It should be noted, however, that by no means all amino acid-fermenting clostridia are proteolytic organisms. Nonproteolytic species are consequently dependent on the availability of free amino acids as growth substrates.

• The fermentation of nitrogen-containing ring compounds

Some clostridia can obtain energy by the fermentation of ring compounds including purines, pyrimidines, and nicotinic acid. The fermentation of purines (guanine, uric acid, hypoxanthine, xanthine) is carried out by *C. acidiurici* and *C. cylindrosporum,* nutritionally highly specialized species, which are unable to ferment other substrates. The fermentation products consist of acetate, glycine, formate, CO_2, or other precursors (Figure 22.16). Only one mole of acetate per mole of purine fermented can be derived directly from a C_2 fragment, since carbon atoms 4 and 5 are the only contiguous carbon atoms in purines. However, the yield of acetate is often greater than one mole, which shows that it must be formed in part from C_1 precursors. Acetate synthesis from CO_2 is a characteristic of certain other clostridial fermentations, discussed below.

• Carbohydrate fermentations by clostridia which do not yield butyric acid as a product

A number of clostridia utilizing carbohydrates as energy sources dissimilate them by pathways other than the butyric acid pathway. These organisms include cellulose-fermenting clostridia, most of which are highly specialized with respect to substrates; some species can ferment only cellulose. The products include ethanol, formate, acetate, lactate, and succinate.

The species *C. thermoaceticum* ferments glucose and other soluble sugars with the formation of acetate as the sole end product; the formation of this product is virtually quantitative, almost three moles of acetate being produced per mole of glucose decomposed. No known pathway of glucose dissimilation permits a direct formation of acetate from all six carbon atoms of the substrate. In fact, only two-thirds of the acetate produced is directly derived from glucose carbon through the reactions of glycolysis; one-third of the acetate is produced by a complex process of total synthesis from CO_2, involving the participation of tetrahydrofolate and a corrinoid coenzyme (a vitamin B_{12} derivative) as carriers

FIGURE 22.16

Origins of some end products from the clostridial fermentation of xanthine.

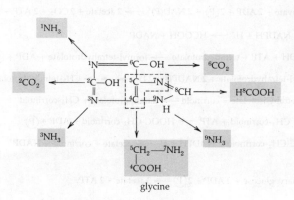

glycine

of the C_1 and C_2 intermediates. The total reaction sequence is outlined in Figure 22.17.

• **The ethanol-acetate fermentation by *Clostridium kluyveri***

The most remarkable of all clostridial fermentations is that performed by *C. kluyveri*. This organism grows only at the expense of a mixture of ethanol and acetate as its energy sources. The main organic products of the fermentation are two higher fatty acids, butyrate and caproate; in addition, some H_2 is produced. If H_2 production is neglected, the fatty acid synthesis can be represented by the two equations:

$$CH_3CH_2OH + CH_3COOH \longrightarrow CH_3CH_2CH_2COOH + H_2O$$

$$2CH_3CH_2OH + CH_3COOH \longrightarrow CH_3CH_2CH_2CH_2CH_2COOH + H_2O$$

The mechanism of net ATP synthesis associated with this fermentation has been discovered only recently and will be outlined for the case of butyrate synthesis.

Ethanol is dehydrogenated in two steps to acetyl-CoA, ferredoxin being the initial acceptor for the second dehydrogenation. Reduced ferredoxin can transfer electrons either to NAD^+ with formation of NADH, or to protons with formation of H_2 (Figure 22.18). As in other clostridia, the synthesis of butyrate then occurs through the cyclic reactions shown in Figure 22.12, represented by the equation:

$$\text{acetyl-CoA} + \text{acetate} + 2NADH + 2H^+ \longrightarrow \text{butyrate} + CoA + 2NAD^+ + H_2O$$

In the absence of H_2 formation, the oxidation of ethanol thus produces the exact quantities of acetyl-CoA and NADH required for butyrate synthesis:

$$\text{ethanol} + CoA + 2NAD^+ \longrightarrow \text{acetyl-CoA} + 2NADH + 2H^+$$

Under these conditions, a net synthesis of ATP cannot occur. However, the formation of H_2 diverts part of the electrons from acetaldehyde oxidation that

FIGURE 22.17

The reactions involved in the conversion of glucose to acetate by *Clostridium thermoaceticum*.

1. $\text{Glucose} + 2\,ADP + 2\,\text{(P)} + 2\,NAD(P)+ \xrightarrow[\substack{\text{Mayerhof} \\ \text{pathway)}}]{\text{(Embden-}} 2\,\text{pyruvate} + 2\,ATP + 2\,NAD(P)H + 2\,H^+$

2. $2\,\text{pyruvate} + 2\,ADP + 2\,\text{(P)} + 2\,NAD(P)^+ \longrightarrow 2\,\text{acetate} + 2\,CO_2 + 2\,ATP + 2\,NAD(P)H + 2\,H^+$

3. $CO_2 + NADPH + H^+ \longrightarrow HCOOH + NADP^+$

4. $HCOOH + ATP + \text{tetrahydrofolate} \longrightarrow \text{formyl-tetrahydrofolate} + ADP + \text{(P)}$

5. $\text{Formyl-tetrahydrofolate} + 2\,NADPH + 2\,H^+ \longrightarrow \longrightarrow \longrightarrow CH_3\text{-tetrahydrofolate} + 2\,NADP^+$

6. $CH_3\text{-tetrahydrofolate} + \text{corrinoid} \longrightarrow \text{tetrahydrofolate} + CH_3\text{-corrinoid}$

7. $CO_2 + CH_3\text{-corrinoid} + ATP \longrightarrow HOOC\text{-}CH_2\text{-corrinoid} + ADP + \text{(P)}$

8. $HOOC\text{-}CH_2\text{-corrinoid} + NADPH + H^+ \longrightarrow \text{acetate} + \text{corrinoid} + NADP^+$

Summary: $\text{glucose} + 2\,ADP + 2\,\text{(P)} \longrightarrow 3\,\text{acetate} + 2\,ATP.$

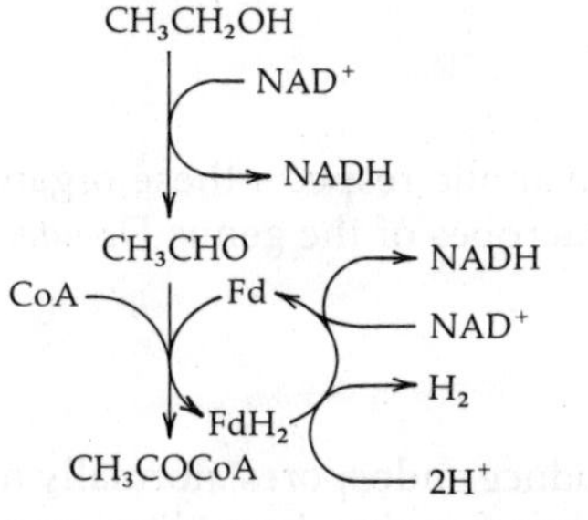

CH$_3$CH$_2$OH

NAD$^+$

NADH

CH$_3$CHO — NADH

CoA — Fd — NAD$^+$

H$_2$

FdH$_2$

CH$_3$COCoA — 2H$^+$

FIGURE 22.18

FIGURE 22.18

The pathway of ethanol oxidation to acetyl-CoA by *Clostridium kluyveri*. Fd denotes ferredoxin.

would otherwise serve for NAD$^+$ reduction, and the molar ratio of NADH:acetyl-CoA produced during ethanol oxidation becomes less than 2:1. As a result, not all the acetyl-CoA formed can be converted to butyrate. Some "surplus" acetyl-CoA is therefore available for ATP synthesis by the reactions:

$$\text{acetyl-CoA} + \text{ⓅP} \longrightarrow \text{acetyl-P} + \text{CoA}$$

$$\text{acetyl-P} + \text{ADP} \longrightarrow \text{acetate} + \text{ATP}$$

The stoichiometry between H$_2$ production and ATP synthesis can be readily calculated. Let us suppose that a moles of H$_2$ are produced per mole of ethanol oxidized. The equation describing ethanol oxidation then becomes:

$$\text{ethanol} + (2 - a)\text{NAD}^+ + \text{CoA}$$

$$\longrightarrow \text{acetyl-CoA} + a\text{H}_2 + (2 - a)\text{NADH} + (2 - a)\text{H}^+$$

and that describing butyrate synthesis becomes:

$$\left(\frac{2 - a}{2}\right)\text{acetyl-CoA} + \left(\frac{2 - a}{2}\right)\text{acetate} + (2 - a)\text{NADH} + (2 - a)\text{H}^+$$

$$\longrightarrow \frac{2 - a}{2}\text{butyrate} + \frac{2 - a}{2}\text{CoA} + (2 - a)\text{NAD}^+ + \frac{2 - a}{2}\text{H}_2\text{O}$$

The "surplus" acetyl-CoA (i.e., that not utilized for butyrate synthesis) is $1 - [(2 - a)/2]$ moles, or $a/2$ moles. Accordingly, for every mole of H$_2$ produced, 0.5 mole of acetyl-CoA becomes available for ATP synthesis. An experimental determination of the balance for this fermentation showed that approximately 0.25 mole of H$_2$ was produced per mole of ethanol used. The ATP yield under these circumstances was therefore roughly $0.25/2 = 0.12$ mole of ATP produced per mole of ethanol oxidized.

THE ANAEROBIC SPOREFORMERS: GENUS *DESULFOTOMACULUM*

The anaerobic sporeformers of the genus *Desulfotomaculum*, which perform anaerobic respiration of organic substrates with SO$_4$$^{2-}$ as a terminal electron acceptor constitute a small group of species, readily distinguishable from the clostridia by several characters: the much higher mean DNA base composition of their DNA; the presence of heme pigments in the cells; and the dependence on SO$_4$$^{2-}$ as a terminal electron acceptor, leading to the formation of large amounts of H$_2$S during growth. The range of oxidizable substrates is limited, lactate being universally used; and the substrate oxidation is always incomplete, resulting in the formation of acetate. In metabolic respects these organisms are analogues of the nonspore-forming strict anaerobes of the genus *Desulfovibrio* (see Chapter 24).

THE ENDOSPORE

The ability to produce endospores, normally never expressed during the vegetative growth of a spore-forming bacterium, constitutes an extremely complex process of differentiation, which is initiated massively in the cellular population as it passes out of exponential growth and approaches the stationary phase. The process leads to the synthesis, within most vegetative cells, of a new type of cell, wholly different from the mother cell in fine structure, chemical composition, and physiological properties. After release from the mother cell, the endospore normally enters a long period of dormancy; however, if subjected to appropriate stimuli, it can germinate and grow out into a typical vegetative cell. The fascinating problems of endospore formation and germination have been intensively and extensively studied by many microbiologists, and only a relatively brief summary of the vast literature on this subject can be presented here. For reasons of experimental convenience, most of the experimental work on endospores has been conducted with members of the genus *Bacillus*; however, both the formation and the germination of endospores in the anaerobic sporeformers appear on present information to involve basically similar events.

• **Endospore formation**

The structural events associated with spore formation have been elucidated by a combination of light and electron microscopic observations. A synthesis based on both types of observations will be presented here (Figure 22.19).

At the shift from logarithmic to linear growth which just precedes the onset of sporulation, each cell contains two nuclear bodies. These bodies coalesce to form an axial chromatin thread. The first definite sign of the onset of sporulation is the formation of a transverse septum near one cell pole, which separates the cytoplasm and the DNA of the smaller cell (destined to become the spore) from the rest of the cell contents. Septum formation is not followed, as in normal cell division, by the development of a transverse wall; instead, the membrane of the larger cell rapidly grows around the smaller cell, which thus becomes completely engulfed within the cytoplasm of the larger cell, to produce a so-called *forespore*. In effect, the forespore is a protoplast, enclosed by two concentric sets of unit membranes, its own bounding membrane, and the membrane of the mother cell which has grown around it. At this stage, the development process becomes irreversible: the cell is said to be "committed" to undergo sporulation. By light microscopy, the forespore appears as a clear, nonrefractile area which is free of granular inclusions.

Once the forespore has been engulfed by the mother cell, there is a rapid synthesis and deposition of new structures that enclose it. The first to appear is the *cortex*, which develops between the inner and outer membranes; shortly afterward, a more electron-dense layer, the *spore coat*, begins to form exterior to the outer unit membrane surrounding the cortex. In the *B. cereus* group an additional, looser and thinner layer, the *exosporium*, forms outside the spore coat (Figure 22.20). Once the spore coat is synthesized, the maturing spore begins to become refractile, although it is not yet heat-resistant. The development of heat resistance

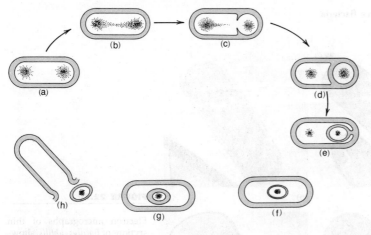

FIGURE 22.19

A diagrammatic representation of the cytological changes accompanying endospore formation in *Bacillus cereus*. (a) Vegetative cell, containing two nuclear bodies (black areas) and granules of reserve material (stippled areas). (b) Condensation of the nuclear material into a rod-shaped element. (c) Beginning of transverse wall formation. (d) Completion of the transverse wall; the forespore with its nuclear material is now cut off from the vegetative cell. (e) Growth of the new wall around the forespore. (f) Completion of the enclosing wall of the spore. (g) Maturation of the spore. (h) Liberated spore, surrounded by a loose outer coat, the exosporium. After I. E. Young and P. FitzJames, "Chemical and Morphological Studies of Bacteria Spore Formation, I," *J. Biophys. Biochem. Cytol.* **6**, 467 (1959).

closely follows two major chemical changes: a massive uptake of Ca^{2+} ions by the sporulating cell and the synthesis in large amounts of dipicolinic acid, a compound absent from vegetative cells. The time sequence of the development of refractility and heat resistance and the synthesis of dipicolinic acid are shown in Figure 22.21. In mature spores the molar ratio of dipicolinate:Ca^{2+} is close to unity, which suggests that dipicolinate occurs as a Ca-chelate. This compound represents 10 to 15 percent of the spore dry weight, and it is located within the spore protoplast. The synthesis of dipicolinate represents an offshoot of the lysine branch of the pathway for the biosynthesis of amino acids of the aspartate family, as shown in Figure 22.22.

The cortex is largely composed of a unique peptidoglycan, containing three repeating subunits (Figure 22.23): a muramic lactam subunit, without any attached amino acids; an alanine subunit, bearing only an L-alanyl residue; and a tetra-peptide subunit, bearing the sequence L-ala-D-glu-meso-DAP-D-ala. They represent, respectively, approximately 55, 15, and 30 percent of the total. There is very little cross-linking between tetrapeptide chains. The distinctiveness of the cortex peptidoglycan is further shown by the fact that *B. subtilis* and *B. sphaericus*, which synthesize chemically different vegetative cell wall peptidoglycans (see Table 22.2), contain essentially similar cortex peptidoglycans.

The outer spore coat, which represents 30 to 60 percent of the dry weight of the spore, is largely composed of protein and accounts for about 80 percent of the total spore protein. The spore coat proteins have an unusually high content of cysteine and of hydrophobic amino acids, and are highly resistant to treatments which solubilize most proteins.

After the completion of spore development, the spore protoplast, accordingly,

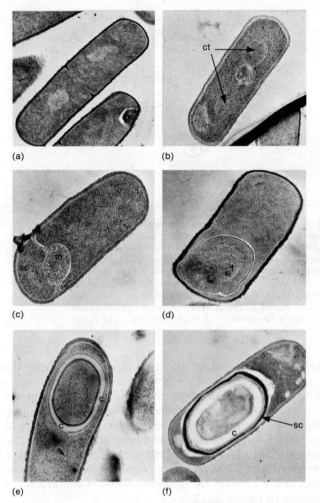

(a)

(b)

(c)

(d)

(e)

(f)

FIGURE 22.20

Electron micrographs of thin sections of *Bacillus subtilis*, showing the sequence of structural changes associated with endospore formation. (a) Vegetative cell in course of exponential growth. (b) Coalescence of the nuclei, to form an axial chromatin thread, ct. (c) Formation of transverse septum (containing a mesosome, m) near one cell pole, delimiting the future spore cell, sc, from the rest of the cell. (d) Formation of a forespore f, completely enclosed by the cytoplasm of the mother cell. (e) The developing spore is surrounded by the cortex, c. (f) The terminal stage of spore development: the mature spore, still enclosed by the mother cell, is now surrounded by both cortex, c, and spore coat, sc. From A. Ryter, "Étude Morphologique de la Sporulation de *Bacillus subtilis*," *Ann. Inst. Pasteur* **108**, 40 (1965).

contains an extremely high content of Ca dipicolinate; and it is enclosed by newly synthesized outer layers of unique chemical structure (the cortex and the spore coat, sometimes also an exosporium), which account for a very large fraction of the spore dry weight. When liberated by autolysis of the mother cell, the mature endospore is highly dehydrated, shows no detectable metabolic activity, and is effectively protected from heat and radiation damage and from attack by either enzymatic or chemical agents. It remains in this cryptobiotic state until a series of environmental triggers initiate its conversion into a new vegetative cell.

• Other biochemical events related to sporulation   Although the synthesis of an endospore is the main enterprise of a sporulating cell, it is by no means the only one. One striking concomitant event, characteristic of *Bacillus thuringensis*, has already been discussed (p. 650), the formation of a

FIGURE 22.21

The increases in the refractility, thermostability of the cells, and dipicolinate content of the population that occur during sporulation in a culture of *Bacillus cereus*. All values are plotted against the age of the culture in hours. After T. Hasimoto, S. H. Black, and P. Gerhardt, "Development of fine structure, thermostability, and dipicolinate during sporogenesis in a bacillus." *Can. J. Microbiol.* **6**, 203 (1960).

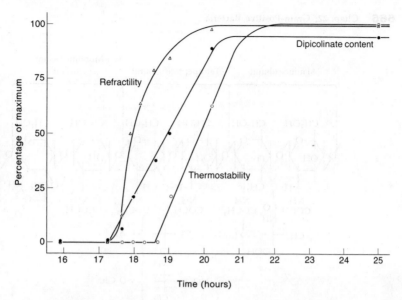

bipyramidal parasporal protein crystal adjacent to each endospore. Various noncrystalline parasporal structures of defined form have been described in other species of *Bacillus* and *Clostridium*.

In many sporeformers, both aerobic and anaerobic, the onset of sporulation is accompanied by the synthesis of a distinctive class of antimicrobial substances: peptides with molecular weights of approximately 1,400 daltons. Many of these peptide antibiotics have been characterized chemically and functionally. They can

FIGURE 22.22

Pathway of biosynthesis of dipicolinic acid showing its relationship to the pathways of biosynthesis of the amino acids: threonine, isoleucine, methionine, and lysine.

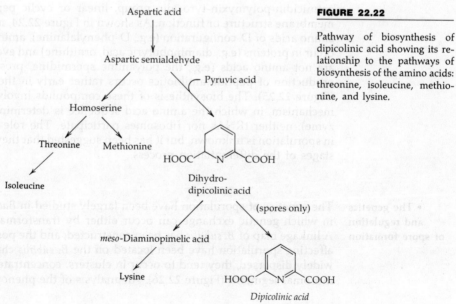

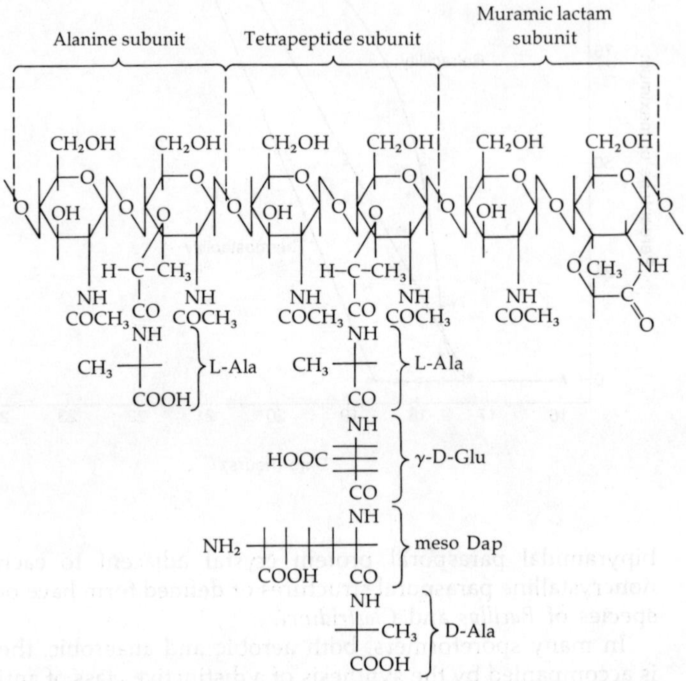

Alanine subunit   Tetrapeptide subunit   Muramic lactam subunit

FIGURE 22.23

Structures of the repeating subunits of the spore cortex peptidoglycan.

be assigned to three classes: edeines, linear basic peptides which inhibit DNA synthesis; bacitracins, cyclic peptides which inhibit cell wall synthesis; and the gramicidin-polymyxin-tyrocidin group, linear or cyclic peptides which modify membrane structure or function. As shown in Figure 22.24, many of them contain amino acids of D-configuration (e.g., D-phenylalanine), amino acids which do not occur in proteins (e.g., diaminobutyric acid, ornithine) and even constituents which are not amino acids (e.g., the polyamine spermidine, present in edeines). The production of peptide antibiotics occurs rather early in the sporulation process (Figure 22.25). The biosynthesis of these compounds involves a novel assembly mechanism, in which the amino acid sequence is determined by a protein (enzyme); neither tRNAs nor ribosomes participate. The role of these compounds in sporulation is unknown, but it has been suggested that they may control various stages of the differentiation process.

• **The genetics and regulation of spore formation**

The genetics of sporulation have been largely studied in *Bacillus subtilis*, a species in which genetic exchange can occur either by transformation or transduction. A linkage map of *B. subtilis* has been constructed, and the positions of many genes affecting sporulation have been located on the *B. subtilis* chromosome. Although widely dispersed, they tend to occur in clusters, concentrated in about five chromosomal segments (Figure 22.26). An analysis of the phenotypic effects of muta-

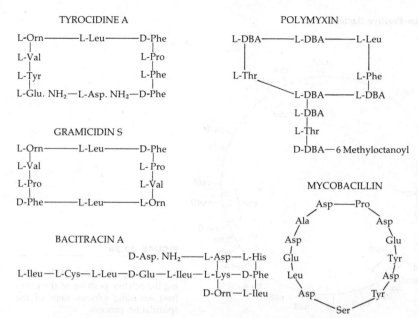

TYROCIDINE A

```
L-Orn————L-Leu————D-Phe
  |                    |
L-Val                L-Pro
  |                    |
L-Tyr                L-Phe
  |                    |
L-Glu. NH₂—L-Asp. NH₂—D-Phe
```

GRAMICIDIN S

```
L-Orn————L-Leu————D-Phe
  |                    |
L-Val                L-Pro
  |                    |
L-Pro                L-Val
  |                    |
D-Phe————L-Leu————L-Orn
```

BACITRACIN A

```
              D-Asp. NH₂—L-Asp—L-His
                                   |
L-Ileu—L-Cys—L-Leu—D-Glu—L-Ileu—L-Lys—D-Phe
                                   |
                     D-Orn—L-Ileu
```

POLYMYXIN

```
L-DBA————L-DBA————L-Leu
  |                    |
L-Thr                L-Phe
  |                    |
      L-DBA————L-DBA
        |
      L-DBA
        |
      L-Thr
        |
      D-DBA—6 Methyloctanoyl
```

MYCOBACILLIN

```
        Asp————Pro
       /           \
     Ala            Asp
      |              |
     Asp            Glu
      |              |
     Glu            Tyr
      |              |
     Leu            Asp
      |              |
     Asp            Tyr
       \           /
         Ser
```

FIGURE 22.24

Structures of certain antibiotics synthesized during the sporulation process by sporeforming bacteria. Abbreviations are: ornithine, orn; valine, val; tyrosine, tyr; glutamine, glu·NH₂; asparagine, asp·NH₂; phenylalanine, phe; proline, pro; leucine, leu; isoleucine, ileu; glutamic acid, glu; lysine, lys; histidine, his; threonine, thr; α-diaminobutyric acid, DBA; aspartic acid, asp; and serine, ser.

tions in these so-called *spo* genes (Figure 22.27) clearly reveals the sequential nature of the process of sporulation. The total number of loci involved in spore formation is still not known with precision, but it is certainly at least 50.

Since the genes governing the sporulation process are not normally expressed during the vegetative growth of spore-forming bacteria, it must be assumed that they are subject to a system of repression, which is relieved when vegetative growth ceases. Massive sporulation can be induced experimentally by removing cells from a growing culture and transferring them to a nitrogen-free medium. For 2 or 3 hours after such a transfer, the provision of a nitrogen source leads to a resumption of vegetative growth. After a certain point, the cells become *committed to sporulation:* addition of a nitrogen source no longer causes a resumption of vegetative growth, the population instead proceeding to sporulate.

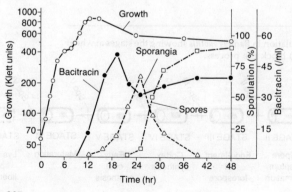

FIGURE 22.25

The time course of growth, bacitracin synthesis, and formation of sporangia and free spores by a culture of *Bacillus licheniformis*.

667

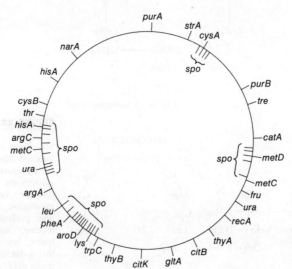

FIGURE 22.26

Linkage map of *Bacillus subtilis* showing the relative position of the genes
(*spo*) encoding various steps of the
sporulation process.

Although repression of sporulation is normally extremely effective during vegetative growth, it can be overcome under certain circumstances. In synthetic media where the growth rate is relatively low, aerobic sporeformers form a small number of spores in the course of growth; these spores are formed at a constant rate, so that their increase in the population parallels the increase in the number of vegetative cells. This behavior suggests that there is a certain probability under such conditions for a vegetative cell to undergo commitment. An even more striking demonstration of the fact that complete arrest of growth is not essential to initiate sporulation has been obtained from experiments on the growth of *B. megaterium* in a chemostat. When glucose is supplied as the rate-limiting nutrient, it is impossible to maintain a vegetative population at a growth rate of less than 0.5 division per hour; instead, mass sporulation occurs, and the population is washed out of the growth chamber.

FIGURE 22.27

Diagrammatic representation of sporulation in *Bacillus subtilis* showing the stages at which various genetic lesions (in *spo* genes) block the process.

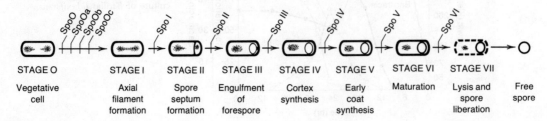

Under special circumstances, sporeformers can be induced to undergo a process known as *microcycle sporogenesis*. After spore germination and outgrowth of vegetative cells, the application of conditions unfavorable for the maintenance of vegetative growth can short-circuit the cell cycle, causing the newly emerged vegetative cell immediately to reenter the sporulation process. This leads to the simplest possible cell cycle:

$$\text{spore} \longrightarrow \text{vegetative cell} \longrightarrow \text{sporangium} \longrightarrow \text{spore}$$

• **Activation, germination, and outgrowth of endospores**

Freshly formed endospores will remain largely dormant even if placed in optimal conditions for germination. The state of dormancy can be broken by a variety of treatments, collectively termed *activation*. Perhaps the most general mechanism for activating spores is *heat shock:* exposure for several hours to an elevated, but sublethal temperature (e.g. 65°C). Heat activation is not accompanied by any detectable change in the appearance of the spores: it simply enables them to germinate when subsequently placed in an environment favorable for this process. Furthermore, heat activation is reversible: if spores are subsequently placed at a lower temperature for some days, their induced germinability declines. A much slower activation takes place upon the storage of spores, even at relatively low temperatures (5°C) or under dry conditions; this activation is irreversible, an increasing fraction of the spore population being capable of germination as storage is prolonged.

When activated spores are placed under favorable conditions, germination can take place. This process is very rapid and is expressed by a loss of refractility, a loss of resistance to heat and other deleterious agents, and the unmasking of metabolic activity, as evidenced by a sudden onset of respiration. These processes are accompanied by the liberation, as soluble materials, of about 30 percent of the spore dry weight. This spore exudate consists largely of Ca-dipicolinate (derived from the spore protoplast) and peptidoglycan fragments (derived from the cortex). The cortex is rapidly destroyed, only the outer spore coat remaining. Germination of activated spores requires a chemical trigger; the specific substances which are active are numerous and varied, including L-alanine, ribosides (adenosine, inosine), glucose, Ca-dipicolinate, and various inorganic anions and cations. The particular germination requirements often differ from species to species, and maximal germination may require a combination of several germinants. Furthermore, germination can be accomplished mechanically by subjecting spores to physical treatments which erode or crack the spore coat.

The process of germination does not appear to be accompanied by significant macromolecular synthesis; the appearance of metabolic activity is a consequence of the unmasking of preexisting but inactive enzymes in the spore protoplast. If germination occurs in a medium which does not contain the nutrients required for vegetative growth, no further change will occur. However, if the nutrients required for macromolecular synthesis are present upon germination, the germinated spore will proceed to grow out into a vegetative cell. Outgrowth involves an initial swelling of the spore within its spore coat, accompanied by a rapid synthesis of a vegetative cell wall, the foundation layer of which may already be present around the spore protoplast at the time of germination. The newly formed vegetative cell then emerges from the spore coat (Figure 22.28), elongates, and proceeds to undergo the first vegetative division.

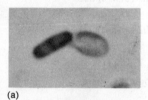

(a)

(b)

FIGURE 22.28

Spore germination in (a) *Bacillus polymyxa* and (b) *B. circulans* (stained preparations, ×5,260). Courtesy of C. F. Robinow and C. L. Hannay.

669

FURTHER READING

I. Spore-forming bacteria

Books   SMITH, N. R., R. E. GORDON, and F. E. CLARK, *Aerobic Spore-forming Bacteria, Agr. Monograph 16.* Washington, D.C.: U.S. Department of Agriculture, 1952.

Reviews and original articles   BARKER, H. A., "Fermentation of Nitrogenous Organic Compounds," in *The Bacteria,* I. C. Gunsalus and R. Y. Stanier (editors), **2**, 151 (1961).

CAMPBELL, L. L., and J. R. POSTGATE, "Classification of the Spore-Forming, Sulfate-Reducing Bacteria," *Bact. Revs.* **29**, 359 (1965).

CUMMINS, C. S., and J. L. JOHNSON, "Taxonomy of the Clostridia: Wall Composition and DNA Homologies in *Clostridium butyricium* and Other Butyric Acid-Producing Clostridia," *J. Gen. Microbiol.* **67**, 33 (1971).

HUNGATE, R. E., "The Anaerobic Mesophilic Cellulolytic Bacteria," *Bact. Rev.* **14**, 1 (1950).

KALTWASSER, H., "Studies on the Physiology of *Bacillus fastidiosus,*" *J. Bact.* **107**, 780 (1971).

KNIGHT, B. C. J. G., and H. Proom, "A Comparative Study of the Nutrition and Physiology of Mesophilic Species in the Genus *Bacillus,*" *J. Gen. Microbiol.* **4**, 508 (1950).

MEAD, G. C., "The Amino Acid-Fermenting Clostridia," *J. Gen. Microbiol.* **67**, 47 (1971).

PERKINS, W. E., et al., "Symposium on Clostridia," *J. Appl. Bact.* **28**, 1–152 (1965).

ROGOFF, M. H., and A. A. YOUSTEN, "*Bacillus thuringensis:* Microbiological Considerations," *Ann. Rev. Microbiol.* **23**, 357 (1969).

SCHOBERTH, S., and G. GOTTSCHALK, "Considerations on the Energy Metabolism of *Clostridium Kluyveri,*" *Arch. Mikrobiol.* **65**, 318 (1969).

WILEY, W. R., and J. L. STOKES, "Effect of pH and Ammonium Ions on the Permeability of *Bacillus pasteurii,*" *J. Bact.* **86**, 1152 (1963).

WOLF, J., and A. N. BARKER, "The Genus *Bacillus:* Aids to the Identification of Its Species," in *Identification Methods for Microbiologists,* B. M. Biffs and D. A. Shapton (editors), IB, 93. New York: Academic Press, 1968.

II. The endospore

Books   GOULD, G. W., and A. HURST (editors), *The Bacterial Spore.* New York: Academic Press, 1969.

HALVORSEN, H. O., R. HANSON, and L. L. CAMPBELL (editors), *Spore, V.* American Society of Microbiology, Washington, D.C., 1972.

Reviews and original articles   COOTE, J. G., and J. MANDELSTAM, "Use of Constructed Double Mutants for Determining the Temporal Order of Expression of Sporulation Genes in *Bacillus subtilis,*" *J. Bact.* **114**, 1254 (1973).

GOULD, G. W., and G. J. DRING, "Mechanisms of Spore Heat Resistance," *Adv. Microb. Physiol.* **11**, 137 (1974).

671 Holt, S. C., and E. R. Leadbetter, "Comparative Ultrastructure of Selected Aerobic Spore-Forming Bacteria: A Freeze-Etching Study," *Bact. Rev.* **33**, 346 (1969).

Hranueli, K., P. J. Piggot, and J. Mandelstam, "Statistical Estimate of the total Number of Operons Specific for *Bacillus Subtilis* Sporulation," *J. Bact.* **119**, 684 (1974).

Kurahashi, K., "Biosynthesis of Small Peptides," *Ann. Rev. Biochem.* **43**, 445 (1974).

Schaeffer, P., "Sporulation and the Production of Antibiotics, Exoenzymes and Exotoxins," *Bact. Revs.* **33**, 48 (1969).

23

GRAM-POSITIVE BACTERIA: THE ACTINOMYCETE LINE

Among the procaryotes, a mycelial growth habit is confined to Gram-positive bacteria, being characteristic of the organisms known as *actinomycetes*. Some of these organisms, the *euactinomycetes,* develop only in the mycelial state, and they reproduce through the formation of unicellular spores, differentiated either singly or in chains at the tips of the hyphae. This group is a large and complex one, containing many genera which are distinguished primarily by their structural and developmental properties. Schematic examples of some of their growth habits are shown in Figure 23.1.

In a second group, known as the *proactinomycetes,* mycelial development is transitory and often limited; specialized spores are not produced, and reproduction occurs primarily by mycelial fragmentation into short, rod-shaped cells. The Proactinomycetes intergrade with a large group of organisms known collectively as *coryneform bacteria,* which are unicellular and reproduce by binary fission. These organisms are characterized by a marked irregularity and variability of cell form; the cells are commonly tapered and club-shaped, in many cases undergoing a transition to a coccoid stage during the developmental cycle.

Lastly, certain Gram-positive, unicellular bacteria which do not form endo-spores, and which are distinguished from the coryneform group by their regular cell shape, appear to be associated with the actinomycete line. They include the lactic acid bacteria and the micrococci.

If motile, members of the actinomycete line bear flagella. Motility is rather rare, although it occurs sporadically in some members of all the groups described above. Most euactinomycetes are permanently immotile, but flagellated spores are formed in a few genera.

For the purposes of the ensuing discussion, it is convenient to divide the members of the actinomycete line into three principal groups (Table 23.1). Group I comprises the unicellular lactic acid bacteria and the micrococci; group II, the

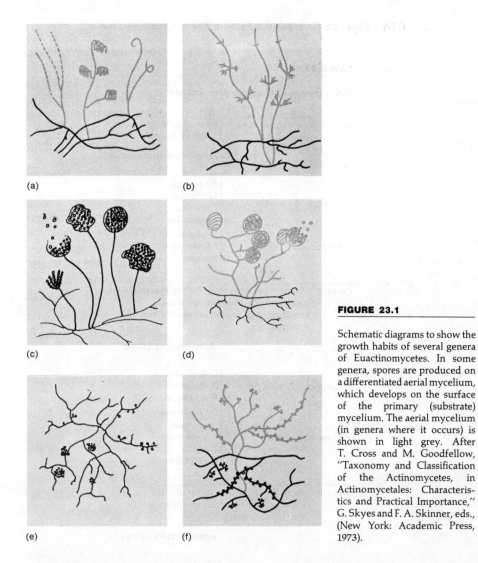

(a)

(b)

(c)

(d)

(e)

(f)

FIGURE 23.1

Schematic diagrams to show the growth habits of several genera of Euactinomycetes. In some genera, spores are produced on a differentiated aerial mycelium, which develops on the surface of the primary (substrate) mycelium. The aerial mycelium (in genera where it occurs) is shown in light grey. After T. Cross and M. Goodfellow, "Taxonomy and Classification of the Actinomycetes, in Actinomycetales: Characteristics and Practical Importance," G. Skyes and F. A. Skinner, eds., (New York: Academic Press, 1973).

coryneform bacteria and proactinomycetes; group III, the euactinomycetes. Not all the genera currently recognized can be described in the confines of this chapter; those selected for discussion are listed in Table 23.1.

• Mean DNA base composition

The overall range of DNA base composition in the actinomycete line is wide (Table 23.2). However, within each group the ranges are, with a few notable exceptions, considerably narrower. Most members of group I have DNA of low GC content, except for *Micrococcus*, in which the values lie near the top of the GC scale. In group III, however, the GC content is characteristically very high; the sole exception is *Thermoactinomyces*, which also differs from all other euactinomycetes in its mode of sporulation (see p. 701). The members of group II have DNA of relatively high GC content, although the span is somewhat wider than that characteristic of group III.

TABLE 23.1

Major groups of Gram-positive bacteria[a]

GROUP I. UNICELLULAR RODS OR SPHERES, OF REGULAR CELL
FORM: REPRODUCTION BY BINARY FISSION

REPRESENTATIVE GENERA

Lactic acid bacteria	Streptococcus Leuconostoc Pediococcus Lactobacillus
Micrococci	Staphylococcus Micrococcus Sarcina

GROUP II. UNICELLULAR, WITH TENDENCY TO VARIABLE OR
IRREGULAR CELL FORM, OR MYCELIAL,
REPRODUCING BY FRAGMENTATION OF THE
MYCELIUM

REPRESENTATIVE GENERA

Coryneform bacteria	Corynebacterium Arthrobacter Propionibacterium
Proactinomycetes	Bifidobacterium Actinomyces Mycobacterium Nocardia Geodermatophilus

GROUP III. VEGETATIVE DEVELOPMENT EXCLUSIVELY
MYCELIAL; REPRODUCTION BY FORMATION OF
SPECIALIZED SPORES

REPRESENTATIVE GENERA

Euactinomycetes	Streptomyces Micromonospora Actinoplanes Streptosporangium Thermoactinomyces

[a] Exclusive of the unicellular endosporeformers.

• **Peptidoglycan structure**

In nearly all Gram-negative procaryotes the cell wall peptidoglycan has the same basic chemical structure, shown schematically in Figure 23.2. This prevalent peptidoglycan type contains *meso*-diaminopimelic acid (*meso*-DAP) in position 3 of the peptide chains; and adjacent chains are cross-linked directly between the free amino group of *meso*-DAP and free carboxyl group of D-alanine (3:4 linkage). The same peptidoglycan occurs in some members of the actinomycete line; but, in addition, a very large number (almost 60) of other peptidoglycan types have

TABLE 23.2

Ranges of mean DNA base composition among members of the actinomycete line

		MOLES PERCENT G + C
GROUP I		
Lactic acid bacteria	Streptococcus	33–44
	Leuconostoc	38–44
	Pediococcus	34–44
	Lactobacillus	35–51
Micrococci	Staphylococcus	30–40
	Micrococcus	66–75
	Sarcina	29–31
GROUP II		
Coryneforms	Corynebacterium	57–60
	Arthrobacter	60–72
	Propionibacterium	59–66
Proactinomycetes	Bifidobacterium	57–64
	Actinomyces	60–63
	Mycobacterium	62–70
	Nocardia	60–72
GROUP III		
Euactinomycetes	Streptomyces	69–73
	Micromonospora	71–73
	Actinoplanes	71–76
	Streptosporangium	69–71
	Thermoactinomyces	44–55

been found in these organisms. The principal variations concern the nature of the diaminoacid in position 3; the presence, number, and nature of additional amino acids which form interpeptide bridges; and the position of the cross link between peptide chains. The distribution of these major variant peptidoglycan structures among the organisms of the actinomycete line is summarized in Table 23.3. Certain taxonomic correlations are evident. Except for a few *Lactobacillus* spp.,

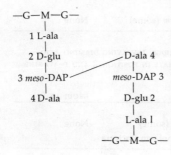

FIGURE 23.2

The primary structure of type A peptidoglycan, which occurs in the walls of certain organisms of the actinomycete line. Abbreviations G, N-acetylglucosamine; M, N-acetylmuramic acid; L (D)-ala, L (D)-alanine; D-glu, D-glutamic acid; *meso*-DAP, *meso*-diaminopimelic acid. The principal variations on this structure that occur in peptidoglycans produced by members of the actinomycete line concern the nature of the diamino acid in position 3; the presence or absence of an interpeptide bridge; and the position of cross-linking between adjacent tetrapeptides (either 3 : 4 or 2 : 4). See Table 23.3 for details.

TABLE 23.3

Distributions of some major peptidoglycan types in organisms
of the actinomycete line

A. *Meso*-DAP IN POSITION 3; ADJACENT PEPTIDE CHAINS WITH DIRECT 3:4
CROSS LINK

GROUP I	GROUP II	GROUP III
Lactobacillus (some)	*Corynebacterium*	*Micromonospora*
	Mycobacterium	*Actinoplanes*
	Nocardia	*Thermoactinomyces*
	Arthrobacter (some)	
	Propionibacterium (some)	
	Geodermatophilus	

B. L-LYSINE IN POSITION 3; INTERPEPTIDE BRIDGE BETWEEN POSITIONS
3 AND 4

GROUP I	GROUP II	GROUP III
Streptococcus	*Bifidobacterium* (some)	None
Pediococcus	*Arthrobacter* (some)	
Leuconostoc		
Staphylococcus		
Micrococcus		
Lactobacillus (some)		

C. L,L-DAP IN POSITION 3; INTERPEPTIDE BRIDGE BETWEEN POSITIONS
3 AND 4

GROUP I	GROUP II	GROUP III
None	*Arthrobacter* (some)	*Streptomyces*
	Propionibacterium (some)	

D. L-ORNITHINE IN POSITION 3; INTERPEPTIDE BRIDGE BETWEEN
POSITIONS 3 AND 4

GROUP I	GROUP II	GROUP III
Lactobacillus (some)	*Bifidobacterium* (some)	None

E. POSITION 3 VARIABLE; INTERPEPTIDE BRIDGE, INVOLVING DIAMINO ACID,
BETWEEN CARBOXYL GROUPS OF D-GLUTAMATE IN POSITION 2 AND D-ALANINE
IN POSITION 4

GROUP I	GROUP II	GROUP III
None	*Arthrobacter* (some)	None
	Other coryneform bacteria	

type B peptidoglycans are characteristic of group I. The euactinomycetes synthesize peptidoglycans of either type A or type C. In group II the variation of peptidoglycan structure is exceptionally large, many different peptidoglycan types occurring among the members of the coryneform group.

• Dissimilatory metabolism and relations to oxygen

The members of the actinomycete line are chemoheterotrophs, which perform aerobic respiration or fermentation (Table 23.4). Aerobes with a purely respiratory metabolism occur in the three subgroups, this metabolic pattern being characteristic of nearly all euactinomycetes. The capacity for fermentation occurs in many representatives of groups I and II. Carbohydrates are the principal fermentable substrates; in most cases (with the exception of *Sarcina*), H_2 is not formed, and either lactic acid or propionic acid accumulates as a major organic end product. Among these fermentative organisms, the relations to O_2 vary. In group I, *Staphylococcus* comprises facultative anaerobes, which can synthesize ATP either by aerobic respiration or by fermentation; the same physiological pattern exists in *Corynebacterium* (group II). In group I the lactic acid bacteria lack a respiratory transport system and are thus entirely dependent on fermentation for the synthesis of ATP, but they can grow well in the presence of air. Other members of both groups I and II with a purely fermentative metabolism (*Sarcina, Propionibacterium, Bifidobacterium, Actinomyces*) can grow only in the absence of air or in the presence of very low partial pressures of oxygen. For all these organisms, molecular oxygen is growth-inhibitory, but not toxic. Consequently, their cultivation does not require the use of special techniques to exclude even brief exposure of the cells

TABLE 23.4

Modes of ATP generation and relations to oxygen in members of the actinomycete line

| | ATP SYNTHESIZED BY | | |
	RESPIRATION	FERMENTATION	RELATIONS TO OXYGEN
GROUP I			
Lactic acid bacteria	−	+	Facultative anaerobes
Staphylococcus	+	+	Facultative anaerobes
Micrococcus	+	−	Strict aerobes
Sarcina	−	+	Aeroduric anaerobes
GROUP II			
Corynebacterium	+	+	Facultative anaerobes
Arthrobacter	+	−	Strict aerobes
Propionibacterium	−	+	Aeroduric anaerobes
Bifidobacterium	−	+	Aeroduric anaerobes
Actinomyces	−	+	Aeroduric anaerobes
Mycobacterium	+	−	Strict aerobes
Nocardia	+	−	Strict aerobes
GROUP III			
Euactinomycetes	+	−	Strict aerobes

to air, of the type essential for the cultivation of the strictly anaerobic nonspore-formers discussed in Chapter 24, and for the cultivation of many clostridia. The anaerobic organisms that belong to the actinomycete line thus represent a special physiological category: *aeroduric* (or aerotolerant) anaerobes.

THE LACTIC ACID BACTERIA

The lactic acid bacteria are immotile, rod-shaped or spherical organisms (Figure 23.3), united by an unusual constellation of metabolic and nutritional properties. The name derives from the fact that ATP is synthesized through fermentations of carbohydrates, which yield lactic acid as a major (and sometimes as virtually the sole) end product.

The lactic acid bacteria are all facultative anaerobes, which grow readily on the surface of solid media exposed to air. However, they are unable to synthesize ATP by respiratory means, a reflection of their failure to synthesize cytochromes and other heme-containing enzymes. Although they can perform limited oxidations of a few organic compounds, mediated by flavoprotein enzymes, either oxidases or peroxidases, these oxidations are not accompanied by ATP formation. The growth yields of lactic acid bacteria are, accordingly, unaffected by the presence and absence of air, the fermentative dissimilation of sugars being the source of ATP under both conditions.

One consequence of the failure to synthesize heme proteins is that the lactic acid bacteria are catalase negative, and hence cannot mediate the decomposition

FIGURE 23.3

The form and arrangement of cells in three genera of lactic acid bacteria: (a) *Lactobacillus;* (b) *Streptococcus;* (c) *Pediococcus* (phase contrast, ×2,180).

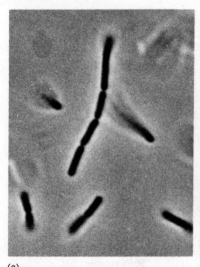
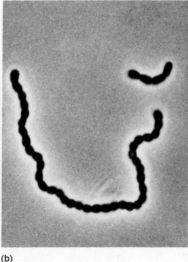
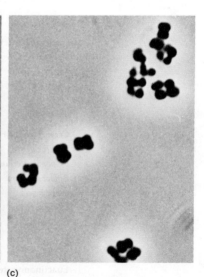

(a)   (b)   (c)

of H_2O_2 according to the reaction:

$$2H_2O_2 \longrightarrow 2H_2O + O_2$$

The absence of catalase activity, readily demonstrated by the absence of O_2 formation when cells are mixed with a drop of dilute H_2O_2, is one of the most useful diagnostic tests for the recognition of these organisms, since they are virtually the only bacteria devoid of catalase that can grow in the presence of air.

The inability of lactic acid bacteria to synthesize heme proteins is correlated with an inability to synthesize hemin, the porphyrin which is the prosthetic group of these enzymes. However, certain lactic acid bacteria acquire catalase activity when grown in the presence of a source of hemin (e.g., on media containing red blood cells). Such species appear to synthesize a protein that can combine with exogenously supplied hemin to produce an enzyme with the properties of catalase.

The inability to synthesize hemin is only one manifestation of the *extremely limited synthetic abilities* characteristic of the lactic acid bacteria. All these organisms have complex growth-factor requirements; they invariably require B vitamins and with one exception (*Streptococcus bovis*), a considerable number of amino acids. The amino acid requirements of many lactic acid bacteria are considerably more extensive than those of higher animals. As a result of their complex nutritional requirements, lactic acid bacteria are usually cultivated on media containing peptone, yeast extract, or other digests of plant or animal material. These must be supplemented with a fermentable carbohydrate to provide an energy source.

Even when growing on very rich media, the colonies of lactic acid bacteria (Figure 23.4) always remain relatively small (at most, a few millimeters in diameter). They are never pigmented; as a result of the absence of cytochromes, the growth has a chalky white appearance which is very characteristic. The small colony size of these bacteria is attributable primarily to low growth yields, a consequence of their exclusively fermentative metabolism. Some species can produce unusually large colonies when grown on sucrose-containing media, as a result of the massive synthesis of extracellular polysaccharides (either dextran or levan) at the expense of this disaccharide; in this special case, much of the volume of the colony consists of polysaccharide. Since dextrans and levans are synthesized only from sucrose, the species in question form typical small colonies on media containing any other utilizable sugar. In the isolation of the spherical lactic acid bacteria, which can grow in media that have an initial pH of 7 or above, the incorporation of finely divided $CaCO_3$ in the plating medium is useful, since the colonies can be readily recognized by the surrounding zones of clearing, caused by acid production (Figure 23.5).

Another distinctive physiological feature of lactic acid bacteria is *their high tolerance of acid*, a necessary consequence of their mode of energy-yielding metabolism. Although the spherical lactic acid bacteria can initiate growth in neutral or alkaline media, most of the rod-shaped forms cannot grow in media with an initial pH greater than 6. Growth of all lactic acid bacteria continues until the pH has fallen, through fermentation, to a value of 5 or less.

The capacity of lactic acid bacteria to produce and tolerate a relatively high concentration of lactic acid is of great selective value, since it enables them to eliminate competition from most other bacteria in environments that are rich in nutrients. This is shown by the fact that lactic acid bacteria can be readily enriched from natural sources through the use of complex media with a high sugar content. Such media can, of course, support the growth of many other chemoorganotrophic

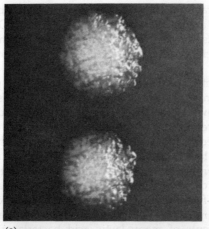

(a)

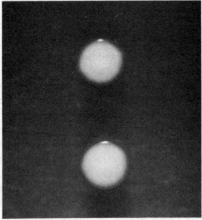

(b)

FIGURE 23.4

Colonies of lactic acid bacteria: (a) *Lactobacillus plantarum* and (b) *Streptococcus lactis* (×9.6).

bacteria, but the competing organisms are almost completely eliminated as growth proceeds by the accumulation of lactic acid, formed through the metabolic activity of the lactic acid bacteria.

As a result of their extreme physiological specialization, the lactic acid bacteria are confined to a few characteristic natural environments. Some live in association with plants and grow at the expense of the nutrients liberated through the death and decomposition of plant tissues. They occur in foods and beverages prepared from plant materials: pickles, sauerkraut, ensilaged cattle fodder, wine, and beer. A lactic fermentation of the sugar initially present occurs during the preparation of pickles, sauerkraut, and ensilage. In fermented beverages the lactic acid bacteria are potential spoilage agents, which sometimes grow and produce an undesirable acidity.

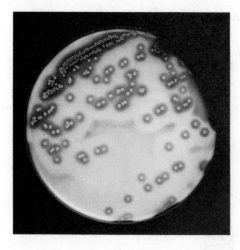

FIGURE 23.5

Colonies of *Streptococcus* growing on an agar medium containing a suspension of calcium carbonate. Lactic acid production has dissolved the calcium carbonate, producing clear zones around each colony.

Other lactic acid bacteria constitute part of the normal flora of the animal body and occur in considerable numbers in the nasopharynx, the intestinal tract, and the vagina. These forms include a number of important pathogens of man and other mammals, all belonging to the genus *Streptococcus*.

A third characteristic habitat of the lactic acid bacteria is milk, to which they gain access either from the body of the cow or from plant materials. The normal souring of milk is caused by certain streptococci, and both rod-shaped and spherical lactic acid bacteria play important roles in the preparation of fermented milk products (butter, cheeses, buttermilk, yogurt).

Because of their activities in the preparation of foods and as agents of human and animal disease, the lactic acid bacteria are a group of major economic importance.

• **Patterns of carbohydrate fermentation in lactic acid bacteria**

It was shown by S. Orla Jensen about 1920 that lactic acid bacteria can be divided into two biochemical subgroups, distinguishable by the products formed from glucose. *Homofermenters* convert glucose almost quantitatively to lactic acid; *heterofermenters*, to an equimolar mixture of lactic acid, ethanol and CO_2. The mode of glucose fermentation can be most simply determined by the detection of CO_2 production. However, since the amount produced by heterofermenters is small (one mole per mole of glucose fermented), its detection requires a special procedure: growth in a well-buffered medium of high sugar content, sealed with an agar plug to trap the CO_2 formed (Figure 23.6).

The metabolic explanation of the dichotomy in fermentative patterns among lactic acid bacteria was discovered many years later. Homofermenters dissimilate glucose through the Embden-Meyerhof pathway (Figure 23.7). However, heterofermenters cannot utilize this pathway, since they lack a key enzyme, fructose-diphosphate aldolase, which mediates sugar-phosphate cleavage:

fructose-1,6-diphosphate

$$\longrightarrow \text{glyceraldehyde-3-phosphate} + \text{dihydroxyacetone phosphate}$$

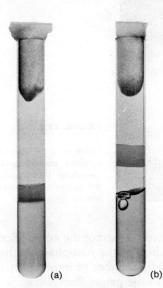

(a)   (b)

FIGURE 23.6

The demonstration of CO_2 production by lactic acid bacteria in tubes of a sugar-rich medium with agar seals: (a) *Streptococcus lactic;* (b) *Leuconostoc mesenteroides.*

These organisms dissimilate glucose through the oxidative pentose phosphate pathway (Figure 23.8). The initial oxidation of glucose-6-phosphate to CO_2 and ribulose-5-phosphate is followed by a C_3-C_2 cleavage of the latter intermediate yielding glyceraldehyde-3-phosphate and acetyl-phosphate. The triose-phosphate is converted to lactic acid through a sequence of reactions identical with those of the homofermentative pathway, while acetyl-phosphate is reduced to ethanol. This reduction involves two successive hydrogenations, which balance the two dehydrogenations involved in the conversion of glucose-6-phosphate to pentose-phosphate and CO_2. As a consequence of its biochemical mechanism, this fermentation necessarily results in a strictly equimolar ratio of the three end products: lactic acid, ethanol, and CO_2. A major difference between the two pathways is their net ATP yields: two moles per mole of glucose fermented by the homofermentative pathway and only one mole by the heterofermentative pathway.

Two heterofermentative lactobacilli, *L. brevis* and *L. buchneri*, cannot grow anaerobically with glucose, because they are not able to effect the reduction of acetyl-phosphate to ethanol, essential to maintain overall redox balance. They can, however, ferment glucose aerobically, reoxidizing NADH at the expense of O_2 by means of a flavoprotein enzyme. The overall reaction for glucose fermentation under these conditions becomes:

$$\text{glucose} + O_2 \longrightarrow \text{lactate} + \text{acetate} + CO_2$$

Both these species can grow anaerobically at the expense of another hexose, fructose, because they possess a mannitol dehydrogenase, which mediates reduction of this ketosugar to the polyalcohol, mannitol:

$$\text{fructose} + \text{NADH} + H^+ \rightleftharpoons \text{mannitol} + \text{NAD}^+$$

This reduction maintains redox balance under anaerobic conditions, the overall equation for fructose fermentation being:

$$3 \text{ fructose} \longrightarrow \text{lactate} + \text{acetate} + CO_2 + 2 \text{ mannitol}$$

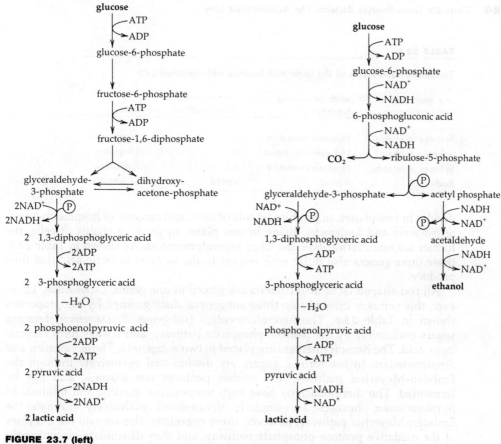

FIGURE 23.7 (left)

The pathway of glucose dissimilation by homofermentative lactic acid bacteria.

FIGURE 23.8 (right)

The pathway of glucose dissimilation by heterofermentative lactic acid bacteria.

Many other heterofermenters of the genera *Lactobacillus* and *Leuconostoc* also contain mannitol dehydrogenase and produce mannitol as a product of fructose fermentation, with a concomitant formation of some acetate in place of ethanol.

Lactic acid bacteria differ with respect to isomers of lactic acid that they produce. This is determined by the stereospecificity of the lactic dehydrogenases which mediate pyruvate reduction:

$$CH_3COCOOH + NADH + H^+ \rightleftharpoons CH_3CHOHCOOH + NAD^+$$

Some species contain only D-lactic dehydrogenase, and hence form the D-isomer; others contain only L-lactic dehydrogenase, and hence form the L-isomer. Certain species contain two lactic dehydrogenases of differing stereospecificity and form racemic lactic acid.

• **Subdivision of the lactic acid bacteria** Lactic acid bacteria with spherical cells are placed in three different genera, distinguished in part on structural and in part on biochemical grounds. *Pediococcus*

TABLE 23.5

Taxonomic subdivision of the lactic acid bacteria with spherical cells

CELL SHAPE AND ARRANGEMENT	MODE OF GLUCOSE FERMENTATION	CONFIGURATION OF LACTIC ACID	GENUS
Spheres in chains	Homofermentative	L-	*Streptococcus*
Spheres in chains	Heterofermentative	D-	*Leuconostoc*
Spheres in tetrads	Homofermentative	DL-	*Pediococcus*
Rods	Varied	Varied	*Lactobacillus*

divides in two planes, to produce tetrads of cells, and consists of homofermenters. *Streptococcus* and *Leuconostoc* divide in one plane, to produce chains of cells; the former are homofermenters, the latter heterofermenters. As shown in Table 23.5, these three genera also differ with respect to the isomers of lactic acid that they produce.

All rod-shaped lactic acid bacteria are placed in one genus, *Lactobacillus*. However, this genus is divided into three subgenera, distinguished by the properties shown in Table 23.6. The heterofermenters (subgenus *Betabacterium*) ferment sugars exclusively by the pentose phosphate pathway, and always form racemic lactic acid. The homofermenters are placed in two subgenera, *Thermobacterium* and *Streptobacterium*. In the former, sugars are dissimilated exclusively through the Embden-Meyerhof pathway, and neither pentoses nor gluconic acid can be fermented. The thermobacteria have high temperature maxima and minima. In *Streptobacterium*, hexoses are similarly dissimilated exclusively through the Embden-Meyerhof pathway; however, these organisms also contain the enzymes of the oxidative pentose-phosphate pathway, and they dissimilate gluconic acid and pentoses through this metabolic route. They are, accordingly, *facultative* homofermenters, in contrast to the *obligate* homofermenters of the subgenus *Thermobacterium*. Both temperature minima and maxima are lower in *Streptobacterium* than in *Thermobacterium*. Although constant for each species, the isomer of lactic acid formed is not a characteristic property of each subgenus.

TABLE 23.6

Distinguishing properties of the three subgenera of *Lactobacillus*

	FERMENTATION OF SUGARS AND PATHWAY EMPLOYED		ISOMER OF LACTIC ACID PRODUCED	TEMPERATURE RANGE	
	HEXOSES	GLUCONATE, PENTOSES		15°C	45°C
Thermobacterium	Embden-Meyerhof	Not attacked	L-, D-, or DL-	−	+
Streptobacterium	Embden-Meyerhof	Pentose-phosphate	L- or DL-	+	−
Betabacterium	Pentose phosphate	Pentose-phosphate	DL-	V[a]	V[a]

[a] Variable from one species to another.

THE MICROCOCCI

In addition to lactic acid bacteria with spherical cells, the Gram-positive cocci include three other genera, distinguished by a combination of physiological, metabolic, and structural characters. The cells of *Staphylococcus* and *Micrococcus* are relatively small (approximately 1 µm in diameter), and do not occur in regular groups. The cells of *Sarcina* are much larger (2 to 3 µm in diameter), and occur in regular, cubical packets (Figure 23.9).

The members of the genus *Staphylococcus* are facultative anaerobes and ferment sugars with the formation of lactic acid as one of the major products. Their DNA base composition (30 to 40 mole percent G + C) is in the same range as that of many spherical lactic acid bacteria. However, they can be readily distinguished from these organisms by several criteria: possession of catalase and other heme pigments; capacity for respiratory metabolism; and much less restricted requirements for carbon and energy (growth will occur on complex media in the absence of carbohydrates). Many also produce carotenoid pigments, absent from all lactic acid bacteria. These organisms are typical members of the normal microflora of the skin, and some are potential pathogens causing either infections or food poisoning.

Micrococcus closely resembles *Staphylococcus* in cell structure, but its members are all strict aerobes and have a wholly different DNA base composition (66 to 72 mole percent G + C). The normal habitat of these bacteria is obscure; they are common in air, and also occur in milk and dairy equipment.

The two species of *Sarcina* are aeroduric anaerobes and vigorous sugar fermenters. Both dissimilate glucose through the Embden-Meyerhof pathway, but with the formation of different end products (Table 23.7). *S. maxima* carries out a typical butyric acid fermentation, whereas the fermentation carried out by *S. ventriculi* yields a mixture of ethanol, lactic acid, acetic acid, and considerable amounts of H_2.

An unusual structural property of *S. ventriculi* is the synthesis of a cell wall which has a thick outer layer of cellulose (Figure 23.10); apart from *Acetobacter xylinum* (see p. 605), it is the only procaryote known to synthesize this polysaccharide.

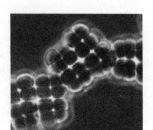

FIGURE 23.9

Sarcina maxima (phase contrast, ×1,630). From S. Holt and E. Canale-Parola, "Fine Structure of *Sarcina maxima* and *Sarcina ventriculi*," J. Bacteriol. **93**, 399 (1967).

TABLE 23.7

Products of glucose fermentation by *Sarcina*[a]

	Sarcina ventriculi	Sarcina maxima
Ethanol	100	0
Lactic acid	10	0
Acetic acid	60	40
Butyric acid	0	77
CO_2	190	197
H_2	140	223

[a] Moles/100 moles of glucose fermented.

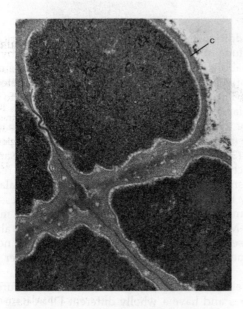

FIGURE 23.10

Electron micrograph of a thin section of a group of cells of *Sarcina ventriculi,* showing the heavy cellulose layer (c) that encloses each cell ($\times$27,900). Courtesy of S. Holt and E. Canale-Parola.

GROUP II: *CORYNEBACTERIUM, MYCOBACTERIUM, NOCARDIA*

The members of group II (coryneform bacteria and proactino-mycetes) present many difficult taxonomic problems; the structure of this group is more unclear and controversial than that of almost any other major bacterial group.

These organisms include several major human or animal pathogens, which were the first representatives to be described and characterized: *Corynebacterium diphtheriae,* the causal agent of diphtheria; *Mycobacterium tuberculosis,* the causal agent of tuberculosis; and *Nocardia farcinica,* the causal agent of the disease of horses known as glanders. Other human or animal parasites or pathogens more or less closely related to each of these major pathogens were subsequently discovered.

Corynebacterium diphtheriae is a normal inhabitant of the respiratory tract. Most strains are nonpathogenic, but acquire the ability to cause diphtheria when infected by a specific phage which confers toxigenicity on the host cell (see Chapter 15).

C. diphtheriae and related animal parasites or pathogens are unicellular, immotile organisms with a characteristic cell shape and arrangement. Just after division, the daughter cells (which are club-shaped, tapering toward the outer poles) undergo a sudden "snapping" movement (Figure 23.11), which brings them into a characteristic angular relationship (Figure 23.12). This appears to result from the fact that the cell wall consists of two layers, only the inner layer invaginating to form the septum. After septum formation is completed, a localized rupture of the outer layer occurs (Figure 23.13). The mechanism of snapping post-fission movement is shown diagrammatically in Figure 23.14.

Corynebacteria are facultative anaerobes capable of both respiration and fermentation. Sugar fermentation by *C. diphtheriae* yields propionic acid as a major end product. The nutritional requirements are complex, but are not known in detail for most species: *C. diphtheriae* requires several B vitamins.

Time (minutes)

0   15   45

FIGURE 23.11

Postfission movement in *Arthrobacter* illustrated by time-lapse pictures of a group of cells growing on agar (phase contrast, ×1,720). From M. P. Starr and D. A. Kuhn, "On the Origin of V-forms in *Arthrobacter atrocyaneus*," *Arch. Mikrobiol.* **42,** 289 (1962).

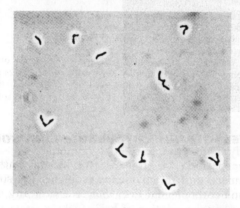

FIGURE 23.12

The typical angular arrangement of dividing cells of a coryneform bacterium, brought about by "snapping" post-fission movement. Phase contrast, ×1,400. From T. A. Krulwich and J. L. Pate, "Ultrastructural Explanation for Snapping Post-Fission Movements in *Arthrobacter crystallopoietes*," *J. Bacteriol.* **105,** 408 (1971).

FIGURE 23.13

Electron micrographs of thin sections of dividing cells of a coryneform bacterium, showing the two-layered structure of the cell wall. (a) Cell in which the transverse wall has been almost completed. cw, cell wall; pm, cell membrane. (b) Cell in which transverse wall formation is complete, showing the separation between the inner (il) and outer (ol) wall layers. (c) Cell which has just undergone snapping post-fission movement, showing the separation (s) between the two parts of the previously continuous outer wall layer. From T. A. Krulwich and J. L. Pate, "Ultrastructural Explanation for Snapping Post-Fission Movements in *Arthrobacter crystallopoietes*," *J. Bacteriol.* **105,** 408 (1971).

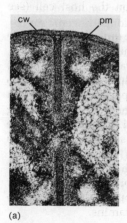
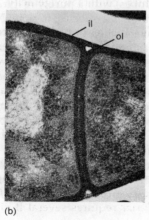
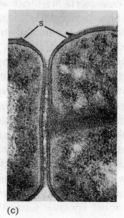

(a)   (b)   (c)

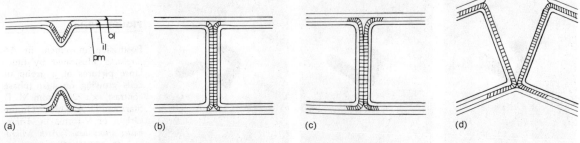

FIGURE 23.14

A diagrammatic representation of the mechanism of transverse wall growth and snapping post-fission movement in a coryneform bacterium. pm, cell membrane; il, inner wall layer; ol, outer wall layer. Only the inner wall layer (cross-hatched) participates in transverse wall formation. (A) initiation of transverse wall formation. (B) completion of transverse wall; (C) extension of inner wall layer, placing tension on outer layer; (D) unilateral rupture of outer wall layer, causing snapping movement. From T. A. Krulwich and J. L. Pate, "Ultrastructural Explanation for Snapping Post-Fission Movements in *Arthrobacter crystallopoietes*," *J. Bacteriol.* **105**, 408 (1971).

The *Mycobacterium* group can form a rudimentary mycelium, which is unstable, and fragments early in growth into slender, immotile rods, sometimes branched. In contrast to corynebacteria, the cells are not tapered and do not show snapping postfission movements. A distinctive though often variable property of mycobacteria, which also occurs in some nocardias, is *acid-fastness*. Cells that have been stained with a hot phenolic solution of basic fuchsin retain this red dye through subsequent treatment with a mineral acid (dilute H_2SO_4 or HCl); all other bacteria stained in this manner are rapidly decolorized by the acid treatment. The property of acid-fastness is associated with a very high content of complex lipids in the cell wall, which makes the cells of mycobacteria and some nocardias waxy and strongly hydrophobic. As a result, colonies have a dry, rough wrinkled surface (Figure 23.15); and in liquid cultures (unless they are grown in the presence of a detergent), the cells cohere to form a tough surface pellicle and a film of growth attached to the wall of the culture vessel.

The mycobacteria are strict aerobes, with a purely respiratory mode of metabolism. Growth, particularly of pathogenic species, is slow. Nonpathogenic mycobacteria, which occur in soil, have higher growth rates. Nonpathogenic mycobacteria do not require growth factors, whereas the nutritional requirements of the pathogenic species are complex.

Many mycobacteria produce yellow or orange carotenoid pigments; and in certain of these species—the so-called *photochromogens*—the synthesis of carotenoids is specifically induced by exposure of cultures to light.

The *Nocardia* group differs from mycobacteria primarily by its greater tendency to mycelial growth. In the initial stages of development, an abundant mycelium is produced, and subsequently fragments into short rods (Figure 23.16). Some of the nocardias are animal pathogens; others are nonpathogenic soil organisms. In nutritional and metabolic respects, they resemble the mycobacteria.

Recent chemical studies have shown that the members of the genera *Corynebac-*

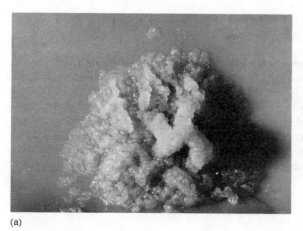

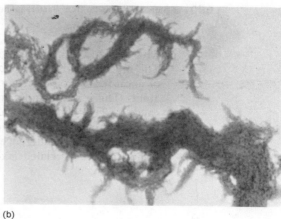

(a)   (b)

FIGURE 23.15

The characteristic appearance of cultures of *Mycobacterium tuberculosis*. (a) Colony growing on the surface of an agar plate, ×7. (b) Cordlike aggregations of cells from a liquid culture, ×345. Courtesy of Professor N. Rist, Institut Pasteur, Paris.

terium, Mycobacterium, and *Nocardia* share a distinctive cell wall composition, unique to the members of these three genera. The wall peptidoglycan is of type A (direct cross linking between *meso*-diaminopimelic acid and D-alanine). Covalently linked to the peptidoglycan is polysaccharide made up of arabinose and galactose; the presence of this wall arabinogalactan confers immunological cross reactivity on cells of the three genera. Furthermore, the wall in all three genera has a high content of lipid, including a distinctive class of lipids known as *mycolic acids*. These are branched, β-hydroxy acids of the general structure:

$$R_1\text{—CH—CH—COOH}$$
$$\phantom{R_1\text{—}}\text{OH}\ \ R_2$$

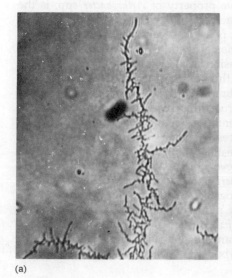

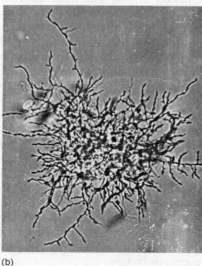

(a)   (b)

FIGURE 23.16

Young surface colonies on agar plates of (a) *Mycobacterium fortuitum* and (b) *Nocardia asteroides* (×648). Courtesy of Ruth Gordon and H. Lechevalier.

where R_1 and R_2 are alkyl groups; they are of relatively high molecular weight. The mycolic acids of mycobacteria contain from 79 to 85 carbon atoms, those of nocardias 48 to 58 carbon atoms, and those of corynebacteria 32 to 36 carbon atoms. These wall components are joined through ester linkages to the arabino-galactan. Thus, a unique wall structure of great molecular complexity is shared by the members of all three genera, a fact that suggests that despite their structural and physiological differences, the corynebacteria, mycobacteria, and nocardias comprise an interrelated taxonomic cluster within group II of the actinomycete line.

GROUP II: AEROBIC CORYNEFORM BACTERIA

The genus *Corynebacterium* is a relatively small and well-defined assemblage, if confined to *C. diphtheriae* and the related animal parasites and pathogens that are facultative anaerobes and possess cell walls of the structure described above. However, many other nonspore-forming Gram-positive bacteria display the cell shape and the ability to undergo snapping post-fission movements that are characteristic of *Corynebacterium*. These so-called *coryneform bacteria* are diverse with respect to physiological, nutritional and metabolic properties, and they occur in a wide variety of habitats.

Strictly aerobic coryneform bacteria are abundant in soil and in milk products; some are pathogenic for plants. Some of these organisms have been attached to the genus *Corynebacterium*, others placed in a variety of special genera. The best-characterized representatives are the soil coryneforms of the genus *Arthrobacter*. These organisms constitute a large fraction of the aerobic chemoheterotrophic population of soil bacteria, and are important agents for the mineralization of organic matter in soil. The most distinctive property of *Arthrobacter* spp. is the succession of changes in cell form that accompany growth (Figure 23.17). In cultures that have entered the stationary phase the cells are spherical and of uniform size, resembling micrococci. When growth is reinitiated, these cells elongate into rods, which undergo binary fission, accompanied by typical snapping post-fission movements. From these rods, thinner outgrowths may develop near one or both poles of the cell, producing branched forms which resemble early developmental stages of mycobacteria or proactinomycetes. Return to the coccoid state may occur either by multiple fragmentation (as in nocardias) or by a progressive shortening of the rods through successive binary fissions. This remarkable growth cycle thus includes features suggestive not only of corynebacteria, but also of mycobacteria and nocardias. However, the arthrobacters are completely different from the *Corynebacterium-Mycobacterium-Nocardia* group in cell wall composition. They synthesize neither arabinogalactan nor mycolic acids, and the various species produce peptidoglycans of numerous chemical types.

In their nutritional properties the *Arthrobacter* group show interesting analogies to aerobic pseudomonads; most species can utilize a wide and varied range of simple organic compounds as principal carbon and energy sources. The majority require growth factors.

691

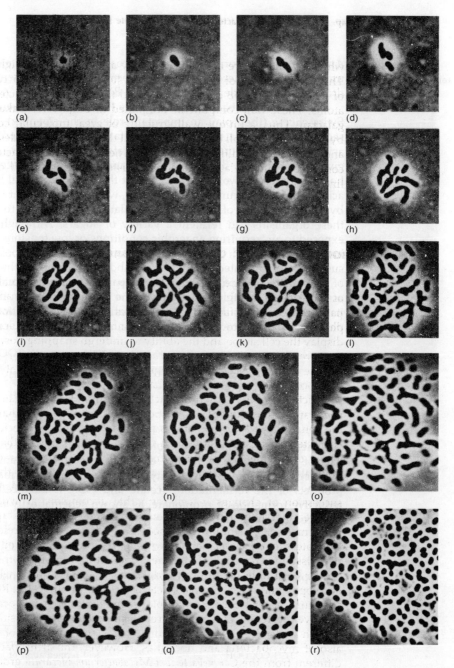

FIGURE 23.17

The cellular life cycle of *Arthrobacter*, shown by successive photomicrographs of the growth of a microcolony from a single coccoid cell on agar over a period of 40 hours. Phase contrast, ×1,020. From H. Veldkamp, G. Van den Berg, and L. P. T. M. Zevenhuizen, "Glutamic Acid Production by *Arthrobacter globiformis*," *Antonie van Leeuwenhoek* **29**, 35 (1963).

• The aeroduric anaerobes of group II

Among the coryneform bacteria and proactinomycetes, several genera are made up of aeroduric anaerobes, which perform distinctive fermentations of carbohydrates. These include *Propionibacterium, Bifidobacterium,* and *Actinomyces.*

The coryneform bacteria of the genus *Propionibacterium* were first isolated from Swiss cheese (they play an important role in the ripening). They develop as a secondary microflora, fermenting the lactate initially produced in the curd by lactic acid bacteria, with formation of propionate, acetate, and CO_2. The two fatty acids give this cheese its distinctive flavor, and the CO_2 produces the characteristic holes. Subsequent work has shown that the primary natural habitat of propionic acid bacteria is the rumen of herbivores, where they ferment the lactate produced by other members of the rumen population. In addition to fermenting lactate, these organisms can ferment a variety of sugars. Although they cannot grow exposed to air, requiring anaerobic conditions or low tensions of O_2, they contain heme pigments, both cytochromes and catalase. Their metabolism is fermentative; sugars are dissimilated through the Embden-Meyerhof pathway, with formation of propionate, acetate, and CO_2, accompanied by some succinate. The formation of succinate is strongly influenced by the content of CO_2 in the growth medium; its formation occurs through a carboxylation of the glycolytic intermediate phosphoenol pyruvate, to yield oxaloacetate, subsequently reduced to succinate:

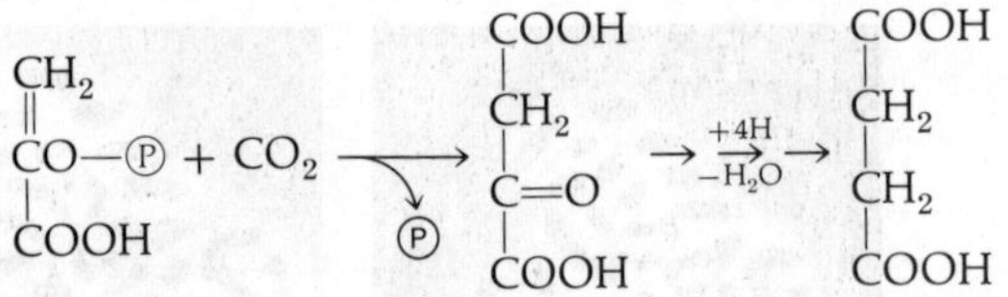
CH2
CO—Ⓟ + CO2
COOH
Ⓟ
COOH
CH2
C=O
COOH
+4H
−H2O
COOH
CH2
CH2
COOH

When lactate is the fermentable substrate, it is initially oxidized to pyruvate. Part of the pyruvate is further oxidized to acetyl-CoA and CO_2, ATP being produced in the conversion of acetyl-CoA to acetate. The formation of the oxidized products of the fermentation, acetate and CO_2, is balanced by a concomitant reductive formation of propionate, as shown schematically in Figure 23.18. The balanced overall equation is:

$$3 \text{ lactate} \longrightarrow 2 \text{ propionate} + \text{acetate} + CO_2$$

Propionate formation from pyruvate occurs through a complex series of reac-

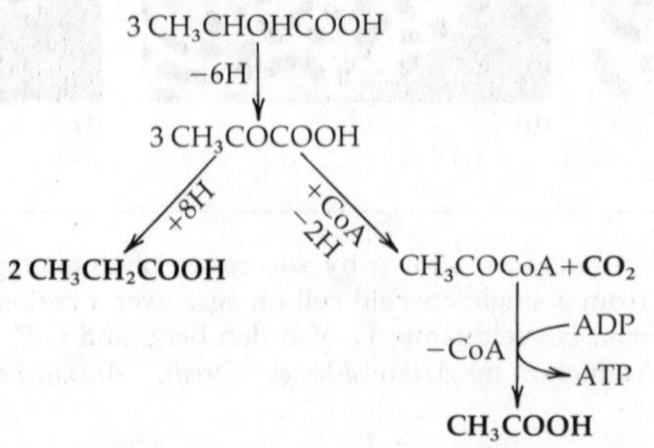
3 CH3CHOHCOOH
−6H
3 CH3COCOOH
+8H
+CoA
−2H
2 CH3CH2COOH
CH3COCoA+CO2
−CoA
ADP
ATP
CH3COOH

FIGURE 23.18

A diagrammatic representation of the propionic acid fermentation of lactic acid.

tions, the successive intermediates being:

pyruvate $\longrightarrow$ oxaloacetate $\longrightarrow$ malate $\longrightarrow$ fumarate
$\longrightarrow$ succinate $\longrightarrow$ succinyl-CoA $\longrightarrow$ methylamalonyl-CoA
$\longrightarrow$ propionyl-CoA $\longrightarrow$ propionate

These steps involve both CoA transfers and transcarboxylations between intermediates in the pathway. The component reactions are:

1. pyruvate + methylmalonyl-CoA $\xrightarrow{\text{(transcarboxylation)}}$ oxaloacetate + propionyl-CoA

2. oxaloacetate + 2NADH + 2H$^+$ $\xrightarrow{\text{(reduction)}}$ succinate + 2NAD$^+$ + H$_2$O

3. succinate + propionyl-CoA $\xrightarrow{\text{(CoA transfer)}}$ succinyl-CoA + propionate

4. succinyl-CoA + propionyl-CoA $\xrightarrow{\text{(transcarboxylation)}}$ propionyl-CoA + methylmalonyl-CoA

The reactions associated with propionate formation are cyclic, as shown in Figure 23.19.

The aeroduric anaerobes of the genus *Bifidobacterium*, which constitute a major fraction of the intestinal microflora of breast-fed infants, resemble the lactic acid bacteria in several respects. They are catalase negative and have complex nutritional requirements; and they ferment sugars, with the formation of lactic acid as a major end product. However, the overall equation is a distinctive one, corresponding neither to a homolactic nor to a typical heterolactic fermentation:

2 glucose $\longrightarrow$ 3 acetate + 2 L-lactate

The biochemical route of this fermentation is unique (Figure 23.20). As in the Embden-Meyerhof pathway, glucose is initially phosphorylated in the 6 position, and is then converted to fructose-6-phosphate. Fructose-6-phosphate undergoes a C$_2$-C$_4$ split, accompanied by an uptake of inorganic phosphate. A complex series of sugar phosphate interconversions is then initiated by a reaction between erythrose-4-phosphate and fructose-6-phosphate, from which two moles of acetyl phosphate and two moles of glyceraldehyde-3-phosphate are ultimately produced.

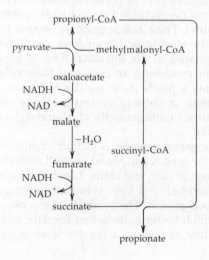

FIGURE 23.19

The cyclic mechanism for the formation of propionate from pyruvate by propionic acid bacteria.

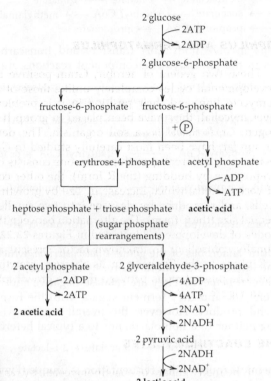

2 glucose
↓ 2ATP → 2ADP
2 glucose-6-phosphate

fructose-6-phosphate   fructose-6-phosphate
                        + Ⓟ

erythrose-4-phosphate   acetyl phosphate
                         ↓ ADP → ATP
                         acetic acid

heptose phosphate + triose phosphate
(sugar phosphate rearrangements)

2 acetyl phosphate   2 glyceraldehyde-3-phosphate
↓ 2ADP → 2ATP   ↓ 4ADP → 4ATP
2 acetic acid   ↓ 2NAD⁺ → 2NADH

                      2 pyruvic acid
                      ↓ 2NADH → 2NAD⁺
                      2 lactic acid

FIGURE 23.20

The pathway of glucose fermentation by *Bifidobacterium;* end products are shown in boldface.

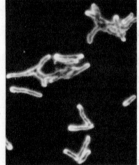

FIGURE 23.21

Actinomyces israelii from a broth culture, showing branched cells and short mycelial fragments (dark field illumination, ×1,120). From J. M. Slack, S. Landfried, and M. A. Gerencer, "Morphological, Biochemical and Serological Studies on 64 Strains of *Actinomyces israelii,*" J. Bacteriol. **97**, 873 (1969).

Acetyl-phosphate is converted to acetate; triose-phosphate is converted via pyruvate to L-lactate. Energetically, this fermentation is slightly more favorable than the homolactic fermentation, since it yields five moles of ATP for every two moles of glucose fermented.

The cells of bifidobacteria are typically swollen, irregular, and branched. The complex nutritional requirements of these organisms include a requirement for N-acetylglucosamine or a β-substituted disaccharide containing this amino sugar (e.g., N-acetyllactosamine). These compounds are present in milk, which accounts for the fact that this is the most favorable medium for bifidobacteria and probably explains their predominance in the intestinal flora of breast-fed babies. When cultivated in a medium containing an excess of N-acetylglucosamine, the cells of bifidobacteria assume a much more regular rod form. Hence the branched, swollen cells characteristic of these organisms probably reflect the fact that they are usually grown with a limiting supply of N-acetylglucosamine, an essential peptidoglycan precursor.

The members of the genus *Actinomyces* include organisms which are members of the normal flora of the mouth and throat, as well as several pathogenic species which produce infections in man and cattle. Like bifidobacteria, they are catalase negative aeroduric anaerobes, and have complex nutritional requirements. CO_2 is often required for good growth. The products of glucose fermentation are more varied than those of bifidobacteria, including formate and succinate, as well as lactate and acetate. They also show a greater tendency than bifidobacteria to mycelial growth in young cultures, although the mycelium is fragile and breaks up readily into rod-shaped or branched fragments (Figure 23.21).

GEODERMATOPHILUS AND DERMATOPHILUS

These two genera of aerobic, Gram-positive bacteria are characterized by developmental cycles completely unlike those of any other members of the actinomycete line, and their position in it is problematical. Since their growth is never mycelial, they have been placed in group II. *Dermatophilus* is an animal pathogen; *Geodermatophilus* a soil organism. The developmental cycles, which appear similar, have been most carefully studied in *Geodermatophilus*. This organism has two very different growth phases: one consists of motile, flagellated rods, which reproduce by budding (the R form); the other consists of nonmotile aggregates of coccoid cells, which increase in size by growth and division of the constituent cells, and reproduce by cleavage into two smaller aggregates of approximately equal size (the C form). The alternation between these two completely dissimilar modes of development, illustrated in Figures 23.22 and 23.23, appears to be nutritionally controlled: an unknown factor, present in certain peptones, converts the R phase to the C phase; in its absence, growth occurs exclusively in the R phase. The interrelations between the two growth phases are illustrated schematically in Figure 23.24.

GROUP III: THE EUACTINOMYCETES

Of the many genera of euactinomycetes, by far the largest is *Streptomyces*. These organisms are abundant in soil, the characteristic odor of damp soil being attributable to a volatile substance which they produce. Since 1945 this bacterial group has acquired great economic importance, as a result of the fact that a large number of therapeutically useful antibiotics are formed as secondary metabolites by streptomycetes (see Chapter 31 for further discussion of this aspect). The intensive search for strains able to produce new antibiotics has led to the description of hundreds of new species, frequently on very slender grounds.

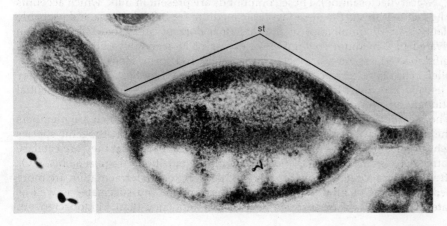

FIGURE 23.22

The R form of *Geodermatophilus*. Electron micrograph of a thin section of a budding cell with two polar stalks (st), from one of which a daughter bud is being formed. ×47,600. Insert: phase contrast photomicrograph of budding-R form cells. From E. E. Ishiguro and R. S. Wolfe, "Control of Morphogenesis in *Geodermatophilus*: Ultrastructural Studies," *J. Bacteriol.* **104,** 566 (1970).

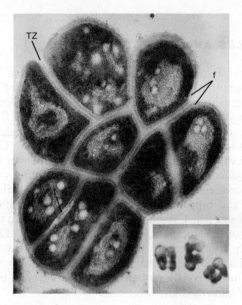

FIGURE 23.23

The C form of *Geodermatophilus*. Electron micrograph of a thin section of a mass of coccoid cells. The cell membrane of each cell is enclosed by a complex wall, composed of a transparent inner layer (t) and fibrous outer layer (f), adjacent cells being separated by a transparent zone (TZ). ×21,600. Insert: phase contrast photomicrograph of the C form. From E. E. Ishiguro and R. S. Wolfe, "Control of Morphogenesis in *Geodermatophilus*: Ultrastructural Studies." *J. Bacteriol.* **104,** 566 (1970).

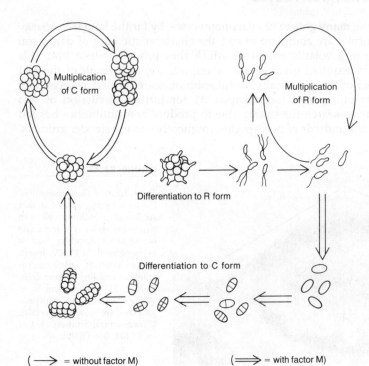

FIGURE 23.24

A diagrammatic representation of the two developmental phases of *Geodermatophilus*, showing the influence of the unknown factor (factor M) present in peptone on the transition between them. From E. E. Ishiguro and R. S. Wolfe, "Control of Morphogenesis in *Geodermatophilus*: Ultrastructural Studies," *J. Bacteriol.* **104,** 566 (1970).

• The Streptomyces developmental cycle

Germination of the small, rod-shaped or oval immotile spores occurs by hyphal outgrowth, 1 to 4 separate hyphae, 0.5 to 1 μm wide, being produced from each spore (Figure 23.25). Elongation and branching of the hyphae produce a firm, compact colony, made up of a substrate mycelium with a smooth, moist surface. A looser aerial mycelium then arises over the surface of the colony. The hyphae of which it is composed are surrounded by an additional wall layer, termed a *sheath*, which makes them slightly thicker and more refractile than those of the substrate mycelium. The sheath material is hydrophobic, so that the aerial mycelium resists wetting. Its color is also generally different from that of the substrate mycelium. During vegetative growth few cross walls are formed, the colony being largely coenocytic. Spores arise exclusively on the aerial mycelium, through fragmentation of the hyphal tip within its sheath, to produce chains of spores still enclosed by a common sheath. The arrangement of sporulating hyphae differs in various members of the genus; two patterns of arrangement are shown in Figure 23.26. At maturity, the spores become separated from one another, each enclosed in a section of the sheath layer, and detach readily from the hyphae. A useful differential character, determinable by electron microscopic examination, is the surface structure of the spores, which reflects the nature of the enclosing sheath material. In some species, spores have a smooth surface; in others, they bear protuberances of various shapes (Figure 23.27).

The streptomycetes are all strict aerobes. Their nutrition is simple; growth factors are not required. Although the range of utilizable carbon and energy sources has not been systematically examined, it appears to be relatively wide, and includes, in addition to simple organic compounds, a number of biopolymers which are hydrolyzed by extracellular enzymes. One natural polymer attacked uniquely by streptomycetes is rubber latex. Many of these organisms also produce several different enzymes capable of hydrolyzing peptidoglycans, and are consequently able to lyse other bacteria through destruction of the peptidoglycan layer of their walls.

• Streptosporangium

The members of this genus occur, like streptomycetes, in soil. Spore germination leads, as in *Streptomyces,* to the formation of a substrate and an aerial mycelium.

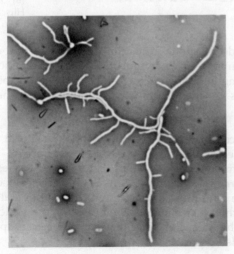

FIGURE 23.25

Germinating spores of a *Streptomyces,* showing early stages in the development of the substrate mycelium. Nigrosin mount (×1,580). The field also contains several spores that have not germinated. Courtesy of C. F. Robinow.

697

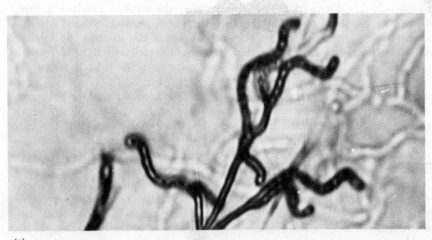

(a)

(b)

FIGURE 23.26

Portions of the aerial mycelium of two streptomycetes, illustrating two different kinds of arrangement of the sporulating hyphae. Photograph (a) ×2,940, courtesy of H. Lechevalier; (b) ×2,550, courtesy of Peter Hirsch.

However, the mode of spore formation differs. The tips of the aerial hyphae become enlarged into vesicles as much as 20 μm wide, enclosed by the sheath, within which nonmotile spores are formed by the septation of a spiral, unbranched hypha. The spores are then liberated by rupture of the sheath.

• **Actinoplanes**   In these actinomycetes an aerial mycelium is not produced. Prior to sporulation, a layer of sheathed hyphae develops on the surface of the brightly colored substrate mycelium, and the tips enlarge into spore vesicles, similar in width to those of *Streptosporangium* (Figure 23.28). Spores arise by septation of branched and coiled hyphae within each vesicle, and they are liberated by rupture of the surrounding sheath, which makes up the wall of the vesicle. In this group the spores are motile by means of polar flagella. *Actinoplanes* species are common in fresh water and also occur in soil.

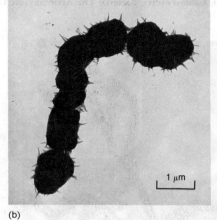

(a)

(b)

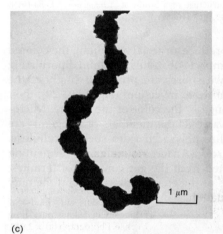

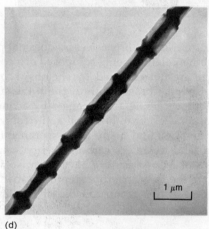

(c)

(d)

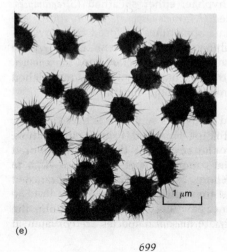

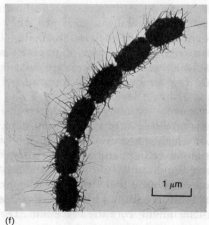

(e)

(f)

FIGURE 23.27

Electron micrographs of the spores of six different *Streptomyces* species, which illustrate various types of surface structure and ornamentation: (a) *S. cacaoi*, showing smooth spores; (b) *S. hirsutus*, showing spiny spores with obtuse spines; (c) *S. griseoplanus*, which has warty spores; (d) *S. aureofaciens*, of a smooth but special "phalangiform" type; (e) *S. fasciculatus*, showing spiny spores with long acute spines; (f) *S. flavoviridis*, showing hairy spores. Courtesy of H. D. Tresner, Lederle Laboratories; reproduced in part from E. B. Shirling and D. Gottlieb, "Cooperative Description of Type Cultures of *Streptomyces. II.* Species Descriptions from First Study," *Intern. J. Syst. Bacteriol.* **18,** 69–189 (1968).

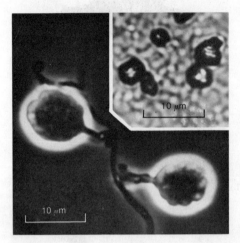

FIGURE 23.28

Spore vesicles of *Actinoplanes*: (a) (inset) A group of mature spore vesicles viewed on the surface of a colony (bright field illumination). (b) Two mature spore vesicles attached to a hypha, mounted in water (phase contrast illumination). From H. Lechevalier and P. E. Holbert, "Electron Microscopic Observation of the Sporangial Structure of a Strain of *Actinoplanes*," *J. Bacteriol.* **89**, 217 (1965).

• Micromonospora

Like *Actinoplanes, Micromonospora* does not produce an aerial mycelium, the colored vegetative colonies being composed of a compact substrate mycelium. Sporulating hyphae are not enclosed by sheaths; brown spores are produced singly at the tips of hyphal branches throughout the substrate mycelium (Figure 23.29), their development being accompanied by a change in the color of the colony. Micromonosporas occur in both soil and fresh water. These members of the genus are strict aerobes, and many of them can utilize polysaccharides such as cellulose, chitin, and xylan as carbon energy sources. Anaerobic, cellulose-fermenting species of *Micromonospora* have been isolated from the gut contents of termites; they are the only anaerobic representatives of the euactinomycetes, and the only members of this group capable of fermentation.

• Properties of the spore of euactinomycetes

In nearly all euactinomycetes, including the four genera described above, spores arise by septation of the distal part of hyphae, either sheathed (*Streptomyces, Streptosporangium, Actinoplanes*) or unsheathed (*Micromonospora*). When spores are produced within a sheathed hypha, they are sometimes enclosed by a layer of hydrophobic sheath material following liberation: this is characteristic of the *Streptomyces* spore. However, in genera such as *Streptosporangium* and *Actinoplanes*, when the spores develop in a vesicular terminal enlargement of the sporulating hypha, they are liberated by a rupture of the enclosing sheath, and thus do not possess a sheath layer.

None of the actinomycete spores formed by hyphal septation is heat-resistant; their thermal tolerance is no greater than that of the vegetative mycelium. However, *Streptomyces* spores can remain viable for long periods in the dry state. Neither the internal cell structure nor the chemical composition of the spores appears to differ significantly from that of the mycelium, and a relatively limited number of studies indicate that the spore enzymes are similar in nature and in level to those of the mycelium. These spores appear to have a reproductive role, following their detachment from the mycelium. They are disseminated either by dispersal through the air or in water.

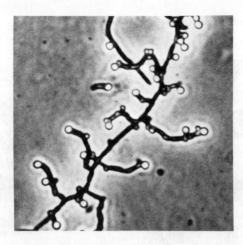

FIGURE 23.29

Micromonospora chalcea, showing spherical spores borne singly at the tips of hyphae (phase contrast, ×2,300). Courtesy of G. M. Luedemann and the Schering Corporation.

• *Thermoactinomyces, an endospore-forming euactinomycete*

Composts of decaying plant materials and manure heaps often undergo spontaneous heating, as a result of the intense metabolic activity of the microbial populations that they contain; the internal temperature may rise to as high as 80°C. Thermophilic actinomycetes are abundant in these habitats, and several different genera have been described. Among these thermophilic actinomycetes are members of the genus *Thermoactinomyces*, which have long been known to produce extremely heat-resistant spores. These spores survive exposure to a temperature of 100°C for as much as an hour, whereas the maximum growth temperature of *Thermoactinomyces* is 65° to 68°C.

Thermoactinomyces forms both a substrate and an aerial mycelium, on both of which spores are produced. They are highly refractile and develop singly at intervals over the entire length of the hyphae (Figure 23.30). Electron micrographs of thin sections (Figure 23.31) show that the immature spores are completely enclosed *within* the hypha, thus resembling the forespores of *Bacillus* and *Clos-*

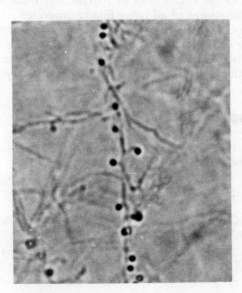

FIGURE 23.30

Phase contrast light micrograph of a portion of the mycelium of *Thermoactinomyces*, showing the refractile endospores, borne on short side-branches, ×1,360. From J. Lacey, *"Thermoactinomyces sacchari* sp. nov., a Thermophilic Actinomycete Causing Bagassosis," *J. Gen. Microbiol.* **66,** 327 (1971).

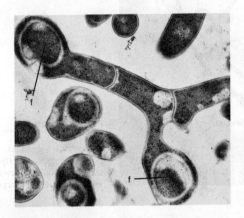

FIGURE 23.31

Electron micrograph of a thin section of a sporulating hypha of *Thermoactinomyces*, showing two forespores (f), enclosed by the cytoplasm of enlarged hyphal tips, ×20,000. From J. Lacey, "*Thermoactinomyces sacchari* sp. nov., a Thermophilic Actinomycete Causing Bagassosis," *J. Gen. Microbiol.* **66**, 327 (1971).

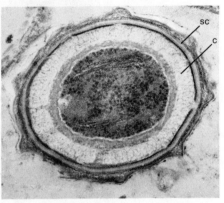

FIGURE 23.32

Electron micrograph of a thin section of a mature endospore of *Thermoactinomyces*, showing the cortex (c) and elaborate spore coat (sc) which surround the spore cell. ×50,600. From J. Lacey, "*Thermoactinomyces sacchari* sp. nov., a Thermophilic Actinomycete Causing Bagassosis," *J. Gen. Microbiol.* **66**, 327 (1971).

tridium; they are released, following maturation, by autolysis of the mycelium. The fine structure of these spores (Figure 23.32) is characteristic of an endospore: they are surrounded by a thick, ridged spore coat which encloses a wide cortex. Furthermore, they have a high content of dipicolinic acid. There can be no doubt, accordingly, that the spores of *Thermoactinomyces* are homologous, both in structure and in function, to the endospores of *Bacillus* and *Clostridium*, and are totally dissimilar from the spores of all other Euactinomycetes.

Moreover, the G + C content of the DNA of *Thermoactinomyces* is far lower than that of other euactinomycetes, lying within the range of many *Bacillus* spp. Consequently, it is possible that this mycelial organism is more closely related to the members of the genus *Bacillus* than to the euactinomycetes.

FURTHER READING

Books   PRAUSE, H. (editor), *The Actinomycetales*, Jena International Symposium on Taxonomy. Jena: G. Fischer, 1968.

SYKES, G., and F. A. SKINNER (editors), *Actinomycetales: Characteristics and Practical Importance*. New York: Academic Press, 1973.

Reviews and original articles   ATTWELL, R. W., T. CROSS, and G. W. GOULD, "Germination of *Thermoactinomyces vulgaris* Endospores," *J. Gen. Microbiol.* **73**, 471 (1972).

BAIRD-PARKER, A. C., "Methods for Classifying Staphylococci and Micrococci," in B. M. Gibbs and F. A. Skinner (editors), *Identification Methods for Microbiologists*, 1A, 59. New York: Academic Press, 1966.

703   BARKSDALE, L., "*Corynebacterium diphtheriae* and Its Relatives," *Bact. Revs.* **34,** 378 (1970).

CANALE-PEROLA, E., "Biology of the Sugar-Fermenting Sarcinae," *Bact. Revs.* **34,** 82 (1970).

GOREN, M. B., "Mycobacterial Lipids: Selected Topics," *Bact. Revs.* **36,** 33 (1972).

HETTINGA, D. H., and G. W. REINBOLD, "The Propionic Acid Bacteria, a Review," *J. Milk Food Tech.* **35,** 295; **35,** 358; **35,** 436 (1972).

ISHIGURO, E., and R. S. WOLFE, "Control of Morphogenesis in *Geodermatophilus*: Ultrastructural Studies," *J. Bact,* **104,** 566 (1970).

KRULWICH, T. A., and J. L. PATE, "Ultrastructural Explanation for Snapping Post-Fission Movements in *Arthrobacter crystallopoietes*," *J. Bact.* **105,** 408 (1971).

SCHLEIFER, K. H., and O. KANDLER, "Peptidoglycan Types of Bacterial Cell Walls and Their Taxonomic Implications," *Bact. Revs.* **36,** 407 (1972).

SHARPE, M. E., and T. F. FRYER, "Identification of the Lactic Acid Bacteria," in B. M. Gibbs and F. A. Skinner (editors), *Identification Methods for Microbiologists,* 1A, 65. New York: Academic Press, 1966.

VELDKAMP, H., "Saprophytic Coryneform Bacteria," *Ann. Rev. Microbiol.* **24,** 209 (1970).

Strict anaerobes that do not form endospores comprise a large and diverse physiological category of bacteria. The technical difficulties involved in their isolation and cultivation are considerable, and, as a result, until recently this group was rather poorly known. A series of new techniques, largely developed by R. E. Hungate during work on the bacterial flora of the rumen, has facilitated the study of these organisms. Their main features (see Chapter 2, p. 41) include the use of prereduced media of low redox potential, isolation via roll tubes, and manipulation of cultures in a stream of O_2-free gas. Unlike the clostridia, these organisms do not form oxygen-insensitive resting cells (endospores) and are therefore rapidly killed by O_2 at all stages of development. It is now evident that some supposed pure cultures isolated before the development of these new techniques were in fact impure; one example of the erroneous conclusions derived from work with such cultures is described on p. 706.

The nonspore-forming anaerobes comprise three distinct physiological subgroups: the *methanogenic bacteria;* the *sulfate-reducing bacteria* of the genus *Desulfovibrio;* and a variety of organisms which ferment sugars, lactate, or amino acids. The two latter subgroups have physiological counterparts among sporeformers [genera *Desulfotomaculum* and *Clostridium,* respectively (see Chapter 22)]. No methanogenic bacteria form endospores.

THE METHANOGENIC BACTERIA

The biological formation of methane (CH_4) is a geochemically important process that occurs in all anaerobic environments in which organic matter undergoes decomposition: swamps, lake sediments, and the digestive tract of animals. It results from the metabolic activity of a small and highly specialized

TABLE 24.1

Subgroups of methanogenic bacteria

GENUS	CELL STRUCTURE	MOTILITY	SUBSTRATES FOR METHANE FORMATION
Methanobacterium (4 spp.)	Rods, sometimes curved	Variable	$H_2 + CO_2$; formate in most spp.
Methanococcus (2 spp.)	Cocci	Variable	$H_2 + CO_2$; formate
Methanosarcina (1 spp.)	Cocci in cubical packets	Absent	$H_2 + CO_2$; methanol; acetate
Methanospirillum (1 spp.)	Helical	Polar flagella	$H_2 + CO_2$; formate

bacterial group, which are terminal members of the food chain in these environments; they convert fermentation products formed by other anaerobes (notably CO_2, H_2, and formate) to methane. Since this gas has a low solubility in water, it escapes from the anaerobic environment and is eventually reoxidized under aerobic conditions by obligate methylotrophs (see Chapter 18). Methanogenesis in the rumen of cattle and sheep has been intensively studied, since it is an important component of the complex microbial activities associated with ruminant digestion (see Chapter 28). In waste disposal plants where sewage is subjected to anaerobic treatment, the growth of these organisms results in the formation of large amounts of methane, which is often collected and used to generate power.

The methanogenic bacteria are Gram-variable organisms, divided into four genera on structural criteria (Table 24.1, Figure 24.1). The universal substrates

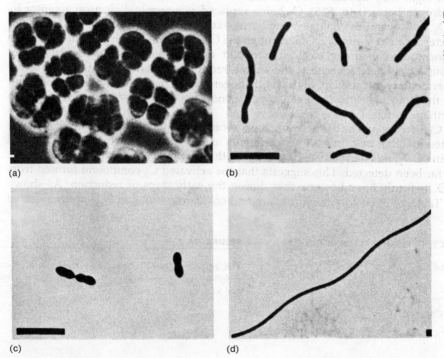

(a) (b) (c) (d)

FIGURE 24.1

Phase-contrast photomicrographs of several methanogenic bacteria. (a) *Methanosarcina barkeri.* (b) *Methanobacterium thermoautrophicum.* (c) *Methanobacterium ruminantium.* (d) *Methanospirillum* sp. In each case, the bar indicates 5 μm. From J. G. Zeikus and V. G. Bowen, "Comparative Ultrastructure of Methanogenic Bacteria," *Can. J. Microbiol.* **21**, 121 (1975).

are $H_2 + CO_2$; some also utilize formate, and one species utilizes methanol and acetate. Ammonia is the preferred nitrogen source. Most of the strains so far studied do not require growth factors; however, some strains from the rumen require fatty acids (acetate, 2-methylbutyrate) and a recently discovered growth factor, coenzyme M (mercaptoethanesulfonic acid).

Much of the early biochemical work on these organisms was conducted on a strain known as "*Methanobacillus omelianskii*," which was apparently able to couple a partial oxidation of primary and secondary alcohols with the reduction of CO_2 to methane, as exemplified by the reaction:

$$2CH_3CH_2OH + CO_2 \longrightarrow 2CH_3COOH + CH_4$$

However, a bacteriological reexamination of this supposedly pure culture has shown that it contains two strict anaerobes: a *Methanobacterium*, which forms methane from CO_2 and H_2, and an unidentified bacterium (organism "S") which oxidizes alcohols, with the formation of H_2 (Figure 24.2). In pure culture neither bacterium can grow at the expense of ethanol and CO_2, since alcohols cannot be attacked by the *Methanobacterium*, and the accumulation of H_2 inhibits growth of organism "S" with alcohols. Hence, the two-membered culture of these bacteria constitutes a well-balanced nutritional symbiosis, which had been maintained in the laboratory ever since its isolation as "*Methanobacillus omelianskii*" over 20 years ago.

• **The metabolism of methanogenic bacteria**

Since the oxidizable substrate most commonly used by these organisms is H_2, at first sight they appear to be the anaerobic counterparts of the strictly aerobic hydrogen bacteria (p. 575), O_2 being replaced as a terminal electron acceptor by CO_2. However, this analogy is not borne out by closer analysis: the methanogenic bacteria do not assimilate CO_2 through the Calvin cycle, and they do not possess a cytochrome-containing electron transport chain, which does occur in another group of obligate anaerobes, the sulfate-reducing bacteria. Many aspects of the intermediary metabolism of the methanogenic bacteria remain obscure, and work on their biochemistry has been largely concerned with the mechanism by which methane is formed.

Cell-free extracts of these organisms can form methane from the substrates listed in Table 24.2. In tracer experiments in which cells or extracts are furnished with $^{14}CO_2$, no intermediates less reduced than methyl compounds (CH_3—R) have so far been detected. This suggests that the activated C_1 compound formed from CO_2 is firmly bound to enzymes during the early steps of reduction. As shown in Table 24.2, both tetrahydrofolate and cobalamin (vitamin B_{12}) are implicated

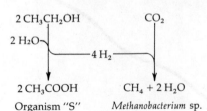

FIGURE 24.2

The respective roles of the two organisms composing "*Methanobacillus omelianskii*" in the conversion of ethanol and CO_2 to acetate and methane.

Substrates from which methane can be formed by cell-free extracts of methanogenic bacteria[a]

$\overset{*}{C}O_2$; $CH_3CO\overset{*}{C}OOH$; $\overset{*}{C}H_2OHCHNH_2COOH$; $H\overset{*}{C}OOH$; $\overset{*}{C}H_3OH$; $H\overset{*}{C}HO$:

$5,\overset{*}{C}H_3$-tetrahydrofolate; $5,10,\overset{*}{C}H_2$-tetrahydrofolate; $\overset{*}{C}H_3$-cobalamin

(vitamin B_{12})

---

[a] The asterisk in each case indicates the carbon atom that becomes a methane carbon.

as methyl group carriers in the terminal reactions leading to methane formation. When supplied with H_2 and ATP, extracts convert the methyl substituent of methyl-cobalamin to methane. Coenzyme M (mercaptoethanesulfonic acid) is involved as a methyl carrier in this reaction. This substance has been detected as a cell constituent only in methanogenic bacteria, and is a growth factor for some strains. The terminal reactions that lead to methane formation are shown schematically in Figure 24.3.

The mechanism of ATP synthesis in methane bacteria is unknown. The intracellular pool of ATP rapidly diminishes when cells are exposed either to chlorinated hydrocarbons (e.g., chloroform), which competively inhibit methane production, or to compounds known to inhibit oxidative phosphorylation (e.g., 2,4-dinitrophenol).

THE GENUS *DESULFOVIBRIO*

In all organisms that use sulfate as a sulfur source, its reduction to sulfide is a normal biosynthetic reaction. This process of *assimilatory sulfate reduction* (Chapter 7, p. 196) is different both in enzymatic mechanism and in

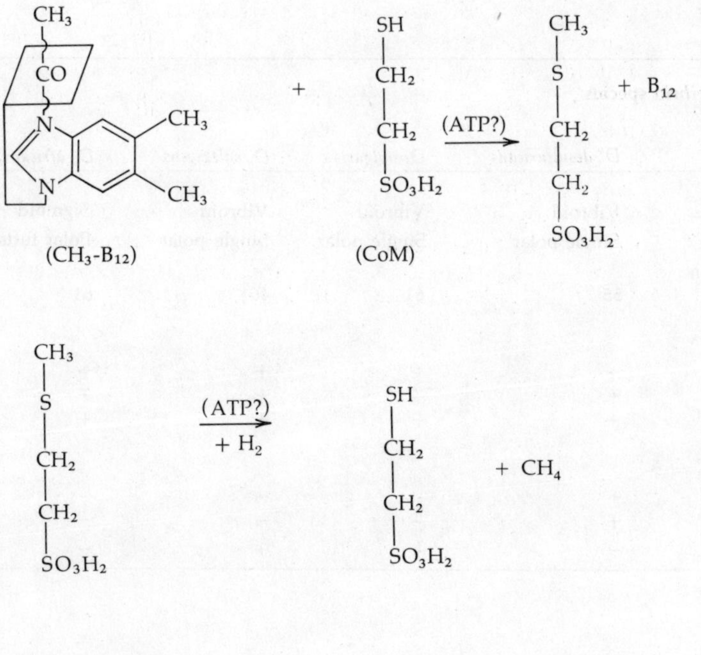

FIGURE 24.3

Terminal steps in the formation of methane.

metabolic function from the use of sulfate as a terminal electron acceptor (*dissimilatory sulfate reduction*). The latter process is confined to two groups of strict anaerobes: sporeformers of the genus *Desulfotomaculum* (Chapter 22, p. 661) and *Desulfovibrio*.

The typical habitats of the dissimilatory sulfate reducers are anaerobic sediments which contain organic matter and sulfate; the activities of these organisms result in a massive generation of H_2S. This often leads to development in the overlying water layer of purple and green sulfur bacteria, which utilize the H_2S produced by the sulfate reducers as a photosynthetic electron donor, reoxidizing it under anaerobic conditions in the light to sulfate. The combined activities of sulfate reducers and anoxygenic photosynthetic bacteria therefore drive an anaerobic sulfur cycle.

The nonspore-forming sulfate reducers of the genus *Desulfovibrio* are Gram-negative, polarly flagellated, curved rods (Table 24.3). These organisms cannot oxidize acetate, which is consequently the endproduct formed through the oxidation of organic substrates. Both malate and lactate are widely utilized; their oxidation to acetate is accompanied by a substrate level phosphorylation via acetyl-CoA (Figure 24.4). Some of these organisms can also ferment pyruvate, malate, or fumarate, in the absence of sulfate: under these conditions, acetate is formed as shown in Figure 24.4, and redox balance is maintained by reduction of part of the substrate, pyruvate being converted to lactate, and malate and fumarate being converted to succinate.

Some *Desulfovibrio* strains can reduce sulfate at the expense of molecular hydrogen. However, they are unable to grow in a mineral medium furnished with H_2 and CO_2, since CO_2 cannot serve as the sole carbon source. Growth requires acetate in addition. The synthesis of cell carbon at the expense of acetate and CO_2 probably involves a ferredoxin-mediated reductive synthesis of pyruvate, known

TABLE 24.3

Properties of *Desulfovibrio* species

	D. desulfuricans	*D. vulgaris*	*D. salexigens*	*D. africanus*	*D. gigas*
Cell form	Vibroid	Vibroid	Vibroid	Sigmoid	Spiral
Flagellation	Single polar	Single polar	Single polar	Polar tufts	Polar tufts
DNA base composition (moles % G + C)	55	61	46	61	60
Substrates utilized					
(a) With sulfate					
Lactate	+	+	+	+	+
Malate	+	−	+	+	−
Pyruvate	+	+	+	+	+
Ethanol					
(b) Without sulfate					
Pyruvate	+	−	−	−	−
Fumarate	+	−	−	−	+
NaCl required	−	−	+	−	−

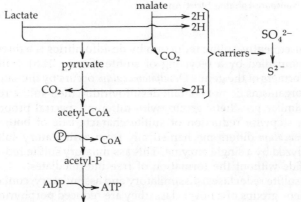

FIGURE 24.4

Pathways of lactate and malate oxidation coupled to sulfate reduction in desulfovibrios.

to occur in certain clostridia and in purple and green bacteria:

$$acetyl\text{-}CoA + CO_2 + FdH_2 \longrightarrow pyruvate + CoA + Fd$$

Strictly speaking, therefore, the *Desulfovibrio* group are not chemoautotrophs, since a substantial part of their cell carbon is always derived from an organic carbon source. Ammonia can serve as sole nitrogen source for these organisms; many are also nitrogen fixers.

• **The mechanism of dissimilatory sulfate reduction**

The reduction of sulfate to sulfide involves the transfer of four pairs of electrons. In dissimilatory sulfate reduction the reduction of sulfate to sulfite is mediated by a complex of three enzymes:

$$SO_4^{2-} + ATP \rightleftharpoons APS + P\text{—}P \quad (ATP\ sulfurylase)$$

$$P\text{—}P + H_2O \longrightarrow 2\circledP \quad (pyrophosphorylase)$$

$$APS + 2e^- \rightleftharpoons AMP + SO_3^{2-} \quad (APS\ reductase)$$

Mechanistically, this conversion is similar to the oxidation of SO_3^{2-} to SO_4^{2-} by thiobacilli and purple sulfur bacteria; in all three groups, the key intermediate is adenylphosphosulfate (APS). Bacterial assimilatory sulfate reduction, however, proceeds via a different activated intermediate, adenylsulfate (see p. 197).

The pathway for the reduction of sulfite to sulfide has been established only recently, primarily as a result of the discovery of the catalytic function of the green protein, desulfoviridin. This protein, abundant in desulfovibrios, was isolated almost 20 years ago, but no function could then be ascribed to it. Desulfovibrin has now been identified as a sulfite reductase, which catalyzes the reaction:

$$3SO_3^{2-} + 2e^- \longrightarrow S_3O_6^{2-}$$

Two other enzymes (trithionate and thiosulfate reductases) convert trithionate to sulfide, each with the formation of sulfite:

$$S_3O_6^{2-} + 2e^- \longrightarrow S_2O_3^{2-} + SO_3^{2-}$$

$$S_2O_3^{2-} + 2e^- \longrightarrow S^- + SO_3^{2-}$$

Consequently, sulfite reduction by desulfovibrios is a three-step process, which is accompanied by a recycling of sulfite (Figure 24.5). Sulfite reduction by the sporeformers of the genus *Desulfotomaculum* occurs by the same pathway; however, these organisms do not contain desulfoviridin; their sulfite reductase has a chemically similar prosthetic group, with different spectral properties.

The stepwise reduction of sulfite characteristic of both *Desulfovibrio* and *Desulfotomaculum* differs mechanistically from assimilatory sulfite reduction, which is catalyzed by a single enzyme. This assimilatory sulfite reductase converts sulfite to sulfide without the formation of free intermediates.

All sulfite reductases, dissimilatory and assimilatory, contain chemically similar prosthetic groups of a novel class: they are reduced porphyrins known as *sirohydrochlorins*, which are not known to occur in any other type of enzyme. The differing spectral properties of sulfite reductases, as exemplified by the enzymes of *Desulfovibrio* and of *Desulfotomaculum*, are related to differences with respect to the chelating metal in the prosthetic group.

• **Electron transport and ATP synthesis**

It was shown by J. Postgate in 1954 that the desulfovibrios, unlike other strictly anaerobic chemoheterotrophs then known, contain a cytochrome of the c type (cytochrome c_3). This was the first indication that sulfate-linked anaerobic respiration might involve the operation of an electron transport chain analogous to that of aerobic respiration. Subsequently, other components of this anaerobic chain have been identified in desulfovibrios: additional c type cytochromes, a cytochrome of the b type, a quinone, a flavoprotein, and several electron carriers containing nonheme iron (ferredoxin, rubedoxin). It is thus evident that the desulfovibrios contain elements of a complex and specialized electron transport chain, coupling substrate oxidation with the various steps of sulfate reduction. There is no doubt that ATP synthesis is associated with electron transfer through this chain; it has been detected in cell-free extracts during H_2 oxidation with sulfite as acceptor, and is abolished by such uncoupling agents as 2,4-dinitrophenol. However, details of the operation of this system are still unknown.

• **Natural activities of sulfate-reducing bacteria**

In suitable anaerobic environments, development of sulfate-reducing bacteria generates large quantities of H_2S, which is often sequestered as insoluble metal sulfides (e.g., FeS). It is now generally accepted that these bacteria have been responsible for the formation of metal sulfide ores and, indirectly, for natural deposits of elemental sulfur (formed by a secondary, probably nonbiological, oxidation of sulfides).

In waterlogged (and hence largely anaerobic) soils the production of H_2S by sulfate reducers may cause damage to plants: this is sometimes a serious economic problem in the cultivation of rice, which is sown in flooded fields (rice paddies).

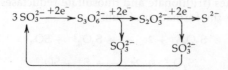

FIGURE 24.5

The pathway of sulfide formation from sulfite in sulfate-reducing bacteria.

NONSPORE-FORMING ANAEROBES WITH A FERMENTATIVE METABOLISM

The members of this physiological category are major constituents of the microflora of the body cavities of animals, areas which provide a series of different oxygen-free ecological niches. The properties of relatively aerotolerant fermentative anaerobes related to actinomycetes were discussed in Chapter 23. The strict anaerobes are principally Gram-negative rods and cocci, although they also include a few Gram-positive cocci. Characteristics of the principal genera are summarized in Tables 24.4 and 24.5. The frequency of the rumen as a source is probably misleading; many of the rumen bacteria may well occur in other regions of the gastrointestinal tract of animals. However, the rumen microflora has been studied in great detail, primarily because of its important role in ruminant nutrition (see Chapter 28). Some species of *Bacteroides*, *Fusobacterium*, *Veillonella*, *Peptococcus*, and *Peptostreptococcus* have been implicated in human or animal infections. However, most of these bacteria appear to be nonpathogenic components of the normal microflora of animals. As shown in Tables 24.4 and 24.5,

TABLE 24.4

Stricly anaerobic cocci with a fermentative mode of metabolism

	GRAM-NEGATIVE			GRAM-POSITIVE		
	Veillonella	*Acid-aminococcus*	*Megasphaera*	*Peptococcus*	*Pepto-streptococcus*	*Ruminococcus*
FERMENTABLE SUBSTRATES						
Cellulose, xylan	–	–	–	–	–	–
Soluble sugars	–	–	+	V[b]	V	+
Lactate	+	–	+	–	–	–
Amino acids	–	+	–	+	+	–
FERMENTATION PRODUCTS						
CO_2	+	+	+	+	V	+
H_2	+	+	Tr[a]	+	V	+
Formate	–	–	–	+	+	+
Acetate	+	+	+	+	+	+
Propionate	+	–	+	+	+	–
Butyrate	–	+	+	+	+	–
Valerate	–	–	+	–	+	–
Caproate	–	–	+	–	+	–
DNA base composition (moles % G + C)	40–44	57	54	36–37	33	40–45
Source	Mouth, intestinal tract	Intestinal tract	Rumen	Intestinal tract	Intestinal tract	Rumen

[a] Tr denotes trace.
[b] V denotes variable.

TABLE 24.5

Strictly anaerobic Gram-negative rods with a fermentative mode of metabolism

	Bacteroides	*Fusobacterium*	*Selenomonas*	*Lachnospira*	*Butyrivibrio*
Structure	Straight rods	Straight rods	Crescent-shaped	Curved rods	Curved rods
Flagellar insertion	Petritrichous or immotile	Peritrichous or immotile	Tuft on concave side	Lateral flagella	Polar monotrichous
Fermentable substrates					
Sugars	+	+	+	+	+
Amino acids	+	+	−	−	−
Fermentation products	Highly varied, often including formate, acetate, propionate, succinate, lactate	Varied; butyrate a major product; also acetate, lactate	Acetate, propionate, CO_2, lactate	Formate, acetate, lactate, ethanol, CO_2	Formate, butyrate, lactate, H_2, CO_3
DNA base composition (moles % G + C)	40–55	26–34	53–61		
Source	Intestinal tract, mouth, rumen,	Mouth, intestinal tract	Rumen, mouth	Rumen	Rumen

both the fermentable substrates and the fermentation products vary from group to group. Apart from the rumen species, very few of these organisms have been fully characterized. The presence of a b type cytochrome has been reported in some *Bacteroides* species, which suggests that in this group ATP synthesis may occur in association with electron transport, as well as at the substrate level, during the fermentation of organic substrates.

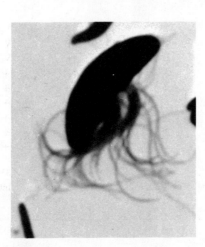

FIGURE 24.6

Selenomonas: flagella stain, showing the crescent-shaped cell and the tuft of flagella inserted laterally on the concave surface of the cell (×3,600). Photo courtesy of C. F. Robinow.

The members of the genus *Selenomonas* have a distinctive cell structure (Figure 24.6). The cells are crescent-shaped and bear a tuft of up to 10 flagella, inserted near the equator on the concave side of the cell.

FURTHER READING

Book   HUNGATE, R. E., *The Rumen and Its Microbes.* New York: Academic Press, 1966.

Reviews and original articles   KINGSLEY, V. V., and J. F. M. HOENIGER, "Growth, Structure and Classification of *Selenomonas*," *Bact. Revs.* **37**, 479 (1973).

LEE, J. P., J. LEGALL, and H. D. PECK, "Isolation of Assimilatory—and Dissimilatory—Type Sulfite Reductases from *Disulfovibrio vulgaris*," *J. Bacteriol.* **115**, 529 (1973).

LEGALL, J., and J. R. POSTGATE, "The Physiology of Sulfate-Reducing Bacteria," *Advan. Microbiol Physiol.* **10**, 82 (1973).

MURPHY, J. J., and L. M. SIEGEL, "Sisoheme and Sirohydrochlorin: The Basis of a New Type of Porphyrin-Related Prosthetic Group Common to Both Assimilatory and Dissimilatory Sulfite Reductases," *J. Biol. Chem.* **248**, 6911 (1973).

WOLFE, R. S., "Microbial Fermentation of Methane," *Advan. Microb. Physiol.* **6**, 107 (1971).

MICROORGANISMS AS GEOCHEMICAL AGENTS

25

The current chemical state of the elements on the outer surface of the earth is, to a considerable extent, a consequence of the chemical activities of living organisms. This fact is dramatically illustrated by the changes that have occurred in the earth's atmosphere. Before life evolved the gases of the atmosphere were highly reduced: nitrogen was present in the form of ammonia (NH_3), oxygen as water (H_2O), and carbon as methane (CH_4). Now, they exist in oxidized form: nitrogen and oxygen as elemental gases (N_2 and O_2) and carbon as carbon dioxide (CO_2). The quantity of these and many other compounds found on the earth's surface represents the net balance between their rates of formation and utilization in biological and geological processes. Such transformation occurs in all regions of the earth that contain living organisms, collectively known as the *biosphere*. The oceans, freshwater, land surface of the continents, and lower portion of the atmosphere comprise the biosphere. This thin film of life on the earth's surface exists in a more or less steady state, maintained by a cyclic turnover of the elements necessary for life, powered by a continuous input of energy from the sun.

The various steps in the cyclic turnover of elements are brought about by different types of organisms. Thus, the continued existence of any particular group of organisms depends on the chemical transformation carried out by others. A break in the cycle at any point would preclude all life. All the major elements necessary for life (carbon, oxygen, nitrogen, sulfur, and phosphorus) are transformed cyclically.

The cyclic nature of transformations in the biosphere can be summarized as follows. Through solar-energy conversion by photosynthesis, CO_2 and other inorganic compounds are withdrawn from the environment and are accumulated in the organic constituents of living organisms. The major producers of organic matter through photosynthesis are unicellular algae (principally diatoms and dinoflagellates) in the ocean and seed plants on land. Organic material thus

715 accumulated provides, either directly or indirectly, the energy sources for all other forms of life.

Insofar as photosynthetic organisms serve as food sources for animals or microorganisms, the elements of major biological importance remain, at least in part, in the organic state, during the transformations which lead to their incorporation into the cells and tissues of the primary consumers. The primary consumers may themselves provide food sources for other organisms so that these elements may persist in organic food chains, made up of many types of nonphotosynthetic organisms. Before they can be again utilized by photosynthetic organisms they usually must be converted once more to inorganic form. This conversion, known as *mineralization*, is brought about largely by the decomposition of plant and animal remains and excretory products by microorganisms, principally fungi and bacteria. It is estimated that 90 percent of the mineralization of organic carbon atoms (i.e., their conversion to CO_2) is the result of the metabolic activities of these two groups of microorganisms. The remaining 10 percent results from the metabolism of all other organisms, as well as the combustion of fuels and other materials. The overwhelming contribution of microorganisms to this process reflects the ubiquity of microorganisms, their significant contribution to the total bulk of living material (their biomass), their high rates of growth and metabolism, and their collective ability to degrade a vast variety of naturally occurring organic materials.

THE FITNESS OF MICROORGANISMS AS AGENTS OF GEOCHEMICAL CHANGE

• **The distribution of microorganisms in space and time**

The omnipresence of microorganisms throughout the biosphere is a consequence of their ready dissemination by wind and water. Surface waters, the floors of oceans over the continental shelves, and the top few inches of soil are teeming with microorganisms that are ready to decompose organic matter that may become available to them. It has been estimated that the top 6 in. of fertile soil may contain more than 2 tons of fungi and bacteria per acre. Any handful of soil contains many different kinds of microbes, presenting at different times microscopic ecological niches for different types to develop. Even on a single soil particle, the conditions may change from hour to hour and from facet to facet.

Let us consider what happens upon the death of a microscopic root hair or a worm in the soil. The organic compounds of the dead tissue are attacked by microorganisms that are capable of digesting and oxidizing these compounds. As oxygen is consumed, conditions may become anaerobic in the immediate proximity of the dead tissue and fermentative organisms develop. The products of fermentation then diffuse to regions in which oxygen is still present or they may be oxidized anaerobically by organisms capable of reducing nitrates, sulfates, or carbonates. Ultimately, the organic compounds will be completely converted to CO_2 or assimilated, the condition will again become fully aerobic, and autotrophs will develop at the expense of such reduced inorganic products as ammonia, sulfide, and hydrogen. Thus, the inorganic products of the decomposition of the plants or animals are eventually completely oxidized. This sequence of events, which occurs on a microscopic scale on a particle of soil, can be observed on a macroscopic scale in nature. When a tree falls into a swamp or a whale decomposes on a beach, the eventual chemical results are essentially the same. Seasonal and climatic conditions may retard or accelerate the cyclic turnover of matter.

In cold climates, decomposition is most rapid in the early spring; in semiarid areas, it is largely restricted to the rainy season.

In nature, only those microorganisms that are favored by the local and temporary environment reproduce, and their growth ceases when they have changed their environment. Most of them are eventually consumed by such everpresent predators as the protozoa, but a few cells of each type persist to initiate a new burst of growth when conditions again become favorable for their development.

• **The metabolic potential of microorganisms**

The relatively enormous catalytic power of microorganisms contributes to the major role they play in the chemical transformations occurring on the earth's surface. Because of their small size, bacteria and fungi possess a large surface–volume ratio compared with higher animals and plants. This permits a rapid exchange of substrates and waste products between the cells and their environment.

Per gram of body weight, the respiratory rates of some aerobic bacteria are hundreds of times greater than that of man. On the basis of the known metabolic rates of microorganisms, one can estimate that the metabolic potential of the microorganisms in the top 6 in. of an acre of well-fertilized soil at any given instant is equivalent to the metabolic potential of some tens of thousands of human beings.

An even more important factor influencing the chemical role that microorganisms play in nature is their high rate of reproduction in favorable environments.

• **The metabolic versatility of microorganisms**

The remarkable ability of microorganisms to degrade a vast variety of organic compounds has led to a widely held conviction that has been termed the principle of microbial infallibility, a principle that was clearly stated by E. F. Gale in 1952: "It is probably not unscientific to suggest that somewhere or other some organism exists which can, under suitable conditions, oxidise any substance which is theoretically capable of being oxidised." With the increasing production of plastics as well as synthetic insecticides, herbicides, and detergents, it has become clear that some substances are remarkably resistant to microbial attack because they persist and accumulate in nature. Even certain naturally occurring organic compounds are somewhat resistant; they accumulate and constitute the organic fraction of soil known as *humus* which confers the deep brown or black color to fertile soils. Because of the importance of humus to agriculture, this complex mixture of persistent organic compounds has been studied extensively. In large degree, it appears to consist of degradation products of a particularly stable component of woody plants known as *lignin*. The remarkable stability of humus has been demonstrated by radiocarbon dating; humus from certain soils is thousands of years old.

These exceptions aside, most organic compounds that are no longer a part of a living organism are rapidly mineralized by microorganisms in the biosphere.

Although some nonphotosynthetic microorganisms (e.g., the *Pseudomonas* group) can attack many different organic compounds, the metabolic versatility of the microbial world *as a whole* is not primarily a reflection of the metabolic versatility of its individual members. Any single bacterial species is only a limited agent of mineralization. Highly specialized physiological groups of microorganisms play important roles in the mineralization of specific classes of organic compounds. For example, the decomposition of cellulose, which is one of the most abundant constituents of plant tissues, is mainly brought about by organisms that are highly specialized nutritionally. Among the aerobic bacteria capable of decomposing cellulose, the gliding bacteria that belong to the *Cytophaga group* (p. 636) are perhaps the most important. The cytophagas can rapidly dissolve and oxidize this insoluble compound, but cellulose is the only substance they can use as carbon source.

It will be recalled that the autotrophic bacteria, responsible for the oxidation of reduced inorganic compounds in nature, are also highly specific. Each type of autotroph is capable of oxidizing only one class of inorganic compounds and, in some cases (the nitrifying bacteria), only one compound.

THE CYCLES OF MATTER

The turnover of the elements that compose living organisms constitute the *cycles of matter*. All organisms participate in various steps of these cyclic conversions, but the contribution of microorganisms is particularly important, both quantitatively (as discussed previously) and qualitatively. For example, certain steps in the nitrogen cycle are exclusively brought about by procaryotes.

THE PHOSPHORUS CYCLE

Considered from a chemical point of view, the phosphorus cycle is simple, because phosphorus occurs in living organisms only in the +5 valence state, either as free phosphate ions (PO_4^{3-}) or as organic phosphate constituents of the cells. Most organic phosphate compounds cannot be taken into the living cell; instead, phosphorus requirements are met by the uptake of phosphate ion. Organic phosphate compounds are then synthesized within the cell, and upon the death of the organism, phosphate ion is rapidly released by hydrolysis.

In spite of the rapid functioning of the phosphorus cycle and the relative abundance of phosphates in soils and rocks, phosphate is a limiting factor for the growth of many organisms because much of the earth's supply of phosphates occurs as insoluble calcium, iron, or aluminum salts. Freshwater often contains phosphate ions in mere trace amounts, being available to animals only after it has been concentrated by the phytoplankton.

Soluble phosphates are constantly being transferred from terrestrial environments to the sea as a consequence of leaching, a transfer which is largely unidirectional. Only small quantities are returned to the land, principally by the deposits of guano by marine birds. Thus, the availability of phosphate for terrestrial forms of life depends on the continued solubilization of insoluble phosphate deposits, a process in which microorganisms play an important role. Their acidic metabolic products (organic, nitric and sulfuric acids) solubilize the phosphate of calcium phosphate, and their production of H_2S dissolves ferric phosphates.

THE CYCLES OF CARBON AND OXYGEN

The cyclic conversions of carbon and oxygen are brought about primarily by two processes—*oxygenic photosynthesis* on the one hand, and *respiration* and *combustion* on the other. It is mainly through the process of oxygenic photosynthesis that the oxidized form of carbon (CO_2) is converted to the reduced state in which it occurs in organic compounds, and that the reduced form of oxygen (H_2O) is oxidized to molecular oxygen (O_2). Although other autotrophs can reduce CO_2 to organic material while oxidizing compounds other than water (NO_2^-, NO_3^-, H_2, Fe^{+2}, and reduced forms of sulfur), the contribution of these processes to the total fixation* of CO_2 is minor.

Heterotrophic metabolism coupled directly or indirectly with the reduction of molecular O_2 completes the cycle by regenerating the major nutrients of oxygenic photosynthesis: CO_2 and water. Algae and plants, as well as the animals that feed on them, contribute their share to this process through their respiratory activities. The bacteria and fungi, however, oxidize the bulk of organic material. Thus, through oxygenic photosynthesis on the one hand and aerobic respiration on the other, the cyclic transformations of carbon and oxygen are obligately linked (Figure 25.1).

Air contains approximately 0.03 percent CO_2 by volume, this concentration being maintained relatively constant as a result of the dynamic balance between photosynthesis and mineralization. When dissolved in slightly alkaline water, bicarbonate (HCO_3^-) and carbonate (CO_3^{2-}) ions are formed:

$$CO_2 + OH^- \rightleftharpoons HCO_3^-$$

$$HCO_3^- + OH^- \rightleftharpoons H_2O + CO_3^{2-}$$

Therefore, bicarbonate serves as the reservoir of carbon for photosynthesis in aquatic environments. The bicarbonate concentration of ocean waters ($\sim$0.002 M) acts as a reservoir for CO_2 for the atmosphere; the oceans trap a large fraction of the CO_2 produced on land, keeping its concentration at a relatively low and constant level.

The importance of the carbon cycle can best be emphasized by the estimate that the total CO_2 contained in the atmosphere, if it were not replenished, would be completely exhausted in less than 20 years at the present rate of photosynthesis. This estimate does not appear too radical when it is realized that the carbon contained in a single giant redwood tree is equivalent to that present in the atmosphere over an area of approximately 40 acres. On land, seed plants are the principal agents of photosynthetic activity. A minor contribution is made by the algae. In the oceans, however, it is the unicellular photosynthetic organisms that play the most important role. The large plantlike algae (seaweeds) are confined in their development to a relatively narrow coastal strip. Since light of photosynthetically effective wavelengths is largely filtered out at a depth of about

---

*A process in which a gaseous compound is converted to a solid one is called *fixation*.

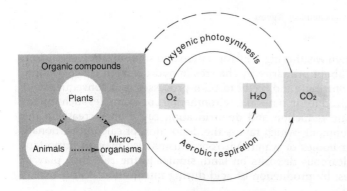

FIGURE 25.1

The carbon and oxygen cycles: oxidations of carbon and oxygen are shown as solid arrows, reductions as broken arrows, and reactions with no valence change as dotted arrows.

50 ft, these sessile algae cannot grow in deeper waters. Because they are free-floating, the microscopic algae of the ocean (known as the phytoplankton) are capable of developing in the surface layers wherever the environment is favorable. Their growth is largely limited by the relative scarcity of two elements: phosphorus and nitrogen. Where these elements are made available as phosphates and nitrates by the runoff of rain water from continents and subsequent distribution by ocean currents, profuse development of phytoplankton occurs. According to one estimate the total annual fixation of carbon in the oceans amounts to approximately 1.2×10^{10} tons, whereas that on the land is about 1.6×10^{10} tons.

Although oxygenic photosynthesis is by far the most important means of reducing CO_2 to organic matter, other processes contribute to a small extent. These include photosynthesis by the purple and green bacteria, CO_2 reduction by the chemoautotrophs, and traces of CO_2 fixed in the metabolism of most organisms.

● **The mineralization process: carbon dioxide formation and the reduction of oxygen**

The biological conversion of organic carbon to CO_2 with the concomitant reduction of molecular oxygen involves the combined metabolic activity of many different kinds of microorganisms. The complex constituents of dead cells must be digested, and the products of digestion must be oxidized by specialized organisms that can use them as nutrients. Many aerobic bacteria (pseudomonads, bacilli, Actinomycetes), as well as fungi, carry out complete oxidations of organic substances derived from dead cells. However, it should be remembered that even those organisms that produce CO_2 as the only waste product of the respiratory decomposition of organic compounds usually use a large fraction of the substrate for the synthesis of their own cell material. In anaerobic environments, organic compounds are decomposed initially by fermentation, and the organic end products of fermentation are then further oxidized by anaerobic respiration, provided that suitable inorganic hydrogen acceptors (nitrate, sulfate, or CO_2) are present.

● **The sequestration of carbon: inorganic deposits**

The carbonate ions in the oceans combine with dissolved calcium ions and become precipitated as calcium carbonate. Calcium carbonate is also deposited biologically in the shells of protozoa, corals, and mollusks. This is the geological origin of the calcareous rock (limestone) that is an important constituent of the surface of continents. Calcareous rock is not directly available as a source of carbon for photosynthetic organisms, and hence its formation causes a depletion of the total carbon supply available for life. Nevertheless, much of this carbon eventually reenters the cycle through weathering. The formation and solubilization of calcium carbonate are brought about primarily by changes in hydrogen ion concentration, and microorganisms contribute indirectly to both processes as a consequence of pH changes that they produce in natural environments. For example, such microbial processes as sulfate reduction and denitrification cause an increase in the alkalinity of the environment, which favors the deposition of calcium carbonate in the ocean and other bodies of water. Microorganisms also play an important role in solubilizing calcareous deposits on land, similar to the role they play in solubilizing phosphates, by production of acid during nitrification, sulfur oxidation, and fermentation.

• **The sequestration of carbon: organic deposits**

A high moisture content, causing oxygen depletion and the accumulation of acidic substances, is sometimes particularly favorable for the accumulation of humus. This phenomenon is most pronounced in *peat bogs,* where, in the course of time, deposits of undecomposed organic matter known as *peat* accumulate. These deposits may extend for hundreds of feet below the surface of the bog. In the course of geological time the compression of peat deposits, probably aided by other physical and chemical factors, has resulted in the formation of coal. Much carbon has thus been sequestered from the biosphere in the form of peat and coal deposits. A second kind of sequestration of carbon in organic form has occurred in deposits of natural oil and gas (methane).

Since the Industrial Revolution, man's exploitation of the stored deposits of organic carbon in the earth's crust has resulted in their very rapid mineralization. Although substantial deposits still remain to be exploited, it is estimated that at current rates of consumption most of the petroleum and natural gas will be used up within a few decades.

THE NITROGEN CYCLE

Although molecular nitrogen (N_2) is abundant, constituting about 80 percent of the earth's atmosphere, it is chemically inert and therefore not a suitable source of the element for most living forms. All plants and animals, as well as most microorganisms, depend on a source of combined, or fixed, nitrogen in their nutrition. Combined nitrogen in the form of ammonia, nitrate, and organic compounds is relatively scarce in soil and water, often constituting the limiting factor for the development of living organisms. For this reason, the cyclic transformation of nitrogenous compounds is of paramount importance in supplying required forms of nitrogen to the various nutritional classes of organisms in the biosphere. The main features of the nitrogen cycle are illustrated schematically in Figure 25.2.

• **Nitrogen fixation**

The turnover of nitrogen through its cycle is estimated to be between 10^8 and 10^9 tons per year. The vast supply of nitrogen gas (N_2) in the atmosphere and

721

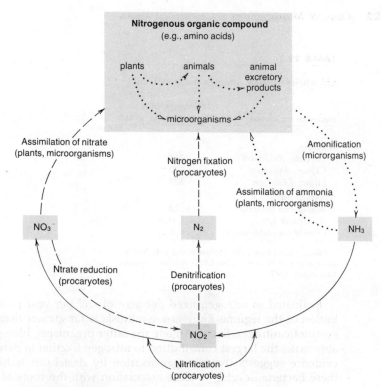

Nitrogenous organic compound
(e.g., amino acids)

plants   animals   animal
                             excretory
                          ⋯> products

⋯> microorganisms ⋰

Assimilation of nitrate
(plants, microorganisms)

Amonification
(micrgorganisms)

Nitrogen fixation
(procaryotes)

Assimilation of ammonia
(plants, microorganisms)

NO₃⁻   N₂   NH₃

Ntrate reduction
(procaryotes)

Denitrification
(procaryotes)

NO₂⁻

Nitrification
(procaryotes)

FIGURE 25.2

The nitrogen cycle: the oxidations of nitrogen are shown as solid arrows, the reductions as broken arrows, and reactions with no valence change as dotted arrows.

the relative scarcity of combined nitrogen on the earth's surface suggest that the process of nitrogen fixation is the rate limiting step. This process is largely a biological one, and bacteria are the only organisms capable of causing it (see Chapter 7). Some nitrogen is fixed by lightning, ultraviolet light, electrical equipment, and the internal combustion engine, but these nonbiological processes are quantitatively insignificant, together accounting for only about 0.5 percent of nitrogen fixation. Even industrial manufacture of fertilizer by the Haber process contributes only about 5 percent. Thus, over 90 percent of all nitrogen fixation is brought about by the metabolic activities of certain bacteria.

Biological nitrogen fixation in nature is mediated in part by free-living bacteria (*nonsymbiotic nitrogen fixation*) and in part by bacteria which exist in mutualistic partnership with plants (*symbiotic nitrogen fixation*).

The most important agents of symbiotic nitrogen fixation are the bacteria of the genus *Rhizobium*, which invade the root hairs of leguminous plants and develop in nodules produced on the roots, where nitrogen fixation occurs. The enzymatic machinery for the fixation of nitrogen is synthesized by the bacterium; the host plant provides an environment which favors the expression of this property (see Chapter 27).

The most important agents of nonsymbiotic nitrogen fixation are heterocyst-forming blue-green bacteria such as *Anabaena* and *Nostoc*. A wide variety of other bacteria are also capable of fixing nitrogen; these include both aerobic bacteria (*Azotobacter, Beijerinckia,* and *Bacillus polymyxa*) and anaerobic bacteria (photosynthetic bacteria and *Clostridium* spp.).

TABLE 25.1

Efficiencies of some nitrogen-fixing systems

ORGANISM OR SYSTEM	POUNDS OF N FIXED/ACRE/YEAR[a]
SYMBIOTIC	
Lucerne: *Rhizobium*	>264
Clover: *Rhizobium*	220
Lupin: *Rhizobium*	132
NONSYMBIOTIC	
Blue-green bacteria	22
Azotobacter spp.	0.26
Clostridium pasteurianum	0.22

[a] Recalculated from E. N. Mishustin and V. K. Shil'nikova, *Biological Fixation of Atmospheric Nitrogen.* London: Macmillan, 1971.

Evaluated as nitrogen fixed per acre of soil per year (Table 25.1), the contribution of the legume-*Rhizobium* symbiosis is far greater than that made by nonsymbiotic nitrogen fixers. Among the latter organisms, blue-green bacteria probably make the largest contribution to nitrogen fixation in nature. However, recent evidence suggests that nitrogen fixation by *Azotobacter* is highly effective when these bacteria develop in close association with the roots of plants; their contribution to nitrogen fixation may therefore have been hitherto underestimated.

Because of the critical agronomic importance of fixed nitrogen, the current world food crisis, and the fact that manufacture of nitrogen fertilizers by the Haber process requires large expenditures of energy, biological nitrogen fixation has become an intensive subject of investigation. Among the important goals of these investigations are the development of new plants capable of harboring nitrogen-fixing symbionts. Although the current range of such plants is wide, it does not include the world's major food crops, wheat and rice, nor the major forage crop, grass.

• The utilization of fixed nitrogen by photosynthetic organisms

Algae and plants assimilate nitrogen either as nitrate or ammonia. If nitrate is the form in which nitrogen is assimilated, it must be reduced in the cell to ammonia. This process of nitrate reduction proceeds only to the extent to which nitrogen is required for growth; ammonia is not excreted. It is this feature in particular that distinguishes *nitrate assimilation* by plants (and also by microorganisms) from *nitrate reduction,* a process of anaerobic respiration which is limited to procaryotes (Figure 25.2).

• The transformations of organic nitrogen and the formation of ammonia

The organic nitrogenous compounds synthesized by algae and plants serve as the nitrogen source for the animal kingdom. During their assimilation by animals, the complex nitrogenous compounds of plants are hydrolyzed to a greater or lesser extent, but the nitrogen remains largely in reduced organic form. Unlike plants, however, animals do excrete a significant quantity of nitrogenous compounds in the course of their metabolism. The form in which this nitrogen is excreted varies from one group of animals to another. Invertebrates predominantly excrete ammonia; but among vertebrates, organic nitrogenous excretion products make their appearance as well. In reptiles and birds, uric acid is the major form in which nitrogen is excreted; in mammals, urea is the principal form. The urea and uric acid excreted by animals are rapidly mineralized by special groups of microorganisms, with the formation of CO_2 and ammonia.

Only part of the nitrogen stored in organic compounds through plant growth is converted to ammonia by animal metabolism and the microbial decomposition of urea and uric acid. Much of it remains in plant and animal tissues and is liberated only on the death of these organisms. Whenever a plant or animal dies, its body constituents are immediately attacked by microorganisms, and the nitrogenous compounds are decomposed with the liberation of ammonia. Part of the nitrogen is assimilated by the microorganisms themselves and thus converted into microbial cell constituents. Ultimately, these constituents are converted to ammonia following the death of the microbes.

The first step in this process of *ammonification* is the hydrolysis of the proteins and nucleic acids with the liberation of amino acids and organic nitrogenous bases, respectively. These simpler compounds are then attacked by respiration or fermentation.

Protein decomposition under anaerobic conditions (*putrefaction*) usually does not lead to an immediate liberation of all the amino nitrogen as ammonia. Instead, some of the amino acids are converted to amines. The putrefactive decomposition is characteristically brought about by anaerobic spore-forming bacteria (genus *Clostridium*). In the presence of air the amines are oxidized by other bacteria with the liberation of ammonia.

• Nitrification   Through all the transformations that nitrogen undergoes from the time of its reductive assimilation by plants until its liberation as ammonia, the nitrogen atom remains in the reduced form. The conversion of ammonia to nitrate (*nitrification*) is brought about in nature by two highly specialized groups of obligately aerobic chemoautotrophic bacteria. Nitrification occurs in two steps: In the first step, ammonia is oxidized to nitrite; in the second, nitrite is oxidized to nitrate. As a result of the combined activities of these bacteria, the ammonia liberated during the mineralization of organic matter is rapidly oxidized to nitrate. Thus nitrate is the principal nitrogenous material available in soil for the growth of plants. The practice of soil fertilization with manure depends on the microbial mineralization of organic matter and results in the conversion of organic nitrogen to nitrate through ammonification and nitrification. Irrigation with dilute solutions of ammonia, which is one of the modern methods used for fertilization, is an even more direct means by which the nitrate content of soil is increased. Ammonia, which can be synthesized chemically from molecular nitrogen, is the most concentrated form of combined nitrogen available because it contains about 82 percent nitrogen by weight. Nitrates are very soluble compounds and are therefore easily leached from the soil and transported by water; hence, a certain amount of combined nitrogen is constantly removed from the continents and carried down to the oceans. In some special localities, notably in the semiarid regions of Chile, deposits of nitrate have accumulated in the soil as a result of the runoff and evaporation of surface water. Such deposits are a valuable source of fertilizer, although their importance has diminished greatly in the course of the last 50 years as a result of the development of chemical methods for making nitrogen compounds from atmospheric nitrogen.

Nitrates have played an important role not only in the development of agriculture but also in the destructive activities of man. Gunpowder, which was the only explosive used for war before the invention of nitroglycerine (dynamite), is a mixture of sulfur, carbon, and saltpeter (KNO_3). During the Napoleonic wars, largely as a result of the British blockade, a shortage of nitrate for gunpowder production occurred in France. This led to the development of "nitrate gardens," in which nitrate was obtained by the mineralization of organic matter. A mixture of manure and soil was spread on the surface of the ground and frequently turned to permit aeration. After the manure had decomposed, nitrate was extracted from the residue.

• **Denitrification**   Many aerobic bacteria can use nitrate in place of oxygen as a final hydrogen acceptor if conditions are anaerobic.

Thus, whenever organic matter is decomposed in soil or water and oxygen is exhausted as a result of aerobic microbial respiration, certain of these aerobes will continue to respire the organic matter if nitrate is present, i.e., by anaerobic respiration. As a consequence, nitrate is reduced. Some bacteria (e.g., *Escherichia coli*) are only able to reduce nitrate to the level of nitrite; others (e.g., *Pseudomonas aeruginosa*) are able to reduce it to nitrogen gas. By this latter process, called *denitrification*, combined nitrogen is removed from soil and water releasing N_2 gas to the atmosphere.

Denitrification is a process of major ecological importance. It depletes the soil of an essential nutrient for plants, thereby decreasing agricultural productivity. Such losses are particularly important from fertilized soils. Although precise values are not available, under certain conditions, the amount of fixed nitrogen fertilizer lost through denitrification may approach 50 percent.

Nevertheless, not all the consequences of denitrification are detrimental. Denitrification is vital to the continued availability of combined nitrogen on the land masses of the earth. The highly soluble nitrate ion is constantly leached from the soil, and it is eventually carried to the oceans. Without denitrification, the earth's supply of nitrogen, including N_2 of the atmosphere, would eventually accumulate in the oceans, precluding life on the land masses except for a fringe near the oceans. Denitrification also maintains the potability of freshwaters, because high concentrations of nitrate ions may be toxic.

THE SULFUR CYCLE

Sulfur, an essential constituent of living matter, is abundant in the earth's crust. It is available to living organisms principally in the form of soluble sulfates or reduced organic sulfur compounds. Reduced sulfur in the form of H_2S also occurs in the biosphere as a result of microbial metabolism and, to a limited extent, of volcanic activity. Except under anaerobic conditions, however, its concentration is low because it is oxidized rapidly in the presence of oxygen, either spontaneously or by bacteria.

The turnover of sulfur compounds is referred to as the *sulfur cycle*. The biological aspects of this cycle are shown in Figure 25.3. In many respects, it shows a striking resemblance to the nitrogen cycle already described.

In addition to the biological sulfur cycle, important nonbiological transformations of gaseous forms of sulfur occur in the earth's atmosphere. It is estimated that some 90 million tons of sulfur in the form of biologically generated H_2S are released to the atmosphere annually; an additional 50 million tons are contributed in the form of SO_2 by the burning of fossil fuels; and about 0.7 million tons in the form of H_2S and SO_2 come from the earth's volcanic activity. In the atmosphere, H_2S is rapidly oxidized by atomic oxygen (O), molecular oxygen (O_2), or ozone (O_3) to SO_2 which might dissolve in water to form sulfurous acid (H_2SO_3); or be oxidized by a second and slower series of reactions (requiring hours or days) to SO_3. When dissolved in water, SO_3 becomes sulfuric acid (H_2SO_4). Some sulfuric acid is neutralized by the small quantities of ammonia in the atmosphere, but much of it returns along with unoxidized H_2SO_3 to the earth's surface in acid form where it causes considerable damage to stone structures and sculptures. The rate of generation of acidic sulfur compounds increases as more fossil fuels are being burned. The problem is particularly acute in areas of high population density, and even now it is causing the rapid destruction of much stone sculpture.

• **The assimilation of sulfate**

Sulfate is almost universally used as a nutrient by plants and microorganisms. The assimilation of sulfate resembles the assimilation of nitrate in two respects. First, like the nitrogen atom of nitrate, the sulfur atom of sulfate must become reduced in order to be incorporated into organic compounds, because in living organisms, sulfur occurs almost exclusively in reduced form as —SH or —S—S—

FIGURE 25.3

The sulfur cycle: the oxidations of the sulfur atom are shown as solid arrows, the reductions as broken arrows, and reactions with no valence change as dotted arrows.

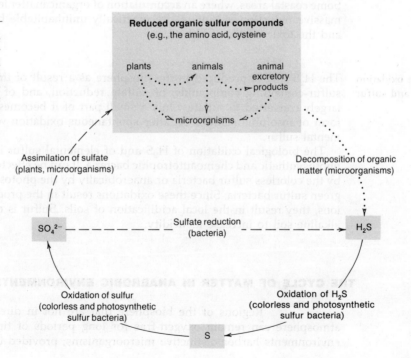

groups. Second, in both cases, only enough of the nutrient is assimilated to provide for the growth of the organism, no reduced products being excreted into the environment.

• **The transformation of organic sulfur compounds and formation of H₂S**

When sulfur-containing organic compounds are mineralized, sulfur is liberated in the reduced inorganic form as H_2S. This process resembles ammonification, in which nitrogen is liberated from organic matter in its reduced inorganic form as ammonia.

• **The direct formation of H₂S from sulfate**

The utilization of sulfate for the synthesis of sulfur-containing cell constituents and the subsequent decomposition of these compounds results in an overall reduction of sulfate to H_2S. H_2S is also formed more directly from sulfate through the activity of the sulfate-reducing bacteria. These obligately anaerobic bacteria oxidize organic compounds and molecular hydrogen by using sulfate as an oxidizing agent. Their role in the sulfur cycle may therefore be compared to the role of the nitrate-reducing bacteria in the nitrogen cycle. The activity of the sulfate-reducing bacteria is particularly apparent in the mud at the bottom of ponds and streams, in bogs, and along the seashore. Since seawater contains a relatively high concentration of sulfate, sulfate reduction is an important factor in the mineralization of organic matter on the shallow ocean floors. Signs of the process are the odor of H_2S and the pitch-black color of the mud in which it occurs. The color of black mud is caused by the accumulation of ferrous sulfide. Some coastal areas, where an accumulation of organic matter leads to a particularly massive reduction of sulfate, are practically uninhabitable because of the odor and the toxic effects of H_2S.

• **The oxidation of H₂S and sulfur**

The H_2S that is produced in the biosphere as a result of the decomposition of sulfur-containing compounds, of sulfate reduction, and of volcanic activity is largely converted to sulfate. Only a small part of it becomes sequestered in the form of insoluble sulfides or, after spontaneous oxidation with oxygen, as elemental sulfur.

The biological oxidation of H_2S and of elemental sulfur is brought about by photosynthetic and chemoautotrophic bacteria. It can be effected either aerobically by the colorless sulfur bacteria or anaerobically by the photosynthetic purple and green sulfur bacteria. Since these oxidations result in the production of hydrogen ions, they result in the local acidification of soils. Sulfur is commonly added to alkaline soil to increase its acidity.

THE CYCLE OF MATTER IN ANAEROBIC ENVIRONMENTS

Regions of the biosphere that are not in direct contact with the atmosphere can remain oxygen-free for long periods of time. Such anaerobic environments harbor distinctive microorganisms; provided that light can penetrate, these microorganisms are able to bring about an almost completely closed anaerobic cycle of matter.

The primary synthesis of organic material under these conditions is mediated by photosynthetic bacteria. The purple and green sulfur bacteria convert CO_2 to cell material, using H_2S as a reductant; the purple nonsulfur bacteria perform an almost complete assimilation of acetate and other simple organic compounds. Upon the death of the photosynthetic bacteria, their organic cell constituents are decomposed by clostridia and other fermentative anaerobes, with the formation of CO_2, H_2, NH_3, organic acids, and alcohols. The hydrogen and some of the organic fermentation products are anaerobically oxidized by sulfate-reducing and methane-producing bacteria. The anaerobic oxidations performed by the sulfate reducers result in the formation of H_2S and acetate, both utilizable in turn by photosynthetic bacteria. The metabolism of the methane bacteria results in the conversion of CO_2 and of some organic carbon (e.g., the methyl group of acetic acid) to methane. Present information suggests that methane cannot be further metabolized under anaerobic conditions. Much of it escapes, however, to aerobic regions, where it is oxidized by *Methanomonas* and other aerobic methane oxidizers. The loss of methane constitutes the only significant leak from this anaerobic cycle of matter. The anaerobic sulfur cycle is completely closed, since sulfate and H_2S are interconverted by the combined activities of sulfate-reducing and photosynthetic bacteria. The anaerobic nitrogen cycle is also closed and is chemically very simple, relative to the cycle in aerobic environments; the nitrogen atom does not undergo valence changes, alternating between ammonia and the amino groups ($R—NH_2$) in nitrogenous cell materials.

Since the participants in this anaerobic cycle of matter are all microorganisms, the cycle does not require much space. In fact, such a cycle can be established in the laboratory, in a closed bottle that contains the appropriate nutrients and that has been inoculated with water and mud from an anaerobic pool. Provided that the bottle is sufficiently illuminated, microbial development will continue in it over a period of years.

THE CYCLE OF MATTER THROUGH GEOLOGICAL TIME

The integration of the various reactions that constitute the cycle of matter results in a balanced production and consumption of the biologically important elements in the biosphere. It is probable that the cycle of matter as we know it today has operated without significant change for at least a billion years. However, there are good reasons to believe that the cycle of matter was considerably different at an early period in the history of the earth, when biological systems first developed on the planet (Figure 25.4).

As discussed at the beginning of this chapter, prior to the emergence of life the elements which are the principal constituents of living organisms were probably present on the primitive earth in their reduced forms. Molecular oxygen was absent from the atmosphere. It is now generally believed that the emergence of living systems was preceded by a long period of chemical organic synthesis involving reactions between the reduced components of the atmosphere catalyzed by ultraviolet light and lightning. The products of these reactions accumulated and underwent further reactions in the primitive oceans. These reactions have been reproduced in the laboratory. If the presumed components of the primitive earth's atmosphere (methane, ammonia, and water) are exposed to irradiation by

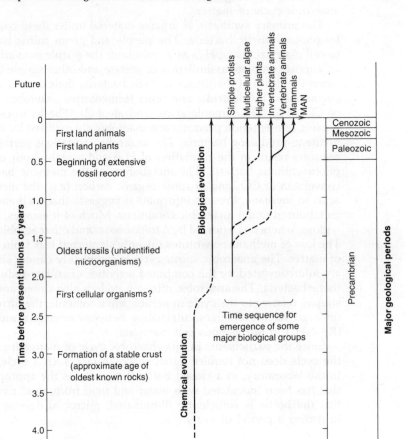

Future

0

0.5 — First land animals
First land plants
Beginning of extensive
fossil record

1.0

1.5

Oldest fossils (unidentified
microorganisms)

2.0

First cellular organisms?

2.5

3.0 — Formation of a stable crust
(approximate age of
oldest known rocks)

3.5

4.0

Formation of planet earth

Time before present billions of years

Chemical evolution

Biological evolution

Simple protists
Multicellular algae
Higher plants
Invertebrate animals
Vertebrate animals
Mammals
MAN

Cenozoic
Mesozoic
Paleozoic

Precambrian

Major geological periods

Time sequence for
emergence of some
major biological groups

FIGURE 25.4

The time scale of terres-
trial evolution.

ultraviolet light or to electrical discharge, a vast array of organic compounds are formed. Among the organic compounds are the constituents of living matter, including sugars, amino acids, purines, and pyrimidines. If phosphate is present in the mixture, nucleotides (including ATP) are formed. If H_2S is present, sulfur-containing amino acids are formed. Under the reducing conditions that prevailed and in the absence of living organisms, these compounds would have accumulated on the primitive earth.

The formation of ever more complex organic compounds and molecular aggregates as a result of chemical interactions eventually resulted in the development of self-duplicating systems, and biological evolution began its course. The first living systems probably had very limited synthetic abilities and depended on fermentative reactions for the generation of energy. With their growth and expansion, the preexisting store of organic raw materials gradually became depleted, favoring the emergence of organisms with an increasing degree of synthetic ability. At a relatively early stage in this primary biochemical evolution, the supply of energy-rich organic compounds must have become limiting, and the further course of biological evolution therefore depended on the acquisition by some organisms of the ability to use light as an energy source. The development of a mechanism for the performance of photosynthesis was therefore one of the earliest and most important steps in biochemical evolution. The first photosynthetic organisms must have been anaerobes, with modes of photosynthetic metabolism analogous to those of the contemporary purple and green bacteria.

The early evolution of photosynthetic organisms culminated in the emergence of forms such as the blue-green bacteria that were able to use water as a reductant for the photosynthetic assimilation of CO_2. Once this point had been attained, the oxidized product of water, molecular oxygen, began to accumulate in the atmosphere, creating the conditions necessary for the evolution of organisms that obtain energy by aerobic respiration. As a consequence of the presence of molecular oxygen, the oxidized forms of nitrogen and sulfur (nitrate and sulfate) became predominant in the biosphere, and the stage was at last set for the establishment of the cycle of matter as we know it today.

THE INFLUENCE OF MAN ON THE CYCLE OF MATTER

The emergence of man as a member of the biological community did not at first significantly affect the cycle of matter on earth. However, the rapid increases in the total size and local density of human populations that have occurred since the Industrial Revolution, coupled with the ever-increasing power of the human species to modify its environment, have begun to change the picture. Within the past century these factors have led to local environmental changes comparable in scale to those produced by major geological upheavals in the past history of the earth. The spread of agriculture, the denudation of forests, the mining and burning of fossil fuels, and the pollution of the environment with human and industrial wastes have profoundly affected the distribution and growth of other forms of life.

• Sewage treatment As a result of the concentration of the human population in large cities, which has proceeded at an ever-increasing pace during the past 150 years, the disposal of organic wastes, both domestic and industrial, has become a major ecological problem. The discharge of untreated urban wastes into adjoining rivers and lakes presents two hazards: the contamination of drinking water by microbial agents of enteric diseases and the depletion of the dissolved oxygen supply as a result of the microbial decomposition of organic matter, leading to the destruction of animal life. For reasons of public health and of conservation, man has been forced to develop methods of *sewage treatment,* which result in the mineralization of the organic components of sewage prior to its discharge into the natural environment.

In a typical sewage treatment plant, the sewage is first allowed to settle. The precipitate, or *sludge,* undergoes a slow anaerobic decomposition, in which methane bacteria play an important part. The soluble organic compounds in the supernatant liquid are mineralized under aerobic conditions. This is sometimes achieved by spraying the liquid on a bed of loosely packed rocks, over which it trickles by gravity. In the *activated sludge process,* air is forced through the sewage, and a floc or precipitate is formed, the particles of which contain actively oxidizing microorganisms. After the organic matter has been largely oxidized, the sludge is allowed to settle. The sludges produced both by anaerobic decomposition and in the activated sludge process consist largely of bacteria, which have grown at the expense of nutrients in the sewage. These residues are eventually dried and used as fertilizer, either directly or after being ashed.

In all processes of sewage treatment the goal is to produce a final liquid effluent in which the biologically important elements have been restored to the inorganic state. Sewage treatment accordingly involves intensive operation of a substantial segment of the cycles of matter under more or less controlled conditions. Even an effluent in which mineralization is complete may, however, produce undesirable ecological effects. If it is discharged into a lake, the consequent enrichment of lake water with nitrates and phosphates can cause an enormous increase in algal productivity, so that the water becomes colored and turbid at certain times of the year. If algal growth is sufficiently massive, the subsequent decomposition of algal organic matter may deplete the dissolved oxygen supply in the lake, with catastrophic effects on its animal life. Such progressive biological degradation of freshwater environments, first encountered in relatively small lakes near urban communities, has now become serious in some of the Great Lakes, particularly Lake Erie.

The threat to inland waters presented by the discharge of mineralized sewage, along with the realization that water containing nitrate ions is dangerous to human health, has prompted the U.S. government to issue new water quality standards. Drinking water must contain not more than 10 mg of nitrate per liter.

• The dissemination of synthetic organic chemicals

In recent decades the chemical industry has produced an ever-increasing variety of synthetic organic chemicals that are being used on an ever-increasing scale as textiles, plastics, detergents, insecticides, herbicides, and fungicides. Some of these, such as textiles and plastics, are virtually completely resistant to microbial decomposition; in nature they tend to remain as permanent unsightly litter. The use of insecticides, herbicides, and fungicides requires their distribution in nature. Some of them are also remarkably resistant to microbial decomposition. Representative examples of their persistence in soil are shown in Table 25.2. In other environments, such as more anaerobic ones, for example, the bottoms of lakes, they persist for even longer periods.

Many of these compounds are toxic for forms of life other than those they are designed to control, and the long-term ecological effects of their dissemination are difficult (if not impossible) to predict, but it is already clear that their accumulation in nature presents very real hazards to many species.

It is now recognized as desirable that any synthetic organic compound widely disseminated in the natural environment should be susceptible to microbial decomposition. During the 1950s alkylbenzene sulfonates (Figure 25.5) became major ingredients of household detergents. The side chains (R) of the compounds were branched aliphatic residues rendering the entire molecule remarkably refractory to microbial decomposition. Accordingly, they passed through sewage treatment plants largely unaltered, and subsequently contaminated supplies of potable water causing it to foam. During the early 1960s the manufacturing process was altered in order to synthesize benzene sulfates with linear aliphatic side chains

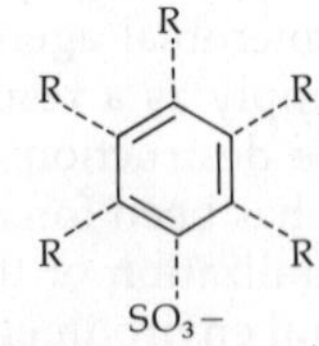

FIGURE 25.5

Generalized structure of alkylbenzene sulfonates. The alkyl R group (either branched or linear) can be at any of the positions indicated on the dotted bonds.

TABLE 25.2

Persistence of pesticides in soil

COMMON NAME	CHEMICAL FORMULA	PERIOD OF PERSISTENCE
INSECTICIDES		
Aldrin	1,2,3,4,10,10-hexachloro-1,4,4a,5,8,8a hexahydro-endo-1,4-exo-5,8-dimethano-napthalene	>9 years
Chlordane	1,2,4,5,6,7,8,8a octachloro 2,3,3a,4,7,7a-hexahydro-4,7-methanoindene	>12 years
DDT	2,2-*bis* (*p*-chlorophenyl)-1,1,1-trichlorethane	10 years
HCH	1,2,3,4,5,6-hexachloro-cyclohexane	>11 years
HERBICIDES		
Monuron	3-(*p*-chlorophenyl)-1,1 dimethylurea	3 years
Simazene	2-chloro-4,6-*bis*(ethylamino)-s-triazine	2 years
FUNGICIDES		
PCP	Pentachlorophenol	>5 years
Zineb	Zinc ethylene-1,2-*bis*-dithiocarbamate	>75 days

(linear alkylbenzene sulfonates); since these are highly susceptible to microbial decomposition, the problem was largely solved.

FURTHER READING

Books ALEXANDER, M., *Microbial Ecology.* New York: Wiley, 1971.

BERNAL, J. D., *The Origin of Life.* New York: World, 1967.

BROCK, T. D., *Principles of Microbial Ecology.* Englewood Cliffs, N. J.: Prentice-Hall, 1966.

STARR, M. P. (editor), *Global Impacts of Applied Microbiology.* New York: Wiley, 1964.

WOOD, E. J. F., *Microbiology of Oceans and Estuaries.* Amsterdam: Elsevier, 1967.

Reviews ALEXANDER, M., "Biodegradation Problems of Molecular Recalcitrance and Microbial Fallibility," *Adv. Appl. Microbiol.* **7**, 35 (1965).

GLEDHILL, W. E., "Linear Alkylbenzene Sulfonates: Biodegradation and Aquatic Interactions," *Adv. Appl. Microbiol.* **17**, 265 (1974).

HOROWITZ, N. H., and J. S. HUBBARD, "The Origin of Life," *Ann. Rev. Genetics* **8**, 393 (1974).

KELLOGG, W. W., R. D. CADIE, E. R. ALLEN, A. L. LAZRUS, and E. A. MARTELL, "The Sulfur Cycle," *Science* **175,** 587 (1972).

PORGES, N., "Newer Aspects of Water Waste Treatment," *Adv. Appl. Microbiol.* **2,** 1 (1960).

POSTGATE, J. R., "New Advances and Future Potential in Biological Nitrogen Fixation," *J. Appl. Bact.* **37,** 185 (1974).

WUHRMANN, K., "Microbial Aspects of Water Pollution Control," *Adv. Appl. Microbiol.* **6,** 119 (1964).

26

SYMBIOSIS

Each group of organisms has had to adapt itself during its evolution not only to the nonliving environment, but also to the other organisms by which it is surrounded. Adaptation to the environment sometimes involves the acquisition of special metabolic capacities which endow their possessor with the unique ability to occupy a particular physicochemical niche. The nitrifying bacteria, for example, can grow in a strictly inorganic environment with ammonia or nitrite as the oxidizable energy source; in the absence of light, no other living organisms are capable of developing in this particular environment, and the nitrifying bacteria are thus freed from biological competition. Withdrawal into a unique physicochemical niche is one means, and a highly effective one, of meeting the challenge of biological competition. A second method, however, which has been adopted by large numbers of microorganisms, has been to meet the challenge *by adapting to existence in continued close association with some other form of life.* This is the biological phenomenon known as *symbiosis.*

TYPES OF SYMBIOSES

The symbiotic associations which microorganisms form with plants and animals, as well as with other microorganisms, vary widely in their degree of intimacy. In terms of the closeness of the association, symbioses may be roughly divided into two categories: *ectosymbioses* and *endosymbioses.* In ectosymbioses the microorganism remains external to the cells of its host;* in endosymbioses the microorganism grows within the cells of its host. The distinction, however, is not always clear-cut; in lichens, for example, the fungal partner forms a projection

*The term *host* refers to the larger of two symbionts.

733

which penetrates the cell wall, but not the cell membrane, of its algal partner (see Figure 27.13).

Symbioses also differ with respect to the relative advantage accruing to each partner. In *mutualistic symbioses* both partners benefit from the association; in *parasitic symbioses* one partner benefits, but the second gains nothing and often suffers more or less severe damage. It is sometimes difficult to determine whether a given symbiosis is mutualistic or parasitic. The degree to which each partner is benefited or harmed can only be evaluated by comparing the fitness of the two members when living independently with their fitness when living in association. Furthermore, the nature of a particular symbiosis can shift under changing environmental conditions, so that a relationship that starts out as mutualistic may become parasitic, or vice versa.

The fact that two organisms have evolved a symbiosis implies that at least one partner derives some advantage from the relationship. The extent to which this partner depends on symbiosis for its existence, however, varies considerably. At one extreme are the microorganisms which populate the *rhizosphere*—the region that includes the surface of the roots, together with the soil immediately surrounding the root hairs, of higher plants. These microorganisms live successfully in other regions of the soil, but they attain higher cell densities in the rhizosphere, where they derive advantages from their proximity to the root hairs. At the other extreme are the obligate parasites, which have never been successfully cultivated outside their hosts.

Thus, symbioses vary with respect to the degree of intimacy (ectosymbiosis vs. endosymbiosis), the balance of advantage (mutualism vs. parasitism), and the extent of dependence (facultative vs. obligate symbiosis).

• **Mutualistic symbioses**

In the following two chapters we shall describe a variety of symbioses in which microorganisms have established mutually advantageous associations with other microorganisms, with plants, and with animals. Here we will confine our discussion to a few examples, to illustrate the varying types of relationships that may properly be regarded as mutualistic.

In mutualistic endosymbioses the microbial symbionts live within the cells of their hosts. In many such associations the microorganism leads a permanently intracellular existence and is passed from one generation of the host to the next in the cytoplasm of the egg. In other associations the microorganism remains intracellular through only a part of the life cycle of the host; at some stage it is liberated into the extracellular environment, from which the next generation of host becomes infected.

Among the mutualistic ectosymbioses we can discern two broad types of associations. First, there are the types in which the microbial symbiont lives on the external surface of its host. Some photosynthetic bacteria, for example, attach themselves to the surfaces of other, nonphotosynthetic bacteria (Figure 26.1) and certain flagellated protozoa which inhabit the termite gut bear on their surface a mantle of spirochetes (Figure 26.2).

Second, many microbial symbionts inhabit the body cavities of their hosts. These associations are still considered as ectosymbioses, because the body cavities, although internal to the whole organism, are external to the tissues and are

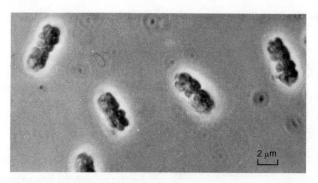

FIGURE 26.1

The ectosymbiosis of green bacteria with a larger, colorless bacterium (the so-called "chlorochromatium" association). Each object is a rod-shaped colorless bacterium coated with the regularly arranged smaller cells of a *Chlorobium* (phase contrast, ×2,190). Courtesy of Norbert Pfennig.

FIGURE 26.2

Spirochetes attached to posterior end of termite flagellate, *Glyptotermes* sp. From H. Kirby, Jr., *Univ. Calif. Publ. Zool.* **45**, 247 (1945).

continuous with the external surfaces of the host. The most familiar examples are the microorganisms that inhabit the digestive tract of mammals; to the same category belong the luminous bacteria that populate the light-emitting organs of some fishes and mollusks (see Figures 26.7 and 26.8).

Of all the ectosymbioses, none is more remarkable than the cultivation of fungi by certain insects, notably the higher termite, *Termes*, and the wood-boring "ambrosia beetles." The termites have special chambers in their nests which contain "fungus gardens," on which the young nymphs and larvae browse. The ambrosia beetles have evolved elaborate methods of infecting their tunnels with fungal spores (Figure 26.3), so that the tunnels become lined with mycelium, on which the beetles feed. These practices closely parallel the cultivation by man of food plants, and in fact all these activities are symbiotic.

The picture of one symbiont literally devouring its partner may at first glance appear to contradict our concept of symbiosis as "existence in continued close association." When one of the partners is a microorganism, however, we are dealing with a *population*, not with an individual. Thus, the cultivated fungus garden of the termite benefits as a population, although a fraction of the population at any given moment is being devoured. In such ectosymbioses, as well as in all endosymbioses, the relationship is one of *reciprocal exploitation,* involving a dynamic balance between offensive and defensive activities of the two members. In the root nodule endosymbiosis, for example, the nodule bacteria infect the plant root and exploit it, the plant ultimately digesting most (but not all) the bacteria. The net result is a superficially peaceful, mutualistic symbiosis.

• Parasitic symbioses

We have defined the parasitic symbioses as those in which one of the partners does not profit from the association and may suffer more or less severe damage. In the case of microbial symbioses with plants and animals, it is usually a simple matter to demonstrate the advantage accruing to the microorganism, but it is often difficult or impossible to evaluate the effect on the host. *Infectious disease,* in which the host is progressively weakened and may eventually die, is obviously a parasitic symbiosis. When however, there is no overt damage to the host, it is difficult to tell whether the relationship is a parasitic or a mutualistic one. For example, the "normal flora" of the mammalian intestinal tract was long assumed to be parasitic, only the microorganisms benefiting from the association. Work with *germ-free animals,* however, has revealed a number of subtle but important benefits which the intestinal flora confers on the host (see page 783).

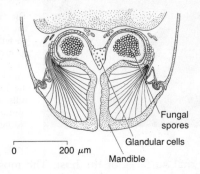

Fungal
spores

Glandular cells

0   200 μm

Mandible

FIGURE 26.3

Section through the head of the ambrosia beetle, *Xyleborus monographus*, showing the special pockets for the storage of fungal spores. After Schedl.

• **Parasitism as an aspect of ecology**

The central problem of ecology is to discover the factors that ultimately determine the survival of a species. Other than the reproductive potential of a species, that is, the number of viable offspring produced per parent per unit of time, the factors that affect survival all fall into one of two categories: those that affect the available food supply and those that affect the rate of destruction of individuals. For any species other than man, who has invented extraordinary ways of destroying himself, the possible fates of an individual are restricted. Animals, other than a few domestic ones, rarely die of old age or from accidental mishaps. Most are either eaten by their natural predators or destroyed by their pathogenic parasites. The distinction between a predator and a pathogenic parasite is a fine one, however, for both satisfy their nutritional needs at the expense of their victim. In the final analysis, the principal difference is that the victim of a predator is killed immediately, whereas the victim of a pathogenic parasite remains alive for an extended period of time.

Since the normal fate of most individuals is to be killed either by predator or by parasite, each species constitutes a link in a biological *food chain*. Microbial cells, for example, are food for many species of plankton, the minute plants and animals that drift in the oceans in great abundance. The plankton serve as the major food source for many marine invertebrates and fishes, which are fed upon in turn by larger fishes and some marine mammals. The largest animals, which seemingly stand at the ends of the food chains, must eventually die and be devoured by microorganisms, so that the food chains are actually *cyclic* in nature.

The fact that every organism is part of a food chain means that—for any heterotroph—the best source of nutrients is another organism. The factors that affect the available food supply of a species are numerous and complex, whether the organism feeds as a predator or as a parasite. For example, an excessively large population may exhaust the food supply to such an extent that the next generation will find an extreme shortage and hence will survive in much smaller numbers. The food supply may also be affected by a change in climate, by long-term geochemical evolution, or by changing competition with other predators or parasites.

The ability of a species to resist being eaten by others depends both on its own defense mechanisms (e.g., protective coloration, armor, ability to fly or burrow, immunity to disease) and on the properties of its predators or parasites.

A new predator or parasite may appear on the scene, for example, or an existing one may evolve more efficient feeding habits.

The living world is thus organized into a large number of intersecting food chains. Each chain consists of a number of species, the populations of which have reached equilibrium in terms of rate of reproduction and destruction. The equilibrium size of a population may shift abruptly if any one of its complex determinants changes. In modern times, the most significant factors affecting this ecological equilibrium have been the activities of man. By building dams, destroying forests, polluting streams, slaughtering game, spraying poisons, and transporting parasites, he has exterminated many species and changed the ecology of many others.

THE FUNCTIONS OF SYMBIOSIS

A symbiont substitutes for part or all of the nonliving environment which free-living organisms occupy; among the myriads of symbioses that have evolved we can find examples in which almost every known environmental function is furnished by one or another symbiont for its partner. For convenience, we shall discuss these functions under four headings: protection, provision of a favorable position, provision of recognition devices, and nutrition.

To determine the functions fulfilled by the partners in a symbiosis it is necessary—in all but the most obvious cases—to separate the partners and study their requirements in isolation. In many cases, this has not yet been achieved, either because the symbionts cannot be separated without damaging them or because the isolated partners cannot be cultivated. Symbionts that have defied attempts to cultivate them in isolation are said to be *obligate symbionts*. The classification of a symbiont as "obligate" is always provisional, since it is always possible that identification of the function performed by its partner will permit its eventual cultivation.

• Protection Endosymbionts, as well as those ectosymbionts which live in the body cavities of animals, are protected from adverse environmental conditions. These habitats protect the symbionts from dessication and—in the case of warm-blooded hosts—from extremes of temperature.

The microbial symbionts of plants and animals also perform functions which protect their hosts. Most notable is the protection which the normal flora of vertebrates offers against invasion by pathogenic (disease-producing) microorganisms; germ-free animals are much more susceptible to infection than their normal counterparts, as we shall describe in Chapter 28. The removal of toxic substances is another function which many microbes perform for their symbiotic partners; in some insects, for example, bacteria harbored in the excretory organs break down uric acid and urea to ammonia, which the bacteria themselves assimilate.

• Provision of a favorable position A symbiotic association may provide one partner with a position that is favorable with respect to the supply of nutrients. Many of the marine ciliated protozoa, for example, are found only on the body surfaces of crustacea, where the host's respiratory and feeding currents assure the microbe of a constant supply of food (Figure 26.4). No less spectacular is the favorable position provided by many marine invertebrates for their photosynthetic algal symbionts. Some of these hosts are phototactic, carrying their photosynthetic partners toward the light. Others, such as the tridacnid clams, house their algal symbionts in special organs which act as lenses to gather light.

The tridacnid clams (family Tridacnidae) have several unique anatomical features, the most prominent of which is the location and thickening of the mantle, the epithelial tissue that lines the shell. Unlike the mantle of all other clams, the mantle of the Tridacnidae is greatly extended along the dorsal, or open, part of the shell, the visceral mass being moved to a more ventral position. The mantle, olive-green in color because of its dense population of algal symbionts, is so thick that it prevents the shell from closing, and its surface is covered with conical protuberances (Figure 26.5).

Sections through the mantle tissue reveal the nature and function of its conical protuberances. Each protuberance contains one or more lenslike structures, the *hyaline organs,* made up of transparent cells. Each hyaline organ is surrounded by a dense mass of algae (Figure 26.6); the function of the lenslike hyaline organ is to permit light to penetrate deeply into the mass of algae.

The tridacnid clams have thus evolved a highly specialized system for cultivating algae within their own tissues. The algae grow extracellularly in haemal channels which lie perpendicular to the exposed surface of the mantle. Senescent and dead algal cells are eventually digested by phagocytic cells of the host.

There is much anatomical evidence to suggest that the tridacnid clams rely heavily on their algae as a source of food. The digestive system is reduced, for example, and the feeding organs are so altered that they screen out all but the most minute particles. Finally, the kidneys are vastly increased in size, presumably to handle the excretion of products formed in the phagocytes by digestion of the algae. The tridacnids thus represent an extreme example of an evolutionary response to symbiosis.

• **Provision of recognition devices**

Bioluminescence is widespread in the animal kingdom, occurring in such diverse groups as jellyfish, earthworms, fireflies, squid, and fish. The emission of light by these animals very often appears to be a recognition device, promoting schooling, mating, or the attraction of prey. In most cases the luminescence is produced by the tissues of the animal itself, but in some species of squid, and in certain fishes, it is produced by luminous bacteria living ectosymbiotically in special glands of the host.

Among the squids (mollusks belonging to the class Cephalopoda), symbiotic luminous bacteria have been identified in a number of species of one suborder, Myopsida. The myopsid squids, also called cuttlefishes, are characterized by their strongly calcified shell. Figure 26.7 shows a male of the genus *Euprymna,* the light organs of which are quite typical of the myopsid squids. The luminous glands are embedded in the ink sac and are partially enclosed in a layer of reflective tissue; just above the glands are lenses made up of hyaline cells which transmit light. In some squids the animal can control the emission of light by a muscular contraction which squeezes the ink sac, pushing it between the light source and

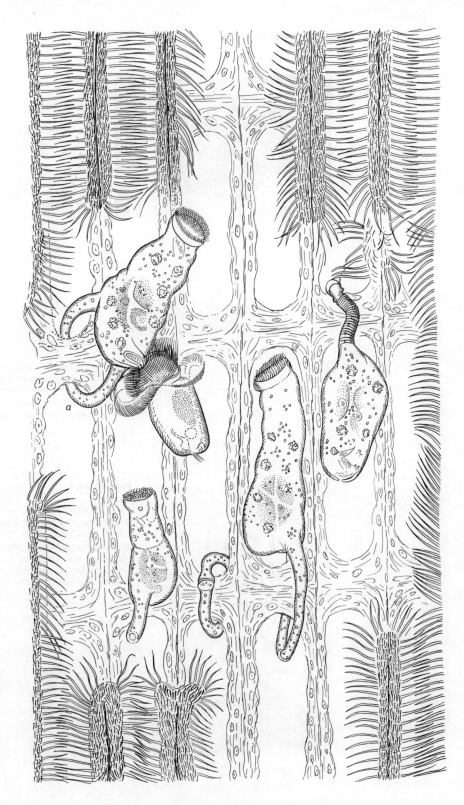

FIGURE 26.4

The ciliate protozoon, *Ellobiophyra donacis*, "padlocked" to the gills of the bivalve, *Donax vittatus*. In (a) *Ellobiophyra* is seen in the process of reproduction by budding. From E. Chatton and A. Lwoff, "*Ellobiophyra donacis* Ch. et Lw., peritriche vivant sur les branchies de l'acephale *Donax vittatus* da Costa*," *Bull. Biol. Belg.* **63**, 321 (1929).

FIGURE 26.5

Underwater photograph of *Tridacna maxima,* showing characteristic exposure of mantle to sunlight. From P. V. Fankboner, "Intracellular Digestion of Symbiotic Zooxanthellae by Host Amoebocytes in Giant Clams (*Bivalvia: Tridacnidae*), with a note on the Nutritional Role of the Hypertrophied Siphonal Epidermis," *Biol. Bull.* **141,** 222 (1971).

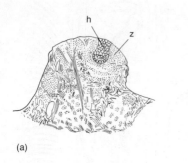

(a)

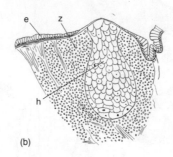

(b)

FIGURE 26.6

Endosymbiotic algae of the clam, *Tridacna crocca:* (a) section through a protuberance on the inner fold of the dorsal mantle edge; (b) enlarged view of a hyaline organ. e, epithelium; h, hyaline organ; z, zooxanthellae. After M. J. Yonge.

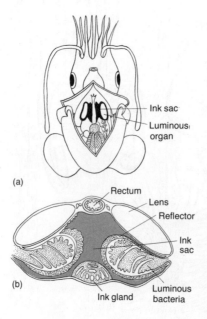

(a)

Ink sac

Luminous organ

(b)

Rectum

Lens

Reflector

Ink sac

Ink gland

Luminous bacteria

FIGURE 26.7

The cuttlefish, *Euprymna morsei.* (a) Male with opened mantle, showing the luminous organs embedded in the ink sac. (b) Cross section through the luminous organs, showing the reflectors, lens, and the open chambers containing the luminous bacteria. After T. Kishitani, "Studien über Leuchtsymbiose von Japanischen Sepien," *Folia Anat. Japan.* **10,** 315 (1932).

741   the lens. Note in Figure 26.7 that the luminous organs are open to the exterior; all evidence suggests that young animals must be infected externally and that intracellular transmission via the egg does not occur.

A large number of unrelated species of fish also possess light organs which consist of open glands containing luminous bacteria. In a few species the organ is provided with a reflecting layer of tissue. The most complex organs are found in the two closely related genera, *Photoblepharon* and *Anomalops*, both of which harbor luminous bacteria in special pouches under the eyes (Figure 26.8). What makes these forms particularly spectacular is their ability to control the emission of light; in *Photoblepharon*, this is accomplished by drawing up a fold of black tissue over the pouch like an eyelid, while in *Anomalops* the light organ itself can be rotated downward against a pocket of black tissue.

The complexity of the organs that have been evolved to control the symbiotic light emission implies that luminescence has great adaptive value for the host; in most cases, it is believed to serve as a recognition device. The functions performed by the host on behalf of the luminous bacteria are undoubtedly those of providing nutrients and protection.

• Nutrition   By far the most common function of symbionts is to provide nutrients for their partners. The provision of nutrients may be indirect, as in the case of fungi that

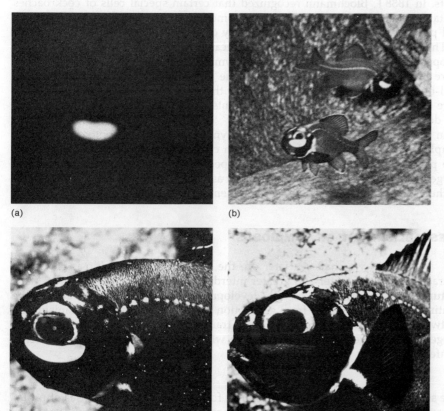

(a)

(b)

(c)

(d)

FIGURE 26.8

The flashlight fish, *Photoblepharon palpebratus*, photographed at night along the reefs in the Gulf of Eilat, Israel. (a) *Photoblepharon* as it appears at night on the reef photographed by the light emission from its own luminescent organ. (b) A pair of *Photoblepharon* in their intertidal territory. (c) Close up of *Photoblepharon* with the lid of the luminescent organ open. (d) Same fish with the lid closed. Parts (b), (c), and (d) are photographs taken with an underwater strobe light. The reflective areas of the lateral line, edges of the fin rays, and operculum are not luminescent. Courtesy of James G. Morin *et al.*, "Light for All Reasons: Versatility in the Behavioral Repertoire of the Flashlight Fish," *Science* **190**, 74–75 (1975).

infect plant roots and thereby increase the water-absorbing capacity of the root system. Usually, however, the nutritional support is direct, the symbiont furnishing one or more essential nutrients to its partner.

The most dramatic and extensively studied example is *nitrogen fixation*. As previously discussed (Chapter 25), the capacity to fix nitrogen is exclusively procaryotic, but many groups of eucaryotes—both plants and animals—have entered into symbiotic associations with nitrogen-fixing bacteria. As an example, nitrogen fixation by root-nodule bacteria will be discussed in some detail in Chapter 27.

Cellulose, as a major plant constituent, provides the principal carbon and energy source for grazing animals as well as for wood-boring insects. Some of these animals are incapable of digesting cellulose; in the ruminants and in at least one group of insects, the termites and wood-eating roaches, cellulose digestion is performed on behalf of the host by symbiotic bacteria and protozoa. The digestion of other complex carbohydrates is often carried out by microbial symbionts living in the digestive tracts of animals.

One of the most intriguing detective stories in biology has been the discovery and elucidation of the endosymbioses between microorganisms and their insect hosts. In 1888 F. Blochmann recognized that certain special cells of cockroaches contain symbiotic bacteria, and soon entomologists discovered bacterial, fungal, and protozoan endosymbionts in a variety of other insects.

The significance of these symbioses became clear when techniques were developed for ridding the insects of their symbionts; such animals require one or more B vitamins to develop normally. The importance of the symbionts to the well-being of their hosts is emphasized by the elaborate mechanisms that insects have evolved for transmitting the symbionts to their young. These mechanisms are described below.

In Chapter 27 we shall describe some symbioses between photosynthetic and nonphotosynthetic partners. In many such cases, the metabolic functions of the two partners are complementary with respect to the metabolism of carbon and oxygen; in effect, therefore, these symbiotic associations carry out a complete cycle of the two elements, according to the scheme described in Chapter 25.

THE ESTABLISHMENT OF SYMBIOSES

As we shall see later on, the evolution of a symbiosis is usually characterized by a greater and greater interdependence of the two partners. This in turn places a premium on the development of mechanisms to ensure the continuity of the symbiosis from generation to generation. Such mechanisms are of two kinds: those in which the host transmits its symbionts directly to its progeny at each generation and those in which each new generation is freshly reinfected.

• **Direct transmission**   The simplest type of direct transmission is found in the endosymbioses of protozoa with algae. The protozoan and its intracellular algal symbiont divide at more or less the same rate, so that each daughter cell of the host receives a proportionate share of algal cells. In some instances cell division is precisely regulated: the host cell, containing two algal symbionts, divides to yield two daughter cells, each containing one symbiont. The symbiont then divides, restoring the number per cell to two (see Figure 27.11).

In the sexually reproducing animals, direct transmission may be accomplished by infection of the egg cytoplasm. This may require an extremely elaborate sequence of host cell movements, morphological changes, and interactions. In certain insects, for example, the microbial symbionts are contained within specialized cells, the *mycetocytes,* which make up organs called *mycetomes.* The mechanism for transferring the symbionts from the mycetome to the egg varies from family to family and may be very complex. A common sequence of events is the following. The symbionts are liberated from the mycetocytes and—being nonmotile in all cases—are passively transported to the ovary by way of the lymph. At the ovary, the symbionts are taken up by special epithelial cells and from these are ultimately transferred to the egg cells.

In other cases, the germ cells become infected early in the embryonic development of the insect. If the insect matures as a male, the germ cells become testes and the symbionts disintegrate. If the insect matures as a female, however, the germ cells become ovaries and the symbionts multiply, so that each egg contains large numbers of them.

The cycle is completed during embryogenesis of the progeny, when events occur which lead to the formation of the symbiont-filled mycetome. The way in which mycetomes are formed varies from family to family, being particularly complicated in those insects which carry several different symbionts, each of which must eventually be housed in its own special type of mycetocyte or mycetome. In every case, the development of the mycetome involves a process of differentiation comparable to that which leads to the formation of any other animal organ. Differentiation is initiated by a series of regulated nuclear divisions in the region of the egg that contains the symbiont mass, and culminates in the formation of the mycetome of the adult insect.

• **Reinfection** Again, it is among the insects that we find the most elegant examples of mechanisms designed to ensure infection of the progeny with symbionts from the mother. Each group of insects has evolved its own set of specialized devices: in some insects a direct anatomical connection between the intestine and vagina guarantees that intestinal symbionts will be copiously smeared on the surface of the eggs as they pass through the ovipositor. When the young larvae hatch, they infect themselves immediately by eating part or all of the eggshell.

Two groups of flies are viviparous: *Glossina* (the tsetse flies) and a large group called *Pupipara,* which are themselves ectosymbionts of mammals and birds. The larvae of these flies are retained in the uterus, where they are nourished by the secretions of greatly developed accessory glands, the "milk glands." The symbiotic microorganisms are localized in these glands and are delivered to the larvae during feeding.

Two mechanisms of transmission are particularly intriguing because they involve a stereotyped, genetically determined *behavior pattern* of the newly hatched larvae. In one, the hatchlings suck up drops of bacterial suspension which exude from the mother's anus during the period of brood care. In the other, discovered in the plant-juice sucking insect *Coptosoma,* the female deposits a bacterium-filled "cocoon" or capsule between each pair of eggs. When the eggs hatch, each larva

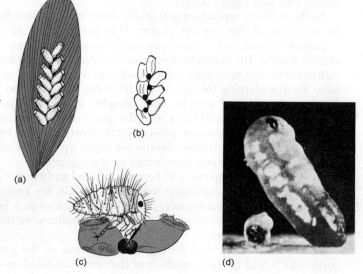

FIGURE 26.9

Coptosoma scutellatum. (a) Eggs deposited on a vetch leaf. (b) The eggs seen from below, showing the symbiont-filled cocoons lying between each pair of eggs. (c) A newly hatched larva sucking the symbiont suspension from a cocoon. (d) Enlarged view of egg and cocoon. (After H. J. Müller.)

sinks its proboscis into a cocoon and sucks up a supply of symbionts (Figure 26.9).

Equally complex are the mechanisms that have evolved in the plant kingdom for the initiation of root nodules when the bacterium, *Rhizobium*, infects its leguminous host. The plant root excretes a number of substances, among which is tryptophan. The bacterial cells in the soil convert tryptophan to the plant growth hormone, indoleacetic acid, and also produce an extracellular polysaccharide capsular material which induces the plant root to excrete the enzyme polygalacturonase. The bacteria then commence penetration of those root hairs which have been induced to grow abnormally by the indoleacetic acid; it is possible that polygalacturonase plays some role in mediating the penetration. The process of nodulation will be discussed in greater detail in Chapter 27.

In general, the more interdependent the symbiotic partners, the more we can expect to find that evolution has produced means for ensuring their continued association. In contrast, the formation of loose associations often seems to depend entirely on chance, and both partners may also be free-living.

THE EVOLUTION OF SYMBIOSES

Natural selection acts on symbiotic associations as well as on individual organisms; symbioses thus have their own phylogenies. In the absence of fossil evidence, such phylogenies are necessarily speculative, but certain trends can nevertheless be deduced from the nature of contemporary symbioses.

It seems inescapable that symbioses evolve in the direction of increasing intimacy. Starting with a loose association, in which one or both organisms finds an optimal environment in the vicinity of the other, an ectosymbiosis may gradu-

745 ally develop. At a later stage in its evolution the relationship may become endo-symbiotic, the small organism penetrating the host tissues and ultimately the host cells.

Once a symbiosis has been established, selection operates to increase its efficiency. An increased degree of adaptation to one highly specialized environment necessarily implies, however, a decreased degree of adaptation to all other environments. The result is a high degree of specialization; the symbiont not only loses its ability to live freely, but it also becomes increasingly *specific* with respect to its choice of partner. Today we find many extreme cases, particularly in the endosymbioses, where neither partner can grow without the other.

At this point we must ask ourselves whether a distinction can be drawn between a totally interdependent pair of symbionts, such as a protozoon that carries an intracellular algal symbiont, and a eucaryotic cell with its contained organelles. Some years ago a distinction might have been proposed in genetic terms, since the coexistence of protozoon and alga involves two distinct genomes, whereas it was assumed that all parts of the eucaryotic cell are controlled by a single genome. Today this distinction can no longer be made, because it is now known that such organelles as mitochondria and chloroplasts contain DNA.

Indeed, as discussed in Chapter 3, it is entirely conceivable that the eucaryotic cell originally arose as an endosymbiosis between two (or more) primitive cell types. The similarities between chloroplasts and endosymbiotic blue-green bacteria, for example, are numerous (Figure 26.10), as are the similarities between mitochondria and endosymbiotic bacteria.

If cell organelles that possess DNA are to be considered as obligate symbionts for the purposes of this discussion, then what can be said of the *plasmids* which

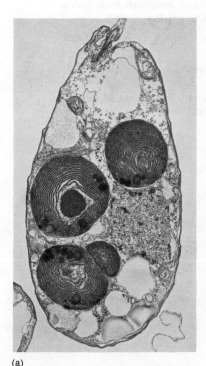 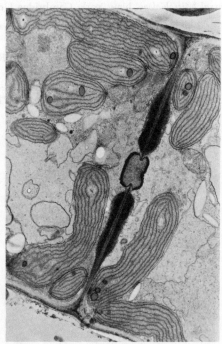

(a)   (b)

FIGURE 26.10

(a) Electron micrograph of a thin section of the flagellated protozoon, *Cyanophora paradoxa*, containing several endosymbiotic blue-green bacteria (×4,370). Courtesy of William T. Hall. (b) Electron micrograph of a thin section of *Pseudogloiophloea confusa*, a red alga, containing several chloroplasts. (The dark structure in the center is a cross-septum of a multicellular filament, ×7,200.) Courtesy of J. Ramus.

were described in Chapter 15? For that matter, what of the *viruses*, which also behave as intracellular foreign genetic elements? All these elements could be classed as obligate symbionts, usually parasitic but under some circumstances entering into mutualistic associations with their host cells. Whether we so classify them depends entirely on whether or not we wish to restrict our definition of symbiosis to relationships between cells (or cell-like organelles). Plasmids and viruses in the intracellular state are neither cellular nor cell-like; they consist of genetic material, either DNA or RNA, and are analogous to the procaryotic chromosome. Although it is obviously true that living systems involve interdependencies at all levels—molecular, organellar, and cellular—it is probably more useful to restrict the term *symbiosis* to *interdependent systems in which the two partners are at least theoretically capable of autonomous existence.*

FURTHER READING

Books   BURNET, F. M., and D. O. WHITE, *Natural History of Infectious Disease*, 4th ed. New York: Cambridge University Press, 1972.

CHENG, T. C. (editor), *Aspects of the Biology of Symbiosis*. Baltimore: University Park Press, 1971.

HENRY, S. M. (editor), *Symbiosis*, Vols. I and II. New York: Academic Press, 1966, 1967.

NUTMAN, P. S., and B. MOSSE, *Symbiotic Associations: 13th Symposium of the Society for General Microbiology*. New York: Cambridge University Press, 1963.

Symbiosis: 29th Symposium of the Society for Experimental Biology. New York: Cambridge University Press, 1975.

TRAGER, W., *Symbiosis*. New York: Van Nostrand Reinhold, 1970.

Review article   PREER, J. R., "Extrachromosomal Inheritance: Hereditary Symbionts, Mitochondria, Chloroplasts," *Ann. Rev. Genetics* **5**, 361 (1971).

SYMBIOTIC ASSOCIATIONS BETWEEN PHOTOSYNTHETIC AND NONPHOTOSYNTHETIC PARTNERS

27

In a great number of mutualistic symbioses, one of the partners is a photosynthetic organism. The function served by the nonphotosynthetic partner varies widely: in some cases, such as the nitrogen-fixing root-nodule bacteria, it is the provision of nutrients; in other cases, such as the fungal partners of lichens, it appears to be protection; and in still other cases, such as the tridacnid clams which house algae, it appears to be the provision of a favorable position.

The contribution of the photosynthetic partner, however, is always the provision of nutrients, i.e., the carbohydrates formed by the fixation of carbon dioxide. The movement of carbohydrate from one symbiont to the other has been studied by allowing ^{14}C-labeled CO_2 to be assimilated in the light, following with time the incorporation of label into metabolites of each partner. In symbiotic associations of algae with invertebrate animals, as well as in associations of algae with fungi (lichens), isotope studies have revealed a number of important adaptations that facilitate the *unidirectional transport of carbohydrate* from the photosynthetic to the nonphotosynthetic partner. For example, the symbiotic algae excrete a much greater proportion of their fixed carbon than do related free-living algae; in many cases this excretion ceases soon after the symbiotic alga is isolated, indicating a specific stimulatory effect by the nonphotosynthetic partner.

The excreted carbohydrate is usually different from the major intracellular carbohydrates of the alga: in most cases it is a carbohydrate which the nonphotosynthetic partner, but not the alga itself, can utilize. For example, the green algae of lichens excrete polyols, such as ribitol, which are not metabolizable by the algae themselves but which are rapidly utilized by the fungal components of the lichens (Table 27.1). This phenomenon explains the unidirectional flow of excreted carbohydrate: the utilization of the excreted material by the nonphotosynthetic partner creates a *concentration gradient*, such that carbohydrate must flow steadily from the alga to its partner. In some cases, the excreted carbohydrate is one that

TABLE 27.1

Carbohydrate movement from photosynthetic to nonphotosynthetic symbiont[a]

PHOTOSYNTHETIC DONOR			NONPHOTOSYNTHETIC RECIPIENT	
ORGANISM	CARBOHYDRATE RELEASED		IMMEDIATE FATE OF CARBOHYDRATE	ORGANISM
Zoochlorellae	Maltose, glucose	$\longrightarrow$	Glycogen, pentoses	Marine invertebrates
Zooxanthellae	Glycerol	$\longrightarrow$	Lipids, proteins	Marine invertebrates
Lichen algae				
Chlorophyceae	Polyols	$\longrightarrow$	Polyols	Lichen fungi
Cyanophyceae	Glucose	$\longrightarrow$	Mannitol	Lichen fungi
Higher plants	Sucrose	$\longrightarrow$	Trehalose, glycogen, polyols	Mycorrhizal fungi

[a] Modified, with permission, from Table 8 in D. Smith, L. Muscatine, and D. Lewis, "Carbohydrate Movement from Autotrophs to Heterotrophs in Parasitic and Mutualistic Symbioses," *Biol. Rev.* **44**, 17 (1969).

may be metabolized by the alga also; in such cases, the unidirectional flow is maintained by the rapid conversion of the carbohydrate in the fungus to a form that only the latter can utilize.

In *mycorrhizas* (associations between fungi and the roots of higher plants), carbohydrate is again found to move from the photosynthetic to the nonphotosynthetic partner. Here the transported carbohydrate appears to be sucrose, which is the form in which carbohydrate is also translocated within the plant. Movement to the fungus thus represents a *diversion of the translocation stream;* in part, this can be accounted for by the rapid conversion of sucrose to fungal carbohydrates such as trehalose and polyols, which the plant cannot utilize. It is possible, however, that the diversion is brought about through the release of plant hormones, many of which are known to be produced by fungi.

SYMBIOSES IN WHICH THE PHOTOSYNTHETIC PARTNER IS A HIGHER PLANT

Microorganisms are found in a number of different symbiotic associations with higher plants. As ectosymbionts, they inhabit the surfaces of leaves (the *phyllosphere*), as well as the soil immediately surrounding the roots (the *rhizosphere*). As endosymbionts, fungi invade the roots to form the associations known as *mycorrhizas*, and certain bacteria invade the roots to form *nitrogen-fixing nodules*.

• **The rhizosphere** The regions of the soil immediately surrounding the roots of a plant, together with the root surfaces, constitute that plant's *rhizosphere*. Operationally, it can be defined as the region, extending a few millimeters from the surface of each root, in which the microbial population of the soil is influenced by the chemical activities of the plant. The major effect observed is a quantitative one: the numbers of bacteria in the rhizosphere usually exceed the numbers in the neighboring soil by a factor of 10 and often by a factor of several hundred.

There is also a qualitative effect. Short Gram-negative rods predominate in the rhizosphere, while Gram-positive rods and coccoid forms are less numerous in the rhizosphere than elsewhere in the soil. No specific association of a particular bacterial species with a particular plant has, however, been established.

The reason for the relative abundance of bacteria in the rhizosphere must certainly be the excretion by plant roots of organic nutrients, which selectively favor certain nutritional types of bacteria. However, no clear-cut nutritional relationships have been discovered, although many organic products excreted by plant roots have been identified. Our state of knowledge concerning the effects of the microbial population of the rhizosphere on the plant is even less satisfactory; despite numerous claims, it remains to be established that the plant benefits from the association. Many free-living soil bacteria, however, perform functions essential for plants, such as nitrogen fixation and the mineralization of organic compounds, so it seems reasonable to assume that some plants do profit from the proximity of some microorganisms.

• **Mycorrhizas**

The roots of most higher plants are infected by fungi. As in so many symbioses, a dynamic condition of mutual exploitation results, both partners benefiting as long as a balance between invasive and defensive forces is maintained. As a result of the infection, the plant root is structurally modified in a characteristic way. The composite root-fungus structure is called a *mycorrhiza*.

The formation of a mycorrhiza begins with the invasion of the plant root by a soil fungus; growth of the fungus toward the root is stimulated by the excretion into the soil of certain organic compounds by the plant. The fungal mycelium penetrates the root cells by means of projections called *haustoria*, and develops intracellularly. In some mycorrhizas the fungus forms intracellular branching structures called *arbuscules* (Figure 27.1); in others it forms characteristic coils.

Depending on the host, the fungus either maintains its intracellular state or undergoes digestion. In the latter case the fungal mycelium persists mainly in the form of *intercellular* hyphae. In all mycorrhizas, however, a large fraction of the mycelium remains in the coil, the intercellular forms tending to produce a compact sheath around the root.

With few exceptions, mycorrhizas are not species specific. A given fungus may be associated with any of several plant hosts, and in most cases a given plant

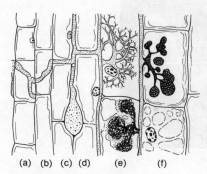

(a) (b) (c) (d)   (e)   (f)

FIGURE 27.1

Drawing showing the penetration of the root of *Allium* by a mycorrhizal fungus. In the first two cell layers (a and b) the fungal mycelium is intracellular. In the third and fourth layers (c and d) it is intercellular; a vesicular storage organ is shown between these layers. In the fifth and sixth layers (e and f) the fungus has formed intracellular branching structures (arbuscules). In (f) the arbuscules are undergoing digestion by the host cells. From F. H. Meyer, "Mycorrhiza and Other Plant Symbioses," in *Symbiosis*, Vol. I (S. M. Henry, editor). New York: Academic Press, 1966.

may form mycorrhizas with any of a number of soil fungi. One species of pine tree, for example, has been found to associate with any of 40 different fungi. A great many free-living soil fungi are capable of forming mycorrhizas. In an experiment performed with pure cultures of free-living fungi and sterile plant roots, over 70 fungal species were found to form mycorrhizas, and many times that number are undoubtedly capable of doing so in nature.

A typical mycorrhiza is shown in Figure 27.2. The stocky, club-shaped appearance results from several effects of the fungus on the root: cell volumes increase but root elongation is inhibited, and lateral root formation is stimulated by *auxins* (plant growth hormones) produced by the fungi.

The mutualistic nature of the mycorrhiza symbiosis can be readily demonstrated in many cases. The fungi that participate are characteristically those which are unable to use the complex polysaccharides which are the principal carbon sources for microorganisms in forest soils and humus. By invasion of plant roots, these fungi avail themselves of simple carbohydrates such as glucose. In fact, the auxins excreted by the fungi induce a dramatic flow of carbohydrate from the leaves to the roots of the host plant.

The plant also benefits from the association. Many forest trees become stunted and die when deprived of their mycorrhiza. Stunted trees can be restored to health by the introduction of suitable mycorrhizal fungi into the soil. The fungus seems to facilitate the absorption of water and minerals from the soil; the absorbing surface of the plant's root system is increased manyfold by the fungal hyphae. The function of a mycorrhiza as an absorbing organ has been confirmed by comparing the uptake of minerals from the soil by plants with and without mycorrhiza. Pines, for example, absorb two to three times more phosphorus, nitrogen, and potassium when mycorrhiza are present than when they are absent.

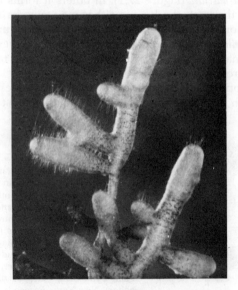

FIGURE 27.2

Mycorrhiza of *Fagus sylvatica*, showing the club-shaped apices of roots and hyphae radiating from the surface. From F. N. Meyer, "Mycorrhiza and Other Plant Symbioses," in *Symbiosis*, Vol. I, S. M Henry (editor). New York: Academic Press, 1966.

• Root nodule bacteria and leguminous plants

It has long been known that the fertility of agricultural land can be maintained by a "rotation of crops." If a given plot of soil is sown year after year with a grass, such as wheat or barley, its productivity begins to decline but can be restored by interrupting this annual cycle with a crop of some leguminous plant such as clover or alfalfa. Roman writers on agriculture recognized that leguminous plants possess this ability to restore or maintain soil fertility which is not shown by other types of plants. It was also known that the leguminous plants have peculiar nodular structures on their roots (Figure 27.3). The plant anatomists of the seventeenth and eighteenth centuries, who examined these nodules in some detail, interpreted them as pathological structures analogous to the galls formed on the shoots of some plants as a result of infestation by insects.

About the middle of the nineteenth century, a new interpretation of the nature of root nodules was offered. At this time, the development of chemical methods enabled scientists to start analyzing the problems of soil fertility and plant growth in chemical terms, and one of the early results of these studies was the elucidation of the role that leguminous plants play in the maintenance of soil fertility. It was found that most plants are limited in their growth by the amount of combined nitrogen in the soil but that leguminous plants are not. Furthermore, by total nitrogen analyses it could be shown that when leguminous plants are grown on nitrogen-poor soil, there is a net increase in the amount of fixed nitrogen in the soil. Since the only possible source of this extra nitrogen is the atmosphere, such experiments demonstrated that leguminous plants, unlike other higher plants, can fix atmospheric nitrogen. Hence, the growth of a crop of legumes on a nitrogen-poor soil results in an increase in the total fixed nitrogen content of the soil, particularly if the crop is plowed under. This is the chemical basis for the long-established practice of crop rotation.

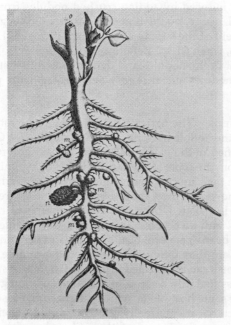

FIGURE 27.3

A seventeenth-century drawing by Malpighi of the root of a leguminous plant, showing the root nodules (m). The large dark object (n) is the coat of the seed from which the plant has developed.

751

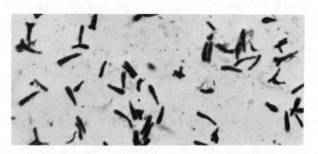

FIGURE 27.4

A stained smear of the contents of a root nodule, showing bacteroids ($\times$1,050). Courtesy of H. G. Thornton and the Rothamsted Experimental Station, United Kingdom.

Once these facts had been established, the question naturally arose as to whether the peculiar nodulations on the roots of leguminous plants had any connection with their ability to fix nitrogen. Occasionally, leguminous plants fail to form nodules, and analyses showed that such plants do not fix nitrogen. When the contents of nodules were examined microscopically, they were found to contain larger numbers of "bacteroids": small, rod-shaped, or branched bodies similar in size and shape to bacteria (Figure 27.4). These facts suggested that the nitrogen-fixing ability of leguminous plants is not a property of the plants as such but results from infection of their roots by bacteria in the soil, such infection leading to the formation of nodules. About 1885 the correctness of this hypothesis was established by showing that if seeds are treated with chemical disinfectants so as to sterilize their surface without impairing their capacity to germinate, and then grown in pots of sterile soil, they will never form nodules. The growth of such plants is strictly limited by the supply of combined nitrogen in the soil. Nodulation can be induced by adding crushed nodules from plants of the same species to the soil. Once nodulation has occurred, the growth of the plants becomes independent of the supply of combined nitrogen (Figure 27.5). The final proof came in 1888, when M. W. Beijerinck succeeded in isolating and cultivating the bacteria present in the nodules and demonstrated that sterile seeds produced the characteristic nodules once more when treated with pure cultures of the isolated bacteria.

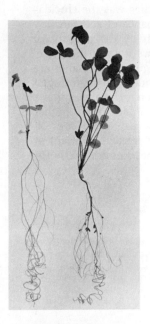

FIGURE 27.5

The effect of nodulation on plant growth. Two red clover plants grown in a medium deficient in combined nitrogen. The one at left, without nodules, shows very poor growth as a result of nitrogen deficiency. The plant at right, with nodules, shows normal growth. Courtesy of H. G. Thornton and the Rothamsted Experimental Station, United Kingdom.

THE NODULE BACTERIA   The agricultural importance of nitrogen fixation led to extensive work on the nodule bacteria. These organisms are Gram-negative motile rods that are classified in the genus *Rhizobium*. It was soon found that the nodule bacteria isolated from the roots of the various kinds of leguminous plants resemble one other closely in their morphological and cultural properties. When inoculated back into plants, however, they show a considerable degree of host specificity. The nodule bacteria isolated from the roots of lupines cannot evoke nodule formation on peas, and vice versa. In contrast, the nodule bacteria from peas, lentils, and broad beans can evoke nodulation in every member of this group of legumes. There are thus differences between the nodule bacteria of peas and those of lupines, which can be detected in terms of their host specificity. The nodule bacteria can be classified into a series of cross-inoculation groups. Strains of any one group have the same host range, which differs from those of the other groups.

The nodule bacteria are normally present in soil. Their numbers are very variable, depending on the nature of the soil and on its previous agricultural treatment. Hence, it not infrequently happens that a leguminous crop will develop poorly in a given plot of soil as a consequence of the fact that the nodule bacteria specific for it are either absent or present in such small numbers that effective nodulation does not occur. Nodulation can be ensured by inoculating the seed with a pure strain of nodule bacteria belonging to the correct cross-inoculation group. Bacterial cultures of proved effectiveness were first made commercially available at the beginning of the twentieth century, and seed inoculation is now a routine agricultural operation. This is by far the most important contribution that the science of soil bacteriology has made to agricultural practice.

Whereas the soil under a nonleguminous crop, such as wheat, may have fewer than 10 *Rhizobium* cells per gram, the same soil will contain between 10^5 and 10^7 *Rhizobium* cells per gram following the development of a flourishing legume crop. The ability of legume plants to stimulate the growth of *Rhizobium* in the soil extends as far as 10 to 20 mm from the roots. The effect is highly specific: bacteria other than *Rhizobium* show little or no stimulation, and growth of the species of *Rhizobium* able to infect that particular leguminous plant is stimulated more than the growth of other species of *Rhizobium*. The substances responsible for this stimulation have not been identified. It has been experimentally established that the high number of *Rhizobium* cells in the rhizosphere of legumes represents stimulation of free-living cells rather than their liberation from nodules, by showing that the increase occurs in the absence of active nodulation.

The number of nodules formed on the roots of the legumes is directly proportional to the density of *Rhizobium* in the soil, up to about 10^4 cells per gram. Above this number, no further increase takes place, and nodule formation may even decline. When the number of *Rhizobium* cells is limited, so that fewer nodules are formed, the size of the nodules is proportionately larger. The result is that the total *volume* of nitrogen-fixing tissue remains fairly constant per acre of leguminous plants.

THE PROCESS OF NODULE FORMATION   In Chapter 26 we summarized what is known about the interactions occurring between the legume root and the free-living bacterium in the soil, interactions which serve to initiate infection. The infection itself begins with the penetration of a root hair by a group of rhizobial cells, and involves the invagination of the root hair membrane. A tube is formed containing bacteria and lined with cellulose produced by the host cell. This tube is called the *infection thread* (Figure 27.6). The infection thread penetrates the cortex of the root, passing through the cortical cells rather than between them.

As the thread passes through a cell, it may branch to produce vesicles that contain bacteria; the walls of the thread and vesicles are continuous with the host cell membrane. The bacteria are finally liberated into the cytoplasm of the host cell; electron micrographs of thin sections of legume roots show that the bacteria are enclosed, either singly or in small groups, in a membranous envelope (Figure 27.7).

Development of the nodule itself is initiated when the infection thread reaches a tetraploid cell of the cortex. This cell, along with neighboring diploid cells, is stimulated to divide repeatedly, forming the young nodule. The rhizobial cells invade only tetraploid plant cells, the uninfected diploid tissue becoming the cortex of the nodules (Figure 27.8). In young nodules the bacteria occur mostly as rods but subsequently acquire irregular shapes, becoming branched, club-shaped, or spherical (the typical bacteroids). At the end of the period of plant growth, the bacteria have often disappeared completely from the nodules; they die, and their cell materials are absorbed by the host plant.

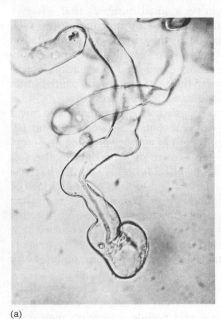

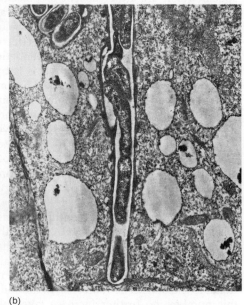

(a)

(b)

FIGURE 27.6

(a) A newly infected root hair. The bacterial infection thread can be seen passing up a root hair, which has curled at the tip as a result of infection. Courtesy of H. G. Thornton and the Rothamsted Experimental Station, United Kingdom. (b) Infection thread crossing a central tissue cell of a nodule aged 1 to 2 days. From D. J. Goodchild and F. J. Bergersen, "Electron Microscopy of the Infection and Subsequent Development of Soybean Nodule Cells," *J. Bacteriol.* **92,** 204 (1966).

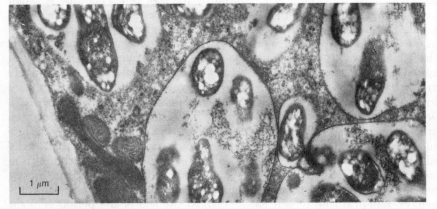

FIGURE 27.7

Mature nodule cell with large membrane envelopes containing four to six bacteroids. No further bacterial growth occurs. From D. J. Goodchild and F. J. Bergersen, "Electron Microscopy of the Infection and Subsequent Development of Soybean Nodule Cells," *J. Bacteriol.* **92,** 204 (1966).

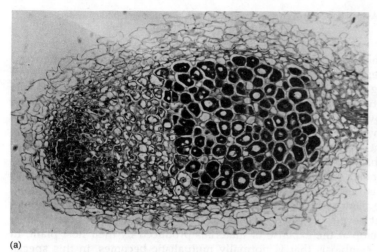
(a)

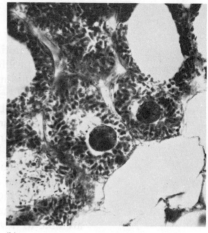
(b)

FIGURE 27.8

(a) Section of a root nodule. The dark cells are filled with bacteria. (b) Section of a nodule at high magnification, showing the individual bacteria in the infected cells. Courtesy of H. G. Thornton and the Rothamsted Experimental Station, United Kingdom.

SYMBIOTIC NITROGEN FIXATION   Until 1975, all attempts to detect significant nitrogen fixation by rhizobial cells isolated and grown in pure culture failed: fixation appeared to require growth within the nodule of the host plant. In 1975, three different laboratories simultaneously reported that several strains of *Rhizobium* would fix nitrogen in pure culture if provided with certain nutrients. The requirements can be met by arabinose, xylose or galactose as carbon source; a dicarboxylic acid such as succinate; and a source of combined nitrogen, such as glutamine or asparagine.

The capacity to fix nitrogen, detected as nitrogenase activity, was found only in cells growing on the surface of solid media; it appears that the conditions necessary for nitrogen-fixation occur only within a dense mass of cells which have previously developed at the expense of combined nitrogen. The necessary conditions may include partial or complete anaerobiosis, since nitrogenase activity in cell-free extracts of nitrogen-fixing bacteria is inhibited by oxygen; the interior of a cell mass growing on agar or in the plant nodule may be sufficiently anaerobic to prevent this inhibition.

It is a curious fact that healthy nodules contain hemoglobin, which neither the plant nor the bacteria can synthesize when grown in isolation. Hemoglobin has two well-known properties: it can be reversibly oxidized and it can bind oxygen avidly. In view of the sensitivity of the nitrogen-fixing activity to inhibition by oxygen, it is possible that the interior of the plant nodule provides the necessary regulation of redox potential for bacteroid activity and that hemoglobin plays some role in this regulation.

THE RELATION BETWEEN MUTUALISM AND PARASITISM IN THE LEGUME SYMBIOSIS   In discussing the formation of nodules we spoke of the "infection" of the root hair by bacteria present in the soil. In many respects, the establishment of the root nodule symbiosis greatly resembles the infection of plants by pathogenic bacteria;

there is an initial destruction of the surface of the root hairs, followed by penetration and proliferation of the bacteria, which then provoke abnormal growth of the surrounding plant tissues. Nevertheless, one does not normally think of the relationship between leguminous plants and root nodule bacteria as a parasitic one, for the obvious reason that the plant gains much from the association. In recent years, however, it has been found that the association does not necessarily benefit the plant. If inoculation studies are conducted with many different strains of leguminous bacteria belonging to one cross-inoculation group, all will give rise to nodulation upon a susceptible plant. With some strains, however, the symbiosis so established is "ineffective" (i.e., it does not permit active nitrogen fixation). As a result, the plant receives little or no benefit from the association, although the bacteria that have established themselves in its roots still profit because they are able to develop at the expense of materials produced by their host plant. In other words, a symbiosis that is normally mutualistic becomes, in this special case, parasitic. It is true that the parasitism is a mild one, since damage to the plant is highly localized in its root system and does not lead to its death. Nevertheless, the situation differs only in degree, not in kind, from the one that occurs when a frankly pathogenic bacterium invades a plant and establishes itself within the plant tissues. This is a good example of the fact, mentioned in Chapter 26, that the dividing line between mutualism and parasitism is a shifting one which can be crossed in either direction.

• **Root nodule bacteria and nonleguminous plants**

The formation of root nodules is not confined to the legumes. There are ten genera of nonleguminous plants, in each of which one or more species is characterized by the presence of root nodules. These genera, of which *Alnus* (the alder) is typical, are listed in Table 27.2. They are not particularly closely related to each other, being distributed among seven families; all but one of these includes nonnodulated genera as well.

TABLE 27.2

Nitrogen-fixing genera of nonleguminous plants[a]

Order: Coriariales	Order: Rhamnales
Family: Coriariaceae	Family: Elaeagnaceae
Genus: *Coriaria*	Genera: *Hippophaë*
Order: Myricales	*Shepherdia*
Family: Myricaceae	*Elaeagnus*
Genus: *Myrica*	Family: Rhamnaceae
Order: Fagales	Genera: *Ceanothus*
Family: Betulaceae	*Discaria*
Genus: *Alnus*	Order: Urticales
Order: Casuarinales	Family: Ulmaceae
Family: Casuarinaceae	Genus: *Trema*
Genus: *Casuarina*	

[a]Taxonomy according to J. Hutchinson, *The Families of Flowering Plants,* 3rd ed. New York: Oxford University Press (Clarendon), 1973.

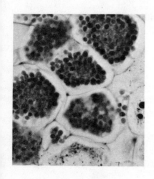

As early as 1896 it was shown that alder plants (genus *Alnus*) can grow well in a medium free of combined nitrogen only if nodulated; by 1962 similar experiments had been done with members of eight of the ten genera listed in Table 27.2. By analogy with the legumes, it was supposed that the nodules were fixing atmospheric nitrogen, and this was eventually confirmed in all cases by the use of ^{15}N. Other parts of the plant do not fix nitrogen.

The fixation process resembles that of legumes in many respects. In both legumes and nonlegumes fixation is inhibited by carbon monoxide and by high levels of oxygen or hydrogen, and is poor when the plants are deficient in cobalt or molybdenum. Finally, hemoglobin has been detected spectroscopically in the nodules of three groups of nonlegumes, *Alnus, Myrica,* and *Casuarina.*

Microscopic examination of stained sections of nonleguminous nodules always reveals structures that suggest symbiotic microorganisms (Figure 27.9). Most workers have interpreted them as actinomycetes, on morphological grounds. Actinomycetes have been isolated from such material, but it is not yet clear whether these isolates represent the nitrogen-fixing symbionts.

Nodulation can be readily induced in most genera by applying suspensions of crushed nodules to the roots; untreated roots show little or no nodulation in control experiments. Whatever the nature of the microbial symbionts, cross-inoculation experiments using crushed nodules show that they possess group specificity. Cross inoculation is often possible between species of a given genus of host plant, but usually not between species of different genera, with the exception of the three genera *Elaeagnus, Hippophaë,* and *Shepherdia,* which belong to one family. Such specificities are reminiscent of the cross-inoculation groups of rhizobia.

The nodule symbiont of one nonleguminous plant, *Trema aspera,* has recently been shown to be a *Rhizobium* species; upon isolation, it can reinfect not only *Trema* but also a number of legumes. The nodules on *T. aspera* resemble legume nodules rather than the actinomycete-carrying nodules of other nonlegumes; *Trema* is the first known case of a *Rhizobium* host which is not a legume.

FIGURE 27.9

Cortical region of a transverse section of a root nodule of *Alnus glutinosa,* showing the dark-stained "vesicles," which are presumed to be the nitrogen-fixing endosymbionts. From G. Bond, "The Root Nodules of Non-leguminous Angiosperms," in *Symbiotic Associations: Thirteenth Symposium of the Society for General Microbiology.* New York: Cambridge University Press, 1963. Section prepared by E. Boyd; micrograph by W. Anderson.

SYMBIOSES IN WHICH THE PHOTOSYNTHETIC PARTNER IS A MICROORGANISM

• Endosymbionts
   of protozoa:
   zoochlorellae and
   zooxanthellae

Many protozoa of the Ciliophora and of the Rhizopoda are hosts to endosymbiotic algae. In freshwater forms the algae are generally green types belonging to the Chlorophyta; in the marine forms the algae are generally yellow or brown types belonging to the dinoflagellates. The two groups of symbionts are called *zoochlorellae* and *zooxanthellae,* respectively.

Figure 27.10 shows zooxanthellae liberated by crushing a foraminiferan protozoon. As found in their hosts, both zoochlorellae and zooxanthellae are invariably coccoid. When cultured free of their hosts, however, zooxanthellae are sometimes observed to form swarming zoospores, which are typical dinoflagellates. Each protozoan cell harbors from 50 to several hundred algae; maintenance of the symbiosis is ensured by similar growth rates of the two partners. Endosymbionts resist digestion by the host. This resistance is undoubtedly related to their location in the host cytoplasm. It should be recalled that microorganisms are taken into the cells of phagotrophic protozoa by phagocytosis and localized inside food vacuoles formed by invaginations of the cell membrane. Their digestion is effected by hydrolytic enzymes liberated into these food vacuoles from lysosomes. The

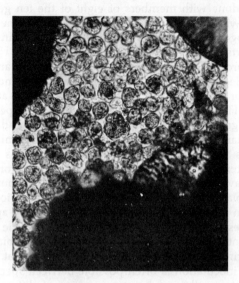

FIGURE 27.10

Zooxanthellae escaping from a crushed forami-niferan. From J. McLaughlin and P. Zahl, "En-dozoic Algae," in *Symbiosis*, Vol. I, S. M. Henry (editor). New York: Academic Press, 1966. Pho-tograph made by J. J. Lee and H. D. Freudenthal.

endosymbionts in the cytoplasm are not contained in food vacuoles, and thus are isolated from the digestive enzymes of the lysosomes.

As we discussed in Chapter 26, symbioses between photosynthetic and non-photosynthetic partners are particularly successful because together the two organisms can carry out a full carbon cycle and a full oxygen cycle. The photo-synthetic partner uses light energy to convert carbon dioxide to organic products, while liberating O_2 from water; the nonphotosynthetic partner uses the O_2 to respire the organic products, producing carbon dioxide as a by-product. This is presumably the basis for the extremely common occurrence of algal-protozoan endosymbioses.

The intimate nature of the relationship is dramatically demonstrated by the fact that many protozoan hosts exhibit *phototaxis* when they harbor a photo-synthetic endosymbiont. In paramecia it has been shown that the alga is the photoreceptor; the movements of the protozoon seem to be controlled by the intracellular concentration of oxygen produced by algal photosynthesis, since phototaxis is exhibited only when the external supply of oxygen is limiting.

• **Endosymbionts of protozoa: cyanellae**

Blue-green bacterial symbionts are called *cyanellae*. They are found in a few genera of freshwater protozoa (e.g., the flagellates *Cyanophora* and *Peliaina*, and the ame-boid rhizopod, *Paulinella*).

Peliaina contains from one to six cyanellae. The symbiosis is maintained by balanced cell division, but this mechanism sometimes fails and a cyanella-free protozoan cell is formed. In *Cyanophora* and *Paulinella*, however, the symbiotic association is perfectly regulated: the protozoan host usually contains two cyanel-lae, each daughter protozoon receiving one cyanella at cell division. The symbiont then divides, restoring the number of two per host.

The cyanellae were discovered as symbionts of protozoa by A. Pascher in 1929.

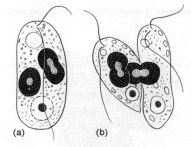

FIGURE 27.11

The protozoon, *Cyanophora*, with its endosymbiotic blue-green bacteria. (a) One of the bacterial cells has begun to divide. (b) At the time of host cell division, each daughter cell receives one symbiont. Compare with Figure 26.10(a). After A. Pascher.

One of Pascher's drawings of *Cyanophora* is shown in Figure 27.11. In recent years *Cyanophora* has been reinvestigated using the techniques of electron microscopy and nucleic acid analysis. Figure 26.10(a) shows a section of *C. paradoxa* as seen in the electron microscope. The endosymbiont has the typical fine structure of a blue-green bacterium, except that it lacks a cell wall, having only a cell membrane as its outer layer. The nucleic acids of the host and symbiont can be separated from each other and can be shown to be typical of a eucaryote and of a procaryote, respectively.

• Symbioses of algae with fungi: the lichens

A lichen is a composite organism, consisting of a specific fungus, usually an ascomycete, living in association with one—or sometimes two—species of algae or blue-green bacteria.* The symbionts form a vegetative body, or *thallus*, of which both the gross structure and the fine structure are characteristic for each lichen "species."

In terms of gross structure, the lichen thalli are divided into three types. The *crustose* lichens adhere closely to their substrate (either rocks or the bark of trees). The *foliose* lichens are leaflike and are more loosely attached to the substrate. The *fruticose* lichens form pendulous strands or upright stalks. Figure 27.12 shows a representative of each type, together with a cross section showing the internal organization of the thallus. The bulk of the thallus is made up of fungal hyphae. In most species these are differentiated into distinct tissues: a closely packed *cortex*, a loosely packed *medulla*, and (in the foliose lichens) attachment regions or *rhizinae*. The algal cells are usually found in a thin layer just below the cortex; in a few species of lichens, however, the fungal hyphae and algal cells are distributed at random throughout the thallus.

Electron micrographs of thin sections show that in most lichens each algal cell is penetrated by one or more fungal haustoria. In some lichens the haustoria penetrate deeply into the algal cells, the membrane of the algal cell invaginating to form a sheath around the haustorium (Figure 27.13). In all cases, the haustoria penetrate only the algal cell wall. The fungi in a few lichens do not have haustoria; instead, there is an intimate contact between the algal and fungal cell walls, which in these species are very thin.

Most lichens propagate by the liberation of *soredia*: small fragments, composed of algal cells and fungal hyphae [Figure 27.14(a)]. In addition, lichens liberate fungal spores [Figure 27.14(b)]. There is considerable evidence, as will be dis-

---

*In discussing the lichen symbioses the general term "alga" will be used to refer to the photosynthetic partner.

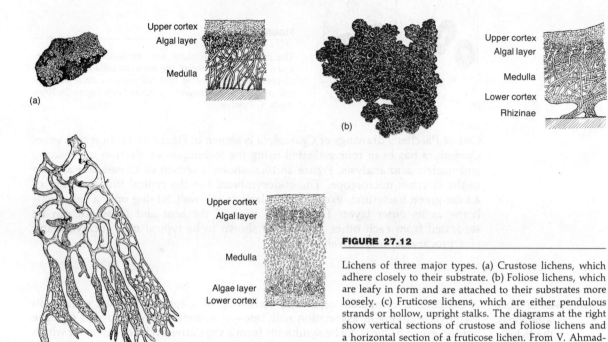

FIGURE 27.12

Lichens of three major types. (a) Crustose lichens, which adhere closely to their substrate. (b) Foliose lichens, which are leafy in form and are attached to their substrates more loosely. (c) Fruticose lichens, which are either pendulous strands or hollow, upright stalks. The diagrams at the right show vertical sections of crustose and foliose lichens and a horizontal section of a fruticose lichen. From V. Ahmadjian, *The Lichen Symbiosis*. Waltham, Mass.: Blaisdell, 1967.

cussed later, that the hyphae produced when these spores germinate make contacts with free-living algal cells and initiate the formation of new lichen thalli.

THE MORPHOLOGICAL CONSEQUENCES OF SYMBIOTIC EXISTENCE   It is relatively easy to separate the symbiotic partners of a lichen and to grow them in pure culture, making possible a comparison of the morphology of each partner as a free-living organism and as a symbiont.

The photosynthetic partner may be severely modified by "lichenization." Certain filamentous blue-green bacteria, for example, fail to form normal filaments when in the thallus; instead, each cell is separated and surrounded by fungal tissue. When isolated from the thallus, the bacterium regains its filamentous growth habit. The green algae found in lichens are so modified that they never produce their characteristic zoospores while part of the lichen thallus.

The fungal partner forms fruiting structures (ascospores and asexual spores) when it is lichenized, but with rare exceptions does not do so when it is isolated and cultivated in the free-living state. In the free-living state it is also incapable of forming a thallus with cortex, medulla, or other tissues. Thus, each partner affects the morphology of the other in a highly specific way.

SPECIES AND SPECIFICITY OF THE SYMBIOTIC PARTNERS   The association of fungus with alga in a lichen is not specific. Thus, a given algal species may be found associated with any one of a variety of lichen fungi, and—conversely—a given fungus may be found associated with any one of a variety of algae. Altogether,

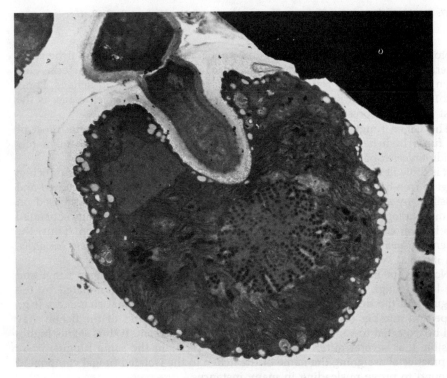

FIGURE 27.13

Electron micrograph of a section through the lichen, *Lecanora rubina,* showing the penetration of an algal cell (*Trebouxia*) by a fungal haustorium. The haustorium has penetrated the outer layer of the cell wall but not the inner layer or the membrane of the algal cell. From J. B. Jacobs and V. Ahmadjian, "The Ultrastructure of Lichens. I. A General Survey," *J. Phycol.* **5**, 227 (1969).

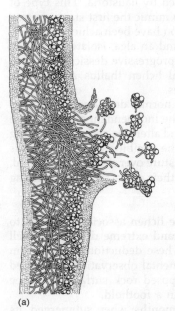
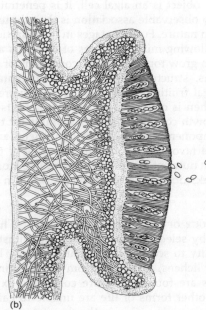

FIGURE 27.14

Lichen reproduction: (a) by the liberation of soredia, made up of fungal threads and hyphae; (b) by the liberation of fungal spores (in this case, ascospores). On germination, a fungal mycelium will develop, which may form a lichen if it comes into contact with an alga cell. From V. Ahmadjian, "The Fungi of Lichens," *Sci. Am.* **208**, 122 (1963).

(a)   (b)

photosynthetic symbionts belonging to 26 genera have been found in lichens: 17 genera of green algae, 1 of yellow-green algae, and 8 of blue-green bacteria. One green alga, *Trebouxia*, is found in more than half of the described lichens. Some 5 to 10 percent of all lichens contain blue-green bacteria, of which *Nostoc* is the most common representative.

The lichen fungi have been classified into several hundred genera; most of the fungi in lichens are Ascomycetes, but a few imperfect fungi and a few Basidiomycetes have also been found.

It is difficult to make any statement about the taxonomic relationships of lichen fungi to free-living Ascomycetes. Historically, the lichens have been studied and classified by specialists who have given them a unique set of names based on the morphology of the composite plant. When a lichen is experimentally separated into its two components, the fungus retains the name of the lichen. Thus, the lichen *Cladonia cristatella* is said to be composed of the fungus, *Cladonia cristatella*, and its algal symbiont, *Trebouxia erici*.

No attempt has been made to integrate the taxonomies of lichen fungi and free-living fungi. Although the morphology of the lichen is dominated by the fungus, and the description of the lichen includes many fungal characters (e.g., shape and number of ascospores), some descriptive features of the lichen may well be expected to vary according to which alga is present. It thus seems highly possible that several different lichen "species" contain the same fungus; if so, the practice of assigning the name of the lichen to the isolated fungal component is bound to prove misleading in many instances.

THE FORMATION AND MAINTENANCE OF THE SYMBIOSIS   Many lichen fungi have been isolated and grown in pure culture. Under conditions of low nutrient supply, the hyphae encircle almost any rounded object of appropriate size which they encounter. If the object is an algal cell, it is penetrated by haustoria. This type of experimentally observable association is thought to mimic the first stages of true lichenization in nature. Further stages in lichenization have been achieved experimentally by allowing mixed cultures of a fungus and an alga, isolated from the same lichen, to grow together under conditions of progressive dessication. After several months, structures typical of the parental lichen thallus are formed, including fungal fruiting bodies and asexual spores.

When a lichen is cultivated, it will continue its normal development only as long as the growth conditions are unfavorable for the independent growth of the individual components. A low supply of nutrients and alternate periods of wetting and drying are favorable for the maintenance of the symbiosis. If the lichen is placed on a rich nutrient medium with adequate moisture over a prolonged period, the union breaks down, and the algae grow out in their characteristic free-living form.

THE PHYSIOLOGY OF THE COMPOSITE ORGANISM   The lichen association seems to have evolved by selection for the ability to withstand extreme drought as well as for the ability to scavenge essential minerals. These deductions follow from the ecology of lichens, as well as from the experimental observations discussed above. Lichens are found in nature colonizing exposed rock surfaces and tree trunks where other forms of life are unable to gain a foothold.

A lichen can remain viable in the dry state for months; when submerged, its water content can change from 2 percent of its dry weight to 300 percent of its dry weight in 30 seconds. Its ability to scavenge minerals is probably related to its production and excretion of *lichen acids,* organic compounds that have the ability both to dissolve minerals and to chelate them. Chelation, the process of binding metal atoms to organic ligands, undoubtedly plays an important role in the solubilization and uptake of minerals by lichens.

More than 100 different lichen acids have been described. Most of them contain two or more phenylcarboxylic acid substituents, with aliphatic side chains. They are not produced by isolated lichen algae, nor—with a very few exceptions—by isolated lichen fungi. A few nonlichen fungi produce similar compounds, however, and it seems likely that the biosynthesis of these compounds in the composite organism is mediated by the fungus.

The production of the lichen acid is thus a creative manifestation of symbiosis. The acids are often excreted in large quantities, crystallizing on the surface of the lichen. In addition to their role as chelating agents, described above, it has been suggested that they inhibit the growth of other microorganisms. Many of the lichen acids do possess strong antibiotic activity, and one of them—usnic acid—is widely used in some European countries as a chemotherapeutic drug for external application.

Lichens grow extremely slowly; an annual increment in radius of 1 mm or less is typical. The range of growth rates is wide, however, and in some species may average 2 to 3 cm/year. Despite their slow growth, lichens form a significant part of the vegetation in some areas; in fact, they are the primary source of fodder for reindeer and caribou in arctic regions.

The ability of lichens to scavenge nutrients at very low concentrations, normally advantageous, becomes injurious to these organisms in regions of industrial air pollution. In such regions the lichen population is greatly reduced or even totally eliminated.

THE SIGNIFICANCE OF THE LICHEN SYMBIOSIS   Both partners of the lichen symbiosis are capable of free-living existence, as shown by the fact that under conditions favorable for the growth of the free-living forms, the symbiosis breaks down. The association is thus of mutual benefit *only in very special ecological situations* (i.e., in environments where nutrients are extremely scarce and where extremes of wetting and drying occur).

The benefit to the fungus of the symbiosis under such conditions is clear: it depends on the alga for its source of organic nutrients. Tracer experiments have confirmed that carbon dioxide fixed by the alga passes rapidly into the fungal mycelium. In those lichens that contain blue-green bacteria as symbionts, the fungus also benefits directly from the atmospheric nitrogen fixed by the bacteria.

The contribution of the fungus to the association is less clear, but there is good reason to believe that it facilitates the uptake of both water and minerals and may also protect the alga from dessication as well as from excess light intensities. Free-living algae, however, are able to grow to a limited extent in the ecological niche inhabited by lichens, and the algal partner may thus be thought to benefit less than does the fungus.

• **Endosymbioses of algae with aquatic invertebrates**   Endosymbiotic algae have been recorded in over 100 genera of aquatic invertebrates, particularly in the coelenterates (jellyfish, corals, sea anemones, hydra), the platyhelminths (flatworms, principally the planarians), the Porifera (sponges), and the mollusks (clams, squid). They are often found in the cytoplasm of cells concerned with digestion (e.g., in the amebocytes of sponges or the phagocytic blood cells of certain clams).

Some of these animal hosts acquire their symbionts with their food, either directly, as in the case of sponges which feed on algae, or indirectly, as in the case of carnivorous animals which find the algae in the tissues of their prey. Once infection has taken place, however, a permanent symbiosis is usually established in which intracellular growth of the algae is restricted.

In most coelenterates, and in certain other invertebrates, the algae are transmitted to the next generation through the cytoplasm of the egg. In such cases it has not been possible to obtain symbiont-free animals, so the significance of the symbiosis is not known. Nevertheless, experiments with isotopically labeled CO_2 show that organic compounds photosynthesized by the algae are utilized by the tissues of the host, and it can be shown that the molecular oxygen generated by algal photosynthesis is several times more than that which would be necessary to provide for the respiratory needs of the host-alga complex. Although the environment of the animal provides dissolved oxygen as well as organic material (principally as plankton), the mechanisms that have evolved for ensuring symbiosis suggest that it is of great ecological significance.

In certain mollusks the photosynthetic symbionts are not algal cells, but *intact, surviving chloroplasts* which are liberated when the algal cells undergo digestion by the host. The animal cell thus becomes photosynthetic, by acquiring a plant organelle. In a sense, a symbiotic relationship can be said to exist between the animal cell and the chloroplast; although the latter does not grow and divide in its new "host," it may persist and function for several months.

The algae found in aquatic invertebrates are of very few types. They are either green algae or dinoflagellates: *zoochlorellae* and *zooxanthellae*, respectively. The algae are presumed to benefit by their intracellular habitat, which supplies a rich supply of essential nutrients. An example of this type of symbiosis (the tridacnid clams) was described in Chapter 26.

FURTHER READING

Books   AHMADJIAN, V., and M. E. HALE (editors), *The Lichens*. New York: Academic Press, 1973.

MARKS, G. C., and T. T. KOZLOWSKI (editors), *Ectomycorrhizae, Their Ecology and Physiology*. New York: Academic Press, 1973.

Review articles   AHMADJIAN, V., "The Lichen Symbiosis: Its Origin and Evolution," *Evolutionary Biology*, Vol. 4, T. Dobzhansky, M. K. Hecht, and W. C. Steere (editors), p. 163. New York: Appleton-Century-Crofts, 1970.

BARNETT, H. L., and F. L. BINDER, The Fungal Host-Parasite Relationship," *Ann. Rev. Phytopathol.* **11**, 273 (1973).

BUSHNELL, W. R., "Physiology of Fungal Haustoria," *Ann. Rev. Phytopathol.* **10**, 151 (1972).

MARX, D. H., "Ectomycorrhizae as Biological Deterrents to Pathogenic Root Infections," *Ann. Rev. Phytopathol.* **10**, 429 (1972).

765

Mosse, B., "Advances in the Study of Vesicular-Arbuscular Mycorrhiza," *Ann. Rev. Phytopathol.* **11**, 171 (1973).

Muscatine, L., and R. W. Greene, "Chloroplasts and Algae as Symbionts in Molluscs," *Int. Rev. Cytol.* **36**, 137 (1973).

Slankis, V., "Soil Factors Influencing Formation of Mycorrhizae," *Ann. Rev. Phytopathol.* **12**, 437 (1974).

Smith, D., L. Muscatine, and O. Lewis, "Carbohydrate Movement from Autotrophs to Heterotrophs in Parasitic and Mutualistic Symbiosis," *Biol. Rev.* **44**, 17 (1969).

Taylor, D. L., "Algal Symbionts of Invertebrates," *Ann. Rev. Microbiol.* **27**, 171 (1973).

Original articles

Kurz, W. G. W., and T. A. La Rue, "Nitrogenase Activity in *Rhizobia* in the Absence of Plant Host," *Nature* **256**, 407 (1975).

McComb, J. A., J. Elliott, and M. B. J. Dilworth, "Acetylene Reduction by *Rhizobium* in Pure Culture," *Nature* **256**, 409 (1975).

Pagan, J. D., J. J. Child, W. R. Scowcraft, and A. H. Gibson, "Nitrogen Fixation by *Rhizobium* Cultured in a Defined Medium," *Nature* **256**, 406 (1975).

Trinkick, M. J., "Symbiosis between *Rhizobium* and the Non-Legume, *Trema aspera*," *Nature* **244**, 459 (1973).

SYMBIOTIC ASSOCIATIONS BETWEEN TWO NONPHOTOSYNTHETIC PARTNERS

28

In many symbioses involving microorganisms neither partner is photosynthetic. With one exception, the examples to be discussed in this chapter are mutualistic; the exception is *Bdellovibrio*, which parasitizes and kills other bacteria.

In the mutualistic nonphotosynthetic symbioses a microorganism may be associated with another microorganism (e.g., the bacterial endosymbionts of protozoa) or with a metazoan partner. In some cases, the microbial symbiont provides its host with a metabolic function, such as the synthesis of a growth factor or the digestion of a complex carbohydrate; in other cases, the microbial symbiont protects its host from invasion by pathogenic parasites. The host in turn furnishes its symbiont with protection, a favorable position, nutrients, or a combination of these functions.

SYMBIOSES IN WHICH BOTH PARTNERS ARE MICROORGANISMS

• *Bdellovibrio bacteriovorus*

The bdellovibrios, of which *Bdellovibrio bacteriovorus* is the type species (see Figure 5.50), are very small, Gram-negative bacteria bearing a single, polar flagellum. They attack and kill other Gram-negative bacteria, multiplying inside the host's cell wall. As isolated from nature they are obligate parasites; rare, host-independent variants can be selected, however, from cultures of host-grown cells. These host-independent strains can grow *in vitro* on peptone media supplemented with B vitamins.

The life cycle of *Bdellovibrio* is unique. It begins in a violent collision with a host cell: the speed of the *Bdellovibrio* cell is so great (as high as 100 cell lengths per second) that a host cell many times its size is carried a considerable distance by the momentum of impact. The parasite immediately attaches to the host cell

766

wall by its nonflagellated end, and rotates about its long axis at speeds exceeding 100 revolutions per second. Shortly thereafter, the host cell rounds up (Figure 28.1), a pore appears in the cell wall at the site of parasite attachment, and the *Bdellovibrio* cell enters the space between the host's cell wall and cell membrane.

The entry step appears to require enzymatic action of the parasite, which liberates proteases, lipases, and a lysozymelike muramidase. In addition, the rapid rotation of the *Bdellovibrio* cell may contribute a mechanical drilling effect to the overall penetration process. The pore that is formed in the host's cell wall is of smaller diameter than that of the parasite, which thus must contract during penetration.

The penetration process is extremely rapid, being completed in a few seconds. There is a lapse of 5 to 10 minutes, however, between attachment and the initiation of penetration, during which period changes occur in the host's cell wall and perhaps in the parasite as well. The rounding up of the host cell, which may occur before penetration is apparent, suggests that the muramidase of the parasite has converted it into a spheroplast; indeed, a host cell can be lysed without penetration, if attacked by many bdellovibrios simultaneously. The rounded host cells are not osmotically sensitive, however, so the term spheroplast may not be entirely accurate.

The bdellovibrio, which loses its flagellum during the penetration process, commences growth in the space between the cell wall and cell membrane. The membrane, although never penetrated by the parasite, becomes porous and leaks cell constituents which serve as nutrients for the parasite. The bdellovibrio

FIGURE 28.1

Sequence of phase contrast photomicrographs illustrating the attack of *Bdellovibrio* on *Pseudomonas tabaci*. The time in minutes after mixing suspensions of host and parasite is indicated on each photograph. After 15 minutes, several small *Bdellovibrio* cells are attached to the host cell, which still has its normal shape. Thereafter, the host cell undergoes progressive distortion, culminating in complete lysis after 275 minutes. From M. P. Starr and N. L. Baigent, "Parasitic Interaction of *Bdellovibrio bacteriovorus* with Other Bacteria," *J. Bacteriol.* **91**, 2006 (1966).

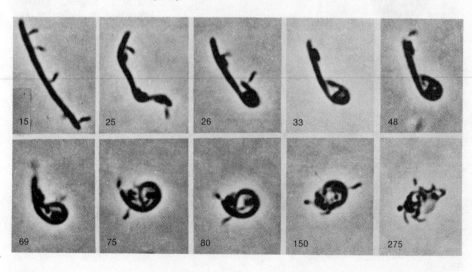

elongates into a filament several times its original length, depending on the size of the host cell, and then segments into a proportional number of flagellated progeny cells; the entire process of multiplication takes about 4 hours. By this time, the host cell wall has undergone further decomposition, and the bdellovibrio progeny are readily liberated (Figure 28.2).

On the surface of a plate covered with a growth of host cells, *Bdellovibrio* will form lytic plaques, superficially similar to those produced by a phage infection (Figure 28.3). In fact, this remarkable bacterium, discovered by H. Stolp only in 1962, was isolated during a search for phages active against bacterial plant pathogens. Stolp observed that plaques developed on his plates after several days of incubation, long after any phage plaques should have appeared. Material isolated from the late-appearing plaques was found to contain a large number of rapidly moving vibrioid cells, which constituted the first isolation of *Bdellovibrio*. The genus name was taken from the Latin word *bdellus*, a leech, since the first observation of its behavior by light microscopy suggested that it remains attached to the outside of its host, sucking it dry of its contents.

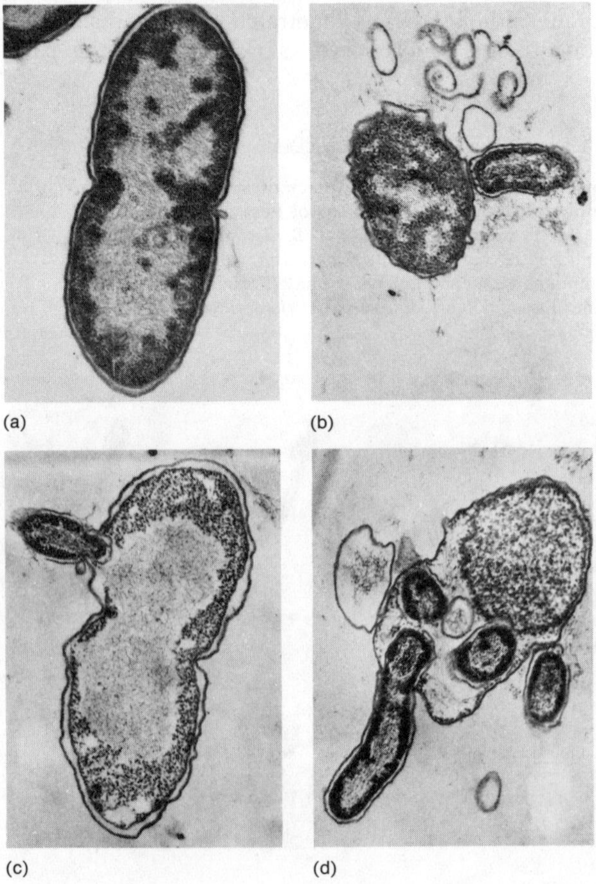

(a)

(b)

(c)

(d)

FIGURE 28.2

Sequence of electron micrographs of thin sections illustrating the interaction of *Bdellovibrio* with a susceptible host bacterium, *Erwinia amylovora* (×28,300). (a) Uninfected host cell. (b) Attachment of *Bdellovibrio*. (c) Penetration of *Bdellovibrio* through cell wall of host. (d) A late state of infection, showing liberation of *Bdellovibrio* progeny. From M. P. Starr and N. L. Baigent, "Parasitic Interaction of *Bdellovibrio bacteriovorus* with Other Bacteria," *J. Bacteriol.* **91**, 2006 (1966).

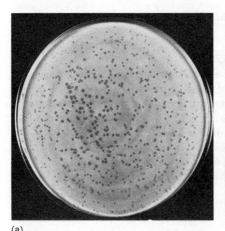

(a)

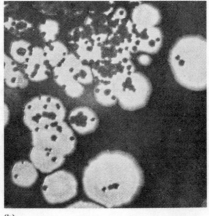

(b)

FIGURE 28.3

Macroscopically visible lysis of susceptible bacteria by *Bdellovibrio*. (a) Plaque formation in a lawn of *Pseudomonas putida* (poured plate). (b) Partial lysis of surface colonies of *Escherichia coli* on a nutrient agar plate streaked with a mixture of host and parasite. From H. Stolp and M. P. Starr, "*Bdellovibrio bacteriovorus* gen. et. sp. n., a Predatory, Ectoparasitic, and Bacteriolytic Microorganism," *Antonie van Leeuwenhoek* **29**, 217 (1963).

Bdellovibrio proved to be a unique group of Gram-negative bacteria, characterized by their very small size (0.3 to 0.45 μm in diameter), the presence of a sheathed flagellum (a property they share with some vibrios), obligate aerobiosis, and obligate parasitism on other Gram-negative bacteria. The last is a key taxonomic property of the group, even though host-independent strains can arise as rare variants.

By screening for organisms which form late-appearing plaques on lawns of Gram-negative bacteria, it has been possible to detect bdellovibrios in a wide variety of natural materials. They have been found in soil samples from many parts of the world, in numbers ranging from 10^2 to 10^5 per gram, as well as in sewage and—in much lower numbers—in pond water and seawater. Most of the strains isolated from nature form a closely related homogeneous group, as indicated by the ability of their DNAs to hybridize with each other, and have been assigned to the species *B. bacteriovorus*. Two other groups of strains have been recognized, based on patterns of enzyme electrophoresis as well as on DNA homologies. They have been assigned to the species *B. stolpii* and *B. starrii* in honor of H. Stolp and M. Starr, two pioneers in the study of bdellovibrios.

The nutrition and physiology of the bdellovibrios have been studied in axenic cultures of host-independent variants. They have been found to lack the ability to catabolize carbohydrates but to be strongly proteolytic, apparently using peptides and amino acids as carbon and energy sources. They are obligate aerobes, respiring substrates via the tricarboxylic acid cycle. The host-dependent strains derive all of their nutrients from the host rather than from the medium.

In Chapter 9 it was stated that most bacteria give a value for Y_{ATP} (the yield of cell mass in grams dry weight per mole of ATP) of about 10. In contrast, bdellovibrios give Y_{ATP} values of 20 to 30; this remarkable efficiency has been shown by S. Rittenberg and coworkers to be due, in part, to the ability of *Bdellovibrio* to assimilate nucleotides directly from the host, conserving the energy-rich phosphate bonds. They are also able to assimilate the host's fatty acids, incorporating some directly into lipids and converting others to their own specific types.

M. Shilo and his collaborators have discovered that host-dependent bdellovibrios can be cultivated in a host-free medium containing an extract of host cells. Growth under these conditions can be rapidly stopped at any point during the cell cycle by the addition of ribonuclease or by washing the cells. These observations suggest that a host-derived RNA factor acts on *Bdellovibrio* without penetrating the cell membrane and that it plays a regulatory (rather than a nutritional) role; the host-independent variants are presumably constitutive mutants which have lost the requirement for an exogenous inducer of an essential function.

The above description of the *Bdellovibrio* life cycle illustrates the fine line that separates the concepts of predator and virulent parasite. In the final analysis, the distinction rests on the viability of the host during the process of attack. If it could be shown, for example, that the host is killed *before* the bdellovibrio begins its multiplication phase, the relationship would better be described as that of predator and prey.

• Bacterial
endosymbionts
of protozoa

Bacterial endosymbionts are extremely widespread in protozoa; they have been described in amebas, flagellates, ciliates, and sporozoa. None of them has been cultivated outside its host, but their bacterial nature has been clearly established on the basis of their morphology, staining properties, and mode of cell division. Some multiply in the nucleus of the host and others in the cytoplasm.

In most cases, the contribution which the bacterium makes to the symbiosis is unknown. In one case, however, its contribution is clear: the bacterial endosymbiont provides its host with amino acids and other growth factors that most protozoa require as exogenous nutrients. The infected host, a trypanosomatid flagellate named *Crithidia oncopelti*, can grow in a simple synthetic medium containing glucose as carbon source together with adenine, methionine, and several vitamins as growth factors. In contrast, another species of *Crithidia* requires not only the above nutrients but also 10 other amino acids (including lysine), hemin, and several additional vitamins. Radioisotope studies showed that in *C. oncopelti* lysine is synthesized via the diaminopimelic acid pathway, characteristic of bacteria. Final proof of the role of the endosymbiotic bacterium found in this protozoon was obtained by fractionating the *Crithidia* cells and showing that diaminopimelic acid decarboxylase, the last enzyme of the biosynthetic pathway leading to lysine, is located in the fraction consisting of the cells of the endosymbiont.

Perhaps the most fascinating, and certainly the most extensively studied, protozoan symbiosis is that of *Paramecium aurelia* and its endosymbiont, *kappa*. In the first of a series of investigations extending over 20 years, T. M. Sonneborn and collaborators showed that most strains of *P. aurelia* fall into two general classes: killers and sensitives. The former liberate toxic particles to which killers are immune but which are lethal for sensitive strains. The ability to liberate toxic particles is genetically controlled by the cytoplasm of the host, rather than by its nucleus; at conjugation, when cytoplasm is exchanged, a sensitive cell mated with a killer is itself converted to a killer.

In attempts to identify the genetic material in the cytoplasm, J. Preer used X rays to inactivate it. Surprisingly, the data yielded a calculated target size for the genetic element so large that it should be visible with the light microscope. Staining experiments were then performed, and the feulgen stain—which is specific for DNA—revealed that the genetic element responsible for liberation of toxic particles is a bacteriumlike endosymbiont which divides by binary fission in the cytoplasm of the paramecium.

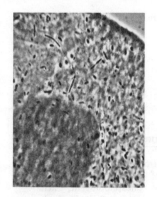

FIGURE 28.4

Stained preparation of unsectioned *Paramecium* containing kappa symbionts (dark rod-shaped bodies) in its cytoplasm. The dark area is the host nucleus; dark phase contrast, ×540. From G. H. Beale, A. Jurand, and J. B. Preer, Jr., "The Classes of Endosymbionts of *Paramecium aurelia*," *J. Cell. Sci.* **5**, 65 (1969).

Kappa, as the endosymbiont was designated, has the morphological and chemical properties of a small bacterium, and can be eliminated from its host by a variety of physical and chemical agents, including many antibiotics. Its loss is irreversible, and the host continues to propagate normally without it. Kappa can be transmitted to sensitive paramecia through extracts prepared from killers, but so far it has not been cultivated outside its host.

Kappa contains DNA and can undergo mutations, including mutation to antibiotic resistance. Its reproduction is dependent on the presence in the host nucleus of a particular gene called K. *Paramecium aurelia* is a diploid organism; the K gene can mutate to the recessive allele, k, and hence a cell may have the genotype KK, Kk, or kk. When a cross between two Kk killers produces a kk segregant, kappa can no longer reproduce and is diluted out during ensuing divisions of the kk host cell. Ultimately, the kk cell gives rise to a clone of sensitive paramecia.

Cells which are infected with kappa harbor several hundred to a thousand of these endosymbionts in their cytoplasm (Figure 28.4). When preparations of purified kappa cells are observed by phase contrast microscopy, some of the cells are found to contain refractile (R) bodies (Figure 28.5). Kappa cells containing R bodies are called *brights*, and those lacking them are called *nonbrights*.

Preer has shown that the toxic particles liberated into the medium are whole, bright kappa cells, which have lost the ability to reproduce further. When brights are fractionated, the toxic activity is found associated with the R bodies, which are seen in the electron microscope to be tightly rolled ribbons of protein (Figure 28.6). It is not clear whether the toxin is the R body itself or a second protein associated with it. The latter seems more likely, in view of the fact that the R body is very stable, while the toxin is very unstable.

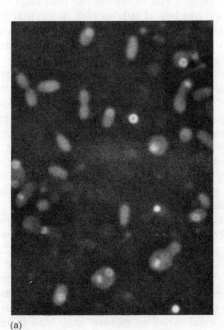

(a)

(b)

FIGURE 28.5

(a) Unfixed, purified preparation of kappa. The rods with uniform color are nonbright kappas; those containing a light spherical refractile body are bright kappas; bright phase contrast, ×4,700. (b) Electron micrograph showing longitudinal section through a bright kappa. Note dark-staining spherical phagelike structures inside the coiled refractile body. Surrounding the refractile body and extending beyond it on either side is a fine membrane, the sheath. (×32,100.) (a) Courtesy of J. Preer; (b) from J. R. Preer, Jr. and A. Jurand, "The Relation between Virus-like Particles and R Bodies of *Paramecium aurelia*," *Genet. Res.* **12**, 331 (1968).

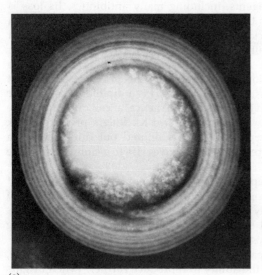

(a)

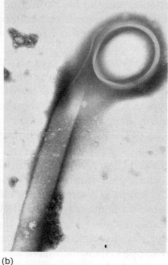

(b)

FIGURE 28.6

Electron micrographs of R bodies of kappa, negatively stained with phosphotungstic acid. (a) Intact, coiled R body, ×119,000. (b) Unrolling R body, ×33,800. (a) From J. R. Preer, Jr., L. B. Preer, and A. Jurand, "Kappa and Other Endosymbionts in *Paramecium aurelia,*" *Bacteriol. Rev.* **38**, 113 (1974); (b) from J. R. Preer, Jr., et al. "The Classes of Kappa in *Paramecium aurelia,*" *J. Cell Sci.* **11**, 581 (1972).

Both the toxin and the R body of kappa have been found to be produced as a consequence of the induction of a defective prophage which is present in the genome of all kappa cells. Cells in which the prophage has been induced are found to contain phage heads and tails as well as circular DNA molecules; they are not lysed, but cease to reproduce further. The phage heads are always found in close contact with R bodies, and it is very possible that R bodies and toxin are coded by phage genes.

In addition to kappa, a number of other bacterial endosymbionts have been found in killer stocks of *Paramecium aurelia* isolated from nature. These have also been designated by Greek letters. One of them, called alpha, has been shown to be a long, spiral gliding organism with strong affinities to *Cytophaga;* it reproduces mainly in the nucleus of its host. The others are eubacteria for which three new genera have been proposed: *Caedobacter*, containing nonflagellated cells, includes the endosymbionts kappa, mu, gamma, and nu; *Lyticum*, containing large, heavily peritrichously flagellated cells, includes lambda and sigma; and *Tectobacter*, containing sparsely peritrichously flagellated cells, includes only delta. Some representatives of these groups are shown in Figure 28.7.

One basis of the mutualistic relationship between *Paramecium* and its bacterial endosymbionts has been clarified by the discovery that one such endosymbiont synthesizes the folic acid required by its host. The equilibrium between the host and endosymbiont is a precarious one, however, and may shift in favor of one or the other. Thus, when bearers of endosymbionts are first cultivated in axenic medium, there is often an unbalanced increase in the reproduction of the endosymbiont leading to the death of the host. Conversely, the rapid growth of the protozoon, once it is established in culture, may lead to the loss of the slower growing endosymbiont by dilution.

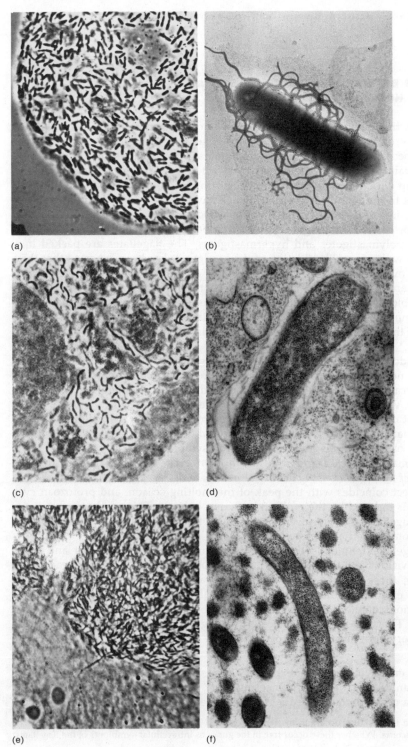

(a)

(b)

(c)

(d)

(e)

(f)

FIGURE 28.7

Some representative bacterial endosymbionts of *Paramecium aurelia*. (a) Lambda in host cytoplasm, stained unsectioned preparation, dark phase contrast, ×750. (b) Isolated lambda, negatively stained electron micrograph showing flagella, ×11,100. (c) Sigma in host cytoplasm, stained unsectioned preparation, dark phase contrast, ×729. (d) Sigma, electron micrograph of section through host cytoplasm showing symbiont and flagella, ×22,000. (e) Alpha in host macronucleus, stained unsectioned preparation, dark phase contrast, ×870. (f) Alpha, electron micrograph of thin section of host macronucleus, ×27,000. From G. H. Beale, A. Jurand, and J. B. Preer, Jr., "The Classes of Endosymbionts of *Paramecium aurelia*," *J. Cell Sci.* **5**, 65 (1969).

SYMBIOSES BETWEEN MICROORGANISMS AND METAZOAN HOSTS

• Ectosymbioses of protozoa with insects: the intestinal flagellates of wood-eating termites and roaches

The woody tissue of trees, consisting mainly of cellulose and lignin, is unavailable as a source of food for most animals; in general, animals do not possess the enzymes necessary to degrade these polymers. Nevertheless, many species of insects obtain the bulk of their food from wood by virtue of an ectosymbiotic relationship with cellulose- and lignin-digesting microorganisms.

Both the termites and cockroaches, which have evolved from a common ancestral group, include some species that eat wood. All the wood-eating species of both groups harbor in their gut immense numbers of flagellated protozoa belonging to the polymastigotes and hypermastigotes. The flagellates are packed in a solid mass within a saclike dilation of the hindgut; it has been reported that they constitute over one-third of the body weight of the insect in some cases. The flagellates are responsible for cellulose digestion, of which the insects themselves are incapable. The flagellates, in turn, are themselves hosts to extracellular spirochetes (Figure 26.2) and to intracellular bacteria, and it is possible that some—if not all—of the cellulases produced by the flagellates derive from their intracellular symbionts.*

The mode of transmission of the flagellated symbionts from one insect generation to the next differs in the two groups. The newly hatched nymphs of termites feed on fecal droplets that exude from the adults; the droplets are laden with symbionts, which infect the young insects. The newly hatched nymphs of cockroaches eat dry fecal pellets that are excreted by the adults; the pellets are laden with flagellate *cysts*, which are able to withstand dessication. The cysts germinate in the gut of the nymphs, reestablishing the symbiosis.

One remarkable feature of the transmission cycle in cockroaches is that *the encystment of the flagellates is regulated by hormones of the insect*. The hatching of eggs in this insect coincides with the peak of the molting season, and protozoan cyst formation is induced by the molting hormone, *ecdysone*. This mechanism ensures that the flagellates will survive dessication in the fecal pellets and be available for infection of the hatching nymphs.

The flagellates enter a sexual cycle following encystment, nuclear and cytoplasmic divisions giving rise to one male and one female gamete from each cyst. Ultimately, these fuse to form a zygote. In an extensive series of studies, L. Cleveland established that sexuality in flagellates is induced by ecdysone, at concentrations of the hormone well below those required to induce molting of the insect. The adaptive significance of this regulation is not clear. It may reflect an obligatory coupling of gametogenesis with encystment.

• Endosymbioses of fungi and bacteria with insects

Microbial endosymbioses are extremely widespread among insects. P. Buchner, the German biologist whose pioneering work on symbiosis has spanned more than half a century, discovered a striking correlation between the diet of insects

---

*Nitrogen fixation also occurs in the termite gut, and is assumed to reflect the activity of nitrogen-fixing bacteria. Whether these occur free in the gut or as intracellular symbionts of the flagellates, is unknown.

and the presence of symbionts: symbionts are never found in insects that have a nutritionally complete diet, but are present in all insects that have a nutritionally deficient diet during their developmental stages. Thus, no carnivorous insect has symbionts, whereas insects that live on blood or on plant sap all contain symbionts. The main function of the symbiont is thus to provide the host with one or more growth factors that are lacking in the insect's diet.

Certain apparent exceptions prove this rule. Mosquitoes, for example, contain no symbionts, although they suck blood. It is only the adult female, however, which takes a blood meal; the larvae and pupae have a nutritionally complete diet consisting of microorganisms and organic debris. Conversely, the granary weevil, *Sitophilus granarius*, contains symbionts although it feeds on nutritionally rich grains. This genus, however, inherits its symbionts from its wood-eating ancestors; it is able to survive and reproduce if freed of its symbionts, *provided that it is fed a nutritionally rich diet*. Without symbionts its choice of food is severely restricted.

THE MICROBIAL ENDOSYMBIONTS   The microbial endosymbionts of insects include both bacteria and yeasts. Most of these have been identified as such solely on the basis of their appearance and mode of reproduction in the host. A few, however, have been successfully isolated and grown in pure culture. For example, one of the symbionts of *Rhodnius*, a kissing bug, has been isolated and identified as an actinomycete of the genus *Nocardia*. Other isolated insect symbionts have proved to be coryneform bacteria or Gram-negative rods. Some yeasts have also been successfully isolated, notably from the long-horned beetles (*Cerambycidae*) and the deathwatch beetles (*Anobiidae*). Although most insects are monosymbiotic, it is not uncommon for a particular species to harbor two or more different microorganisms. The relationship between insects and their endosymbionts appears to be highly specific; an insect species can often be identified reliably by observing the nature of its symbionts.

THE LOCALIZATION OF THE ENDOSYMBIONTS   The microbial endosymbionts are housed within specialized cells of the insect (Figure 28.8). These are called *mycetocytes* when they harbor yeasts and *bacteriocytes* when they harbor bacteria. Some authors refer to both as mycetocytes, and we will use this terminology here.

In some insects the mycetocytes are scattered randomly throughout a normal tissue, such as the wall of the midgut or the *fat body*, a loose, discontinuous tissue lining the body cavity. In many insects, however, the mycetocytes are restricted to special organs called *mycetomes*, the only function of which is to house the endosymbionts. It is possible to trace an evolutionary series of steps between ectosymbiosis, in which the symbionts develop in the lumen of the insect gut, and endosymbiosis in mycetomes. Figure 28.9 shows schematically the principal parts of the insect digestive tract. Figure 28.10 illustrates the localization of endosymbionts in out-pocketings or *blind sacs* of the insect midgut. Figure 28.11 shows how, in a series of species of anobiid beetles, the blind sacs have evolved to become more and more independent of the midgut. In the most primitive endosymbioses the symbionts are found both extracellularly in the gut lumen and intracellularly in the blind sacs. In the most advanced forms the symbionts are completely isolated, and the blind sacs have evolved into independent organs, or mycetomes.

In some insects the mycetocytes are localized in the Malpighian vessels, the excretory organs of the insect. In certain genera of the *Curculionidae* (the family that includes weevils, snout beetles, and curculios), two of the six Malpighian

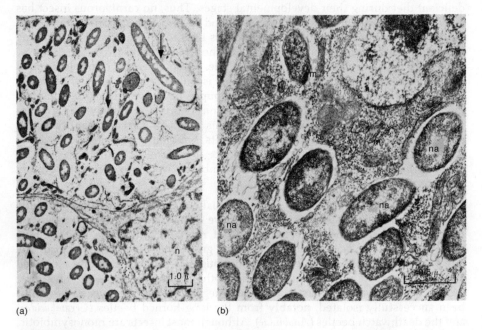

(a)   (b)

FIGURE 28.8

Mycetocytes of the insect, *Sitophilus granarius;* electron micrographs of thin sections.
(a) Low magnification; arrows indicate bacterial endosymbionts, ×3,330. (b) High
magnification, ×19,500. m, mitochondria; n, mycetocyte nucleus; na, nuclear area of
endosymbiotic bacterium; r, ribosomes. From I. Grinyer and A. J. Musgrave, "Ultrastructure and Peripheral Membranes of the Mycetomal Microorganisms of *Sitophilus
granarius* (L.) (Coleoptera)," *J. Cell Sci.* **1,** 181 (1966).

vessels have become anatomically specialized for this purpose and have evolved
into club-shaped mycetomes (Figure 28.12). In a number of other insects the
mycetomes are detached from the gut, forming essentially independent structures
in the body cavity.

THE SIGNIFICANCE OF THE INSECT ENDOSYMBIOSES   The essential role played by the
endosymbionts in the nutrition of the host can be demonstrated by artificial
elimination of the symbionts and study of the behavior of the symbiont-free

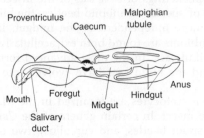

FIGURE 28.9

Schematic diagram of the digestive tract of the insect.

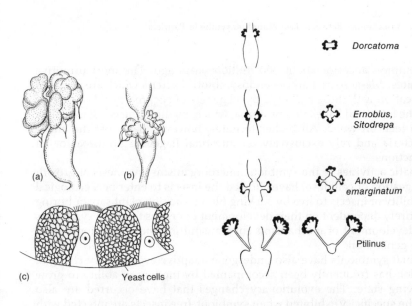

Dorcatoma

Ernobius,
Sitodrepa

Anobium
emarginatum

Ptilinus

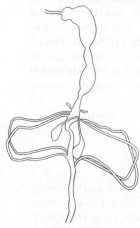

FIGURE 28.10 (left)

Blind sacs of the midgut of *Sitodrepa panicea*, an anobiid beetle: (a) larva; (b) adult; (c) epithelium of the blind sac of the larval midgut, showing yeast-filled mycetocytes separated by sterile cells with brush borders. After A. Koch.

FIGURE 28.11 (right)

Blind sacs of the midgut of a series of anobiid beetles, showing evolutionary development of the blind sacs as independent organs. Left column: longitudinal sections. Right column: cross sections. After A. Koch.

(a)   (b)

(c)   Yeast cells

FIGURE 28.12

The adult gut of *Apion pisi*, showing the transformation of two of the six Malpighian tubules into mycetomes. After A. Koch.

insects. Elimination has been accomplished by a variety of ingenious methods. In insects that smear their eggs with symbionts, the egg surface can be sterilized. In insects with well-defined and isolated mycetomes, such as the stomach disc of *Pediculus*, the louse, the mycetome can be surgically removed. Some insects can be freed of their symbionts by the use of high temperatures or of antibiotics. In some cases, growth of symbiont-free insects is severely retarded, and the adult stage may not be reached (Figure 28.13). In other cases, the principal effect is to disturb the reproductive system: the female organs may be damaged, or their formation may be completely blocked.

In many such experiments the loss of the symbionts can be totally compensated for by the provision of vitamins, particularly the B vitamins. In cockroaches (family Blattidae) it has also been shown that the symbionts provide the host with some essential amino acids. Feeding the young insects ^{14}C-labeled glucose led to the appearance of labeled tyrosine, phenylalanine, isoleucine, valine, and arginine in symbiotic, but not in symbiont-free, individuals. The injection of ^{35}S-labeled sulfate similarly showed that the methionine and cysteine of the cockroach are synthesized by the symbiotic bacteria.

The bacterial symbionts of some insects also appear to aid the host in the breakdown of nitrogenous waste products (uric acid, urea, and xanthine).

THE EVOLUTION OF THE INSECT-MICROBE SYMBIOSES   The evolutionary relationship between a specific symbiosis and the diet of the host can be clearly traced in the termites. The fossil record shows that the termites and the cockroaches split

FIGURE 28.13

The effect of symbiont loss on the growth of larvae of *Sitodrepa panicea*: (a) symbiont-free larva on normal diet; (b) symbiont-free larva on normal diet plus 25 percent dried yeast; (c) normally infected larva on normal diet without supplementation. After A. Koch.

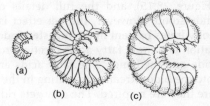

(a)   (b)   (c)

off from a common ancestor about 300 million years ago. The most primitive group of termites, *Mastotermes,* harbors endosymbiotic bacteria which are identical in type and location with those harbored by all genera of cockroaches. *Mastotermes* also harbors the intestinal flagellated protozoa, which, by digesting cellulose, allow their hosts to feed on wood. All higher termites, however, have lost the endosymbiotic bacteria and rely exclusively on intestinal flagellate protozoa for all symbiotic functions.

The enzymatic activities of the symbiotic microorganisms (synthesis of growth factors and digestion of cellulose) have allowed the insects to enter new ecological niches. The ability of insects to live by sucking blood, sucking plant sap, or boring in wood is entirely dependent on their development of organs to house symbionts and to their development of mechanisms for transmitting these symbionts from generation to generation.

The microbial symbionts have also undergone adaptive evolutionary changes. Such adaptation has frequently been accompanied by the loss of ability to grow in the free-living state. The evolutionary changes that have occurred are also indicated by the specificity exhibited when symbiont-free insects are infected with "foreign" symbionts. The anobiid beetle *Sitodrepa,* for example, can be reinfected either with its normal symbiotic yeast or with a foreign yeast. The former infects only the normal mycetocytes, whereas the latter infects all the epithelial cells of the blind sacs and midgut. Furthermore, the foreign yeast is not transmitted to the adult stage during morphogenesis. It is of particular interest that a beetle which harbors its normal symbiont *cannot* be infected with the foreign yeast that easily infects symbiont-free individuals; the normal symbiont appears to confer on its host an immunity to infection by related microorganisms, a phenomenon frequently observed in symbiotic relationships of vertebrates as well as of invertebrates.

• **The ruminant symbiosis**

The ruminants are a group of herbivorous mammals that includes cattle, sheep, goats, camels, and giraffes. Ruminants, like other mammals, cannot make cellulases. They have evolved an ectosymbiosis with microorganisms, however, which enables them to live on a diet in which the major source of carbon is cellulose.

The digestive tract of a ruminant contains no less than four successive stomachs. The first two, known as the *rumen* (Figure 28.14), are essentially vast incubation chambers teeming with bacteria and protozoa. In the cow the rumen is a bag with a capacity of about 100 liters. The plant materials ingested by the cow are mixed with a copious amount of saliva and then passed into the rumen, where they are rapidly attacked by bacteria and protozoa. The total microbial population is enormous, and the population density is of the same order as that of a heavy laboratory culture of bacteria (10^{10} cells per milliliter). Many different microorganisms are present (Figure 28.15), and the full details of their biochemical activities are not yet understood. However, the net effect is clear: the cellulose and other complex carbohydrates present in the ingested fodder are broken down with the eventual formation of simple fatty acids (acetic, propionic, and butyric) and gases (carbon dioxide and methane). The fatty acids are absorbed through the wall of the rumen into the bloodstream, circulating in the blood to the various tissues of the body where they are respired. The cow gets rid of the gases formed

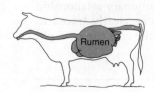

FIGURE 28.14

Diagram of the intestinal tract of the cow, to show the rumen.

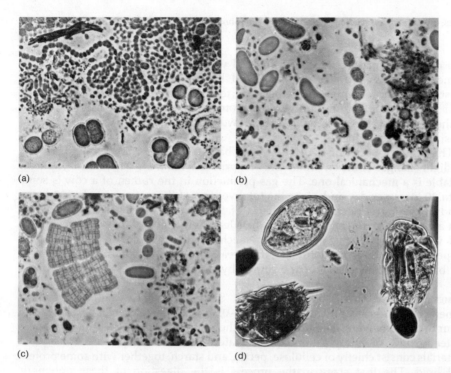

(a) (b)

(c) (d)

FIGURE 28.15

Some microorganisms from the rumen of the sheep. (a), (b), and (c) Bacteria (ultraviolet photomicrographs, ×732). From J. Smiles and M. J. Dobson, "Direct Ultraviolet and Ultraviolet Negative Phase-contrast Micrography of Bacteria from the Stomachs of the Sheep," *J. Roy. Micro. Soc.* **75**, 244 (1956). (d) Ciliate protozoa (×9). Courtesy of J. M. Eadie and A. E. Oxford.

in the rumen by belching at frequent intervals. The microbial population of the rumen grows rapidly, and the microbial cells pass out of the rumen with undigested plant material into the lower regions of the cow's digestive tract. The rumen itself produces no digestive enzymes, but the lower stomachs secrete proteases, and as the microbial cells from the rumen reach this region they are destroyed and digested. The resulting nitrogenous compounds and vitamins are absorbed by the cow. For this reason, the nitrogen requirements of the cow and other ruminants are much simpler than those of other groups of mammals. Whereas man or the rat requires many amino acids (the so-called essential amino acids) preformed in the diet, the ruminant can grow on ammonia or urea, which are excretion products in most mammals. These simple nitrogenous compounds are built up into microbial proteins by the rumen population.

The evolution of the rumen has involved both structural and functional modifications of the gastrointestinal tract. The principal structural modification is the development of a complex stomach, of which the largest compartments are essentially fermentation vats. The functional modifications that ruminants have undergone are even more profound. In the first place, the salivary glands do not secrete enzymes, the saliva being essentially a dilute salt solution (principally sodium bicarbonate and sodium phosphate) that provides a suitable nutrient base for the microbes of the rumen. In the second place, the lower fatty acids have very largely replaced sugar as the primary energy-yielding substrate. This, in turn, has led to changes in the enzymatic makeup of nearly all the tissues in the body, which respire fatty acids far more rapidly than do the tissues of nonruminants. Finally, the source of amino acids and vitamins has become very largely internalized (microorganisms instead of ingested food materials).

For the microorganisms that have taken up residence in the rumen, the situation also offers advantages; they are provided with an environment always rich in fermentable carbohydrates, well buffered by the saliva, and maintained at a constant favorable temperature, the body temperature of the cow. Individually, their ultimate fate is to fall prey to the proteolytic enzymes in the lower regions of the digestive tract; for the species, however, the rumen provides a safe and constant ecological niche.

The rumen association is in a delicately balanced equilibrium, easily disturbed by slight changes of the environment. The principal failure to which this symbiosis is liable is a mechanical one. The gas production in the rumen of a cow is some 60 to 80 liters per day and, since the total volume of the rumen is only 100 liters, steady belching is necessary to get rid of the accumulating gases. For reasons that are not fully understood, certain diets lead to foaming of the rumen contents, and when this happens the belching mechanism of the cow fails to function properly. This causes a painful and, if untreated, eventually fatal affliction known as "bloat," (i.e., distention of the rumen by the trapped gases).

METABOLIC ACTIVITIES OF RUMEN BACTERIA   Since the redox potential (E_0') of the rumen contents is steadily maintained at -0.35 V, all the microbial processes that occur in the rumen are anaerobic ones. As the ruminant grazes, the rumen receives a steady flow of finely ground plant materials mixed with saliva. The plant materials consist chiefly of cellulose, pectin, and starch, together with some protein and lipid. The first stage in the process is the digestion of these polymeric macromolecules. A great deal of attention has been given to the identification of the microorganisms responsible for the digestion of cellulose, since this is the major digestive process in the rumen. Between 1 and 5 percent of the bacterial cells in the rumen have been found to be cellulolytic; they produce an extracellular cellulase that hydrolyzes cellulose, glucose appearing as the final product of digestion.

The cellulose-digesting bacteria of the rumen, like the other principal ruminant microorganisms, are all strict anaerobes. Several different species of cellulose-digesting bacteria have been isolated and described: *Bacteroides succinogenes*, *Ruminococcus flavofaciens*, *R. albus*, and *Butyrovibrio fibrisolvens*. Of these, *B. succinogenes* is a Gram-negative organism which should probably be classified as a myxobacterium, since it shows gliding movement. It forms succinic acid and lesser amounts of acetic acid. *Ruminococcus* produces principally succinic acid, and *Butyrovibrio* produces principally butyric acid.

The great bulk of the bacterial population, however, is noncellulolytic. These organisms rapidly utilize the glucose and cellobiose produced by the cellulolytic species; their great efficiency in scavenging these molecules presumably accounts for their predominance over the cellulolytic forms. Furthermore, many of the rumen bacteria (including some of the cellulolytic species) are capable of digesting starch, pectin, proteins, and lipids. Indeed, only the lignin of the ingested plant material escapes digestion by the rumen flora.

The products of digestion of polysaccharides, proteins, and lipids are fermented by the rumen bacteria. In the rumen these metabolic activities lead to the accumulation of the gases CO_2 and methane and the fatty acids acetic, propionic, and butyric acids. When the predominant rumen bacteria are isolated and studied as pure cultures, however, not only are the above listed products formed but also hydrogen gas, together with large amounts of formic, lactic, and succinic acids.

By adding radioactive isotopes of these compounds to rumen contents, the reasons for their failure to accumulate in the rumen have been discovered. Thus, hydrogen gas is quantitatively combined with carbon dioxide to form methane, by the organism *Methanobacterium ruminantium;* the lactate is fermented to acetate plus smaller amounts of propionate and butyrate, by such organisms as *Peptostreptococcus elsdenii;* and the formate is converted first to carbon dioxide and hydrogen, by a variety of bacteria, and then to methane by *M. ruminantium.* Finally, the succinate is rapidly decarboxylated to propionate by *Veillonella alcalescens* and other bacteria.

The net result is the formation of carbon dioxide, methane, and acetic, propionic, and butyric acids, in remarkably constant proportions. Carbon dioxide accounts for 60 to 70 percent of the gases, methane accounting for the remainder; acetic acid represents 47 to 60 percent, propionic acid represents 18 to 23 percent, and butyric acid represents 19 to 29 percent of the fatty acids, respectively.

THE RUMEN PROTOZOA Protozoa were seen microscopically in rumen contents as early as 1843, but almost 100 years elapsed before they were successfully isolated and cultivated *in vitro.* Representatives of several genera have now been cultivated, notably the oligotrichous ciliates *Diplodinium, Entodinium, Epidinium, Metadinium,* and *Ophryoscolex,* and the holotrichous ciliates *Isotricha* and *Dasytricha.* Unfortunately, attempts to grow these protozoa in axenic (bacteria-free) culture have not yet been successful.

All such cultures to date have been contaminated with bacteria, intracellular as well as extracellular, so a final conclusion cannot be drawn from the enzymological and metabolic experiments that have been reported. Nevertheless, some tentative conclusions have been reached, on the basis of experiments in which the protozoa were maintained alive for extended periods of time in the presence of high levels of bactericidal antibiotics. These experiments have implicated species of *Diplodinium* and *Metadinium* in the digestion of cellulose and species of *Entodinium* and *Epidinium* in the digestion of starch. Many of these protozoa are active predators on the rumen bacteria.

• Ectosymbioses of microorganisms with birds: the honey guides

The honey guides, a group of birds belonging to the genus *Indicator,* are found in Africa and India. Their name accurately describes their behavior: they literally guide honey badgers, as well as humans, to the nests of wild bees, where they wait for their follower to break open the hive. When the badger (or human) has departed, the honey guide proceeds to feed on the remnants of honeycomb that have been left exposed.

This behavior became all the more remarkable when it was discovered that these birds do not possess enzymes for digesting beeswax. Instead, they harbor in their intestines two microorganisms which carry out the digestion for them: a bacterium, *Micrococcus cerolyticus,* and a yeast, *Candida albicans.* The micrococcus is a highly specialized symbiont, depending on a growth factor that is produced in the small intestine of the honey guide.

• Ectosymbioses of microorganisms with mammals: the "normal flora" of the human body

Both the skin and the mucous membranes of the body are directly accessible to the external environment; soon after an infant is born these surfaces become populated by a characteristic flora. The skin becomes contaminated during passage through the birth canal; the mucous membranes may be sterile at birth but become contaminated within hours. Of the many microorganisms that reach these surfaces, only those that are particularly suited to growth in such environments become established; these constitute the "normal flora," which remains remarkably constant. The bacteria of the normal flora do not cause disease unless accidentally introduced into normally protected regions of the human body or as a result of physiological changes within the host. For example, a radical change in diet or infection with a virus may alter the conditions within the body in such a fashion that a member of the normal bacterial flora can become pathogenic.

Each region of the body provides a distinctive ecological niche which selects for the establishment of a characteristic flora. The bacteria that populate the skin, for example, include mainly corynebacteria, micrococci, nonhemolytic streptococci, and mycobacteria. Moist regions of the skin harbor yeasts and other fungi. The skin flora is selected both by nutritional conditions and by antibacterial agents, such as the fatty acids secreted by the skin. The microorganisms of the skin flora are so firmly entrenched in the sweat glands, sebaceous glands, and hair follicles that no amount of bathing or scrubbing can totally remove them.

The throat and mouth support a variety of microorganisms, including representatives of most of the common eubacterial groups. Gram-positive cocci, including micrococci, pneumococci, and *Streptococcus salivarius*, are common throat inhabitants. The Gram-negative cocci are represented by members of both the aerobic genus *Neisseria* and the anaerobic genus *Veillonella*. The mouth and throat also harbor large numbers of both Gram-positive and Gram-negative rods. The former include mainly lactobacilli and corynebacteria, while the latter include members of the genera *Bacteroides* and *Spirillum*. Spirochetes (*Treponema dentium*), yeasts (*Candida albicans*), and Actinomycetes (*Actinomyces israelii*) are also common mouth inhabitants. All these organisms are normally harmless, but if the mucous membranes are injured, many can invade the body tissues and produce disease.

The same types of organisms inhabit the nasopharynx and are at least potentially able to reach the lungs. However, a series of protective mechanisms keep the trachea and bronchi relatively free of live bacteria. First, most of the bacteria adhere to the mucous lining of the nasopharynx and cannot readily move to the lungs because the ciliated epithelial cells that line the trachea constantly sweep the mucus upward. Second, the lungs are the site of extremely active phagocytosis, a mechanism whereby foreign particles are engulfed and destroyed by special amebalike cells.

The stomach and small intestine are unsuitable environments for bacterial development, but the large intestine harbors an extremely large resident flora. In the intestines the nature of the flora changes with diet and with age. Breast-fed infants, for example, have always a predominant intestinal population of *Bifidobacterium*, an organism not commonly found elsewhere. This organism disappears soon after weaning.

In adults the predominant bacteria of the intestinal tract are *Bacteroides* spp., *Escherichia coli*, and *Streptococcus faecalis*. Organisms of the genus *Bacteroides* are strictly anaerobic, nonmotile, Gram-negative rods; they are the most characteristic organisms of the mammalian intestinal tract and are seldom found elsewhere. Their numbers in human feces exceed 10^9 cells per gram, whereas the density of *E. coli* rarely exceeds 10^8 cells per gram. Other genera, such as viridans streptococci and micrococci, are present in lower numbers. The yeasts are represented by *Candida* and *Torulopsis*, and protozoa by several genera: *Balantidium*, a ciliate found only in man; *Entamoeba*; and flagellates of the genus *Trichomonas*. The proportions of the different microbial types are largely dependent on diet. Oral chemotherapy with such agents as antibiotics or sulfonamides can cause striking changes in the intestinal flora.

The principal bacteria of the vagina are the lactobacilli, which maintain a low pH as a result of their fermentative activity. This acidity is responsible for preventing the establishment of other forms. When the lactobacilli of the vagina are reduced in numbers during chemotherapy with antibacterial drugs, or during the use of oral contraceptives, vaginal infections by bacteria and by yeasts commonly occur.

In hospital practice today many serious infections are seen with bacteria that normally reside harmlessly in the host. Probably the most common source of such "endogenous infections" are the intestinal contents, which normally contain 10^{10} to 10^{11} bacteria per gram of stool. In persons with diseases that interfere with host defense mechanisms or in circumstances when drug treatment has the same effect, it is common for bacteria that reside normally in the bowel to invade the bloodstream. Examples of diseases that may interfere with host defense mechanisms include leukemia, which may produce severe lowering of circulating leukocytes, and multiple myeloma, which is a disease of antibody-producing cells. Drugs like cortisone may also increase susceptibility to invasion by organisms of the normal intestinal flora.

The normal flora of the mammalian body provides another example of a symbiotic relationship which can shift from mutualism to parasitism and back again. In the absence of circumstances which permit them to invade the tissues, organisms of the normal flora benefit the host by preventing the establishment of virulent pathogens to which the host is often exposed.

• **Germ-free animals**

By the use of complex equipment and elaborate techniques, it has been possible to deliver germ-free animals by caesarean section and to rear them in a germ-free environment. Adult animals in such an environment will mate and produce germ-free litters, so that colonies of germ-free animals can be maintained.

The chambers in which the animals are delivered and reared are equipped to prevent the entry of organisms such as bacteria and fungi, thus permitting experimentation on the role of the normal flora in the growth and development of the host, as well as on resistance to infection by virulent pathogens.

The development of the germ-free animal is abnormal in several respects. The cecum of the germ-free animal is greatly enlarged, the lymphatic system is poorly developed, and the germ-free animal makes much less immunoglobulin than does the normal animal.

Comparisons of normal and germ-free animals have not revealed any effects of the normal flora on the nutritional requirements of the host. However, striking effects have been observed on the *resistance or susceptibility of the host to infectious diseases*. For example, germ-free rats do not develop dental caries, whereas control animals do. Cavities in the teeth appear, however, when the germ-free rats are infected with streptococci. A more complicated role of the normal flora in producing disease is revealed in the case of infection by the protozoan pathogen, *Entamoeba histolytica*, the agent of amebic dysentery. This organism cannot produce disease in germ-free guinea pigs, but it does so if the animals are first infected with *Escherichia coli* or *Enterobacter aerogenes*. Thus, organisms typical of the normal flora potentiate the virulence of a pathogen which, by itself, cannot survive in the intestine.

In contrast, the presence of the normal flora protects the host against some infectious diseases. For example, normal rats develop resistance to *Bacillus anthracis* spores at an early age, the number of spores in a dose required to kill 50 percent of the animals in a given experiment rising from 10^4 to 10^9 very soon after birth. Germ-free rats never develop this resistance, which can thus be attributed to the presence of the normal flora.

FURTHER READING

Books   BUCHNER, P., *Endosymbiosis of Animals with Plant Microorganisms.* New York: Wiley, 1965.

HENEGHAN, J. B. (editor), *Germ-Free Research (Biological Effects of Gnotobiotic Environments).* New York: Academic Press, 1973.

HUNGATE, R. E., *The Rumen and Its Microbes.* New York: Academic Press, 1966.

ROSEBURY, T., *Microorganisms Indigenous to Man.* New York: McGraw-Hill, 1962.

SKINNER, F. A., and J. A. CARR (editors), *The Normal Microbial Flora of Man.* New York: Academic Press, 1974.

Review articles   GORDON, H. A., and L. PESTI, "The Gnotobiotic Animal as a Tool in the Study of Host-Microbial Relationships," *Bacter. Rev.* **35**, 390 (1971).

PREER, J. R., L. B. PREER, and A. JURAND, "Kappa and Other Endosymbionts in *Paramecium aurelia,*" *Bact. Rev.* **38**, 113 (1974).

STARR, M. P., and J. C.-C. HUANG, "Physiology of the Bdellovibrios," *Advan. Microbial Physiol.* **8**, 215 (1972).

STARR, M. P., and R. J. SEIDLER, "The Bdellovibrios," *Ann. Rev. Microbiol.* **25**, 649 (1971).

STOLP, H., "The Bdellovibrios: Bacterial Parasites of Bacteria," *Ann. Rev. Phytopathology* **11**, 53 (1973).

Original articles   BRYANT, M. P., "Normal Flora-Rumen Bacteria," *Am. J. Clin. Nutrition* **23**, 1440 (1970).

HOROWITZ, A. T., M. KESSEL, and M. SHILO, "Growth Cycle of Predacious Bdellovibrios in a Host-Free Extract System and Some Properties of the Host Extract," *J. Bacteriol.* **117**, 270 (1974).

RITTENBERG, S. C., and R. B. HESPELL, "Energy Efficiency of Intraperiplasmic Growth of *Bdellovibrio bacteriovorus,*" *J. Bacteriol.* **121**, 1158 (1975).

29

MICROBIAL PATHOGENICITY

The term pathogenicity denotes the ability of a parasite to cause disease. Pathogenicity is a taxonomically significant attribute, being the property of a species; thus, the bacterial species *Corynebacterium diphtheriae* is said to be pathogenic for man. The individual strains of a bacterial species may, however, vary widely in their ability to harm the host species, and this relative pathogenicity is termed *virulence*. Virulence is accordingly an attribute of a strain, not a species; one may speak of a highly virulent, a weakly virulent, or even an avirulent strain of *C. diphtheriae*.

In general, the virulence of a strain of a pathogenic species is determined by two factors: its *invasiveness*, or ability to proliferate in the body of the host, and its *toxigenicity*, or ability to produce chemical substances—*toxins*—that damage the tissues of the host. It is characteristic of bacterial toxins that they are capable of damaging or killing *normal* host cells (i.e., the cells of a host that has not previously been exposed to the infectious agent in question). Certain pathogenic microorganisms, however, cause damage to the vertebrate host by a mechanism that is more indirect and does not come into play unless or until the host has previously experienced specific infection. This mechanism is known as *hypersensitivity*, or *allergy*, and involves an immune response by the already sensitive host to a cell component of the parasite which is *nontoxic* for a normal host.

The role played by invasiveness in damaging the host varies widely. Some pathogens are so toxigenic that an extremely localized infection may result in the production and diffusion through the host of sufficient toxin to cause death. The classical example of such a disease is diphtheria, in which the pathogen, *Corynebacterium diphtheriae*, multiplies in the throat and produces a diffusible toxin that affects virtually all the tissues of the body. At the other extreme are pathogens that must invade and multiply extensively within the body of the host, in order to produce enough toxin to cause damage or to enable the toxin to reach specifically susceptible tissues. The classical example of such a disease is anthrax, in which the pathogen, *Bacillus anthracis*, is present in enormous numbers in the bloodstream in the terminal stages of infection.

Microbial parasites are transmissible from one individual of the host species to another; the establishment of a pathogen in the body of a fresh host is called *infection;* if damage to the host is produced, the process is called *infectious disease.* Infectious microorganisms reach their hosts by a variety of routes: in contaminated food or drink, in airborne droplets, by direct contact with an infected individual, by animal bite, or by the introduction of contaminated material into a wound. The unbroken skin is an impregnable barrier, and organisms which lodge there penetrate no further. The mucous membranes, however, are readily penetrated by virulent microorganisms; thus, all microbial diseases (other than those transmitted by animal bite or wound contamination) begin as infections of the mucosal membranes of the respiratory, intestinal, or genitourinary tracts.

The ability of a pathogen to develop in its host depends on the balance between the growth-promoting and growth-inhibiting factors which the host environment provides. The growth-promoting factors are the nutrients required by the microorganism; the growth-inhibiting factors include a variety of cellular and chemical *host defense mechanisms* which the pathogen must overcome if it is to survive and proliferate. Variations in these factors among animal species, and among the tissues of a given species, determine the host and tissue specificities of each pathogenic microorganism.

There are three general modes of infection among the various pathogens: some multiply only on the surface of the mucosal epithelium; some penetrate and multiply in the epithelial cells; and others pass through the epithelium, entering the deeper tissues of the body and the circulatory system. In the following sections we shall discuss these modes of infection, describing in each case the host defense mechanisms that the pathogen must overcome. First, however, we shall survey the toxins by means of which microbial parasites damage their hosts.

MICROBIAL TOXINS

A microbial toxin may cause a highly specific type of tissue damage in its host. When a wound is infected by *Clostridium tetani*, for example, the tetanus toxin diffuses up regional motor nerves to the central nervous system where it binds to neuronal surface receptors and suppresses the normal synaptic inhibition of nerve impulses. The toxin of *C. botulinum* produces specific damage of a different type to the nervous system: it inhibits the release of acetylcholine at myoneural junctions, so that the nerve fiber no longer stimulates the muscle fiber to which it is attached.

In many cases, however, the damage represents a secondary, *nonspecific* host response to the primary action of the toxin. For example, the regulation of the passage of ions and water from the tissues into the bowel, which is mediated by the bowel epithelial cells, can be disrupted by many different agents. One agent is the toxin of *Vibrio cholerae*, which causes severe and often fatal fluid loss. Fluid loss is thus a nonspecific host response to the binding of cholera toxin by the cells of the intestinal epithelium.

Another type of nonspecific damage which reflects a response of the host tissues to infection is *inflammation,* a complex series of cellular changes described in detail later in this chapter. The nonspecific character of the inflammatory response makes it extremely difficult to discover whether the microbial product which initiates the response is a true toxin or an *allergen*—an antigenic product of the microorganism which induces an inflammatory allergic response in a previously sensitized host. Even in experiments involving germ-free animals, it is almost impossible to rule out a previous, sensitizing exposure to an antigen introduced in food or on airborne dust.

• **The nature of toxins**

Most of our knowledge of microbial toxins has come from work on pathogenic bacteria. The search for bacterial toxins began shortly after the discovery of the role of bacteria as etiological agents of human disease. By 1890 the toxins of two important human pathogens, *Corynebacterium diphtheriae* and *Clostridium tetani*, had been discovered. In each case, the discovery was made in the same manner; the bacterium was grown *in vitro* in a culture medium, and a sterile filtrate prepared from the fully grown culture was observed to cause death when injected into experimental animals. Furthermore, autopsies revealed that these animals showed the characteristic lesions associated with the specific natural infection. The toxic substances proved to be heat-labile and are now known to be proteins. Because they were present in the medium, not associated with the bacterial cells, they were termed *exotoxins.*

A number of other pathogenic bacteria have been subsequently shown by comparable methods to produce exotoxins that have specific effects, clearly significant in the causation of the specific disease. However, filtrates prepared from cultures of many important pathogens failed to show toxicity. This led to the examination of the bacterial cells themselves, killed by heat, as possible toxic agents. Such experiments showed that the *cells of nearly all Gram-negative pathogenic bacteria are intrinsically toxic;* furthermore, heat-killed cells of many *nonpathogenic* Gram-negative bacteria show similar toxic effects. The heat-stable toxins associated with the cells of Gram-negative bacteria came to be known as *endotoxins.* As we shall describe later, the endotoxins are relatively nonspecific, all producing much the same clinical and pathological symptoms when injected into experimental animals. Many years of intensive study were required to reveal their nature and cellular origin; it is now known that *endotoxins are lipopolysaccharide-protein complexes, derived from the outer layers of the cell walls of Gram-negative bacteria.*

The names "exotoxin" and "endotoxin" to designate these two classes of toxic substances can be misleading, since there is now good evidence to show that many "exotoxins" are associated with the bacterial cells during growth and are liberated only after death and lysis of the bacteria. Exotoxins can, however, be distinguished from endotoxins by their chemical nature. The former are simple proteins, whereas the latter are molecular complexes which contain protein, lipid, and polysaccharide. Nevertheless, these names are now so firmly entrenched that they are not likely to be abandoned, and we shall use them in the subsequent discussion.

The examination of the cells and culture filtrates of pathogenic bacteria grown *in vitro* led to the recognition of a number of microbial products that damage the host. There remained, however, many important bacterial pathogens, including the causative agents of anthrax and plague, for which this approach failed to reveal any significant toxic product. The environmental conditions in a laboratory culture are always different from those provided in the body of an infected animal, and the recognition of this obvious but previously overlooked fact led to *the search for bacterial toxins that are produced by pathogens in the animal body*. Such work, conducted mainly by H. Smith and his collaborators, was centered originally on anthrax and led to the discovery of the specific exotoxin produced by *Bacillus anthracis*. Later, the toxin of *Yersinia pestis*, the agent of plague, was demonstrated for the first time by comparable experiments.

The toxins of both the anthrax and plague organisms were found to be complexes of two or more substances, each of which was nontoxic by itself but which together acted synergistically to produce a toxic effect. Such knowledge permitted the refinement of the assay systems for the toxins to such an extent that—in each case—it became possible to establish the production of toxin by cultures of bacteria *in vitro*.

The failure to discover the toxin of a virulent organism may often result from the lack of a suitable assay system, particularly in the case of organisms specifically pathogenic for man. For example, the toxin of *Vibrio cholera*, the agent of cholera, escaped detection for many years; neither culture filtrates nor cell extracts exhibited toxicity when injected into experimental animals. The toxin was discovered, however, when filtrates of cultures of *V. cholerae* were injected into ligated intestinal loops of rabbits; such filtrates produced the gross fluid loss and mucosal damage characteristic of the natural disease. Using this assay system, it was possible to purify the cholera toxin and characterize it as a heat-labile protein. The same assay system has permitted the detection and isolation of the toxins of virulent strains of *Escherichia coli*.

• The ecological significance of toxins

For an understanding of pathogenesis, it is not sufficient to isolate a toxic substance from a pathogenic bacterium. If such a substance is to be implicated as a determinant of virulence, it must be shown to produce one or more of the specific symptoms of the disease. Furthermore, the site of action and the effective concentration must be ones that could plausibly obtain in the course of a natural infection. These criteria of ecological significance are extremely difficult to satisfy and have been fully satisfied in relatively few cases. Two other criteria are often used: a correlation between toxin production and virulence in different strains of a pathogenic species, and the ability of antiserum directed against the toxin to protect animals from infection.

All the above mentioned criteria have been satisfied for the toxins of botulism, tetanus, and diphtheria. These exotoxins cause all the symptoms of the respective diseases and are produced only by virulent strains. Antisera against these toxins fully protect individuals from natural disease. Many other microbial products identified as toxins are, without doubt, also ecologically significant, but in each case one or more of the criteria mentioned above is lacking.

• Bacterial exotoxins

Many virulent bacteria liberate toxic substances into the medium when grown *in vitro*. Some of these appear to be ecologically significant toxins; they are listed in Table 29.1. Others do not seem to be related to the diseases produced by the pathogens *in vivo*. Gram-positive bacteria of the genera *Streptococcus*, *Staphylococcus*, and *Clostridium*, for example, produce numerous cytolytic toxins when grown *in*

TABLE 29.1

Some ecologically significant exotoxins

ORGANISM	EXOTOXIN	TARGET TISSUE	MODE OF ACTION
Clostridium botulinum	Neurotoxin	Myoneural junctions	Inhibits release of acetylcholine
Clostridium tetani	Neurotoxin (tetanospasmin)	Central nervous system	Suppresses synaptic inhibition
Clostridium perfringens	α toxin	General (local wound tissue)	Lecithinase activity (cytolysis)
Corynebacterium diphtheriae	Diphtheritic toxin	General (disseminated)	Inhibits protein synthesis
Staphylococcus aureus	α toxin	General	Cytolytic
	"Enterotoxin"	Nerve cells	Unknown
Shigella dysenteriae	Enterotoxin	Intestinal epithelium	Interferes with regulation of electrolyte transport
Vibrio cholerae	Enterotoxin	Intestinal epithelium	Interferes with regulation of electrolyte transport
Yersinia pestis	"Guinea pig toxin"	General	Unknown

vitro. These toxins, many of which appear to act enzymatically, bring about the lysis of a variety of types of mammalian cell (including erythrocytes, thus the designation of some of them as *hemolysins*). The extent, if any, to which the cytolytic toxins contribute to the pathology of the host is unclear, however, with the possible exception of the staphylococcal α toxin.

THE PRODUCTION AND LIBERATION OF EXOTOXINS   The toxic proteins produced by the Gram-positive bacteria appear to be true exotoxins: they are liberated into the medium by intact cells during logarithmic growth. These proteins are not found in the cytoplasm of the bacteria, and it seems probable that they are synthesized by membrane-associated, rather than cytoplasmic, ribosomes. Gram-positive bacteria are able to excrete extracellular enzymes as well as exotoxins, but the mechanism by which these proteins traverse the cell membrane and cell wall is unknown.

The extracellular toxins of the Gram-negative bacteria, however, are not activity excreted into the medium; they accumulate intracellularly and are liberated when the cells die and undergo autolysis.

The function served by toxin production is not apparent, and many nontoxigenic strains develop just as well as their toxigenic counterparts in the host. Indeed, the genes that determine bacterial exotoxins have been found in many cases to be carried by dispensable plasmids or prophages harbored by the bacterium, and not to be part of the bacterial genome. To date, the diphtheria toxin, the erythrogenic toxin of *Streptococcus pyogenes*, enterotoxin A and the α toxin of *Staphylococcus aureus*, and the toxin of *Clostridium botulinum* type D have all been found to be determined by bacteriophage genes, while several toxins produced by strains of *Escherichia coli* have been found to be determined by plasmid genes.

In each of these cases, the loss of the prophage or plasmid renders the cell nontoxigenic, and toxin production is regained when the prophage or plasmid is reintroduced into the cell.

The synthesis of some bacterial exotoxins is strongly influenced by the concentrations of certain metal ions in the medium. High concentrations of iron, for example, inhibit the production of diphtheria toxin, tetanus toxin, the α toxin of *Clostridium perfringens*, and the enterotoxin ("neurotoxin") of *Shigella dysenteriae*, while zinc ions are *required* for the production of α toxin by *C. perfringens*. The mechanism of these effects is not yet known, but it has been speculated that the toxin genes may form parts of operons that are subject to genetic regulation; if so, iron and zinc may act by binding to the products of regulatory genes, either repressors or activators. Such hypotheses can be subjected to direct test, since it is now possible to produce diphtheria toxin in a cell-free protein-synthesizing system programmed with DNA of the toxinogenic phage.

The toxins of *C. botulinum* types A and B, and the ε toxin and ι toxin of *C. perfringens*, are formed by the proteolytic cleavage of larger, nontoxic polypeptides. The detection of such toxins requires that proteolytic enzymes either be liberated into the culture medium by the bacteria or be present in the fluid or cells of the assay system.

THE MODES OF ACTION OF EXOTOXINS   Toxins exert their effects either by destroying specific cellular components or by interfering with specific cellular functions; in a few cases, it has been possible to demonstrate such effects using purified toxins and preparations of sensitive cells. It is difficult, however, to be certain that the effect observed *in vitro* occurs during the natural infection, or—if it does occur—that it represents the *primary* effect of the toxin. A number of exotoxins, for example, are strongly hemolytic, but the lysis of red blood cells is not of major significance in any of the diseases in which these toxins have been implicated.

Some exotoxins have been shown to act as *hydrolytic enzymes*, degrading essential components of host cells or tissues. The α toxin of *Clostridium perfringens*, for example, is a lecithinase. Lecithin is an important lipid constituent of cell membranes and mitochondrial membranes; its hydrolysis by the α toxin results in the destruction of the membrane of many types of cells and could be a primary cause of the tissue damage that occurs in gas gangrene. *Clostridium perfringens* produces a number of other exotoxins, including the k toxin, a collagenase.

The diphtheria toxin, produced by *Corynebacterium diphtheriae*, is an example of a toxin which acts by interfering with a normal cell function rather than by destroying a cell component. When diphtheria toxin is added to mammalian cells in culture, the syntheses of DNA, RNA, and protein are almost instantly inhibited. The *primary* effect of the toxin, however, appears to be the inhibition of protein synthesis, since diphtheria toxin also inhibits the incorporation of amino acids into polypeptides in cell-free systems. Using such systems, it has been found that the primary site of toxin action is the enzyme, transferase II, which catalyses one step in the transfer of the growing polypeptide chain from one tRNA molecule to another on the surface of the ribosome (see p. 238). The toxin enzymatically catalyses the transfer of the ADP-ribose moiety of NAD to transferase II, which is consequently inactivated.

The study of the mode of action of diphtheria toxin has been complicated by the fact that the toxin, as liberated from the bacterium, has no enzymatic activity. Proteolytic digestion in the presence of a thiol reducing agent cleaves the molecule into two fragments, called A and B. Fragment A has full enzymatic activity, while B has none; A and B must be linked together, however, in order for A to enter the mammalian cell, so that neither A nor B is toxic by itself. It is not known whether the cleavage of the toxin with liberation of the enzymatically active fragment occurs at the mammalian cell membrane or in the cytoplasm.

A number of pathogens produce *enterotoxins*, which act specifically on the epithelial cells of the intestine. They include *Vibrio cholerae* and certain other vibrios, *E. coli, C. perfringens,* and *Shigella dysenteriae.* The *Shigella* enzyme is usually referred to as the "Shiga neurotoxin," but it is now recognized that its effects on the central nervous system (seen in the rabbit) are secondary to the vascular damage which it causes upon injection. (Conversely, the "enterotoxin" of *Staphylococcus aureus,* which causes nausea and vomiting, is actually a neurotoxin which acts on the central and autonomic nervous systems.)

The true enterotoxins have been found to act by binding to specific receptors on the membranes of sensitive cells. The bound toxin activates a membrane adenyl cyclase, causing a sharp rise in the intracellular concentration of cyclic AMP; the high concentration of cyclic AMP, in turn, interferes with the normal regulatory processes of the cell and causes an abnormally high rate of electrolyte transport out of the tissues and into the lumen of the intestine. The electrolytes are accompanied by water, and the resulting loss of fluid from the tissues leads to acidosis and frequently shock. Death often occurs from shock, unless the fluid and electrolytes lost from the circulation are replaced.

• Bacterial endotoxins

Endotoxins are lipopolysaccharide-protein complexes derived from the cell walls of Gram-negative bacteria. They are antigenically active and are identical with the somatic, or O, antigens of the whole cell. When a culture of Gram-negative bacteria is allowed to age and autolyze, the endotoxins are liberated in soluble form. They can also be extracted from cell suspensions by appropriate procedures.

Endotoxins have been isolated from all pathogenic Gram-negative bacteria; the best known are those of the enteric bacteria belonging to the genera *Salmonella, Shigella,* and *Escherichia.* The endotoxins of these organisms exhibit two distinct activities: pyrogenicity (fever production) and toxicity.

An endotoxin complex can be separated into a lipopolysaccharide fraction, which is pyrogenic and toxic, and a protein fraction. The protein fraction is neither pyrogenic nor toxic but confers antigenicity on the whole complex. (The role of proteins in antigenicity will be discussed in more detail in Chapter 30.) The isolated lipopolysaccharides are highly active: 10^{-6} g will produce fever in a horse weighing 700 kg.

The purified endotoxins of virulent, as well as of avirulent, enteric bacteria cause many of the symptoms of disease when injected into animals. They are *inflammatory agents,* which increase capillary permeability and produce cellular injury. The injured host cells release additional inflammatory agents, which probably contribute significantly to damage.

The body temperature of the mammal is regulated by mechanisms that are controlled by certain centers in the brain; a number of different substances can, upon introduction into the bloodstream, interfere with the regulatory mechanisms and cause fever. As fever accompanies many different types of inflammation, a search was made in injured tissues for pyrogenic substances. It was eventually found that all pyrogenic preparations contained contaminating bacterial endotoxins.

The analysis of pyrogenicity then focused on the endotoxins, which were found to have extremely high activities. Endotoxins do not act directly on the thermoregulatory brain centers; they cause the release of an *endogenous pyrogen* from polymorphonuclear leucocytes. This substance, the chemistry of which is still unknown, is directly responsible for the fever.

The inflammatory and pyrogenic effects of the endotoxins are *nonspecific*. Although they undoubtedly contribute to the general pathology of the infection, they are not responsible for the specific symptoms of disease caused by the Gram-negative pathogens. It is also important to emphasize that fever and toxicity accompany severe infections with Gram-positive organisms that do not contain endotoxins.

The endotoxins of enteric bacteria cause transient changes in nonspecific resistance to infection when injected into animals in minute doses. Figure 29.1 shows what happens when mice are injected with a purified lipopolysaccharide fraction of an endotoxin and then challenged at successive time intervals by injection with a virulent strain of *E. coli*. Immediately after the administration of the lipopolysaccharide, the mice show greatly increased susceptibility to bacterial infection. The LD_{50} (the number of bacterial cells per animal required to kill 50 percent of the treated animals) is reduced a thousandfold. Within 24 hours, however, their susceptibility to infection diminishes markedly, the LD_{50} rising until it reaches a value about 100,000 times greater than that of the controls. The increase in resistance lasts for 3 to 5 days and then disappears. The role of endotoxin in increasing the resistance of animals to infection is nonspecific; a purified lipopolysaccharide prepared from one species of Gram-negative bacterium confers resistance to infection by other species of Gram-negative bacteria. The mechanism of this effect is not known. It may play a very important role in nature, since all mammals (including man) are constantly exposed to the endotoxins of enteric bacteria.

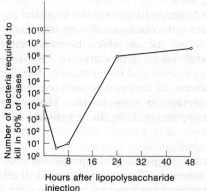

FIGURE 29.1

The effect of endotoxin lipopolysaccharide on resistance to infection (see text for explanation). Replotted from data of O. Westphal.

INFECTIONS OF THE MUCOSAL EPITHELIUM

Almost all microbial infections begin at the surfaces of the mucosal membranes. They can be divided into two groups: those which progress to the subepithelial tissues of the body and those which remain localized at the surface. In the latter the toxins produced by the microorganisms may cause only local damage, or they may diffuse throughout the body and produce lesions in the "target" tissues, i.e., those containing specifically sensitive cells.

• **Attachment without penetration**   The epithelia of the respiratory tract, intestinal tract, and vagina carry a dense coat of microorganisms known collectively as the "normal flora" (see Chapter 28). Each of these epithelial regions has its own, typical normal flora, made up of microorganisms which adhere firmly to the epithelial cells. For example, in the mouth, *Streptococcus salivarius* adheres to the epithelial cells of the tongue and the cheek, but is not present on the surface of the teeth.

Another streptococcus, *S. mutans,* which is a component of the normal flora of the mouth, adheres specifically to the dental enamel and is the etiologic agent of caries. Its adhesion is mediated by a group of capsular polysaccharides, formed only when this bacterium grows at the expense of sucrose. Attack on the dental enamel is caused by intense and localized acid production. These facts explain the long-recognized correlation between the frequency of dental caries and the level of sucrose consumption in man.

The normal flora constitutes a strong barrier to the establishment of invading pathogens, by competing for attachment sites and nutrients and by producing inhibitory substances such as organic acids. The mucous membranes themselves secrete antimicrobial substances, including long-chain fatty acids and lysozyme. In the respiratory tract the epithelial cells beat their cilia constantly and rhythmically, sweeping the film of secreted mucus and loosely adhering microorganisms toward the outer portals of the body.

All these defense mechanisms must be overcome in order for a pathogen to establish itself on the mucosal epithelium; how this is accomplished is not clear, but the ability to compete with the normal flora is an absolute requirement. As discussed in Chapter 28, studies on germ-free animals have confirmed the protective role of the normal flora.

Some examples of pathogens which remain localized on the epithelial surface are *Bordetella pertussis* and *Corynebacterium diphtheriae,* which infect the respiratory tract of mammals, and *Vibrio cholerae,* which infects the intestinal tract. *B. pertussis* is the agent of whooping cough; it produces an exotoxin which acts locally to produce a subepithelial necrosis and inflammation. The exotoxin of *C. diphtheriae* diffuses through the submucosa and is disseminated throughout the body by the bloodstream, causing damage to many tissues. The exotoxin of *V. cholerae* acts locally on the intestinal epithelial cells, as described earlier in this chapter.

• **Penetration of the epithelial cells**   *Shigella dysenteriae* is an example of a pathogen which enters the epithelial cells, causing an erosion and ulceration of the mucosal epithelium of the intestinal tract. The mechanism of penetration appears to involve the diffusion of an extracellular

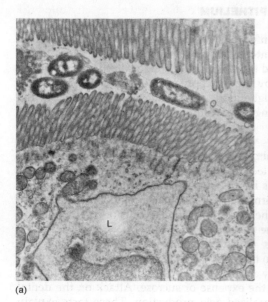
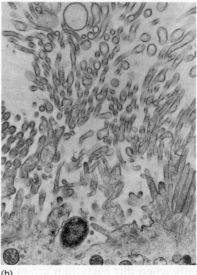

(a)   (b)

FIGURE 29.2

Electron micrographs, thin sections of absorptive epithelial cells of the guinea pig ileum,
12 hours after challenge with *Salmonella typhimurium*. (a) Several organisms are close
to the brush border, which is still intact. L, lymphocyte migrating between epithelial
cells, ×8,400. (b) Degeneration of the brush border and the apical cytoplasm occurs
near an adsorbed bacterium, ×5,000. From A. Takeuchi, "Electron Microscope Studies
of Experimental *Salmonella* infection. I. Penetration into the intestinal epithelium by
Salmonella typhimurium," *Am. J. Pathol.* **50**, 109 (1967).

bacterial product which induces a phagocytosislike activity of the epithelial cells:
the microvilli of these cells can be seen in electron micrograph thin sections to
degenerate when the bacteria are about 300 nm away (Figure 29.2), and the
bacteria pass into the cells within inverted cytoplasmic membranes, ultimately
killing the cells which they enter.

INFECTIONS OF THE SUBEPITHELIAL TISSUES

Once past the mucosal epithelium, the invading microorganism
encounters several new and extremely effective host defense mechanisms, as well
as a considerable variation from tissue to tissue in the biochemical milieu. The
invasiveness of the microorganism depends on its ability to multiply in the
environment provided by the host tissues and fluids and on its ability to overcome
the host defenses.

• The biochemical   During their evolution, the invasive parasitic microorganisms have adapted to
     milieu   grow in one or another tissue of the host. The high degree of tissue specificity

exhibited by many of them may be inferred to reflect differences between tissues in the biochemical milieu which they provide, at least in those instances in which no difference in defense activity is apparent. To date, however, only one such biochemical difference has been clearly documented, and that is the case of erythritol. Erythritol is the preferred carbon source of several species of *Brucella*, which cause abortions in ungulates; it is found in high concentrations in the placenta of ungulates, but not in other ungulate tissues or in the placenta of various species (man, rat, rabbit, guinea pig) resistant to placental infection by these organisms.

In some infectious diseases, notably tuberculosis, the limiting factor for microbial growth is the *availability of free iron*. Both the vertebrate host and the microbial parasite depend on chelating agents, which they excrete, for the transport of iron into their cells. Mammals liberate iron-chelating proteins called *transferrins* into the bloodstream. Bacteria excrete a variety of iron chelators of low molecular weight; the specific chelator excreted by tubercle bacilli is called *mycobactin*.

The interactions between chelators and cells are highly specific. Thus, mammalian cells, but not tubercle bacteria, can utilize transferrin-bound iron, but the opposite is true for mycobactin-bound iron. The result is a *battle for iron*, in which the outcome depends on the relative binding strengths and concentrations of the chelating agents excreted by host and parasite. The injection of iron or an appropriate microbial chelating agent into an infected host favors the parasite, while the injection of agents which depress free iron levels protects the host against infection.

The effect of iron on the course of some microbial infections may be complicated by its influence on toxigenicity: high concentrations of available iron depress exotoxin production by such pathogens as *Corynebacterium diphtheriae* and *Clostridium tetani*, while favoring their invasiveness.

• **Constitutive host defenses**

Vertebrates have evolved numerous mechanisms that keep microbial parasites in check. Some of these mechanisms are *constitutive*: they are properties of the normal host. Others are *inducible*: they appear only in hosts that have undergone an induction process such as prior exposure to the infectious organism or its products.

The principal constitutive defenses encountered by microorganisms which penetrate beyond the epithelial layers are the *phagocytic cells* of the body, which engulf and digest foreign particulate matter; certain *antimicrobial substances* in the tissues and circulating fluids; and a complex of reactions known as *inflammation*.

• **The phagocytic cells**

Phagocytic cells are found both in the circulatory system and in the tissues. The white blood cells, or leucocytes, are described in Table 29.2 and Figure 29.3. They include two types which are actively phagocytic: *monocytes* and *granulocytes*.* Granulocytes owe their name to the presence of granules in their cytoplasm; they are further differentiated, on the basis of the staining properties of the granules, into *eosinophils* (granules stainable by the acid dye eosin), *basophils* (granules stainable by the basic dye methylene blue), and *neutrophils* (granules stainable by a mixture of acidic and basic dyes). All the granular leucocytes are capable of phagocytosis,

---

*Granulocytes are also known as *polymorphonuclear leucocytes*, because of the irregular shape of their nuclei.

TABLE 29.2

The leucocytes

CELL TYPE	LOCATION AND DEVELOPMENT	PRINCIPAL FUNCTION IN DEFENSE
Monocytes	Circulate in the bloodstream; move by chemotaxis into inflamed tissues, developing into tissue macrophages	Phagocytosis
Granulocytes	Circulate in the bloodstream; move by chemotaxis into inflamed tissues	Phagocytosis (principally in the neutrophilic granulocytes)
Fixed macrophages	Lining the sinuses of the liver, spleen, bone marrow, and lymph nodes (the reticuloendothelial system)	Phagocytosis
Lymphocytes	Circulate in the bloodstream; move by chemotaxis into inflamed tissues. Antigenic stimulation of one type (B cells) causes enlargement and development into plasma cells	Secretion of antibody (plasma cells); cellular immune functions (T cells) (see Table 29.3)

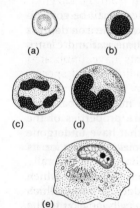

(a) (b)

(c) (d)

(e)

FIGURE 29.3

Some types of blood cells: (a) an erythrocyte; (b) a small lymphocyte; (c) a granulocyte; (d) a monocyte; (e) a macrophage.

but only in the neutrophils is phagocytosis sufficiently active to play a significant role in host defense.

When tissue infection occurs, chemotactic substances are liberated which attract both neutrophils and monocytes to the site of injury. During this process, which we shall describe in more detail in the section on inflammation, the monocytes enlarge and develop into cells called *macrophages*, which have enhanced phagocytic activity. Macrophages also occur as fixed cells in certain organs: the liver, spleen, bone marrow, and lymph nodes. Since they are for the most part reticular (forming a supporting network) or endothelial (lining the sinuses) in these organs, the fixed macrophages are collectively referred to as the *reticuloendothelial system*. The reticuloendothelial system constitutes a major line of defense in the animal body, actively filtering out microorganisms that have penetrated outer barriers.

THE PHAGOCYTIC PROCESS   The process of phagocytosis occurs in two steps. In the first step the microorganism becomes *attached to the membrane of the macrophage*. Attachment then triggers the ingestion process: pseudopods extend on all sides of the microorganism, fusing with one another so as to enclose it within a membrane-limited vacuole (Figure 29.4). The ingestion process requires metabolic energy, most of which is derived from glycolysis. The lactic acid produced by glycolysis diminishes the pH in the vacuole to 5.5 or below.

THE DESTRUCTION OF PHAGOCYTIZED MICROORGANISMS   Once a phagocytic vacuole has formed in the phagocyte, it fuses with lysosomes, and the enclosed bacteria are killed and digested through the action of the complex mixture of substances released from the lysosomes. These include several microbicidal substances: two enzymes, lysozyme and myeloperoxidase; basic polypeptides; and a protein termed *phagocytin*. A variety of lysosomal hydrolases with low pH optima digest the macromolecular constituents of the killed bacterial cells. Of these components,

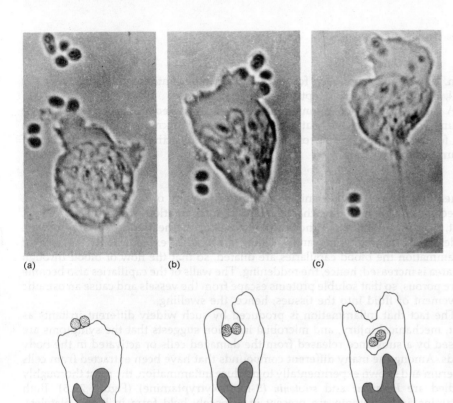

(a) (b) (c)

(d) (e) (f)

FIGURE 29.4

The ingestion of pneumococci by a phagocyte in the presence of antibody. (a) Two pneumococci are in contact with a pseudopodium of the phagocyte. (b) The same two pneumococci are inside the phagocyte, one on each side of a lobe of the nucleus; a group of four pneumococci are in the process of being ingested. (c) Six of the eight pneumococci have been engulfed. (d), (e), (f) Schematic diagram of the stages in phagocytosis, showing the formation of the phagocytic vacuole by an inversion of the cell membrane. (a), (b) and (c), from W. B. Wood, Jr., M. R. Smith, and B. Watson, "Studies on the Mechanism of Recovery in Pneumococcal Pneumonia: IV. The Mechanism of Phagocytosis in the Absence of Antibody," *J. Exptl. Med.* **84,** 402 (1946).

myeloperoxidase appears to have a particularly important role as a lethal agent: the phagocytes of individuals who are deficient in the myeloperoxidase system are relatively ineffective in destroying ingested microorganisms, and these individuals are highly susceptible to infectious diseases.

THE RESISTANCE OF VIRULENT MICROORGANISMS TO PHAGOCYTIC DESTRUCTION   Many pathogens owe their invasiveness, and thus their virulence, to their ability to resist phagocytosis or intracellular destruction. In some cases, this resistance reflects the excretion by the microorganism of substances which block the phagocytic process; these substances, along with other microbial products that promote the invasive spread of pathogens in the host, are known collectively as *aggressins*. Some aggressins inhibit the chemotactic response of blood phagocytes; some (notably the capsule) block attachment and ingestion; and some inhibit the intracellular destruction of phagocytized cells. In addition, many pathogens excrete substances, called *leucocidins*, which kill phagocytes.

• Antimicrobial
substances in body
tissues and fluids

We have seen that phagocytes contain a variety of antimicrobial substances. One of these, lysozyme, is present in many other parts of the body: it has been detected in tears, nasal secretions, saliva, mucus, and extracts of various organs including skin. The tissues have been found to contain other antimicrobial substances as well, including basic polypeptides and polyamines.

A different group of antimicrobial substances, called *beta lysins,* occur in the serum. The beta lysins are heat-stable substances of unknown composition, which are formed by the clotting of platelets. They are active against a number of Gram-positive bacteria.

• **Inflammation**

When a tissue of a higher animal is subjected to any of a variety of irritations, it becomes "inflamed." The characteristics of inflammation—reddening, swelling, heat, pain—are familiar. Although the causes of the heat and pain are not well understood, the reddening and swelling are readily explained. In the area of inflammation the blood capillaries are dilated, so that the flow of blood through the area is increased; hence, the reddening. The walls of the capillaries also become more porous, so that soluble proteins escape from the vessels and cause an osmotic movement of fluid into the tissues; hence, the swelling.

The fact that inflammation is produced by such widely different irritants as heat, mechanical injury, and microbial infection suggests that the symptoms are caused by a substance released from the damaged cells or activated in the body fluids. Among the many different compounds that have been extracted from cells or serum and shown experimentally to produce inflammation, the most thoroughly studied are *histamine* and *serotonin* (5-hydroxytryptamine) (Figure 29.5). Both histamine and serotonin are present in a loosely held form in blood platelets, as well as in the cells of many tissues, and are released in response to a variety of stimuli.

As the inflammatory response develops, a striking change occurs in the behavior of the granulocytes. At first, they adhere to the inner walls of the capillaries. Next, they push their way between the cells of the capillary walls and enter the tissues, a process that can take as little as 2 minutes [Figure 29.6(a-d)]. If the inflammation has been initiated by a bacterial infection, the granulocytes move toward the focus of infection in response to chemotactic substances liberated by the bacteria. As the inflammation progresses, the phagocytes release lysosomal enzymes that damage and eventually destroy neighboring tissue cells.

In the later stages of inflammation the granulocytes that have accumulated at

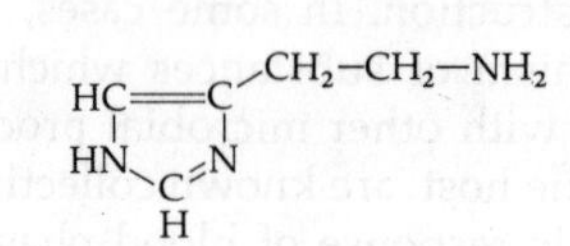

FIGURE 29.5

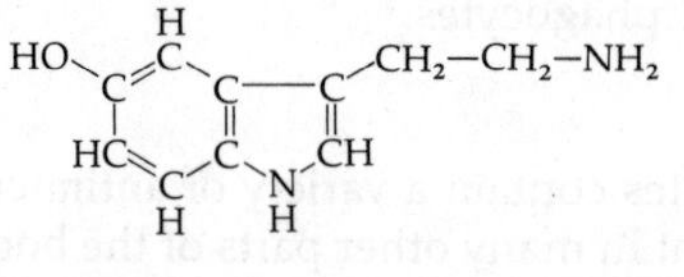

The structures of histamine and serotonin. Histamine is the decarboxylation product of histidine; serotonin is the decarboxylation product of 5-hydroxytryptophan.

histamine

serotonin

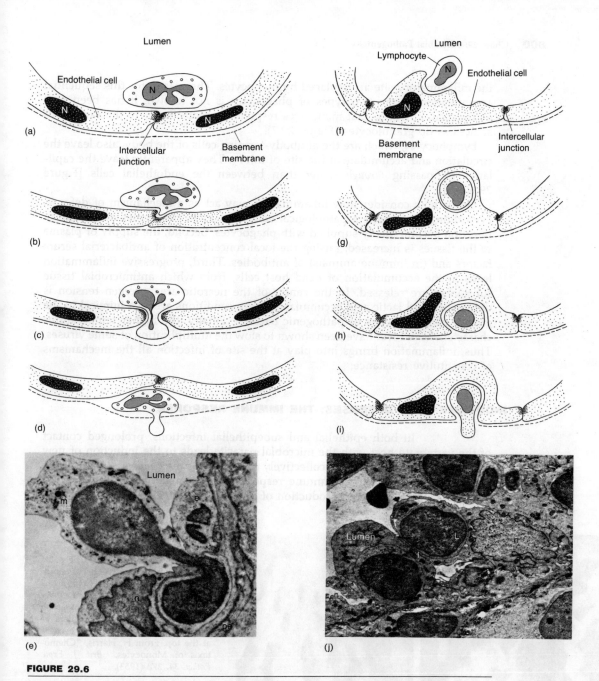

FIGURE 29.6

(a to d) Stages in the migration of a granulocyte through the venule wall. The cell penetrates an intercellular function and remains extracellular at all times. (e) Part of an inflamed venule. Cell m is a monocyte, which is penetrating an intercellular junction by the same mechanism; e, endothelium; n, nucleus of an endothelial cell; pe, periendothelial sheath. (f to i) Stages in the migration of a small lymphocyte through an endothelial cell of a venule. The lymphocyte is totally intracellular at one stage in the passage. (j) Part of a venule from a normal lymph node; a lymphocyte (l) is completely enclosed by the cytoplasm of an endothelial cell; n, nucleus of the endothelial cell. From V. T. Marchesi and J. L. Gowans, ''The Migration of Lymphocytes through the Endothelium of Venules in Lymph Nodes: An Electron Microscope Study,'' *Proc. Royal Soc. B,* **159,** 283 (1964).

the inflammatory site are replaced by monocytes. The reason for this sequential accumulation of the two types of phagocytes is not known; it has been well established, however, that monocytes respond chemotactically to the same substances as do granulocytes (Figure 29.7).

Lymphocytes, which are the antibody-forming cells of the body, also leave the circulation and accumulate at the site of injury. They apparently leave the capillaries by passing *through* rather than between the endothelial cells [Figure 29.6(f–i)].

Let us now consider how inflammation may act as a mechanism of defense, even though it is itself a pathological condition. First, the tissues at the site of infection become richly supplied with phagocytes. Second, the supply of plasma to the tissues is increased, raising the local concentration of antibacterial serum factors and (in immune animals) of antibodies. Third, progressive inflammation leads to the accumulation of dead host cells, from which antimicrobial tissue substances are released. In the center of the necrotic area, oxygen tension is diminished and lactic acid accumulates; these conditions are also inimical to the growth of many types of pathogenic bacteria. Finally, the higher temperatures characteristic of fever have been shown to slow the multiplication of some viruses. Thus, inflammation brings into play at the site of infection all the mechanisms of constitutive resistance.

INDUCIBLE HOST DEFENSES: THE IMMUNE RESPONSE

In both epithelial and subepithelial infections, prolonged contact of the vertebrate host with the microbial parasite leads to the induction of new and specific defense reactions collectively referred to as the *immune response*. There are two distinct classes of immune response: the production of circulating or secreted *antibodies*, and the production of specifically *sensitized cells*. The immune

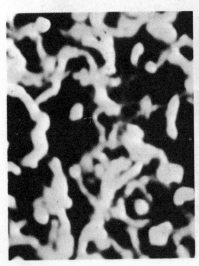 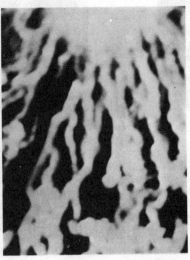

(a)   (b)

FIGURE 29.7

Chemotaxis of monocytes. The paths of the monocytes have been photographically recorded as light areas. Photograph (a) shows random movement of the monocytes; photograph (b) shows the directed movement of monocytes toward a clump of bacteria at the top. From H. Harris, "Chemotaxis of Monocytes," *Brit. J. Exptl. Pathol.* **34**, 278 (1953).

response usually reduces the toxigenicity and invasiveness of the parasite to the point where the constitutive host defenses can eliminate it. In some cases, however, the pathogen possesses virulence factors that interfere with the immune response or confer resistance to it. Furthermore, in some infectious diseases the immune response itself seriously damages the host tissues by the reaction known as *hypersensitivity* (*allergy*).

• **Antibody-mediated immunity**

The term *antibody* refers to a group of related globulin proteins which are capable of specific, noncovalent binding to the molecules which induce their formation. Such inducing molecules are called *antigens*; practically all proteins are antigenic, as are many polysaccharides. In addition, many types of molecules which are not intrinsically antigenic become antigenic when attached to proteins; this class includes lipids, nucleic acids, carbohydrates, and numerous small molecules.

There are several classes of antibody molecules, or *immunoglobulins* as they are called, but they all have certain structural features in common. The basic unit of each antibody molecule is a complex of four polypeptide chains, two identical heavy chains and two identical light chains, folded in such a manner as to produce on its surface two identical regions, called Fab, which contain the *antigen-binding sites*. There is also a third region, called Fc, which gives the antibody additional specificity such as the ability to bind to the membranes of certain cells (Figure 29.8).

The antigen-combining sites are shallow cavities in the globulin molecule which fit, as a lock fits a key, complementary projections on the surfaces of antigens. Such a projection is called an *antigenic determinant,* and the precise portion which fits the antibody's combining site is called a *haptenic group*.

The antigenic specificity of a protein can be changed if the protein is coupled chemically to a small molecule which by itself is wholly devoid of antigenicity.

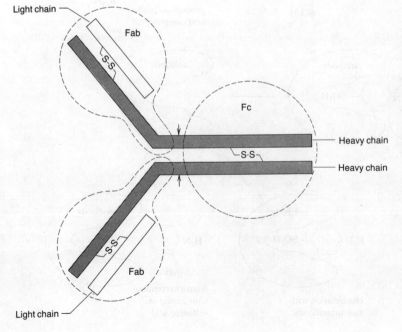

Light chain

Fab

S-S

Fc

S-S

Heavy chain

Heavy chain

S-S

Fab

Light chain

FIGURE 29.8

Schematic diagram of an immunoglobulin. The molecule consists of two heavy polypeptide chains and two light polypeptide chains, held together by disulfide bridges as shown. The molecule can be hydrolysed with papain, which cleaves it at the site indicated by the arrows to produce three fragments: two identical fragments, called Fab, each of which contains one antigen-binding site, and one crystallizable fragment, called Fc, which contains sites that determine the binding of antibody to specific host cells.

The antibodies that are formed against the modified protein may not combine with the original protein but may be specific in their ability to combine with the modified protein. Furthermore, these antibodies can combine with the small molecule itself, unattached to any protein. For example, the molecule sulfanilic acid can be coupled to a protein, as shown in Figure 29.9, and used to immunize an animal. The ability of the resulting antibody to combine with free sulfanilic acid is shown by the fact that free sulfanilic acid will specifically interfere with the combination between antibody and the protein to which sulfanilic acid is bound. The specificity of the antibody is remarkable; for example, an isomer of sulfanilic acid, in which the —SO$_3$H group is in the *ortho* position relative to the amino group, will not combine with antibody formed against the protein-sulfanilate in which the —SO$_3$H group is *para* to the amino group.

FIGURE 29.9

The coupling of sulfanilate to protein, and the specificity of the antibody formed against the complex.

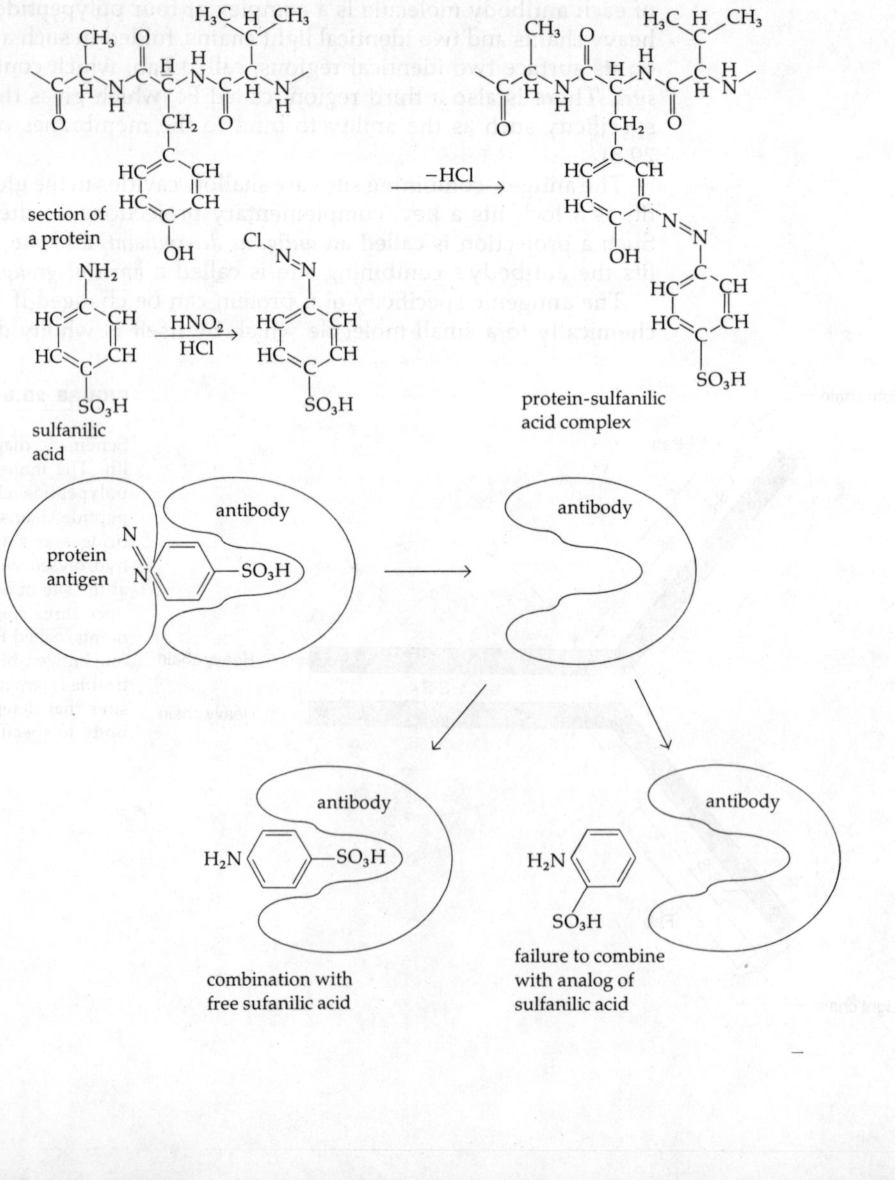

protein-sulfanilic acid complex

section of a protein

sulfanilic acid

combination with free sulfanilic acid

failure to combine with analog of sulfanilic acid

Sulfanilic acid thus behaves as a *hapten*. A hapten may be defined as a non-immunogenic molecule which, when coupled to a protein, determines the antigenic specificity of the complex; in its free form a hapten will combine with the antibody, the specificity of which it has determined.

An individual vertebrate is capable of forming antibodies against many thousands of different antigens. Since antibodies are so highly specific, this means that an individual can produce thousands of different heavy and light chains, each with a unique set of amino acid sequences. It is known from molecular genetics that differences in amino acid sequence reflect differences in the nucleotide sequence of DNA. The inescapable conclusion is that the genome of the vertebrate animal contains many thousands of genes coding for different immunoglobulin polypeptides, each determining a different antigen-binding specificity, or—alternatively—that the genetic differences arise by somatic mutation and/or recombination during the development of the antibody-producing cells of the individual.

THE PRODUCTION OF ANTIBODY   Antibodies are produced by a class of white blood cells called *lymphocytes*. There are two types of lymphocytes, which play very different roles in the immune response: "B cells," which actively secrete antibody, and "T cells," which mediate cellular immunity and also play a role in the process by which antigens induce antibody synthesis by B cells (Table 29.3).

During the differentiation of B cells, one (and only one) set of immunoglobulin polypeptide genes is activated in each cell. Activation is random, different B cells producing different immunoglobulins; successive cell divisions of an activated B

TABLE 29.3

Some properties of B cells and T cells

PROPERTY	B CELLS	T CELLS
Antigen-binding receptors on cell surface	Specific antibodies	Specific receptor molecules: relation to antibodies uncertain
Response to binding of cognate antigen	Enlarge, multiply, develop into plasma cells which secrete antibodies	Enlarge, multiply, liberate the products described below
Function in antibody production	Synthesize and liberate antibodies	Stimulate antibody production by B cells by (1) liberation of factors which stimulate B cell development (2) direct interaction with B cells and macrophages (in response to certain classes of antigen)
Cytotoxic activity	None	Antigen-stimulated T cells kill antigen-bearing target cells on contact
Effects on macrophages	None	Liberate factors which (1) stimulate phagocytic activity of macrophages (2) attract macrophages (3) inhibit further movement of macrophages

cell thus give rise to a *clone* of specific antibody-forming cells. At birth, the lymphocytes of an individual constitute a mixed population of such clones, the total population forming in aggregate many thousands of different antibodies. T cells undergo a parallel process of diversification, the cells of a given clone containing a unique *antigen receptor* in their cell membranes. The two cell types enter the circulation as *small lymphocytes.*

The antibody molecules produced by small B lymphocytes are not secreted in detectable amounts, but they are present in abundance on the surface of the cell membrane. Small B lymphocytes rarely divide, except when one or more specific antigen molecules combine with the membrane-bound antibody. This causes the cell to grow and divide rapidly; its numerous progeny mature into a new, larger type of cell (the *plasma cell*) which synthesizes and secretes antibody at a high rate. T cells are also stimulated to proliferate and enlarge as a result of contact with specific antigen; the specific products which they liberate include some which potentiate the response of related B cells and others which function in the processes of cellular immunity (see below).

The events described above account for many of the phenomena which are observed during the infectious process. The antigens on the surface of the microorganism or liberated from it bind to the surfaces of those small lymphocytes, both B and T, which bear the cognate antigen receptors. These lymphocyte clones are then greatly amplified by proliferation (a process called *clonal selection*), and the B cell progeny (plasma cells) begin actively producing their specific class of antibodies. Ultimately, the infection terminates and no more antigen is produced; antibody synthesis declines, and many of the enlarged lymphocytes revert to the small, resting type. The individual now possesses vastly enlarged clones of *memory cells* which are capable of producing, at the next stimulus, antibody of the specific type. Thus, an individual who has recovered from a microbial disease will, on reexposure, produce antibodies against that microorganism much more rapidly than during the initial infection: this is the so-called *anamnestic response.*

The requirement of B cells for T cell cooperation in the process of antibody formation varies with the type of antigen involved. In those cases in which T cells must be present, there is also a requirement for an interaction with *macrophages* which have antigen bound, nonspecifically, to their surfaces. Thus, in many cases, antibody formation by B cells requires a three-way cooperation between B cells, T cells, and macrophages.

ANTIMICROBIAL CONSEQUENCES OF ANTIGEN-ANTIBODY BINDING Both virions and microbial toxins must bind to receptors on host cell membranes if they are to damage the host. The combination of antibody with a virion or toxin molecule may be sufficient to block such binding, and thus to protect the host completely. This is termed *neutralization.*

The binding of antibody to the surface of microbial cells has little direct effect on microbial growth or toxin production. However, it makes all bacterial cells much more *susceptible to phagocytosis.* Furthermore, antibody-binding makes the cells of Gram-negative bacteria susceptible to lysis by a set of normal blood proteins called *complement.* Finally, the combination of antibody with soluble microbial antigens triggers a greatly increased *inflammatory response* by the host.

THE LYSIS OF ANTIBODY-COATED CELLS BY COMPLEMENT   In 1894 R. Pfeiffer and his coworkers discovered that certain Gram-negative bacteria were lysed by antiserum prepared against them. The serum lost its lytic activity when heated to 56°C for a few minutes; the activity was restored by the addition of fresh *normal* serum. Lysis of the bacterial cells thus requires both specific antibody and a nonspecific component of normal serum; the latter is named *complement*. Complement is now known to be a group of proteins rather than a single substance; in human serum, complement comprises 11 different proteins. Antibody-coated Gram-negative bacteria are generally susceptible to lysis by complement, whereas Gram-positive bacteria are totally resistant to this host defense mechanism. The basis of this difference has not been discovered.

Red blood cells, which have bound antibodies directed against their surface, are also lysed by complement. The mechanism of this *complement fixation reaction* has been studied in detail and can be summarized as follows.

The first step is the binding of antibody to the surface of the red cell. Antibody that is so bound is altered, perhaps by an allosteric transition, so that in turn it binds the component of complement called C'1.* C'1, which has been discovered to be a complex of three different proteins, has no enzymatic activities. When bound to the antigen-antibody complex, however, it is converted to a form that has both esterase and protease activity and reacts with C'4, cleaving it to form a fragment (C'4b) that binds to the complex also. The complex now consists of cell-antibody-C'1-C'4b; this complex now binds C'2, which is cleaved by C'1 to form a bound fragment called C'2a.

In a further series of steps the complex successively adds C' components until it consists of cell-antibody-C'1, 4b, 2a, 3b, 5b, 6, 7, 8, and 9. At the C'5 step, the surface of the red cell membrane becomes covered with lesions (Figure 29.10); lysis does not occur, however, until C'8 and C'9 have been bound.

ANTIBODY-FACILITATED PHAGOCYTOSIS (OPSONIZATION)   As already mentioned, microbial cells coated with antibody are more susceptible to phagocytosis. Their susceptibility is further enhanced by the binding of C'1a, 4, 2a, and 3. The process by which antibodies and complement molecules render cells susceptible to phagocytosis is called opsonization (Gr. *opsonein*, to prepare food for). Opsonizing antibodies act by combining with, and thus neutralizing, antigenic components of microbial cells which block phagocytosis.

Agglutinating antibodies also promote phagocytosis. Since antibodies are bivalent, and the microbial cell has many haptenic groups on its surface, the combination of antibody with cells in appropriate proportions leads to the formation of clumps of cells, which are more readily phagocytized than single cells.

INDUCTION OF THE INFLAMMATORY RESPONSE   Complement is fixed not only by antibody-coated cells, but also by the molecular aggregates that form when multivalent soluble antigens react with antibodies. Antigen-antibody complexes that have bound C'1a, 4, 2a, 3, and 5 release polypeptide fragments of C'3 and C'5 that cause mast cells (basophil-like cells associated with capillaries and connective tissues) to release histamine; the complexes also serve as chemotactic

*Complement is abbreviated C'. The components of complement have been numbered, the numbering system reflecting in part the historical order of their discovery.

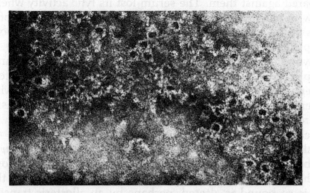

FIGURE 29.10

Sheep red cell, sensitized with rabbit antisheep red cell antibody and lysed with complement, ×233,000. Courtesy of M. J. Polley.

attractants for some phagocytes. Antigen-antibody complexes, through their activation of complement, thus increase the intensity of the inflammatory response over that induced by the infectious process alone.

HOST DAMAGE CAUSED BY ANTIBODY-MEDIATED HYPERSENSITIVITY   We have seen that a primary contact with antigen may produce the immune state, in which a greatly amplified population of specific small lymphocytes persists as "memory cells," ready to respond to subsequent antigen challenge by further proliferation and antibody production.

Under some circumstances, however, the primary contact with antigen may leave the host not only with lymphocytic memory cells, but also with *specific antibodies attached to the membranes of mast cells and basophils*. If antigen molecules of the original type are reintroduced into the body at a later date, they combine with the cell-bound antibody and trigger the release of massive amounts of histamine and serotonin. These compounds mediate excessive inflammatory reactions, and sometimes trigger a fatal response called *anaphylactic shock*.

A different type of inflammation is produced when antigen-antibody aggregates accumulate in the blood and lymphatic vessels, fixing complement and inducing an inflammatory response so severe as to cause more host damage than host protection. The damage in such cases is mainly produced by lysosomal enzymes released from neutrophils, and is characterized by the destruction of small blood vessels (vasculitis). The lesions which are seen are called *Arthus reactions*, after their discoverer; they are seen in some bacterial infections, in which case bacterial antigens are believed to be responsible.

Two serious diseases occur as sequels to localized, superficial infections by Group A streptococci: acute glomerulonephritis, in which kidney tissue is damaged, and rheumatic fever, in which heart valve tissue is damaged. In both cases, antibodies to streptococcal antigens cross react with host tissue antigens, producing the specific damage: a phenomenon known as *autoimmunity*. Further damage is produced every time the individual is reexposed to streptococcal antigens.

The pathological results of immune sensitization are called *hypersensitivity*, or *allergy*. Sensitivity mediated by circulating antibody molecules is referred to as "immediate type," since the inflammatory response to the secondary introduction of antigen is very rapid. Sensitivity which results from the accumulation of specifically sensitized T cells is referred to as "delayed type," since the inflammatory response to the secondary introduction of antigen is much slower to appear. The latter process is discussed below, in the section on cell-mediated immunity.

MICROBIAL RESISTANCE TO ANTIBODY-MEDIATED HOST DEFENSES We have seen that the coating of bacteria by antibody and complement factors C'1 through C'3 renders them susceptible to phagocytosis and—in the case of Gram-negative bacteria—to lysis by the further actions of complement factors C'5 through C'9. Some Gram-negative pathogens, however, produce surface components which—although antigens themselves—block the processes of opsonization and lysis.

The best known of these substances are the K antigen of *Escherichia coli* and the Vi antigen of *Salmonella typhi*. Their mechanism of action was determined by experiments in which red blood cells were coated with K or Vi, as well as with antired blood cell antibody, and then treated with complement. The presence of K or Vi was found to block one or more steps in the sequence of C' interactions.

The antigenic determinants of the K and Vi antigens are acidic polysaccharide groups of the Gram-negative cell wall. Antibody may be formed against these groups; such antibodies do not bring about complement-mediated lysis, although they remain opsonic. A good correlation has been found between the K antigen level of *E. coli* pathogens and their relative virulence in systemic infections. The correlation is not as clear in enteric infections, in which other aggressins may be more important.

• Cell-mediated immunity

A large number of immune responses have been found to be mediated not by antibodies secreted by B cells, but rather by the secretions and direct cell contact activities of immune (antigen-specific) T cells. This relationship is dramatically revealed in individuals who are genetically deficient in T cell production: they are highly susceptible to infections by protozoa, fungi, viruses, and intracellularly proliferating bacteria, and they are deficient in tumor rejection and graft rejection. However, the activities of sensitive T cells in the immunized normal individual may, upon subsequent exposure to antigen, produce inflammatory responses of pathological severity. This is the condition of "cell-mediated hypersensitivity," or "delayed-type" hypersensitivity.

PROTECTION AGAINST MICROBIAL INFECTION When a T cell binds its cognate antigen, it acquires *cytotoxicity*: direct contact between the T cell and a "target cell" bearing the specific antigen leads to the death of the target cell. The cytotoxic activities of T cells have been mainly observed in experimental systems in which the target cells (carrying the specifically stimulatory antigen) have been vertebrate cells from another species. T cell cytotoxicity probably plays an important role in suppressing viral infections, by the killing of infected host cells that express viral antigens on their surfaces.

Another modification of T cells which results from antigen binding, and which is of major significance in host protection against bacterial infections, is the liberation of a substance which activates macrophages. The activated macrophages, anthropomorphically referred to as "angry killers" by some authors, have greatly increased phagocytic activity and much higher levels of lysosomal enzymes. Their increased ability to kill ingested microorganisms is *nonspecific*. Thus, a host who has recovered from a microbial infection possesses a population of T cells which, on later exposure to the antigens of the same microorganism, will activate macrophages to become angry killers not only of this microorganism but of other microorganisms as well.

HOST DAMAGE CAUSED BY CELL-MEDIATED HYPERSENSITIVITY   Antigen-activated T cells not only activate macrophages, but also attract them and subsequently inhibit their movement. Since T cells themselves are attracted to loci of infection, the net result is the concentration of both activated T cells and macrophages at the site of an infection. The liberation of soluble toxins (lymphotoxins) by the T cells and of lysosomal enzymes by macrophages produces extensive tissue necrosis. Such necrosis is dramatically evident in the lesions which form around tubercle bacilli, and the very reactions which serve to contain the bacteria (cell-mediated immunity) also produce the characteristic pathology of the disease (cell-mediated hypersensitivity).

IMMUNOSUPPRESSIVE ACTIVITIES OF MICROBIAL PATHOGENS   Some microbial pathogens produce substances which *suppress the immune response.* These substances represent a subclass of aggressins. For example, a substance can be extracted from the cytoplasm of Group A streptococci that inhibits antibody formation against sheep red cells in mice, and polysaccharide extracts of *Bordetella pertussis* will diminish the delayed hypersensitivity response to tuberculin (an antigenic fraction from tubercle bacilli). In the latter case, it is probable that the aggressins inhibit the proliferation and/or activities of lymphocytes.

FURTHER READING

Books   AJL, S., et al., (editors), *Microbial Toxins,* Vols. 1–8. New York: Academic Press, 1970–1972.

DAVIS, B. D., R. DULBECCO, H. N. EISEN, H. S. GINSBERG, W. B. WOOD, JR., and M. MCCARTY, *Microbiology,* 2nd ed. New York: Harper & Row, 1973.

ROSE, N. R., F. MILGRAM, and C. J. VAN OSS (editors), *Principles of Immunology.* New York: Macmillan, 1973.

SMITH, H., and J. H. PEARCE (editors), *Microbial Pathogenicity in Man and Animals: 22nd Symposium of the Society for General Microbiology.* New York: Cambridge University Press, 1972.

Review articles   BURROWS, W., "Cholera Toxins," *Ann. Rev. Microbiol.* **22,** 245 (1968).

COLLIER, R. J., "Diphtheria Toxin: Mode of Action and Structure," *Bacteriol. Rev.* **39,** 54 (1975).

GALLY, J. A., and G. M. EDELMAN, "The Genetic Control of Immunoglobulin Synthesis," *Ann. Rev. Genetics* **6,** 1 (1972).

HIRSCH, J. G., "Phagocytosis," *Ann. Rev. Microbiol.* **19,** 339 (1965).

PATIL, S. S., "Toxins Produced by Phytopathogenic Bacteria," *Ann. Rev. Phytopathol.* **12,** 259 (1974).

PIERCE, N. F., W. B. GREENOUGH, and C. C. J. CARPENTER, JR., "*Vibrio cholerae* Enterotoxin and Its Mode of Action," *Bacter. Rev.* **35,** 1 (1971).

809

Schwab., J. H., "Suppression of the Immune Response by Microorganisms," *Bact. Rev.* **39**, 121 (1975).

Spector, W. G., and D. A. Willoughby, "The Inflammatory Response," *Bact. Rev.* **27**, 117 (1963).

Original articles   Gill, D. M., A. M. Pappenheimer, Jr., and T. Uchida, "Diphtheria Toxin, Protein Synthesis and the Cell," *Fed. Proc.* **32**, 1508 (1973).

Klebanoff, S. J., "Myeloperoxidase-Halide-Hydrogen-Peroxide Antibacterial System," *J. Bacteriol.* **95**, 2131 (1968).

Krinsky, N. I., "Singlet Excited Oxygen as a Mediator of the Antibacterial Action of Leukocytes," *Science* **186**, 363 (1974).

Murphy, J. R., A. M. Pappenheimer, Jr., and S. T. deBorms, "Synthesis of Diphtheria *Tox*-Gene Products in *Escherichia coli* Extracts," *Proc. Natl. Acad. Sci. U.S.* **71**, 11 (1974).

Weinberg, E. D., "Iron and Susceptibility to Infectious Disease," *Science* **184**, 952 (1974).

30

MICROBIAL DISEASES OF MAN

Man is host to a variety of pathogenic bacteria, protozoa, and viruses. In addition, certain members of the normal microbial flora that inhabit the skin and mucous membranes may become invasive and produce disease when the mechanisms of host immunity are suppressed.

The properties of the parasite which cause damage to the host vary among the major groups. As discussed in the previous chapter, toxins are responsible for the pathological consequences of many bacterial infections; they are absent or rare in infections by fungi and protozoa, however, most of which owe their pathogenicity to the induction of hypersensitivity reactions. Hypersensitivity also plays a role in many viral diseases, along with the cell damage caused directly by intracellular viral growth.

Differences are also observed in the inducible mechanisms of host resistance by means of which the different microbial groups are kept in check. Circulating antibodies play a major role in the defense against many bacterial pathogens, but are probably of little significance in other types of infections. Cell-mediated immunity, however, is of major importance in host resistance to fungi and viruses.

These conclusions have emerged from observations of persons exhibiting different states of *immunodeficiency,* either genetic or induced by immunosuppressive drugs. Thus, persons who are deficient in the production of circulating antibodies are highly susceptible to respiratory infections by Gram-positive bacteria; persons who are deficient in T cell functions, however, tend to succumb to infections by fungi and viruses, as well as to bacteria which grow predominantly intracellularly (e.g., tubercle bacilli and brucellae). There are not yet sufficient data concerning protozoan infections to evaluate the relative contributions of antibody-mediated and cell-mediated immunity.

In this chapter we shall document some of the most important microbial diseases of man, and in so doing illustrate the variety of mechanisms which underlie host-pathogen relationships.

BACTERIAL DISEASES

The obligate bacterial parasites of man depend for their survival on their transmission from one individual to another. They have thus evolved, in each case, a characteristic *portal of exit, mode of transmission,* and *portal of entry.* In the following sections we shall group the bacterial diseases according to their modes of transmission, because such a classification tends to link ecologically related pathogens.

• **Diseases transmitted by fecal contamination of food and drink** The intestinal tract is the natural habitat of many kinds of bacteria, most of them harmless under ordinary conditions. A number of intestinal inhabitants are serious pathogens, however; these include the causative agents of typhoid and paratyphoid fevers, dysentery, cholera, and the *Salmonella* infections incorrectly referred to as "bacterial food poisoning." Some do their damage locally, whereas others spread from the intestinal tissues to other parts of the body. All, however, have two important attributes in common: they leave the host in excreted fecal matter and must enter the next host via the mouth to reach the intestines once again.

The *enteric diseases,* as they are called, are thus acquired principally by swallowing food or drink contaminated with feces. Before the introduction of modern sanitation, water supplies were constantly subject to direct contamination from privies and faulty sewers. Today, however, contamination by these means has become rare, and other modes of transmission have become relatively more important. The common housefly is an effective agent of transmission because it visits both feces and food indiscriminately. Furthermore, there are far more healthy carriers of enteric pathogens than frank clinical cases, so that anyone who handles food is a potential source of contamination. Hence, only the strictest attention to personal hygiene on the part of food handlers can prevent the spread of enteric disease.

Many animals, including cattle and fowl, may be naturally infected with members of the genus *Salmonella,* which cause the enteric infections known as bacterial food poisoning. It is, consequently, possible to become infected by eating contaminated meat or eggs.

Some important bacterial diseases transmitted by the fecal contamination of food and drink are described in Table 30.1.

• **Diseases transmitted by exhalation droplets** The transmission of disease by the respiratory route is called *droplet infection* because in such cases the pathogenic organisms are carried from person to person in microscopic droplets of saliva. In countries that practice modern methods of sanitation, droplet infection is by far the most important route by which disease is spread. Every time a person sneezes, coughs, or even speaks loudly, he exhales a cloud of tiny droplets of saliva. Each droplet contains a little dissolved protein, as well as varying numbers of the microorganisms that inhabit the mouth and respiratory tract; the droplets quickly evaporate, leaving in the air great numbers of minute flakes of protein that bear living bacteria. A person suffering from a

TABLE 30.1

Some human diseases transmitted by fecal contamination

DISEASE	ETIOLOGIC AGENT	PATHOGENESIS
Typhoid fever	*Salmonella typhi* (*S. typhosa*), a Gram-negative, peritrichously flagellated rod. Facultatively anaerobic; mixed-acid fermentation. (See Chapter 20.)	The organisms first multiply in the gastrointestinal tract. Invasion of the bloodstream via the intestinal lymphatics and thoracic duct leads to dissemination throughout the body. Heavy growth occurs in the biliary tract, from which the bowel is invaded. Foci in the lungs, bone marrow, and spleen.
Enteric fevers, gastroenteritis, Salmonella septicemias ("bacterial food poisoning")	*Salmonella typhimurium, S. schottmülleri, S. choleraesuis;* Gram-negative, peritrichously flagellated rods. Facultatively anaerobic; mixed-acid fermentation. (See Chapter 20.)	Enteric fevers are diseases characterized by wide dissemination of the organism throughout the body. Enteric fever caused by salmonellae other than *S. typhi* are milder than typhoid fever and are called "paratyphoid fevers"; *S. schottmülleri* is the most common cause of enteric fevers in the U.S. Gastroenteritis is a salmonellosis in which the organism remains localized in the gastrointestinal tract; *S. typhimurium* is the most common cause of salmonella gastroenteritis in the U.S. Salmonella septicemias are most commonly caused by *S. choleraesuis.*
Cholera	*Vibrio cholerae,* a Gram-negative, polarly flagellated, curved rod. Facultatively anaerobic; mixed-acid fermentation. (See Chapter 20.)	The organism multiplies extensively in the small intestine. An exotoxin acts on the mucosal cells; water and electrolyte loss leads to shock.
Bacterial dysentery	*Shigella dysenteriae,* a Gram-negative, immotile rod. Facultatively anaerobic; mixed-acid fermentation. (See Chapter 20.)	Lesions are formed in the terminal ileum and colon. Abdominal cramps, diarrhea, and fever are produced.

respiratory infection will certainly contaminate every other person in whose presence he coughs, sneezes, or speaks. The only way to prevent such spread would be to require that all individuals wear masks equipped with filters. Such an extreme measure has rarely been found practicable or enforceable. The result is that in a crowded city, a highly infective respiratory pathogen such as the influenza virus can spread from one person to several million in as short a time as 6 or 8 weeks.

Many highly important diseases involve infection of the respiratory tract and are spread by droplet exhalation. Some important bacterial respiratory diseases are described in Table 30.2.

• Diseases transmitted by direct contact   There are a small number of pathogens for which the portal of entry is the skin or the mucous membranes and which depend on direct contact for transmission. This group includes the causative agents of the venereal diseases, syphilis and gonorrhea. The responsible organism in each case cannot survive for long outside the host and requires direct contact of mucous membranes for transmission. Sexual intercourse is thus one of the chief means of spreading these diseases, although syphilis can also be acquired before birth, and gonorrhea during birth, from an infected mother.

TABLE 30.2

Some human diseases transmitted by exhalation droplets

DISEASE	ETIOLOGIC AGENT	PATHOGENESIS
Diphtheria	*Corynebacterium diphtheriae*, a Gram-positive, immotile rod; tends to be club-shaped. Postfission movements result in characteristic "palisade" arrangements. Facultatively anaerobic; propionic acid fermentation. (See Chapter 23.)	The organism establishes itself in the throat and remains localized in the upper respiratory tract. An exotoxin is produced which is disseminated by the bloodstream to all parts of the body. Local inflammation causes an exudate that forms a "diphtheritic pseudomembrane," blocking the air passages.
Tuberculosis	*Mycobacterium tuberculosis*, a pleomorphic, immotile, acid-fast rod. Obligately aerobic. (See Chapter 23.)	The bacteria multiply both intracellularly and extracellularly within lesions of the lungs, called tubercles. An enlarging tubercle may discharge into a bronchus, promoting spread of the disease to other parts of the lung and (via exhalation droplets) to other individuals. Less frequently, organisms are disseminated via the bloodstream to set up secondary (metastatic) lesions in other organs. The toxins of *M. tuberculosis* have not been identified; at least some of the host damage results from hypersensitivity reactions of the delayed type.
Plague	*Yersinia pestis*, a Gram-negative, immotile, small rod. Facultatively anaerobic; mixed-acid fermentation. (See Chapter 20.)	A disease of domestic and wild rodents, plague is transmitted to man by flea bite. The disease is called bubonic plague when it is characterized by enlarged, infected lymph nodes ("buboes"). In severe cases the organism is disseminated to other organs; when the lungs become infected, the disease becomes directly transmissible from man to man by droplet infection. This form of the disease is called *pneumonic plague* and is highly contagious.
Meningococcal meningitis	*Neisseria meningitidis* (meningococcus), a Gram-negative, immotile coccus forming pairs of cells. (See Chapter 19.)	Meningococci are carried harmlessly in the nasopharynx by 25 percent or more of the population. For unknown reasons, it occasionally invades the bloodstream and then localizes in the meninges (the membranes surrounding the spinal cord).
Streptococcal infections	*Streptococcus pyogenes*, a β-hemolytic, Gram-positive coccus growing in chains. Lactic acid homofermentation. (See Chapter 23.)	The organism develops first in the pharynx. Strains harboring a particular bacteriophage form erythrogenic toxin: if the individual is sensitive to the toxin, a skin rash appears and the disease is called *scarlet fever*. Further spread of the organism may lead to mastoiditis, peritonitis, puerperal sepsis, cellulitis of the skin, or erysipelas. Rheumatic fever is a common sequel to recurrent streptococcal pharyngitis and is correlated with high titers of antibody to streptococcal antigens. No organisms can be isolated from the diseased heart tissue, which appears to be damaged by an immunological reaction.
Pneumococcal pneumonia	*Streptococcus pneumoniae* (pneumococcus), a Gram-positive coccus typically forming diplococci. Lactic acid homofermentation. See Chapter 23.)	Between 40 and 70 percent of the adult population carry pneumococci in their throats. The organisms reach the lungs when normal barriers are malfunctioning (e.g., during viral respiratory infections).
Other respiratory infections	*Hemophilus influenzae; Bordetella pertussis*. Small, Gram-negative, nonmotile rods. *Hemophilus* species require hematin and nicotinamide nucleoside for growth.	*H. influenzae* causes respiratory infections in children; it is also the most common cause of bacterial meningitis in children. *B. pertussis* causes whooping cough; it rarely penetrates the mucosa of the respiratory tract and does not spread to other parts of the body.

In the tropics there are several diseases caused by organisms closely related to the agent of syphilis that are normally not transmitted by sexual intercourse. They all start as skin infections and require direct contact for transmission. Yaws is an example of this group.

Three other nonvenereal diseases that are transmissible by direct contact are anthrax, tularemia, and brucellosis. All are diseases of animals that can be transmitted to man. Brucellosis, a disease of goats, cattle, and swine, constitutes a severe occupational hazard for animal handlers, including veterinarians, meat packers, and dairy workers. Tularemia, a disease of wild rodents, is often contracted by hunters or butchers who handle the carcasses of wild game. The principal bacterial diseases transmitted by direct contact are described in Table 30.3.

• **Diseases transmitted by animal vectors**

Certain pathogens have become adapted during their evolution to existence in two or more alternative hosts. The plague bacillus, for example, can multiply in rats, fleas, and man; the flea carries it from rodent to rodent or from rodent to man, and it never has to survive in environments unsuitable for growth.

The agents of plague and tularemia are the only eubacteria in this category, but there are many viral, rickettsial, protozoan, and spirochetal diseases transmitted by animal vector. Epidemics of malaria, yellow fever, rabies, typhus, and plague spread in this fashion have radically altered the course of human history. The very property of having alternative hosts, which is such an advantage to parasites, has also led to their eventual control by man. By eliminating either the *vector* (the species that transmits the pathogen) or the *reservoir of infection* (the species from which the vector derives the infection), man has been able to eradicate such diseases from large areas.

The chief diseases transmitted by animal bite, together with their vectors and reservoirs, are listed in Table 30.4.

• **Diseases acquired by ingestion of bacterial toxins**

Serious diseases are caused by the ingestion of food containing the toxin of *Clostridium botulinum* or of *Staphylococcus aureus*. Although the disease is not further transmitted by the victim, an outbreak affecting many people can occur when a common food source becomes contaminated. Before the introduction of a strict canning code, the food canning industry was responsible for many deaths each year from botulism (ingestion of *C. botulinum* toxin). Food handlers who have open staphylococcal skin lesions are still a common cause of outbreaks of staphylococcal food poisoning.

• **Wound infections**

Whenever unsterilized foreign material penetrates a wound, microorganisms will certainly be introduced. If conditions in the wound are suitable for the growth of one or more contaminating microbes, an infection results that may eventually spread through the tissues or circulatory system.

Introduction into wounds cannot be considered a "natural" route of transmission, being too irregular and infrequent to ensure perpetuation of a parasitic species. Most often infected wounds are found to harbor ordinary soil-dwelling bacteria, such as the clostridia. The clostridia are obligate anaerobes that do not

TABLE 30.3

Some human diseases transmitted by direct contact

DISEASE	ETIOLOGIC AGENT	PATHOGENESIS
Anthrax	*Bacillus anthracis*, a Gram-positive, spore-forming, immotile rod. Obligately aerobic. (See Chapter 22.)	Anthrax is a natural disease of domestic and wild animals, including mammals, birds, and reptiles. Man acquires the disease primarily by direct contact, the organism gaining entry through a minor abrasion of the skin. A "malignant pustule" forms at the site of infection and may be followed by a fatal septicemia.
Tularemia	*Francisella tularensis*, a short, immotile, Gram-negative rod.	Tularemia (discovered in Tulare County, California) is a natural disease of wild rodents. Man usually acquires the disease through the handling of infected carcasses and skins, although it can also be transmitted by arthropod bite (ticks, deer flies). The organism spreads throughout the body via the lymphatics and bloodstream; lesions develop in the lungs, liver, spleen, and brain. Growth is primarily intracellular.
Brucellosis	*Brucella melitensis, B. abortus, B. suis*: small, Gram-negative, immotile rods.	All three species are capable of infecting a wide range of mammals, although each has a preferred host: *B. melitensis*, goats; *B. abortus*, cattle; *B. suis*, swine. In cattle, localization of the organism in the pregnant uterus (which often causes abortion) results from a specific growth stimulation by erythritol, which is present only in the vulnerable host tissues. In man, the organism is widely disseminated in the body and multiplies primarily within host phagocytic cells. The disease may undergo periodic remissions, in which case it is called *undulant fever*.
Gonorrhea	*Neisseria gonorrhoeae* (gonococcus), a Gram-negative coccus, forming pairs of cells. (See Chapter 19.)	The organism is transmitted by sexual contact; newborn infants may acquire serious eye infections from passage through an infected birth canal. Following sexual contact, the organism penetrates the mucous membranes of the genitourinary tract; infection usually is restricted to the reproductive organs, although septicemia may occur.
Syphilis	*Treponema pallidum*, a spirochete. (See Chapter 5.)	The organism is transmitted by sexual contact; it may also be transmitted to the fetus during pregnancy. Following sexual contact, the organism penetrates the mucous membranes of the sexual organ, forming a local primary lesion, or "chancre." Secondary lesions develop several weeks later in the eyes, bones, joints, or central nervous system. If untreated, the disease may progress several years later with the formation of tertiary lesions of the heart valves, central nervous system, eyes, bones, or skin. The toxins responsible for the disease are unknown; delayed hypersensitivity is believed to account for part or all of the damage occurring in the tertiary stage.

grow in healthy tissues. Deep wounds, however, form an ideal environment, since dead (necrotic) tissues are present, air is excluded, and tissue oxygenation is reduced as the result of impaired circulation. Clostridia spores are so ubiquitous in nature that any deep wound into which clothing or soil is introduced has a high probability of being contaminated with one or another species of *Clostridium*. Many of these organisms produce potent exotoxins that kill the surrounding host tissues. One species, *C. tetani*, produces a toxin that affects the nerves and causes muscle spasms; infection, if not treated, is almost invariably fatal. This disease is called tetanus, or "lockjaw." Other clostridia cause severe local damage (gangrene) at the site of infection.

TABLE 30.4

Some human diseases transmitted by animal bite

ECOLOGY, WITH RESPECT TO MAN	DISEASE	TYPE OF MICRO-ORGANISM	VECTOR	RESERVOIR
Man serves as accidental host, not as a reservoir	Plague	Bacterium	Flea	Rat, other rodents
	Tularemia	Bacterium	Tick	Wild rodents, tick[a]
	Rabies	Virus	Dog, jackal, bat, etc.	Same as vector
	Endemic typhus[b]	Rickettsia	Flea	Rat
	Rocky Mountain spotted fever	Rickettsia	Tick	Wild rodents
Man is one of two or more reservoirs	African sleeping sickness	Protozoon	Tsetse fly	Man, wild mammals
	Yellow fever	Virus	Mosquito	Man, monkeys
	Relapsing fever	Spirochete	Louse, tick	Man, tick,[a] rodents (?)
Man is sole reservoir	Malaria	Protozoon	Mosquito	Man
	Epidemic typhus[b]	Rickettsia	Louse	Man
	Dengue fever	Virus	Mosquito	Man

[a] Ticks act both as a vector and as a reservoir, because microorganisms multiply in the body of the tick and are transmitted through ovary and egg from one tick generation to the next.
[b] The two types of typhus fever are caused by closely related strains of rickettsiae.

Although the clostridia are the most dangerous wound pathogens, many other bacteria may become established in wounds. Common wound contaminants include staphylococci, streptococci, enterobacteria, and pseudomonads.

Leptospirosis, which begins as a wound infection, is an occupational disease among workers who are exposed to frequent contact with polluted water. The leptospirae are parasites of pigs, dogs, and rodents and are excreted in the urine of infected animals. They can infect man only through minor wounds or breaks in the skin, and the disease is thus most common among men who work in wet places, such as sewers, fish markets, wet fields, or canals. The common wound infections are described in Table 30.5.

• **Rickettsial diseases**

The rickettsias are extremely small, obligately parasitic bacteria; all are *intracellular* parasites (see Chapter 5).

One of the outstanding features of the rickettsias is their parasitic relationship to arthropods (lice, fleas, ticks, and mites). These are their natural hosts, in which they usually live without producing disease. The rickettsias have also become adapted to mammalian hosts, to which they are transmitted by arthropod bite. Thus, arthropod–mammal–arthropod chains of transmission are common. In most

TABLE 30.5

Some common wound infections

DISEASE	ETIOLOGIC AGENT	PATHOGENESIS
Tetanus	*Clostridium tetani*, a Gram-positive, spore-forming, peritrichously flagellated rod. Obligately anaerobic. (See Chapter 22.)	*C. tetani* is a common soil organism; it is also of common occurrence in the feces of animals (but not of humans). It produces disease when accidentally introduced into wounds. Spore germination and growth require anaerobic conditions, which obtain when the wound is necrotic and vascular damage is severe. A neurotoxin is produced which is transported via the bloodstream and the peripheral nerves to the spinal cord, where it acts principally on the anterior horn cells.
Gas gangrene	*Clostridium perfringens, C. novyi, C. septicum*: Gram-positive, spore-forming, peritrichously flagellated rods. Obligately anaerobic. (See Chapter 22.)	Development of the organism in anaerobic wounds is accompanied by accumulation of hydrogen gas produced by fermentation. A variety of soluble toxins are produced, including the α toxin of *C. perfringens*, a lecithinase.
Leptospirosis	*Leptospira icterohaemorrhagiae, L. canicola, L. pomona*: spirochetes. (See Chapter 5.)	The leptospirae cause mild, chronic infections of rodents and domestic animals, which shed them continuously in the urine. Man acquires the disease through contact with urine-contaminated water; the portal of entry is the skin. A disseminated infection results, during which the organisms can be cultured from the blood.

cases, man is only an accidental host, not forming a part of a transmission chain; the one exception is louse-borne typhus. The principal rickettsial diseases of man are described in Table 30.6.

• Chlamydial diseases The chlamydias, like the rickettsias, are obligate intracellular parasites of birds and mammals (see Chapter 5). In their natural hosts they tend to produce prolonged, latent infections; overt disease is more characteristic of an infection acquired from a different host species.

Four human diseases are caused by chlamydias: psittacosis (ornithosis); lymphogranuloma venereum; and two diseases of the eye, trachoma and inclusion conjunctivitis. The last two are caused by closely related organisms, both of which are classified as *Chlamydia trachomatis*. The agent of inclusion conjunctivitis, however, normally inhabits the human genital tract, from which it occasionally is spread to the eye; the agent of trachoma normally inhabits the tissues of, and surrounding, the eye itself.

The principal chlamydial diseases of man are described in Table 30.7.

FUNGAL DISEASES

The fungal diseases of man are either *mycoses*, caused by true infection, or *toxicoses*, caused by the ingestion of toxic fungal metabolites.

• The mycoses A small number of fungi are capable of causing human disease by actual infection. For most of these, invasion of host tissue is accidental, their normal habitat being

TABLE 30.6

Some representative rickettsial diseases of man

DISEASE	RESERVOIRS, VECTORS, AND TRANSMISSION TO MAN	REMARKS
Epidemic typhus	The rickettsia is carried from man to man by the body louse.	High mortality rates in man.
Endemic typhus	The rickettsia is normally a parasite of rats and fleas. It is maintained in nature by rat–flea–rat chains of transmission. Mild epidemic, involving man–louse–man chain of transmission, may follow bite by flea.	Milder disease than epidemic typhus. Epidemic and endemic typhus are caused by different, but closely related, rickettsias.
Scrub typhus (tsutsugamushi fever)	The true reservoir is the mite; the rickettsias are passed from one mite generation to the next via the eggs. They are also transmitted to rats by arthropod bite, and the rat is thus a secondary reservoir. Man is infected by bite of a mite or flea.	Confined to the Far East.
Spotted fevers	The rickettsias are tick parasites and are passed from one tick generation to the next via the egg. They are also transmitted to a variety of mammalian hosts, including rodents and domestic animals, which thus form secondary reservoirs. Man is infected by tick bite.	There are several closely related diseases of this type, with differing geographical distributions (e.g., Rocky Mountain spotted fever, Mediterranean fever, South African tick-bite fever).
Rickettsialpox	The reservoirs are the house mouse and its mites; man is infected by mite bite.	The rickettsia is antigenically related to the spotted fever group. Rickettsialpox has only been observed in urban areas.
Q fever	The rickettsia is a parasite of numerous wild animals and ticks; the latter transmit it to goats, sheep, and cattle. Man acquires the disease by inhalation of infected dust, drinking infected milk, or by direct contact with animals or animal products.	This is a unique rickettsial disease, in that it is transmissible to man by means other than arthropod bite.

the soil. The exceptions are the *dermatophytes*, which inhabit the epidermis, hair, and nails; these are transmissible from person to person or from animal to person.

No toxins have yet been discovered to play a role in the pathogenesis of the mycoses, and it is believed that most of the host damage seen in fungal infections arises from hypersensitivity reactions, particularly of the delayed (cell-mediated) type. Localized lesions, resembling the classical delayed hypersensitivity responses of tuberculosis, are common, and represent areas where slow, hyphal growth is taking place. Dissemination throughout the body usually involves the persistence of unicellular yeast forms within phagocytes.

The fungal diseases are generally grouped according to their depth of penetration. We will consider them as forming three such groups: the *dermatomycoses;* the *subcutaneous mycoses;* and the *deep,* or *systemic mycoses.*

• **The dermatomycoses**

The scaly, annular skin lesions caused by dermatophytes are called *tinea* (Latin, worm or insect larva), as they were originally thought to be caused by worms or by lice. They are generally classified according to the affected part of the body: *tinea pedis* is better known as athlete's foot, *tinea capitis* as ringworm of the scalp, and *tinea corporis* as ringworm of the nonhairy skin of the body. Most tineas are caused by members of three genera of fungi: *Trichophyton, Microsporum,* and

Chlamydial diseases of man

DISEASE	AGENT	PRINCIPAL TRANSMISSION ROUTE	PATHOGENESIS
Psitticosis (ornithosis)	*Chlamydia psittaci*	Inhalation of feces from infected birds (many species)	Inflammation of the lungs; fever. Fatality rate may be as high as 20 percent in untreated cases.
Lymphogranuloma venereum	*C. trachomatis* (rarely, *C. psittaci*)	Sexual intercourse	Skin lesions and enlarged lymph nodes in genital regions of the body.
Trachoma	*C. trachomatis*	Mechanical spread by fingers or contaminated objects	Lesions form in tissues of and surrounding the eyes; may progress to blindness.
Inclusion conjunctivitis	*C. trachomatis*	Newborn infants may be infected from the genital tract of the mother. Adults acquire the disease by sexual contact and subsequent finger-to-eye transfer.	Inflammation of the conjunctiva.

Epidermophyton. Trichophyton can grow in hair, skin, and nails; *Microsporum* can grow in hair and skin only; and *Epidermophyton* can grow in skin and occasionally in nails.

These organisms are transmitted by direct contact or by contact with infected hair clippings or epidermal scales. Animals form an additional reservoir; for example, over 30 percent of dogs and cats in the United States carry *M. canis*, an organism that can cause ringworm of the scalp in humans.

• **The subcutaneous mycoses**

These infections are initiated when certain soil-inhabiting fungi are introduced beneath the skin by thorns or splinters, or as contaminants of wounds. The diseases that develop are grouped into three categories: *sporotrichosis*, characterized by ulcerating skin lesions, is caused by a yeastlike fungus called *Sporotrichum schenkii; chromoblastomycosis*, characterized by skin lesions containing dark brown yeast cells, may be caused by any of several fungal species; and *maduramycosis*, characterized by a generalized destruction of the tissues of the foot or hand, may be caused by any of several fungi, as well as by several Actinomycetes.

• **The systemic mycoses**

A small number of fungal species produce deep lesions of the infected organ or widely disseminated lesions in the body. They include four soil inhabitants: *Blastomyces dermatitidis, Histoplasma capsulatum, Coccidioides immitis,* and *Cryptococcus neoformans.* They also include normally harmless inhabitants of the body such as *Candida albicans,* which become invasive only when the individual's normal antimicrobial defenses have been disturbed. For example, *Candida* will cause disease when the normal flora has been suppressed by antibiotic therapy, when immuno-

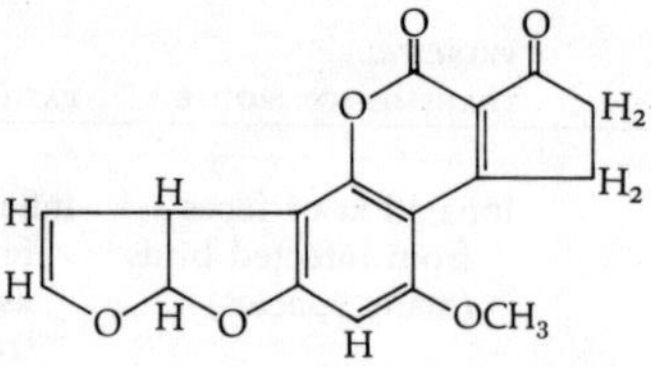

Aflatoxin B1

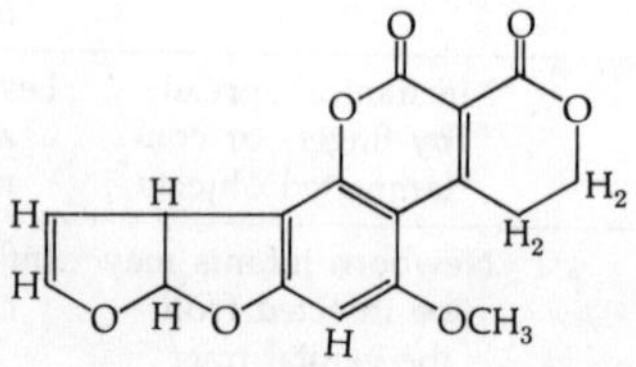

FIGURE 30.1

Aflatoxin G1   The structures of some aflatoxins.

suppressive treatment is being given, or when the individual is severely debilitated by another disease.

The major systemic mycoses are described in Table 30.8.

• **The toxicoses**   Many fungi produce poisonous substances, called *mycotoxins*, which cause serious—sometimes fatal—diseases if ingested. They also produce a variety of *hallucinogens*, such as lysergic acid. The mycotoxins of importance to man include the toxins produced by the poisonous mushrooms, the toxins of *Claviceps purpurea* (ergot, a pathogenic parasite of rye), and the *aflatoxins*.

The aflatoxins are formed by *Aspergillus flavus*, which grows in a variety of plant materials. Such crops as peanuts and grains, if not properly dried, may contain sufficient levels of aflatoxins to cause severe liver damage. In the United States rigid standards of food processing and storage, combined with the imposition of maximum permissive levels of toxin in foodstuffs, have effectively prevented diseases caused by aflatoxins. In India and many parts of Africa, however, aflatoxins are a serious problem.

The structures of some aflatoxins are shown in Figure 30.1. They have been found to bind to DNA and to inhibit RNA synthesis; they are also potent carcinogens, causing hepatomas in experimental animals. Indeed, a high correlation has been observed between human liver cancer and aflatoxin intoxication.

PROTOZOAN DISEASES

Of the many thousands of protozoan species, only about 20 cause disease in man. Their impact on worldwide human health, however, is far out of proportion to their numbers, and it has been estimated that at any given moment a quarter of mankind is afflicted with severely debilitating protozoan diseases. Malaria alone accounts for over 100 million cases a year, a million of

TABLE 30.8

Some important systemic mycoses

ORGANISM	MORPHOLOGY	PATHOGENESIS	EPIDEMIOLOGY
Candida albicans	In tissues and exudates: Oval, budding yeast cells, 2–6 μm. On agar: Submerged growth in the form of a pseudomycelium made up of adhering, long, budding cells. Surface growth in the form of budding yeast cells.	*Candida* is a harmless member of the normal flora of the mucous membranes in the respiratory, gastrointestinal, and female genital tracts. In debilitated patients it may produce systemic progressive disease or localized lesions of the skin, mouth, vagina, or lungs.	Most individuals harbor the organism, so transmission is not a factor in the disease. Prevention requires maintenance of the normal host defenses, including the normal flora. *Candida* infections often follow disturbances of the normal flora by antibiotic therapy.
Cryptococcus neoformans	In tissues and spinal fluid: Spherical or oval, budding, encapsulated yeast cells, 5–12 μm. On agar: Yeast cells only; no mycelium.	Infection occurs via the respiratory tract. Most common clinical manifestation is chronic meningitis, which may be accompanied by lesions of the skin and lungs. Untreated cases are ultimately fatal.	Bird feces are the main source of infection; the disease is not transmissible from person to person.
Blastomyces dermatitidis	In tissues: Thick-walled, budding, spherical forms, 8–15 μm. On agar: Spherical forms at 37°C, filamentous at 25°C.	The infection may be limited to the lungs, or it may be widely disseminated to the skin, bones, viscera, and meninges. Lesions consist of small abscesses and tuberclelike granulomas (masses of macrophages).	*Blastomyces* appears to inhabit the soil and to be acquired by inhalation of dust. It is not transmissible from person to person.
Histoplasma capsulatum	In tissues: Oval, budding yeast cells 2–4 μm. On agar: Yeast cells at 37°C; filamentous at 25°C, with tuberculate spores.	The disease usually remains localized in the lungs, but in a minority of cases it becomes widely disseminated throughout the body. The tissue lesions resemble tubercles.	The organism occurs in the soil and grows abundantly in bird feces (chicken houses) and bat guano (caves). It is acquired by the inhalation of contaminated dust; it is not transmissible from person to person.
Coccidioides immitis	In tissues or pus: Thick-walled sphericles, 15–75 μm, filled with endospores. On agar: Mycelial, with cask-shaped arthrospores.	Respiratory infection, characterized by an influenzalike illness. Rarely, the disease progresses to a disseminated form closely resembling tuberculosis, with lesions in all organs and the central nervous system.	*Coccidiodes* is endemic in the soil and in rodents in the southwestern United States. It is acquired by inhaling contaminated dust; it is not transmissible from person to person.
Geotrichum candidum	In sputum: Rectangular arthrospores, 5 × 10 μm, or thick-walled, oval yeast cells. On agar: Mycelial, with arthrospores.	Chronic bronchitis or lesions in the mouth.	The natural habitat is not known, but it may be the mouth and intestinal tract of man.

which are fatal. In addition to the human diseases which they cause, the protozoan parasites are indirectly responsible for widespread malnutrition in Africa, as a result of the diseases which they produce in domestic cattle.

Pathogens are found in each of the four main groups of protozoa; the principal ones are summarized in Table 30.9. In the following sections the protozoan diseases are grouped according to their modes of transmission, which are as varied as they are in the bacteria.

TABLE 30.9

Some important protozoan diseases of man

GROUP	ORGANISM	DISEASE	PRINCIPAL TRANSMISSION ROUTE
Flagellates	*Giardia lamblia*	Flagellate diarrhea	Ingestion of cysts (fecal contamination)
	Trichomonas vaginalis	Genitourinary tract infections	Sexual intercourse
	Trypansoma gambiense ⎫ *T. rhodesiense* ⎭	(African) sleeping sickness	Tsetse flies (Glossinidae)
	T. cruzi	(American) Chagas' disease	Triatomid bugs
	Leishmania donovani	Kala-azar (infection of spleen, liver, lymph nodes, bone marrow)	Sandflies (Phlebotomus)
	L. tropica	Cutaneous lesions	Sandflies (Phlebotomus)
	L. braziliensis	Nasopharyngeal infections	Sandflies (Phlebotomus)
Amebas	*Entamoeba histolytica*	Amebic dysentery; infections of liver, spleen, brain	Ingestion of cysts (fecal contamination)
	Naegleria sp.	Amebic meningoencephalitis	Penetration of mucous membranes
Ciliates	*Balantidium coli*	Balantidial dysentery	Ingestion of cysts (fecal contamination)
Sporozoans	*Plasmodium falciparum* *P. vivax* ⎫ *P. malariae* ⎬ *P. ovale* ⎭	Malaria, hemoglobinuria ⎫ Malaria ⎭	Female *Anopheles* mosquito
	Toxoplasma gondii	Disseminated infections, particularly of the reticuloendothelial system	Transplacental infection of the fetus (congenital infection); ingestion of cysts; inhalation (?).
	Isospora belli ⎫ *I. huminis* ⎭	Intestinal infections	Ingestion of cysts

• **Diseases transmitted by the ingestion of cysts**   Within each of the main groups of protozoa there are pathogens that are transmitted by the ingestion of cysts (thick-walled, resting cells). Four of these—*Entamoeba histolytica, Giardia lamblia, Balantidium coli,* and *Isospora* species—are parasites of the gastrointestinal tract; they have relatively simple life cycles, involving a proliferative state (*trophozoites*) and a cyst stage. The cysts are passed in the feces to the external environment, where they survive and contaminate food and water.

The cysts of *Toxoplasma gondii* may be transmitted in different ways. They are found in the skeletal muscle tissue of sheep and pigs, so that ingestion of incompletely cooked meat may be a source of infection. Toxoplasma cysts have also been seen in lung alveoli, so that inhalation of contaminated dust may be a second means of transmission. Domestic cats have recently been found to form another reservoir of this organism and to shed cysts in their feces. The ingestion or inhalation of cysts from this source is probably an important cause of infections in man.

Two novel mechanisms of host damage have been discovered in the intestinal protozoan parasites. *Giardia lamblia* has a "suction disc" on its ventral surface, by means of which it attaches itself to the intestinal epithelium of the host (Figure 30.2). In severe infections the inner surface of the upper small intestine may be totally covered by attached parasites, causing the mechanical blockage of fat absorption—a deficiency responsible for some of the pathological effects of infection.

The second novel mechanism occurs in *Entamoeba histolytica*. As its name implies, this organism causes the lysis of host cells, particularly leucocytes. Leucocytolysis requires direct contact between the leucocyte and the ameba; recent electron microscopic studies have shown that *Entamoeba* possesses *surface lysosomes*, each with a vermiform appendage which appears to serve as a trigger (Figure 30.3). Contact between the trigger and a host cell may cause eversion of the lysosome, with destruction of the leucocyte by the lysosomal enzymes.

• Diseases transmitted by insect vectors

In three groups of pathogenic protozoa—the trypanosomes, the leishmanias, and the malarial parasites of the genus *Plasmodium*—parts of the life cycle occur in an insect host. The insect serves as the vector for transmission of the pathogen from human to human.

There are two groups of trypanosomes: the African species, *T. gambiense* and *T. rhodesiense*, which are transmitted by the tsetse fly (*Glossina* spp.) and cause sleeping sickness; and the American species, *T. cruzi*, which is transmitted by triatomid bugs and causes Chagas' disease. A typical trypanosome is illustrated in Figure 4.10(a) (p. 100).

The leishmanias are transmitted by sandflies (*Phlebotomus* spp.), and *Plasmodium* is transmitted by the *Anopheles* mosquito. With the exception of the triatomid bugs, all the insect vectors pump the parasites into the host's bloodstream in a stream of saliva while taking a blood meal; only the female *Anopheles* mosquito transmits its parasite, for the male does not suck blood. The triatomid bugs do not inject the trypanosome parasites, but instead deposit them in feces on the host skin while feeding. The parasites are rubbed into the puncture wound or into the ocular conjunctiva when the insect bite is scratched. The organisms then develop within macrophages and ultimately within muscle cells.

The trypanosomes multiply in the bloodstream, the American forms multi-

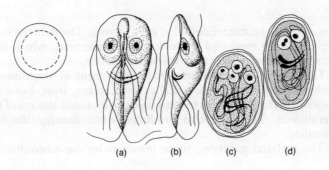

(a) (b) (c) (d)

FIGURE 30.2

Giardia intestinalis. (a) "Face" and (b) "profile" of vegetative forms; (c) and (d) cysts (binucleate and quadrinucleate stages). Circle represents red blood cell for size comparison. From Jawetz, E., J. L. Melnick and E. A. Adelberg, *Review of Medical Microbiology*, 12th edition. (Lange Medical Publications, Los Altos, Calif., 1976.)

(a)

(b)

(c)

FIGURE 30.3

Surface lysosomes of *Entamoeba histolytica*. (a) Scanning electron micrograph of *E. histolytica* trophozoite, showing seven lysosomes in surface view, three of which have the trigger device in view, ×2,130. (b) Close-up view of a surface lysosome with a protruding trigger, ×6,800. (c) Electron micrograph of thin section of a surface lysosome, ×36,600. From R. D. P. Eaton, E. Meerovitch, and J. W. Costerton, "The Functional Morphology of Pathogenicity in *Entamoeba histolytica*," *Ann. Trop. Med. and Parasitol.* **64**, 299 (1970).

plying in the reticuloendothelial cells as well. The damage they cause to the host appears to be mediated by toxins or by allergens, which affect principally the central nervous system and the heart muscle.

Leishmania donovani (Figure 30.4), the agent of the disease called Kala-azar, multiplies in reticuloendothelial cells of spleen, liver, bone marrow, and lymph nodes. Other leishmanias tend to produce localized lesions of the skin and mucous membranes. The microbial products which damage the host have not been identified.

The malarial parasites, upon injection by the mosquito, develop first within

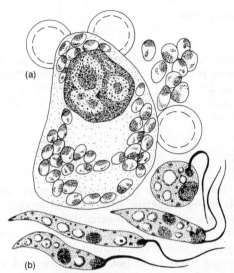

FIGURE 30.4

Leishmania donovani. (a) Large reticuloendothelial cell of spleen with amastigotes. (b) Promastigotes as seen in sandfly gut or in culture. Circles represent red blood cells for size comparison. From Jawetz, E., J. L. Melnick and E. A. Adelberg, *Review of Medical Microbiology,* 12th edition. (Lange Medical Publications, Los Altos, Calif., 1976.)

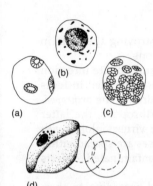

FIGURE 30.5

Plasmodium falciparum in red blood cells. (a) Young trophozoites (triple infection); (b) mature trophozoite; (c) mature schizont; (d) mature gametocyte. Circles represent uninfected red blood cells for size comparison. From Jawetz, E., J. L. Melnick and E. A. Adelberg, *Review of Medical Microbiology,* 12th edition. (Lange Medical Publications, Los Altos, Calif., 1976.)

liver parenchymal cells and later within red blood cells (Figure 30.5). The pathological effects of infection are produced mainly during the latter phase. Anemia develops as a consequence of red cell destruction, but the anemia seen in malaria is not sufficient to account for the disease symptoms. Rather, it appears that factors are liberated during the red cell cycle that cause damage to bone marrow, spleen, kidney, and other organs. Whether these factors are products of the parasite or of the host is not yet known.

The diseases caused by insect transmitted parasites are restricted to regions of the world that represent the normal habitats of their insect vectors. Thus, sleeping sickness is found only in parts of Africa, and Chagas' disease is found only in parts of Central and South America. Malaria, however, may occur in almost any part of the world, following the widespread distribution of the *Anopheles* mosquito. Control or eradication of *Anopheles* has been accomplished in many regions of the temperate zones, however, so that today malaria is generally limited to the tropics and subtropics.

Within a given geographical region, the insect transmitted diseases occur in patterns that reflect both the local habitats of the insects (e.g., the stagnant water breeding areas of *Anopheles*) and seasonal variations in their abundance.

• **Venereal disease**   *Trichomonas vaginalis* infects the male urethra and prostate as well as the female vagina. The organisms, an example of which is shown in Figure 4.10(b) (p. 100), do not form cysts and cannot survive for long outside the body; sexual intercourse is therefore the only means of transmission. The parasites, and the damage which they cause, are restricted to the genitourinary tracts; in females, the infections often go undetected.

• **Diseases caused**
  by free-living
  soil amebas   Certain free-living soil amebas have been found to cause meningoencephalitis in man, apparently by their accidental introduction into the nasal mucosa. The organisms that have been isolated from the infected tissues have so far been

classified in the genus *Naegleria;* members of a related genus, *Hartmannella,* have been found capable of producing a similar meningoencephalitis in experimental animals. All cases to date have been traced to infections acquired during swimming in freshwater rivers and lakes; extensive brain damage is produced, but the factors responsible are as yet unidentified.

VIRAL DISEASES

The viruses damage their hosts either by destroying the cells in which they multiply or by triggering hypersensitivity reactions.

The *respiratory viruses* are liberated in exhalation droplets and initiate infection in the surface tissues of the respiratory tract. The *enteric viruses* are shed in feces and are acquired by the ingestion of contaminated food and drink. The *arthropod-borne viruses* (*arboviruses*) multiply in both vertebrates and arthropods, the latter serving as the vectors of their transmission; they inject the viruses along with saliva through the skin when taking a blood meal. Other viruses are spread either by *direct contact* between individuals or by contact with contaminated objects (*fomites*).

The principal viral agents of human disease are grouped according to their transmission routes in Table 30.10, which also indicates their major target organs (i.e., those in which the most damage is produced by viral replication). Animal viruses are sometimes classified according to these target organs: for example,

TABLE 30.10

Viruses of human disease, grouped according to their transmission routes

PRINCIPAL TRANSMISSION ROUTE	VIRUS	MAJOR TARGET ORGANS
Respiratory	Influenza	Respiratory tract
	Parainfluenza	Respiratory tract
	Respiratory syncytical	Respiratory tract
	Measles	Respiratory tract, skin
	Mumps	Parotid glands, testes, meninges
	Adenoviruses	Respiratory tract
	Rhinoviruses	Respiratory tract
	Coxsackieviruses (some)	Respiratory tract, many others
	Coronaviruses	Respiratory tract
Enteric	Polioviruses	Intestinal mucosa, lymph nodes, central nervous system
	Coxsackieviruses (some)	Many
	Echoviruses	Gastrointestinal tract; occasionally disseminated
	Hepatitis virus	Liver (also kidneys and spleen)
Direct contact, fomites	Herpes simplex	Oral mucous membranes
	Poxviruses	Skin, many others
	Rubella	Skin, many others
Animal bite	Arboviruses	Many
	Rabies	Central nervous system

those which multiply primarily in the central nervous system are called *neurotropic viruses*, and those which produce prominent skin lesions are called *dermatropic viruses*. It should be emphasized that the subdivision of viruses in terms of their modes of transmission or their target organs is not correlated with the taxonomic subdivision of viruses in terms of their physicochemical properties, as presented in Table 12.1.

The ability of viruses to multiply in—and thus damage—certain organs but not others reflects a high degree of *tissue specificity*. In a number of cases this specificity has been traced to the presence or absence of *viral receptors* on the cell surface; for example, homogenates of susceptible organs can be shown to bind virions, but homogenates of nonsusceptible organs do not.

Viruses also show marked *host specificities*, being able to infect some animal species but not others; the presence or absence of viral receptors has been shown to be the determining factor in certain cases of host specificity also. For example, primate cells, but not rodent cells, have receptors for polio virions; poliovirus RNA, however, infects both types of cells with low efficiency.

• Host defenses: phagocytes The phagocytic cells of the body constitute a primary line of defense against viruses as they do against cellular pathogens. This barrier is breached by the virulent viruses that multiply within macrophages and are released in increased numbers to infect other types of cells.

• Host defenses: interferon Practically all animal cells, but particularly bone marrow cells, spleen cells, and macrophages, are induced by viral infection to synthesize and secrete a protein that interferes with viral multiplication. This protein, called *interferon*, interferes with the multiplication of all viruses, not only with the type which induced its formation. Different animal species produce different interferons, which much more effectively protect cells of the same animal species than cells of a different species.

Animal cells can be induced to synthesize interferon by *double-stranded RNA*, but not by any other form of nucleic acid. Since all RNA viruses produce double-stranded intermediates during their replication, their ability to induce interferon synthesis is easily explained. DNA viruses may also cause some double-stranded RNA to be produced in infected cells, although it is not known to be required for their replication or maturation. Such double-stranded RNA has actually been found in cells infected with vaccinia virus (a DNA virus) and to be effective in inducing interferon.

Interferon does not act directly to inhibit viral multiplication; rather, it induces the formation of a second protein, called *antiviral protein*, which is the true inhibitor. Interferon induces antiviral protein synthesis not only in the cell in which it is itself synthesized, but also in the surrounding cells to which it diffuses. Its sphere of activity in host defense is thus greatly enlarged.

Antiviral protein appears to inhibit viral multiplication by blocking the translation of viral messenger RNA into viral proteins, which could account for its ability to protect against both RNA and DNA viruses.

• Host defenses: immune responses Most viruses are good antigens and induce both neutralizing antibodies and cell-mediated immune mechanisms. Cellular immunity appears to play a more important role than humoral antibodies in protection against certain viruses, such as the poxviruses and the enveloped viruses. It also serves to eliminate host cells which have acquired new surface antigens as the result of harboring latent viruses or the genomes of tumor viruses.

The cell-mediated immune response brings into play not only an increased activity of macrophages, but also an increased level of interferon production, since antigen-activated lymphocytes release interferon along with the other factors described in Chapter 29. To be highly virulent, a virus must thus have the following properties: (1) it must not be neutralized by antibodies directed against it (i.e., it must bind to host cells and penetrate them even when coated with antibody), or it must be a poor inducer of antibody synthesis; (2) it must be able to multiply, or at least resist destruction, within phagocytic cells; and (3) it must either be able to resist interferon action or be a poor inducer of it.

• **Mechanisms of cell damage: immune responses**

Many viruses trigger hypersensitivity reactions of the delayed type, producing nonspecific tissue damage as described in Chapter 29. Some viruses possess antigens which cross react with host cell antigens, i.e., the antibodies induced by the virus may react with a particular type of host cell, producing an auto-immune disease. This is particularly likely to occur in infections by viruses which acquire an envelope of modified host membrane during the liberation process (see Chapter 12). These virions induce antibodies which are active not only against themselves, but also against uninfected as well as infected cells. In other words, the association of viral antigens with the host membrane may cause normal membrane constituents to become antigenic.

• **Mechanisms of cell damage: cytopathic effects**

The symptoms of viral disease reflect, in many cases, the pathological changes undergone by the infected cells themselves. Such changes, which often terminate in the death of the cell, are of two types: morphological, in which the membrane and internal structures of the cell are visibly disrupted, and biochemical, in which the synthetic processes and other physiological activities of the cell are impaired.

The virus-induced synthesis of new proteins appears to be necessary for most of these cytopathic effects; such proteins, which thus behave as intracellular, nondiffusible toxins, may or may not be incorporated into the mature virion. The toxic effects of virus-induced proteins can be demonstrated by extracting them from infected cells and adding them to cultures of uninfected cells at high concentrations. For example, of two viral coat proteins which can be extracted from adenovirus infected cells, one causes morphological damage to uninfected cells and the other causes the depression of macromolecular syntheses.

The precise nature of the biochemical lesions induced by viral products is, in most cases, not known. In phagocytic cells much of the cell damage appears to result from the disruption of lysosomes, which discharge their lytic enzymes into the otherwise protected cytoplasm of the cell. In other types of cells, the virus-induced proteins appear to block host cell metabolism at a number of points, leading to the shutoff of normal macromolecular syntheses and the diversion of metabolic precursors into the pathways of viral synthesis. In some viral infections, large masses of virions or unassembled viral subunits accumulate in the nucleus

829   or in the cytoplasm; these *inclusion bodies,* as they are called, may become large enough to cause mechanical damage to the cell.

• **Latent viral infections**   Some viruses are able to *persist within host tissues* for long periods of time without producing overt disease. Such *latent infections* may eventually be converted into acute diseases by external factors which upset the equilibrium between host and parasite.

Latent infections of two different types have been observed, depending on the particular virus involved. In some cases, a *carrier state* is established at the cellular level: at any given time only a few cells are infected and are liberating virus; antibodies and other host defenses keep the extracellular virus population in check but do not eradicate it completely. In other cases, viral replication occurs at a moderate rate without harming the host cells; viral genomes are transmitted to daughter cells *intracellularly* at the time of cell division.

Eventually, some change in the physiology of the host upsets the balance, and the resulting increase in viral multiplication and maturation produces an episode of acute disease. Such events appear to explain the recurrent episodes of *herpes simplex* infections in a given individual, which are provoked by such factors as sunlight, fatigue, or fever.

A number of formerly unexplained diseases (e.g., subacute sclerosing panencephalitis and kuru) have now been found to follow prolonged periods (sometimes many years) of latent viral infection. These diseases, called *slow virus infections,* apparently result from the slow development of humoral and cell-mediated hypersensitivity responses over the long period of latent infection.

FURTHER READING

Books   BAKER, J. R., *Parasitic Protozoa.* London: Hutchinson University Library, 1969.

BARUA, D., and W. BURROWS, *Cholera.* Philadelphia: W. B. Saunders, 1974.

CLUFF, L. E., and J. E. JOHNSON, III. *Clinical Concepts of Infectious Diseases.* Baltimore: Williams & Wilkins, 1972.

COHEN, J. O. (editor), *The Staphylococci.* New York: Wiley, 1970.

GOLDBLATT, L. A. (editor). *Aflotoxin: Scientific Background, Control, and Implications.* New York: Academic Press, 1969.

JAWETZ, E., J. L. Melnick, and E. A. ADELBERG, *Review of Medical Microbiology,* 12th ed. Los Altos, Calif.: Lange Medical Publications, 1976.

PURCHASE, I. F. H. (editor). *Mycotoxins.* New York: Elsevier Scientific, 1974.

STORZ, J., *Chlamydia and Chlamydia-induced Diseases.* Springfield, Ill.: Thomas, 1971.

VILCEK, J., *Interferon.* New York: Springer, 1969.

YOUMANS, G. P., P. Y. Paterson and H. M. Sommers, *The Biologic and Clinical Basis of Infectious Diseases.* Philadelphia: W. B. Saunders, 1975.

Review articles   BARKSDALE, L., "*Corynebacterium diphtheriae* and Its Relatives," *Bact. Rev.* **34,** 378 (1970).

BRAY, R. S., "*Leishmania,*" *Ann. Rev. Microbiol.* **28,** 189 (1974).

BRENER, Z., "Biology of *Trypanosoma cruzi,*" *Ann. Rev. Microbiol.* **27,** 347 (1973).

CULBERTSON, C. G., "The Pathogenicity of Soil Amebas," *Ann. Rev. Microbiol.* **25,** 231 (1971).

DE CLERCQ, E., and T. C. MERIGAN, "Current Concepts of Interferon and Interferon Induction," *Ann. Rev. Med.* **21,** 17 (1970).

FUCCILLO, D. A., J. E. KURENT, and J. L. SEVER, "Slow Virus Diseases," *Ann. Rev. Microbiol.* **28,** 231 (1974).

MIROCHA, C. J., and C. M. CHRISTENSEN, "Fungus Metabolites Toxic to Animals," *Ann. Rev. Phytopathol.* **12,** 303 (1974).

SMITH, H., "Mechanisms of Virus Pathogenicity," *Bact. Rev.* **36,** 291 (1972).

LINSELL, C. A., and F. G. PEERS. "The Aflotoxins and Human Liver Cancer," *Recent Results Cancer Research* **39,** 125 (1972).

Original articles   MACLEAN, D. J., J. A. SARGENT, I. C. TOMMERUP, and D. S. INGRAM, "Hypersensitivity as the Primary Event in Resistance to Fungal Parasites," *Nature* **249,** 186 (1974).

PAPPENHEIMER, A. M., Jr., and D. M. GILL, "Diphtheria," *Science* **182,** 353 (1973).

31

THE EXPLOITATION OF
MICROORGANISMS BY MAN

The role of microorganisms in the transformations of organic matter was not recognized until the middle of the nineteenth century. Nevertheless, microbial processes have been used by man since prehistoric times in the preparation of food, drink, and textiles; in many cases, these processes became controlled and perfected to an astonishing degree by purely empirical methods. The outstanding examples of *traditional microbial processes* are those used in the production of beer and wine; the pickling of certain plant materials; the leavening of bread; the making of vinegar, cheese, and butter; and the retting of flax. The rise of microbiology, which revealed the nature of these traditional processes, led not only to great improvements in many of them, but also to the development of entirely new industries based on the use of microorganisms which had previously not been exploited by man.

THE USE OF YEAST BY MAN

Yeasts traditionally and still play an important role from a technical and industrial standpoint. Although many genera and species of yeast exist in nature and many are used industrially, the yeasts of greatest technical importance are strains of *Saccharomyces cerevisiae*. They are used in the manufacture of wine and beer and in the leavening of bread.

The manufacture of alcoholic beverages was already well established in early civilizations, most of which had myths about the origin of wine making that attributed the discovery to divine revelation. This fact suggests that even in very ancient times the beginnings of the art were already shrouded in prehistoric darkness. The use of yeast as a leavening agent for bread originated in Egypt about 6,000 years ago, and spread slowly from there to the rest of the western world.

The discovery that alcohol can be distilled, and so concentrated, originated either in China or the Arab world. Distilleries began to appear in Europe in the middle of the seventeenth century. At first the alcohol manufactured was used only for human consumption, but with the industrial revolution, the demand for alcohol as a solvent and chemical raw material developed and the distilling industry grew very rapidly.

• The making of wine

The making of wine involves the fermentation of the soluble sugars (glucose and fructose) of the juices of grapes into CO_2 and ethyl alcohol. After the grapes are harvested, they are crushed to yield a raw juice or *must*, a highly acidic liquid containing 10 to 25 percent sugar by weight. In many parts of the world the mixed yeast flora on the grapes serves as the inoculum. In such a natural fermentation a complex succession of changes in the yeast population occurs; in the later stages the so-called true wine yeast, *Saccharomyces cerevisiae* var. *ellipsoideus*, predominates. In other areas, California, for example, the must is first treated with sulfur dioxide, which virtually eliminates the natural yeast flora; it is then inoculated with the desired strain of wine yeast. The fermentation proceeds vigorously, usually being completed in a few days. Often it is necessary to control the rate of the fermentation or to cool the fermenting mixture, in order to prevent a rise of temperature which would affect quality or even kill the yeast. Must from both red and white wine grapes (*Vitis vinifera*) is white and results in a white wine. Since the color of red grapes is in the skin, red wines are made by fermentation in the presence of the skins. The alcohol developed during fermentation extracts the color into the wine. Following fermentation the new wines must be clarified, stabilized, and aged to produce a satisfactory final product. These processes require months, and for high quality red wines, even years. During the first year, many wines (particularly red) undergo a second spontaneous fermentation, the *malo-lactic fermentation*, which can be caused by a variety of lactic acid bacteria (*Pediococcus, Leuconostoc,* or *Lactobacillus*). This fermentation converts malic acid, one of the two major organic acids of grapes, to lactic acid and CO_2, thus converting a dicarboxylic acid to a monocarboxylic acid and reducing the acidity of the wine. Although the malo-lactic fermentation proceeds spontaneously, slowly, and undramatically (sometimes even without the winemaker's knowledge), it is absolutely vital to produce red wines of good quality from grapes grown in cool districts, which yield wines with too high an initial acidity to be palatable.

Certain special types of wine undergo additional microbial transformations. Sparkling wines (champagne types) undergo a second alcoholic fermentation under pressure at the expense of added sugar, either in the bottle or in bulk; the CO_2 thus produced carbonates the wine. The secondary fermentation is conducted with varieties of wine yeast which readily clump following fermentation and are consequently easily removed. Sherries (wines of the type produced in the Jerez district of Spain) are fortified with alcohol to about 15 percent, exposed to air, and allowed to develop a heavy surface growth of certain yeasts which impart the unique sherry flavor to the wine.

Some European sweet wines, notably those from the Sauternes district of

France, undergo even more complex microbial transformations. Prior to picking, the grapes become spontaneously infected with a fungus, *Botrytis cinerea*. This infection causes water loss, thus increasing sugar content, and destruction of malic acid, thus decreasing the acidity of the grapes. Certain favorable changes of flavor and color occur. The resulting very sweet must from these infected grapes is fermented by so-called *glucophilic yeasts,* i.e., yeasts that rapidly ferment the glucose leaving residual fructose (the sweeter of the two sugars). The product is a sweet dessert-type wine.

Although the high alcohol content and low pH ($\sim$3.0) of wines make them unfavorable substrates for growth of most organisms, they are subject to microbial spoilage. The problem of the "diseases" of wines was first scientifically explored by Pasteur, whose descriptions of the responsible organisms and recommendations for preventing their development are still valid today. The most serious spoilage problems are those that occur if wines are exposed to air. Film-forming yeasts and acetic acid bacteria grow at the expense of the alcohol, converting it to acetic acid, thus souring the wine. Serious diseases can also be caused by fermentative organisms in the absence of air. Rod-shaped lactic acid bacteria can grow anaerobically at the expense of residual sugar and impart a "mousy" taste to the wine. Wine yeasts can grow in sweet wines even after bottling; although such growth does not alter the flavor, the wine becomes cloudy and hence less attractive. Wine spoilage can be prevented by pasteurization, but this somewhat diminishes quality. Wines of low alcohol content that also contain sugar are particularly subject to spoilage, now commonly prevented either by chemical additives such as sulfur dioxide or by sterilization through filtration. The roles of certain microorganisms in the manufacture and spoilage of wines are summarized in Table 31.1.

• **The making of beer**

Beers are manufactured from grains which, unlike grape or other fruit musts, contain no fermentable sugars. The starch of the grains must be *saccharified* (hydrolyzed to the fermentable sugars, maltose, and glucose) prior to fermentation by yeasts. Three principal grains were traditionally used for the production of beer by man: barley in Europe, rice in the Orient, and corn in the Americas. In each case, a different solution to the saccharification of starch was found. In the case of barley, starch-hydrolyzing enzymes (amylases) of the grain itself were used. Barley seeds contain little or no amylase, but upon germination, large amounts of amylase are formed. Hence, barley is dampened, allowed to germinate, and is then dried and stored for subsequent use. Such dried, germinated barley, called *malt,* is dark in color as a result of the exposure to increased temperatures during drying and has more flavor than untreated barley seeds. The starch of barley remains largely unaffected by the malting process. Hence, the first step in beer making is the grinding of malt and its suspension in water to allow hydrolysis of the starch. Malt itself is sometimes used as the total source of starch, or, if a lighter beer is desired, unmalted barley or some other cereal grain is added to the saccharifying mixture. In the United States large quantities of rice are used in the manufacture of beer. Concomitantly with the hydrolysis of starch, other enzymatic processes occur, including the hydrolysis of proteins. After saccharification has reached the desired stage, the mixture is boiled to stop further enzymatic changes and it is then filtered. Hops (the pistillate inflorescence bracts of the vine, *Humulus lupus*) are added to the filtrate (*wort*) and contribute a soluble resin, which imparts the characteristic bitter flavor of beer and which also acts as a preservative against the growth of bacteria. The use of hops is a relatively recent modification

TABLE 31.1

Partial list of microorganisms involved in the production or spoilage of wines

ORGANISM	ROLE(S) IN WINE MAKING OR SPOILAGE	CHEMICAL OR PHYSICAL CHANGE EFFECTED
Saccharomyces cerevisiae var. *ellipsoideus*	1. Primary alcoholic fermentation 2. Carbonation of sparkling wines by a secondary fermentation 3. Clouding of sweet wines	Glucose and/or fructose $\longrightarrow$ ethanol + CO_2
Pediococcus, Leuconostoc, and *Lactobacillus*	Malo-lactic fermentation	1. Malic acid $\longrightarrow$ lactic acid + CO_2 2. Flavor enrichment
Flor sherry yeasts (*Saccharomyces beticus, cheresiensis,* and *fermenti*)	Grow as heavy surface layer (Flor) to produce sherry flavor	1. Oxidize ethanol to acetaldehyde 2. Produce flavor components
Botrytis cinerea	Grows in certain regions (e.g., Sauternes) on the surface of grapes used to produce sweet wines	1. Desiccates grapes 2. Oxidizes malic acid to CO_2 and H_2O 3. Adds flavor and color
Acetic acid bacteria and film-forming yeasts	Spoil wine exposed to air	Oxidize ethanol to acetic acid
Lactic acid bacteria, notably *Lactobacillus trichodes*	Spoil wine anaerobically	Produce "mousy" flavor

of the art of beer making, having been introduced about the middle of the sixteenth century; even today, unhopped beer is made in some countries. After filtration, the hopped wort is ready for fermentation.

In contrast to wine fermentations, beer fermentations are always heavily inoculated with special strains of yeast derived from a previous fermentation. The fermentation proceeds at low temperatures for a period of 5 to 10 days. All yeasts used in making beer are *Saccharomyces cerevisiae*, but not all strains of *S. cerevisiae* can be used to make good beer. During the course of time, special strains with desirable properties have been selected, known as *brewer's yeast*. Before Pasteur, the selection and maintenance of good yeast strains was an empirical art. The success of a brewer depended largely on his ability to obtain a suitable strain and propagate it from batch to batch without its becoming too heavily contaminated by undesirable microorganisms. Good brewer's yeasts were developed over a period of centuries; they cannot be found in nature. Like cultivated higher plants, they are a product of human art, and in recognition of this fact, the brewer refers to other yeasts (including other strains of *S. cerevisiae*) as "wild yeasts."

Since Pasteur's time the recognition, testing, and maintenance of good strains of brewer's yeast have been placed on a scientific basis. The pioneer of this work was E. C. Hansen, who worked in the Carlsberg Brewery in Copenhagen. Strains of brewer's yeast fall into two principal groups known as *top* and *bottom* yeasts. Top yeasts are vigorous fermenters, acting best at relatively high temperatures

(20°C), and are used for making heavy beers of high alcoholic content, such as English ales. Their name derives from the fact that during fermentation they are swept to the top of the vat by the rapid evolution of CO_2. In contrast, bottom yeasts are slow fermenters, act best at lower temperatures (12°C to 15°C), and produce lighter beers of low alcoholic content of the type commonly made in the United States. Their name derives from the fact that the slower rate of CO_2 evolution allows them to settle to the bottom of the vat during fermentation.

The diseases of beer, like those of wine, were first scientifically studied by Pasteur. They occur most commonly following fermentation, either during maturation or following bottling. One agent is a wild yeast, *Saccharomyces pasteurianus*, which imparts a disagreeable bitterness to beer. Lactic acid bacteria sometimes make beer acidic and cloudy. They develop principally when the temperature becomes too high during maturation and storage. Acetic acid bacteria may at times also cause souring, particularly in barreled beer exposed to air. The principal methods of avoiding spoilage are the use of pure yeast strains as starters and pasteurization of the final product. Some beers are now sterilized by filtration prior to bottling, a process which avoids the slight damage to flavor that results from pasteurization.

Although wines from grapes and beer from barley are the characteristic fermented beverages of the western world, rice serves as the source of most fermented beverages (e.g., sake) in the Orient. The problem of hydrolyzing rice starch as a preliminary to fermentation has been solved by the use of amylases from molds, principally *Aspergillus oryzae*. In the manufacture of sake the first step is the preparation of a culture of the mold. Mold spores, saved from a previous batch, are sown on steamed rice and are allowed to grow until the mass of rice is thoroughly permeated with mycelium. This material (*koji*), which serves both as a source of amylase and as an inoculum, is added to a larger batch of steamed rice mixed with water. Hydrolysis of the starch proceeds, and when sufficient sugar has accumulated, a spontaneous alcoholic fermentation begins. Lactic acid bacteria as well as yeasts are present in koji, so lactic acid is produced in addition to alcohol and CO_2. The production of alcoholic beverages from grain in the Orient thus differs from the Occidental process in two respects: saccharification is effected by microorganisms, and saccharification proceeds simultaneously with fermentation.

In the Americas as well as in certain regions of the Middle East, yet a third agent of saccharification is used—human saliva, which contains amylases. Indians in Central and South America prepare a corn beer by chewing the grains and spitting the mixture into a vessel, where it is allowed to undergo a spontaneous alcoholic fermentation.

• **The making of bread**

An alcoholic fermentation by yeasts is an essential step in the production of raised breads; this process is known as the leavening of bread (after the old word for yeast, "leaven"). The moistened flour is mixed with yeast and allowed to stand in a warm place for several hours. Flour itself contains almost no free sugar to serve as a substrate for fermentation, but there are some starch-splitting enzymes present that produce sufficient sugar to support leavening. In the highly refined flours, commonly used in the United States, these enzymes have been destroyed and sugar must be added to the dough. The sugar is rapidly fermented by yeast. The carbon dioxide produced becomes entrapped in the dough, causing it to rise, while the alcohol produced is driven off during the baking process.

Yeast produces other more subtle changes in the physical and chemical properties of the dough, which affect the texture and flavor of the bread. This fact became evident when J. von Liebig, a German chemist of the last century, invented baking powder, a mixture of chemicals which produces carbon dioxide when moistened. Liebig anticipated that baking powder would replace yeast. Although Liebig's invention is widely used in other forms of baking, it did not supplant yeast as a leavening agent for bread.

The yeasts used in bread making all belong to the species *Saccharomyces cerevisiae* and have been derived historically from strains of top yeasts used in brewing. Until the nineteenth century, yeasts for bread making were obtained directly from the nearest brewery. The commercial production of compressed yeast by industry was greatly stimulated by the application of mass production techniques to bread making. A large modern bakery may use many hundreds of pounds of yeast daily, for about 5 pounds of yeast are required to leaven 300 pounds of flour. Much of the bakers' yeast manufactured today is dried under controlled conditions that maintain viability of the yeast cells, a treatment which facilitates shipment and storage.

MICROBES AS SOURCES OF PROTEIN

Because of their rapid growth, high protein content, and ability to utilize organic substrates of low cost, microorganisms are potentially valuable sources of animal food. The growth of the science of animal nutrition has led to the development of a new industry, based on the cultivation of yeast for use as a supplement in animal feeds. Since the goal is to obtain cell material, the organisms are always grown under forced aeration to maximize the growth yield. Even if the oxygen supply is maintained at a high level, there is the risk that when carbohydrates are used as substrates part of the substrate will be diverted to alcohol. Consequently, strictly aerobic yeasts of the genus *Candida* are used in preference to fermentative yeasts.

• **Production of yeasts from petroleum**

The cost of raw material is a factor of paramount importance in the production of microorganisms for use as food, and cheap sources of carbohydrate (e.g., whey, molasses, sulfate waste liquor) were initially used for growth of food yeasts. However, since aerobic growth conditions are used, all compounds that can support respiratory metabolism may serve. This led to the development of processes which utilize petroleum as a substrate. Petroleum is still very cheap, compared with other possible substrates, and since hydrocarbons are the most highly reduced of organic compounds, growth yields at their expense are extremely high.

The British Petroleum Corporation has built an industrial unit in France for the cultivation of *Candida lipolytica* in an aqueous emulsion of crude petroleum. This yeast can oxidize aliphatic, unbranched hydrocarbons of chain length C_{12} to C_{18}, compounds that comprise part of the complex mixture of alkanes present in crude petroleum. Their selective removal by the growth of *Candida lipolytica* produces a dewaxed petroleum that is much more easily refined. The economic feasibility of the British Petroleum process depends on its twofold function: simplification of refining and protein production.

Considerable efforts are also being made to develop processes for the production of microbial protein at the expense of other cheap substrates. One potential substrate is the gas methane, a major petrochemical product. However, many problems remain to be solved before an effective industrial process can be based on its use. The most important of these are the relatively low growth rates of methane-oxidizing bacteria and their tendency to produce large quantities of extracellular slime.

• Production of specific amino acids

The great potential value of microorganisms as foods or feed supplements lies in their high protein content. This makes them the best agents for the rapid and efficient conversion of other more readily available organic compounds into protein, of which the world is becoming critically short. This point becomes evident when protein production by cattle and by yeast is compared. A bullock weighing 500 kg produces about 0.4 kg of protein in 24 hours. Under favorable growth conditions, 500 kg of yeast produce over 50,000 kg of protein in the same period.

Many plant foods contain sufficient protein to supply the quantitative needs of mammals, but they cannot serve as sole sources of dietary protein because their proteins are deficient in certain specific amino acids required by mammals. Wheat protein is low in lysine, rice protein in lysine and threonine, corn protein in tryptophan and lysine, bean and pea protein in methionine. The addition of the deficient amino acid(s) to diets that contain a single source of vegetable protein will render them adequate. The practicality of fortifying diets of vegetable protein with individual amino acids has been amply demonstrated in numerous experiments with both animals and humans. Thus, the world shortage of certain specific amino acids—notably, lysine, threonine, and methionine—is more critical than the shortage of total protein. The microbial production of *specific amino acids* has therefore been intensively studied.

Since the metabolism of microorganisms is precisely regulated (Chapter 9), microorganisms normally synthesize quantities of amino acids just sufficient to meet their growth requirements. However, naturally occurring and mutant strains of some microorganisms have defective mechanisms for the regulation of specific biosynthetic pathways and, as a consequence, excrete large amounts of certain amino acids into the medium. Methods for the microbial production of nutritionally important amino acids are now available and are constantly being improved.

THE USE OF ACETIC ACID BACTERIA BY MAN

When wine and beer are freely exposed to the air, they frequently turn sour. Souring is caused by the oxidation of alcohol to acetic acid, mediated by the strictly aerobic acetic acid bacteria. The spontaneous souring of wine is the traditional method of manufacturing vinegar. The word vinegar is derived from the French "vinaigre" which literally means "sour wine."

The manufacture of vinegar still remains largely empirical. The principal modifications introduced during the past century concern the mechanical rather than the microbiological aspects of the process. In the traditional *Orleans Process*, which is still used in France, wooden vats are partially filled with wine, and the acetic acid bacteria develop as a gelatinous pellicle on the surface of the liquid. The conversion of ethanol to acetic acid takes several weeks—the rate of the process being limited by the slow diffusion of air into the liquid. The survival of this slow and inefficient method is attributable to the high quality of the product.

When the taste of the product is not of primary importance, vinegar is made by more rapid methods from cheaper raw materials (e.g., diluted distilled alcohol and cider). These methods are designed to accelerate oxidation by improved aeration and regulation of temperature, but they remain microbiologically uncontrolled. The oldest such method, developed early in the nineteenth century, utilizes a tank that is loosely filled with wooden shavings through which the liquid is circulated. The liquid is trickled into the tank and air is blown countercurrent to the liquid flow. The acetic acid bacteria develop as a thin film on the wooden shavings, thus providing a large surface of cells which are simultaneously exposed to the medium and to air. Once a bacterial population has become established on the shavings, successive batches of vinegar can be produced quickly; solutions initially containing 10 percent alcohol can be converted to acetic acid in 4 or 5 days. Much vinegar is still made by this method, but modern, stirred, deep tank fermentors, similar to those used to produce antibiotics, are now being introduced.

The oxidation of ethanol to acetic acid is an example of the *incomplete oxidations* carried out by acetic acid bacteria. Certain other incomplete oxidative conversions by acetic acid bacteria are industrially important. Gluconic acid, which is used by the pharmaceutical industry, is made by oxidation of glucose by acetic acid bacteria. Many sugar alcohols are converted to sugars by acetic acid bacteria. One such reaction in commercial use is the production of sorbose from sorbitol. Sorbose is used as a suspending agent for certain pharmaceuticals, and it is an intermediate in the manufacture of L-ascorbic acid (vitamin C).

THE USE OF LACTIC ACID BACTERIA BY MAN

Lactic acid bacteria produce large amounts of lactic acid from sugar. The resulting decrease in pH renders the medium in which they have grown unsuitable for the growth of most other microorganisms. Growth of lactic acid bacteria, therefore, is a means of preserving food; in addition, they produce flavor components.

• **Milk products**   The manufacture of such milk products as butter, cheese, and yogurt involves the use of microorganisms, among which the lactic acid bacteria are particularly important. The discovery of the roles played by microorganisms in the preparation of these foods has led to the development of a special branch of bacteriology known as *dairy bacteriology*.

Many lactic acid bacteria occur normally in milk and are responsible for its spontaneous souring. Milk souring provides a means of preserving this otherwise highly unstable foodstuff, and the manufacture of cheese and other fermented milk products undoubtedly began largely as a means of preservation.

The manufacture of cheese involves two main steps: *curdling* the milk proteins to form a solid material from which the liquid is drained away; and the *ripening* of the solid curd by the action of various bacteria and fungi, although certain fresh cheeses are essentially unripened.

The curdling process may be exclusively microbiological, since acid production of lactic acid bacteria is sufficient to coagulate milk proteins. However, an enzyme known as *rennin* (extracted from the stomachs of calves) which curdles milk is also often used for this purpose.

The subsequent ripening of the curd is a very complex process, and is highly variable, depending on the kind of cheese being made. The ripening process is chemically variable. In the young cheese, all nitrogen is present in the form of insoluble protein, but as ripening proceeds, the protein is progressively cleaved to soluble peptides and ultimately to free amino acids. The amino acids can be further decomposed to ammonia, fatty acids, and amines. In certain cheeses, protein breakdown is restricted. For example, in Cheddar and Swiss cheese, only 25 to 35 percent of the protein is converted to soluble products. In soft cheeses, such as Camembert and Limburger, essentially all the protein is converted to soluble products. In addition to changes in the protein components, ripening involves considerable hydrolysis of the fats present in the young cheese. The enzymes present in the rennin preparation contribute somewhat to the ripening process, but microbial enzymes in the cheese play the major role. The types of microorganism involved are varied. Hard cheeses are ripened largely by lactic acid bacteria, which grow throughout the cheese, die, autolyze, and release hydrolytic enzymes. Soft cheeses are ripened by the enzymes from yeasts and other fungi that grow on the surface.

Some microorganisms play highly specific roles in the ripening of certain cheeses. The blue color and unique flavor of Roquefort cheese is a consequence of the growth of a blue-colored mold, *Penicillium roqueforti*, throughout the cheese.* The characteristic holes in Swiss cheese are formed by carbon dioxide, a product of the propionic acid fermentation of lactic acid by species of *Propionibacterium*.

Butter manufacture is also in part a microbiological process, since an initial souring of cream, caused by milk streptococci, is necessary for subsequent separation of butterfat in the churning process. These organisms produce small amounts of acetoin which is spontaneously oxidized to *diacetyl*, the compound responsible for the flavor and aroma of butter. Since streptococci differ markedly in their ability to produce acetoin, it has become common practice to inoculate pasteurized cream with pure cultures of selected strains.

In many parts of the world milk is allowed to undergo a mixed fermentation by lactic acid bacteria and yeasts which produces a sour, mildly alcoholic beverage (e.g., kefir and kumiss).

The roles of microorganisms in the manufacture of milk products are summarized in Table 31.2.

• The lactic fermentation of plant materials

Certain lactic acid bacteria are found characteristically on plant materials. These organisms are responsible for the souring that occurs in the preparation of pickles, sauerkraut, and Spanish-style olives. In these lactic acid fermentations, sugars initially present in the plant materials serve as the fermentable substrates. The lactic acid produced imparts flavor to the product as well as protecting it from further microbial attack.

---

*In the United States a white mutant of *Penicillium roqueforti* is sometimes used to produce a mold-ripened cheese for those who find the flavor desirable but the color objectionable.

TABLE 31.2

Microbiology of milk products

PRODUCT	PROCESS	MICROORGANISMS[a]
Buttermilk	Lactic acid fermentation	*Lactobacillus bulgaricus*
Yogurt	Lactic acid fermentation	*L. bulgaricus* + *Streptococcus thermophilus*
Kefir	Alcoholic and lactic acid fermentations	*Streptococcus lactis* + *L. bulgaricus* + lactose-fermenting yeasts
Cheeses (in general)	Initial lactic acid fermentation temperature, 35°C	*S. lactis* or *S. cremoris*
	fermentation temperature, 42°C	Various thermophilic lactic acid bacteria, principally lactobacilli
Hard cheeses (e.g., Cheddar and Swiss)	Proteolysis and lipolysis	Various lactic acid bacteria within the cheese
Soft cheeses (e.g., Camembert, Brie, and Limburger)	Proteolysis and lipolysis	Surface growth, initially of fungi (*Geotrichum candidum* and *Penicillium* spp.), sometimes followed by *Bacterium linens* and *B. erythrogenes*
Swiss cheese	Propionic acid fermentation	*Propionibacterium* spp.
Roquefort	Lipolysis and production of blue mold pigment	*Penicillium roqueforti*

[a] Microorganisms generally associated with the process.

The preservative value of a lactic acid fermentation is also exploited in the ensilaging of green cattle fodder. After plant materials have undergone fermentation in a silo, they may be kept indefinitely without risk of decomposition.

• Dextran production

Some lactic acid bacteria belonging to the genus *Leuconostoc* produce large amounts of an extracellular polysaccharide known as dextran when grown with sucrose. Dextran is a polyglucose of high but variable molecular weight (15,000 to 20,000,000); the average molecular weight varies with the strain employed. These lactic acid bacteria first came to the attention of industrial microbiologists for their nuisance value; they occasionally develop in sugar refineries, and the large amounts of gummy polysaccharide produced may literally clog the works.

Dextran is now produced industrially, following the discovery that dextran derivatives which have been chemically cross linked to make them insoluble in water can act as molecular sieves. Columns of such modified dextrans (marketed largely under the trade name of *Sephadex*) retard the passage of small molecules, and thus permit the physical fractionation of solutes which differ in molecular weight. Sephadex columns can be used for molecular weight determinations in the range of 700 to 800,000 daltons, after calibration with compounds of known molecular weight.

Another class of microbial polysaccharides now being produced industrially are the chemically complex extracellular polysaccharides synthesized by aerobic pseudomonads of the *Xanthomonas* group. These substances have the physical property of forming thixotropic gels, and in addition, are stable at relatively high temperatures. As a result, they have a wide variety of industrial uses, notably as lubricants in the drilling of oil wells and as gelling agents in paints with a water base.

THE USE OF BUTYRIC ACID BACTERIA BY MAN

• **The retting process**

Retting is a controlled microbial decomposition of plant materials designed to liberate certain components of the plant tissue. The oldest retting process, which has been used for several thousand years, is the retting of flax and hemp to free the bast fibers used in the making of linen. These fibers, made up of cellulose, are held together in the plant stem by a cementing substance, pectin; their physical separation is difficult. The goal of retting is to bring about decomposition of the pectin, thus freeing the fibers without simultaneous decomposition of the fibers themselves. The plant stems are immersed in water; the plant stems become water logged, microbial decomposition begins. Initially, aerobic microorganisms develop and use up the dissolved oxygen, making the environment suitable for the subsequent development of the anaerobic butyric acid bacteria. These organisms rapidly attack the plant pectin, freeing the fibers. If retting is unduly prolonged, cellulose-fermenting bacteria will also develop and destroy the fibers. An analogous retting process is used in the preparation of potato starch. Its purpose is to free the starch-containing cells in the potato tuber from the pectin in which they are embedded.

• **The acetone-butanol fermentation**

In the past 50 years, certain *Clostridium* spp. have been used on a very large scale for the production of the industrial solvents, acetone and butanol. Many clostridia carry out a fermentation of sugars with the formation of carbon dioxide, hydrogen, and butyric acid. Some carry out further reactions, converting the butyric acid to butanol and the acetic acid to ethanol and acetone. The commercial development of the so-called *acetone-butanol fermentation* mediated by *Clostridium aceto-butylicum* began in England just before World War I and expanded rapidly during the war because acetone was needed as a solvent in the manufacture of explosives. After World War I, the demand for acetone diminished, but the process survived because another major product of the fermentation, butanol, found a use as a solvent for the rapid drying of nitrocellulose paints in the growing automobile industry. A by-product of the fermentation, the vitamin, riboflavin, also helped to maintain its commercial feasibility.

Today, this industry has virtually disappeared as a result of the development of competing methods, only in part microbiological, for the synthesis of the major products. Both acetone and butanol are produced in large amounts from petroleum; a microbiological process based on the use of yeasts is the principal source of riboflavin.

The acetone-butanol fermentation made important technological contributions to industrial microbiology. It was the first large-scale process in which the exclusion of other kinds of microorganisms from the culture vessel was of major importance to the success of the operation. The medium used for the cultivation of *Clostridium acetobutylicum* is also favorable for the development of lactic acid bacteria; if these organisms begin to grow, they rapidly inhibit the further growth of the clostridia through lactic acid formation. An even more serious problem is infection with bacterial viruses, to which clostridia are highly susceptible. Thus, the acetone-butanol fermentation can be operated successfully only under conditions of careful microbiological control. The establishment of this industry led to the first successful use of *pure culture methods on a mass scale,* which were later improved and refined in connection with the industrial production of antibiotics.

THE MICROBIAL PRODUCTION OF CHEMOTHERAPEUTIC AGENTS

The period since World War II has seen the establishment and extremely rapid growth of a major new industry, the use of microorganisms for the synthesis of chemotherapeutic agents, particularly antibiotics and hormones. The development of this industry has had a dramatic and far-reaching social impact. Nearly all bacterial infectious diseases which were, prior to the antibiotic era, major causes of human death have been brought under control by the use of these drugs. In the United States, bacterial infection is now a less frequent cause of death than suicide or traffic accidents.

• **The rise of chemotherapy**

The importance of acquired immunity as a means of protection against specific bacteriological diseases was recognized shortly after the discovery of the role of microorganisms as the etiological agents of infectious diseases (see Chapter 1). For several decades thereafter control of infectious disease was based exclusively on the use of antisera and vaccines, and was largely preventative; usually, little could be done to cure infections after they had appeared.

A different kind of approach to the control of infectious disease was developed by the German chemist Paul Ehrlich, who initiated an empirical search for synthetic chemicals that possess *selective toxicity* for pathogenic microorganisms. He coined the word *chemotherapy* to describe this approach to the control of infectious disease. Ehrlich's efforts produced one limited success: in 1909 he discovered synthetic organic compounds containing arsenic which were effective in the treatment of syphilis and other spirochetal infections, but which had severe side effects.

The next significant advance in chemotherapy was also made empirically. Large numbers of aniline dyes were screened for antibacterial chemotherapeutic activity and one substance of this class, *prontosil,* was found to be effective. However, prontosil possessed no antibacterial action *in vitro.* Its antibacterial activity in infected animals was then shown to be attributable to a colorless breakdown product, *sulfanilamide,* formed in the animal body. Sulfanilamide possesses antibacterial activity both *in vitro* and *in vivo.* D. D. Woods observed that the inhibition of bacterial growth by sulfanilamide can be reversed by a structural analogue, *p*-aminobenzoic acid (Figure 31.1).

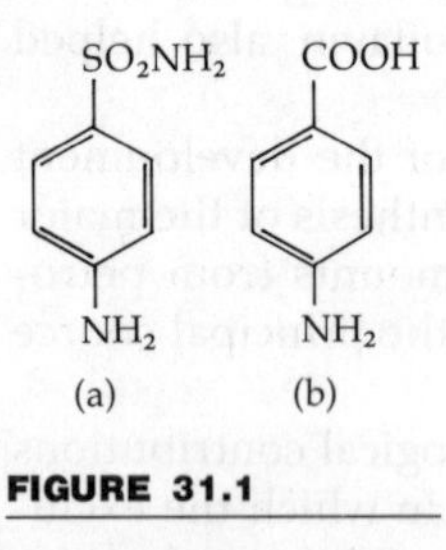

Wood then made a brilliant series of deductions: that *p*-aminobenzoic acid is a normal constituent of the bacterial cell; that it has a coenzymatic function; and that this function is blocked by sulfanilamide as a result of its steric resemblance to *p*-aminobenzoic acid. In fact, *p*-aminobenzoic acid proved to be not a coenzyme, but a biosynthetic precursor of the coenzyme folic acid; sulfanilamide blocks its

FIGURE 31.1

The structures of (a) sulfanilamide and (b) *p*-aminobenzoic acid.

SO₂NH₂   COOH

NH₂   NH₂
(a)   (b)

conversion to this end product. Sulfanilamide is selectively toxic because most bacteria must synthesize folic acid *de novo*, whereas mammals obtain it from dietary sources.

Wood's work appeared to offer a *rational approach to chemotherapy* through the synthesis of analogues of known essential metabolites. In succeeding years, thousands of structural analogues of amino acids, purines, pyrimidines, and vitamins were synthesized and tested; but very few useful chemotherapeutic agents were discovered.

The great modern advances in chemotherapy have come from the chance discovery that many microorganisms synthesize and excrete compounds which are selectively toxic to other microorganisms. These compounds, called *antibiotics*, have revolutionized modern medicine.

The first chemotherapeutically effective antibiotic was discovered by Alexander Fleming in 1929. He observed, as many before him had done, that on a plate culture of bacteria which had become contaminated by a mold, bacterial growth in the vicinity of the mold colony was inhibited. He reasoned that the mold was excreting into the medium a chemical that prevented bacterial growth. Sensing the possible chemotherapeutic significance of his observation, he isolated the mold, which proved to be a species of *Penicillium,* and established that culture filtrates contained an antibacterial substance which he called *penicillin.*

Penicillin proved to be chemically unstable, and Fleming was unable to purify it. Working with impure preparations, he demonstrated its remarkable effectiveness in inhibiting the growth of many Gram-negative bacteria, and he even used it with success for the local treatment of human eye infections. In the meantime, the chemotherapeutic effectiveness of sulfanilamides had been discovered, and Fleming, discouraged by the difficulties of penicillin purification, abandoned further work on the problem.

Ten years later, a group of British scientists headed by H. W. Florey and E. Chain resumed the study of penicillin. Clinical trials with partly purified material were dramatically successful. By this time, however, England was at war; and the industrial development of penicillin was undertaken in the United States, where an intensive program of research and development was begun in many laboratories. Within 3 years penicillin was being produced on an industrial scale, an astonishing achievement in view of the many difficulties which had to be overcome. Penicillin remains one of the most effective chemotherapeutic agents for treatment of many bacterial infections.

Penicillin proved to be effective primarily against infections caused by Gram-positive bacteria. Its startling success in the treatment of such infections prompted intensive searches both at universities and in industry for new antibiotics. A second clinically important antibiotic, *streptomycin,* which is effective against both Gram-negative bacteria and *Mycobacterium tuberculosis,* was discovered by A. Schatz and S. Waksman.

Streptomycin was the first example of an antibiotic possessing a *broad spectrum of activity,* effective against many Gram-positive and Gram-negative bacteria. Other antibiotics with even broader spectra of activity (for example, the tetracyclines) have been subsequently discovered. Antibiotics have proved to be less useful in the treatment of fungal infections: antifungal antibiotics such as nystatin and amphotericin B are considerably less effective therapeutically than their antibacterial counterparts, at least in part because their toxicity is far less selective. Good antiviral antibiotics are yet to be found.

Since 1945, thousands of different antibiotics produced by fungi, Actinomycetes, or unicellular bacteria have been isolated and characterized. A small fraction of these are of therapeutic value; about 50 are currently produced on a large scale for medical and veterinary use. Their nomenclature is complicated, one antibiotic often being sold under several different names. There are two reasons for this proliferation of names. First, many antibiotics are members of a group of compounds, all of which possess similar structures; a name is required for the *class of compounds,* as well as for each *individual representative.* Second, each manufacturer of an antibiotic assigns to it for marketing purposes a *trade name* which, by law, only he can use. To protect a trade name for exclusive use, the law requires that another name, available for general use, be also assigned to the antibiotic in question; this is called the *generic name.* The generation of multiple names can be illustrated by the example of an antibiotic which in the United States is given the generic name, *rifampin.* The generic name of the same compound in Europe is *rifampicin.* Its class name is rifamycin. It is sold under the trade names *Rifactin* and *Rifadin,* among others.

The generic names, sources, uses, and mode of action of some antibiotics are shown in Table 31.3.

TABLE 31.3

Properties and uses of certain antibiotics

CHEMICAL CLASS	GENERIC NAME	BIOLOGICAL SOURCE	EFFECTIVE CHEMOTHERAPEUTICALLY AGAINST	MODE OF ACTION
β-lactams	Penicillins	*Penicillium* spp.	Gram-positive bacteria	
	Cephalosporins	*Cephalosporium* spp.	Gram-positive and gram-negative bacteria	Inhibit synthesis of bacterial cell wall (peptidogylcan)
Macrolides	Erythromycin	*Streptomyces erythreus*	Gram-positive bacteria	Inhibit 50S ribosome function
	Carbomycin	*S. halstidii*		
Aminoglycosides	Streptomycin	*S. griseus*	Gram-positive and gram-negative bacteria	Inhibit 30S ribosome function
	Neomycin	*S. fradiae*		
Tetracyclines	Tetracycline[a]	*Streptomyces aureofaciens*	Gram-positive and gram-negative bacteria; rickettsias; bedsonias	Inhibits binding of aminoacyl-t RNAs to ribosomes
Polypeptides	Polymyxin G	*Bacillus polymyxa*	Gram-negative bacteria	Destroys cytoplasmic membrane
	Bacitracin	*B. subtilis*	Gram-positive bacteria	Inhibits synthesis of bacterial cell wall (peptidoglycan)
Polyenes	Amphotericin B	*S. nodosus*	Fungi	Inactivate membranes containing sterols
	Nystatin	*S. nouresii*	Fungi	
—	Chloramphenicol[b]	*S. venezuelae*	Gram-positive and gram-negative bacteria; rickettsias	Inhibits translation step of ribosome function

[a] Made microbiologically and by chemical dehydrochlorination of chlorotetracycline.
[b] Now made by chemical synthesis.

Antibiotics are exceedingly varied in chemical structure. Examples of some of the various chemical classes are shown in Figure 31.2.

• **Mode of action of antibiotics**

The search for new antibiotics remains an empirical enterprise, and their physiological significance for the microorganisms that produce them is obscure. However, the reasons for their selective toxicity are in many cases now known. In general, antibiotics owe their selective toxicity to the fundamental biochemical differences between procaryotic and eucaryotic cells, their toxic effect being the consequence of their ability to inhibit one essential biochemical reaction specific either to the procaryotic or to the eucaryotic cell (see Chapter 3).

• **The production of antibiotics**

The antibiotics were the first industrially produced microbial metabolites which were not *major* metabolic end products. The yields, calculated in terms of conversion of the major carbon source into antibiotic, are low and are greatly influenced by the composition of the medium and by the other cultural conditions. These facts have encouraged intense research directed toward improving yields. For this purpose, genetic selection has proved remarkably successful. The wild type strain of *Penicillium chrysogenum* first used for penicillin production yielded approximately 0.1 gram of penicillin per liter. From this strain a mutant was selected which produced 8 grams per liter under the same growth condition, a sixtyfold improvement in yield. Subsequent strain selection following chemical mutagenesis has led to the development of new strains with even greater capacity for antibiotic production. By such *sequential genetic selection,* improvements of antibiotic yield as great as a thousandfold have often been obtained. Most genetic improvement has been empirical; large numbers of mutagenized clones are evaluated for their abilities to produce larger quantities of the antibiotic. However, with increased knowledge of the pathways of biosynthesis of antibiotics more rational approaches are being exploited. It is now possible to select strains in which control of the synthesis of known precursors of an antibiotic has been altered by mutation. Such strains produce larger amounts of the precursor, and sometimes also larger amounts of the antibiotic end product.

The synthesis of antibiotics begins only after growth of the organisms that produce them has virtually ceased (Figure 31.3). They belong to a class of microbial products called *secondary metabolites,* because their synthesis is not associated with growth. The control mechanisms that trigger the synthesis of secondary metabolites as growth ceases are a fascinating but almost completely unexplored aspect of biochemical regulation.

Although the microorganisms used to produce antibiotics are all aerobes and are grown under conditions of vigorous aeration, the production process is generally referred to in the technical literature as a "fermentation." Antibiotics are produced by so-called *submerged cultivation methods,* using deep stainless steel tanks which must be subjected to continuous forced aeration and rapid mechanical agitation. The provision of adequate aeration is of great importance to yield, and the energy expended for aeration contributes appreciably to the cost of production.

When a microorganism is grown aerobically in tanks with capacities of tens of thousands of gallons containing a rich, nonselective medium, the maintenance of a pure culture poses numerous special engineering problems. For the successful production of antibiotics, pure cultures are indispensable. This fact was first revealed during penicillin production. Many bacteria produce an enzyme, peni-

streptomycin (aminoglycoside)

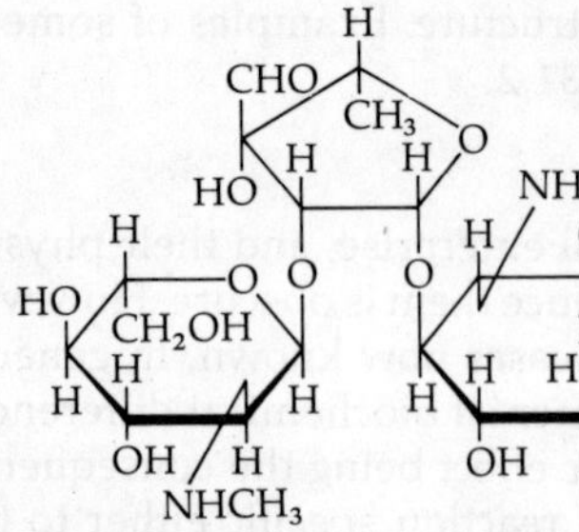

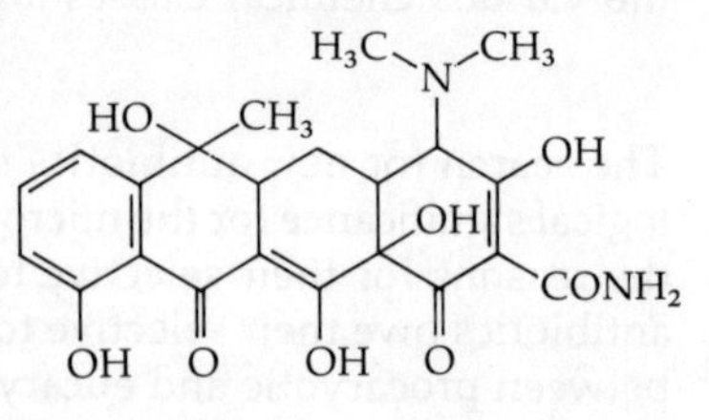

tetracycline

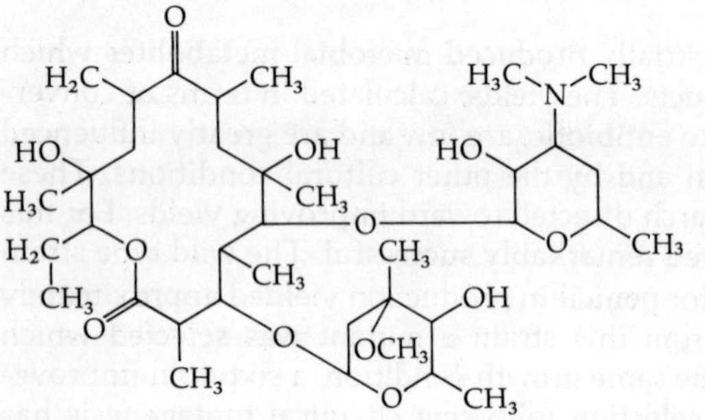

erythromycin (macrolide)

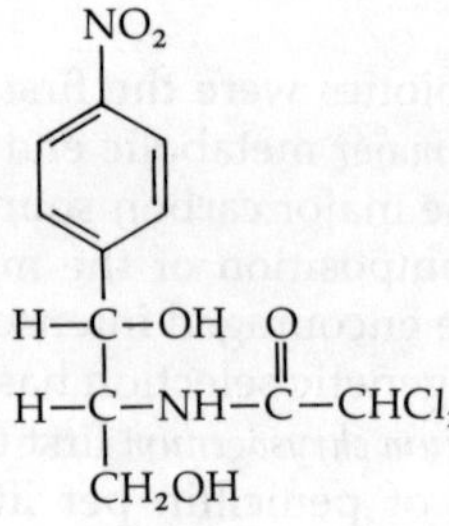

chloramphenicol

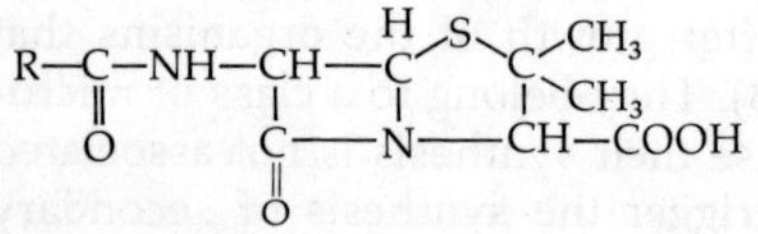

penicillins β(-lactam)
(R—group variable.)

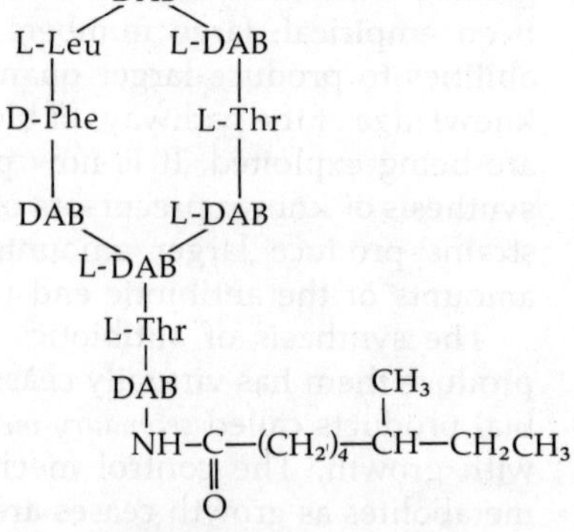

polymyxin B
(polypeptide)

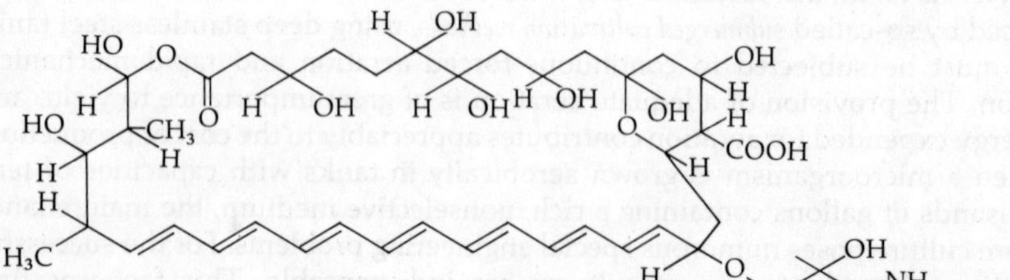

DAB
L-Leu   L-DAB
D-Phe   L-Thr
DAB   L-DAB
L-DAB
L-Thr
DAB

filipin

FIGURE 31.2

Structures of some antibiotics illustrating the wide diversity of the chemical classes. Polymyxin B is a cyclic polypeptide made up of the amino acid residues: leucine (leu), phenylalanine (phe), threonine (thr), and α,γ-diaminobutyric acid (DAB).

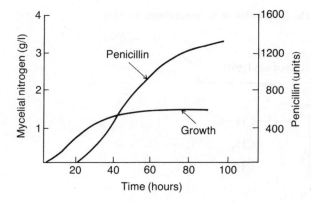

FIGURE 31.3

Temporal relationship between growth of *Penicillium chrysogenum* and its production of penicillin. After W. E. Brown and W. H. Peterson, "Factors Affecting Production of Penicillin in Semi-pilot Plant Equipment," *Ind. Eng. Chem.* **42**, 1769 (1950).

cillinase, which catalyzes the hydrolytic cleavage of the four membered β-lactam ring of penicillin, with resulting loss of antibiotic activity. Consequently, contamination of a fermentor by penicillinase-producing bacteria can result in complete destruction of the accumulated penicillin.

In the manufacture of antibiotics the microbial product is sometimes subjected to subsequent chemical modification. One example is the chemical substitution of the acyl group of natural penicillins (this group is designated as R- in the structural formula shown in Figure 31.2) to produce a large variety of semi-synthetic penicillins (Figure 31.4). Another example is the catalytic dehydrochlorination of chlorotetracycline to produce the more effective substance, tetracycline.

• Microbial resistance to antibiotics

The antibiotic era of medicine began abruptly some 30 years ago. How long it will last has become an open question. Although the search for new antibiotics continues undiminished, the rate of their discovery has declined sharply; most of the really effective antibiotics have probably already been discovered. Furthermore, strains of pathogens resistant to antibiotics have begun to develop at an alarming rate. Although most strains of staphylococci were sensitive to penicillin G when it was first introduced into medical practice, essentially all hospital acquired staphylococcal infections are now resistant to this antibiotic. A problem of even greater concern is the appearance of bacterial strains that are simultaneously resistant to several antibiotics, the so-called *multiply-resistant strains*. Between 1954 and 1964 the frequency of multiply-resistant strains of *Shigella* in Japanese hospitals rose from 0.2 percent to 52 percent.

Bacterial resistance to an antibiotic is sometimes acquired by the mutation of a chromosomal gene, which modifies the structure of the cellular target. A good example is mutationally acquired streptomycin resistance. This antibiotic deranges bacterial protein synthesis by attachment to one of the proteins in the 30S subunit of the ribosome. Some mutations of the gene that encodes this ribosomal protein destroy the ability of the protein to bind streptomycin, but they do not substantially affect ribosomal function; the cell in which such a mutation occurs consequently becomes streptomycin-resistant.

Resistance can also be acquired as a result of infection of the bacterial cell by a plasmid belonging to the class of *resistance factors* (R factors, see Chapter 15). These plasmids often confer simultaneous resistance to several antibiotics. They carry genes which encode enzymes that catalyze chemical modifications of

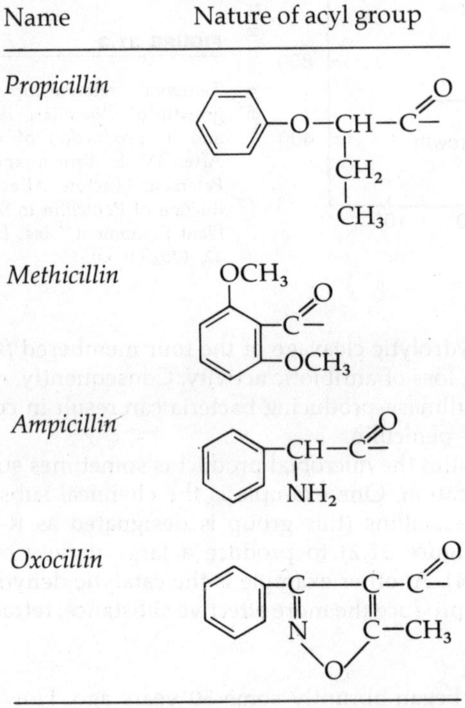

Name   Nature of acyl group

Propicillin

Methicillin

Ampicillin

Oxocillin

FIGURE 31.4

Some semisynthetic penicillins now in chemical use, showing the chemically introduced acyl substituents. (See Figure 31.2 for the general structure of penicillins.)

antibiotics, converting them to derivatives without antibiotic action. For example, streptomycin resistance conferred by an R factor may be caused either by adenylation, phosphorylation, or acetylation of the antibiotic; all these chemical modifications result in the loss of antibiotic activity. Multiply-resistant bacterial strains almost always owe their resistance to the presence of an R factor. Since these plasmids have wide host ranges, often being readily transferable between different bacterial species, their increasing dissemination in natural bacterial populations is by far the most serious aspect of the problem of microbial resistance to antibiotics. Unless some solution to this problem can be found, the future therapeutic effectiveness of antibiotics is in jeopardy.

• Microbial   Cholesterol (Figure 31.5) and chemically related steroids are structural components
transformations   of eucaryotic cellular membranes and therefore universal chemical constituents
of steroids   of eucaryotes. During the evolution of vertebrates, special pathways were evolved for the conversion of these universal cell constituents to new and functionally specialized steroids: the *steroid hormones*, which are potent regulators of animal development and metabolism. Steroid hormones are formed in specialized organs, through the secondary metabolism of cholesterol, a C_{27} steroid. The adrenocortical hormones are synthesized in the adrenal gland and are all C_{21} compounds, such as *cortisone* (Figure 31.5); the sex hormones are synthesized in the ovary or the testis and are C_{18} or C_{19} compounds (Figure 31.5). Accordingly, by relatively slight chemical modifications of the basic steroid structure, vertebrates have evolved

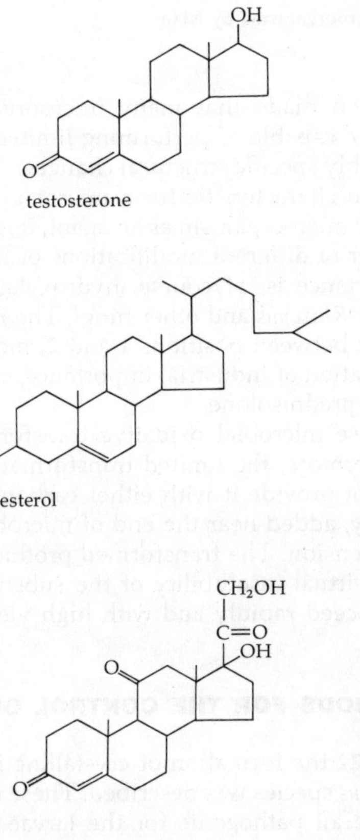

OH

testosterone

O

cholesterol

HO

CH_2OH
$C=O$
OH

O

O

cortisone

FIGURE 31.5

The structure of cholesterol, a C_{27} steroid, and of two mammalian steroid hormones for which cholesterol is a biosynthetic precursor—cortisone (C_{21}) and testosterone (C_{19}).

two new subclasses of steroid molecules with highly specific physiological functions and of great potency.

The elucidation of the structures and general functions of mammalian steroid hormones was completed about 30 years ago, but it was only in 1950 that possible chemotherapeutic uses for them became apparent, with the discovery that cortisone treatment can relieve dramatically the symptoms of rheumatoid arthritis. Today, cortisone and its derivatives are very widely used to treat a variety of inflammatory conditions, and additional uses for steroid hormones have emerged in the treatment of certain types of cancer and as oral contraceptives. The production of these compounds has now become a major industry.

Since the steroid hormones are produced by mammals in very small quantities, it was evident that their isolation from animal sources could not supply the clinical needs. Accordingly, the chemists turned their attention to the synthesis of those substances from plant sterols, which are abundant and can be cheaply prepared. One major chemical obstacle soon became apparent. All adrenocortical hormones are characterized by the insertion of an oxygen atom at position 11 on the ring system (Figure 31.6), by an organ-specific enzymatic hydroxylation of the biosynthetic precursor in the adrenal gland. Although it is easy to hydroxylate the steroid nucleus chemically, it is extremely difficult to insert a hydroxyl group at a specific position, and the specific 11-hydroxylation essential to the successful synthesis of cortisone from cheaper steroids became a major stumbling block to the development of a successful industrial process.

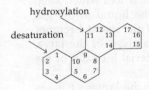

hydroxylation

desaturation

FIGURE 31.6

The ring system of steroids, showing the numbering of the carbon atoms and the specific sites of two commercially important chemical modifications that are mediated by microorganisms.

849

The discovery was then made that many microorganisms—fungi, Actinomycetes, and bacteria—are capable of performing limited oxidations of steroids, which cause small and highly specific structural changes. The positions and nature of these changes are often characteristic for a microbial species, so that by the selection of an appropriate microorganism as an agent, it is possible to bring about any one of a large number of different modifications of the steroid molecule. Of particular practical importance is, of course, hydroxylation at the 11 position, which can be mediated by *Rhizopus* and other fungi. The introduction of a double bond by dehydrogenation between positions 1 and 2, mediated by a *Corynebacterium*, is another transformation of industrial importance, essential in the synthesis of a cortisone derivative, prednisolone.

The substrates for these microbial oxidative transformations are essentially insoluble in water. Furthermore, the limited transformation which the microorganism can effect does not provide it with either carbon or energy. The steroid substrates are, accordingly, added near the end of microbial growth, in the form of a finely dispersed suspension. The transformed products are released into the medium. In spite of the virtual insolubility of the substrates in water, many of these transformations proceed rapidly and with high yields.

MICROBIOLOGICAL METHODS FOR THE CONTROL OF INSECTS

In Chapter 22 the formation of crystalline inclusions in the sporulating cells of certain *Bacillus* species was described. These bacilli (*Bacillus thuringensis* and related forms) are all pathogenic for the larvae (caterpillars) of certain insects, specifically, a very wide range of insects belonging to the Lepidoptera (butterflies and related forms). Following the isolation of the crystalline inclusions from sporulating bacterial cells, it was shown that all the primary symptoms characteristic of the natural disease of insects could be reproduced by feeding larvae on leaves coated with the purified crystals. The crystals consist of a protein which is insoluble in water under neutral or mildly acid conditions but which can be dissolved in dilute alkali. The gut contents of larvae are, in general, alkaline, and when the ingested crystals reach the gut, they are dissolved and partially hydrolized. This modified protein attacks the cementing substances which keep the cells of the gut wall adherent, and as a consequence, the liquid in the gut can diffuse freely into the blood of the insect. The blood of the insect becomes highly alkaline, and this change in pH induces a general paralysis of the larva. Death, which ensues much later, appears to result from bacterial invasion of the body tissues.

The protein crystals possess a highly specific toxicity for the larvae of many Lepidoptera but are wholly nontoxic for other animals (including all the vertebrates) and for plants. They thus provide an ideal agent for the control of many serious insect pests which damage plant crops. Recognition of this fact has led recently to the development of a new microbiological industry: the large-scale production of the toxic protein, for incorporation in dusting agents that can be used to protect commercial crops from the ravages of caterpillars. In industrial practice, the protein itself is not chemically isolated. Instead, the crystal-producing bacilli are grown on a large scale, harvested after the onset of sporulation with its accompanying crystal production, dried, and incorporated in a dusting powder.

THE PRODUCTION OF OTHER CHEMICALS BY MICROORGANISMS

The widespread use of microorganisms in the chemical and pharmaceutical industries has come about because of the recognition that it is often cheaper to use a microorganism for the synthesis of a complex organic compound (for example, an antibiotic) than to synthesize it chemically. Microbial syntheses also have distinct advantages in the preparation of optically active compounds, since chemical synthesis leads to racemic mixtures which must subsequently be resolved.

As previously discussed, the microbial production of acetone and butanol, once the major source of these chemicals, has now been largely superseded by chemical synthesis. Nevertheless, the microbial production of many relatively simple and cheap organic compounds remains competitive with chemical methods of synthesis. These compounds include gluconic acid, produced by *Aspergillis niger* and acetic acid bacteria, and citric acid, produced by *A. niger*.

The production of two vitamins, vitamin B_{12} and riboflavin, provides an instructive lesson in the economics of industrial microbiology. Both are now produced commercially by microbial means. Vitamin B_{12} is produced by certain *Pseudomonas* spp. Although the yields are very low, this process remains competitive because of the very high price of the product: the structural complexity of vitamin B_{12} virtually precludes a commercially feasible chemical synthesis. Riboflavin (vitamin B_2) is a much simpler compound, which can be readily prepared by chemical synthesis. It is still produced microbiologically, as a result of the discovery that certain plant pathogenic fungi (*Ashbya gossypii* and *Eremothecium ashbyi*) overproduce this vitamin and excrete the excess into the medium. By further genetic selection and improvement of culture methods, strains have been developed which produce so much riboflavin that the vitamin crystallizes in the culture medium.

THE PRODUCTION OF ENZYMES BY MICROORGANISMS

The production of microbial enzymes, either pure or partly purified, is an important aspect of industrial microbiology. The uses of microbial enzymes in medicine and industry (Table 31.4) are remarkably diverse.

THE USE OF MICROORGANISMS IN BIOASSAYS

Microorganisms are extensively used for the performance of bioassays, to determine the concentration of certain compounds in complex chemical mixtures. Such assays can be used both for the quantitative determination of growth factors and for the quantitative determination of antibiotics and other specific growth inhibitors.

The principle of bioassays of growth factors is very simple. A medium is prepared which contains all nutrients required for growth of the test organism, except the substance to be assayed. If that substance is added to the medium

TABLE 31.4

Partial list of microbial enzymes produced industrially

NAME OF ENZYME	MICROBIAL SOURCE	USES	REACTION CATALYZED
Diastase	Aspergillus oryzae	Manufacture of glucose syrups; digestive aid	Hydrolysis of starch
Acid-resistant amylase	A. niger	Digestive aid	Hydrolysis of starch
Invertase	Saccharomyces cereviseae	Candy manufacture (prevents crystallization of sugar)	Hydrolysis of sucrose
Pectinase	Sclerotina libertina	Clarification of fruit juices	Hydrolysis of pectin
Protease	A. niger	Digestive aid	Hydrolysis of protein
Protease	B. subtilis	Removal of gelatin from photographic film to recover silver	Hydrolysis of protein
Streptokinase	Streptococcus sp.	Promotes healing of wounds and burns	Hydrolysis of proteins
Collagenase	Clostridium histolyticum	Promotes healing of wounds and burns	Hydrolysis of protein (collagen)
Lipase	Rhizopus sp.	Digestive aid	Hydrolysis of lipids
Cellulase	Trichoderma konigi	Digestive aid	Hydrolysis of cellulose

in limiting amounts, the growth of the test organism will be proportional to the amount added. The first step in developing a bioassay is, accordingly, to measure the relationship between the amount of growth obtained and the amount of the limiting nutrient added (i.e., determine the yield coefficient, see Chapter 8). An example of the relationship obtained is shown in Figure 31.7.

If samples of a material suspected to contain the nutrient in question are added to the basal medium, their quantity present can be determined from the amount of growth that occurs. The lactic acid bacteria are frequently used for bioassays because of their very extensive growth factor requirements. By modifying a single basal medium so that different nutrients are growth limiting, it is possible to assay a large number of different amino acids and vitamins with a single test organism. Another advantage of lactic acid bacteria as test organisms is that growth can

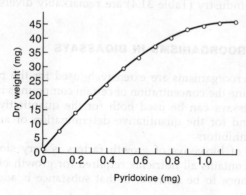

FIGURE 31.7

Microbial bioassay curve for the vitamin pyridoxine. The curve expresses the relationship between the amount of pyridoxine supplied and the amount of growth (dry weight) obtained. The organism used was a pyridoxine-requiring mutant of the mold Neurospora. From J. L. Stokes, Alma Larsen, C. R. Woodward, Jr., and J. W. Foster, "A Neurospora Assay for Pyridoxine," J. Biol. Chem. 150, 19 (1943).

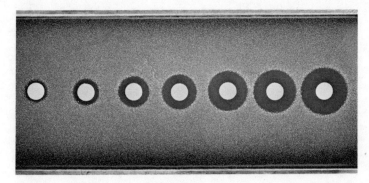

FIGURE 31.8

A bioassay for the antibiotic penicillin. A tray of nutrient agar was seeded with the test bacterium, and differing amounts of penicillin were placed on the paper discs. The amount of penicillin increases from 10 units on the left-hand disc to 10,000 units on the right-hand disc. From the areas of the zones in which bacterial growth is inhibited, a curve relating penicillin concentration to the extent of inhibition may be derived.

be estimated indirectly by determining the amount of lactic acid produced. The vitamin content of foods is still determined by bioassay, but amino acids are now largely determined by chemical methods. Bioassay is also an indispensable tool in the detection and purification of new growth factors.

The quantitative determination of antibiotics and other antimicrobial compounds is conducted by the so-called *cup plate* assay originally developed for the assay of penicillin. An agar medium is densely seeded with the test bacterium, and a number of glass cylinders are placed on its surface. A known dilution of a solution of an antibiotic is added to each cup, and the plate is incubated until growth has occurred. The antibiotic diffuses into the surrounding agar during incubation and produces a zone of inhibition. The diameter of this zone is a function of the initial concentration of the antibiotic in the solution which was added to the cup. A standard curve, relating the diameter of the zone of inhibition to antibiotic concentration, permits the assay of solutions containing unknown concentration of the antibiotic. A slight modification is the substitution of paper discs soaked in solutions of the antibiotic for the glass cylinders (Figure 31.8).

FURTHER READING

Books CASIDA, L. E., *Industrial Microbiology*. New York: Wiley, 1968.

FOSTER, E. E., R. E. NELSON, M. L. SPECK, R. N. DOETSCH, and J. C. OLSON, *Dairy Microbiology*. Englewood Cliffs, N.J.: Prentice-Hall, 1957.

FREED, M., *Methods of Vitamin Assay*. New York: Interscience Publishers, 1966.

FREITAS, Y. M., and F. Fernandes, *Global Impacts of Applied Microbiology*. Bombay: The Examiner Press, 1971.

ZÄHNER, H., and W. K. MAAS, *Biology of Antibiotics*. New York: Springer-Verlag, 1972.

Reviews AMERINE, M. S., and R. C. KUNKEE, "Microbiology of Wine Making," *Ann. Rev. Microbiol.* **22**, 323 (1968).

ROGOFF, M. H., "Crystal-Forming Bacteria as Insect Pathogens," *Adv. Appl. Microbiol.* **8**, 29 (1966).

INDEX

A

A (see Absorbancy)
AAA pathway, 208
Aberration:
    chromatic, 50
    spherical, 50
Abiogenesis, 4
Abortion:
    infectious, 815
    phage infection, 387
Absorbance of bacteria, effect of wavelength on, 281
Absorbancy, 280
    relationship to bacterial cell mass, 281
Acetic acid, end product of fermentation, 170
Acetic acid bacteria, 49, 590, 604–606
    use by man, 837
Acetoacetyl-S-ACP, 218
Acetobacter, 590, 591, 604
    pasteurianum, 185
    xylinum, 336, 337, 605
Acetohydroxyacid synthetase, 212
α-Aceto-α-hydroxybutyrate, 212
Acetoin, 170, 628, 839
α-Acetolactic acid, 170, 212
Acetone, end product of fermentation, 170
Acetone-butanol fermentation, 654
N-Acetylglucosamine, 243
N-Acetylglutamic acid, 207
N-Acetylglutamic acid synthetase, 262
N-Acetyl-γ-glutamyl phosphate, 207
N-Acetylmuramic acid, 243
N-Acetylornithine, 207
O-Acetylserine, 211
Acholeplasma, 126

blastoclosticum, 307
Achromatium, 569, 570
Acidaminococcus, 711
Acid-fastness, 688
Acinetobacter, 590, 610
    regulation of β-ketoadipate pathway in, 265–267
ACP (see Acyl carrier protein)
Acrasieae, 116
Acridines, 403, 409
Actinomyces, 674, 675, 677, 692
    antibioticus, 328
    israelii, 694, 782
    streptomycini, 328
Actinomycete line, 672–702
    modes of ATP generation, 677
    principal groups of, 672, 674
Actinoplanes, 676, 698
    spore vesicles of, 700
Activated sludge process, 729
Activation energy, 305
Active transport, 295
Acute glomerulonephritis, 806
Acyl carrier protein, 216
Acyl-S-ACP, 218
Adanson, Michael, 506
Adenine, structure of, 199
Adenosarcoma, 394
Adenosine monophosphate, synthesis of, 201
Adenosine triphosphate:
    anhydride linkage in, 155
    chemical-coupling hypothesis for generation of, 184
    chemiosmotic hypothesis for generation of, 184, 185
    electron donors and acceptors in various modes of generation, 180

ester linkage in, 155
    mechanism of generation by electron transport, 184
    role in coupling catabolism to biosynthesis, 156
    structures of, 155
    yield, 284
Adenovirus, 368, 372, 390, 826
Adler, J., 343
Aerobe, definition of, 33
Aerobic pseudomonads, 590, 591
Aerobic sporeformers, 646–652
Aeroduric (aerotolerant) anaerobes, 678
Aeromonas, 613, 618, 620, 621, 625
    hydrophila, 617, 619, 625
    punctata, 616, 625
    shigelloides, 625
Aerotaxis, 343
Aerotolerant anaerobes, 309
Aflatoxin B$_1$, structure of, 820
Aflatoxin G$_1$, structure of, 820
Agar, a solidifying agent, 15
Age of cells at division, 285
Aggressins, 797
Agmatine, 214
Agrobacterium, 590, 591, 597–599
    tumefaciens, 336, 599
AICAR (see Aminoimidazole carboxamide ribotide)
Akinetes, 138, 140
D-Alanine, 243
Alcaligenes, 590, 597–599
    eutropha, 134, 574, 576
    paradoxus, 134, 574
Alder tree, nitrogen fixation by, 756–757
Aldrin, 731
Ale, 834

of metabolic intermediates, 159
role in dehydrogenase reactions, 158
structure of, 158
Nicotinamide adenine dinucleotide
phosphate, role in dehydrogenase
reactions, 158
Nicotinamide ribotide, structure of, 158
Nicotinic acid, 32, 279
fermentation of, 659
nif genes, 623
Nitrate reduction, assimilatory, 194–195
Nitrification, 723
Nitrifying bacteria, 16, 565–568
Nitrite reductase, assimilatory, 195
Nitrobacter, 566
winogradskyi, 568
Nitrococcus, 566
mobilis, 568
Nitrogen, role in microbial nutrition, 29, 31
Nitrogen cycle, 720–724
Nitrogen fixation, 195
discovery of the role of microorganisms,
17
by *Enterobacter*, 623
Nitrogen-fixing systems, efficiencies of, 722
Nitrosococcus, 566
Nitrosocystis oceanus, 567
Nitrosoguanidine (*see* N-Methyl-N'-nitro-
N-nitrosoguanidine)
Nitrosolobus, 566
multiformis, 567
Nitrosomonas, 124, 566
europaea, 320, 567
Nitrosospira, 566
brienses, 568
Nitrospina, 566
gracilis, 568
Nitrous acid, 403, 407, 409
Nitrous acid-induced mutations, 408
Nocardia, 674–677, 686
asteroides, 689
farcinica, 686
group, 688–689
opaca, 574, 575
Noctiluca, 98, 99
Nodulation:
effect on plant growth, 752
ineffective, 756
Nodule:
cell, 754
process of formation, 753–755
Nomenclature, binomial system of, 504
Nonhalophiles, 300, 301
Nonheme iron, 174
Nonsense codon, 415
Nonspore-forming anaerobes, 704–707
Nostoc, 762
Novick, R., 483
Nu symbiont of Protozoa, 772
Nuclear division, 82
Nucleic acid hybridization, 512–517
Nucleocapsid (*see* Capsid)
Nucleolar organizer, 67
Nucleolus, 67
Nucleoplasm, 78
Nucleoside diphosphokinase, 202
Nucleosides, structure of, 199
5'-Nucleotidase, 304, 305
Nucleotides, synthesis of, 198–205

Nucleus:
bacterial, 356–362
interphase, 67
Numerical aperture, 51
Nutrient agar, 16
Nutrient broth, 16
Nutrition of microorganisms:
principles of, 27
role of elements, 28, 29
Nystatin, 85, 843, 844
use in obtaining pure cultures, 23

O

O antigens, 622, 791
O side-chain, 323
Obligate anaerobes, techniques for
cultivation of, 41
Oceanospirillum, 590, 608
Ochiai, K., 480
Ochre codon, 415
Octadecenoic acid, 309
OH-Spheroidenone, 532
Okazaki, R., 232
Okazaki fragments, 232
Okenone, 532
Okiba, T., 480
Olives, Spanish style, 839
OMP (*see* Orotidine monophosphate)
Oncogene theory, 398
Oncogenesis, 395
Oncogenic viruses, 395–400
One-gene-one-enzyme hypothesis, 188
One-step growth curve, 375
Operator genes, 419
Operon, 420
regulation of, 422
Ophryoscolex, 781
Opsonein, 805
Opsonization, 805–806
Order, 505
Organelles, 84
of eucaryotes, possible origin of, 745
Orla Jensen, S., 681
Orleans process, 837
Ornithine, 207, 214
Ornithosis, 817, 819
Orotic acid, 203
Orotidine monophosphate, 203
Oscillatoria, 139, 347, 348, 544
redekei, 636
Osmolarity, 299
Osmophile, 299
Osmoregulation, 299
Osmotic pressure, 299
O-specific chains, 245
Outer membrane, 124, 293
Overoxidizers, 604
Oxalacetic acid, structure of, 167
Oxaloglutaric acid, 208
Oxidase test, 184, 590
Oxidation:
of glucose, energy yield, 161
of inorganic compounds, 171
Oxidation-reduction potential, 180
Oxidative phosphorylation, 161

Oxygen:
control of concentration, 41
effect on bacteria, 309
Oxygenases, 312
Oxygen-sensitive enzymes, 312

P

Palmitic acid, 217, 309
Palmitoleic acid, 217
Pantothenic acid, 32, 206
Papillae of colonies, 448
Papilloma viruses, 397
Papillomas, 394
Papovaviruses, 372, 390, 395
Paracolon group, 620–623
Parainfluenza, 826
Paramecium aurelia, 770
Paramylum, 91
Paramyxoviruses, 372, 390
Parasexuality, 496–497
Parasite symbioses, 735–737
Paratyphoid fever, 812
Pascher, A., 758
Passive diffusion, 294
Pasteur, L., 652, 833, 834
experiments:
discovery of anaerobic life, 9
role of microorganisms in
transformation of matter, 7–9
on significance of fermentation, 9
spontaneous generation, 5, 6
Pasteurella, 612, 621
Pasteurization, 43
Pathogenicity, definition of, 785
Pathogens:
opportunistic, 595
plant, 595
Paulinella, 758
Peat bogs, 720
Pediculus, 777
Pediococcus, 674, 675, 678, 683, 832, 834
halophilus, 300
Peliaina, 758
Pelodictyon, 138, 350, 555
clathratiforme, 556
mesh formation by, 141
Pelonema, 347
Penicillin, 84
discovery of, 843
mode of action, 329
use in obtaining pure cultures, 23
Penicillinase, 303, 305
Penicillins, 844, 846
Penicillium, 110
chrysogenum, 845
roquefortii, 839, 840
Pentachlorophenol, 731
Pentose phosphate pathway, 165
PEP (*see* Phosphoenolpyruvic acid, role in
substrate level phosphorylation)
Peptide antibiotics, 665, 667
Peptide bond, 227
Peptidoglycan, 318–330
of actinomycete line, 674–677
cross-linking, 322
distribution of, 83
of endosporeformers, 646–647
function of, 83, 328–330